HOMEOWNERS' ENCYCLOPEDIA OF HOUSE CONSTRUCTION

HOMEOWNERS' ENCYCLOPEDIA OF HOUSE CONSTRUCTION

Morris Krieger

McGraw-Hill Book Company

New York St. Louis San Francisco

Auckland
Bogotá
Düsseldorf
Johannesburg
London
Madrid
Mexico
Montreal
New Delhi
Panama
Paris
São Paulo
Singapore
Sydney
Tokyo
Toronto

Library of Congress Cataloging in Publication Data

Krieger, Morris, date.
Homeowners' encyclopedia of house construction.

Includes index.
1. House construction—Dictionaries.
2. Building—Dictionaries. I. Title.
TH4811.K73 690'.03 77-19157
ISBN 0-07-035497-9

The editors for this book were Jeremy Robinson and Margaret Lamb, the designer was Edward J. Fox, and the production supervisor was Frank P. Bellantoni. It was set in Optima by University Graphics, Inc.

The pronouns "he" and "his" have been used in a purely generic sense in this book to accommodate the text to the limitations of the English language and avoid awkward grammatical constructions.

Contents

To the Reader

In one of the books I consulted during the writing of this book, the author quotes an old saying that during the course of construction, not one house but three houses are built—

> The one the architect thinks he designed.
> The one the architect actually did design.
> The one which the builder finally erects.

I would like to amend this saying by adding two more houses, one at the beginning and another at the end—

> The one the homeowner imagines he is getting.
> •
> •
> •
> The one the homeowner takes delivery of.

The building, or purchase, of a house is the single greatest expense and responsibility that most families assume during their lifetimes. Unfortunately, most families know nothing whatever about the design and construction of houses. They think in terms of numbers of bedrooms and bathrooms, step-down living rooms, cathedral ceilings, kitchen sizes, quarter and half acres, miles to the nearest train station or shopping center, and so on. When they wander through a house they tend to judge quality by the size of the rooms, on the erroneous theory that the bigger the better. They open and close closet doors. They bang on walls. These gestures are the equivalent of judging the quality of an automobile by kicking the tires and slamming the doors—and automobile manufacturers are expert at building a solid-sounding "thunk" into a closing door; it's a question of acoustics, not strength.

It is of course elemental that a house have the style and living qualities a family wants. But the family also has a right to expect solid construction, sound materials, and good workmanship. That they are disappointed in these expectations more often than not is first, because they are not experts in house construction and, second, because they don't know how to find someone who is. For most families, the soundness of the house they finally buy is like the quality of the medical attention they receive—a complete mystery to them. They pay both their medical bills and mortgage installments on pure faith that they are getting their money's worth.

This book, then, is written for the genus *homeowner,* for that variant form of the genus known as the *do-it-yourselfer* (who has spread like a weed in these times of high labor costs), and for those who aspire, either tentatively or firmly, to becoming homeowners and do-it-yourselfers themselves someday.

For those who intend building their own house, whether or not they hire an architect or general contractor to help and advise them, the articles in this book should help them close the wide, wide gap in quality that often exists between the house they think they are paying for and the house they actually take delivery of. I hope also that the articles will stimulate the would-be-builder to think more imaginatively about the materials and building techniques that are available to him: I hope it becomes a good book to browse in for its educational value.

For those who intend buying, a reading of the appropriate articles on house construction, electrical and heating systems, and plumbing should help them develop a sharper, more critical eye for the construction standards to which any particular house was built, as well as its current condition. They may also learn not to walk away from a house that is going at a discount because of some obvious defect. Perhaps the defect is not as deep-seated or expensive to fix as may at first appear.

As for the do-it-yourselfer, he won't find any instructions on how to build any particular projects or undertake any particular repairs; there are already any number of excellent books on the market that he can consult. What he *will* find here is background information on building methods and building materials that will help him do a more substantial, intelligent job than he might otherwise do, no matter what the project.

This book will also, I hope, serve still another purpose. Most houses are, in the nature of things, old houses. A new house often grows old with surprising rapidity; it was badly built in the first place. As a house ages, troubles develop. The roof leaks, the plumbing blocks up, the heating system conks out. The homeowner calls in a local contractor for an opinion and a price estimate. The price will (naturally) seem too high. The homeowner calls in other contractors and discovers to his dismay that each one thinks something else has gone wrong. Worse, each contractor gives him a different price estimate for fixing the problem. Even when all the contractors agree on the fault and how to fix it, their price estimates are sometimes wildly different. The homeowner doesn't live who hasn't been faced with a dilemma of this sort, and who hasn't felt uneasily that no matter what he decides, he is going to be taken advantage of.

What is the homeowner to do? Well, it is my hope and intention that by reading the relevant articles in this book the innocent homeowner will learn enough about the construction of his house to deal knowledgeably with all the contractors and repairmen he meets. Not, I hasten to add, that he will be able to take off his coat and do the job himself if he wants to, nor will he ever know as much about plumbing systems as a plumber or about roofs as a roofing contractor, and so on. But he *will* know enough about his house to talk sense to these gentlemen and know when they are talking sense to him.

Morris Krieger

Acknowledgments

I would like to acknowledge the generosity of the following individuals (and their companies and associations) for permission to use the illustrations and tables that appear in this book. But I am especially grateful to them for the time they have spent reading the articles for technical accuracy. I have incorporated most of the changes suggested to me, and I have no doubt that as a result the articles are much, much better. Any errors that remain are, of course, my responsibility.

Glenn R. Avery (American Plywood Association), Ken Backman (National Fire Protection Association), Edward C. Benfield (Stanley Works), Dave E. Brackell (Gypsum Association), A. Waldo Brown (National Oak Flooring Manufacturers' Association), Bernard J. Brown (Jackson Water Heater Co.), J. Bruce Burkland (Brick Institute of America), H. B. Carlsen (Gypsum Association), Robert C. Carmody (Copper Development Association), J. M. Carney (American Plywood Association), Calvin L. Cooksey (Cast Iron Soil Pipe Institute), R. H. Edes (Producers Council), K. Eric Ekstrom (National Woodwork Manufacturers' Association), Charles Farley (Brick Institute of America), Jack Finley (General Electric Co.), John M. Foehl (Copper Development Association), Mrs. D. M. Gardner (Western Wood Products Association), Lewis S. Goodfriend (Consultant Engineer), William J. Groah (Hardwood Plywood Manufacturers' Association), Kenneth A. Gutschick (National Lime Association), James Hackett (American Plywood Association), John M. Havas (Summer Rider and Associates), Howard Haynes (Illumination Engineering Society), Paul K. Heilstedt (BOCA International), Patrick J. Higgins (National Association of Plumbing-Heating-Cooling Contractors), Robert A. Holcombe (National Forest Products Association), Paul R. Hollenbeck (West Coast Lumber Inspection Bureau), John E. Kaufman (Illuminating Engineering Society), Dennis Kirchoff (National Sanitation Association), Robert Kleinhaus (Tile Council of America), Howard Lieberman, P.E., Kenneth F. Lelen (Sumner Rider and Associates), Harry C. Leslie (Forest Products Laboratory, USDA), Dr. Nina I. McClelland (National Sanitation Foundation), J. W. Meusel (Atlantic Cement Company), Richard D. Peacock (National Fire Protection Association), Virgil G. Peterson (Red Cedar Shingle and Handsplit Shake Bureau), John Quinlan (The Anaconda Company), Thomas B. Redmond (National Concrete Masonry Association), Richard Reilly (Jares, Baum, Bowles, Inc.), Robert E. Ross (Hydronics Institute), Richard L. Sanderson (BOCA International), Gary Schoenfeld (National Concrete Masonry Association), John G. Shope, P.E., Robertson Sillars (Portland Cement Association), Robert M. Stansel (Adhesives and Sealant Council), Paul Stier (Hill and Inowlton), Alan R. Trellis (National Association of Home Builders), Robert F. Welsh (Stucco Manufacturers' Association), Robert E. Wilde (American Concrete Institute), William H. Winter (Architectural Woodwork Institute), and Alan H. Yorkdale (Brick Institute of America).

Acoustical Materials

This article describes sound-absorbent materials that might be used in the home. For a more general description of how sounds are transmitted, controlled, and measured, see ACOUSTICS.

All materials will absorb at least some of the sound energy that impinges upon them. It may be as little as a sheet of glass, which will absorb about 2½ percent; the rest of the sound energy is either reflected toward the source or transmitted through the glass. At the other extreme is an open window that will, as far as a person inside a room is concerned, "absorb" almost 100 percent of the sound energy "striking" it; an open window is a near-perfect sound absorber.

Between these extremes lies everything else. Ordinary building materials such as wood, brick, stone, plaster, and so on, absorb something less than 5 percent of the sound energy striking them. To be counted as an acoustical material, a substance must absorb more than 50 percent of the impinging sound energy. To do this, the material must have a porous, fibrous structure—that is, it must have a great many convoluted, interconnected passageways exposed to the surface that run deep into the material; and the thicker the material, the better a sound absorber it will be. This is because the air molecules that carry the sound energy can migrate deeply into the interior of the material, where, as they repeatedly strike and bounce off the surrounding fibers, their rapid, vibratory motion is converted by friction into heat.

Mere lightweightness counts for little or nothing as far as the ability to absorb sound is concerned. Cork and foamed plastics, for example, are very lightweight materials. Their structures consist of innumerable cells. However, almost none of these cells lie exposed on the surface nor are they interconnected with each other. Whatever sound-absorptive properties these materials may have is due to the fact that any impinging sounds may be reflected from them in a diffuse fashion because of the irregularity of their surfaces.

Interestingly enough, thin, hard materials like a sheet of metal or plywood will also absorb impinging sound vibrations, as long as they are stiff enough. They do this by vibrating sympathetically with the frequency of the sounds that strike them. The energy of the sound waves is thus dissipated by the vibratory motion they induce in the panels. Panels of this kind absorb mainly low-frequency sounds. High-frequency sounds are largely reflected from or transmitted through the panels.

Noise Reduction Coefficients

Every building material, and each kind of wall and floor/ceiling construction, will absorb (or transmit) sounds according to their frequencies. It depends on whether, and at what frequencies, sympathetic vibrations can be induced in the material or construction. Each material or construction will be most efficient at absorbing sounds having particular frequencies. Materials and constructions, therefore, are tested at a range of frequencies to determine their overall sound-absorptive properties, these frequencies being 125, 250, 500, 1000, 2000, and 4000 cycles per second. The percent of impinging sound absorbed at each frequency is called the *sound-absorption coefficient* for that frequency. Tables of the sound-absorption coefficients for a wide range of materials and constructions are published for use by architects, acoustical engineers, and others. For an example, *see* ACOUSTICS (Table 2).

As a practical matter, knowing how a particular acoustical material reacts to a range of impinging sounds is too complicated for a builder or the ordinary homeowner. Instead, a simplified rating system is used in which each material is assigned a *noise reduction coefficient* (NRC) (*see* ACOUSTICS). For any given acoustical material, its NRC is obtained by averaging the sound-absorption coefficients obtained at 125, 250, 500, 1000, and 2000 cycles. The resulting average is the material's NRC. NRC ratings are widely used, by manufacturers of sound-absorption materials, and most builders and homeowners tend to take these NRC ratings at face value. It should be remembered, however, that these are average figures. Any particular material may absorb either more or less sound at a given frequency than is indicated by its NRC and, therefore, it may be more or less suitable for a particular application than other materials having a similar NRC.

What Acoustical Materials Can and Cannot Do

Although it is true that acoustical materials can reduce the apparent loudness of sounds within a room, it is important to understand how this is accomplished. Otherwise, it will be impossible to understand when acoustical materials should be installed and when not.

When a sound originates within a room, the sound waves reach an occupant of that room in two ways. First, a small proportion of the sound waves travel directly from their source to the occupant. Second, most of the sound waves bounce repeatedly off the walls, ceiling, and floor of the room before they reach the occupant. There is a very slight time lag between the moment when the direct sound waves reach the occupant and when these reflected sound waves reach him. Furthermore, because these reflected sound waves reach the occupant at slightly different times, depending on the size and proportions of the room and on the relative positions of the origin and occupant in the room, the reflected sound waves give a reverberant, echo-like quality to the original sound. It is this reverberant sound that is responsible for the warm, rich quality we associate with indoor sounds.

What effect does the installation of an acoustical material have on the sound within the room? Assume the ceiling has been covered with acoustic tiles. As a consequence, the occupant will have the distinct impression that the room is now much quieter than it was before. The acoustic tiles will not have any effect at all on the intensity of the sound waves traveling directly from the source to the occupant. This should be obvious. The original sound will be just as loud as it was before. As for the reverberant sound waves, however, they will decrease in intensity much more quickly because every time a sound wave bounces off the ceiling, a proportion of its sound energy will be absorbed by the acoustic tiles. How much of the sound energy will be absorbed will depend on the material's NRC.

It is possible to have too much of a good thing, to overdo the amount of acoustical material installed in a room. If, for example, a room not only has acoustic tiles on the ceiling but has wall-to-wall carpeting and walls covered with drapes as well, sounds become deadened. Voices lose their warm, rich quality. Instead, they sound thin, flat, and distant, as if the speakers were standing in an open field some distance from each other. If the room is large enough, it is even possible for voices to become lost altogether. It is necessary, therefore, to strike a balance between excessive noise levels within a room and excessive sound absorption.

Although acoustical materials can do a great deal to reduce the intensity of sounds that originate in a room, they are unable to do much with sounds that originate outside a room and are transmitted into it. (For the reasons, *see* ACOUSTICS.)

It will suffice here to note that outside airborne sounds enter a house by inducing a vibratory motion in the windows, doors, and walls. These structures then transmit these sounds into the house. Acoustical materials cannot do anything at all to prevent this mode of sound transmission. They simply lack the mass and solidity to dampen sound vibrations.

The reader might object that if an acoustical material can absorb sounds originating within a room, why can't it absorb sounds originating outside the room? One part of the answer is that the sounds "absorbed" by an acoustical material consist not only of the sounds that are truly absorbed but also of the sounds that pass completely through the material and are not, therefore, reflected back to the occupant. And a large proportion of the sound that is "absorbed" by an acoustical material is, in fact, transmitted completely through the material because a lightweight material (which an acoustical material is) is inherently incapable of preventing the transmission of sounds through it. (For an explanation, *see* ACOUSTICS.)

Another part of the answer is that acoustical materials absorb mainly high-frequency sounds. An acoustical material absorbs very little of the low-frequency sounds that strike it and most of the irritating outdoor sounds that homeowners object to are such low-frequency sounds as traffic and aircraft noises.

Furthermore, sounds are absorbed within a room only after the sound waves have struck the acoustical material a great many times. When an outside sound enters a room, it passes through the acoustical material only once, with a negligible effect on its intensity. But once an outdoor sound has entered a room its intensity will be reduced quickly by repeated reflection off the wall, floor, and ceiling surfaces. But the original sounds passing through the wall directly to an occupant will be unaffected by this attenuation of the reflected sounds.

Finally, acoustical materials are not the effective sound absorbers that one is led to expect by manufacturers' literature. Innumerable tests by a great many laboratories have shown that the maximum possible amount of sound reduction that can be achieved by the installation of an acoustical material is something like 9 to 10 decibels, with most materials achieving a sound reduction of about 5 decibels. (For a definition of decibels, dB, for short, *see* ACOUSTICS.) Thus, although it is perfectly feasible to install acoustical materials in a room and achieve a reduction in the overall intensity of the sound, one should also remember that these sound levels are comparatively low to begin with. It is quite another kettle of fish to try and reduce the extremely high sound levels originating inside a boiler factory, say, by installing acoustical materials. It can't be done. And if the reader happens to live next to an airport runway where the noise levels may reach an intensity of 100 dB, all that the installation of acoustical materials will do for his peace of mind is reduce the intensity of the noise within his house to 90 or 95 dB, which is trivial.

Acoustics

Acoustics is the science of sound. Insofar as this article is concerned, acoustics is the science of *controlling* or *limiting* or *suppressing* unwanted sounds. That is, we will take acoustics to mean acoustical *noise control*. The subject is more encompassing than might at first appear for it includes: (1) the general design and layout of a house to prevent or limit the transmission of unwanted sounds from one part of a house to another, or from outside the house to inside, (2) methods of construction that will prevent or limit the transmission of unwanted sounds from one part of a house to another part, or from outside the house to inside, and (3) the use of sound-absorbent materials on the walls and ceiling of a room to reduce the level of sound within the room. It is this last item that most people understand to mean by "sound control" or "soundproofing" or "sound insulation." This is, however, a comparatively minor aspect of acoustics.

Acoustics has become much more important since World War II because we are living in an increasingly noisy world. Traffic noises and the roar of jet aircraft outside the house have made life within more uncomfortable. Inside the house there is radio, television, and hi-fi equipment, innumerable electrical gadgets that whine, grind, or throb, and the constant, enervating, low-frequency noise of heating equipment in winter and air conditioners in summer (though, in truth, we seldom complain about the noises we ourselves make). Compounding the problem has been the development of new lightweight materials that, whatever structural and economic advantages they may confer, vibrate easily in response to impinging sounds and thus both amplify and pass all these sounds along.

In this article, then, we shall describe how the intensity of sounds is measured, how sounds are transmitted throughout a house (or from outside the house to inside), how unwanted sounds can be prevented from spreading (that is, how the sounds are *isolated*), and how a room can be *insulated* against unwanted sounds. Note the distinction between the *isolation of* sounds and *insulation against* sounds. The distinction is important and should be kept in mind.

SOUND INTENSITY AND DECIBELS

The faintest possible sound that a person can hear has about one-trillionth (1/1,000,000,000,000) watt per sq meter of energy. A confidential *sotto-voce* conversation will have about 10,000 times more energy than this. An ordinary office or dinner-table conversation will have 100,000 times more energy, and so on, to the loudest sound one can hear—a nearby jet engine operating at full power, say—which would be about a trillion times more intense than the faintest possible sound we could hear, and quite painful.

This is an enormous range of energy. That our ears can hear this range of sound energy is a tribute to their design and construction. How do the ears do it? By compressing the range of sound intensity. It is a peculiarity of the way we hear that we do not hear equal increments of sound energy as equal increments of "loudness." That is, if sound A is 10 times more intense than sound B, as measured objectively by a sound-level meter, we would hear the difference in intensity not as 10 times "louder" but only as twice as "loud." And "loud" is in quotation marks because loudness is a very subjective quality that differs among individuals.

Our hearing system compresses enormously the range of sound intensities that strike our ears, and the more intense the sound, the more drastically is the actual intensity of the sound compressed. In short, we hear differences in the intensity of sounds logarithmically. This is shown in Fig. 1, which compares the logarithmic system of measuring sounds with the decimal system we are all familiar with. Note that the logarithms measure the *ratios* of sound energy, not the actual values. That is, a 100 percent change in the logarithmic scale is equal to a 10-point change on the decimal scale. A 1000 percent change on the logarithmic scale is equal to a 20-point change on the decimal scale. A 10,000 percent change on the logarithmic scale is equal to a 30-point change on the decimal

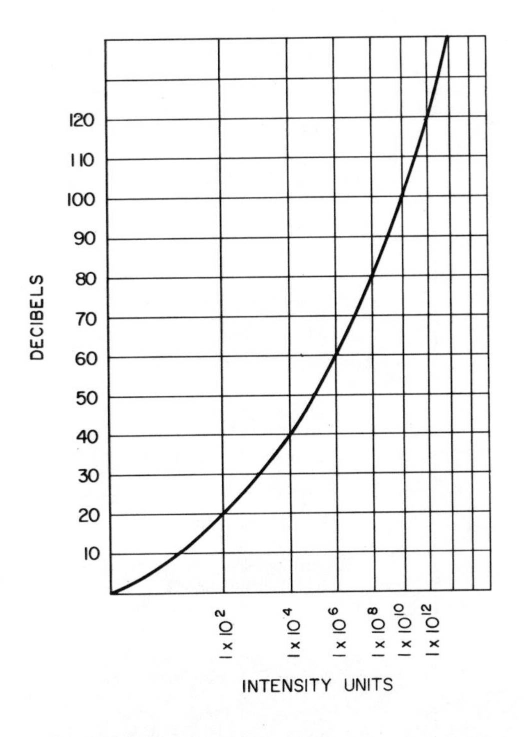

Fig. 1. The relationship between actual sound intensities and their subjective loudness is shown in the diagram. Decibels measure subjective loudness.

scale. And so on, between any two points on this chart within its entire range. What all this means is that if sound A is 100 times more intense than sound B, as measured by a sound-level meter, we will think sound B is only twice as loud. And if sound A is 1000 times more intense than sound B, we will think sound B is only three times as loud, and so on. This relationship between the subjective loudness of sounds and their actual sound intensities is true over the entire range of our hearing.

In acoustics, the relationship between actual sound intensities and their subjective loudness is measured by the *decibel* scale (abbreviated dB), named in honor of Alexander Graham Bell. Thus, on the decibel scale, a 10-point change in the apparent loudness of a sound means that the sound seems to us to become twice as loud (or one-half as loud) as it was before. In fact, of course, the actual intensity of the sound will be 100 times as great as (or less than) it was before. One of the things that the decibel scale does, therefore, is allow us to rate the intensity of sound subjectively. In Table 1, for example, a whisper is shown to be about 10 dB louder than the rustle of leaves; that is, the whisper will sound about twice as loud to us as the leaves do. In fact, of course, the noise made by the whisper is 100 times as loud as the noise made by the leaves. In the same way, Table 1 shows that there is about a 10-dB difference in apparent loudness between the noise of a loud horn in traffic and the sound made by an express train passing close to us at high speed. Again, the actual difference in sound intensity between traffic noises and an express train is enormously greater than the difference between a whisper and the sound of rustling leaves. Nevertheless, the subjective reality is that the train will sound only twice as loud as traffic and rustling leaves will sound only twice as loud as a whisper.

Table 1. The Decibel Scale of Sound Intensities

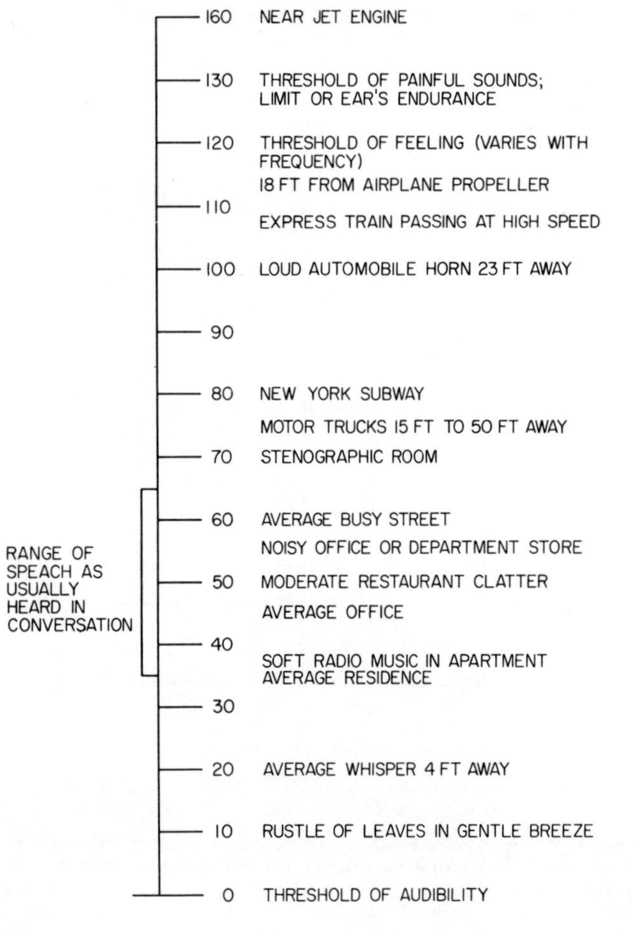

SOUND ISOLATION

With this understanding of what decibels represent firmly in our minds, we can now consider how sounds travel from one room to another within a house and what can be done to *isolate* the sound—to prevent its traveling from one room to another.

Imagine two rooms separated from each other by a solid partition. The construction of the partition is irrelevant for the moment. A sound originating in one of these rooms will travel through the air as alternately compressed and rarefied molecules of air. The force with which these air molecules strike the partition, though extremely weak, is still sufficiently strong to cause the partition to vibrate sympathetically with the original sound. And because this apparently solid partition vibrates, no matter how imperceptibly, the original sound will be transmitted into the second room.

The loudness with which the sound is heard in the second room will be weaker than the original sound for several reasons. First, all sound waves attenuate in strength as they travel away from the source of the sound. For example, whatever the strength of the original sound (and neglecting for the moment the effects of reverberant sound within a room), 10 ft from the source the sound waves will be about one-half as energetic as they were originally; 20 ft from the source they will be about one-quarter as energetic; 30 ft away they will be about one-eighth as energetic, and so on, until the sound becomes too weak to be heard. This conforms to our ordinary experience of how sounds travel. In short, the force with which the original sound impinges on a partition will depend on the distance between the origin of the sound and the partition.

Second, a certain percentage of the impinging sound waves will be reflected back from the partition into the room. The actual percent reflected will depend on the nature of the surface. A smooth, hard plaster surface, for example, reflects more sound than heavy cloth drapes do.

Third, some of the sound will be absorbed by the wall, depending, again, on the nature of the wall material. Materials like plaster, glass, concrete, and plywood absorb very little sound. A soft, porous material like fiberboard absorbs a great deal. The type of porosity in the material is important. A material like fiberboard that has a fibrous structure in which a great many interconnecting internal passageways are exposed on the surface will trap a large percentage of impinging air molecules. The molecules will dissipate their energy in the form of heat as they repeatedly strike the sides of these passageways. The structure of another type of porous material consists of self-contained cells that are not exposed to the air. Cork and styrofoam are examples. These materials are relatively poor sound absorbers.

The proportion of impinging sound energy that is absorbed by any given material is measured by its *absorption coefficient*. This is a number that will lie somewhere between 0 (for complete reflection of the sound) to 1.0 (for complete absorption). An open window, for example, has an absorption coefficient of 1.0, since it can reflect nothing, while a sheet of glass has an absorption coefficient of 0.027, which means it reflects all but 2.7 percent of the sound energy striking it. Elaborate tests have been conducted on a wide range of construction materials to determine their absorption coefficients at different frequencies. Table 2 shows a representative sample of materials and their absorption coefficients.

When one subtracts the sound that is either reflected or absorbed, what remains is the sound energy that actually passes through the partition. The difference, in decibels, between the original amount of sound energy on opposite sides of the partition is the partition's *transmission loss*. A partition that is very efficient in preventing the passage of sound has a high transmission loss; it is an excellent sound

Table 2. Absorption Coefficients of Common Building Materials*

Complete tables of coefficients of the various materials that normally constitute the interior finish of rooms may be found in the various books on architectural acoustics. The following short list will be useful in making simple calculations of the reverberation in rooms.

Materials	Coefficients					
	125 cps	250 cps	500 cps	1000 cps	2000 cps	4000 cps
Brick, unglazed	0.03	0.03	0.03	0.04	0.05	0.07
Carpet, heavy, on concrete	0.02	0.06	0.14	0.37	0.60	0.65
Same, on 40-oz hairfelt or foam rubber	0.08	0.24	0.57	0.69	0.71	0.73
Concrete block, coarse	0.36	0.44	0.31	0.29	0.39	0.25
Concrete block, painted	0.10	0.05	0.06	0.07	0.09	0.08
Fabrics						
Light velour, 10 oz per sq yd, hung straight, in contact with wall	0.03	0.04	0.11	0.17	0.24	0.35
Medium velour, 14 oz per sq yd, draped to half area	0.07	0.31	0.49	0.75	0.70	0.60
Heavy velour, 18 oz per sq yd, draped to half area	0.14	0.35	0.55	0.72	0.70	0.65
Floors						
Linoleum, asphalt, rubber, or cork tile on concrete	0.02	0.03	0.03	0.03	0.03	0.02
Wood	0.15	0.11	0.10	0.07	0.06	0.07
Wood parquet in asphalt on concrete	0.04	0.04	0.07	0.06	0.06	0.07
Glass						
Ordinary window glass	0.35	0.25	0.18	0.12	0.07	0.04
Gypsum board, ½-in. nailed to 2 × 4's 16 in. o.c.	0.29	0.10	0.05	0.04	0.07	0.09
Plaster, gypsum, or lime, smooth finish on tile or brick	0.13	0.15	0.02	0.03	0.04	0.05
Plaster, gypsum, or lime, rough finish on lath	0.02	0.03	0.04	0.05	0.04	0.03
Same, with smooth finish	0.02	0.02	0.03	0.04	0.04	0.03
Plywood paneling, ⅜-in. thick	0.28	0.22	0.17	0.09	0.10	0.11

*This table is reprinted in abridged form through the courtesy of the Acoustical Materials Association. The values reported here are reviewed from time to time as more up-to-date information becomes available from researchers in the field of acoustics. The reader is referred to the annual bulletin of the AMA for future changes.

barrier. On the other hand, a wall that has a low transmission loss is a poor sound barrier.

Innumerable tests of different wall constructions under standardized conditions have been made by a number of manufacturers and testing organizations such as the American Society for Testing and Materials (ASTM), to determine the ability of particular constructions in preventing the transmission of sound. According to ASTM standards E 90 and C 423, these tests are conducted at 18 standard frequencies within the range of 100 to 5000 cycles per second. Table 3 gives the transmission-loss values, in decibels, for common types of wall and floor constructions. The effectiveness of any particular construction shown in Table 3 as a sound barrier may be judged by comparing its transmission loss with the following figures:

25 to 30 dB loss—A conversation conducted in normal tones will be heard through the partition.

30 to 35 dB loss—A loud conversation can be heard through the partition, but much of the conversation will be unintelligible.

35 to 40 dB loss—Normal conversation is inaudible. Loud conversations can be heard, but the words are muffled and unintelligible.

40 to 45 dB loss—Even a loud conversation is inaudible, but singing, musical instruments (especially brass instruments and the piano), and a radio or television set turned up loud can be heard distinctly enough to be annoying.

45 dB loss and up—Even loud sounds will be muffled.

In addition, for single-family dwellings, it is considered that the partition between a bedroom and adjoining rooms should have a transmission loss of 40 to 45 dB; the partitions between a bathroom and adjacent rooms (except between a bedroom and its connecting bathroom) should have a transmission loss of 40 to 45 dB; and the partition between any other two rooms, and between any room and the outdoors, should have a transmission loss of 35 to 40 dB. The party wall of a two-family dwelling separating one dwelling from the other should have a transmission loss of better than 45 dB.

These figures are merely a rough guide to the qualities of any particular construction as a sound barrier. Many other things have to be considered as well. For example, once sounds pass through a partition, the actual intensity of the sounds that reach the ear may be less because drapes and furnishings installed in the room will absorb a portion of the sounds. The effect of the drapes and furnishings will be to increase the actual transmission loss. In addition, other noises may be present that help to mask, or muffle, the sounds coming from the next room. This masking noise gives the illusion that the transmission loss is greater than it actually is. Background noise is always present inside most houses, whether the occupants are aware of it or not, and its presence does help to increase the effective transmission loss of any particular wall construction. The following levels of background noise are typical:

Under 25 dB—A low background-noise level usually found in quiet suburban or residential communities. Traffic noises are absent, and so are the sounds of mechanical equipment.

25 to 35 dB—Average background-noise levels, in which one can hear light or distant traffic noises and perhaps low-level mechanical equipment.

35 dB and up—Prominent street and traffic noises, noisy central air-conditioning equipment and noisy window-mounted air-conditioner units.

In any room in which extreme quiet is required, these background noise levels may themselves become a nuisance that must be dealt with rather than ignored.

Finally, there is the frequency of an offending noise to be considered. The decibel ratings given above are for average conditions in which high-frequency and low-frequency sounds are jumbled together. In general, however, low-frequency sounds are less easy to suppress than high-frequency sounds. In a general way, the sound resistance of a typical wall construction is reduced by about 5 dB for every halving of the frequency. That is, if a wall construction is rated as having a 40 to 45 dB transmission loss at a frequency of 2000 cycles, the

Table 3. Transmission-Loss Values for Common Types of Wall and Floor Constructions

Building construction	Transmission loss (dB) at listed frequencies (cps)						Source*
	125	250	500	1000	2000	4000	
Single walls							
2-in. solid gypsum sand-aggregate plaster (18 lb per sq ft)	31	32	33	38	45	53	1
6-in. hollow-core cinder block, painted both sides (33 lb per sq ft)	29	31	36	40	46	52	1
6-in. hollow-core cinder block, $\frac{5}{8}$-in. sand-aggregate plaster both sides (43 lb per sq ft)	36	33	38	45	50	56	1
$4\frac{1}{2}$-in. solid brick, plastered both sides (45 lb per sq ft)	34	35	40	51	57	60	1
7-in. stone-aggregate concrete, plastered both sides (90 lb per sq ft)	44	42	52	58	66	70	1
2×4 wood studs, $\frac{1}{2}$-in. gypsum board both sides (6 lb per sq ft)	20	30	36	41	43	42	2
2×4 wood studs, $\frac{1}{2}$-in. sand-aggregate plaster on $\frac{3}{8}$-in. gypsum lath both sides (16 lb per sq ft)	27	25	31	44	34	50	1
Double walls							
Two wythes of plastered 3-in. dense concrete, 3-in. airspace between (bridging in airspace and at edges) (85 lb per sq ft)	38	40	51	54	57	65	1
Two wythes of plastered $4\frac{1}{2}$-in. solid brick, 2-in. airspace between (sound-absorbing material in airspace—bridging at edges only) (90 lb per sq ft)	43	50	52	61	73	78	1
Two wythes of plastered $4\frac{1}{2}$-in. solid brick, 12-in. airspace between (wythes *completely* isolated) (90 lb per sq ft)	57	70	83	93	—	—	1
Floor-ceilings							
Typical residential floor-ceiling wood finish; and subfloors on wood joists, gypsum lath and plaster below (about 15 lb per sq ft)	24†	32†	40†	48†	51†	54†	1
Concrete floor slab, $\frac{1}{2}$-in. plaster finish coat below (about 45 lb per sq ft)	43†	40†	44†	53†	56†	58†	1
Doors							
$1\frac{3}{8}$-in. hollow-core wood door, normally hung	5	11	13	13	13	12	2
$1\frac{3}{8}$-in. solid wood door, normally hung	10	13	17	18	17	15	2
$1\frac{3}{8}$-in. solid wood door, fully gasketed	16	18	21	20	24	26	2
Specially constructed $2\frac{5}{8}$-in. wood door, full double gasketing	20	23	29	23	31	37	2

**Sources:* 1. Beranek, L. L. (ed.). *Noise Reduction,* chap. 13, McGraw-Hill Book Company, New York, 1960. 2. Bolt Beranek and Newman Inc., unpublished data.

†Number is not a transmission-loss value but a room-to-room noise reduction value adjusted for a receiving room with a 0.5-sec reverberation time at the listed frequency. The actual transmission-loss value should be within ± 2 dB of the listed noise reduction value.

same wall will have a transmission loss of only 25 to 30 dB for sounds having a frequency of 250 cycles. The higher the sound frequencies, the easier they are to suppress, which is why such low-frequency sounds as traffic noises, jet aircraft, fan rumble, and air-conditioning noises are so difficult to suppress. Fortunately, our sense of hearing makes up for this characteristic of sounds, in part, by being much less sensitive to low-frequency sounds.

Soundproof Wall Construction

There are two surefire methods of building a partition having a high transmission loss: (1) make the wall heavy, (2) build the wall in two completely separate halves with an air space between.

The heavier a partition—that is, the more mass it has per unit volume—the more effective a sound barrier it will be. A solid 6-in.-thick concrete wall or an 8-in.-thick brick wall make much more effective sound barriers than the usual 2×4 in. stud wall in which $\frac{1}{2}$-in-thick gypsumboard is nailed to both sides of the wall. The reason is that the impinging air molecules find it much more difficult to induce a vibratory motion in a massive wall than in a lightweight wall. Weight makes the difference. However, there is a limit to how much one can increase the weight of a wall, and a limit to the effectiveness of the weight increase, as is shown in Fig. 2.

There is a second method, however, of obtaining a high

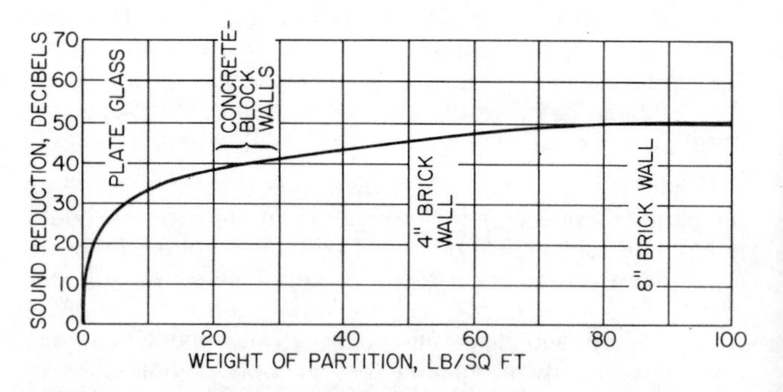

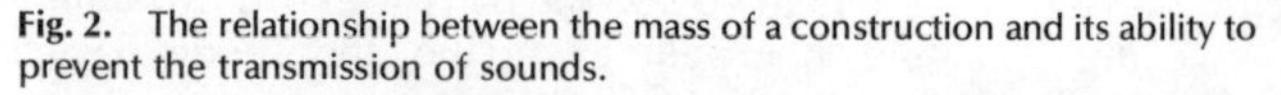

Fig. 2. The relationship between the mass of a construction and its ability to prevent the transmission of sounds.

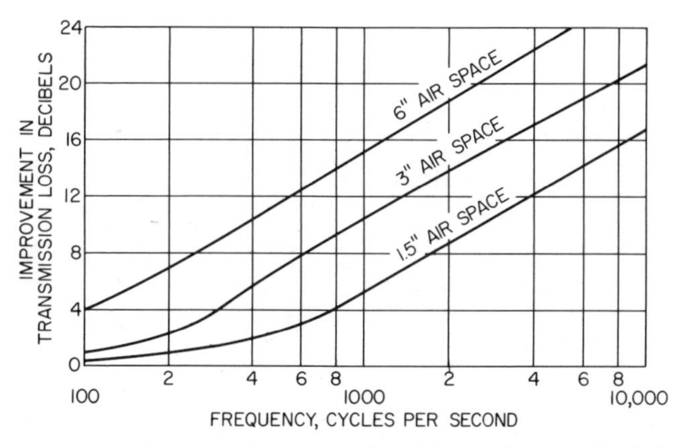

Fig. 3. The wider an air gap within a wall construction, the more effectively will the construction prevent the transmission of sounds through it *(Callender, Time-Saver Standards, McGraw-Hill).*

transmission loss in a wall—building an air cavity within the partition. This is shown in Fig. 3. To be effective, however, the two halves of the wall must not have any interconnection between them whatsoever, since any such interconnection will transmit sounds across the cavity.

It may appear, by analogy with heat insulation, that stuffing the wall with an insulating material such as glass fiber or rock wool will increase the transmission loss of the wall. Experience shows, however, that an insulating material increases the transmission loss by 2 or 3 dB at most. Insulation materials don't work for the same reason that acoustical materials don't—they are too light and porous. It is one of the great fallacies of modern home construction that stuffing an acoustical material in a partition will help prevent the transmission of sound through it. In addition, trying to increase the transmission loss of a wall, or a ceiling, by attaching acoustical tiles, insulation board, drapes, or a similar lightweight, porous material against one of the external surfaces is also a waste of money because the lightweight and porous nature of these materials makes them inherently incapable of preventing the transmission of sounds through themselves, much less through a wall or ceiling. However, it should be added that attaching an insulation material to one of the *interior* surfaces of a wall partition does help increase the transmission loss through the partition, if the installation is done correctly.

An alternative to an air cavity is to support the two sides of a partition on flexible mountings, which are in turn attached solidly to the partition studs. The walls are thus free to vibrate, and to dampen the vibrations of impinging sound waves. Flexible mounts of this type, of which there are several patented designs on the market, are as effective as an air space in increasing the partition's transmission loss.

An even simpler but equally effective method of obtaining a high transmission loss (about 45 to 50 dB) at small expense is to nail fiberboard to both sides of the studs—making sure the nails are driven well below the surface of the fiberboard—and then to attach ⅝-in.-thick gypsumboard panels to the fiberboard with adhesive. For a description of this method, *see* DRYWALL CONSTRUCTION. The fiberboard has a springy quality to it that enables it to absorb sound vibrations just as flexible mounts do, and the gypsumboard adds the mass that is necessary if the impinging sound vibrations are to be dampened. If one wants a plaster wall instead of a gypsumboard wall, then one can attach gypsum lath to flexible mountings, over which a standard plaster coating is applied. Such a plaster wall will have a transmission loss of about 50 dB. (For procedure, *see* PLASTERING.)

In an existing house, it would be difficult, if not impossible, to tear down a partition in order to erect a more sound-resistant partition in its place, not to mention the expense of doing so. There is, however, a simple and quite efficacious method by which the transmission loss of an ordinary 2 × 4 in. stud wall can be vastly improved—the construction of a false wall immediately next to the stud wall with a cavity between them. The construction is shown in Fig. 4. Under no circumstances should the false wall touch the real wall at any point, not even for the installation of electrical outlet boxes, wiring, insulation, or whatever. The two walls must remain completely isolated. If pipes or any other equipment must run through the wall, then care should also be taken to isolate them from the walls to prevent their transmitting vibrations from one wall to the other. A construction of this kind is capable of achieving an STC rating as high as 50 dB (see below).

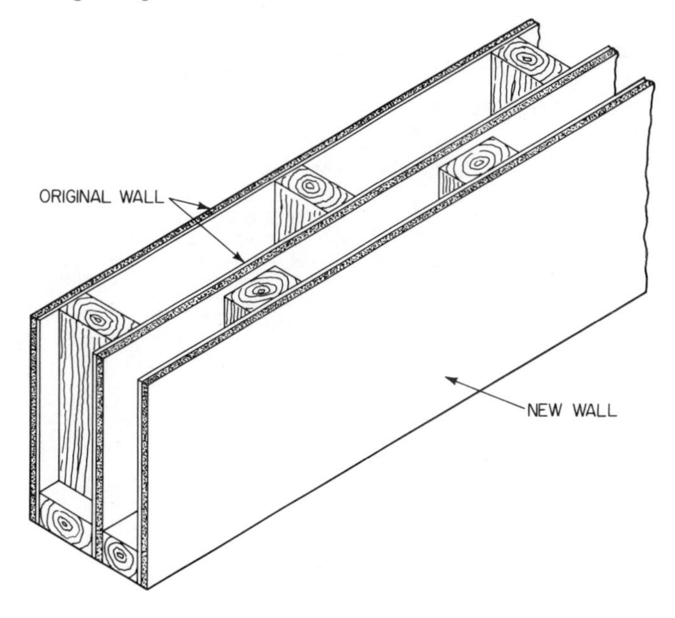

Fig. 4. Building a false wall next to an existing wall greatly increases the transmission loss between two rooms. The false wall should be spaced at least 1 in. from the real wall.

STC Ratings

Figure 5 shows a variety of possible wall constructions. Each type of construction has been given an "STC" rating, the "STC" standing for *sound transmission class.* The STC ratings are an attempt to summarize, in the form of a single numerical value, the results of the transmission-loss tests mentioned above.

The STC rating for a given construction is obtained as follows. First, sound transmission class *contours* are drawn on a graph, as shown in Fig. 6. These contours indicate that subjective response of the ear to differences in apparent loudness. Note that the contours slope upward and to the right as frequency increases, which indicates that high-pitched sounds are subjectively louder to the ear and are, therefore, more disturbing than low-pitched sounds having the same sound intensity. The transmission-loss test results for the construction are now plotted on the graph. Most constructions and individual wall materials vary in their ability to suppress impinging sound vibrations, depending on the frequencies of the vibrations. Figure 6 shows that this particular construction is least capable of preventing the passage of sounds having a frequency of about 1000 cycles. At this frequency, the transmission-loss curve drops to the 30-dB contour. As this is the point

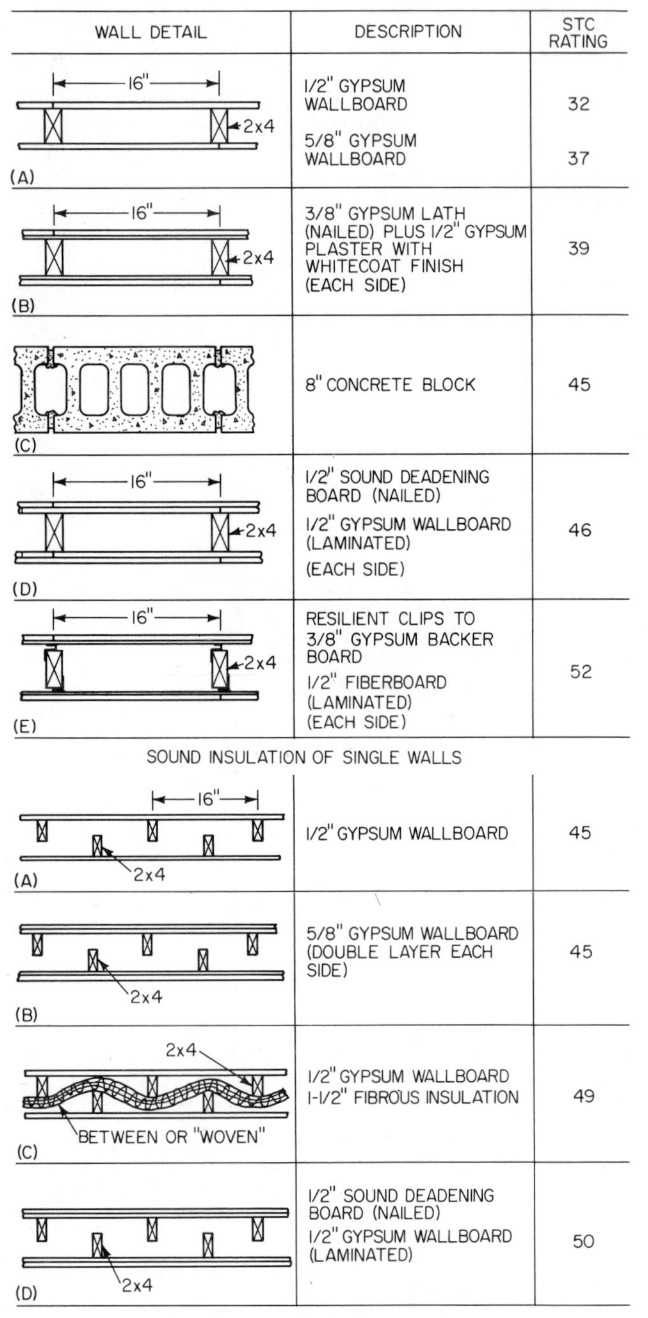

WALL DETAIL	DESCRIPTION	STC RATING
(A) 16" 2x4	1/2" GYPSUM WALLBOARD 5/8" GYPSUM WALLBOARD	32 37
(B) 16" 2x4	3/8" GYPSUM LATH (NAILED) PLUS 1/2" GYPSUM PLASTER WITH WHITECOAT FINISH (EACH SIDE)	39
(C)	8" CONCRETE BLOCK	45
(D) 16" 2x4	1/2" SOUND DEADENING BOARD (NAILED) 1/2" GYPSUM WALLBOARD (LAMINATED) (EACH SIDE)	46
(E) 16" 2x4	RESILIENT CLIPS TO 3/8" GYPSUM BACKER BOARD 1/2" FIBERBOARD (LAMINATED) (EACH SIDE)	52

SOUND INSULATION OF SINGLE WALLS

WALL DETAIL	DESCRIPTION	STC RATING
(A) 16" 2x4	1/2" GYPSUM WALLBOARD	45
(B) 2x4	5/8" GYPSUM WALLBOARD (DOUBLE LAYER EACH SIDE)	45
(C) 2x4 BETWEEN OR "WOVEN"	1/2" GYPSUM WALLBOARD 1-1/2" FIBROUS INSULATION	49
(D) 2x4	1/2" SOUND DEADENING BOARD (NAILED) 1/2" GYPSUM WALLBOARD (LAMINATED)	50

SOUND INSULATION OF DOUBLE WALLS

Fig. 5. STC ratings show how effectively these single and double wall constructions prevent the transmission of sounds *(U.S. Forest Service)*.

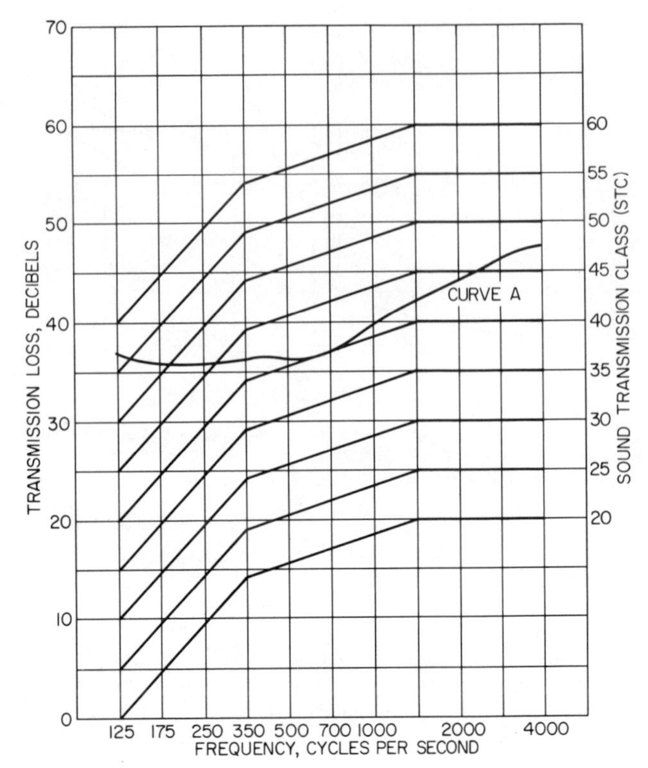

Fig. 6. Determining the STC rating of a particular wall construction. In this case the construction has an STC rating of 30 dB.

at which the construction tests most badly, it is assigned an STC rating of 30 dB.

Doors, Windows, and Cracks

For the sake of simplifying the discussion of how sounds travel from one room to another, we have assumed that all partitions are solid and without openings. This is obviously artificial, since all rooms have doors and most rooms have windows as well. The presence of doors and windows can alter radically the efficiency with which a partition acts as a sound barrier.

And so can cracks. A thin crack 2 or 3 ft long in an otherwise solid brick wall can reduce the transmission loss of the wall by 50 percent. A long crack of this kind (the bottom of every door has one) transmits sounds of all frequencies: a small hole—a keyhole, say—transmits mainly high-frequency sounds. A very small opening can let in a remarkable amount of sound. The author remembers hearing of a New York brownstone that was converted into apartments. When the house was built, ¾-in.-diam. brass speaking tubes had been installed in the walls. They led from the upstairs bedrooms to the kitchen, which was located on the ground floor. These tubes were left in place when the building was converted, which gave the tenant occupying the new apartment in the former kitchen an unparalleled opportunity to listen in on the conversations of the other tenants in the building. And all these sounds came through ¾-in. holes located in the wall.

An ordinary hollow-core door with side panels ⅛-in. thick, which is the cheapest and most widely installed door in present-day dwellings (*see* DOOR, WOOD), has a transmission loss of about 15 dB, a figure that includes the cracks that usually surround the sides and bottom of the door. As we have noted, one can hear a conversation through a partition rated at 25 to 30 dB, so having a poorly mounted hollow-core door connecting two rooms is little better than having no door at all. Installing a rubber seal around the edges of the door will increase the transmission loss to about 20 dB, which will still allow an ordinary conversation to be overheard. Even a solidly constructed, well-hung door with a rubber seal properly installed around the frame will have a maximum transmission loss of 30 dB.

The larger a wall, the less effect a door has on the overall transmission loss. When a door (or window) occupies 10 percent or more of a wall area, its STC rating will control the STC rating of the entire wall. This being the case, it is obviously a waste of money to build a partition with a transmission loss greater than of a door in that wall.

The transmission loss of a doorway can be increased considerably by installing double doors. A double door is usually installed, for example, between a doctor's waiting room and consulting room. One of the doors will be used as an ordinary door with the second door usually being swung back against the wall where it is out of the way. Both doors are closed at the same time only when the conversation between doctor and patient is confidential.

When double doors are installed, the air space between the doors should be as wide as possible, and at least 4 in. wide in any case. The heavier and more solid the door construction the greater its transmission loss, of course, and installing rubber seals around the edges of both doors will increase the transmission loss even more. A double-door installation of this type can have an overall transmission loss of 45 dB and more.

What to do about the windows, if anything, depends on the amount of outside noise, the total area of the windows in the wall, and the amount of sound isolation desired. As with doors, there is no point in constructing a wall having a high transmission loss if sounds are going to come through the windows anyway. Under laboratory conditions, an ordinary ⅛-in.-thick pane of glass has a transmission loss of about 25 to 30 dB. But when this glass is installed in a wood or metal sash that can be opened and closed, the transmission loss drops considerably.

A simple and inexpensive method of increasing the transmission loss of a window installation is to permanently install storm sash. That is, the storm-sash frame should be caulked to the window frame to prevent any sounds leaking past the edges of the sash. A more expensive method of increasing the transmission loss is to install a double window having an air space at least 4 in. wide between the panes. The transmission loss of such a construction in a brick wall will be between 42 and 47 dB.

When windows are permanently sealed, however, another way of ventilating the room must be found. It then becomes necessary to install a forced-air ventilating system or an air-conditioning system of some kind. Whatever method of ventilation is used, the outside air should never be obtained from an inlet installed in the same wall as the windows. This would negate sealing the windows in the first place.

Impact Sounds

The sounds we have discussed so far travel through the air, and the soundproofing techniques we have discussed have dealt with methods of preventing the transmission of these airborne sounds through solid barriers.

There is, however, another class of sounds that is perhaps more irritating and insidious than airborne sound. These are the *impact sounds* that travel through the solid structure of a house, and especially through the floor-ceiling construction. Impact sounds include, for example, the sound of footsteps transmitted directly through a floor-ceiling to the room below, the sounds transmitted by a piano directly to the floor via its legs, water hammering transmitted throughout a house by the water-supply pipes, and fan and compressor noises transmitted directly from an air-conditioning unit to the window ledge on which it is resting, and from there to the rest of the building structure.

Airborne sounds and impact sounds cannot be compared directly, and the STC ratings we have described are of no use in measuring the sound levels of impact noises transmitted through a construction, or the effect of these impact noises upon the occupants of a dwelling. Figure 7, for example, shows several floor-ceiling constructions. Most of these constructions have enough mass and rigidity to suppress effectively any airborne sounds impinging on them, and this is reflected in their high STC ratings; it is rare to hear conversations or other airborne sounds in a room directly above or below the room in which these sounds originate. Figure 7 also shows the *impact noise ratings* (INR) of these constructions. The two methods of measurement have nothing to do with each other.

The assignment of an INR to a floor-ceiling construction is

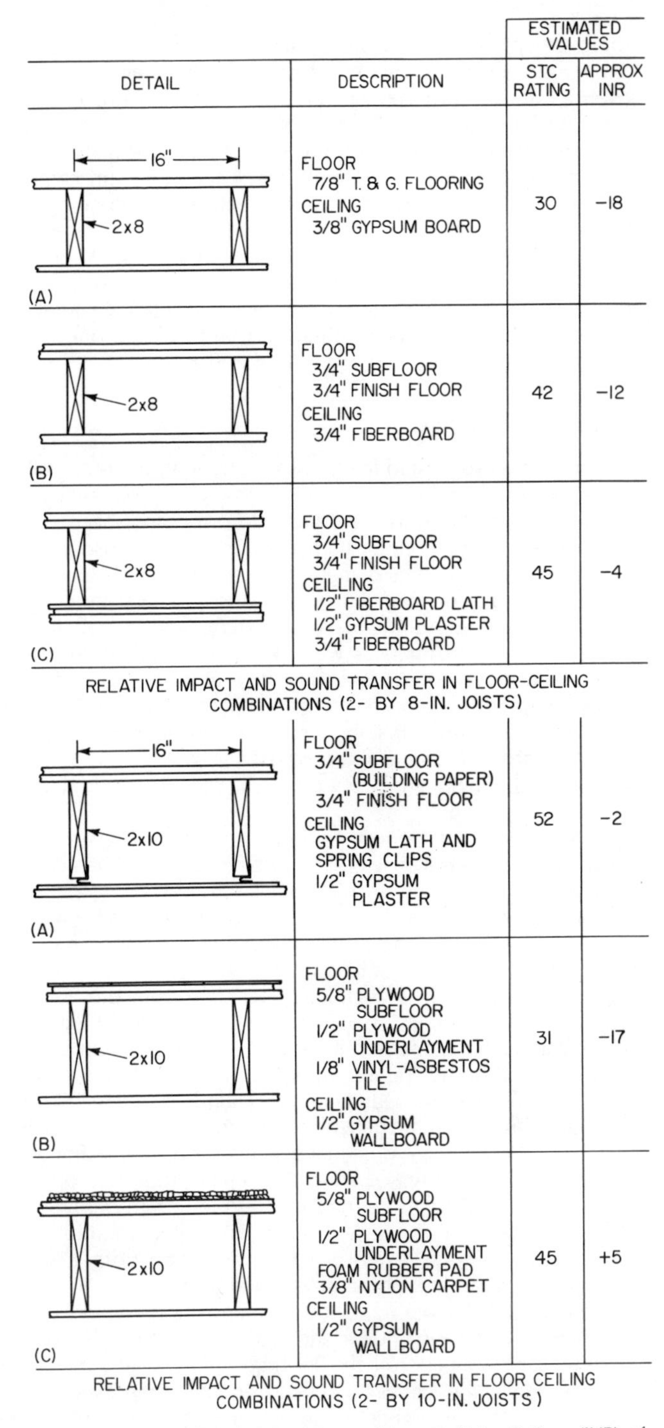

DETAIL	DESCRIPTION	ESTIMATED VALUES STC RATING	APPROX INR
(A) 16" 2x8	FLOOR 7/8" T. & G. FLOORING; CEILING 3/8" GYPSUM BOARD	30	−18
(B) 2x8	FLOOR 3/4" SUBFLOOR 3/4" FINISH FLOOR; CEILING 3/4" FIBERBOARD	42	−12
(C) 2x8	FLOOR 3/4" SUBFLOOR 3/4" FINISH FLOOR; CEILLING 1/2" FIBERBOARD LATH 1/2" GYPSUM PLASTER 3/4" FIBERBOARD	45	−4

RELATIVE IMPACT AND SOUND TRANSFER IN FLOOR-CEILING COMBINATIONS (2- BY 8-IN. JOISTS)

DETAIL	DESCRIPTION	STC RATING	APPROX INR
(A) 16" 2x10	FLOOR 3/4" SUBFLOOR (BUILDING PAPER) 3/4" FINISH FLOOR; CEILING GYPSUM LATH AND SPRING CLIPS 1/2" GYPSUM PLASTER	52	−2
(B) 2x10	FLOOR 5/8" PLYWOOD SUBFLOOR 1/2" PLYWOOD UNDERLAYMENT 1/8" VINYL-ASBESTOS TILE; CEILING 1/2" GYPSUM WALLBOARD	31	−17
(C) 2x10	FLOOR 5/8" PLYWOOD SUBFLOOR 1/2" PLYWOOD UNDERLAYMENT FOAM RUBBER PAD 3/8" NYLON CARPET; CEILING 1/2" GYPSUM WALLBOARD	45	+5

RELATIVE IMPACT AND SOUND TRANSFER IN FLOOR CEILING COMBINATIONS (2- BY 10-IN. JOISTS)

Fig. 7. A comparison of STC ratings and Impact Noise Ratings (INR) of different ceiling/floor constructions *(U.S. Forest Service)*.

an attempt to measure how effectively the construction suppresses impinging impact noises. The ratings are based on the ability of a standard construction to suppress various types of standardized impact sounds. It is assumed that this standard construction will suppress these impact noises by a reasonable amount. This standard construction has been given an INR of zero. The impact noise resistance of other floor-ceiling constructions are now compared to this standard construction and their performances are rated accordingly. If, for example, a particular floor-ceiling construction suppresses impact noises more effectively than the standard construction, it is given a plus (+) rating, the value of which depends on the amount of improvement. If, on the other hand, the construction suppresses impact noises less effectively than the standard, it is given a minus (−) rating, the value of which depends on the amount of deterioration. Typical INRs are shown in Fig. 7.

In an existing house, the only thing that can be done to reduce impact sounds is to install heavy carpeting and carpet pads on the floors. Carpeting is very effective at dampening or muffling the sounds of footsteps and similar impact noises. It will, in fact, improve the INR by +20 to +30 points. (As far as the STC rating of a floor-ceiling construction is concerned, carpeting will have no effect whatsoever on the STC rating.) Pianos and hi-fi equipment should rest on flexible mountings of some kind to reduce the amount of impact noise.

Suspended Ceilings

It is sometimes suggested that a suspended acoustical-tile ceiling of the type found in offices and commercial establishments will absorb impact sounds transmitted from a floor above. In this type of construction, a light metal framework is suspended from the ceiling by wires or metal rods and acoustical tiles are then set in place on this framework. This type of ceiling will absorb impinging sounds to some extent, and it will hide exposed pipes and ducts as well, but it will have almost no effect whatsoever on suppressing any airborne sounds or impact sounds that may be transmitted through the ceiling from the floor above, mainly because of the lightness of the overall construction and also because the ceiling, since it is suspended by rigid wires or rods, will merely transmit the sounds unchanged.

Another type of suspended ceiling often installed to suppress unwanted sounds is a false ceiling suspended below the actual ceiling by wood furring strips. This false ceiling is constructed by analogy with a false wall. This construction suffers from the same basic defect as any light construction—it simply hasn't the mass to absorb the impinging sound vibrations. And whereas a false wall is completely separate from the actual wall, a false ceiling must of necessity be supported by the actual ceiling. Such a false ceiling does have some slight effect in reducing airborne sounds when it is suspended by flexible mounts, but it will not reduce impact sounds.

A more fundamental difficulty with suspended ceilings is that, even when they do reduce sound levels, the impact sounds that are suppressed are only a small proportion of the sounds that continue to be transmitted to the room below via the wall studs and floor joists. It is very difficult, if not impossible, to achieve any degree of sound reduction from a suspended ceiling unless an effort is made at the same time to prevent the passage of impact sounds along the solid pathways afforded by the flooring, joists, and studs. In particular, the wall studs must be isolated from the flooring, and the wall construction must also be heavy and solid enough to prevent the inducement of sympathetic sound vibrations. This is very difficult to achieve in a 2 × 4 in. stud wall.

SOUND INSULATION

We come at last to the subject of *sound absorption,* that aspect of acoustics most people think of when they hear the words "sound control," "soundproofing," or "sound conditioning."

By sound absorption, we mean the ability of the wall and ceiling surfaces of a room to absorb impinging sound waves and prevent their being *reflected back into the room.* This sound absorption is usually achieved by the application of acoustical materials of some kind to the walls and ceiling. These materials are capable of altering the general quality of sound within the room, as well as reducing the overall sound level.

Sound Propagation in a Room

Imagine an ordinary room. Within this room, a short, sharp sound of some kind originates. As the sound waves spread away from the source, they strike the walls, ceiling, and floor and are reflected from these surfaces. If you are sitting in this room and have an extraordinarily acute sense of hearing, you will hear first the original sound and then an intensification of that sound as all these reflections strike your ears. After the original sound has died away, the reflected sounds will continue to bounce about for a brief period; then they, too, will fade away.

These reflected sounds produce what is called *reverberation*—a multiplicity of very weak echoes that impinge too rapidly upon the ear to be distinguished individually but which, in sum, give the original sound a rich, warm quality that it would otherwise lack. One has only to contrast the sound of a voice heard indoors with the same voice heard outdoors where there is no reverberant effect. Outdoors, the voice sounds much smaller and thinner.

The reverberation, however, cannot last too long. If it does, the individual sounds will bounce back and forth an excessive number of times and interfere with each other. The sound acquires an echo-like, booming quality that is unnatural and may make speech unintelligible.

What one does when one applies an acoustical material to the walls and ceiling of a room is alter the reverberant characteristics of that room; and, also, reduce the overall sound levels a slight amount. We shall not concern ourselves with the manipulation of this reverberation. This is a very complex subject and has more to do with maximizing the quality of the music heard at home and in the concert hall than with general noise and sound levels in the home.

Most people who consider installing an acoustical material in their home are usually influenced by their experience with these materials in an office or commercial establishment. In the typical large office, for example, there is the constant clatter of typewriters. Copiers and other kinds of office equipment make their noises. Telephones ring continually. There is the constant buzz of conversation. Yet all these sounds seem to have a subdued quality to them that the observer infers, correctly, is due to the use of an acoustical material on the ceiling. And, indeed, if it weren't for this material, the apparent noise level in most offices would be much higher than it is, and the offices would be much more enervating places to work than they are.

But the analogy between noises in the home and in the office is inexact. In the first place, we can control or limit the noise level within our home to suit our convenience. If a television set or radio is too loud, we can easily turn the volume down to a more comfortable level. If the children are noisy, we can send them outdoors to play or tell them to be quieter. If we are operating an appliance or a piece of electrical equipment, the noise level is rarely loud enough to be irritating. Besides, as we have noted, the noises we make ourselves are rarely objectionable to us, whatever others may think of them.

In the second place, and more important, the noises that *are* objectionable in a home usually originate *outside* the home, or *outside* the room in which we happen to be working or resting. And as we have emphasized several times, acoustical materials are almost useless in preventing the transmission of these outside noises into the room—they simply lack the weight, the solidity, and the mass that are required to dampen or suppress these sound vibrations.

What acoustical materials *can* do is absorb some of the sounds that originate within the room itself. The amount of sound absorbed is limited. Innumerable tests have shown that the most that acoustical materials can achieve in the way of reducing sound levels is 5 to 10 dB, with 5 dB being the usual maximum. If the typical noisy office has a sound level of 50 dB, say, then installing a sound-absorbent material will reduce the apparent sound level to 40 or 45 dB. In addition, the material will also reduce the reverberant effect produced by the sounds bouncing off the walls, ceiling, and furnishings. It is this loss of reverberation that makes the typical office sound much less noisy than a sound-level meter might indicate.

But in a home it is this reverberant quality to a room that gives warmth and color to the human voice. When a sound-absorbent material is used to excess in a room, the voice acquires a flat, thin quality that is unnatural. There is a curiously distant quality to a conversation, the voices seem to come from a distance, and the sound patterns of the words may fall off so quickly that comprehension becomes difficult. It is, for example, the silent, mortuary-like atmosphere that one sometimes encounters in the executive offices of a large corporation, where the ceilings are covered with acoustical tile, the windows with heavy drapes, and the floors with thick carpeting. Silence is golden.

Brick

The emphasis in this article is on brick made from burned clay or shale. For the technique by which a mason or bricklayer lays brick, *see* BRICKLAYING; for a description of the overall construction of one- and two-family brick dwellings, *see* BRICK-MASONRY CONSTRUCTION.

Brick made of clay have been used as a building material for about 10,000 years. Brick have many advantages. The raw material, clay, can be obtained almost everywhere. The technique of converting the clay into small, easily handled brick units is straightforward. Brick are manufactured in a large number of standard sizes (see Fig. 1), and in hundreds of shades and finishes. Joined together with mortar, an art that is easily acquired with practice, brick are used to build walls, floors, foundations, facades, and arches of any size and shape, and in highly decorative patterns as well. And, once built, these structures require almost no maintenance and can last for hundreds of years.

The principal structural defect of brick is its porosity. Brick must be porous, otherwise the units could not be mortared together. But if a brick should be excessively porous, and if it should become soaked with water during cold weather, and if this water should freeze, it will expand within the brick. The expansion will force apart the particles of which the brick are made. Over a period of time, if enough freezing-thawing cycles occur, the brick will gradually disintegrate. Large flakes may also split from the surface of the brick because of frost action, a process called *spalling*. Nevertheless, it is easy enough to manufacture brick that will withstand this weathering effect or, if disintegration begins, to take steps that will prevent any further damage.

BRICKMAKING

Brick are made by a straightforward process in which the original clay or shale is *tempered;* that is, mixed with water, formed into units of the desired size and shape, and fired in a kiln to drive out the water and fuse the particles of clay together.

Not every clay makes a satisfactory brick. If there is too much water in the clay, more than 20 to 30 percent, the molded brick will shrink excessively as it dries out. This shrinkage is undesirable in itself and also because it may cause the brick to crack and warp. A sandy clay shrinks less but the presence of sand reduces the cohesiveness of the clay particles, making a brick that is weak and crumbly. Sandy clays, therefore, are undesirable.

Too much lime is also undesirable in a clay, first because the lime will attract an excessive amount of water, with all the consequences described above, and, second, because a brick made from a limey clay will attract an excessive amount of water after it has been fired. If, in addition, the lime has not been adequately pulverized and small particles remain, some of these particles will absorb enough water that *lime pops* will occur because the pressure exerted by the absorbed water will force chips to spall off the surface of the brick. If one looks closely at one of these pits, one can see at the bottom of the pit the white dot of lime that was responsible for the pop.

For the sake of a consistent product, therefore, manufacturers today take great pains to prepare the clay properly. It is first broken up, stones and other foreign substances are removed, and the clay is pulverized into a fine powder by heavy rollers. Mineral oxides may be added to give the brick a desired color, or batches of clay from different deposits may be mixed together to give the brick certain desired strength characteristics.

The pulverized clay is then tempered with a suitable amount of water and the plastic mass is then *pugged* in a pugg mill, which resembles a concrete mixer and in which the clay and water are mixed to an even consistency. The clay is then ready to be formed into units.

Soft-Mud Process

When brick were made by hand, the clay had to have from 20 to 30 percent water, by volume, if the clay were to have the plasticity required for hand-molding. In the *soft-mud process* of making brick, traditional methods are followed but the process has been mechanized. Today, the clay, untouched by human hand, is forced into molds by a plunger and formed under pressure. To prevent the brick units from sticking to the sides of the molds, the molds are either *sand-struck* or *water-struck.* In sand-striking, the insides of the molds are coated with a thin layer of sand; in water-striking, they are dipped in or sprayed with water. Sand-striking is the more common method in use as brick produced by it are cleaner looking and sharper-edged.

Stiff-Mud Process

The soft-mud process has been largely superseded by the *stiff-mud process,* which is the method by which most brick are made today. In the stiff-mud process, the clay has only enough water—12 to 15 percent, by volume, to give it the necessary cohesiveness. In the brickmaking machine, the stiff clay is forced through a die by an auger, like toothpaste being squeezed from a tube. The length of clay is cut by wires into individual units of the desired length or width as it is extruded from the machine.

Dry-Press Process

A third method of making brick, the *dry-press process,* is used to make very dense, well-formed brick having very high compressive strength. These dry-pressed brick are widely used as *face brick,* which are brick meant to be exposed on the facade of a building.

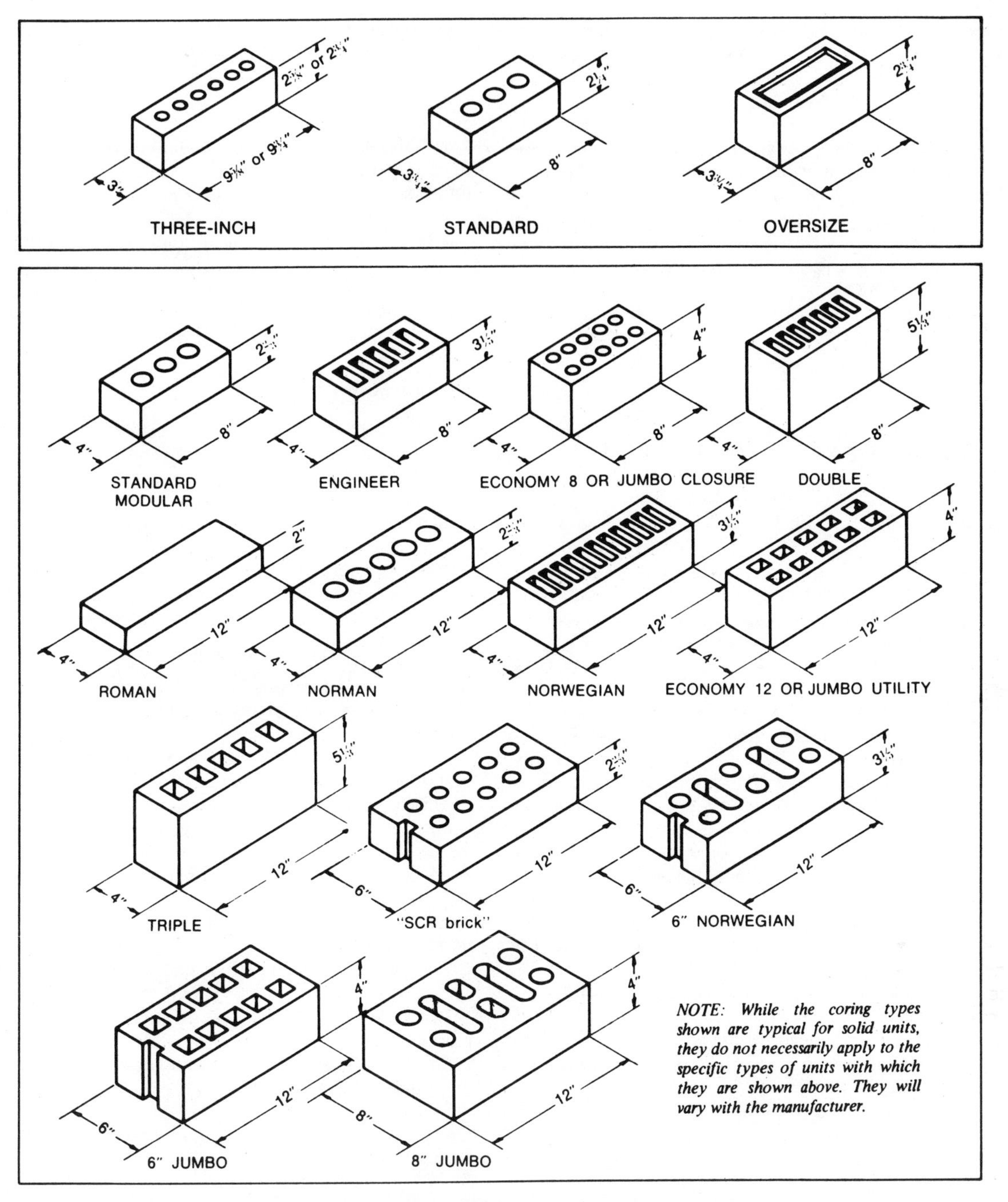

Fig. 1. Brick are made in a variety of shapes and sizes, many of which are known by traditional names *(Brick Institute of America).*

In the dry-press process, the clay contains only 7 to 10 percent water, by volume. The brick units are formed by plungers that push the clay into metal molds. The sides of the molds then come together to compress the clay under high pressure. The pressure is released for a moment, then reapplied with a force of from 500 to 1500 lb per sq in. The plungers then push the completed brick units out of the molds.

Brick Drying

Most of the water in a freshly molded (or *green*) brick unit fills the pore spaces between the clay particles. This water must be removed before the unit is fired, otherwise it will shrink an excessive amount during firing, which will cause it to warp and perhaps crack as well. In the old days, brick were dried in open sheds, a procedure that took anywhere from 1 to 6 weeks to complete, depending on the original water content of the brick. Nowadays, they are dried in special chambers, using the waste heat from the kiln to drive out the water. This drying out usually takes only 8 to 12 hr.

Today, most brick are fired in *tunnel kilns* in which the green brick are loaded on carts that are then drawn slowly from one end of the kiln to the other, the temperature being regulated in the different parts of the kiln to achieve the desired firing. It takes from 40 hr to 6 days for the firing to be completed, depending on the characteristics of the clay, plus an additional 2 to 3 days for cooling.

There are several stages in the actual firing process. As the temperature of the kiln increases from ambient temperature at the beginning of the firing to a temperature of 250 to 350°F, which takes about 12 hr, any remaining free water in the bricks is driven off. This stage is called *water smoking.* As the temperature continues to increase to 800 to 1400°F, the water that is chemically bound to the clay particles is driven off. Somewhere between 1000 and 1800°F, any organic substances in the clay are oxidized and the oxygen, carbon, and sulfur in the clay are consumed. Finally, at some temperature between 1600 and 2400°F, depending on the characteristics of the clay, vitrification begins. That is, the particles of clay become soft enough to stick together. In most brick, complete vitrification is not allowed to take place in order that the brick will retain their porosity. Generally speaking, the higher the final kiln temperature, the greater the shrinkage and the darker the brick become.

BRICK TYPES, SIZES, AND GRADES

Building Brick

Building brick, which are also known as *common brick* and *backing brick,* are used for the more humdrum types of construction, such as backing for an exterior course of *facing brick,* and to construct foundations, retaining walls, and piers. What is important in building brick is not their appearance but their strength and resistance to moisture.

Brick Grades

Building brick are made in three grades, which differ according to the intended use.

Grade SW brick are intended for underground service, where the units may become saturated with water and then be subjected to freezing temperatures. The brick are also intended for outside use above ground in cold, wet climates where there is some likelihood that the units will be subjected to frost action.

Grade MW brick are also intended for outside use in cold, wet climates, but only above ground where it is unlikely the units will be saturated with water but where there is still a chance they may be subjected to frost action.

Grade NW brick are for use everywhere else in a building, that is, in either an interior or exterior wall where the units will back up a course of face brick and there is very little chance they will ever be subjected to frost action, no matter how cold and wet the climate. Grade NW brick can also be used on the outside of a building in a cold climate, as long as the annual precipitation does not exceed 20 in.

Brick Sizes

Building brick are made in the sizes shown in Fig. 1. Note that the sizes of the brick may be nominal or actual. Nominal sizes are used by a bricklayer when he is laying out a job and calculating the quantity of brick he will need. The nominal size is usually ½ in. larger in every dimension than the actual size, to take into account the thickness of the mortar joint between the bricks.

Facing Brick

Facing brick are intended for use wherever appearance is important, either inside or outside a house. Facing brick, therefore, are made in a variety of surface textures and in a wide range of colors and sizes. They may also be glazed to enhance their appearance.

Brick Grades

Facing brick are made in grades SW and MW for use under the same conditions described above for common brick. In addition, three *types* of facing brick are made. The types differ according to the intended purpose of the construction.

Brick Types

Type FBX brick are made to high standards of size, color match, and accuracy of surface finish. The units have a minimum of mechanical imperfections.

Type FBS brick have a greater variation in finished size, color match, and surface finish than Type FBX brick.

Type FBA brick are not selected at all for uniformity of color, size, or surface finish. In fact, an architect or builder will sometimes specify Type FBA facing brick because he *wants* a variety of shades and surface textures in the brickwork.

Surface Finish

The most common surface finishes available in facing brick are shown in Fig. 2. A smooth-surfaced brick is a natural consequence of the brick having been formed in a mold or die having smooth sides. A scored or combed appearance is obtained by using a die having the appropriate projections cut into its sides. The pattern is transfered to the clay as it is forced through the die. A roughened texture is obtained by pressing a revolving wire brush against the sides of the clay as it emerges from the die.

Glazed Brick

A *glaze* is an impervious, fire-resistant, permanent, and glassy finish on the exterior of the brick. Usually only one side of the brick is glazed.

Ceramic-glazed brick are made by brushing or spraying a *slip* on the brick after they have been dried but before they are fired. The slip consists of a mixture of minerals—ball clay, kaolin, flint, and feldspar. The glaze is applied over this slip. As the brick are fired in the kiln, the slip provides the color and the glaze the protective glassy coating. By selecting the proper combination of minerals for the slip, a wide range of bright colors can be obtained. There are, in fact, about four dozen standard ceramic glaze colors from which one can choose.

Ceramic-glazed facing brick are available in two grades and two sizes.

Grade S (Select) brick are intended for use with narrow mortar joints.

Grade SS (Select Sized) brick are intended for use where the dimensions of the individual units must be closely maintained.

Type I brick are glazed on one face only.

Type II brick are glazed on two opposite faces. Type II brick are used where both faces will be exposed.

Secondhand Brick

An architect or builder will sometimes use secondhand brick on the exterior or interior of a house. The brick have a weathered, antique appearance that many people find attractive. Secondhand brick should be used with caution, however. They should never be used structurally, that is, to support a load, because it is impossible to tell by their appearance how strong or durable they are. If the brick have been part of a chimney, they may be impregnated with oily or tarry substances. The danger here is that the pores of the brick may be filled with these substances and the mortar will not be able to grip them. Any wall of which these units are a part might easily collapse for this reason.

Nor is it particularly cheap to use secondhand brick. One might, for example, obtain a free truckload from a demolition site just for the cost of hauling them away, but a bricklayer will have considerable trouble chipping off the old mortar and setting the brick properly. In the end, the cost of laying the brick will probably be the same as if new brick had been used in the first place.

Fig. 2. A few of the most common brick textures *(Brick Institute of America).*

BRICK STRENGTH

Brick are supposed to take crushing loads, which they do very well indeed. In a brick wall, it is always the mortar joints that are the weak points in the construction, never the brick themselves. The tensile strength of a brick unit may be only 5 to 10 percent of its compressive strength, but since brick are never placed in a position where they will ever have to withstand a tensile load, the tensile strength is relatively unimportant.

Tables 1A and 1B show the relationship that exists between the standard grades of brick and their minimum compressive strengths. The compressive strength of a brick depends primarily on the degree of vitrification it has undergone during its firing. The higher the kiln temperature, the greater the vitrification; the greater the vitrification, the more the clay particles will have fused together and the stronger and more durable the brick will be. It is the number of these fused interconnections that basically determines a brick's strength and internal homogeneity. The characteristics of the clay used has only a minor influence on the strength of brick.

There is a limit to the amount of vitrification a brick should undergo because, as we have already noted, a brick must be porous if the mortar is to get a good grip on it. But porosity may also result in frost damage, although no one really understands how this frost damage occurs. All that is known for certain is that, in a general way, the greater a brick's porosity, the greater the likelihood that it will be damaged by frost, if conditions are suitable.

Table 1A. ASTM Physical Requirements for Facing Brick and Solid Masonry Units

Designation	Minimum compressive strength (brick flatwise), gross area, psi (MPa)		Maximum water absorption by 5-hr boiling, percent		Maximum saturation coefficient*	
	Average of 5 brick	Individual	Average of 5 brick	Individual	Average of 5 brick	Individual
Grade SW	3000 (20.7)	2500 (17.2)	17.0	20.0	0.78	0.80
Grade MW	2500 (17.2)	2200 (15.2)	22.0	25.0	0.88	0.90
Grade NW	1500 (10.3)	1250 (8.6)	no limit	no limit	no limit	no limit

*The saturation coefficient is the ratio of absorption by 24-hr submersion in cold water to that after 5-hr submersion in boiling water.

Table 1B. ASTM Physical Requirements for Facing Brick and Solid Masonry Units Made from Clay or Shale

Designation	Minimum compressive strength (brick flatwise), gross area, psi (MPa)		Maximum water absorption by 5-hr boiling, percent		Maximum saturation coefficient*	
	Average of 5 brick	Individual	Average of 5 brick	Individual	Average of 5 brick	Individual
Grade SW	3000 (20.7)	2500 (17.2)	17.0	20.0	0.78	0.80
Grade MW	2500 (17.2)	2200 (15.2)	22.0	25.0	0.88	0.90

*The saturation coefficient is the ratio of absorption by 24-hr submersion in cold water to that after 5-hr submersion in boiling water.

SELECTION OF BRICK

In the selection of ordinary building brick, the principal considerations are grade and delivered price. One should select brick according to their intended structural use, but even within each grade there are a number of things one must watch for. The brick units should be uniform in shape, size, and color. There should not be any broken corners, cracks, pebbles, or small stones visible on the surface, nor should the edges have a twist to them. The surfaces should not be *too* smooth, because this may lead to weak mortar joints; nor should the bricks have a glossy look, as this might indicate excessive vitrification. Sound brick will ring metallically when two units are struck together. Although some variability in color must be accepted, the range of colors should not be too great, since variability in color is an indication of variability in strength. In general, assuming that a group of brick have been made from the same clay, the darker the brick, the stronger and more durable they will be.

As for the selection of facing brick, everything that has been said about building brick applies also to facing brick, with the additional complication that one is now also faced with questions of taste and style. Trying to select a facing brick by examining a sample in an office or salesroom is like trying to decide on a suit material by looking at a 4-in.-square swatch under fluorescent lighting. Some people can do it, some people can't. The color and texture of a brick will vary greatly according to the time of day and the angle of the sun. If it is at all possible, the wisest course is to locate a house that has been built of the same type of brick that you are considering and examine the appearance of the walls before making a final decision.

Bricklaying

This article describes the techniques by which brick units are laid and, by extension, how structural-clay-tile and concrete-blocks units are laid as well, although the latter two types of masonry are not mentioned in this article at all. The existence of mortar, suitable for laying brick is assumed; for a description of the properities of mortar *see* MORTAR. For a description of the various sizes and kinds of brick that are made, *see* BRICK. For a description of the different types of brick structures and the structural properties of brick walls, *see* BRICK-MASONRY CONSTRUCTION.

The first thing that should be understood about bricklaying is that it comes in two standards—ordinary and first class. An ordinary standard of bricklaying is used to build most dwellings, apartment buildings, and commercial structures; that is, wherever price is the important consideration. The bricklayer has won the contract on the basis of a low bid, and he can maximize his profit only by laying the brick as rapidly as possible. Quality of construction, as long as it conforms with the requirements of the local building code, is definitely a secondary consideration.

One finds first-class workmanship, on the other hand, in expensive dwellings, bank buildings, and such "monumental" structures as United States post offices, courthouses, and other government buildings—jobs for which the bricklayer had bid according to detailed written specifications that he must follow exactly. Here, quality comes first. It is interesting that *ordinary* and *first-class* standards are sometimes called *ordinary* and *inspected* standards, or *commercial* or *uninspected* (as against inspected), the implication being that you won't get first-class workmanship unless the bricklayer is closely supervised. And, in fact, in tests conducted by the Brick Institute of America on the strength of brick walls, it was found that walls built to the highest standards of supervised workmanship were about one-third stronger than walls built to ordinary, unsupervised standards of workmanship.

The difficulty for most homeowners is, of course, that first-class workmanship, even when unsupervised, is so much more expensive than ordinary workmanship. It takes time for a conscientious bricklayer to proportion the mortar correctly, mix it thoroughly, spread it carefully, and to set each brick in place just so. It takes time to build a wall that is exactly square, plumb, and level, and to make sure the mortar joints are watertight. As everyone knows, time is money—to a conscientious, diligent bricklayer as well as to a careless, money-hungry one.

This article, then, describes how brick *should* be laid if it is to meet the highest standards of workmanship. Knowing what these standards are, the reader will be in a position to judge the quality of any brickwork he examines, and a do-it-yourselfer who is contemplating building a brick structure by himself will have some idea of what is involved in achieving a first-class standard of workmanship.

GENERAL CONSIDERATIONS

The strength and durability of brickwork depends on three factors: (1) the inherent strength of the brick used, (2) the inherent strength and weather resistance of the mortar used, and (3) the quality of workmanship. We will assume that the strength of both the brick and mortar are adequate. (*See* BRICK; MORTAR.) The third factor, the quality of the workmanship, depends on two things: (1) the kind of *bond*, or pattern, in which the brick are laid, and (2) the way in which the joints between the brick units are made.

Brick Bond

Figure 1 illustrates the types of bond most often used in house construction. A bricklayer can, of course, lay brick in any of a number of very complicated and decorative bonds—it is up to the builder to specify what he wants. Although *bond* is usually synonymous with *pattern*, it can also mean the strength with which the brick grip the mortar. All the bonds illustrated,

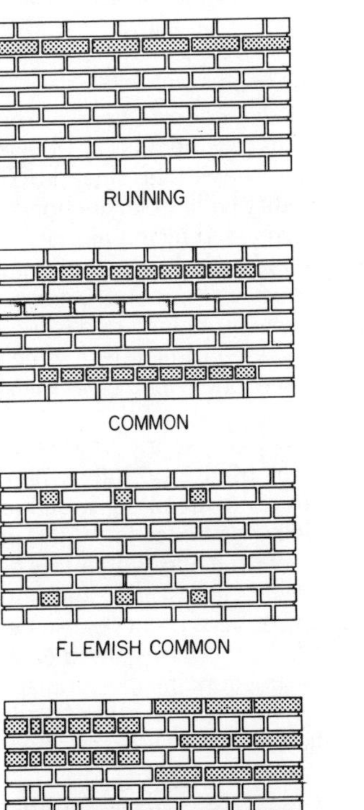

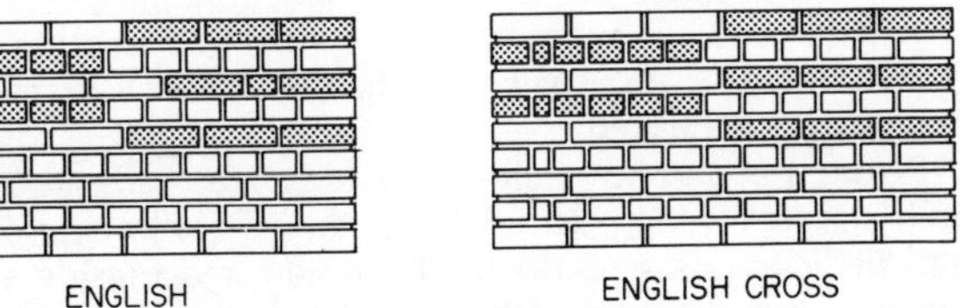

Fig. 1. Types of brick masonry bonds *(U.S. Army).*

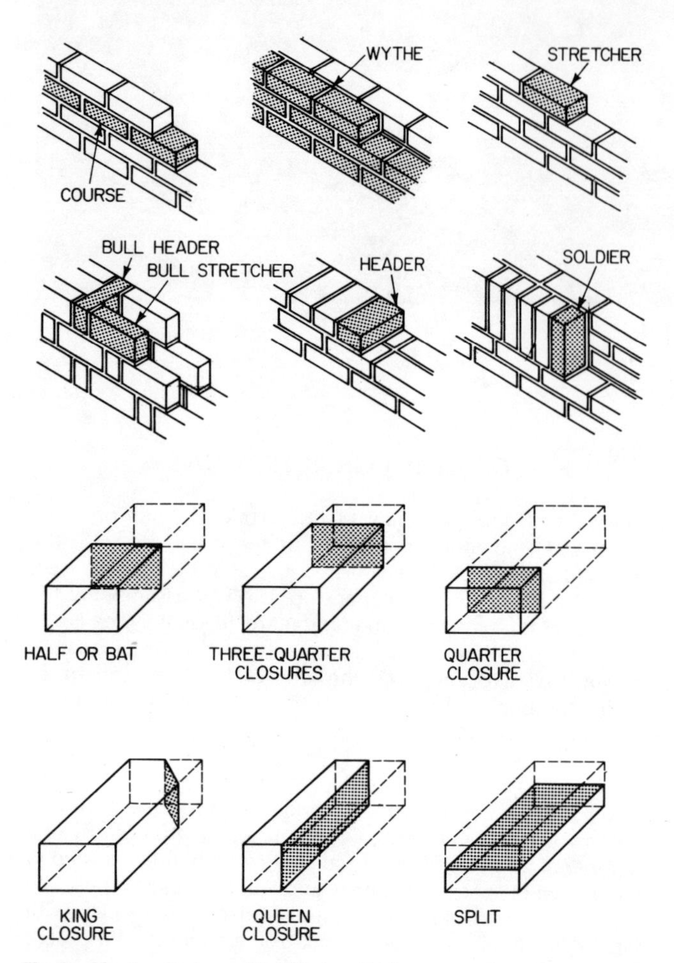

Fig. 2. The terminology of bricklaying. Brick are named according to the way in which they are laid in place in a wall. Individually cut brick also have distinctive names *(U.S. Army).*

except for the *stack bond,* are interlocked to some degree. This interlocking helps to tie the brick units together into a stronger structure. This is obviously not the case with the stack bond, the units of which must be tied together with metal anchors of some kind, in addition to the bond provided by the mortar.

Of the bonds illustrated, the *running bond* is the simplest. It consists of all stretchers and no headers (see Fig. 2 for an explanation of bricklaying terminology). Running bond is used mainly in brick-veneer and cavity-wall construction in which metal ties are used to strengthen the brickwork (*see* BRICK-MASONRY CONSTRUCTION).

Common or *American bond* is the most widely used bond of all. It is the cheapest to lay and is used especially where appearance is of no importance—in commercial construction or to back up decorative face brick. As shown in Fig. 1, it consists of several rows, or *courses,* of stretchers with a row of headers set every fifth, sixth, or seventh course. The headers tie together the inner and outer *wythes,* or tiers, of an 8-in.-wide or 12-in.-wide wall. The closer the header courses are to each other, the stronger the overall construction; header courses spaced six courses apart are most common. The other bonds shown in Fig. 1 are usually specified where a more attractive appearance is desired.

Figure 3 shows how full-size headers are used to construct an 8-in.-thick wall in Flemish bond; this is how brick should be laid in first-class work to tie the inside and outside wythes together. In ordinary work (and in brick-veneer and cavity-wall construction as well), full-size headers are not often used. *Blind, false,* or *snap headers,* as they are called (all referring to the fact that half-size bricks are used), are used instead in each course, full-size headers being used only every fifth or sixth course, if then, to tie the wythes together.

Mortar Joints

The mortar-filled space between two adjacent bricks is called a *joint.* The mortar has several purposes: (1) it binds the bricks together, (2) it provides a base that will support irregularly shaped bricks evenly, thus distributing the weight of the brick units above evenly upon the brick units below, and (3) it makes the construction watertight.

There are no rules that determine how thick a joint should be. It depends on the regularity of the bricks and their porosity. For glazed face brick, which are made to very close tolerances by the dry-press process (*see* BRICK), the joints need be only 1/8 in. thick. For very irregularly surfaced common brick, the joints may be as much as 3/4 or 1 in. thick. The average thickness for ordinary brickwork is, however, between 3/8 in. and 1/2 in. In general, the thinner the joint, the stronger the construction, with 1/4-in.-thick joints making the strongest walls. A thick joint will also shrink down more than a thin joint as the mortar sets, shrinkage that may cause a crack to open up between the mortar and a brick—another reason for having the joints as thin as possible.

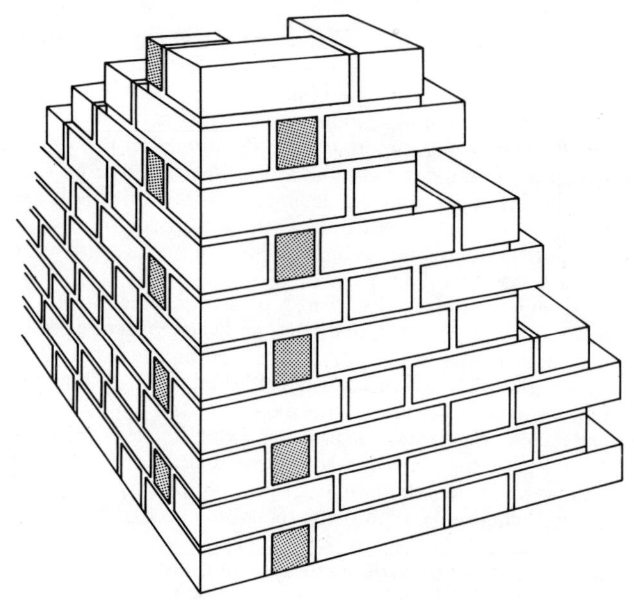

Fig. 3. A wall set in Flemish bond with what is called an English corner.

BRICKLAYING TECHNIQUE

A skilled bricklayer makes bricklaying look deceptively simple. It's like watching a skilled dressmaker or tailor rapidly stitch up a seam by hand. If you pick up the needle and thread yourself and try to emulate her or him, you quickly discover just how difficult it is to make those neat, even stitches.

The bricklayer scoops up as much mortar with his trowel as he thinks necessary from the pile on the mortarboard, usually enough to spread over four or five brick. The trick is to load up the trowel fully with mortar without having the mortar lump together in a pile. The bricklayer then spreads the mortar on the brick in one smooth motion. He slings the mortar gently from the trowel with a snap of his wrist, at the same time turning the trowel from the horizontal to an almost vertical position. The mortar will slide smoothly off the trowel and spread itself evenly on top of the brick. It is all in the wrist. The bricklayer then riffles the mortar from side to side with the flat of his trowel to spread it evenly over the brick. The point of the trowel is depressed slightly as this is done, which leaves a shallow furrow down the center of the mortar. The furrow makes it easier to set the brick units in place.

To place a brick unit, the bricklayer picks up the brick with one hand, at the same time scooping up additional mortar with his trowel. He spreads the mortar thickly on the side of the brick and pushes the brick into position against its neighbor. The excess mortar, both under and at the side of the brick, is squeezed out of the joint as the bricklayer presses the brick into position. The pressure also causes the shallow furrow in the mortar bed to disappear. The excess mortar that has been extruded from the joints is cut off with the edge of his trowel and thrown back onto the pile of mortar.

It is important that all the spaces between the brick units be completely filled with mortar if the joints are to be strong, durable, and watertight. Tests by the National Bureau of Standards have shown that solidly filled joints will result in a brick wall that is from 24 to 109 percent stronger than a wall in which gaps and air pockets have been left in the joints. The less experienced a bricklayer, the more mortar he will lay down to achieve this solid joint: he may make the joint as much as 1 in. thick, which will, for the reasons mentioned above, probably result in a leaky wall. But regardless of how thick, or thin, joints are, an experienced bricklayer will always squeeze *some* excess mortar out of the joints as an indication that they are, in fact, completely filled.

Pointing

Pointing is the process of filling joints with mortar after the bricks have been laid. The bricklayer uses a long, narrow *pointing trowel* that is only ¼ to ½ in. wide. Pointing is a confession that a poor job of bricklaying has been done, since if an adequate amount of mortar had been placed in the joints in the first place, pointing would be unnecessary. A wall may, of course, require pointing after years of exposure to the weather, but that is another story.

Buttered Joints

As has been mentioned, hard, dense brick can be set with very thin joints. Instead of slopping on a bed of mortar for the bricks to rest on, the brick unit is "buttered" with mortar that has been mixed to the consistency of butter—soft but still firm enough to hold its shape. With his trowel the bricklayer butters all four edges of the bottom of the brick, the center of the brick being left dry. The technique is similar to the way that one might clean jam off a knife by wiping the blade along the edges of a slice of bread. The end of the brick is buttered with mortar in the same way and the brick is then set in place. Buttered joints are usually 1/4 in. thick, though they may be thinner if the brick is especially hard and very accurately sized.

One problem encountered in laying hard brick is that they may provide very little grip for the mortar—the brick units tend to slide about on top of the mortar rather than remaining in place. For this reason also, too many courses of hard brick cannot be set on top of each other within a short period of time. The mortar must be given time to harden, otherwise the weight of the bricks will squeeze the mortar out of the bottom joints and the bricks themselves will tend to slide out of position, which will make for a very untidy looking wall.

Types of Finished Joints

Having placed a course, or several courses, of brick, the bricklayer must now finish off the exposed joints neatly before the mortar can set hard. This process of smoothing a joint is called *striking* the joint. The most common methods of doing this are shown in Fig. 4.

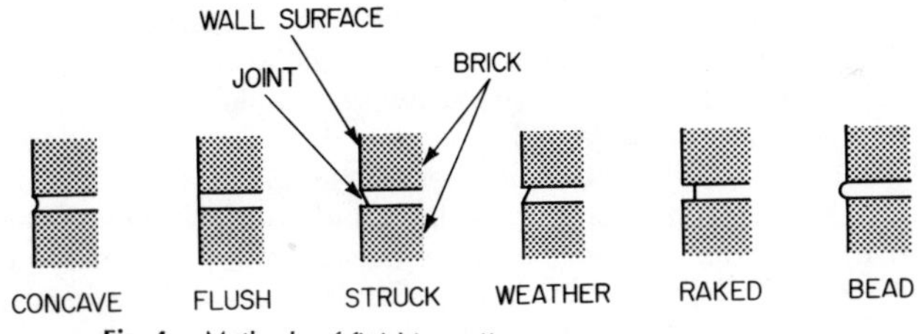

Fig. 4. Methods of finishing off a mortar joint *(U.S. Army)*.

The simplest thing the bricklayer can do to strike a joint is just cut the joint flush with the surface of the wall, using the edge of his trowel as a knife. Cutting a joint, as it is called, in this way tends, however, to drag the mortar along with the trowel, and this may leave a fine crack between the mortar and the brick. The crack will be located at the bottom of the joint if the cut is made in an upward direction, or at the top of the joint if the cut is made in a downward direction. These cracks provide openings through which water may eventually find its way.

The next simplest way in which a bricklayer can finish a joint is to make a *flush joint,* which he does by setting the point of his trowel on the brick units along the bottom edge of a joint and then drawing his trowel along the joint at a slight angle. A *weather joint,* which is similar to a flush joint but sheds water more effectively, is made by setting the point of the trowel on the brick units at the top edge of a joint and then drawing the trowel along the joint at a slight angle. Both flush and weather joints look neater than an ordinary cut joint and, since they also pack the mortar in somewhat, they are also somewhat stronger than a cut joint.

A *raked joint* is a favorite with those architects who like the long, low, horizontal look that strong, parallel lines give to brickwork. An ordinary cut joint is made first and the mortar is then raked out to a depth of ¼ to ⅜ in., using a special tool. The bricklayer then strikes the joint with another special tool called a *jointer* that packs the remaining mortar in tightly. Raked joints are time-consuming and thus expensive to make, and regardless of their architectural effect, they result in an exterior wall that tends to leak. In particular, raked joints should not be made with rough-textured brick, since the wall will be that much more difficult to make watertight. There is, however, no structural reason that raked joints cannot be made in interior brickwork for its decorative effect. Raked joints are sometimes also made in exterior walls that are to be stuccoed over. The deep joints help provide a grip for the stucco.

The *concave joint* is the best and least expensive method of striking a joint and also the most weather-resistant. A concave joint is made with a special tool having a half-round shape. The tool is somewhat larger in diameter than the thickness of the joint. After the mortar has been cut flush, the bricklayer presses the tool as forcibly as he can into the joint, drawing the tool along the joint as he does so. The pressure plus the form of the tool combine to compact the mortar solidly, making a dense and very water-resistant joint. An alternative method is to use a V-shaped tool, which produces a deeper-looking effect.

Having struck the joints, the bricklayer completes the job by going over the brickwork with a stiff brush to brush away any excess mortar that may be clinging to the edges of the joints.

LAYING A BRICK WALL

Plans of an architect-designed brick house will usually indicate not only the size and style of brick to be used but also the joint spacing. The location of the doors and windows and the height of the windows above the foundation will also be specified so that the brick can be laid without any special cutting or fitting. That is, the bottoms of the windows will be flush with the top of one course of brick; and the tops of the windows and doors will be flush with the tops of other courses of brick. In addition, the distances between the windows and doors, as well as the overall lengths and heights of the outside walls, will be some even multiple of the brick size. Carelessness in this respect adds greatly to the cost of laying the brick because the bricklayer will have a considerable amount of cutting and fitting to do.

But whether the house has been designed by an architect or not, a bricklayer always begins a wall by first accurately locating the corners of the building. When there is a crew of bricklayers on the job, the foreman has the responsibility of locating the corners and making sure they are truly square as well.

Once the corners have been located and squared, the foreman will "lay out the bond" to see how well a full course of bricks fit into the overall length of the wall. For a wall being laid in American bond, the first course will always consist of a row of headers. But whatever the bond, the foreman, or bricklayer, must first lay the brick units down accurately on top of the foundation, including the joint spacing between units as well. If he ends up a bit short or a bit over, he must adjust the positions of the bricks slightly until he has obtained an even-looking appearance. If there is an awkward-sized gap to fill, he will try to slip a three-quarter brick into the course, since such a brick disrupts the overall pattern the least.

To build a wall that is straight, plumb, and level, one must start by building the corners straight, plumb, and level. Everything depends on how well and how accurately the corners are made. Usually the best bricklayer on a crew is put on this job because of the skill required. Building up the corners is called *laying the leads.*

The bricklayer will first make a *gauge strip,* which is a length of wood about the size of a yardstick on which he marks the thicknesses of several courses of brick and the thicknesses of the joints between. Another length of wood called the *story pole,* which is usually made from a 2 × 2 and is long enough to reach from the foundation to the second-floor-joist level, is marked to indicate the heights of the window sills, lintels, and second-floor joists, as well as any other important features that will affect the laying of the courses.

Corners are built up in a pyramidal shape (see Fig. 3). The bricklayer takes great pains to lay each brick accurately in place. Nothing is allowed to fit by eye alone, although the bricklayer will constantly check his work by sighting along the brickwork. Each brick unit is leveled with the help of a spirit level. The straightness of each course is checked with a straightedge. The overall squareness of the corner is checked with a large square.

As soon as he begins to lay the second course in the lead, the brick units are checked for plumbness with a plumb rule. The overall height of each course is checked with the gauge strip. After each brick has been set in place, it is positioned by tapping it in or out with the handle of the trowel. All these pains are taken because once a brick has been laid in place and the mortar has begun to set, it is fatal to try and move the brick—the bond will be permanently broken. Everything must be done right the first time, brick by brick, as the corner is built up.

By the time both corners have been completed, the joints struck, and the excess mortar brushed away, the mortar in the bottom courses will have set hard. A nail is now driven into the mortar at each corner. A line is hung over the tops of the nails and positioned about $^{1}/_{16}$ in. away from the wall. The line is held taut by tying half bricks or other weights to the ends of the line. The main portion of the wall is now laid, care being taken to set the brick units level with, and their top edges $^{1}/_{16}$ in. away from, the line. As each course is completed, the line is raised to the next course and another row of bricks laid.

Most residential brick walls are 8 in. wide. Thus, the 4-in. wide outside wythe whose construction we have been describing is backed by an inside wythe that is also 4 in. wide. Whatever the bond or pattern on the outside face of the wall, the inside wythe is almost invariably set in common bond and with cut joints. Figure 5 shows the general construction. There are only a few points to note.

First, all the joints in the inside wythe should be filled with mortar as solidily as the joints in the outside wythe.

Second, the space between the two wythes should also be filled solidly with mortar. This is called a *collar joint,* One way of doing this is to *parge,* or plaster, the back of the outside wythe with mortar at the same time that the inside wythe is being built up. The parging coat should be at least $^{1}/_{2}$ in. thick and, when a brick is pressed into place, mortar should squeeze from between the brick and the parge coat.

Third, the brick units in each inside course should form an interlocking pattern with the brick units in the outside course, as this will increase the strength of the construction.

Fourth, the inside wythe should be built up at the same time as the outside wythe so that the parge coat can form a solid link between then.

The entire wall—including both the inside and outside wythes—is completed up to the top of the corner leads. The bricklayer must then lay out the leads for another five, six, or seven courses, and the next section of the wall is built up.

A good bricklayer is always conscious of the bond, or pattern, of the brickwork, and he will try to set his headers and stretchers in the outside wythe to maintain an even appearance. If the outside wythe has an English or Flemish bond, or a special decorative bond, it is especially important that the brick units be aligned vertically as well as horizontally. The bricklayer may draw vertical pencil marks on the brickwork, using his plumb rule as a guide, to help him set the bond accurately. If the bond is unusual or complicated, he may even make a drawing beforehand which he will then follow.

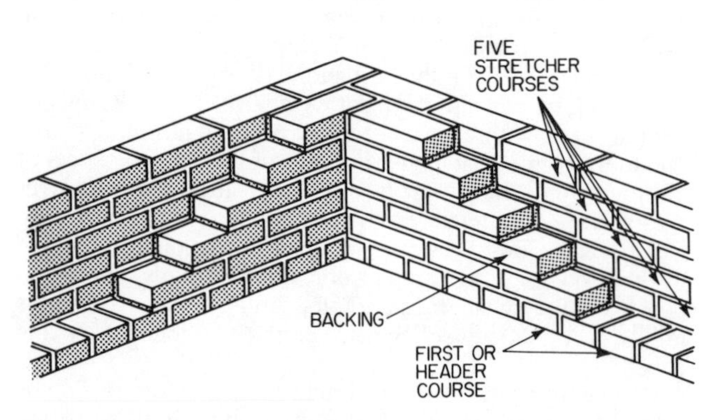

Fig. 5. Laying the inside wythe of a wall after the outside wythe has been constructed to the height of the next header course, which ties the two wythes together.

Brick-Masonry Construction

This article describes the various ways in which a one- and two-family dwelling may be constructed of brick. For a description of the brick itself and the technique by which brick are laid, *see* BRICK; BRICKLAYING. For other types of masonry walls, *see* BRICK-VENEER CONSTRUCTION; CONCRETE-BLOCK CONSTRUCTION; STUCCO.

We all know that, no matter how hard he huffed or how desperately he puffed, the big bad wolf could not blow down the brick house built by the most industrious and prudent of the three little pigs. The story reflects our feeling about brick houses—that they are solid, strong, and long-lasting.

The feeling is accurate. Assume we have a typical one-story brick dwelling. The total dead and live loads supported by the walls will be something on the order of 150 lb per sq ft. Ordinary brick units, of the type used in house construction, will have a compressive strength of around 6000 lb per sq in. It would appear that an 8-in.-thick wall made of these units would have a total compressive strength of 576,000 lb for each and every linear ft of wall (6000 lb per sq in. × 8 in. wide × 12 in. long = 576,000 lb per linear ft)!

In fact, the actual construction will be considerably weaker than this because: (1) the workmanship will be less than perfect and (2) the mortar binding the brick units together is considerably weaker than the brick itself. For ordinary uninspected construction, for example, which is the standard at which most houses are built (*see* BRICKLAYING), the assumed strength of the construction must be reduced by one-third. Therefore, the actual compressive strength of the wall will be ⅔ × 576,000 = 384,000 lb per linear ft.

Furthermore, depending on the type of mortar used in the construction (*see* MORTAR), the overall compressive strength of the wall will be reduced by another four-fifths to two-thirds. If we should be using type N mortar, for example, which results in the weakest construction, the compressive strength will be reduced to one-fifth, or 76,800 lb per linear ft (⅕ × 384,000 = 76,800). This is still considerably greater than the 150-lb-per-sq-ft load that we assumed to be bearing down on the wall.

Most building codes require that brick walls 35 ft or less in height be at least 8 in. thick; this requirement includes the great majority of dwellings built. As an optional requirement, these building codes allow brick walls to be only 6 in. thick in one-story dwellings if the wall is 9 ft or less in height.

These 6-in. and 8-in. dimensions refer to the amount of solid material in a cross-section of the wall, not its overall thickness. Thus, if a wall has a 2-in. air gap (a *cavity*) between the inner and outer wythes, the wall must then be at least 10 in. wide overall to meet the code requirement for an 8-in.-wide wall.

Figure 1 shows the main types of brick wall being built. These walls are divided into *barrier* walls, which are solidly constructed throughout, and *drainage* walls, which have an air space, or cavity, that separates the outer and inner wythes of brick. The reason for calling these constructions *barrier* and *drainage* will become apparent.

BRICKS AND WET WEATHER

Brick is more or less porous. When it rains, the eaves of a house will prevent some of the rain from striking a brick wall but a considerable quantity of rain will still soak into the brick. When the rain stops and the weather turns clear, this moisture will evaporate back into the atmosphere again. A brick wall can absorb an enormous quantity of water without doing any harm either to itself or to the interior of the house. What moisture does penetrate the wall usually evaporates from the inner surface as water vapor; the inside of the house may then feel chilly and damp.

Assume the brick wall we are discussing is well built, which in this context means that all the mortar joints in the wall, and especially the vertical joints, are filled solidly with mortar. If this is not the case, and if the wall should become thoroughly saturated with water during an exceptionally heavy rainfall, the water may find channels within the joints that will enable it to penetrate deeply into the wall. If such a channel should open onto the interior side of the wall, then water stains may appear on the wallpaper or plaster, or water may even run out in a stream, depending on the amount of water that has soaked into the wall and the size of the channel.

Even when a brick wall is very solidly constructed, without any voids or defects in the joints at all, the volume of water that strikes the wall during an exceptionally heavy rainfall may completely saturate the brick. Fortunately for those living in brick houses, in the overwhelming majority of well-constructed brick walls (and even in those walls that are not too well constructed), the possibility of rain soaking completely through is almost nil. Very heavy rain must fall for about 24 hr, accompanied by driving winds of at least 50 mi per hr, before an 8-in.-thick wall will become completely saturated.

Interestingly enough, it is far more probable that a relatively new wall will leak than a wall built 50 years ago and more. The reason is that engineers and manufacturers have been *too* successful in increasing the strength and rigidity of brick walls.

In an old wall, the mortar usually contains a relatively large proportion of lime. The lime gives the mortar a certain "softness," or resiliency; the mortar will give a bit if the wall is subjected to a stress of any kind. In fact, deep within a wall the lime takes such an exceptionally long time to cure that even when a crack does develop, the lime may "heal" the crack by resealing it. The wall is thus able to accommodate itself to the stress without developing cracks either in the mortar or the brick. Indeed, it is rare to find cracks in an old brick wall no matter how thick or thin it may be or how much of a load it has been supporting.

But at about the time of World War II, high-strength brick and high-strength mortars were developed, the purpose being to maintain the overall strength of masonry walls while reducing their overall thickness from 12 in., which was a common construction thickness, to 8 in. The purpose was to reduce

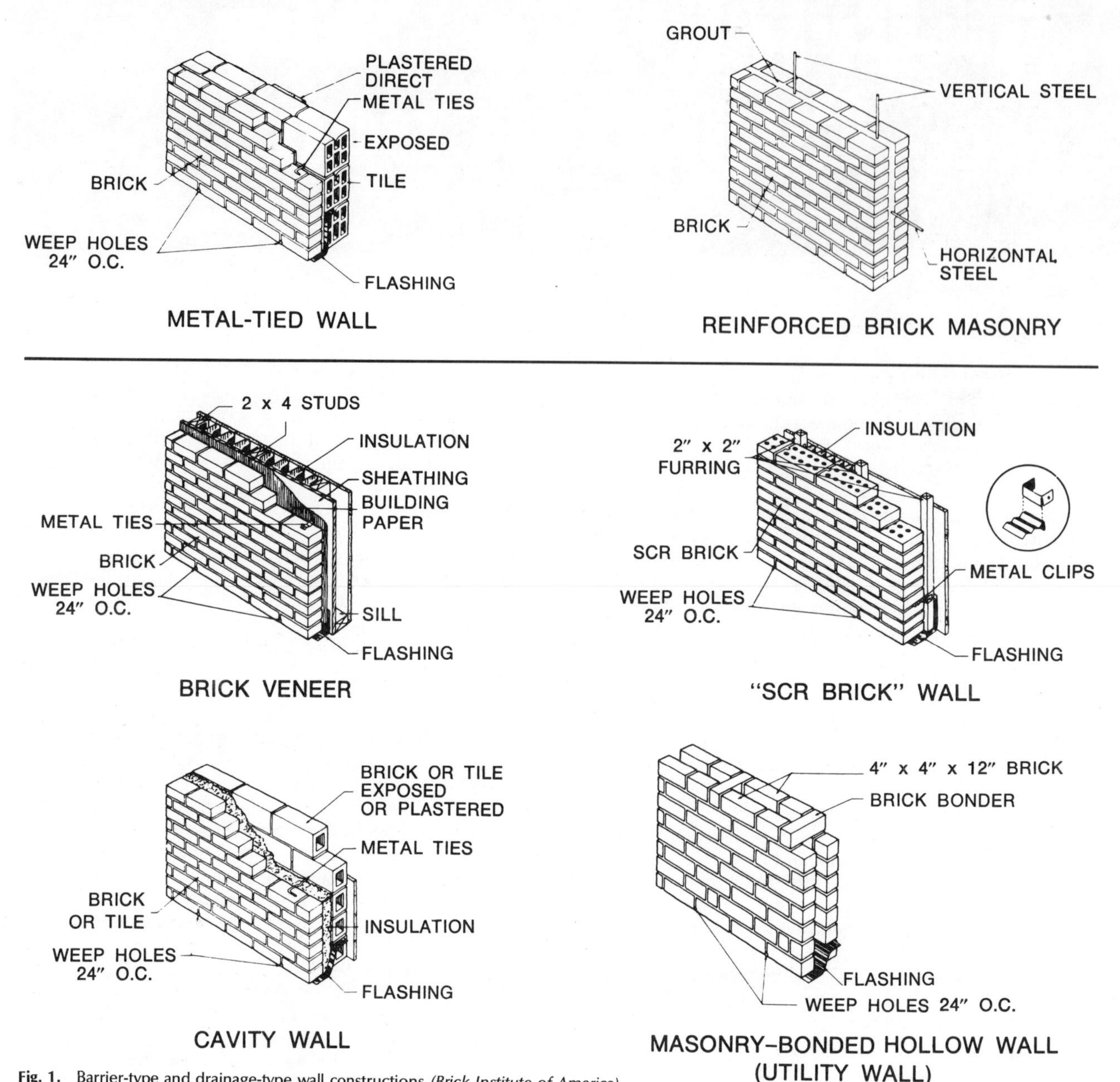

Fig. 1. Barrier-type and drainage-type wall constructions *(Brick Institute of America).*

construction costs while maintaining wall strength. This endeavor was completely successful.

However, high-strength mortar contains a large proportion of portland cement, which shrinks as it sets; the mortar is also far more rigid and unyielding than a mortar having a high lime content. As a consequence, brick walls built with a high-strength mortar almost invariably develop a network of very fine cracks in the mortar joints because the mortar is too rigid to accommodate itself to any stress or expansive force. In freezing weather, for example, this inability to give with a stress makes it much more likely that the mortar will crack and erode over time because of frost action. (It should be noted, however, that these fine cracks have no effect on the overall strength of the wall.)

In addition, high-strength brick is quite dense and has very little porosity. In wet weather very little moisture penetrates the brick. The brick may even be impervious to water. The rain then simply runs down the face of the wall. However, the network of fine cracks in the mortar make it possible for the water to soak into the mortar joints and from there completely through the wall.

If we assume the brick do have some porosity to them, the density of high-strength mortar makes it difficult for this moisture to soak into the brickwork in an even manner—to disperse itself, so to speak. Instead, the water that has been absorbed by the brickwork tends to collect along the tops of the mortar joints. If channels are present in the joints, this water will follow the channels into the interior of the building.

BARRIER-WALL CONSTRUCTION

Having described a rather discouraging set of circumstances, we should now emphasize that the overwhelming proportion of solid brick walls are and remain watertight for their entire lifetimes. But, in recognition that such problems have been raised, the concept of *barrier walls* (see Fig. 1) has been developed to solve them. A barrier wall is a wall so solidly constructed that water cannot possibly make its way through.

For one thing, in a barrier wall there are no *header courses* in the brickwork. (Header courses consist of brick that run completely through a wall from front to back.) Headers make it possible for the water in a saturated wall to soak through the wall even though the wall is free of any construction defects. If a builder wants a decorative bond on the exterior of the wall, then *false headers,* which are merely half-sized bricks, must be used to develop the bond pattern. To make up for the loss in strength, since headers strengthen the overall construction, the inner and outer wythes must then be tied together with metal reinforcement.

Another technique used to prevent the flow of water through a barrier wall is to *parge* the backside of the outside wythe. That is, the surface of the bricks is plastered over with a ½-in.-thick layer of mortar as the wall is built up. Parging is quite an effective waterproofer.

Furthermore, there is no reason why one should use the strongest possible brick or the strongest possible mortar to build a brick wall. As we have pointed out, a well-made brick wall for a dwelling is more than capable of supporting any loads imposed on it. It is more sensible in a dwelling to have a wall that is able to absorb stresses rather than having a wall that tries to meet these stresses head-on.

More important than their individual strength characteristics, brick and mortar should be suited to each other. It is a mistake, for example, to use a high-strength brick with a low-strength mortar. It is an even worse mistake to use a low-strength brick with a high-strength mortar. The best-made and longest-lived walls are built when the strengths of the brick and mortar are matched to each other. For example, brick that has a rather low compressive strength (grade NW brick, for example, which has a compressive strength of around 1500 lb per sq in.) should be used with type O mortar. (For a description of brick grades, *see* BRICK. For a description of mortar types, *see* MORTAR). For medium-strength brick (that is, grade MW brick, which has a compressive strength of around 2500 lb per sq in.), type N mortar should be used. And for high-strength brick (that is, grade SW brick, which has a minimum strength of 3000 lb per sq in.), type S or M mortar should be used.

Finally, in a barrier wall one should always provide an air space of some kind between the interior surface of the wall and the interior wall finish to prevent any moisture in the wall from soaking through the wall finish. The most common method of providing this air space is to attach furring strips to the brickwork (see Fig. 2) to which the plaster lath or gypsum wallboard is then attached. This air space not only prevents any water

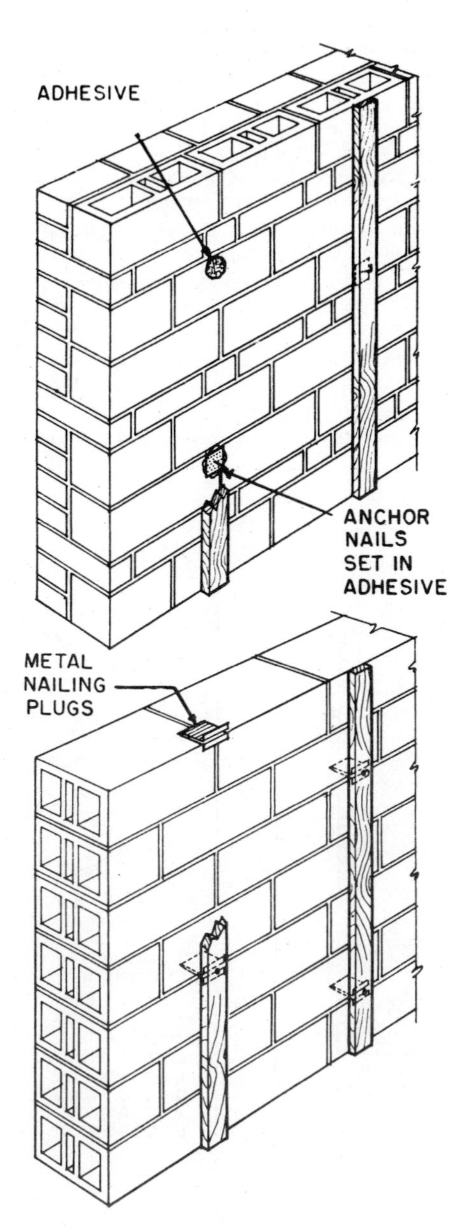

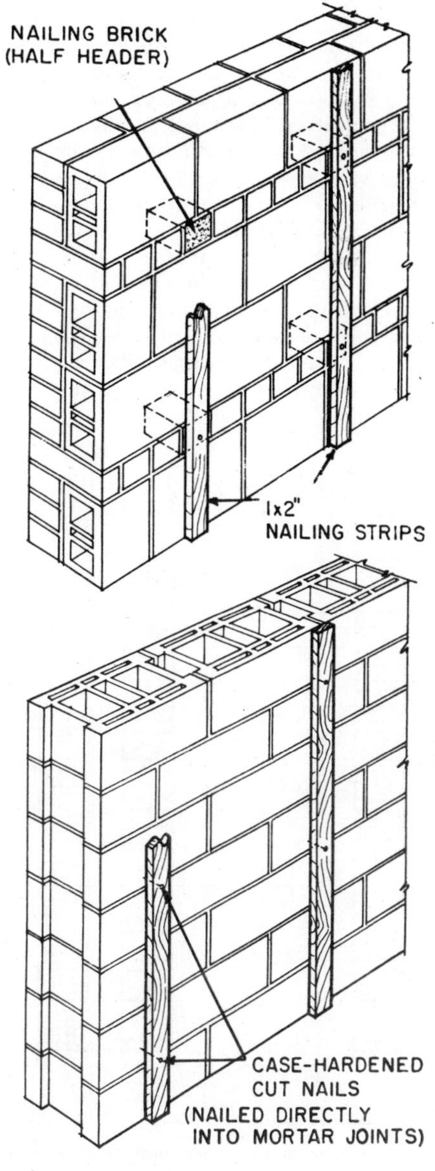

Fig. 2. Methods of attaching furring strips to the inside wall surface of a masonry wall *(Brick Institute of America).*

damage to the interior wall finish, it enables air to circulate at the back of the wall and carry away any water vapor that may otherwise condense on the back of the interior wall finish.

Solid brick walls built in this way are as weatherproof as one could wish, except possibly in those sections of the country in which there are frequent and severe rainstorms accompanied by high winds. In such locales, barrier-type walls may not be able to withstand the elements, and it would be wiser to build a drainage-type wall instead.

DRAINAGE-WALL CONSTRUCTION

The basic characteristic of a drainage (or cavity) wall (see Fig. 3) is the existence of an air gap, or cavity, between the inner and outer wythes of brick. This air cavity is usually 2 in. wide, though it may be as much as 3 in. wide. Apart from Z-shaped metal ties—3/16 in. in diameter and 6 in. long—that join the wythes to each other, the wythes are completely separate. There is no way, therefore, for water that has penetrated the outer wythe to cross over to the inner wythe. The amount of water that might find its way across the wythes by way of the metal ties is negligible.

Cavity walls have several advantages over solid barrier walls—advantages so substantial that almost all solid-brick homes built today have cavity walls, despite the fact that they cost about 20 percent more to build than barrier walls. The cavity is, in fact, an absolute barrier to the passage of water. Cavity walls will remain dry in the most torrential of climates, even when the standard of workmanship is not all that it should be. Cavity walls also have a 25 percent greater insulating value than barrier walls of the same thickness. In the hot, desert climate of the southwestern United States, the 2-in.-wide air gap between the inner and outer wythes of a cavity wall may be all the insulation required to keep the interior of the house at a comfortable temperature during the day. In northern climates, the air gap is equally effective in keeping the heat inside the house during cold weather.

An unanticipated bonus of cavity-wall construction was the discovery that the metal ties between the wythes not only increased the overall strength of the wall, they also gave the construction a useful amount of flexibility. In hot weather, for example, the bricks in the outer wythe tend to expand more than the bricks in the inner wythe. Under similar circumstances, a solid barrier wall might develop cracks. But in a cavity wall the metal ties absorb this stress by expanding slightly, and without reducing the strength of the construction in any way.

Structurally, cavity walls and barrier walls have about the same strength. They also have about the same resistance to sound penetration, both having a sound rating of about 50 to 55 dB (see ACOUSTICS). Both also have a fire-resistance rating of between 4 and 5 hr.

Poor workmanship can, however, reduce the watertightness of a cavity wall, and several points must be kept in mind when such a wall is being built.

First, there should be a number of *weep holes* in the bottom course of the wall. Weep holes are vertical gaps left between the bricks. They are usually spaced 24 in. apart. Any water that has penetrated the outer wythe will run down the wall and be discharged through the weep holes. Obviously, these weep holes must remain clear if the inner wythe is to remain dry. Weep holes are sometimes blocked by careless bricklayers who will allow goblets and drips of mortar to fall into the cavity as they build up the wall.

A bricklayer can do one of two things to prevent the blocking of weep holes. He can make a 45-degree bevel in the mortar bed before he lays a course of brick, as shown in Fig. 4. The bevel edge will prevent excess mortar being squeezed out of the joint as the brick are set in place.

Second, or in addition, the bricklayer can install a 4-to-6-ft-long, 2-in.-wide strip of wood called a *shield* into the cavity. A rope is tied or nailed to each end of the shield. What mortar falls into the cavity will be caught by the shield. When the bricklayer reaches a height at which he must lay metal ties across the wythes, he raises the shield, clears off any mortar, sets the metal ties in place, and then rests the shield on the ties as he continues to build up the wall.

Failure to use such a wood shield (and the failure is com-

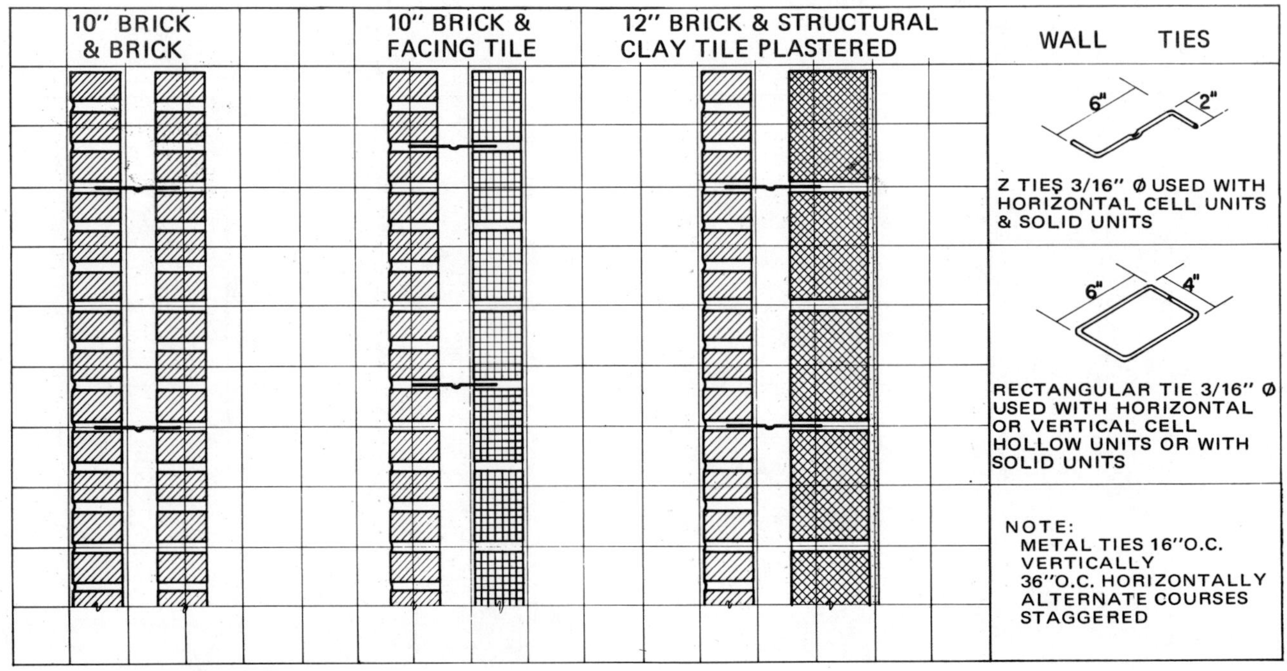

Fig. 3. Typical cavity-wall constructions *(Brick Institute of America).*

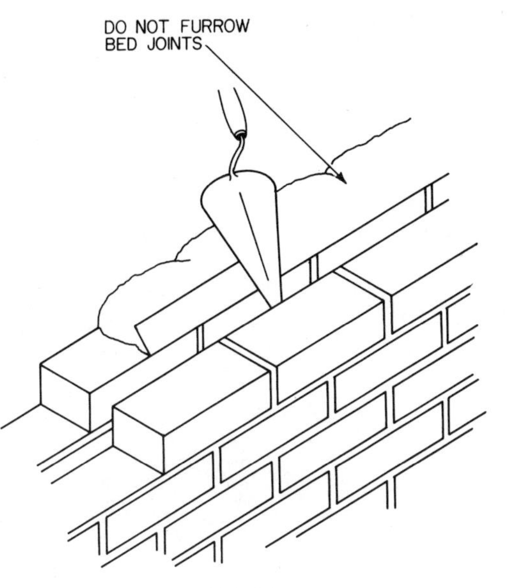

Fig. 4. Beveling a layer of mortar will prevent the mortar falling into the cavity.

mon, bricklayers thinking it a nuisance and a reflection on their skill), might also result in goblets of mortar falling on the metal ties. These goblets form bridges across which water can travel.

Water might also splash across the cavity after striking pieces of brick or mortar that are projecting into the cavity; it is also necessary to be on guard against these potential projections when building the wall.

But these faults are all rather minor compared to the failure to provide adequate *flashing* between the inner and outer wythes wherever windows, doors, vents, or other parts of the house structure are built into or through the wall. Leaks in a cavity wall almost always occur because of inadequate flashing at these points. Flashing should be installed at points of potential leakage. *See* FLASHING.

One should also make sure that there is a *dampproof course,* as it is sometimes called, at the very bottom of the wall. (For a description, *see* FLASHING.) A dampproof course does not protect the wall against rain so much as it prevents dampness in the soil or foundation wall from rising into the brickwork by capillary action.

COMPOSITE-WALL CONSTRUCTION

The expense of laying standard sizes of brick plus the development of specially shaped structural clay tile and concrete blocks has led to *composite-wall* construction, in which the outside wythe of the wall consists of standard-size clay brick while the inner wythe consists of larger, more economical, and easier-to-work-with tile or concrete block (see Fig. 5). One obtains all the benefits of an all-masonry wall without the expense of an all-clay-brick wall.

Everything that has been said about the advantages and disadvantages of solid and cavity walls made from brick alone applies also to these composite walls. Composite walls can be used wherever all-brick walls can be used. They are equally strong and equally watertight. In addition, the cavities in the structural-clay-tile units and concrete-block units give the construction much of the insulating advantages of an all-brick cavity wall.

One unfortunate characteristic of solid composite walls is that brick and concrete blocks expand at different rates when exposed to the hot summer sun, or when they are soaked with water, with the concrete blocks expanding more than the brick. As a result, the mortar joints may crack at the front of the construction. In addition, concrete blocks tend to shrink a bit after a wall has been completed, after the moisture they usually absorb during construction has evaporated. For this reason also, the mortar joints in a solid composite wall are likely to develop cracks over a period of time. There is nothing that can be done about this differential expansion except to make sure that the strengths of the mortar and masonry are matched to each other.

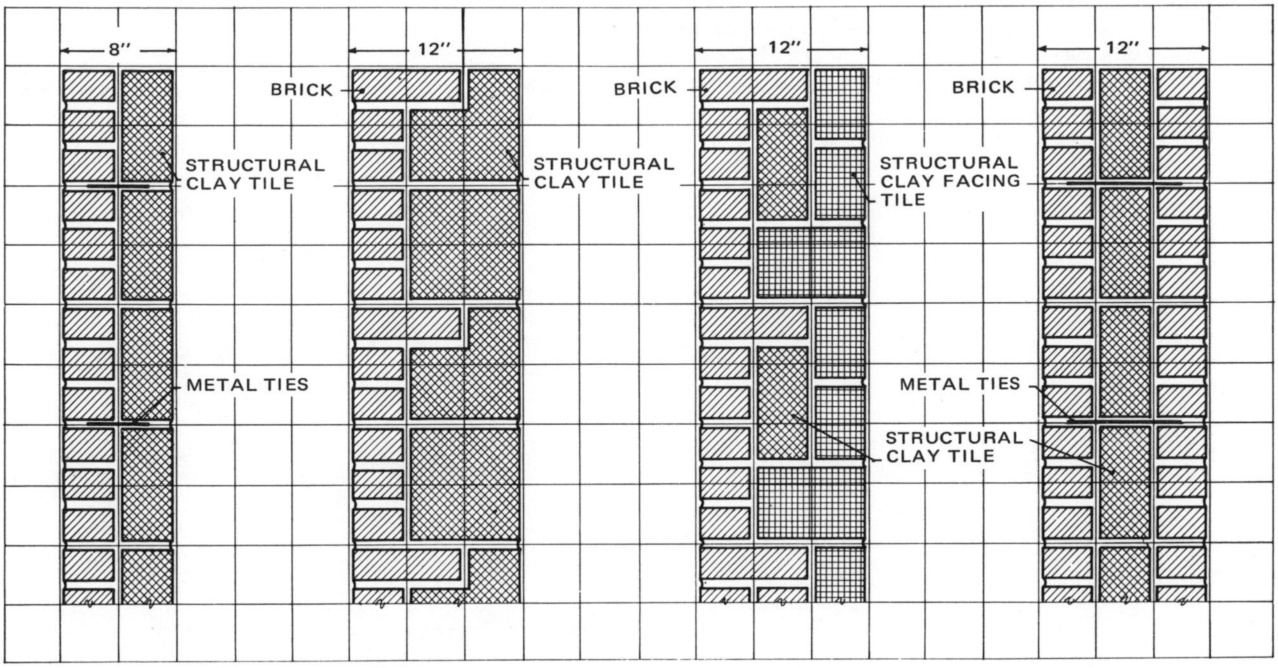

Fig. 5. Typical composite-wall construction *(Brick Institute of America).*

SCR BRICK CONSTRUCTION

In its never-ending battle against the forces of evil, as represented by the many lumber and plywood associations, the Brick Institute of America (formerly the Structural Clay Products Institute) developed in the 1950s SCR (which stands for Structural Clay Research) brick (see Fig. 6). This occurred after the major building-code associations had approved the use of a 6-in.-thick masonry wall in one-story dwellings. SCR brick have reduced considerably the cost of constructing a brick wall, and their use is widespread and increasing. SCR brick are manufactured in all the colors, textures, and structural grades in which ordinary clay brick are manufactured.

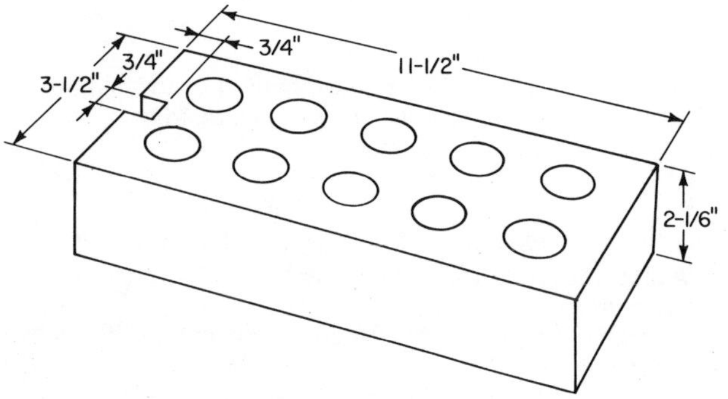

Fig. 6. SCR brick, developed by the Brick Institute of America, for 6-in.-thick wall construction.

SCR brick are made to modular dimensions (*see* MODULAR CONSTRUCTION). Assuming for example, the mortar joints will be $\frac{1}{2}$ in. thick, the nominal size of the brick being $2\frac{2}{3} \times 6 \times 12$ in., three courses of brick with $\frac{1}{2}$-in.-thick joints between them will be exactly 8 in. high. Each brick has 10 holes, which makes it easy to grasp and set in place. These holes also allow the bricklayer to install anchor bolts through over-lapping courses, or to install steel reinforcement bars if the masonry is being made earthquake resistant. One side of each brick also has a notch in one end into which a window or door jamb can fit.

BRICK-AND-PILASTER WALLS

The Brick Institute of America has also developed a 4-in.-thick brick-and-pilaster wall for use in dwellings that reduces the overall cost of brick-masonry construction to that of traditional wood-frame construction (see Fig. 7). The pilasters are located at 4-ft intervals, and their primary function is to stiffen the wall construction. They also act as furring that supports any insulation that may be attached to the wall as well as the usual interior wall finishes. The gaps created by the pilasters also give the wall construction many of the advantages of a traditional cavity wall.

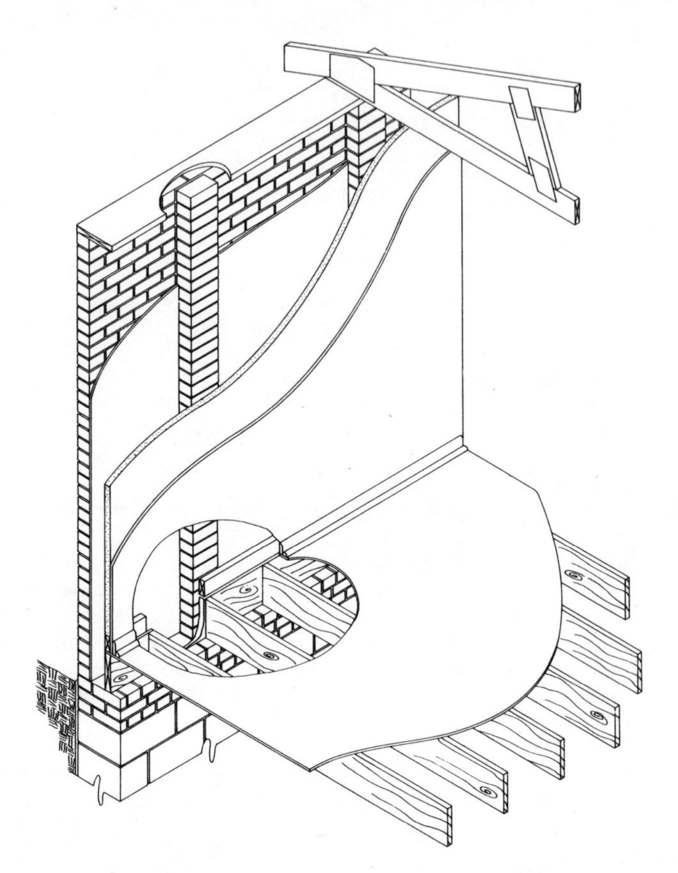

Fig. 7. The brick-and-pilaster wall developed by the Brick Institute of America *(Brick Institute of America).*

Brick-Veneer Construction

In brick-veneer construction, a 4-in.-wide facing of brick is constructed over the sheathing of a wood-frame house to give the walls the appearance of having been made of solid brick, but without incurring the expense of all-brick construction (see Fig. 1). Structurally, brick veneer adds little to the strength of the walls—it is primarily decorative and must, in fact, be supported by the wall sheathing. But apart from its decorative effect, a properly constructed brick-veneer facing adds substantially to the thermal-insulating properties of the walls, and it will pay for itself over the long run in reduced air-conditioning costs during hot summer months and in reduced fuel bills during cold winter months.

If a new house is being designed from the first to have a brick-veneer facing, the foundations can be made strong enough to support the weight of the brick and wide enough to give them adequate bearing. The doors and windows can also be selected especially to fit into the thicker wall sections. But if an old house is being remodeled by the addition of a brick veneer, then it will be necessary to add a 5-in.-wide shelf of concrete to the existing foundation to give the brickwork the support it requires (see Fig. 2). It may also be necessary to install new and wider door and window frames, and perhaps modify the existing exterior finish to make room for the brickwork.

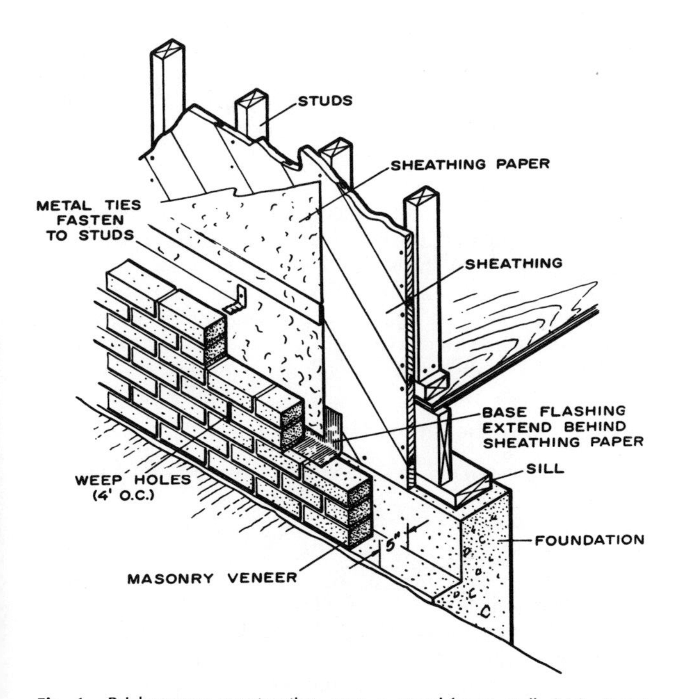

Fig. 1. Brick-veneer construction over a wood-frame wall *(U.S. Forest Service).*

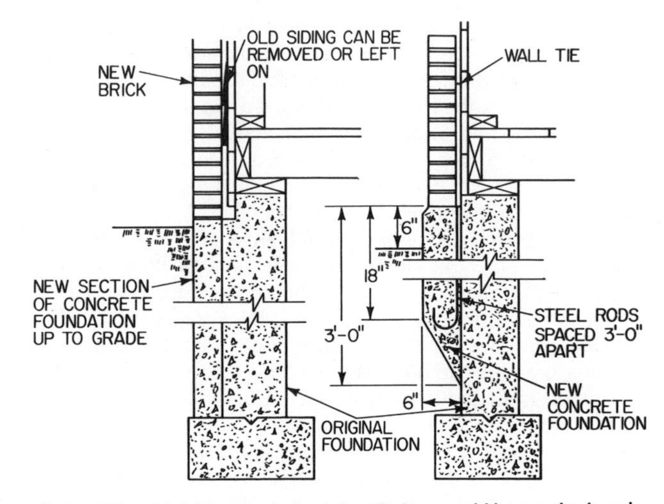

Fig. 2. When a brick-veneer facing is installed on an old house, the foundation walls must be strengthened to support the weight of the brick. Two methods of adding to the existing foundation are shown.

Moisture and Brick-Veneer Construction

In the *platform-framed* house that is so common in the United States (for a description, *see* WOOD-FRAME CONSTRUCTION), the wood members of the framework installed horizontally tend to swell and shrink in height over the course of a year because of annual changes in atmospheric humidity. The amount of swelling and shrinking that takes place will depend on local climatic conditions. It is possible for a platform-frame house located in a part of the country that has cold, dry winters and hot, humid summers to shrink and expand as much as an inch or two over the course of a year.

Brick, however, is entirely unaffected by changes in humidity. If, then, a brick-veneer facing is added to the walls of a platform-framed house, the differential rates of contraction and expansion that occur between the wood and brick may result in damage either to the brick veneer or to the wood framework. Assume, for example, that over a hot, humid summer the wood has absorbed a considerable amount of moisture before the brick veneering is added. The wood sheathing and the brick units will be tied together during the course of construction as described below. When the wood subsequently dries out over the next winter, it will attempt to shrink back down again (because a wood-frame house *always* expands up from and then shrinks back down to the foundation to which it is securely fastened) but it will be unable to do so because it is now firmly tied to the brick veneering. It is conceivable, therefore, that as the wood shrinks the entire wall will be pulled up off its foundation.

On the other hand, assume that the wood is dry when the brick veneering is added but that the wood subsequently absorbs moisture. The wood frame will thereupon expand upward during the next hot, humid summer. As it does, it may very well force the brickwork up with it—brickwork is very weak in tension—with the result that large cracks will open up in the mortar joints.

A *balloon-framed* house (for a description, *see* WOOD-FRAME CONSTRUCTION), however, is almost entirely immune from the effects of these humidity changes because all the lumber in the framework is installed vertically, and lumber does not shrink or swell a significant amount in a lengthwise direction. (For a discussion of wood shrinkage and expansion, *see* WOOD.) Therefore, if brick veneer is to be applied to a new wood-frame house, it is much to be preferred that the builder balloon-frame the structure.

From what has been said, it should be apparent that before a homeowner decides to add brick veneering to an old house, he must have the house examined and ascertain whether the house is platform- or balloon-framed. If the house is platform-framed, he must then find out what the annual rates of humidity change are and what local experience has been in regard to the application of brick veneer over an old platform-framed house. It may be that the dimensional changes in the structure will not be large enough to worry about.

One problem the owner of an old house won't have to worry about is whether his house has been built of green lumber. From all that has been said it should be apparent that the effects of any humidity changes on a brick-veneer installation will be minimized if the moisture content of the wood is in equilibrium with the humidity level that exists locally. That is, the wood must be thoroughly seasoned. Depending on local conditions, seasoned wood will have a moisture content of between 9 and 12 percent. Newly purchased lumber can, however, have a moisture content as high as 19 percent. If brick veneering is to be applied immediately to the walls of a newly constructed house, the builder must first ascertain the moisture content of the wood that is to be used for construction. If it is too high, he will have to defer construction of the brickwork until the wood dries out naturally. This may take a month or two. In the meanwhile, construction must cease. Construction may, however, be completed up to the installation of the wall sheathing.

Brickwork

For an introduction to this subject, *see* BRICK; BRICKLAYING; BRICK-MASONRY CONSTRUCTION. The brick used for veneering and the technique of laying the brick in no way differs from the information contained in these articles. A few points must be emphasized, however.

Most building codes require that brick used for veneering be nominally 4 in. wide (which means it will actually be at least 3½ in. wide), although some codes will allow brick veneer that is to be built only one story high (that is, 8 ft high) to be actually 3 in. wide.

Common, or American bond, in which all the courses are stretcher courses makes the strongest construction and is most often used for brick veneering. A brick-veneer facing can almost always be recognized at a glance—all the courses will be stretcher courses. More decorative bond patterns that require headers as well as stretchers to make up the pattern (for a definition of these terms, *see* BRICKLAYING) will require *false* headers, that is, bricks that are only half size. The construction, therefore, will not be as strong as when common bond is used. A decorative brickwork pattern, therefore, may require that additional reinforcement be used to strengthen the construction; one must check with one's local building code.

A brick wall that is 3 in. to 3½ in. wide and 8 or more ft in height is not self-supporting. The brickwork requires additional support if it is to be prevented from collapsing because of excessive wind pressures, earthquake shocks, or defects in the construction. The required support is provided by metal *ties* or *anchors* that are nailed to the wall studs or wood sheathing. The nails that hold the ties in place should never be driven into sheathing made of insulation board, fiberboard, or gypsumboard. These materials simply do not have the necessary nail-holding ability. If any of these materials should be used for the wall sheathing then care must be taken to drive the nails into the studs *behind* the sheathing.

Ties come in a variety of designs but they are usually either small metal rods with a hook on one end that locks into the mortar and a small circlet on the other end that is held by the nail, or they consist of a corrugated strip of metal about 6 in. long and 1 in. wide. The latter are shown in Fig. 1. The strip is bent 90 degrees, one end being laid in the mortar and the other end being nailed to the sheathing. The best ties are made of copper, which won't corrode in mortar and is unaffected by moisture, although most installations use ties made of galvanized iron, which does corrode but is much cheaper. Whatever the metal, the nails used to hold the ties to the sheathing should be made of the same metal as the ties to prevent corrosion of the ties due to electrolytic action.

An air space at least 1 in. wide but not more than 2 in. wide must be maintained between the brickwork and the sheathing. This air space is necessary to prevent moisture reaching the sheathing, as a 4-in.-thick brick wall is hardly watertight, even when great care is taken with the construction, because most brick is inherently porous and will become saturated with water whenever there is an exceptionally heavy rainfall (*see* BRICK-MASONRY CONSTRUCTION). Even if a dense, relatively nonporous brick is used, small cracks will usually appear in the mortar joints over a period of time because of weathering or because of stresses induced in the brickwork by uneven settlement of the house. The air space, therefore, provides a barrier across which water cannot pass. Instead, the water will run down the inner surface of the brickwork and be discharged through weep holes let into the bottom course for this purpose, as shown in Fig. 1.

During cold, damp winters, water vapor migrating out of the house tends to condense against the cold, inner surface of the brickwork (*see* CONDENSATION). The presence of an air gap then helps to prevent this condensate soaking into the wall sheathing, which is much more susceptible to water damage than the brick are. If the brickwork were mortared flat against the sheathing, and if there were no other surface against which this water vapor could condense in cold weather, it is possible that the sheathing would in time be destroyed by moisture.

Since there is the strong probability that water will some day run down the inner surface of the brickwork, care must be taken when the ties are installed that they do not tip downward toward the sheathing, as this would provide pathways across which the water could reach the wall sheathing.

For the same reason, the bricklayer must take care not to drop mortar into the air space as he is building up the wall. Bits of mortar stuck within the air space will provide bridges across which water can reach the sheathing.

Flashing and Sheathing Paper

Flashing must always be installed over the top of the foundation before a brick-veneer wall is constructed (see Fig. 1). For a description of the method used to keep water out of the interior of a dwelling, *see* FLASHING. The flashing material, which consists either of a sheet of metal or 30-lb asphalt-saturated felt, is laid directly on top of the foundation wall. It should extend over the front of the foundation by about 1 in., with the front edge of the flashing bent down a bit to form a drip, and it should extend up against the sheathing by at least 6 in.

Sheathing paper is then nailed over the entire surface of the sheathing and over the flashing as well, as shown in Fig. 1, to keep moisture from getting behind the flashing. The sheathing paper serves two purposes. Since bricklaying is a very wet job, it will keep the wood framework and sheathing dry as the wall is being built. Thereafter, it will help keep water away from the sheathing. Since, as we have mentioned, there is an excellent probability that some moisture will get behind the brickwork and into the air gap, one should use the heaviest grade of waterproof sheathing paper that is available.

Preparing the Foundation

One sq ft of nominal 4-in.-wide brick weighs about 37 lb. The 8-ft-high brick-veneer wall will, therefore, weigh about 300 lb per linear ft, and twice this for a two-story house. This additional weight must be supported by the foundation.

In a new house, the foundation can be built wide enough to accommodate both the wood framework and the brick veneer. The foundation, therefore, must be built about 4¾ in. wider than usual.

If brick veneering is being added to an old house, the builder has several alternative methods of preparing the foundation. The preferred method is to pour a 5-ft-wide shelf of concrete against the existing foundation wall, with the shelf resting on the footing. Alternatively, he can build an auxiliary foundation wall by laying a course of concrete blocks on top of the existing footings or he can bolt steel shelf angles to the existing foundation, after having first made sure that this construction is approved by the local building code. The idea of supporting the brickwork on a shelf in this way is derived from skyscraper construction in which the exterior wall consists of a thin veneer of stone or brick supported by shelf angles at each floor level. The use of shelf angles for dwellings seems to be an inexpensive and reasonable method of providing a support for the brickwork. The problem with this method is that the strength of the foundation is completely unknown, since it is extremely unlikely that tests were ever made to determine the strength of the concrete at the time the foundation was poured, and it is equally unlikely that the current strength of the foundation can now be ascertained. The foundation might hold; then again, it might not.

Whatever the method used to support the brickwork, the top surface of the foundation must be dead level if the bricks are themselves to be both adequately supported and level. The first course of brick must be laid out and the corners built up square, level, and plumb (*see* BRICKLAYING; that article describes also how a *story pole* is used by the bricklayer to make sure that the brick courses come flush with the tops of the door and window frames and the bottoms of the windows).

Sometimes in an old house with a wood-siding or wood-shingle exterior finish, the builder doesn't bother to remove this exterior finish before he builds the veneer wall. There is no harm in leaving the old finish. If it is watertight, there is an advantage in doing so—and it is certainly cheaper to let it be than try and remove it. However, if the old finish is being left, there are still some parts of the exterior that must be removed—water tables, window sills, decorative cornices, and the like. Any openings that are left in the wall must be covered over with flashing and/or sheathing paper to prevent water penetration into the walls through these openings.

Building Code

A building code is a statute, in the form of detailed regulations, that has been enacted by a municipality or other government entity to ensure that all the buildings constructed within its jurisdiction meet certain minimum standards of health and safety. The building code has its legal justification in the inherent power of any government to protect its citizens from any harm likely to come to them because of unhealthy or unsafe conditions.

Building codes, therefore, are concerned with such things as structural adequacy, the quality and strength of the materials used, sound workmanship, the correct installation of approved electrical wiring and equipment, the correct installation of approved gas-, coal-, or oil-heating equipment and their piping, the correct installation of approved sanitary plumbing fixtures and their piping, the fire resistance of the materials used, and the existence of fire exits.

In most municipalities, the plans for all new construction must be approved by officials of the buildings department before construction begins, and these same officials must have access to the property at all times and be able to inspect all equipment, materials, and workmanship before the building is approved for occupancy. If the equipment, workmanship, or materials do not meet the standards of the building code, these officials have the authority to order that the necessary changes be made before they issue a *certificate of occupancy,* as it is called.

Once a building has been approved and occupied, if the owner should thereafter want to make a basic alteration in the electrical, heating, or plumbing systems, or a basic change in the structure, the buildings department must approve the alterations beforehand and inspect the workmanship and materials after the alteration has been completed. If the job is signed off by a licensed electrician or plumber, adequacy of workmanship and materials is usually assumed, although the buildings department always reserves the right to make a subsequent inspection and order any changes it thinks necessary.

The impulse that led to the development of building codes was humanitarian. Without the legal restraint imposed by a code, nothing prevented a builder from putting up the cheapest, shoddiest, and most densely packed dwellings he could get away with. Perhaps the best-known examples of such jerry-built construction in the United States were the cold-water tenements—the "old law" tenements—that once covered most of the Lower East Side of Manhattan, and much of the other sections of New York City as well. The crowded, unsanitary conditions that existed in these tenements, and the large number of fires that occurred because of these conditions, led to the development and adoption of building codes.

The codes that have been adopted by most large- and medium-size cities in the United States are a mixture of engineering knowledge, local building customs, and an accommodation to local political realities. Originally, building codes were of the *specifications* type, which required that all construction be accomplished using specified materials in a specified way. The builder has very little leeway in the materials he can select or the methods of construction he can use. One curious example of how this specification-type building code has influenced building styles was the requirement in New York City that the rooftop water-storage tanks that are connected to standby fire-sprinkler systems be constructed of wood staves. As a result, anyone who has every looked down on midtown Manhattan from a skyscraper has been struck by the sight of innumerable wooden water tanks with conical roofs perched rather incongruously on top of all that steel and concrete.

Specification-type building codes were probably a necessity in a day when speculative builders tried to get by using the cheapest materials they could buy, assembled in the most slipshod manner. But times changed, and new materials were developed. Since World War II there has been a swing toward building codes of the *performance* type, in which the performance standards of a material or structure are outlined and the builder is free to select whatever materials or building techniques will meet these standards.

For example, a specification-type code for a house sewer will simply specify that cast-iron pipe of a certain quality and size be used and that the pipe be installed in a specified manner. The plumbing contractor has no choice or say in the matter. In a performance-type building code, however, the code will specify that the piping not be affected by any corrosive or harmful substances in the sewage or in the soil in which the pipe is buried, that the pipe meet certain minimum strength requirements, and that the pipe not be affected by temperature changes within a specified range. The plumbing contractor is free to use plastic pipe, cast-iron pipe, or gold pipe if he wants to, as long as he can show the local buildings officials that the pipe does in fact meet their standards.

The spread of performance-type codes has been helped enormously by the materials specifications published by the American National Standards Institute (ANSI; formerly the American Standards Association), by the American Society for Testing and Materials (ASTM), by branches of the U.S. government, especially the Department of Commerce, and by an enormous number of specialized industry groups. To ensure the adequacy of any material used or construction, all a local building code need do is specify, in the case of house sewers, for example, that the polyvinyl chloride pipe used (if a plastic pipe is being used) meets Standard Specification D 2665-73 published by the ASTM.

Small municipalities are not in a position to undertake independent studies of building materials and construction techniques and then publish their own building codes based on their findings. They have come instead to depend on the work being done by four nonprofit organizations whose membership consists of building officials. Those organizations have published *model codes* that all municipalities are free to adopt in whole or in part. The virtues of adopting a model code are that the municipality is assured of obtaining a well-thought-out and up-to-date performance-type building code based on sound construction practices, a code that is updated periodically in the light of changing conditions and the availability of new materials. All the model-code organizations are prepared to help a municipality establish a buildings department complete with all the necessary forms and procedures.

The four organizations and their published codes are as follows:

American Insurance Association (formerly the National Board of Fire Underwriters)—*National Building Code.*
Building Officials and Code Administrators International (formerly Building Officials Conference of America)—*Basic Building Code*
International Conference of Building Officials (formerly the Pacific Coast Building Officials Conference)—*Uniform Building Code.*
Southern Building Code Congress—*Standard Building Code.*

These codes are not intended primarily for one- and two-family dwellings, although they do apply to dwellings also. In 1971 these four organizations joined together to publish a model building code that *is* devoted exclusively to dwellings, *The One and Two Family Dwelling Code,* which is essentially a distillation of all four codes as they apply to dwellings.

There are several other model codes that are of importance to builders of one- and two-family dwellings. The National Fire Protection Association publishes an enormous range of books and pamphlets dealing with the construction of fire-resistant buildings, the use of fire-resistant materials, and fire-fighting procedures and equipment. Among their publications is the *National Electrical Code,* which most municipalities have adopted in toto as a basic part of their own building codes. The *National Electrical Code* contains approved methods of installing electrical wiring and equipment for all types of buildings. Licensed electricians are required by most municipalities to be thoroughly familiar with this code; the electricians must pass a test based on their knowledge of the code before they can receive their licenses.

In 1972 the National Fire Protection Association published an abridgement of the electrical code called the *Electrical Code for One- and Two-Family Dwellings.* Anyone wiring or rewiring a house should install the wiring according to the requirements of this publication.

Three of the model organizations mentioned above have also promulgated model plumbing codes, which are as follows:

BOCA—*Basic Plumbing Code*
ICBO—*Uniform Plumbing Code*
SBCC—*Standard Plumbing Code*

These codes were published because their predecessor, the *National Plumbing Code,* which had been published last in 1944 as a cooperative effort of the United States government and many engineering societies and industry associations, had fallen hopelessly out of date. As with the electrical code, most municipalities have now adopted one of the above model plumbing codes in full as a basic part of their own building code.

Building Loads

One of the less-thought-about facts of house construction, though certainly one of the most important once you do think about it, is that houses stay up once they are put up. They stay up despite the handicap that many houses are built with less than adequate workmanship or with poor materials, and that the houses are subjected to a great many different and often quite severe stresses during their lifetimes. Some of these stresses are inherent in the structure, others are imposed from outside. The inherent stresses are *dead loads.* The stresses imposed from outside include *live loads, wind loads, snow loads,* and *earthquake loads.*

DEAD LOADS

The dead loads include the weights of all the structural parts of the building—the studs, joists, rafters, roofing, flooring, insulation, plastering, and so on. The dead loads also include the weights of all the mechanical equipment permanently installed in the house—the plumbing and heating equipment, ductwork, central air-conditioning equipment, kitchen appliances, and the like. In sum, the dead loads include everything in, on, or attached to the house that cannot possibly be omitted or eliminated if the house is to be considered complete and habitable. Since the weights and locations of all these materials are known, the dead loads can easily be calculated and the structure designed to support them.

LIVE LOADS

Live loads include everything that the inhabitants carry into a completed building to make it habitable, including themselves. Live loads, therefore, are additions to the floor dead loads. Anything that is movable is usually considered as part of the live load, though sometimes live loads shade off into dead loads. The permanent installation of large bookcases with their weight of books, a grand piano, or machine-shop equipment, can be considered either as live loads or as dead loads. In the former case, it is assumed that the joists will be able to support this weight without any additional reinforcement. In the latter case, the builder may find it necessary to reinforce the joists to help them support the additional loads.

In American building codes, the first-floor joists must be strong enough to support a live load of 40 lb per sq ft; the second-floor joists must be strong enough to support a live load of 30 lb per sq ft; and attic joists, if the attic is unoccupied, must be strong enough to support a live load of 20 lb per sq ft. If the attic space is to be converted at a later date into a living space, a bedroom or study, then the attic floor joists must be capable of supporting a live load of 30 lb per sq ft. Staircases are required to support a live load of 100 lb per sq ft.

These requirements are very conservative. The weight of all the furniture in a typical living room, apportioned over the entire floor area, will seldom exceed 5 lb per sq ft. If one were to crowd 20 people into a 12 × 15 ft living room for a party, the live floor loading would not exceed 20 lb per sq ft. Nevertheless, one never knows what future loads a floor may be required to support, and it is undoubtedly wiser to expect the worse possible loading than to have the floor start sagging after a period of use.

Wood joists must be strong enough to support both their own weight and whatever live loads are imposed on them. But joists that are designed *only* for strength will tend, over a period of time, to sag of their own weight as well as the weight of the live loads they are supporting. This sagging will cause the plaster to crack, not to mention what it will do to the peace of mind of the occupants. Joists, therefore, must be large enough to be stiff as well as strong. Building codes require that joists deflect a maximum of 1/360th their unsupported lengths when they are supporting their maximum design loads. For a 20-ft-long joist, this means, for example, that the joist should sag no more than 2/3 in. at the center of its span. Joists will rarely, if ever, be subjected to their maximum design loadings so the actual amount of sag in the typical dwelling will in fact be imperceptible to the occupants.

WIND LOADS

The wind is the most important outside force affecting the structural integrity of a house. The map in Fig. 1 shows the maximum wind pressures that can be expected in different parts of the United States. The contour lines represent the force of the wind in lb per sq ft of *pressure,* not mi per hr of velocity, since it is the pressure pushing against the house that is of interest to us. The figures represent *average* maximum wind pressures; that is, the force exerted by the wind over a 5-min period. Gusts may add perhaps another 50 percent to these values. In addition, the shape of a house will sometimes increase the force with which the wind strikes it. For a rectangular-shaped house, for example, a *shape factor* of 33 percent must be added to the basic wind pressure.

Thus, including both gust loads and a shape factor, an engineer designing a new house usually assumes a wind loading of something around 15 lb per sq ft, which is equivalent to a gust velocity of about 75 to 80 mi per hr (refer to Table 1).

In most parts of the United States, a house capable of withstanding a wind load of 15 lb per sq ft will be more than strong enough. However, along the Gulf and Atlantic coasts as far north as Chesapeake Bay, and 50 miles inland from these coastlines, houses must expect to withstand winds of hurricane force sometime during their existence and the anticipated wind loadings must be increased accordingly. It is improbable, however, that wind loads will ever exceed 40 lb per sq ft.

When calculating the wind pressures acting on a house, an engineer assumes that the wind will act in a horizontal direc-

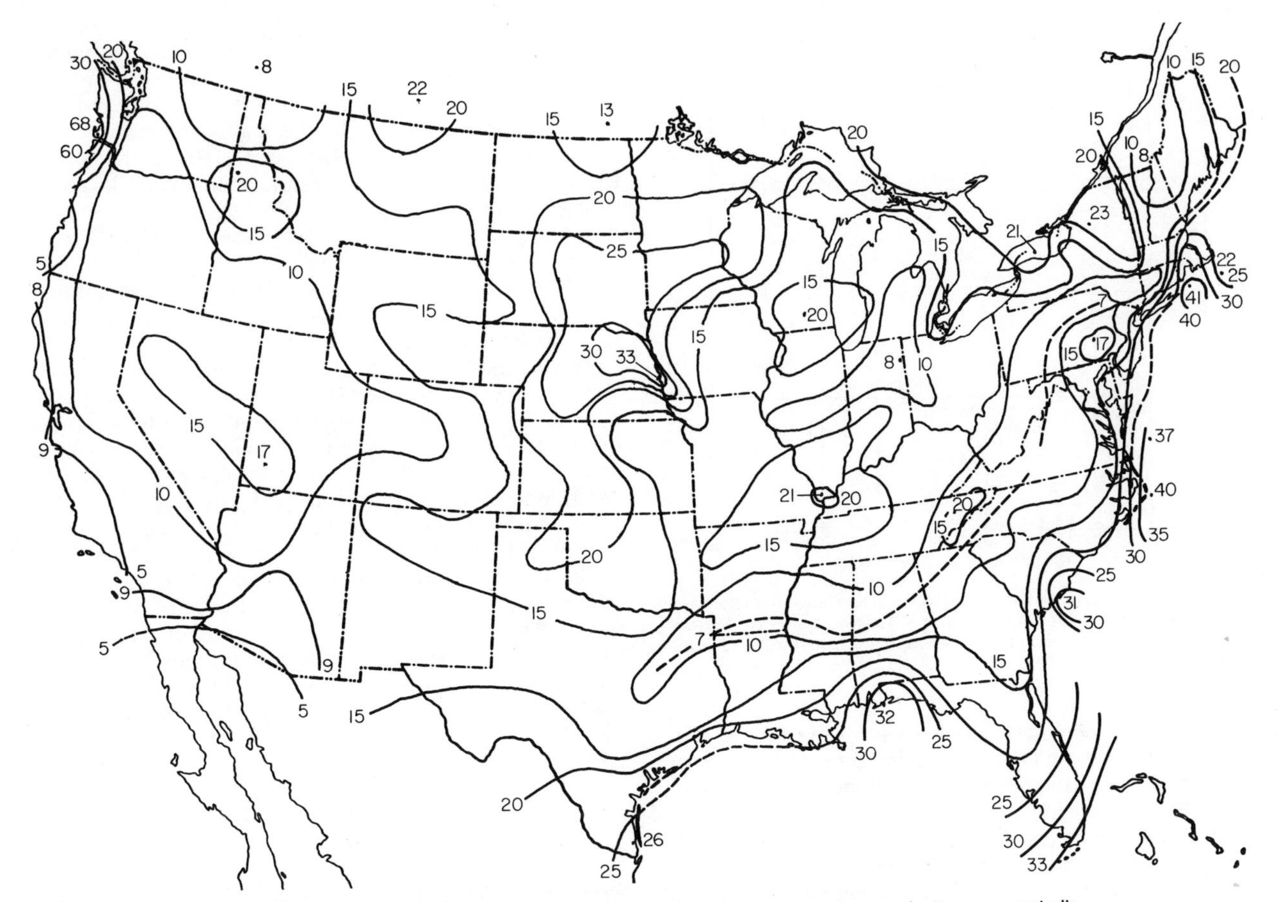

Fig. 1. Wind map of the United States. The contour lines represent maximum velocity pressures in lb per sq ft. Use Table 1 to convert these pressures to equivalent miles per hour *(U.S. Dept. of Commerce).*

Table 1. Wind Speeds and Velocity Pressures

True wind speed, mi per hr	Velocity pressure, lb per sq ft
5	0.064
10	0.256
15	0.575
20	1.023
25	1.600
30	2.302
35	3.133
40	4.092
45	5.179
50	6.394
55	7.737
60	9.208
65	10.810
70	12.530
75	14.390
80	16.370
85	18.480
90	20.720
95	23.080
100	25.580
105	28.200
110	30.950
115	33.830
120	36.830

tion. It doesn't always, of course. Most people are also likely to assume that the wind pushes down against a pitched roof. This assumption is not quite accurate. Imagine a strong wind blowing against a house that has a gable roof (see Fig. 2). When the wind strikes the house, it will be deflected to the sides and upward. As the wind passes over the roof, its velocity must increase. As its velocity increases, its pressure must decrease. Depending on the shape of the roof, there may even be a negative pressure acting against the downwind side of the roof.

The roof may, in fact, be pulled up and away from the walls supporting it. Tiles and shingles that are "blown off" a roof are not pushed off by the wind; they are sucked off by negative air pressures. Everyone has seen photographs of wind-damaged houses that have had their roofs "blown off." What has actually happened is that a suction pressure has lifted the roof from the walls. I don't think anyone has seen a photograph of a house that has had its roof blown *in* during a windstorm.

For the typical gable roof, the negative pressure on the windward side of the roof increases in a regular way as the

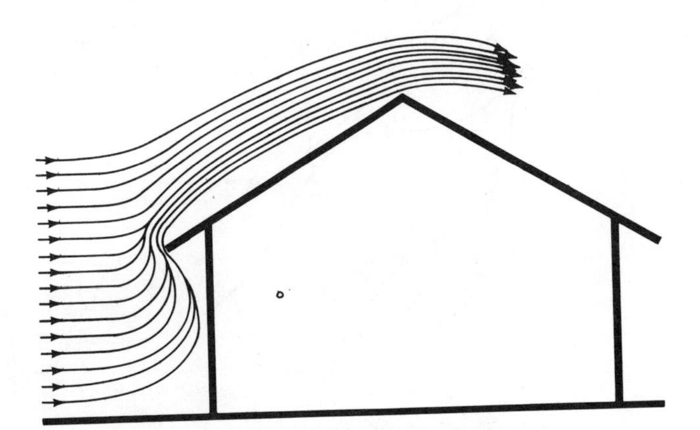

Fig. 2. Wind blowing against a house with a gable roof.

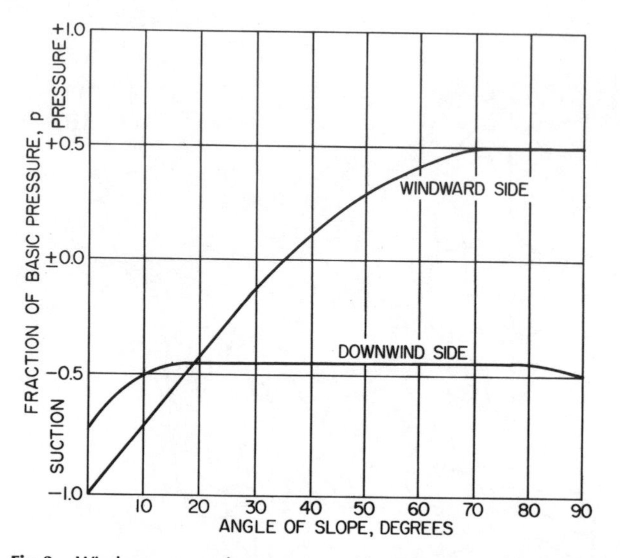

Fig. 3. Wind pressure and suction on a gable roof, as a function of the basic wind pressure. Note that the suction on the downwind side of the roof remains constant at about −0.5 of the basic wind pressure regardless of the roof slope.

pitch of the roof increases. This is shown in Fig. 3. At roof slopes of 30 degrees and more, the wind pressure turns positive and increases as the pitch of the roof increases.

As for the walls, wind pressures acting on the downwind side of a house will be less than those acting on the windward side. Much depends on the basic shape of the house. In sum, however, a house is as vulnerable to suction forces as it is to pressure forces. When the windows rattle and the framework shakes during a high wind, it is a combination of positive and negative pressures acting against the house that are responsible for the shakes and rattles. To be more exact, it is the relative changes in air pressure between the inside and outside of the house that are responsible for the shakes and rattles.

The roof is more vulnerable to wind pressures than the walls for two reasons. The first we have mentioned already—the strong negative forces acting against the roof. The second is that in most dwellings the juncture between the walls and roof rafters is structurally one of the weakest parts of the overall house construction. This is especially true for platform-framed houses in which the rafters are nailed to a *top plate,* which is itself none too securely nailed to the wall studs (*see* WOOD-FRAME CONSTRUCTION). The great difficulty in attaching the rafters to the walls in a wood-framed house is that it is very difficult to build this joint so that it will resist tensile stresses, that is, stresses that tend to pull the nailed joint apart. The joint must be reinforced if it is to resist high wind loads. For a description of reinforcement methods, *see* ROOF FRAMING.

In those parts of the country where tornadoes or hurricanes are common, local building codes often require that the roof, walls, and foundation of a platform-framed house be securely joined together with iron strapping. (Incidentally, there is no point in trying to design a house against a direct hit by a tornado—the wind velocities within the funnel of a tornado have been estimated to be as high as 500 mi per hr, which is equivalent to wind pressures of about 75 lb per sq ft.)

SNOW LOADS

A roof is usually designed to support a dead load of 20 lb per sq ft. This includes the weight of the rafters and roofing material, whether it be asbestos shingles, wood shingles, asphalt roll roofing, clay tiles, or whatever, although the roof is usually

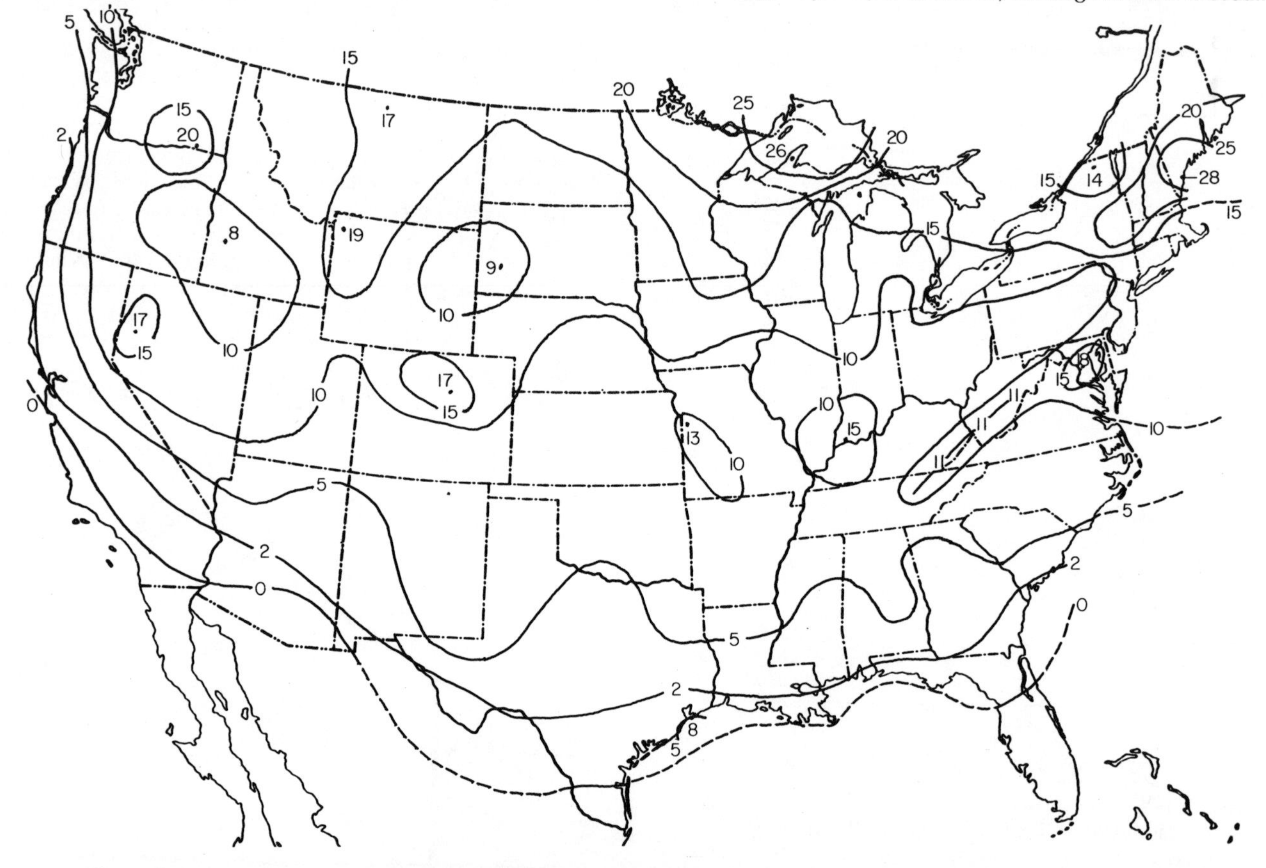

Fig. 4. Snow load map of the United States. The contour lines represent the basic snow load in lb per sq ft *(U.S. Dept. of Commerce).*

built a bit stronger than normal if the rafters are required to support the weight of slate or clay-tile roofing. If a flat roof is to be used as a sun deck or roof garden, its design live load must be increased accordingly. The total dead and live loads on a roof garden are assumed to be 100 lb per sq ft; on a sun deck, 60 lb per sq ft.

In addition, in northern parts of the United States, roofs are required to support the live load of snow for several months of the year. Snow weighs about 8 lb per cu ft when dry and from 10 to 15 lb per cu ft when wet or tightly packed. This snow load is calculated on the basis of the horizontal area covered by the roof, not the actual area of the roof itself, which, of course, will be greater when the surfaces of the roof are slanted. If the total anticipated snowfall for *two months* (the usual method of calculating the total snow load) exceeds 20 lb per sq ft, then it is the snow load that will determine how strong the roof must be constructed, not the weight of the rafters and roofing material.

Figure 4, a map showing the average snowfall in different parts of the United States, indicates in a rough way how strongly a roof must be built in different parts of the country. The requirements of local building codes can be expected to reflect local experiences. One place in the Northwest, for example, has had snow loads greater than 240 lb per sq ft, and it is not uncommon in many parts of the Rockies for the snowfall to reach 25 ft, which is equivalent to a snow load of 300 lb per sq ft.

There is an alternative to strengthing the roof rafters when exceptionally heavy snow loads are anticipated. This is to make the roof steeper. The steeper a roof, the less the maximum anticipated snow load for the obvious reason that snow has difficulty sticking to a steep surface. A roof that slopes 60 degrees or more hasn't any snow-load requirement at all.

The question arises: How does one calculate combined snow and wind loads? The answer is that one never assumes that heavy accumulations of snow and strong winds will occur simultaneously, which seems sensible. If the anticipated snow loads are greater than the anticipated wind loads, the snow loads will determine the roof design; otherwise the wind loads will be the determining factor.

EARTHQUAKE LOADS

The map in Fig. 5 shows the relative frequency and intensity of earthquakes in the United States. In zone 0, no danger what-

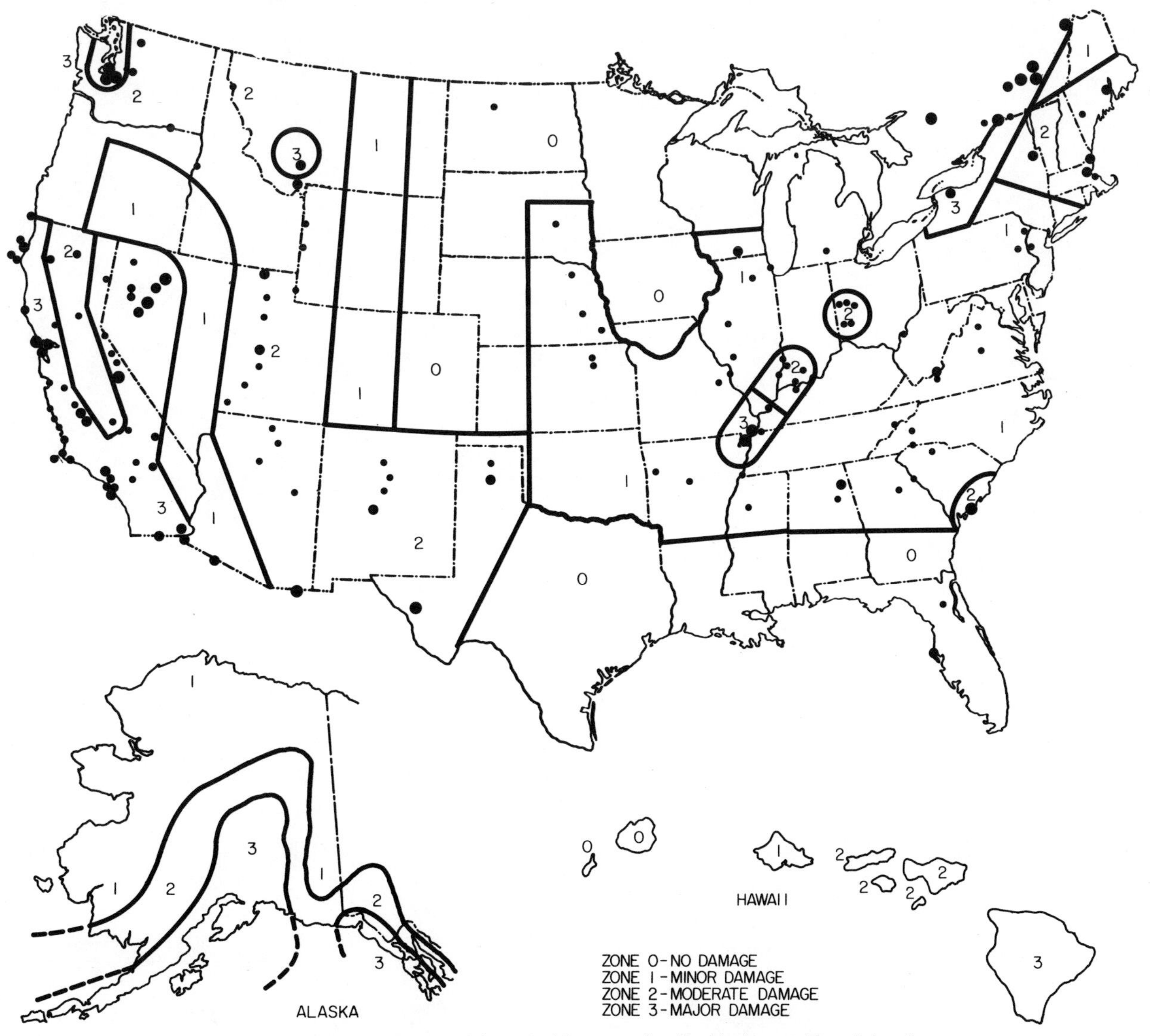

Fig. 5. Earthquake zones of the United States, as described in the text. The relative sizes of the dots indicates the relative intensity of the earthquake shocks experienced.

ever exists from earthquake shocks and no special precautions need be taken against earthquakes in the construction of a house. In zones 1, 2, and 3, the probable amount of damage resulting from an earthquake doubles as the number of the zone increases. Thus, zone 2 is expected to experience twice as much damage from an earthquake, on the average, as zone 1 will; and zone 3 is expected to experience twice as much damage from an earthquake, on the average, as zone 2 will. The positions of the dots indicate roughly the locations of earthquake epicenters; their size indicates the intensity of the earthquake shocks.

An earthquake occurs whenever the earth relieves an internal stress that has built up within it. These stresses occur between two great masses of rock that meet each other along what is called a *fault*. Whatever the reason for the buildup of the stress, it is relieved when these two great masses adjust their positions relative to each other along the fault. It is this suddent adjustment in positions that is felt as an earthquake shock. The shock is immediately followed by several smaller shocks, which are actually vibratory motions as these masses settle down to their new positions.

The shock and the vibrations that follow it result in horizontal movements on the surface of the earth, movements that tend to whip a house back and forth several times in a series of short, sharp, horizontal jolts. Most building codes ignore these earthquake loads in their required design loads for one- and two-family dwellings when the dwellings have a wood framework. The reason is that wood is peculiar among construction materials in that it can absorb extremely large—but momentary—stresses without suffering any damage or failure. Since the loads imposed on a house by an earthquake are basically horizontal loads of the same sort that are imposed by wind loads, though the earthquake loads are much shorter and sharper, of course, a wood-frame house that can resist the usual design wind loads of 15 lb per sq ft is considered to be capable of resisting the shock loads produced by most earthquakes. Therefore, for a wood house, there is no need to take earthquake loads into account in the design of the house.

Masonry houses, on the other hand, are far more likely to require special construction techniques to resist earthquake loads. Brick, concrete, concrete blocks, and stonework are much more brittle materials than wood, and they are held together by rigid mortar joints. Masonry hasn't the resilience of wood to sudden, sharp loadings. An unreinforced masonry wall will very likely suffer some damage in even a moderate earthquake. If the shock is severe, an entire masonry wall may collapse. In earthquake-prone areas, therefore, masonry buildings must be especially reinforced to resist sudden horizontal loads. For a description of the necessary reinforcements, *see* CONCRETE-BLOCK CONSTRUCTION.

Built-up Roofing

Built-up roofing is a method of applying a permanent, waterproof covering to a flat or moderately sloped roof, that is, a roof having a slope not greater than 3 in 12.

HOT-BITUMEN METHOD

Several layers, or *plies,* of felt that have been impregnated with a bituminous compound are laid down, each ply being cemented in place with hot bitumen. The roofing felt is then covered with a layer of gravel or crushed slag that will protect the bitumen and felt against weathering and the deleterious effects of the sun.

Compared to sheet-metal roofing, which is the other method of installing roofing to protect a flat roof from the weather, built-up roofing is much cheaper, if nowhere as long lived. Built-up roofs are described as 10-, 15-, or 20-year roofs, depending on the number of felt plies that are laid down. These lifetimes are guaranteed by the manufacturers of the materials used, as long as their materials are installed according to their specifications and in the presence of their inspectors. (But, sad to say, domestic installations are not guaranteed in this way by manufacturers, although a contractor may do so.) The actual lifetimes of built-up roofing are, however, much longer than the specified lifetimes.

Installation Techniques

Either coal-tar pitch or asphalt can be used as the bituminous coating; it depends mainly on the slope of the roof. (Note that *bituminous* and *bitumen* are generic names that refer to both coal-tar pitch and asphalt. For convenience, *bituminous* or *bitumen* will be used in this article when both tar and asphalt are meant; otherwise the reference will be specifically to either tar or asphalt.) Tar has a lower melting point than asphalt and in hot weather it will flow very slowly into any holes or splits in the roofing, thus maintaining its integrity. Tar is also more water-resistant than asphalt, which makes it a more suitable material for completely flat or very slightly pitched roofs on which standing pools of water are likely to remain after a rain. On a sloped roof, however, tar tends to flow slowly downhill in hot weather and asphalt then becomes the material of choice.

Roofing felt is usually described as being 15-lb or 30-lb material, which is the weight of one layer of the felt when it covers 100 sq ft of area. The felt is sold in 36-in.-wide rolls. The felt must match the bitumen used. That is, if tar is being used as the cement, the felt must be saturated with tar. If asphalt is being used, the felt must be saturated with asphalt. All felts were made originally from rag stock. Those that still are are called rag felts, since felts are also made of asbestos or glass fiber now. Both these materials are fireproof and rotproof, which rag felts are not, and they are used more and more for this reason. Asbestos felts contain about 15 percent rag content, the rag fibers providing a cohesiveness to the felt that asbestos by itself lacks.

Figure 1 shows how the felt plies are laid down and mopped over for 20-year roofing. A 15-year roof would have one ply less, and a 10-year roof would have two plies less. The description that follows is for a 20-year built-up roof.

First, a layer of 5-lb building (or sheathing) paper is nailed to the roof sheathing to protect the interior of the house against any liquid bitumen that might otherwise find its way through cracks or joints in the sheathing (assuming, that is, that the underside of the roof is not covered by plaster, gypsumboard, or some other kind of interior finish). The strips should overlap by about 2 in. If the underside of the roof is not exposed to view, the building paper should be nailed down using galvanized roofing nails that are just long enough to penetrate the sheathing. If the underside of the roof *is* exposed to view, the nails should *not* penetrate the sheathing.

Two "dry" plies of 15-lb felt are then laid down, the strips overlapping by about 16 in., as shown in Fig. 1. They are called dry plies because they are not mopped down with hot bitumen but are held in place by roofing nails. Note that the first strip should extend over the edge of the roof by 18 in. This overhanging strip will be folded back later, just before the gravel stop is installed (as described below) to provide a watertight edge to the roofing.

Additional strips of 15-lb felt are now laid down. Note in Fig. 1 that the first strip, next to the edge of the roof, is only 10⅔ in.

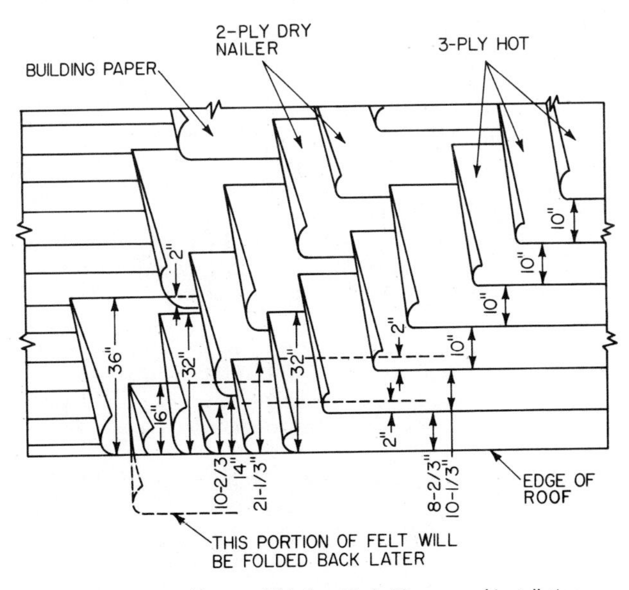

Fig. 1. Application of layers of felt for a 5-ply 20-year roof installation.

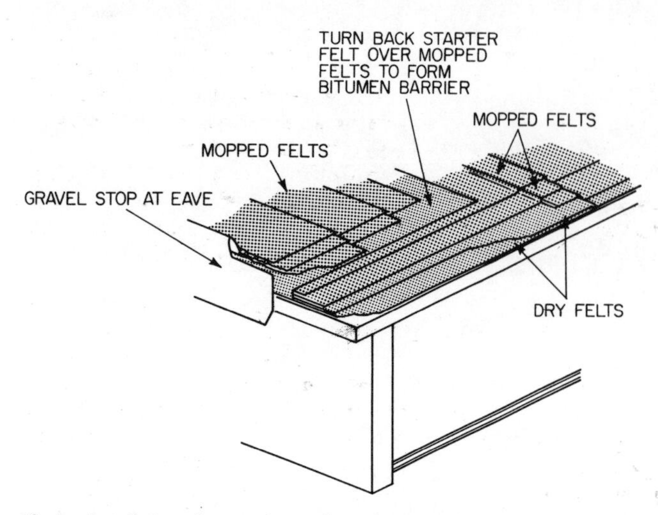

Fig. 2. Installation of a gravel stop along the eave of a roof.

wide; that the second strip of felt is only 12⅓ in. wide and it, too, begins at the very edge of the roof; and that the third strip of felt is 32 in. wide and it, too, begins at the very edge of the roof. Thus, there are three layers of felt abutting the edge of the roof, each of which is mopped down with hot bitumen. Arranging the plies in this way ensures that there will be equal thicknesses of felt over the entire area of the roof. The remainder of the strips are then laid down at 10-in. intervals, each strip being mopped down with hot bitumen.

The basic principle that must be adhered to when mopping down the felts is that one felt should never touch another. It is the bitumen that provides the watertight protection, not the felts. It is, therefore, the integrity of the coatings of bitumen that should be maintained. The function of the felts is primarily to strengthen the installation. Mopping down the hot bitumen with this in mind and making sure also that the felt is adhering firmly to the bitumen are the tricky parts of installing built-up roofing; they are the reasons why it is always best to employ a crew of skilled workmen.

Once all the felts have been mopped down, the 18-in.-wide strip of felt that has been hanging over the edge of the roof is folded back on top of itself and a galvanized steel *gravel stop* (its name is self-descriptive) is installed all along the edge of the roof, as shown in Fig. 2. The gravel stop is first bedded down in a thick layer of *flashing compound,* which is a very thick, cold asphalt emulsion, to make a watertight seal. The gravel stop is then nailed to the roof sheathing (through the mopped felt plies) with corrosion-resistant threaded roofing nails. Two narrow strips of felt, the first 6 in. wide and the second 9 in. wide, are then mopped down on top of the gravel stop with bitumen to seal in the nails and to prevent moisture getting under the gravel stop.

The final step in the installation of built-up roofing is pouring a coat of hot bitumen over the entire roof, at the rate of about 150 lb per 100 sq ft of roof. Crushed gravel or slag is immediately poured into the hot bitumen. The gravel is the size known as *pea.* It ranges between ¼ and ⅝ in. in size and is applied at the rate of 400 lb per 100 sq ft. Crushed slag, which is the same size as the gravel but lighter in weight, is applied at the rate of 300 lb per 100 sq ft. The gravel or slag must be embedded solidly in the bitumen.

If in Fig. 1 you count the total number of dry and mopped plies of felt at any one point, you will note they total five. The 5-ply overlap has occurred because of the way in which the felts were spaced as they were laid down. A 4-ply roof (i.e., a 15-year roof) is installed in exactly the same way as the 5-ply roof just described except that the mopped plies are spaced 17 in. apart. For a 3-ply (i.e., a 10-year) roof, the mopped plies are spaced 33 in. apart. It is this difference in felt spacing that is the fundamental difference between a 10-, a 15-, and a 20-year roof.

Alternative Techniques

Instead of laying down two dry plies of 15-lb felt to cover a bare roof, as shown in Fig. 1, a single ply of 30-lb felt, the strips being spaced 32 in. apart, can be nailed down. The felt will overhang the edge of the roof by about 18 in. as already described. The plies are then mopped down and the overhanging strip is then folded back upon itself, as already described.

Instead of pouring a flood coat of hot bitumen and then embedding crushed gravel or slag in it, 120-lb mineral-surfaced *cap sheets* can be mopped down on top of the 15-lb mopped felts. These cap sheets have had crushed particles of slate or mica embedded into their top surface during their manufacture. The particles serve the same protective function as crushed gravel or slag do; however, cap sheets are not as strong nor do they offer the same degree of protection to the roof as gravel or slag do, the latter two materials being preferred on flat roofs.

Cap sheets are preferred to crushed gravel or slag only when built-up roofing is installed on a roof having a pitch greater than 3 in 12. The cap sheets overlap by only 1 in. The mopped plies of 15-lb felt are spaced as already described for a 10-, 15-, and 20-year roof. As there is no loose gravel or slag, a gravel stop is unnecessary. Instead, a galvanized-steel *drip edge* is installed to prevent water soaking into the roof sheathing and then working its way back under the roofing by capillary action. The drip edge is nailed down on top of the dry plies (which are *not* folded back upon themselves in this type of installation) and both the mopped plies and the cap sheet are then mopped down over the drop edge.

COLD-PROCESS ROOFING

A modified version of built-up roofing uses a cold asphalt emulsion instead of hot asphalt as a cement and waterproofing membrane. This, of course, does away with the need for a kettle in which to melt the asphalt and the need for a large crew of men, and with the smells and combustion danger as well. The asphalt emulsion, which is known as asphalt *roofing cement,* pours very, very thickly. The better the quality, the thicker it is. It is usually applied by scooping it onto the ply and then spreading it about evenly with a serrated trowel, similar to the trowel that is used to apply the mastic for asphalt floor tiles.

The solvent in the emulsion acts as a binder. As the solvent vaporizes, it softens the asphalt that has been saturated into the felt during its manufacture, which allows the felt and roofing cement to adhere together.

Otherwise, the general installation procedure is much the same as when hot asphalt is used. Roofing applied by the cold-process technique is also described as a 20-, 15-, or 10-year roof, depending on the number of plies laid down. However, as there is no final flood coat of asphalt to finish off the job, 120-lb or 80-lb cap sheets are used instead.

The whole point of using the cold-process technique is that it is less expensive, less messy, less bother, and less difficult to do. When a homeowner finds it necessary to redo his roof, and he wants to do the job himself, he will invariably use cold-process materials. Basically, however, the cold-process technique is intended to maintain an existing roof; for patching an area, say, not to lay down an entire new roof. For that, the hot-bitumen method described above is always to be preferred.

Chimney

This article describes the principles underlying the design and construction of chimneys, which are basically air shafts through which smoke and combustion gases can escape from the fireplace, boiler, and/or furnace in which they originate. The air shaft itself is called a *flue,* but a *chimney* includes both the flue and the surrounding construction, which is traditionally of brick or stone but often consists nowadays of an insulated metal pipe. *See also* FIREPLACE.

CHIMNEY DESIGN

A chimney (see Fig. 1) actually has two functions. It is a passageway through which combustion products and smoke are discharged harmlessly into the atmosphere, and it also sets up a draft that draws fresh air into the fireplace or combustion chamber to sustain the combustion. Obviously, the more efficient the withdrawal of combustion gases from the combustion chamber, the more efficiently is additional fresh air provided to sustain the fire.

To produce an adequate draft, the chimney must have height. The hot combustion gases, being less dense than the atmosphere, spiral upward within the flue and are discharged from the top of the chimney. If the outside air is quite calm, the gases may continue to rise in a straight column for several feet, which increases the effective height of the chimney and thus the strength of the draft. Sometimes a breeze will blow across the top of the chimney in such a way as to suck the gases out of the flue, which also effectively increases the height of the chimney; of course, one shouldn't count on these vagaries of the wind to sustain a draft.

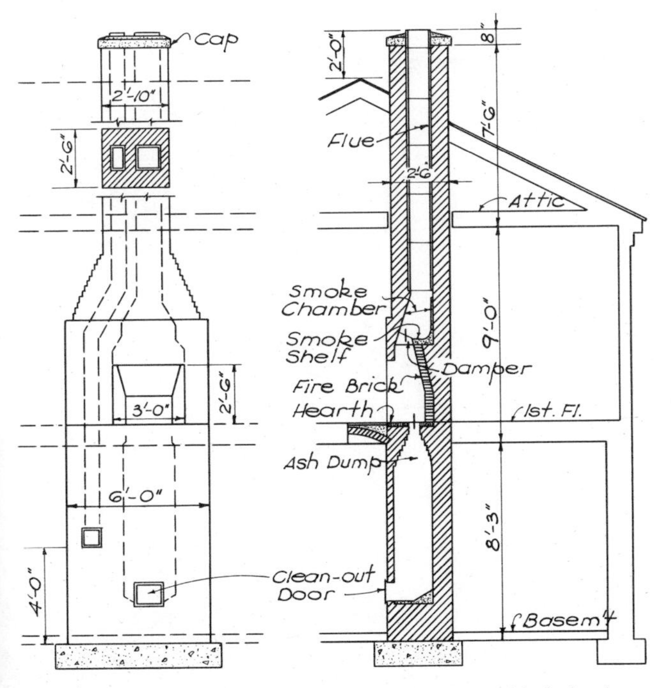

Fig. 1. A typical chimney design for dwellings *(U.S. Dept. of Agriculture).*

An efficient draft requires, for one thing, that the passageway between the combustion chamber and the top of the chimney be completely airtight, since any cracks or openings in the *smokepipe* (described below) or flue will reduce considerably the strength of the draft.

For another thing, an efficient draft requires that the difference in temperature between the atmosphere and the combustion gases be as great as possible. A large temperature differential means there will be a large difference in the density between the atmosphere and the combustion gases, a difference that ensures the gases will rise rapidly through the flue.

An *interior* chimney (one built entirely within the walls of a house) functions more efficiently than does an *exterior* chimney (one which forms part of the exterior wall of a house) because in an interior chimney the rising column of combustion gases is likely to remain hotter longer. An exterior chimney may be affected by cold winds and freezing rains. If cold penetrates the brickwork, it will lower the temperature of the flue. This will in turn increase the density of the combustion gases and, therefore, their rate of ascent will decrease. The result will be a fluttering, smoky fire that refuses to burn strongly. It is a basic principle of sound chimney design, therefore, that the exterior walls of a chimney be thick enough to prevent, or minimize, chilling the column of rising gases. Cold air sucked into the flue through cracks in the masonry or through badly fitted joints in the smokepipe also has a chilling effect on the combustion gases, which is one reason why such openings should be guarded against during construction.

It sometimes happens that a downdraft of air will enter a flue and chill the ascending column of hot gases. Such a downdraft may occur because the configuration of the roof and chimney (or the configuration of nearby houses, trees, or topography with respect to the chimney) is such that winds blowing from a given direction will tend to blow down the chimney. Downdrafts may also occur because the flue is too large for the amount of combustion gases produced by the fire; that is, while the combustion gases are rising up through the center of the flue, cold air is descending along the sides of the flue.

Sizing a flue should never be a problem to a builder. Most manufacturers of boilers, furnaces, gas appliances, or prefabricated fireplaces will specify in their literature the optimum size and minimum height of flue required for the most efficient operation of their equipment. For the typical dwelling, a square flue 9 × 9 in. in size or a round flue 8 in. in diam. is usually considered adequate.

Problems may arise, however, when one type of heating

equipment is replaced by another. When, for example, a coal-burning furnace is replaced by a gas-fired furnace, the flue will immediately become too large for the amount of combustion gases produced by the gas fire, since gas is a much more efficient fuel than coal. This particular problem can be easily dealt with, however, by installing a constriction at the outlet of the furnace.

On the other hand, a great many homes have been built in which the chimney flue is sized for a gas-fired furnace. In these times of natural-gas shortages and increasing natural-gas prices, a family may decide to switch from gas to either oil or coal. If they do, they will discover that the chimney flue is now too small to sustain an efficient draft for the new fuel, and their fuel costs will be higher than anticipated. A family planning a new house today would be well advised, therefore, to size the chimney flues of the house for coal, even if they intend burning oil or gas. They will be able to use the oil or gas efficiently in a flue sized for coal, but the opposite is not true.

Whatever the minimum flue height recommended by a manufacturer, most building codes also have minimum chimney height requirements that must be met. Local codes usually require that the top of a chimney be 2 to 3 ft higher than the top of a gable roof or the top of the parapet on a flat roof. This requirement is meant to ensure that any sparks or burning materials ejected from the chimney will be carried high enough into the atmosphere to be cooled before they can land on the roof, where they might otherwise ignite some combustible material.

It is the general opinion that a chimney must be at least 15 ft high from smokepipe to chimney top if it is to draw well. A 15-ft chimney is sometimes esthetically unsuited to the low, ground-hugging style of architecture favored by many. A low chimney may be prettier but it will never produce as efficient a draft as a high chimney, and the owners of a low-chimneyed house may find they are bothered with smoky fires, excessive soot buildup in the flues, and excessively high fuel costs.

Flues and Flue Liners

There is a very close analogy between the way gases flow through a flue and the way water flows through a pipe. Just as water flows with maximum efficiency in a pipe that is smooth, free of obstructions, and without sharp bends, combustion gases rise in a flue most efficiently when the inner surface of the flue is smooth, free of projections or soot buildup, and without sharp bends or constrictions.

At one time, all chimneys were built without flue liners; the walls of the flue consisted of the bricks of which the chimney was built. In such chimneys, if the mortar joints aren't smoothly made, the flow of combustion gases is likely to be disrupted and the resultant turbulence will reduce the effective cross-sectional area of the flue. In addition, soot tends to catch in the crevices existing in the brickwork. In time, a heavy deposit of soot may build up that further reduces the cross-sectional area of the flue.

There are other problems with flue-linerless chimneys. During a heavy rainstorm with driving winds, water tends to soak into the bricks. As a result, the column of gases in the flue becomes chilled. If an excessive amount of water soaks into the bricks, which is far from improbable given the porous nature of brick and the exposed position of most chimneys, this excess water may find its way through cracks in the mortar and enter not only the flue but the interior of the house as well. If this should happen, the ceiling and walls of the house may be permanently stained by the soot and products of combustion that are washed out of the bricks by the water. It is impossible to remove stains such as these.

Finally, combustion gases contain both water vapor and (possibly) sulfur compounds. These combine to form a mild solution of sulfuric acid. Over a prolonged period of time, this acid will eat away the mortar holding the bricks together. This disintegration of mortar is most likely to occur at the very top of the chimney where the flue gases are most exposed to the weather and have the most opportunity to combine with rain or water vapor in the atmosphere, though it will happen to some extent along the entire length of the flue. The fragile, crumbly appearance of a great many chimney tops on old houses, which is evident to the most inexperienced eye, is caused by this prolonged acid attack.

For all these reasons, building codes today require that all flues be lined, usually with a vitrified-fireclay pipe having walls at least ⅝-in. thick. These flue liners are made in square, rectangular, and round cross sections, and in a wide range of modular and nonmodular sizes; the usual section length is 2 ft.

Flue liners have a great many advantages in the chimney construction. They act as heat insulators to some degree, and this improves the efficiency of the draft. They provide a uniform cross section, which also helps improve the efficiency of the draft, as does the fact that they have a smooth surface. Flue liners also protect the brickwork against excessive heat, which might cause the joints to crumble away in time. They also make the entire construction more airtight.

Since flue liners are made in square, rectangular, and round cross sections, which of them should be selected? Square and rectangular liners are obviously more useful with a brick chimney since a bricklayer finds it easier to lay bricks against a flat flue than against a round one; thus, the cost of installation is less. But a round flue is much more efficient than either a square or rectangular flue because, as we mentioned above, the natural tendency of the flue gases is to rise in a spiral. This is why an 8-in.-diam. round flue and a 9-in.-sq flue are equivalent. In a square-cornered flue, the corners are simply dead spaces that fill up with soot deposits and provide channels along which cold air can flow down the flue and chill the ascending gases.

Smokepipe and Soot Pocket

A *smokepipe* connects the appliance in which the fire is burning (whether kitchen range or stove, fireplace, boiler, or

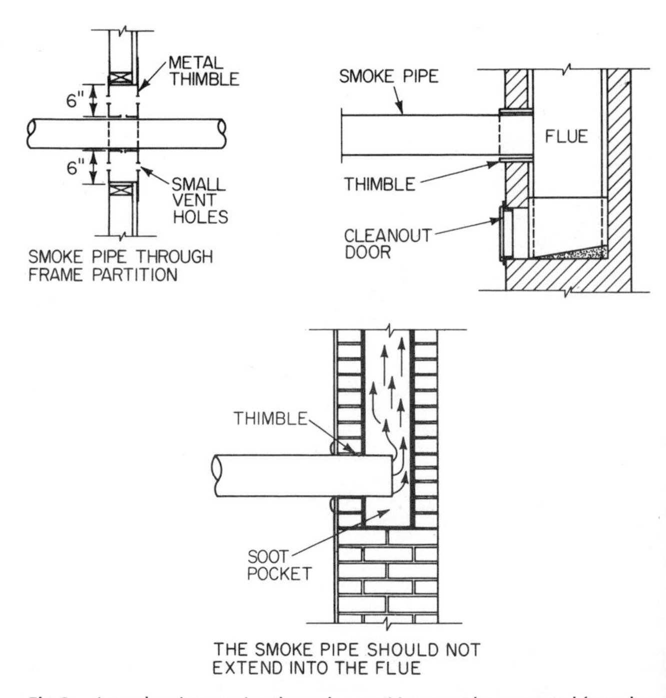

Fig. 2. A smoke pipe passing through a partition must be separated from the wall construction by a metal ventilating shield. The smoke pipe enters a chimney as shown above.

furnace) and the bottom of the flue (See Fig. 2). Each appliance must be connected to a *separate* flue through a *separate* smokepipe if each appliance is to operate as it is supposed to. Furthermore, each smokepipe should be as short as possible—and never more than 10 ft long—which means the appliances must be located as close to the chimney as possible.

The diameter of the smokepipe is usually specified in the manufacturer's literature. The way in which smokepipes are installed is closely regulated by building codes to prevent fires. For example, a smokepipe must never be closer than 9 in. to any wood construction or to any other combustible material. Whenever the smokepipe is less than 18 in. from any such combustible material, it must be wrapped in a fire-resistant material, such as sheet asbestos.

A smokepipe must never pass through an enclosed space, such as through flooring, to reach a flue, because of the danger of temperature buildup. The smokepipe must always enter a flue at the story level it originates on. If the smokepipe must pass through a wall or partition to reach the flue, a hole at least 12 in. larger in diameter than the smokepipe must be cut in the partition or wall and a double-walled, metal *ventilating shield* installed in order to separate the partition or wall from the smokepipe (refer to Fig. 2); either that or brickwork at least 4 in. thick must be built entirely around the smokepipe.

Finally, the smokepipe must enter the flue horizontally, and without actually extending into the flue, as shown in Fig. 2; if it does extend into the flue, the smooth flow of combustion gases up the flue will be disrupted. The hole in the bottom of the chimney through which the smokepipe passes is usually fitted with a metal *thimble* that fits closely around the smokepipe. The space between thimble and smokepipe is packed with *boiler clay* or *boiler putty,* and the space between thimble and chimney is filled with mortar, concrete, or fireclay. If the smokepipe should pass directly through brickwork without a thimble being installed, a close-fitting metal collar should be slipped over the smokepipe and the space between collar and brickwork filled solidly with concrete, mortar, or fireclay.

The space at the bottom of the flue, just below the smokepipe, is called a *soot pocket.* It provides a place where soot can accumulate. Since soot is combustible, it must be removed at periodic intervals—once a year usually. Access to the soot pocket is through a small cast-iron *cleanout door* built into the chimney. The cleanout door must fit tightly to prevent any air leakage into the flue, and each flue must have its own soot pocket and cleanout door. Any attempt to economize on the construction by having a common soot pocket that serves several smokepipes is very poor practice since the interconnection between the flues at the common soot pocket will interfere seriously with the draft in each individual flue. Nor should the soot pockets be made too deep; 10 in. below the bottom of the smokepipe is sufficient. If a soot pocket is made deeper than this, there is a chance that cold air may collect in it, which will chill the entire flue.

CHIMNEY CONSTRUCTION

Foundation

Brick weighs about 120 lb per cu ft. Knowing this figure, and the dimensions of a chimney, it is easy enough to calculate the chimney's overall weight. One must also know the loadbearing capacity of the soil on which the chimney is to be built. Knowing both the weight of the chimney and the loadbearing capacity of the soil, one can then design a footing that is large enough to support the weight of the chimney without the chimney's settling.

If the walls of a house are to be made of masonry, an exterior chimney can be considered as an integral part of the wall

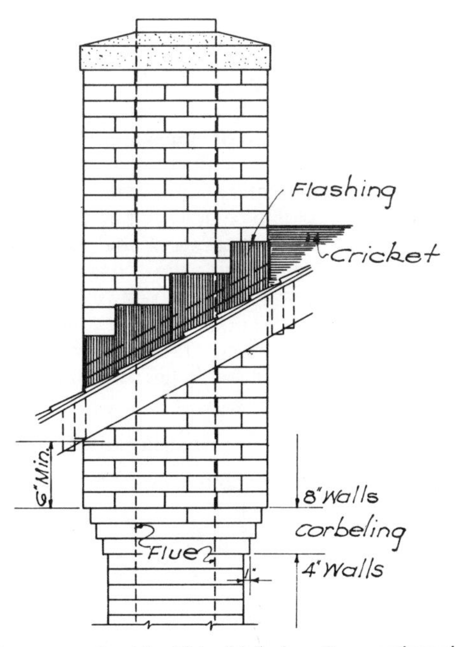

Fig. 3. A chimney must be at least 8 in. thick where it passes through a roof to prevent premature chilling of the hot combustion gases rising up through the flue *(U.S. Dept. of Agriculture).*

construction and the footings designed accordingly.

If, however, the house is of wood-frame construction, then the difference in weight per unit area between the chimney and the rest of the house would be so disproportionate that a separate foundation must be built to support the chimney; otherwise any settling of the chimney would also cause the house to settle. Chimney footings for a one-story house are considered adequate if they are 8 in. thick and extend at least 6 in. beyond the chimney in every direction. For a two-story house, the footings should be 12 in. thick.

Brickwork

The technique of laying brick to construct a chimney is exactly the same as described for building up the walls of a house (see BRICKLAYING). Those parts of the chimney that are also part of

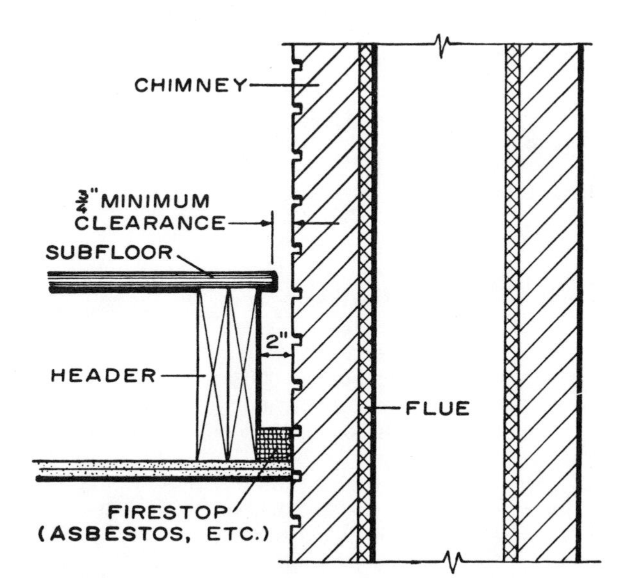

Fig. 4. An air gap at least 2 in. wide must be left between any part of a chimney and the wood framing that surrounds the chimney *(U.S. Forest Service).*

the exterior wall are usually 8 in. thick, whether flue linings are incorporated in the construction or not. The interior parts of a chimney are usually 4 in. thick. Sometimes an attempt is made to reduce the cost of construction by making the exterior side of a chimney only 4 in. thick. This practice is unwise because it increases the likelihood that the flue gases will become chilled during cold weather and so reduce the efficiency of the draft.

It is especially important that those parts of the chimney that extend above the roof line be at least 8 in. thick all around, as shown in Fig. 3, not only to prevent any potential chilling of the flue gases but also to enable the construction to resist any bending stresses that may be imposed on it by high winds. Such winds might easily bend the brickwork in a 4-in.-thick chimney enough to open up cracks in the mortar joints—cracks that will inevitably enlarge over the years through normal weathering.

Firesafe Construction

As already mentioned, building codes require that a chimney be isolated from the wood-frame structure of a house to prevent the wood from catching fire. Not that the bricks themselves will ever become that hot, but because, if there should be cracks in the mortar joints, flames could find their way out of the chimney. In this connection, one should remember that soot is combustible, and in an old chimney the entire flue may be covered thickly with soot. If this soot should ever catch fire, and if there should be any cracks in the construction, there is a serious danger that the flames would spread from the chimney into the rest of the house. This is another reason why flue liners are so strongly recommended for all chimneys. They will protect the house from any fires that might originate in the flue.

Therefore, there is only one circumstance allowed in chimney construction when wood is ever to be placed in direct contact with the brickwork. This is when 1½-in.-wide furring strips are nailed to the *corners only* of a chimney so that lathing or mouldings can be adequately supported. Otherwise, all wood studs, furring, and lathing must be kept at least 2 in. away from the chimney, as shown in Fig. 4.

Concrete

Concrete is a mixture of portland cement, water, sand, and crushed gravel or stone. The water and cement form a *cement paste* in which the sand and stone or gravel are mixed. The sand and stone or gravel together make up the *aggregate* of a concrete mixture. The aggregate serves no structural function. It is merely a filler that adds low-cost bulk to the cement paste; it usually makes up about 75 percent of a given mass of concrete, by volume. Although a poor aggregate can reduce the strength of a batch of concrete considerably, good aggregate adds only slightly to the strength of the cement.

ADVANTAGES AND DISADVANTAGES OF CONCRETE

The two principal advantages of concrete as a construction material are its relative cheapness and the ease with which it can be handled and placed while it is in the plastic state. The principal *structural* advantages of concrete are its great compressive strength and its durability. Concrete can withstand very high compressive loads. This is what makes concrete so suitable for the foundations, walls, and columns of buildings, and for driveways and walks as well.

The principal structural disadvantage of concrete is its poor tensile strength. That is, it cannot withstand pulling or bending loads without cracking or breaking. For this reason, steel rods, or *reinforcement steel,* are often embedded in concrete, the reinforcement steel providing the tensile strength the concrete lacks. Concrete with reinforcement steel embedded in it is *reinforced concrete.*

In addition to its poor tensile strength, concrete, like most construction materials, expands in hot weather and when wet and contracts in cold weather and as it dries out. Unless these movements are allowed for during construction, the concrete will crack.

Table 1. Compressive Strength of Concrete for Different Water-Cement Ratios*

Water-cement ratio, gal per bag of cement	Probable compressive strength at 28 days, psi	
	Non-air-entrained concrete	Air-entrained concrete
4	6000	4800
5	5000	4000
6	4000	3200
7	3200	2600
8	2500	2000
9	2000	1600

*Strengths are based on 6- by 12-in. cylinders moist-cured 28 days. With excessive percentages of entrained air, strength will be less than listed.

Source: ACI 613-54.

And, contrary to common belief, solid concrete is not impervious to water. Some moisture will migrate into the best-made concrete. But if the concrete should be excessively porous, which can happen if too much water has been used in mixing it, moisture can easily enter the concrete after it has cured. If this moisture should be present within the concrete when cold weather comes, the moisture may freeze, which may result in serious frost damage to the structure.

Despite these limitations, concrete is an inherently strong and durable construction material. If the proportions of water, cement, and aggregate are carefully calculated (as described later in this article), and if the concrete is placed and allowed to cure according to simple but definite rules, it is possible to obtain from the concrete all the strength and durability that is inherent in it.

MIXING CONCRETE

Water-Cement Ratio

As shown in Table 1, the ratio of water to cement in a batch of concrete is the principal determinant of the concrete's final strength. At one time the instructions for preparing a batch of concrete would have contained proportions such as 1:2:4, indicating that 1 part of portland cement to 2 parts of sand to 4 parts of gravel by volume were to be mixed together, after which sufficient water was to be added to obtain a workable mixture. This procedure ignored entirely the importance of the water-cement ratio. It also resulted very often in the preparation of a very weak concrete, since the natural tendency is to add enough water to make placement of the concrete as easy as possible—the sloppier the better, as far as the workmen are concerned. This manner of specifying the proportions of concrete is obsolete and should never be followed.

In theory, it takes only 3 gal of water to hydrate completely 1 cu ft of cement. (A sack of cement contains 1 cu ft exactly, and the sack weighs 94 lb.) But this water-cement ratio produces a mixture that is too stiff to be worked. In practice, therefore, additional water, between 4 and 8 gal per sack of cement, is used to obtain a workable mixture.

But, as is shown in Table 1, the greater the proportion of water in a water-cement ratio, the weaker the final concrete will be. The additional water that is necessary to achieve a workable batch will only evaporate from the concrete as the concrete sets, and it will leave behind in the concrete innumerable voids. This is the reason there will always be some porosity in concrete. But when an excessive amount of water has been used, there will be an excessive number of voids, which may cause the concrete to leak badly. If these voids should be filled with moisture when cold weather comes, they will cause the frost damage alluded to above.

As a general rule, therefore, 6 gal of water per sack of cement should be the maximum amount used for making concrete; and the less the amount of water that is used, the stronger the concrete will be. Also included in the 6 gal is whatever surface moisture is contained in the sand that is part of the aggregate, as described below.

Aggregate

Aggregate is divided into *fine* aggregate, which consists of all particles 1/4 in. and less in size, and *coarse* aggregate, which consists of everything else up to roughly 2 1/2 in. in size. The maximum size of the coarse aggregate in any particular mix depends mainly on the width of the forms into which the concrete is to be poured. As a general rule, the largest size particles should not exceed one-fifth the distance across the narrowest part of the forms.

Coarse aggregate (see Fig. 1) is usually obtained by crushing a suitable stone, such as trap rock, granite, or hard limestone. Trap rock is preferred, when it is available, as it is an exceptionally hard, dense, and durable substance. After the rock has been crushed, it is graded as to size by being passed through a series of screens. These piles of segregated coarse aggregate are later mixed together in the desired proportions, as it is an even gradation of particle sizes that makes a good concrete, not mere quantity.

Fine aggregate (see Fig. 2) consists essentially of sand. Quartz sand is preferred because of its hardness and durability. Some sands, such as those consisting of particles of feldspar or mica, should not be used as they tend to decompose upon exposure to air. Like the coarse aggregate, the best fine aggregate for concrete has an even gradation of particle sizes, though with a larger proportion of coarse particles.

Water

As for the water, any water that is good enough to drink is good enough for mixing concrete. The converse is also true. If water isn't good enough to drink, it isn't good enough for mixing concrete. Seawater can be used, but if it is, it will decrease the 28-day strength of the concrete by about 12 percent. One group of substances water should never contain are sulfates. Sulfates will attack the concrete and will gradually reduce its strength.

Fig. 1. The photograph above shows how a well-graded coarse aggregate looks before being separated by size, as shown in the photograph below. The smallest particles are 1/4 in. in size, the largest are 1 1/2 in. *(Portland Cement Association).*

Fig. 2. The photographs show an example of a well-graded sand before (top) and after (bottom) being separated by size. The particles range in size from dust to about 1/4 in. *(Portlant Cement Association).*

Slump Test and Concrete Proportioning

A freshly mixed batch of concrete will slump if it is poured on a flat surface. Experience has shown that concretes having particular slumps are desirable for different types of structures. The amount of slump is thus a useful indication of the workability of the concrete while it is plastic and its potential strength when cured. The amount of slump is measured by a standardized *slump test* in which a truncated metal cone, 12 in. high, is filled with a sample of the concrete. The concrete is tamped down solidly, the cone is raised, and the concrete will then slump. The amount of slump is then measured from the top of the truncated cone, as shown in Fig. 3.

The amount of slump depends in part on the water-cement ratio (which is given and should not be changed), in part on the

Fig. 3. (Left) A workman tamping fresh concrete into a cone in preparation for a slump test. (Right) Measuring the amount of slump from the top of the cone. This is a medium-wet concrete mixture *(Portland Cement Association).*

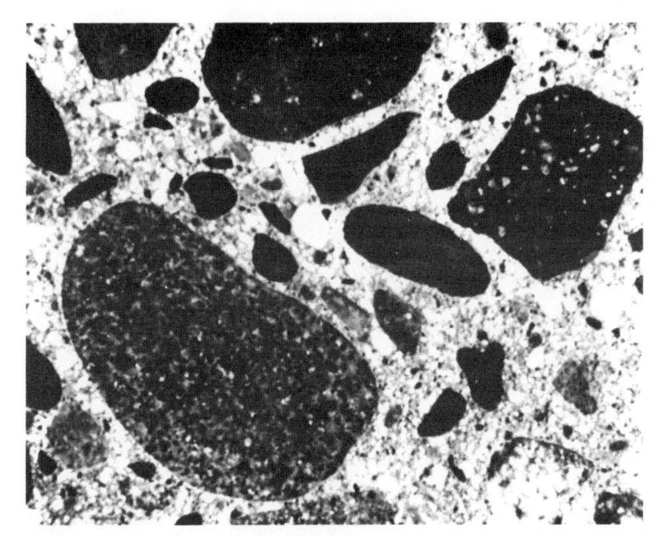

Fig. 4. A polished cross section through a sample of concrete. Each piece of aggregate is completely surrounded by hardened cement paste *(Portland Cement Association).*

total proportion of aggregate in the mixture, and in part on the ratio between the fine and coarse aggregates in the mixture.

Ideally, the aggregate should range evenly in size from the largest pieces of gravel or stone to the smallest particles of sand, as shown in Fig. 4. All the voids between the largest pieces of gravel should be filled by smaller pieces of gravel, and all the remaining voids should be filled by progressively smaller sizes of sand. In an ideal mix, only enough cement paste should be required to fill the voids left between all the particles of aggregate. Only then will one obtain the densest and strongest possible concrete mixture at the lowest possible cost.

On a large engineering project, the concrete can be especially formulated for the application. The physical characteristics of the fine and coarse aggregates can be ascertained and the exact proportions of water, cement, and coarse and fine aggregates can then be mixed together under closely controlled conditions to obtain the characteristics desired in the concrete, which are chiefly a specified 28-day compressive strength and a certain slump.

On a small job, however, such as the pouring of the foundations for a house, while it is easy enough to control the water-cement ratio, it is not so easy to control either the particle sizes or the ratio of fine to coarse aggregate. Instead, one has to use what one has.

As for the water-cement ratio, one must remember to include as part of the ratio the water that has soaked into the sand pile. The amount of water in a pile of sand is easily estimated. As shown in Fig. 5, one squeezes a handful of the sand. If the sand is merely damp, the assumption is that the sand contains about ¼ gal of water per cu ft. If the sand is wet (and most sands are "wet"), it will contain about ½ gal of water per cu ft. Very wet sand will contain at least ¾ gal of water per cu ft and may contain as much as 1¼ gal if the sand is very fine-grained. Whatever the moisture estimate, this quantity of water must be subtracted from the amount of water added separately to the mix.

As for the total amounts of fine and coarse aggregates required in a batch of concrete, the only way to proceed on a small job is by trial and error, as it is impossible to tell without physically segregating the particles just how evenly they are graded in the aggregate. Practically, a builder will refer to tables such as Tables 2 and 3, which are useful primarily as *guides,* not as recipes to be followed exactly. Using the tables, therefore, a builder will work up trial batches which he can then check by the slump test.

Attempting to increase the workability of a batch of concrete by adding more water is a very common and a very bad mistake. Not only is the original water-cement ratio upset, which will reduce the strength of the concrete, but a too-fluid mixture will allow the coarse aggregate to separate from the mix and settle to the bottom of the form. This segregation of the coarse aggregate is one of the main causes of porous concrete that cracks easily and leaks badly.

The method used to check the workability of a batch of concrete is to trowel it, as shown in Fig. 6. Only three or four back-and-forth strokes of a trowel should be necessary to obtain a smooth surface. If the surface is still lumpy after this troweling, it is an indication of insufficient sand in the aggregate. If the batch works too easily and a film of water is brought to the surface, it is an indication of insufficient coarse aggregate.

After the proportions have been determined, sufficient concrete can be mixed to complete the job. If the amount of concrete required is less than about 1 cu yd, this quantity can

Fig. 5. (Left) If squeezed sand falls apart in the hand, it is damp. (Center) If the sand forms a ball, it is wet. (Right) If the sand sparkles and wets the hand, it is very wet *(Portland Cement Association).*

Table 2. Suggested Concrete Mixes with Separated Fine and Coarse Aggregates

Kind of work	Gal of water added to each 1-sack batch if sand is: Damp*	Wet† (average sand)	Very wet‡	Suggested mixture for 1-sack trial batches§ Cement, sacks (cu ft)	Aggregates Fine, cu ft	Coarse, cu ft
Concrete subjected to severe wear, weather, or weak acid and alkali solutions						
5 gal. of water per sack of cement (3/4-in. max. size aggregate)	4½	4	3½	1	2	2¼
Floors (such as home, basement, dairy barn), driveways, walks, septic tanks, storage tanks, structural beams, columns and slabs						
6 gal of water per sack of cement (1-in. max. size aggregate)	5½	5	4¼	1	2¼	3
6 gal of water per sack of cement (1½-in. max. size aggregate)	5½	5	4¼	1	2	3½
Foundation walls, footings, mass concrete, etc.						
7 gal of water per sack of cement (1½-in. max. size aggregate)	6¼	5½	4¾	1	3	4

*Damp describes sand which will fall apart after being squeezed in the palm of the hand.
†Wet describes sand which will ball in the hand when squeezed but leaves no moisture on the palm.
‡Very wet describes sand that has been subjected to a recent rain or recently pumped.
§Mix proportions will vary slightly depending on gradation of aggregates.

Table 3. Materials Required to Make 1 cu yd of Concrete with Different Sizes of Separated Coarse and Fine Aggregates

Maximum aggregate size, in.	Suggested mixture for 1-sack trial batches* Cement, sacks (cu ft)	Aggregates Fine, cu ft	Coarse, cu ft	Materials per cu yd of concrete Cement, sacks	Aggregates Fine cu ft	Fine lb	Coarse cu ft	Coarse lb
¾	1	2	2¼	7¾	17	1550	19½	1950
1	1	2¼	3	6¼	15.5	1400	21	2100
1½	1	2½	3½	6	16.5	1500	23	2300
1½	1	3	4	5	16.5	1500	22	2200

*Mix proportions will vary slightly depending on gradation of aggregates.

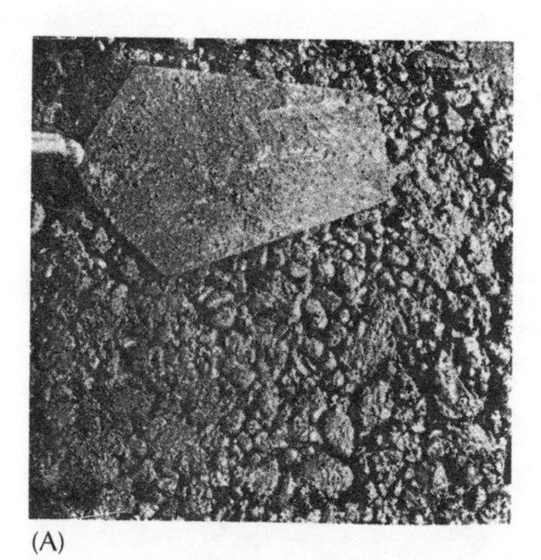
(A)

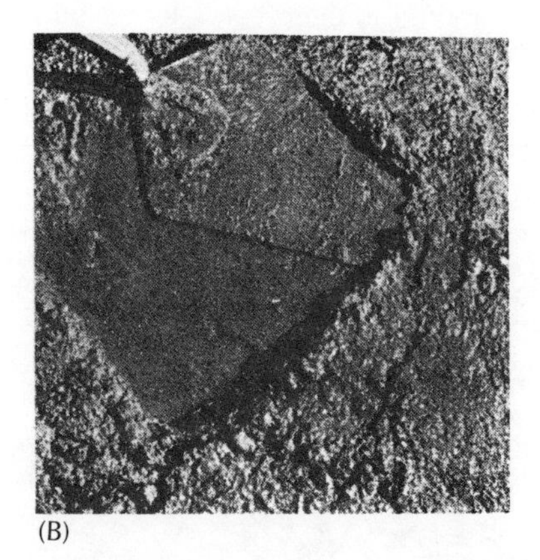
(B)

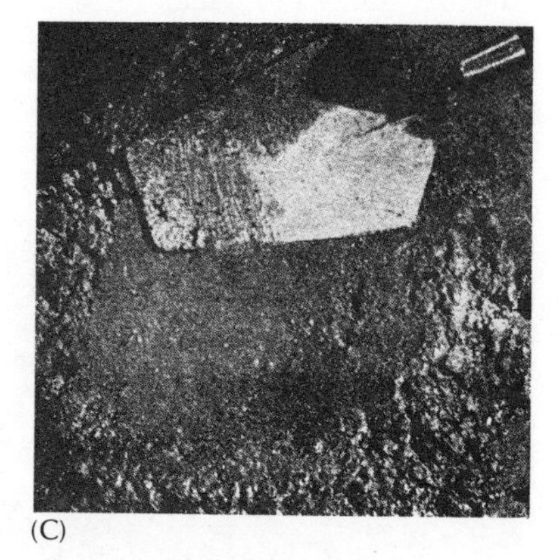
(C)

Fig. 6. The workability of concrete can be checked by working a batch over with a trowel. (*A*) Mixture with an excess amount of coarse aggregate. (*B*) Mixture with an excess amount of fine aggregate. (*C*) Mixture that is correctly proportioned *(Portland Cement Association).*

be mixed by hand in a tub. Working alone, one man can mix about 1 cu yd of concrete in an hour. If a mechanical mixer is being used, the ingredients can be added to the mixer in the same sequence as described above for working up test batches. The mixture should be mixed 1 min for every 1 cu yd, or portion thereof, in the mixer.

PLACING CONCRETE

Preliminary Preparations

Before concrete can be placed into a form, the form must be properly constructed. When placing concrete, one must be especially careful to produce a tight bond between the cement paste, the aggregate, and the sides and bottom of the formwork. That is, the concrete must be well compacted, excess air must be worked out of the concrete, and there should not be any voids or air pockets between the concrete and the formwork. The principal method by which solid, tight concrete is obtained on small jobs is by *spading;* on large projects a mechanical *vibrator* is used.

The single greatest mistake one can make while placing concrete is to allow the coarse aggregate to separate from the mix and settle to the bottom of the formwork. This may happen, for example, if the concrete has too high a slump or if it is spaded or vibrated excessively. In general, concrete has been spaded or vibrated a sufficient amount when the coarse aggregate just disappears below the surface of the concrete. If a film of water or thin mortar paste should appear on the surface of the concrete, the spading or vibration has been excessive.

Segregation of coarse aggregate can also occur if the concrete is allowed to fall freely more than 3 to 5 ft, or if the concrete is allowed to ricochet off the sides of the formwork or bounce off the reinforcement steel. Concrete will also segregate if it is allowed to travel down a sloping chute for any distance. All these faults should be avoided.

Adding Fresh Concrete to Hardened Concrete

Sometimes it is necessary to interrupt the placing of concrete before a form has been completely filled. In addition, it is the usual practice to allow footings and slabs to harden before walls or columns are placed on top of them. Before fresh concrete can be placed on top of hardened concrete, the surface of the hardened concrete must be prepared to make sure that there will be a tight, waterproof joint between the two layers.

The hardened concrete must be level, clean, and damp. If the concrete has dried out, it must be soaked for several hours beforehand to make sure it will not suck water out of the fresh concrete.

Finally, the surface of the hardened concrete must be rough, with a great many stones lying exposed on the surface to help lock the fresh concrete in place. The simplest method of obtaining the necessary roughness is simply not to smooth or spade the concrete before it does harden.

FINISHING CONCRETE

Screeding

The surface of fresh concrete is leveled and smoothed by drawing a long, flat straightedge across the top of the form with a back-and-forth motion. This procedure is called *screeding.* Screeding is usually a two- or three-man job. Two men work the straightedge forward while the third man shovels excess concrete from in front of the straightedge, or he adds concrete in front of the straightedge if there are any low spots.

The screeding should then be repeated since the first screeding will have raised a wave of concrete behind it as the straightedge is moved forward.

Floating

After the screeding has been completed, the surface of the concrete is immediately worked over with a wood *float.* Floating eliminates high spots, it fills in low spots, and it imparts a smooth if coarse texture to the concrete. Concrete should be floated just enough to smooth and level the surface. Excessive floating will bring a thin film of mortar to the surface which may become a source of potential trouble; it will only dust, spall, and crack with the passing of time.

Troweling

If a smooth, hard, dense surface is desired, the concrete must be worked over with a steel trowel after it has been floated and all the water has evaporated from the surface. Troweling should not begin until the surface of the concrete has begun to set. This point has been reached when the surface appears dull and firm, and troweling does not raise any water or mortar to the surface. If it does, the troweling has been started too soon.

It is not good practice to add water or dry cement to the surface in order to obtain a very fine, smooth finish, a technique that is often used with gypsum plaster on the interior walls of a house (*see* PLASTERING). It just won't work with concrete; fine cracks will in time appear on the surface and the surface will then begin to spall.

Brooming

A nonskid surface is usually obtained by drawing a broom having coarse, stiff bristles over the surface of the concrete immediately after it has been floated and while it is still soft.

CURING CONCRETE

When a batch of concrete is first mixed, it forms a plastic mass that can be poured into prepared forms with little effort. In about an hour the concrete sets into a rigid mass that weighs about 150 lb per cu ft. The cement paste is said to *hydrate,* that is, the individual particles of cement absorb the surrounding molecules of water into their molecular structures. As they do, the cement crystallizes into a kind of rigid gel, something like gelatine, that gradually changes with the passing of time into a solid mass of minute, interlocked crystals. The longer hydration continues, the stronger the concrete will become. It is standard engineering practice, therefore, to calculate the final design strength of concrete on the basis of a 28-day curing period, although under exceptional conditions concrete has been known to continue increasing in strength for a quarter of a century and longer.

The time during which concrete hydrates and increases in strength is its *curing period.* Concrete cannot hydrate, or cure, unless there is water present within the concrete. Throughout the curing period, therefore, all the exposed surfaces of the concrete must be kept moist. As long as the concrete is kept moist, curing will continue and the concrete will become increasingly stronger, denser, and more impervious to water.

Once concrete is allowed to dry out completely, however, hydration stops. Usually it is the surface of concrete that is adversely affected by a too-short curing period. For example, the surface of a sample of concrete that has been kept moist for 28 days will be twice as strong as a surface that has been kept moist for only 3 days.

The outside air temperature is an extremely important factor in proper curing. If the air temperature is too high, over 90°F, say, the water in the concrete may evaporate away before hydration can be completed. An excessive internal temperature may also interfere with proper hydration. Concrete that has been mixed and placed during very hot weather is never as strong as concrete that has been placed when the air temperatures are 70°F and below, mainly because of the difficulty of keeping the concrete properly moist. Furthermore, concrete that sets too quickly during hot weather is more likely to crack afterward because it will have shrunk an excessive amount during its curing, and it will not thereafter be able to withstand the stresses imposed on it by large changes in temperature.

Nor can concrete be placed during freezing weather unless special precautions are taken to keep the concrete above 50°F for at least 4 days after it has been placed. If the temperature of the concrete should fall below 50°F, it will never harden properly. And if the concrete is allowed to freeze before it has set, it will be permanently damaged.

To cure concrete properly, the exposed surfaces must be kept continually moist from the moment the concrete first begins to set. The concrete must thereafter be kept continually moist for a minimum of 14 days, and longer if at all possible.

The simplest method of keeping concrete moist is to spray the surface with water at frequent intervals. Sand or burlap can be spread over the surface to help retain moisture. A covering is necessary, in any case, for the first 3 days after placing to protect the concrete from the direct rays of the sun.

Special curing compounds can also be sprayed on the concrete. The spraying should take place as soon as the surface of the concrete has lost its watery appearance. Properly applied, a curing compound will allow the concrete to continue curing even after the concrete has been placed in service, as with a concrete highway, for example. Spraying on a curing compound is often the only practical method of curing concrete that has been poured into an unusual shape.

A third method of curing concrete is to spread a sheet of polyethylene film or building paper over the surface of the concrete. The polyethylene or building paper should overlap the sides of the exposed concrete. If several sheets of film or paper must be laid down to cover the concrete, the sheets should overlap by at least 12 in., and the edges of the sheets should also be weighted down in some way. This covering must remain on top of the concrete for the entire curing period.

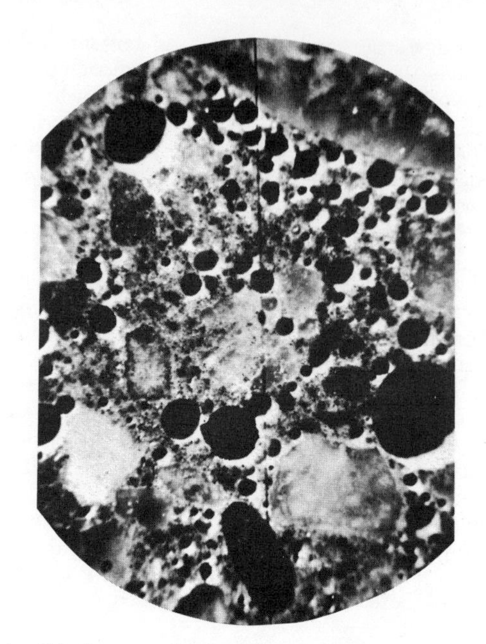

Fig. 7. A polished cross section through a sample of air-entrained concrete, as seen through a microscope *(Portland Cement Association)*.

AIR-ENTRAINED CONCRETE

There are a great many substances that will produce an extremely large number of very small air bubbles under suitable conditions. If one of these *air-entraining agents* is added to cement, the result is *air-entrained* concrete. After such a concrete has been poured and has set, dispersed within every square inch of the concrete will be innumerable extremely small bubbles (i.e., ranging from 0.003 to 0.050 in. in size), as shown in Fig. 7. These bubbles give the concrete the remarkable ability to absorb the stresses imposed on it by large temperature changes or by frost action. That is, air entrainment will make the concrete almost impervious to damage by the elements.

Some of the air-entraining agents added to concrete include derivatives of wood resins, animal or vegetable fats, fatty acids, sulfonated organic compounds such as detergents, hydrogen peroxide, and inorganic substances such as powdered aluminum and zinc. Usually 3 to 7 percent of the air-entraining agent by volume is added to a given quantity of portland cement.

Air-entrained concrete is almost as strong as ordinary concrete and far, far more durable. Its use is strongly recommended for all foundation work, exposed flat surfaces, and wherever freezing conditions or extreme changes in temperatures are anticipated.

Air-entrained concretes require special proportions of sand in the aggregate and the size of the sand particles must also be closely specified. That is, the sand should be of such a size that it will pass through a No. 30 sieve and be retained by a No. 50 sieve. Air-entrained concrete is more workable than ordinary concrete (which allows less water to be used in the water-cement ratio and thus adds to its advantages over ordinary concrete), and it has a somewhat sticky quality when wet.

Concrete Block

This article describes the manufacture and characteristics of concrete blocks. For a description of the techniques by which these blocks are assembled into walls and foundations, *see* CONCRETE-BLOCK CONSTRUCTION.

Concrete blocks, or *concrete masonry units,* as they are more formally called, are the most utilitarian of masonry construction materials. They have all the advantages of a masonry material: great compressive strength, a long, maintenance-free life, a comparatively high heat insulation value, and good sound insulating properties; they are fire-resistant, rot-, rodent- and termite-proof; and their assembly into foundations, walls, and partitions is straightforward and requires no unusual manual skills. Although more expensive than wood, concrete blocks are, for a masonry material, cheap and simple to use.

The principal drawbacks of ordinary concrete blocks are their porosity, which requires that they be waterproofed or protected against rain and damp soil, and their plain appearance. Concrete blocks are associated in the popular mind with warehouses, factories, garages, and the cheapest of public housing projects. They are, therefore, considered undesirable for use in private dwellings, which is a mistake, as they are an excellent construction material for interior bearing walls that will later be covered over with plaster, gypsum wallboards, or wood paneling; for foundations; and as a backing for exterior walls that are to be faced with brick veneer or stucco.

MANUFACTURE AND CURING

The ingredients used for making concrete blocks are portland cement, water, and an aggregate such as gravel or crushed rock, which is exactly what ordinary concrete consists of. Where the manufacture of concrete blocks differs from the mixing of ordinary concrete is that in the making of concrete blocks a smaller amount of water is added to the mix—just enough to enable the cement paste and aggregate to adhere to each other without having the blocks crumble apart after they have been molded.

In the usual manufacturing plant, the ingredients are thoroughly "dry-mixed" and then fed into iron molds that are part of a heavy press. The molds are filled with the damp concrete mix and the molds are then vibrated to consolidate the mix. A plunger descends to press the ingredients firmly together and another plunger then punches out the cores from the solid block. The cored blocks are then removed from the molds. In a large plant this molding process is highly automated and thousands of blocks can be manufactured in a day, by machines that produce 1500 blocks per hr. At the other extreme, small machines are available that operate by hand levers and make only one block at a time; just the thing for the rabid do-it-yourselfer who is determined to do *every*thing by himself.

Once unmolded, the blocks must be cured. For a description of the curing of concrete, and the great importance of curing concrete properly so that it may develop its maximum strength, *see* CONCRETE. It will suffice here to note that it takes about 28 days for concrete to cure naturally and develop most of its final compressive strength, as well as shrink down to its final size, although under suitable conditions concrete will continue to cure and increase in strength for years after it has been mixed and poured.

In a small manufacturing plant, the newly molded blocks may simply be stacked in piles under a shelter of some kind, kept damp for a few days, and then allowed to cure naturally for 2 to 4 weeks. In more modern, highly automated plants, the curing period is accelerated by stacking the blocks in a special curing room and increasing the temperature or pressure, or both, that the blocks are exposed to. In one process, for example, wet steam at a temperature of around 125°F is allowed to circulate around the stacked blocks. Under these conditions, the blocks will be substantially cured in about 24 hr. In another process, the blocks are sealed in an autoclave and high-pressure steam at a temperature of about 370°F is allowed to circulate around the blocks. In this latter process, the blocks will be substantially cured in about 12 hr. This forced curing not only enables the concrete to set extremely fast, but the process also results in blocks that are much stronger and that shrink less than blocks that are cured slowly under atmospheric conditions.

It is important for the builder to know whether the concrete blocks he intends using are completely cured or only partially cured—or not cured at all—as concrete shrinks as it cures. The total amount of shrinkage may not seem critical. Concrete blocks that are left to cure naturally under atmospheric conditions will shrink an amount that is equal to ½ in. to ¾ in. per 100 ft of block. (Steam-cured blocks shrink from ⅜ in. to ½ in. per 100 ft.) If partially cured or completely uncured blocks are used to build a wall, it is probable that tension cracks will open up in the wall as the blocks continue to cure and to shrink.

Cured concrete blocks will also swell if they become wet. Most concrete blocks are quite porous and will absorb a very large amount of water. It is important, therefore, once concrete blocks have been cured, to keep them as dry as possible until they have been assembled into a wall or foundation. The blocks should never be stored directly on the bare ground, as they will absorb moisture from the soil by capillary action. Nor should a pile of concrete blocks ever be left exposed to the weather, where they are likely to be soaked by rain. It may take weeks for the blocks to dry out enough for them to be used, since if damp concrete blocks are used for building a wall, cracks will undoubtedly open up in the completed wall as the blocks dry out and shrink down again to their original size.

Heavyweight and Lightweight Blocks

The weight of a concrete block depends largely on the weight of the aggregate used in the concrete mix. A standard portland-

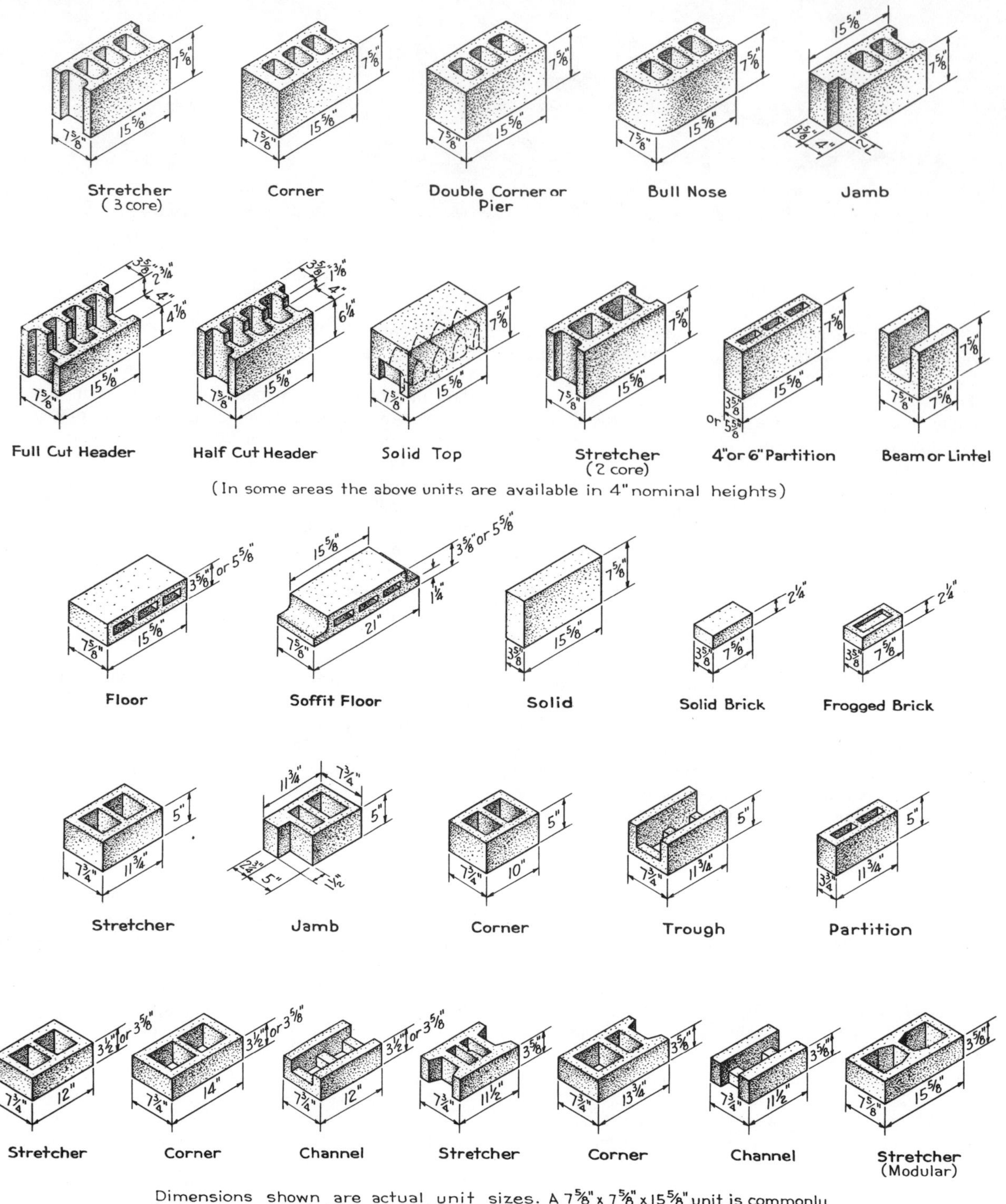

Fig. 1. Typical shapes and sizes of concrete masonry units *(Portland Cement Association).*

cement-sand-and-gravel mix will weigh about 130 lb per cu ft. Individual blocks 8 × 8 × 16 in. in size will weigh between 40 and 50 lb each. Heavyweight blocks are used where high compressive strength and a good resistance to water penetration are necessary, that is, in below-grade foundations and single-wythe walls subject to wind-driven rain.

All other concrete-block construction for dwellings, which includes interior wall partitions and above-grade exterior walls in temperate climates mainly, are almost always made of lightweight concrete blocks, which are blocks in which the aggregate has a porous, cellular nature. Examples of such aggregates include cinders, expanded clay or shale, blast-furnace slag, pumice, perlite, and vermiculite. All these aggregates are crushed, washed, and graded to size before they are used. Concrete blocks made with such aggregates will weigh somewhere between 30 and 40 lb each.

Lightweight blocks are particularly suitable for residential construction, where structural loads are usually quite light.

Their light weight makes it easy for masons and bricklayers to handle them (which makes the job go faster and thus reduces the overall labor costs), and their porous structure gives the blocks excellent thermal-insulating properties. It may appear after what has been said above regarding the tendency of concrete blocks to expand when wet that their porous nature would be a disadvantage, but if such blocks ever do become soaked with water (and there is no reason why they should once they have been properly installed) their porosity ensures they will dry out again just as quickly.

Sizes, Shapes, and Styles

Sizes

The most common size of concrete block is 8 × 8 × 16 in., although blocks are also available in thicknesses of 2, 4, 6, 10, and 12 in., in 4-in. heights, and up to 32 in. long. It should be understood that these are *nominal* sizes; that is, the size of a block *including* the mortar joints between it and its neighbors. Since a mortar joint is *usually* ⅜ in. thick, the *actual* size of the concrete block will be ⅜ in. less all around. That is, the actual size of an 8 × 8 × 16 in. block will be 7⅝ × 7⅝ × 15⅝ in., as shown in Fig 2. This point should be clearly understood because concrete blocks are manufactured with actual dimensions that are short ⅜ in. in their nominal dimensions. If one doesn't know the actual size of the blocks one is using, one won't know how thick the mortar joints should be. And if one makes a mistake in the thickness of the mortar joints, the finished dimensions of the job may differ significantly from the intended dimensions.

Shapes

Figure 1 shows the main shapes and sizes of concrete block available in most sections of the United States. All the shapes shown are made in both full-length and half-length sizes so that a wall can be built with little or no cutting of the blocks. However, the range of shapes actually available in any particular locality may be larger or smaller than what is shown in Fig. 1. It is always prudent for a builder to find out for certain just what is available in his community before he starts construction.

Cores

As far as the builder is concerned, whether a block has two or three cores is largely irrelevant. The main purpose of the cores is to reduce the weight of the block. The presence of cores also reduces the overall compressive strength of the block, of course, but this fact is, or should be, taken into account in the design. If a concrete-block wall must support an especially heavy load, heavyweight blocks can be used or steel reinforcement rods can be installed. For a description of these processes, *see* CONCRETE-BLOCK CONSTRUCTION.

It might be noted that a *solid* concrete block is defined as a block that consists of 75 percent solid matter. A *hollow* block, therefore, by definition has cores that occupy more than 25 percent of the gross volume. In fact, the cores of most hollow blocks usually occupy about 45 percent of their gross volume.

CONCRETE-BLOCK SPECIFICATIONS

All loadbearing blocks are classified as either grade N or grade S blocks. This grading system applies to both solid and hollow blocks. Grade N blocks are intended for general use above and below grade in conditions that may or may not be exposed to water penetration. Grade S blocks are intended for use in exterior locations where they are protected from the weather and in interior locations.

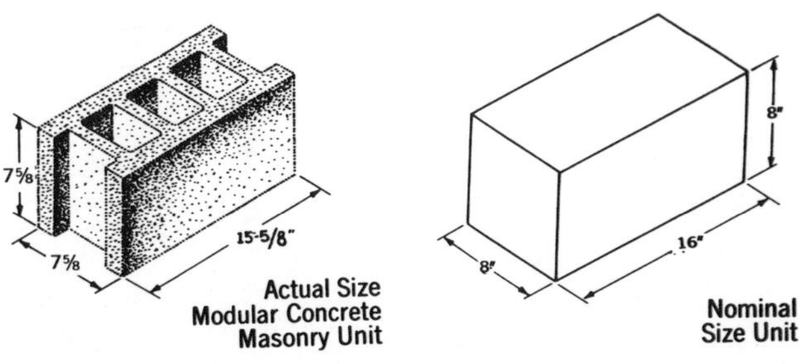

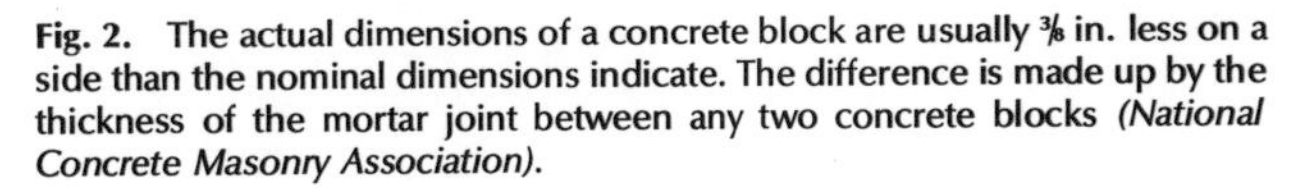

Fig. 2. The actual dimensions of a concrete block are usually ⅜ in. less on a side than the nominal dimensions indicate. The difference is made up by the thickness of the mortar joint between any two concrete blocks *(National Concrete Masonry Association).*

The basic difference between grade N and grade S blocks is in the type of aggregate used in their manufacture. Grade N blocks have denser, harder aggregates than grade S blocks. They are, therefore, better able to withstand heavy compressive loads. The table below shows the minimum required compressive strengths, in pounds per square inch, of both solid and hollow loadbearing blocks and of lightweight blocks, according to ASTM specifications.

	Average of 3 blocks, psi	*Individual unit, psi*
	Gross Area	
Hollow loadbearing blocks		
Grade N	1000	800
Grade S	700	600
Solid loadbearing blocks		
Grade N	1800	1500
Grade S	1200	1000
	Net area	
Hollow nonloadbearing blocks		
All types	600	500

These figures are for concrete blocks under test conditions. The building codes that govern actual construction practices and loadings take into account the fact that mortars are always much weaker than the blocks they bind together and that there is likely to be a wide variation in the quality of the workmanship. Building codes, therefore, always add a margin of safety to these loadings. For example, a wall built of hollow loadbearing concrete blocks with type N mortar can support a maximum compressive load of only 80 lb per sq in., but if type M mortar is used, the same wall can support a maximum compressive load of 100 lb per sq in. (*see* MORTAR.)

Concrete blocks are also classified according to their moisture content. Blocks that meet the maximum moisture-content requirements shown in the following table are classified as type I blocks. If the blocks fail to meet these requirements (which means in fact that the requirements have been ignored by the manufacturer), they are classified as type II blocks. The maximum moisture content of the blocks is in comparison to the moisture content of the blocks when they are saturated with moisture.

Linear shrinkage, percent	*Maximum moisture content, percent**		
	Humid	*Intermediate*	*Arid*
0.03 or less	45	40	35
0.03 to 0.045	40	35	30
0.045 to 0.065 Max.	35	30	25

* Average of 3 blocks, based on average annual relative humidity condition at jobsite.

Concrete-Block Construction

This article describes the methods by which nonreinforced and reinforced concrete-block foundations and walls are built. For a description of the blocks themselves, *see* CONCRETE BLOCK. *See also* BRICKLAYING; BRICK-MASONRY CONSTRUCTION; CONCRETE.

Concrete-block construction is not one of the preferred methods of building a house in the United States. Concrete-block walls have a plain, not to say ugly, appearance and most people tend to associate concrete-block construction with industrial structures and government-sponsored housing projects.

Nevertheless, the use of concrete blocks for home construction has much to commend it. Concrete-block construction is by far the least expensive method of building a masonry house, with all the advantages that masonry confers: it is a solid, stable, and long-lasting material and inherently fire-resistant, rot-, vermin-, and termite-proof. A house made of concrete blocks is also more thermal- and sound-resistant than a wood-frame house. The expenses of operating a heating and/or air-conditioning system in such a masonry house will always be less than for a wood-frame building. Furthermore, the construction techniques are simple. Any man who is handy with tools can, if he wishes, erect foundations and walls made of concrete blocks by himself.

As for appearance, a concrete-block structure is usually finished in stucco or brick veneer (*see* STUCCO; BRICK-VENEER CONSTRUCTION). These not only add to the basic advantages of masonry construction but are themselves both attractive and long-lasting exterior finishes. Whatever the situation in the United States, where wood-frame construction is enormously more widespread than masonry construction, throughout Europe and the rest of the world the structures of most homes and small apartment buildings are now being built basically by the methods described in this article.

As for the disadvantages of concrete-block construction, there is the cost, which in the United States is usually greater than for a wood-frame house of the same size (but not *much* greater). Concrete-block walls also tend to expand and shrink with changes in humidity and temperature. The walls, therefore, tend to crack along the mortar joints unless special control joints are built into the construction (as described below) to absorb this expansion and contraction. The tendency to crack can be an especially severe disadvantage in a below-grade foundation in areas where the soil is wet or where the ground tends to soak up and hold water during rainy weather. The foundation walls must, therefore, be waterproofed if the basement is to remain dry. But if the proper construction techniques are followed, as described in this article, the walls both above- and below-grade will remain crack-free and watertight for the life of the building.

NONREINFORCED CONSTRUCTION

Building Design

The standard concrete block is nominally 8 x 8 x 16 in. in size. Standard half-size blocks 8 in. long are also manufactured. It follows that the most economical method of building a house with concrete blocks is to follow the principles of modular design in which 8 in. is the basic unit of measurement. That is,

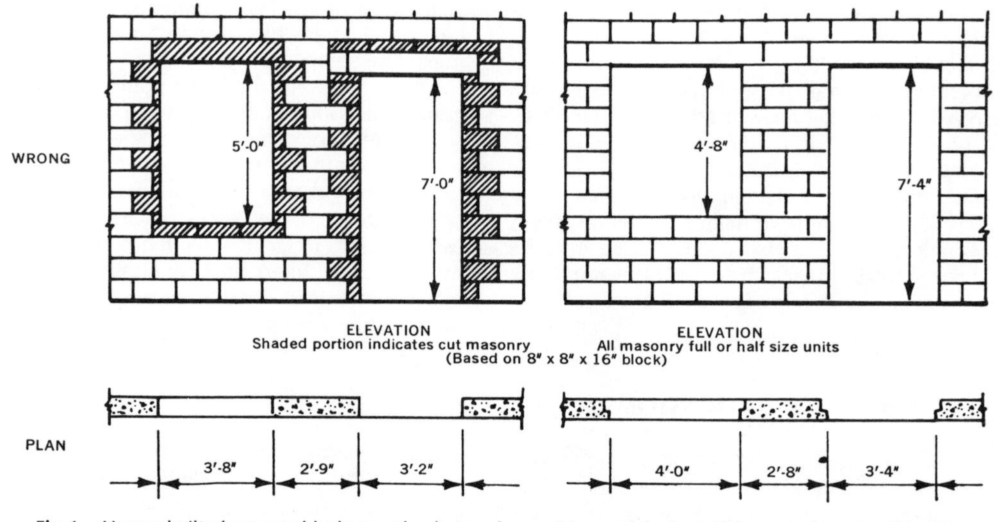

Fig. 1. Houses built of concrete blocks must be designed on an 8-in. modular basis if the cost of construction is to be minimized. Otherwise, a great many of the blocks must be cut to size individually, as shown in the left-hand illustration *(National Concrete Masonry Association).*

all the walls must be some multiple of 8 in. in length, and the doors and windows must be so spaced from each other and from the corners of the building that they are also some multiple of 8 in. apart from each other (see Fig. 1). *See also* MODULAR CONSTRUCTION.

The sizes of the doors and window frames, as well as the overall height of the wall, must also be multiples of 8 in., since the blocks are 8 in. high. In some areas of the country, concrete blocks 4 in. high are available, but the builder must check that blocks this size are available locally before the dimensions of the doors, windows, and walls are set. Failure to keep this principle of modular design in mind will mean that a great many of the blocks may have to be cut by hand to some odd dimension; this will add considerably to the cost of construction.

Local building codes usually specify the maximum allowable stresses permitted on a concrete-block wall. It is very common, for example, to allow a maximum compressive load of 80 lb per sq in. on a foundation wall built of concrete blocks. Most building codes will also specify the minimum thickness of a concrete-block wall, the usual requirement being that the walls of one- and two-story dwellings be 8 in. thick. A dwelling three stories high must have first- and second-story walls that are 12 in. thick, while the third-story walls need be only 8 in. thick.

While concrete-block walls have great compressive strength for loads that bear directly downward, they have very little lateral strength; that is, they cannot resist loads that push sideways against the walls. The construction, therefore, requires that additional support be provided against such lateral loads. This support is usually provided by the joists that span the house at each floor level. Until these joists are in place, the soil that was removed so that the footings and/or foundation walls could be built must not be shoveled back against the walls, otherwise the lateral pressure exerted by the soil might collapse the walls. An unsupported concrete-block wall may also collapse because of the lateral pressures exerted by high winds or an earthquake. Therefore, if a wall must remain standing unsupported for a time, it should always be braced to protect it against an unexpected lateral load, as shown in Fig. 2.

Laying Concrete Blocks

A bricklayer who is unfamiliar with laying concrete blocks will discover the technique is dead simple compared to laying brick. The general principles are the same, but the large size of the blocks and the thickness of the mortar joints make the job much easier (*see* BRICKLAYING). There is one major difference in technique between laying brick and blocks. Brick must be soaked with water before they are laid to prevent the too-rapid absorption of water from the mortar, which will cause the mortar to set prematurely and thus weaken it, but concrete blocks must be set while they are completely dry. The reason is the tendency of concrete blocks to expand when wet and then to shrink back down again as they dry out. To minimize this expansive tendency when the mortar is applied, the blocks must be completely dry. If the blocks should become wet for some reason before they are laid, tension cracks will appear in the construction—usually in the form of a step pattern in the mortar joints—as the blocks dry out.

Once construction has started, a partially completed wall must never be left exposed to the weather. The wall must always be covered by a tarpaulin, building paper, or some other waterproof material, at the end of the workday or whenever it begins to rain. These precautions will help prevent shrinkage in the completed wall.

Mortar and Mortaring

The mortar used will depend on the function of the wall and the loads it is expected to support. Foundations, load-bearing piers or partitions, or exterior walls built in areas where high winds, earthquakes, or severe frosts are likely to occur must be built using type M or S mortar. Elsewhere in the construction, or where extreme weather conditions or earthquakes are unlikely to occur, type N mortar or masonry cement can be used (*see* MORTAR).

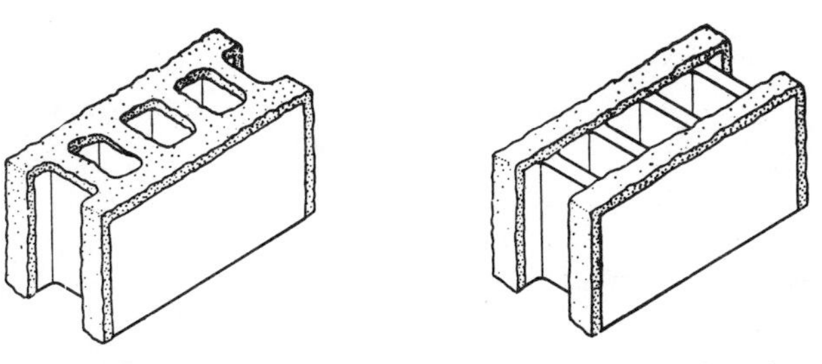

Fig. 3. Examples of full bedding and face shell bedding in the application of mortar *(Portland Cement Association)*.

Fig. 2. A concrete-block masonry wall has very little resistance to lateral loads. The wall must be supported by bracing until it can be supported by the ceiling joists *(Portland Cement Association)*.

There are two ways in which the mortar can be spread on the blocks—full mortar bedding and face-shell bedding. In full mortar bedding, the mortar is spread over both the webs and face shells of the block, as shown in Fig. 3. In face-shell bedding, mortar is spread only over the face shells. Building codes usually specify full mortar bedding on all loadbearing walls, whether above or below grade, with face-shell bedding being permitted elsewhere in the construction.

Heavyweight versus Lightweight Blocks

As for the type of block to use, the choice is between heavyweight blocks and lightweight blocks. Which type is chosen is usually simply a matter of availability. In general, heavyweight blocks should be used for foundation walls and for above-grade walls subject to wind-driven rain. Lightweight blocks are usually used everywhere else in the construction. *See* CONCRETE BLOCK.

Laying the Blocks

Before actual construction begins, the mason must check the accuracy of the layout. He does this by snapping a chalk line on the footing or foundation wall and laying down a row of blocks along the line of the intended wall. The blocks are not mortared in place but simply laid down with a ⅜-in. gap between the blocks. If the line of blocks comes out exactly even with the end points of the wall, good. If it doesn't, something is wrong with the dimensioning somewhere, and the mason must either shift the spacing of the blocks slightly, or, if this won't work, he must find the fault and correct it before construction can begin.

But assuming nothing is wrong, the mason begins construction by spreading a full bed of mortar on top of the footing or foundation wall. Figures 4 to 8 show the general procedure that is followed from the spreading of the mortar to the completion of the first course. We will note only a few points here.

If one examines the webs of a concrete block, one will note that one side of the block has thicker webs than the other. This happens because the dies that compact and form the blocks while they are in the molds have tapered sides to facilitate their punching out the cores. The thicker side of the webs must always be *up* in order to provide as large a bedding surface as possible for the mortar.

The accuracy with which a wall is built, and the overall

Figs. 4–8. Laying the first course of concrete blocks *(Portland Cement Association)*.

squareness of the house as well, depends on the accuracy with which the corner blocks are set in place, while the overall flatness of the wall depends on the accuracy with which the first course is laid from corner to corner. The mason sets the corner blocks in place as shown in Figs. 9 to 13, and using his level as a guide, he adjusts their positions by rapping them with the handle of his trowel until each block is flat, plumb, and level.

Figs. 9–13. Building up the corner of a concrete-block structure. Each block is plumbed and leveled individually and the corner as a whole must be straight, plane, and square *(Portland Cement Association)*.

Fig. 14. When a row of concrete blocks has been laid correctly, the last block to be laid in that row—the closure block—will slip neatly into place without dislodging any of the mortar that has been buttered on its two sides *(Portland Cement Association)*.

The corners having been completed, the mason stretches a line between the corners and lays the first course, checking the alignment of each block with his level as it is set in place. The last block to be set in place in each course is the *closure block*. If the job has been accurately done, the closure block will fit neatly in place with a ⅜-in.-wide mortar joint on either side, as shown in Fig. 14. The height of each course is checked with a *story pole*, or *course pole*, which is a 1 x 2 in. length of wood marked off in 8-in. increments.

Making Watertight Joints

A good-looking, weathertight job will depend in large part on how well the mortar joints are *tooled*. The mortar is allowed to set until it has become so stiff that the mason can press his fingernail into it only with difficulty. The mortar is allowed to reach this stage before it is tooled because mortar shrinks as it sets. As it shrinks it pulls away slightly from the blocks, which opens up minute cracks that may enlarge in time and thus reduce the watertightness of the construction. The mason waits, therefore, for the propitious moment. He then compacts the mortar solidly into the joints by pressing against the joints as forcibly as he can with the jointing tools shown in the illustrations.

The top course of a foundation wall and the top courses of a wall at the first- and second-story levels must be made of solid masonry blocks. The required solidity is obtained either by laying specially made blocks that have solid tops, which are available in many, but not all, communities, or by filling the cores in the top course of blocks with concrete or mortar. Solid blocks prevent the entrance of water into the interior of the wall and they also provide the solid bearing surface that is necessary to support the wood *sill plates* and *top plates* on which the joists and rafters will rest (*see* WOOD-FRAME CONSTRUCTION).

Wall Intersections

If there are any interior walls made of concrete blocks in the construction, these walls must never be joined to the exterior walls with an interlocking bond. Only the corners of an exterior concrete-block wall may interlock. The reason is that the interior and exterior walls may shrink at different rates, or at different times, which might stress the joints enough to open up large cracks in the construction.

Fig. 15. Two intersecting concrete-block bearing walls are tied together by 28-in.-long metal straps, as shown in the illustrations. The ends of the straps are bent over about 2 in. and embedded in concrete-filled cores. The metal straps should be spaced not more than 4 ft apart *(Portland Cement Association).*

Two intersecting *loadbearing* walls are connected by steel tie bars, which are $\frac{1}{4} \times 1\frac{1}{4} \times 28$ in. long, with 2-in.-long right-angle bends at each end (see Fig. 15). The tie bars are embedded in concrete or mortar. These tie bars provide the necessary lateral support for the walls. They should be spaced not more than 4 ft apart in the walls.

Where a *nonloadbearing* interior partition meets an exterior wall, the necessary lateral support is provided by strips of 2-in.-wide expanded-metal lath or hardware cloth, installed as shown in Fig. 16. These strips are cut about 18 in. long and they are embedded in the mortar joints as the exterior wall is constructed, the unsupported ends of the strips being left to dangle until the partition wall is begun. The strips are then embedded in the mortar joints of the partition wall. The strips are required at every second course.

Fig. 16. When a nonloadbearing wall intersects a loadbearing wall, the two are tied together by wire mesh embedded in the mortar joints, as shown. The wire mesh is embedded in every other course of blocks *(Portland Cement Association).*

Control Joints

We have already mentioned the importance of installing control joints in order to prevent shrinkage cracks in the walls. A control joint is nothing more than a weak point deliberately built into the construction. If a wall is expected to develop a crack, which an ordinary concrete-block wall is likely to do, it is only sensible to decide in advance where one would prefer the crack to occur. The control joint (or weak point) can then be placed where its appearance is either minimized or completely hidden. Furthermore, the control joint can be built in such a way that if it does in fact crack, water will still be prevented from penetrating the wall.

A control joint always runs vertically up a wall as shown in Fig. 17. There are several inherently weak points in a concrete-block wall where control joints are especially suitable: (1) where two walls intersect, (2) at the juncture of the wall with a column or pilaster, (3) above or next to doors, windows, or chases, (4) wherever the height of the wall changes. In a very long wall, a control joint should be located at 20- to 30-ft intervals. In a professionally designed building, the architect or engineer will always note on the plans exactly where control joints are to be located.

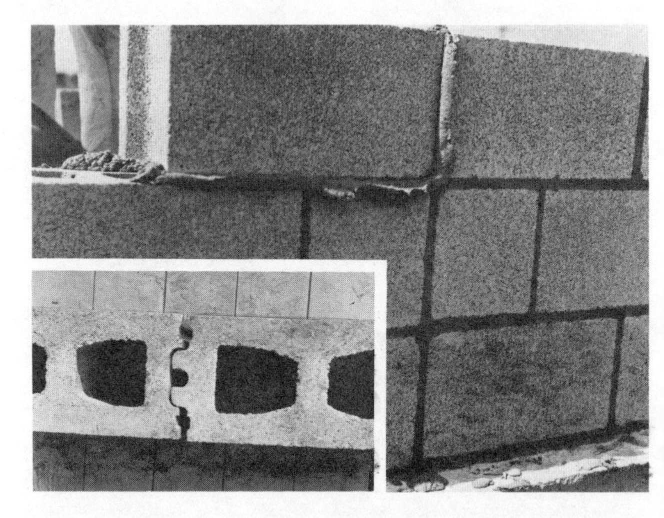

Fig. 17. Control joints are necessary in concrete-block walls to relieve any expansion stresses. Control joints are continuous vertical joints in the wall. They can be made by installing a strip of building paper on one side of the mortar joint or by the use of special blocks as shown (inset). Both types of control-joint construction allow the wall to expand without weakening the overall construction *(Portland Cement Association).*

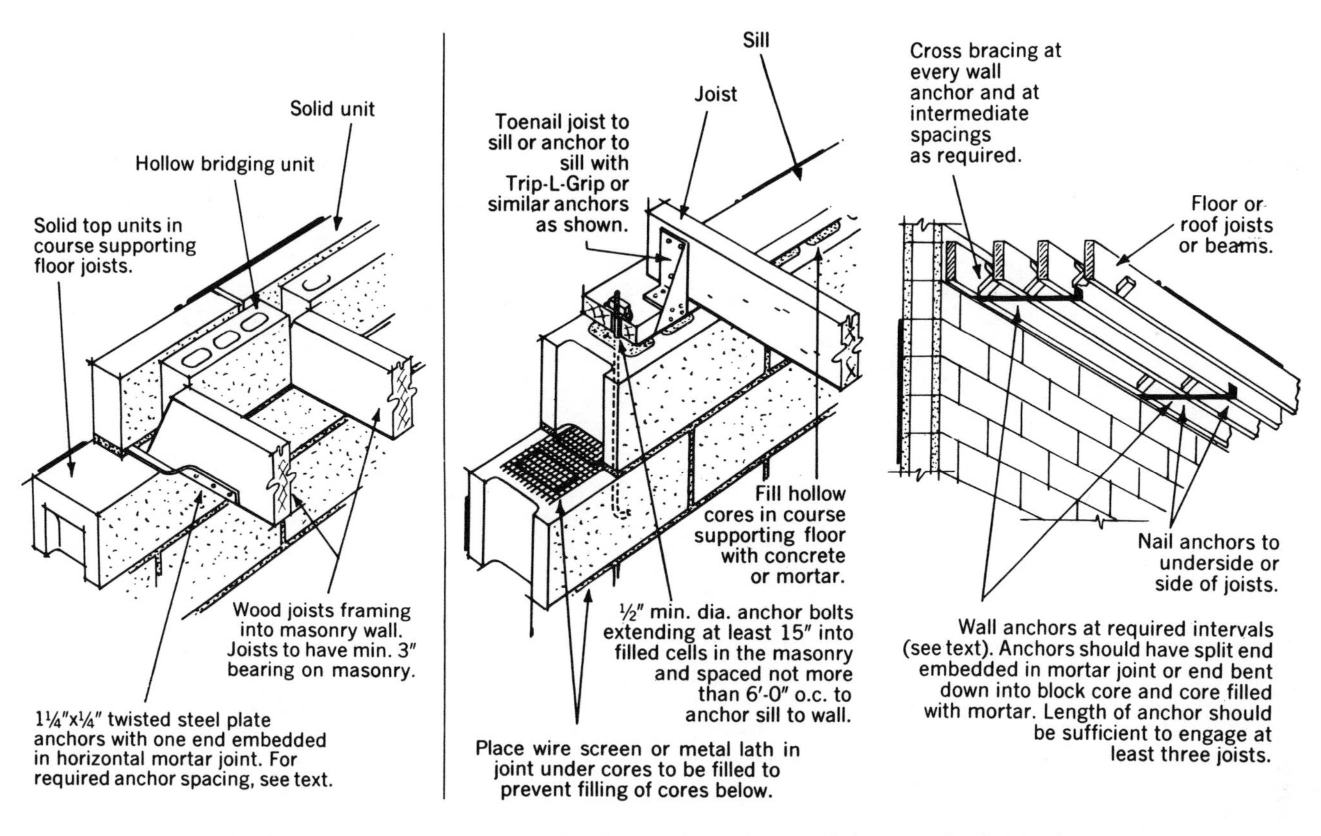

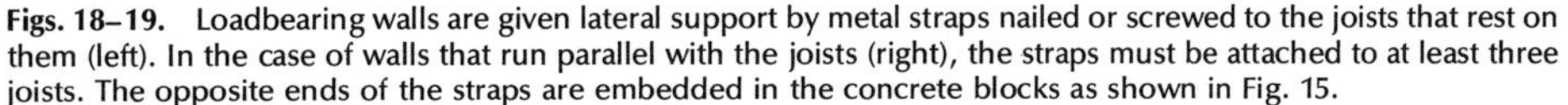
Figs. 18–19. Loadbearing walls are given lateral support by metal straps nailed or screwed to the joists that rest on them (left). In the case of walls that run parallel with the joists (right), the straps must be attached to at least three joists. The opposite ends of the straps are embedded in the concrete blocks as shown in Fig. 15.

Joists and Lateral Support

The lateral support provided the walls by wood joists is not simply a matter of laying the joists in place. They must be anchored securely to the walls by ¼ × 1¼ × 16 in. steel straps, as shown in Fig. 18. Where the ends of joists rest directly on top of a concrete-block wall, the straps are placed a maximum of 6 ft apart. One end of each strap is bent over about 2 in. and embedded in a concrete-filled core, as has already been described. The remaining length of the strap is drilled in several places along its length and the strap is nailed or screwed to the wood.

The same type of strap is used to anchor a wall that runs parallel to the joists, as shown in Fig. 19. The straps are embedded in concrete-filled cores, as already described. The straps are spaced a maximum distance of 8 ft apart. The free ends of the straps must be long enough to reach across three joists, and the end of the strap must also be T-shaped or be bent up so that it can be securely nailed or screwed into the farthermost joist. The strap is nailed or screwed to the intermediate joists as well.

Cavity-Wall Construction

A cavity wall consists of two wythes, or rows, of masonry separated from each other by an air space that is at least 2 in. wide but not more than 3 in. wide. For a general description of cavity walls, their advantages, and their method of construction, *see* BRICK-MASONRY CONSTRUCTION. The reader should refer to that article first. Everything said there regarding brick cavity-wall construction will apply also to concrete-block cavity-wall construction, except as noted below.

The concrete blocks used for a cavity wall need be only 4 in. wide. This size, plus the nominal 2-in. width of the cavity, will make a wall that is 10 in. thick overall (see Fig. 20). The wall

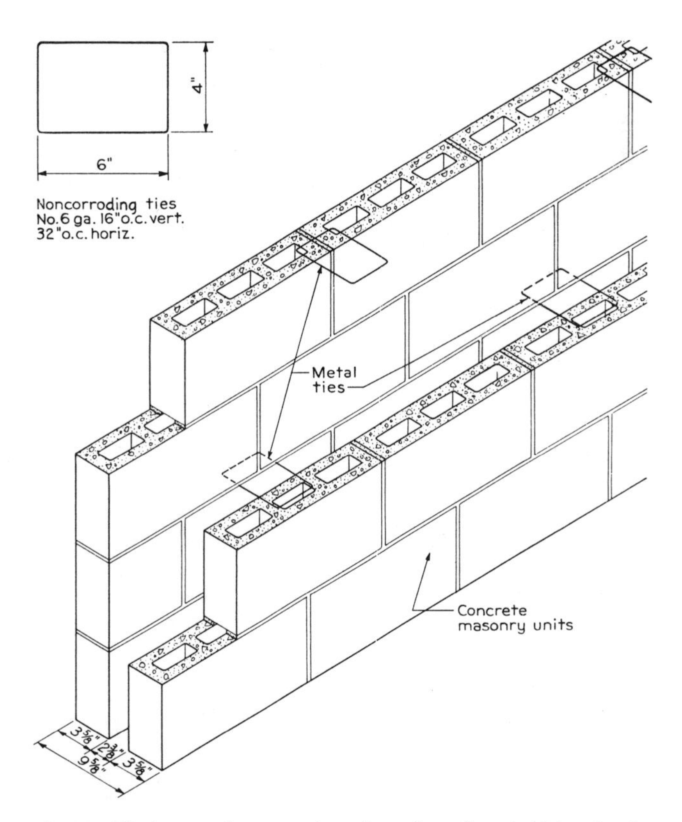

Fig. 20. The inner and outer wythes of a cavity wall are held together by metal ties, as shown *(Portland Cement Association).*

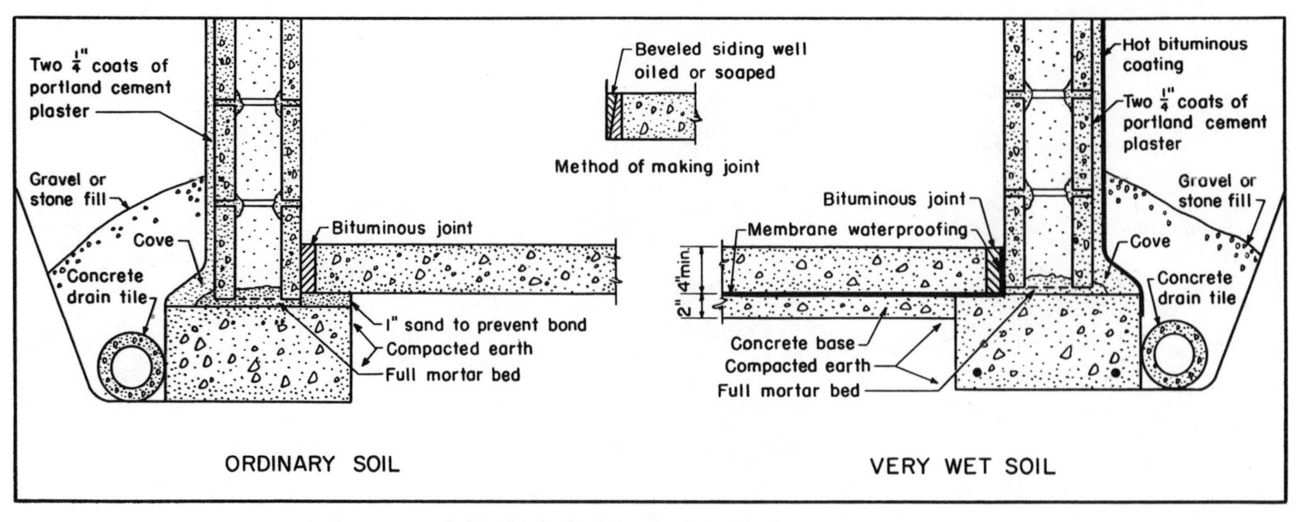

Fig. 21. Methods recommended by the Portland Cement Association to waterproof belowgrade construction in ordinary and in very wet soils *(Portland Cement Association).*

can be made stronger, if necessary, by using 6-in.-wide blocks for the inner wythe.

The two wythes must be tied together to strengthen the wall laterally. In a brick wall, this reinforcement is usually provided by Z-shaped pieces of wire embedded in the mortar. In a concrete-block cavity wall, the wythes are usually tied together by rectangular-shaped pieces of No. 6 wire as shown in Fig. 20.

Brick walls always require flashing over the doors and windows to prevent water leaking through the joints and into the interior of the house. Similar flashing is required for concrete-block cavity walls.

Weep holes are also necessary in a cavity wall. The weep holes are located, every two or three blocks apart, in the mortar joints at the bottom of the first course of blocks. The best way of making a weep hole is to grease a length of rubber or plastic tubing, set the tubing in place at the very bottom of the wall so that it extends up inside the cavity for several inches (to prevent the weep hole's being clogged with mortar that may be dropped into the cavity), mortar the blocks in position in the usual way, and, after the wall has been completed, pull the tubing out.

Waterproofing

The exterior of below-grade foundation walls must always be finished off with two ¼-in.-thick coats of portland-cement plaster to ensure the watertightness of the construction. The application of these finish coats is called *parging.* The mortar can either be the same mortar used for the rest of the construction or it can consist of 1 part portland cement to 2½ parts of sand (by volume), plus the necessary water.

After the first parge coat has been applied, it is allowed to stiffen and it is then roughened with a wire brush to provide a key for the second parge coat. The first parge coat is then allowed to harden for at least 24 hr. Just before the second coat is applied, the first coat should be wet down by spraying it lightly with water. The second coat must be kept damp for at least 48 hr so that it may cure properly and develop its full strength.

If the parging is being done during very hot, dry weather, the wall should be dampened slightly with a spray of water before it is parged. This will kill some of the wall's porosity and will prevent its sucking the water out of the parge coat too quickly, which would weaken the coat. Note that this is the only occasion on which water is ever applied to concrete blocks for any reason whatsoever.

If the soil if normally wet, after the parge coat has hardened, the wall should be coated with a bituminous waterproofing compound such as coal-tar pitch or asphalt. After the parge coat has dried hard, it is primed with either creosote or an asphalt primer. If the waterproofing compound is coal-tar pitch, the primer must be creosote; if it is asphalt, the primer must also be asphalt. The primer is allowed to dry and two coats of the hot bituminous compound are then brushed on top of the parge coat. Each coat must be applied as one continuous layer, with the second coat being brushed crosswise to the direction of the first coat.

In addition, drainage tile should be laid all around the perimeter of the foundation, and gravel should be placed against the foundation wall to facilitate the drainage of water from the soil around the building. The construction is shown in Fig. 21.

REINFORCED CONSTRUCTION

Reinforced-concrete-block construction (see Fig. 22) is the small builder's way of obtaining all the advantages of reinforced-concrete construction without any of the expense. Instead of having to construct an elaborate and expensive system of forms and then hire experienced crews to install the reinforcement steel and pour the concrete, the builder can do everything himself.

The mason must make sure that the cores of the concrete blocks in which the reinforcement steel rods are to be placed are aligned with each other for the full depth of the wall as the wall is built up. After the steel rods have been inserted into the core spaces, the cores are completely filled with a fluid cement grout and, once this grout has hardened, *voila!*—homemade reinforced-concrete-block walls that are almost as strong as traditional reinforced-concrete construction and only a little more laborious and expensive to build than ordinary concrete-block walls.

Vertical Reinforcement

The usual reinforcement steel used for home construction consists of deformed, ⅜-in.-diam. bars. Typically, two reinforcement bars are placed in each corner cavity, in the cavities on both sides of every door and window, and 32 in. apart along the entire length of the wall. This vertical spacing may, however, vary from 16 in. to 48 in., depending on the antici-

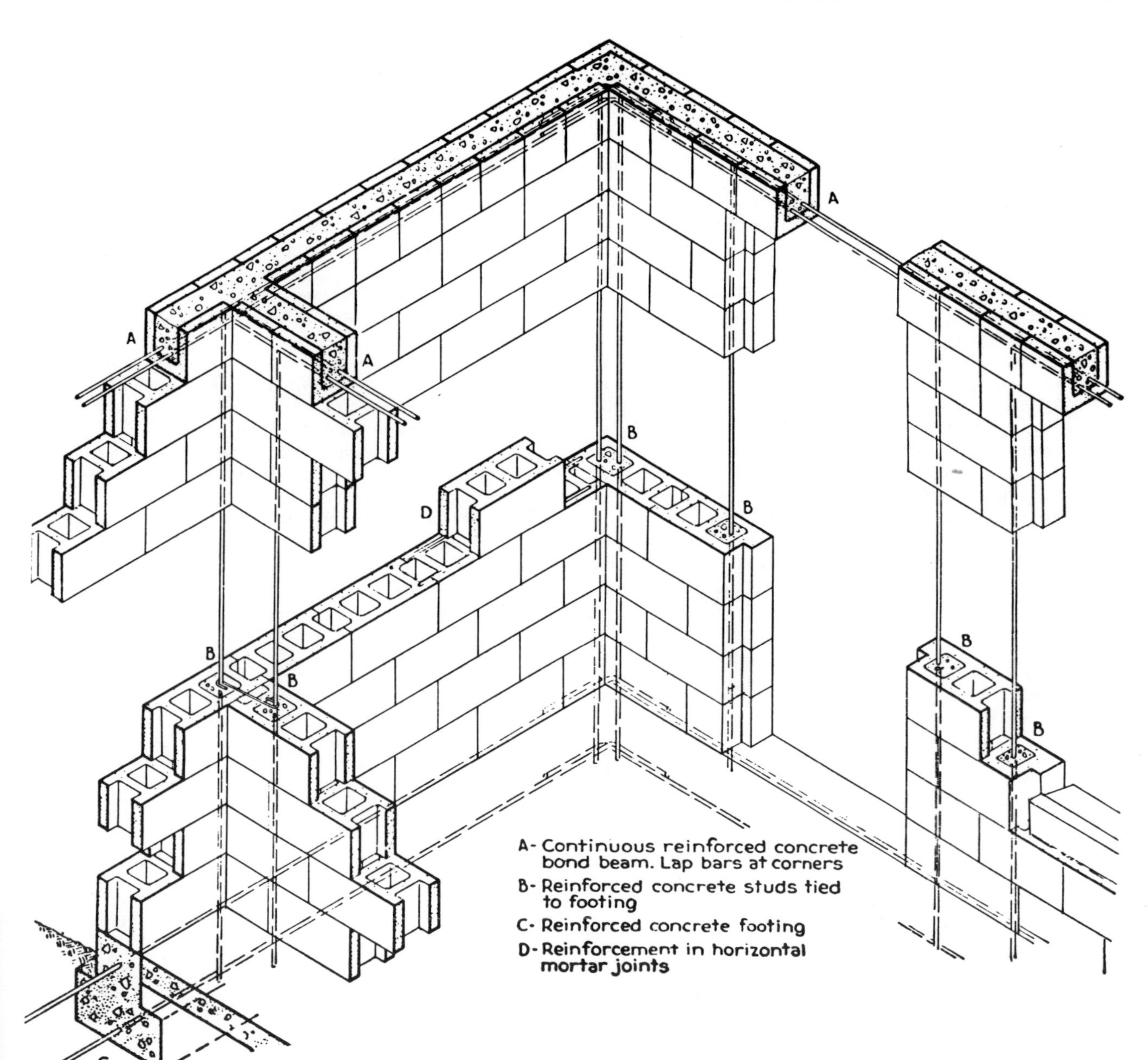

Fig. 22. Reinforcement bars are placed both vertically and horizontally in a concrete-block construction to reinforce it. Such reinforcement is often required in earthquake-prone areas of the United States *(Portland Cement Association).*

pated loads. The cores containing the reinforcement bars must have a clear, unobstructed area of at least 2 × 3 in. that runs from the top of the wall to the bottom. The bars are centered within the core by *spacer ties* installed at the top and bottom of the wall. If the reinforcement bars are 8 ft long, they must also be supported at their centers.

As the walls are built up, it is important that full mortar bedding be used on the blocks that will contain the reinforcement bars; otherwise, when the grout is poured into the core cavities, it will simply leak away into the adjacent cores. It is also important that the mortar in the joints not be extruded into the core cavities as this extruded mortar will interfere with the pouring of the grout, which could result in the formation of air pockets within the cavities. Nor should a full bed of mortar be spread over the top of the foundation wall or footing when the first course of blocks is being laid; the grout should bind with the top of the foundation or footing, not with an intervening layer of mortar.

Horizontal Reinforcement

A concrete-block wall is usually reinforced horizontally by reinforcement bars installed in the top course of the wall, as well as above and below all window openings and above all door openings. If the anticipated loads are likely to be unusually severe for some reason, additional reinforcement bars may be located in a course at the 4-ft-high level of the wall.

To install the reinforcement, two ⅜-in.-diam. bars are laid in *bond beam blocks* (see Fig. 22) that form the top course of the wall. The reinforcement bars should extend around the entire perimeter of the building. Where the ends of two bars meet, they should overlap by at least 1 ft to maintain the continuity of the reinforcement. Once the reinforcement bars have been set in place, grout is poured into the trough formed by the bond beam blocks. It is a basic principle of sound construction that, where vertical and horizontal reinforcement bars meet, the grout should be poured around their juncture at the same time in order to bind the vertical and horizontal reinforcement solidly together.

Grout and Grouting

Grout consists of portland cement and sand to which sufficient water is added to enable the resulting slurry to pour easily into the core cavities. The proportions of the mix are usually 1 part portland cement to 2¼ to 3 parts sand, by volume. If the core cavities are larger than 2 × 3 in. in size, which happens when blocks containing only two cores are used, 1 to 2 parts of *pea gravel,* by volume, can be added to the mixture in order to reduce the overall cost of the grout. The amount of water that should be added can be determined accurately only by experiment. Enough water should be added that the grout will slump 8 to 10 in., as measured by the slump test. (For a description, *see* CONCRETE.) The grout should slump as much as possible without the aggregate's separating. The ingredients should be mixed together in a mixer for at least 5 min. The grout should be poured as soon as possible after it has been mixed, and positively within 1½ hr after the water has been added. If the grout hasn't been used by then, it should be discarded.

The method used to pour the grout is called *low-lift grouting.* In low-lift grouting, the walls are built up to a maximum height of 4 ft before the grout is poured. After the grout has set, the walls are built up another 4 ft. The fact that the walls are built in 4-ft stages allows rather short reinforcement bars to be used, which can be inserted easily into the core cavities as the courses are laid. This facility of installation makes low-lift grouting eminently suitable for home construction.

The points that should be kept in mind with low-lift grouting are that the reinforcement bars should overlap by at least 1 ft wherever they meet, and that the mortar should be allowed to harden for at least 24 hr before the grout is poured. The grout should be poured only to within 1½ in. of the top of the wall after the first 4-ft-high section has been built. This space provides a key within the cores that will help lock the top 4-ft section to the bottom 4-ft section when the final grout is poured.

Immediately after the grout has been poured, it should be *puddled,* that is, a long 1 × 2 in. stick should be thrust into the grout and worked up and down vigorously for several minutes to consolidate the grout and enable any trapped air bubbles to work their way free. The grout is then left undisturbed until it has just begun to set. The grout is then repuddled. This second puddling is necessary because the grout will have shrunk away slightly from the sides of the cavity as it begins to set. Repuddling recompacts the grout and ensures that there will be a solid bond between the concrete blocks, the reinforcement steel, and the grout.

Condensation

We are all familiar with the phenomenon of condensation—the beads of moisture that form on a cold glass of beer in the summertime, for example, or the frost that forms on a window on a cold winter morning.

Less well known is the insidiously destructive role played by condensation within the structure of a house. Condensation within the walls may cause the exterior paint to blister and peel away from the siding; the siding and sheathing behind the paint may even rot away if the condensation is extreme enough. Condensation within an enclosed roof may rot the roof sheathing, and the condensed moisture will stain the ceiling below as it slowly drips, drips, drips. Condensation collecting on cold-water pipes buried within the walls can drip on and discolor the wall finish, and it may be the cause of decay as well. In a house with an unventilated or poorly ventilated crawl space, condensation under the flooring may cause both the flooring and the joists to rot, and in a house with a full basement, condensation may result in a summer-long film of moisture on both the walls and floor; puddles of water may even cover the floor. Obviously, condensation—its cause and prevention—is a subject homeowners should know something about.

Interestingly enough, condensation as a destructive agent is a fairly recent problem, a consequence primarily of the increasing use of insulation in homes built since World War II. Uninsulated houses rarely suffer from condensation problems. To understand why this is so, it will first be necessary to review the basic physical principles underlying the condensation of water vapor within the home.

Water vapor (the quantity of which is usually measured in grains of water per lb of dry air) is a gas, an integral part of the mixture of gases that make up the atmosphere. These molecules of water vapor are smaller than molecules of either nitrogen or oxygen, the principal constituents of the atmosphere, and they can, therefore, easily pass through barriers that are considered "airtight," such as wood-framed walls, masonry walls, and even concrete.

The actual amount of water vapor in any given volume of the atmosphere varies continuously. There is no lower limit to the amount of water vapor that may be present in the atmosphere (and the amount, therefore, may be zero), but there *is* an upper limit. This upper limit depends on the temperature of the air. The higher the temperature, the greater the volume of water vapor the atmosphere can hold—but only up to a given maximum. When the atmosphere is holding as much water vapor as it possibly can *at a given temperature,* the air is said to be *saturated* with water vapor, or its relative humidity is said to be 100 percent. Air that is saturated with water vapor at a given temperature is also said to be at its *dew-point temperature.*

If the temperature of air saturated with water vapor should increase, and if no more water vapor enters the atmosphere, the air will no longer be at its dew-point temperature, and an additional quantity of water vapor can then enter the atmosphere. On the other hand, if the temperature of saturated air should fall, some of the water vapor in the air will—indeed, *must*—condense on the nearest solid object.

Water vapor moves with the current of air of which it is a part, of course. But in addition, water vapor has a motion of its own that is independent of the movement of the air as a whole. This motion occurs because water vapor exerts a pressure that is independent of the pressure exerted by the atmosphere as a whole. This *vapor pressure,* as it is called, increases with any increase in the water-vapor content of the atmosphere. When two adjacent areas of the atmosphere have different water-vapor pressures, the water vapor will always flow from the area of high pressure to the area of low pressure. To put the same fact another way, water vapor will always flow from an area of greater saturation to an area of lesser saturation. And, since warm air almost invariably has a greater water-vapor content than cool air, it follows that water vapor usually flows from a warm area to a cool area.

What are the practical consequences of these facts? During cold weather, the air inside a house is much warmer and also much more humid than the air outside the house. The water vapor within the house, therefore, tends to migrate from the interior of the house to the exterior. To do so, it must pass through the walls of the house, the materials of which (plaster, various wallboards, wood, and brick) are all more or less permeable to water vapor. The only effect these wall materials will have on the movement of water vapor out of the house will be to slow it down, each according to its permeability.

Walls and Water Vapor

We are now in a position to understand how water vapor can enter a wall and either condense on the inner surface of the wall, within the wall construction, or pass entirely through the wall without condensing at all.

Figure 1a shows a cross-section of a very common type of uninsulated wood-frame wall construction. The air-temperature line shows how, on a cold winter day, the air temperature drops from, say, 70 to 10°F at various points within the wall. As the air temperature drops, the dew-point temperature—which is, you will recall, a measure of the maximum amount of water vapor air can hold at any given temperature—also drops.

In the situation represented in Fig. 1a, the water vapor does not condense within the wall because the air temperature never falls to the dew-point temperature. The water vapor, therefore, remains in the gaseous state and it will eventually diffuse completely through the wall.

This is not, however, the situation that exists in a house having insulated walls. There is no doubt as to the great value of insulation as a conserver of heat and a reducer of fuel costs. But there is also no doubt that, unless the proper precautions are taken during the installation of insulation, the presence of the insulation will increase the probability that any water vapor within the wall will condense there. Figure 1b shows why this is so.

Note how in Fig. 1b the insulation produces a much steeper drop in the air temperature across the wall space. This is only as it should be, since it is the function of insulation to act as a temperature barrier. As a result, the temperature of the outside wall surface will be much lower in an insulated wall than it would be in an uninsulated wall. Because of this steep drop in air temperature, the air-temperature line and the dew-point-temperature line now meet within the wall space, with the consequence that some of the water vapor in the air will condense on the nearest solid surface, which happens to be the inner face of the sheathing. The actual amount of water vapor that will condense will depend on the temperature differential across the wall and on the amount of water-vapor present within the wall space.

Unfortunately for the point we are making, the general improvement in the American standard of living that has occurred since World War II has led to the widespread use of appliances—clothes washers, clothes driers, and dehumidifiers especially—that have increased greatly the amount of water vapor present within a house. In addition, the gradual but continuous improvement in the efficiency of home heating equipment, plus the tendency of many people to keep their houses at a temperature that is higher than necessary for good health, enables the air inside a house to support these increased levels of water vapor. Finally, the widespread use of weather stripping around the doors and windows, plus the installation of storm sash, has stopped up many of the pathways by which cold air once infiltrated a house and provided an unintended ventilation that, among other things, reduced the humidity of the air within the house.

For all these reasons, the typical insulated house of today usually contains a heavy load of water vapor that must be got rid of. It follows that there is a considerable movement of water vapor into the walls. If this water vapor condenses within the walls, and if the amount of condensation is great enough, the sheathing will in time become soaked with water. The water may gradually soak into the exterior siding as well. If it does, the paint will peel off the siding, since the best paint in the world can't stick to a wet surface. If the condensation continues for a prolonged period, the insulation will become soaked with water also, which will reduce considerably its effectiveness as a heat barrier. Over a long enough period of time, the lower sections of the sheathing and studs may decay—and all this will happen unbeknown to the homeowner until some unusually severe stress to the house causes the decayed wood to collapse.

It should be noted that it is not the condensation of water vapor per se that is responsible for wood decay. The presence of moisture is, however, the indispensable condition that makes it possible for molds to grow and spread within the wood. It is these molds that destroy the wood.

Vapor Barriers and Vapor Traps

For some years after condensation problems began to show up, builders and architects were baffled as to the mechanism by which condensation occurred, but there is now no doubt that the mechanism is as described in the preceding paragraphs. Once the cause of the problem had been recognized, the solution was equally obvious. This was to install vapor barriers in the walls that prevented the movement of water vapor into the walls in the first place.

A *vapor barrier* is a material that is completely impermeable to water vapor. The principal materials used include aluminum foil, polyethylene films, and building paper, which consists of paper or felt impregnated with asphalt. Nowadays all insulated homes are automatically provided with vapor barriers, at least by reputable builders.

There is only one place in a wall where a vapor barrier should be installed and that is on the *warm* side of the wall, as close to the plaster or wallboard as possible. Figure 1c shows how a vapor barrier installed in this location in an insulated wall lowers the dew-point temperature within the wall to a level that the air temperature cannot possible fall to, even in the coldest weather.

This statement will be true, however, only if the vapor barrier is correctly installed; that is, without any significant gaps, splits, or holes through which water vapor can pass. To be effective, the barrier must be as absolute as possible. A few small gaps in the barrier, which occur usually at the studs where the edges of the material meet, may seem a trivial matter, as indeed it is if one is thinking in terms of weeks, months, or even a year or two. But a house is supposed to last at least 50 years (according to the depreciation period allowed by our tax laws, anyway), and a considerable amount of water vapor can enter a wall and condense there. And it doesn't take much dampness for rot to set in.

Although condensation will not be a problem in an uninsulated wall, and although it might become a serious problem in an insulated wall, it most certainly will cause trouble in a wall in which the homeowner or builder has created, through ignorance, a *vapor trap* that prevents the diffusion of water vapor out of the wall entirely. A vapor trap is an impermeable barrier installed on the *cold* side of a wall. It can be built into an exterior wall in a number of different ways. The builder may, for example, create a vapor trap by using a building paper in back of the sheathing that is impermeable to water vapor instead of a permeable building paper, that is, one that can "breathe."

Or a homeowner may replace the wood siding in an old house with vinyl or aluminum siding, and the contractor may install the siding in such an airtight manner—with all the joints well-sealed and without any vent plugs in the panels—that a vapor trap is created where once the water vapor could migrate freely through gaps between the old siding. Not that there is anything wrong with vinyl or aluminum siding; it is just that air must be allowed to circulate through a wall.

The same sort of vapor trap may be created when an oil-base paint is applied to wood siding with such thoroughness and care that an impermeable barrier is painted onto the surface of the wall. Perversely enough, an extremely well-painted wood siding may be the cause for the ultimate failure of the paint film. To prevent any condensation within the walls, therefore,

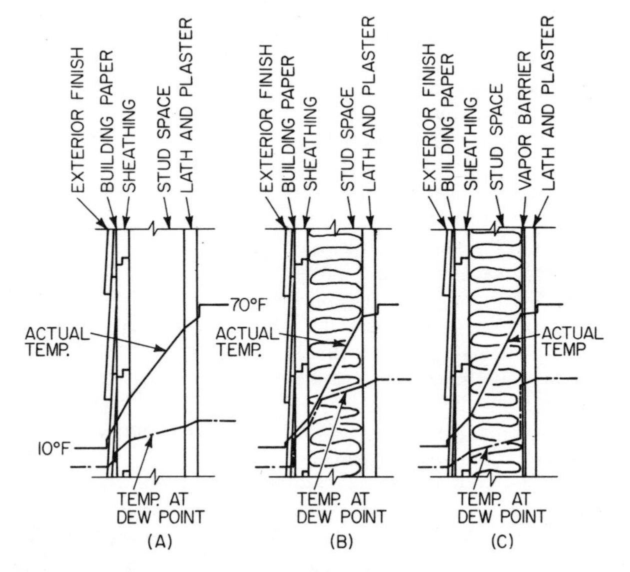

Fig. 1. The relationship between temperature, dew point, and condensation within a wood-frame wall.

both vinyl and aluminum siding, and painted wood siding as well, should always have vent plugs installed under the eaves to allow outside air to circulate through the walls.

"Blown-In" Insulation

Condensation is always a possibility when insulation is sprayed into, or "blown in," a wall through holes cut in the siding between the studs. This method of insulating an old wood-frame house is quite common, but the situation created within the walls is that of Fig. 1b. However simple it may be to install loose insulation in an old house, installing a vapor barrier within such a wall is almost impossible. What to do?

The solution turns out to be quite simple. First, vent plugs should be installed in the siding to allow air to circulate freely within the wall space. Second, the interior surface of the wall should be made impermeable. This can be accomplished by painting the wall with aluminum paint, which will provide the necessary impermeable barrier. A coat or two of an ordinary interior wall paint over the aluminum paint will provide the desired decorative finish. Alternatively, the homeowner can paste vinyl wallpaper to the wall, which will at the same time provide both the necessary vapor barrier and the decorative finish.

Condensation on Windows and Wall Surfaces

Another type of condensation problem may occur in a house having uninsulated walls. The humidity level within the house may be so high that, during the night, after the heating system has been turned down, or off, water vapor condenses on both the window surfaces and interior wall surfaces. It is, of course, the frost on the windows that is immediately apparent—Jack Frost has been here. But most homeowners don't realize that the wall surfaces will be covered by an invisible film of condensed water vapor as well. If this condensation is severe enough, moisture may in time stain the wall surface or wall-paper, and it may also soak into and ruin any draperies, furniture, framed pictures, or carpeting that may be touching the wall. Heavily frosted windows are always an indication of excessive humidity levels within a house and the homeowner should take warning at its appearance.

One remedy is to increase the night-time temperatures in the house to prevent the condensation, but this will not, of course, remove the cause. Another is to improve the ventilation, which will remove excess water vapor from the house. Increasing the circulation of air will also prevent condensation, which is why a fan is often directed against a cold window, as in automobile defrosters. Insulating the walls (if it is possible) will, of course, prevent the walls becoming cold in the first place.

ROOF CONDENSATION

Figure 2 shows a cross section through a typical wood-frame roof construction in which the underside of the rafters is enclosed by a plaster or wallboard ceiling. During the winter, conditions within such a roof are more favorable for the formation of condensation than they are in a wall of similar construction. During the day, the warm, moist air within the house rises toward the roof. The roofing material, made hot by the sun's radiations, even on the coldest day, encourages this upward movement of air. Any water vapor that diffuses into the roof space is trapped there because the roofing material is usually impermeable, the main exception to this being wood shingling.

As the sun sets, the roof temperature falls rather abruptly below the temperatures existing inside the house, and any trapped water vapor will condense on the roof sheathing. In below-freezing weather, the condensate will freeze, only to melt and drip down on the ceiling below when the sun rises the next day. The result will be large, disfiguring water stains on the ceiling. Sometimes roofing nails project through the sheathing. This occurs especially when the roofing consists of asphalt shingles. These nails conduct cold with much greater efficiency than the asphalt does and therefore provide focal points upon which water vapor can condense. In freezing weather, therefore, the condensate will freeze as tiny balls upon the points of the nails. The next day, as the sun heats the nails, the ice melts. Over a prolonged period water dripping from these nails will produce on the ceiling below a multitude of small, rust-colored stains.

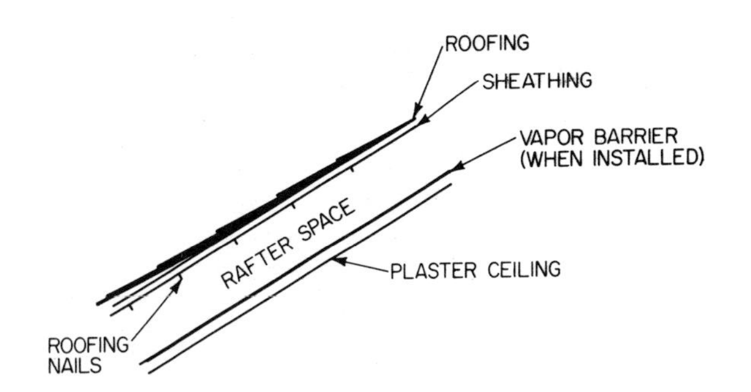

Fig. 2. A typical roof installation in which the rafter space is enclosed.

The practice of installing insulation between the rafters, however useful it is for retaining heat within a house, only encourages the condensation of moisture upon the roof sheathing, for the same reasons mentioned above for walls. Condensation within the roof space can be prevented by installing a vapor barrier under the rafters, but this solution only raises another problem: what to do about the buildup of moisture within the house generally, which now has no place to escape? The best long-term solution is to prevent the accumulation of excess moisture in the house in the first place; but since this solution is probably impossible for most families, the next best solution is to vent the air from the kitchen and bathrooms directly to the outdoors.

Water vapor may also condense in the ceiling below the floor of an unfinished attic if, for some reason, there is no natural circulation of air through such an attic. But if the attic is ventilated, as it should be, any water vapor that diffuses through the ceiling and into the attic will be removed by the natural circulation of air through the attic. If this natural ventilation is inadequate for any reason, it may be necessary to install an exhaust fan in the downwind side of the attic to help remove humid air from the house.

BASEMENT CONDENSATION

The condensation problems discussed so far are winter or cold-weather problems. Condensation in a basement is a summer problem. In most full basements, the floor consists of concrete that has been placed directly on the soil. The summer temperatures of this soil will range from 40 to 45°F in the northernmost parts of the United States to something around 70°F in the southernmost parts of the United States. In hot, humid weather, water vapor entering the basement will condense on the concrete floor, which will be at the same temperature as the soil below. If the concrete is porous enough, the condensed water will quickly diffuse through the concrete and be absorbed eventually into the soil, assuming the soil is dry enough to absorb it, as it usually is under a full basement.

But if the surface of the concrete has been troweled to a hard, dense finish, this finish will hamper the diffusion of water through the concrete. If the floor has been covered with vinyl tiles, linoleum, or a similarly impermeable flooring material, or if an impermeable vapor barrier such as polyethylene film has been laid under the concrete to keep moisture from entering the basement from *below,* the condensate must perforce remain on the basement floor while the owner of the house thinks dark thoughts about cracked walls and shoddy construction.

The way to prevent a wet basement in a humid climate is to construct the basement floor properly in the first place. This means that the concrete must be placed over a 6-in.-deep bed of coarse gravel or rock, in which one or several rows of drain tile have been laid, the drain tile emptying into a sewer that will carry away any water that happens to collect in the gravel bed. Since any water that may be in the soil cannot possibly migrate *up* through a gravel bed but will instead drain away, the soil under the concrete floor will remain reasonably dry. This, in turn, means that the underside of the concrete floor will have a lower vapor pressure than exists on the top of the floor. This being so, any water vapor in the basement will tend to diffuse through the concrete and condense on the gravel bed below.

When condensation occurs in an existing concrete-floored basement, and if the soil below the basement is reasonably dry, as it probably will be, getting rid of the condensate means providing a pathway through the concrete that will allow the condensate to diffuse out of the basement. If an impermeable flooring material has been installed over the concrete, it must be removed. If the surface of the concrete has a hard, dense finish, this finish must be removed, either with a hammer and chisel or by jackhammering the concrete. The surface can then be made smooth again by troweling a layer of porous concrete over the surface.

CONDENSATION ON SLAB-ON-GRADE FLOORS

In humid weather, water vapor will condense on a concrete floor that has been placed on or just below grade level for the same reasons it will condense in a full basement. In fact, condensation is more likely to occur on a slab floor since a great many of these floors are covered with vinyl tiles or a similarly impermeable flooring material laid directly on the concrete, whereas in most full basements the concrete is left unfinished. A rug laid upon a slab-on-grade floor is just as likely to result in condensation since the rug will retain heat, which will increase the rate of condensation in the concrete directly under the rug.

The solution recommended for a full basement is unsuitable for a slab-on-grade floor in communities where the building code requires that a vapor barrier be placed under the slab to prevent its absorbing moisture from the soil below. In effect, the requirement for a vapor barrier is perpetuating the very condition it is intended to prevent, since it will prevent any condensate that collects on top of the slab from diffusing through the concrete into the soil below. These codes seem not to recognize that in humid weather condensation is more likely to occur on top of a floor slab rather than that moisture will migrate through the floor slab from below. A vapor barrier is often required even when a coarse gravel bed has been laid down to prevent water collecting under the slab.

The only thing the occupant of such a house can do to dry out his floor during humid weather is to provide adequate ventilation throughout the house.

CONDENSATION IN CRAWL SPACES

The soil under a basementless house may release as much as 100 lb of water vapor a day into the air. If there is poor ventilation in the crawl space under such a house, the water vapor will make its way into the house, which will then feel continually damp and clammy, and it will also condense on the cool undersides of the house—on the girders, joists, and flooring. A substantial amount of damage can be done to the house by wood decay before the decay is discovered by the homeowner.

To prevent the buildup of water vapor within a crawl space, two things are necessary: (1) adequate ventilation must be provided, and (2) a vapor barrier, which usually consists of a sheet of polyethylene film, must be laid down on top of the soil in such a way that water vapor cannot make its way into the air past the overlapped edges of the film strips.

Door, Wood

A door is a movable barrier that closes off a *doorway,* which is an opening in a wall through which people and their possessions must pass in order to enter or leave a house or go from one room to another. Most doors swing on hinges, or *butts,* fastened to one side of the door. A few doors—mainly closet and wardrobe doors in modern houses, but also large living-room and dining-room doors in older or more expensive dwelling—slide sideways, on special tracks, to open and close.

A door fits closely into an especially constructed frame that forms the sides and top of the doorway. The construction of this framework is much the same for all doors, both hinged and sliding, and regardless of the size or quality of the door (see Fig. 1). The two sides of the frame are the *side jambs;* in a hinged door, they are sometimes called the *butt* and *lock jambs,* according to function. The top of the frame is the *head jamb,* or simply the *head.* The jambs are surrounded by *casing,* which is decorative wood trim that sets off the doorway and also covers the gaps that exist between the jambs and the adjacent wall construction.

An interior hinged door also has a projecting strip of wood, a *doorstop,* nailed all around the perimeter of the doorway against which the door abuts. The door must have something to knock up against when it closes, otherwise it will simply keep on swinging, which will pull the butts out of the jamb after a while. In an *exterior* door, the doorstop consists of a rabbet cut about ½ in. deep all around the side and head jambs.

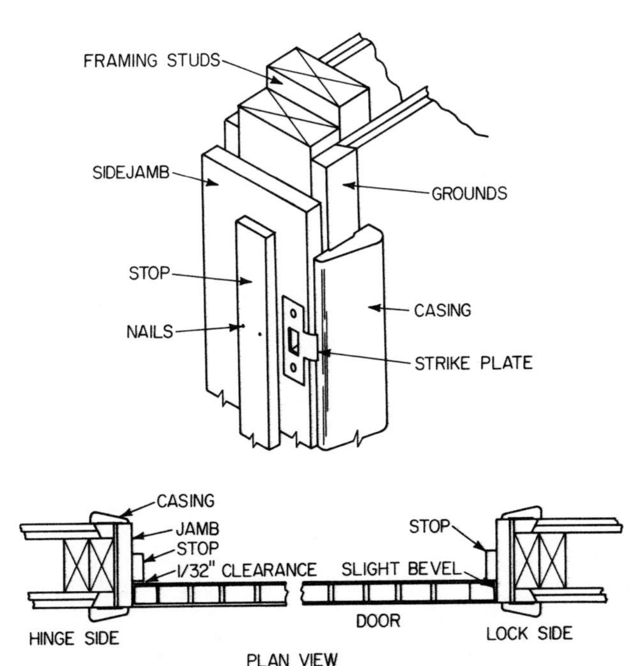

Fig. 1. Detail of an interior door installation, showing doorstop *(U.S. Forest Service).*

An exterior door will also have a *sill* installed at the bottom of the doorway. The sill is a raised, sloping piece of wood, stone, or concrete that is about as wide as the jambs. Its function is to prevent rain entering the building past the bottom of the door. An interior doorway leading into a bathroom will usually have a sill installed also, especially if the bathroom floor is tiled, to prevent any water that overflows onto the floor from running out the door. Most exterior doors will also have a *threshold* attached to the sill. This is a thin strip of wood or metal that is about as wide as the door is thick that further closes off the gap at the bottom of the door. So much for basic nomenclature.

DOOR CONSTRUCTION

An enormous range of door styles and sizes exists. These doors are made from a wide variety of woods and wood products that vary greatly in quality, in the way they are manufactured, and in the way they are fastened to each other to form the finished door. The price range of doors is correspondingly wide. What follows, therefore, can be only a broad introduction to the principal types of doors that are available.

Doors are basically of two types—panel and flush. *Panel* doors consist of a wood framework in which are installed thinner wood panels. A panel door can have any number of panels, which are arranged in standard patterns (see Fig. 2). *Flush* doors are just that, perfectly flat on both sides, without any inset paneling or raised decorative trim to break the surface. Panel doors are used mainly in houses built in a traditional style, and flush doors are usually installed in dwellings built in a modern style.

Panel Doors

Figure 2 also shows the construction of a typical panel door. The door has a vertical member on each side called a *stile.* The stiles run the full length of the door. The stile to which the butts are attached is known variously as the *hinge, butt,* or *hanging* stile. The other stile is the *closing* or *lock* stile. There can also be one or more intermediate stiles, but if there are, they are generally smaller and they never run the length of the door.

The horizontal members are called *rails.* There are always at least two rails, one at the top and the other at the bottom of the door, the bottom rail usually being much wider than the top rail. There can also be one or more intermediate rails, which run from one stile to the other and join them together more securely. These are known as *cross* rails. The cross rail in which the lock is installed is also called the *lock* rail. The greater the number of rails, the less tendency the door will have to sag as it gets older.

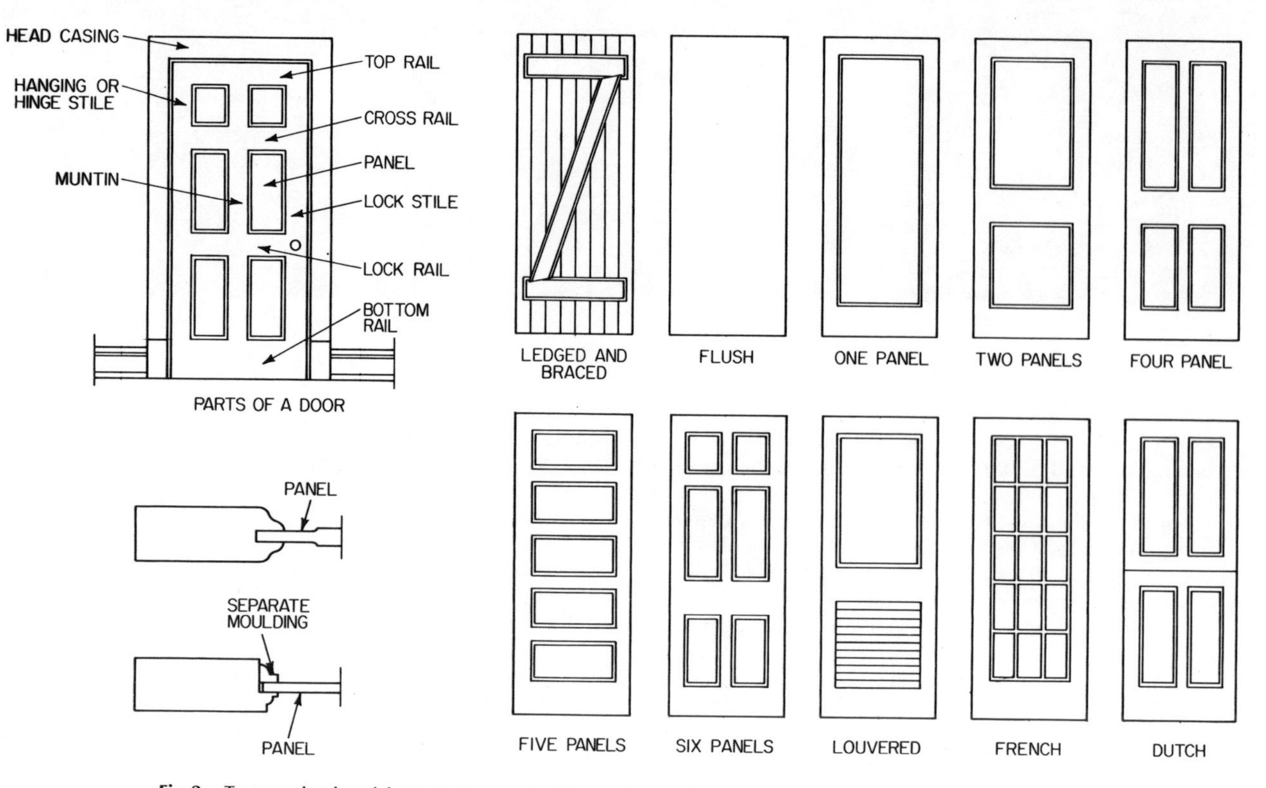

Fig. 2. Types and styles of doors. In the paneled doors, the panels can be held in place either by grooves cut into the stiles or by separate moldings nailed to both the stiles and rails.

The open spaces between the stiles and rails are filled by thin wood panels. These panels may be held in place in either of two ways. The edges of the stiles and rails can be plowed (that is, grooves are cut into them) to receive the panels, the entire door then being assembled together at the same time; or the panels can be held in place by mouldings nailed to the already assembled stiles and rails. These mouldings are called *sticks* in the trade, and their installation is called *sticking* or *stickering*.

The stiles and rails of panel doors may be constructed of solid pieces of lumber or veneered plywood. Paneled doors made of solid wood are not considered to be of the highest quality. They are used, rather, when the door is to be painted. Better-quality doors intended be stained or varnished are made from hardwood veneers. It is possible to make a panel door completely from solid pieces of hardwood, of course, but such doors have a bad tendency to warp. Veneered plywood is an excellent material as it combines the attractive appearance of hardwood with the strength and dimensional stability of plywood.

The stiles of a veneered plywood door are edged with matching hardwood strips that are ½ to ¾ in. thick. The presence of these strips allows the carpenter to plane down the sides of the door to obtain a close fit in the doorway without destroying the appearance of the door.

As for the panels, they may also be made either from a solid piece of wood or from veneered plywood. Again, veneered plywood is preferred because it is dimensionally more stable than solid wood, which has a tendency to split or shrink in its frame. But, if the panels are to have raised faces or beveled edges, and if the door is to be stained or varnished, the panels must be made from a solid piece of wood.

Paneling is always set loosely within its frame. It is never glued, nailed, or otherwise fastened to the frame. To do so would not allow the frame or the paneling to swell or shrink in response to normal changes in humidity, and it is likely that the panels would warp or crack in time.

Flush Doors

Flush doors are constructed basically much like ordinary plywood panels (*see* PLYWOOD). The doors, however, have much thicker center cores. There are two main types of core used—those that are of solid construction and those that are not (see Fig. 3).

Hollow-Core Doors

Hollow-core flush doors are used inside a house, where the temperatures are equitable and the humidity stable throughout the year. Under these conditions, there is not much danger that these lightly-constructed doors will encounter any unusual stresses that will cause them to warp.

A hollow-core flush door will have a framework of stiles and rails made usually of ponderosa pine. On less expensive doors meant to be painted, the stiles will remain exposed. On better-quality doors meant to be stained or varnished, the stiles will have veneer strips glued over them that match the face veneer used on the side panels. On the best-quality doors, solid hardwood strips ½ to ¾ in. thick are glued to the stiles.

Hollow-core flush doors also require solid blocking into which mortises for the door locks can be cut. These *lock blocks* are glued to the stiles and are anywhere from 20 to 27 in. long and 4 to 4⅝ in. wide—plenty of room in which to locate a lock. Less expensive doors will have a lock block built into only one edge of the door, and the carpenter had better not bore into the wrong edge. More expensively made doors will have a lock block installed along each of the edges of the door.

The center core space may be filled with any of a wide variety of lightweight materials in themselves flimsy but which, when glued to the facing panels, increase the strength of the door tremendously. The same principle is used in modern aircraft construction. Some of the most common of these fillers are shown in Fig. 3.

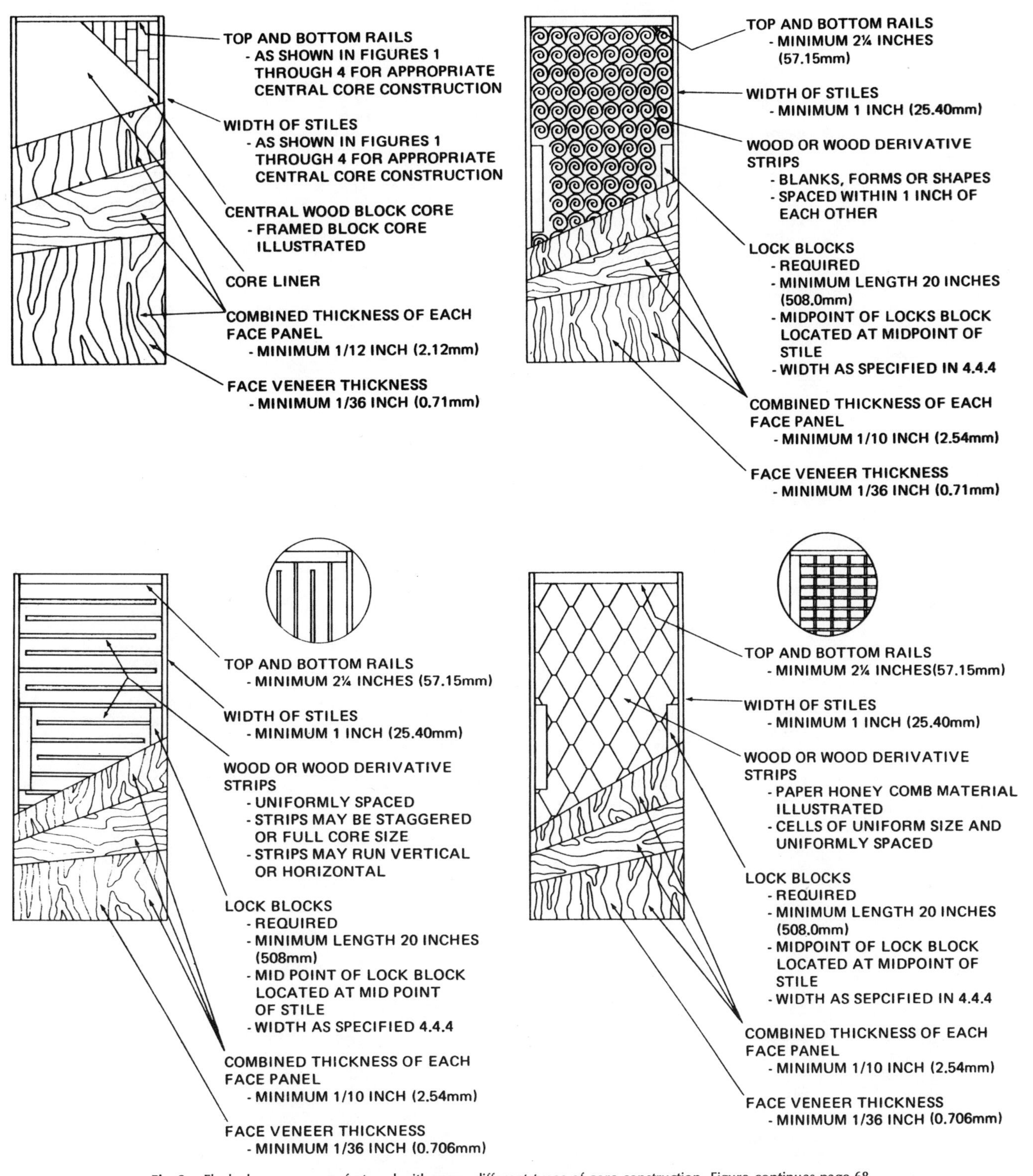

Fig. 3. Flush doors are manufactured with many different types of core construction. Figure continues page 68 *(National Woodwork Manufacturers Association)*.

Solid-Core Doors

Solid-core flush doors are used mainly for the exterior doors of a house, as there will be times when a considerable difference in humidity and/or temperature will exist on opposite sides of the door, conditions that may cause a hollow-core door to warp. Solid-core doors are also much less likely to transmit cold into the house during the winter. Solid-core doors may also be used in the interior of a house where there is a likelihood that a stressful condition may arise, such as on opposite sides of a bathroom door, for example, or on opposite sides of an attic door.

Solid-core doors usually have a framework of wood strips that are ½ to ¾ in. wide. These strips are nonstructural equivalents of the stiles and rails in a panel door. This light framework is filled with small blocks of wood (i.e., *core blocks*) all planed down to the same thickness as the framework. Sometimes these core blocks will be glued together into one solid block,

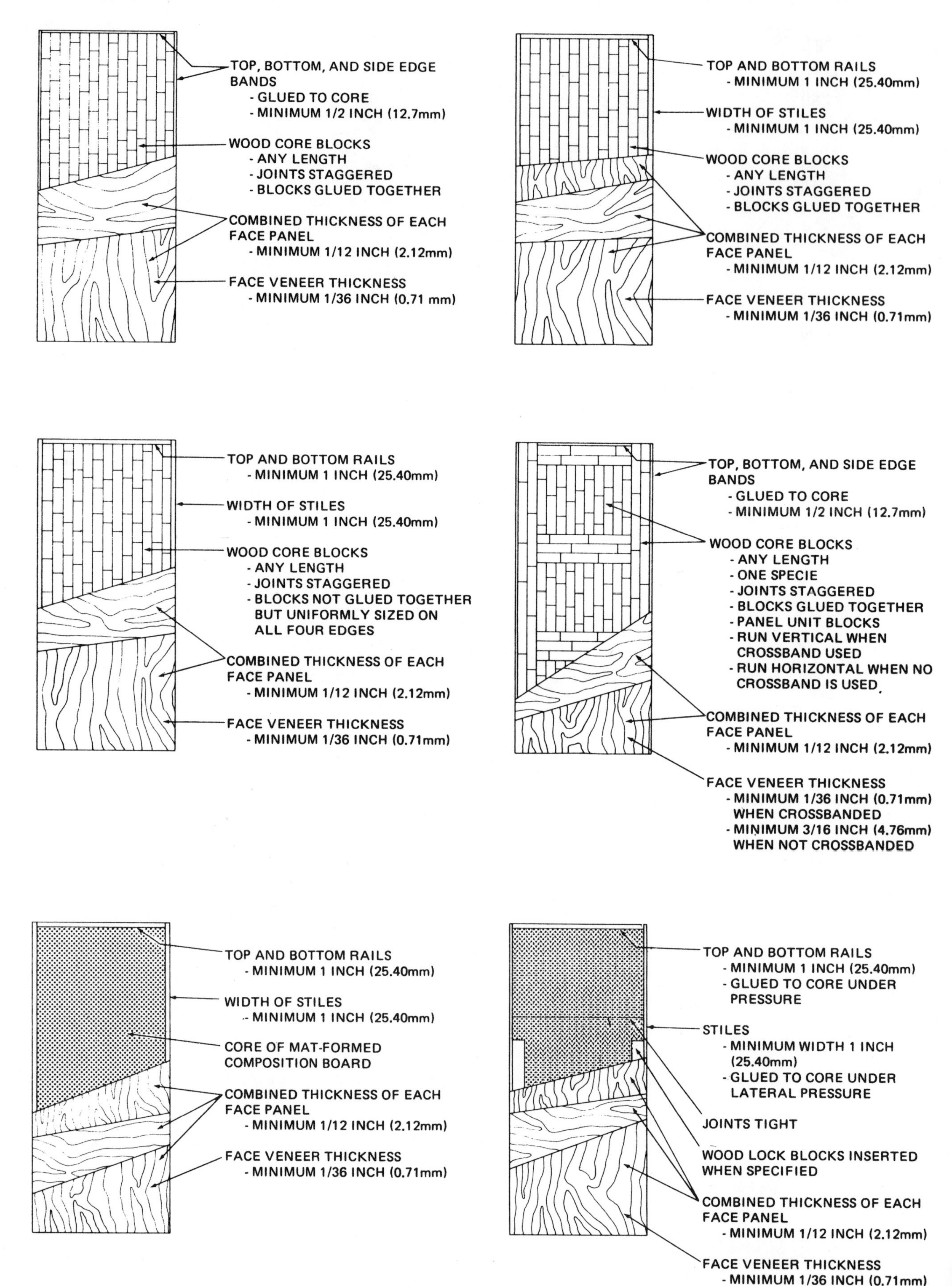

Fig. 3 (continued).

which is then planed down as a unit to its final thickness. In other doors, the core blocks are merely laid in place and glued not to each other but to the face panels. White pine was once widely used for core blocks because of its lightness and dimensional stability but ponderosa pine is used nowadays, as it has the same qualities.

Each side of the door is faced with a one- or two-ply hardwood veneer. If one-ply panels are used, they will be $\frac{1}{8}$ to $\frac{1}{4}$ in. thick. If two-ply panels are used, the inner ply will usually be $\frac{1}{8}$ in. thick, and the outer ply will be from $\frac{1}{16}$ to $\frac{1}{8}$ in. thick. Laminates printed with a simulated wood pattern or in bright colors are being used increasingly for the face plies.

DOOR SIZES

Occasionally an architect may design an elaborate entrance door for an expensive dwelling, which must be especially made, but apart from this rarity doors nowadays are *stock* doors, that is, they are mass produced in standard sizes and installed by a carpenter "as is." Table 1 shows the wide range of stock-sized flush doors that are made; one can find in the table almost any combination of width, thickness, and height one would ever need. If a customer should want an odd-sized door, most mills would make the door to order but would charge a premium, of course. Despite the complexity of Table 1, most doors are made in only a few popular sizes, which are as follows.

Interior doors are usually $1\frac{3}{8}$ in. thick, the major exception being small closet doors, which are often only $1\frac{1}{8}$ in. thick. Interior door heights are usually 6 ft 8 in., for both first-floor and upper-floor doors, though sometimes the upper-floor doors are only 6 ft 6 in. high. Interior door widths are usually 2 ft 6 in., except for bathroom doors (which may be 2 ft 4 in. wide) and clothes-closet and linen-closet doors (which may be 2 ft 0 in. wide).

Exterior doors are usually $1\frac{3}{4}$ in. thick, 6 ft 8 in. high, and 3 ft 0 in. wide, although service-entrance doors may be only 2 ft 8 in. wide. These are the usual minimum dimensions. Many main entrance doors will be larger than this—doors 7 ft 0 in. high and from 2 to $2\frac{1}{4}$ in. thick are common.

In fact, the sizes of the doors in a dwelling may be taken as a reliable indication of the general construction standards followed by the builder. If the door sizes are made to the standard dimensions noted above, one can assume that the rest of the construction follows customary standards as well. If the doors are smaller and/or thinner than the sizes mentioned above, it is likely that the builder cut corners in the rest of the construction as well. If the doors are larger and heavier than ordinary, it is a good indication that the builder tried to build a house of better than average quality.

Table 1. Standard Sizes of Wood Flush Doors

Interior Flush Doors Solid-Core

$1\frac{3}{8}$"	$1\frac{3}{4}$"
1'6" × 6'6"	2'6" × 6'0"
6'8"	6'6"
2'0" × 6'0"	6'8"
6'6"	7'0"
6'8"	2'8" × 6'0"
7'0"	6'6"
2'4" × 6'0"	6'8"
6'6"	7'0"
6'8"	3'0" × 6'8"
7'0"	7'0"
	3'4" × 6'8"
	7'0"

Table 1. Standard Sizes of Wood Flush Doors (continued)

Interior Flush Doors Hollow-Core

$1\frac{3}{8}$"	$1\frac{3}{4}$"
1'6" × 6'6"	2'4" × 6'8"
6'8"	6'10"
6'10"	7'0"
7'0"	2'6" × 6'0"
1'8" × 6'6"	6'6"
6'8"	6'8"
6'10"	6'10"
7'0"	7'0"
1'10" × 6'6"	2'8" × 6'0"
6'8"	6'6"
6'10"	6'8"
7'0"	6'10"
2'0" × 6'0"	7'0"
6'6"	2'10" × 6'0"
6'8"	6'6"
6'10"	6'8"
7'0"	6'10"
2'2" × 6'0"	7'0"
6'6"	3'0" × 6'0"
6'8"	6'6"
6'10"	6'8"
7'0"	6'10"
2'4" × 6'0"	7'0"
6'6"	

Exterior Flush Doors Hollow-Core

$1\frac{3}{4}$"	
2'8" × 6'8"	3'0" × 6'8"
6'10"	6'10"
7'0"	7'0"
2'10" × 6'8"	
6'10"	
7'0"	

Exterior Flush Doors Solid-Core

$1\frac{3}{4}$"	$2\frac{1}{4}$"
2'4" × 6'8"	3'0" × 6'8"
7'0"	7'0"
2'6" × 6'8"	3'4" × 6'8"
7'0"	7'0"
2'8" × 6'8"	3'6" × 6'8"
7'0"	7'0"

Flush Dwarf Doors

$1\frac{1}{8}$"	
1'6" × 4'0"	2'4" × 4'0"
4'6"	4'6"
5'0"	5'0"
5'6"	5'6"
2'0" × 4'0"	2'6" × 4'0"
4'6"	4'6"
5'0"	5'0"
5'6"	5'6"

(When so specified, doors can be furnished $1\frac{3}{8}$" thick.)

DOOR INSTALLATION

Doorframe Installation

There is a difference of opinion regarding just when the doorframes should be installed. One group of builders will set the doorframes in place immediately after the house frame has been completed and the walls and roof installed but before the finish flooring or the plaster or wallboard are installed. A second group of builders will wait until construction of the house has been almost completed and the finish flooring has been laid down. That is, the first group considers the doorframe installation to be part of the rough carpentry, and the second group considers the doorframe installation to be part of the finish carpentry.

As far as the homeowner is concerned, the advantages to him are with the second group of builders. The point of installing the doorframes as soon as possible, from the point of view of a builder, is that the edges of the doorframes will form *grounds,* or straightedges, that the plasterers can use to level the plaster (*see* PLASTERING). This technique is certainly faster and cheaper than installing separate grounds, which are shown in Fig. 1, and the builder is assured that the doorframes (which he will have bought ready-made) and the wall finish come out exactly flush with each other, which will make it easier to install the casing.

When doorframes are installed this soon, however, the danger always exists that the wood will become saturated with moisture as the plasterers and others work around it, with the consequence that the doorframe may warp as the wood dries out. (The doorframe will probably have been preprimed by the manufacturer to minimize this danger, but the danger still exists.) The edges of the doorframe may also be nicked and gouged as the workmen move themselves, their tools, and their scaffolding about. Finally, the carpenters may leave unsightly gaps along the edge of the floor where they butt the ends of the finish flooring against the installed jambs. None of these problems can occur if the doorframes are considered part of the finish trim and are installed after the rest of the house has been completed.

Doors and doorframes are widely sold as completely finished, prepackaged units, the doors being already fitted and hung within the assembled doorframe, all ready for installation in the rough opening. Doors and doorframes are also sold as complete but dissassembled kits, with the jambs precut to size, the wood routed out in both jambs and door for the butts and lock, and the entire package requiring only assembly into the rough door opening.

In these prefabricated door assemblies, the doorframes are already trimmed to one of several predetermined widths. Which width the builder orders will depend on the overall interior-wall thickness, which is usually the actual width of the 2 × 4 in. stud plus ½ in. to ¾ in. on each side of the stud, depending on the thicknesses of the finish wall. Plastered walls are usually 5¼ in. thick overall, and ½-in.-thick gypsum wallboard walls are usually 4⅝ in. thick overall. Prefabricated doorframes are also sold in which the widths of the jambs can be adjusted slightly to accommodate the overall width of a particular wall opening.

Rough door openings are usually 3 in. higher and 2½ in. wider than the size of the doors that will be fitted into them. Before the wall is constructed, therefore, the carpenter must know what the door sizes will be, information that he will obtain from the house plans. When constructing the rough opening, the carpenter must make sure that the side studs (or *bucks*) and the lintel are cut from lengths of straight lumber and that this lumber is installed as plumb, level, and square as possible, otherwise the finish carpenter will have trouble installing the doorframes so *they* are plumb, level, and square.

It is conventional to determine the "hand" of a door, that is, the direction in which the door swings on its butts when it opens, as if one were standing outside a room looking in. Thus, for an interior door one would be standing in a corridor, while for an exterior door one would be standing outside the house. If neither of these conditions apply, then the "hand" is determined when the butts are not visible from the side on which one is standing.

Therefore, when one is standing outside a house or room, if the butts are on one's right, the door is a right-hand door. If the butts are on one's left, the door is a left-hand door. As for the locks, if, when standing outside a house or room, the door opens away from one (the usual direction of swing), the lock is a *regular* bolt type. If the door opens toward one, the lock is a *reverse* bolt type.

Door Hanging

Doors must always be handled and stored carefully. If possible, they should never be delivered to the site until the house has been enclosed so they can be stored in a dry place. Doors should never lean against a wall in such a way that only one edge of the door is resting on the wall—the door might develop a twist. It is better for both doors and walls if the doors are laid flat until they are ready to be installed.

A well-fitted door will have a 1/16-in. clearance on top and along both sides, and a ⅜-in. to ½-in. clearance at the bottom (see Fig. 4). If carpeting is to be laid under the door, the thickness of this carpeting (plus the thickness of any pad) must be known in advance, if possible, so that the appropriate amount of space can be allowed for. If a threshold is to be installed at the bottom of the door, the final clearance need be only 1/16 in.

To fit the door, it is first wedged tightly into the doorframe in such a way that it bears solidly against the butt jamb. The carpenter then checks the clearance between the door and the lock jamb and makes whatever alterations may be necessary by planing the edge of the door. The fit between the head jamb and the top of the door is checked in the same way, the top of the door being planed, if necessary, to obtain a good fit. Finally, the fit of the butt jamb and the door is checked and this edge is planed, if necessary. Once an accurate fit has been achieved, the lock side of the door is beveled slightly with a plane to permit the inner edge of the door to clear the jamb as the door swings open. If the door were left with a perfectly square edge, this edge would scrape against the jamb every time the door opened and closed.

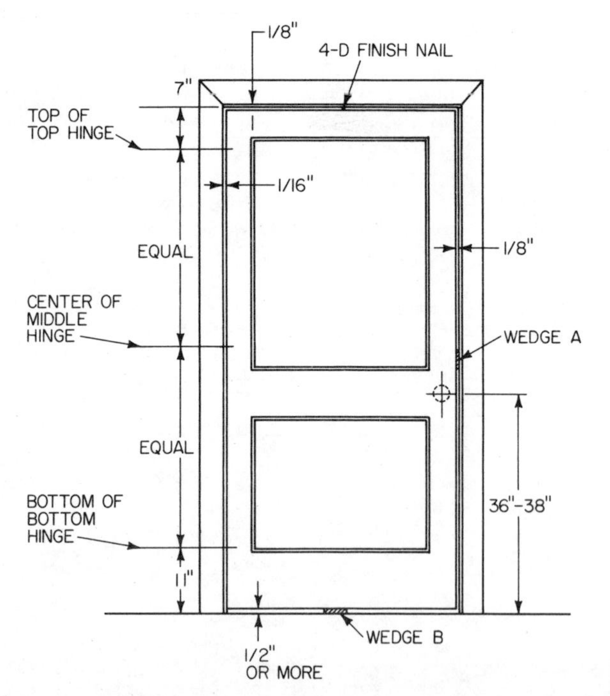

Fig. 4. Laying out the door hinges and obtaining the proper fit of a door within its frame.

Butt Installation

The number and size of butts installed on any particular door will depend on the weight and size of the door. Interior doors usually have two butts installed, exterior doors have three, as do the doors leading into an unheated attic, as three butts will counteract any tendency the attic door may have to warp.

The butts used for residential doors are usually of the type called *loose-pin* butt hinges (see Fig. 5). That is, the pin holding the two halves (or *leaves*) of the butt together merely drops into the *knuckle,* i.e., the round, central portion of the butt. If the pin is held in place by a screw located at the bottom of the knuckle, the butt is call a *fast-joint* butt hinge.

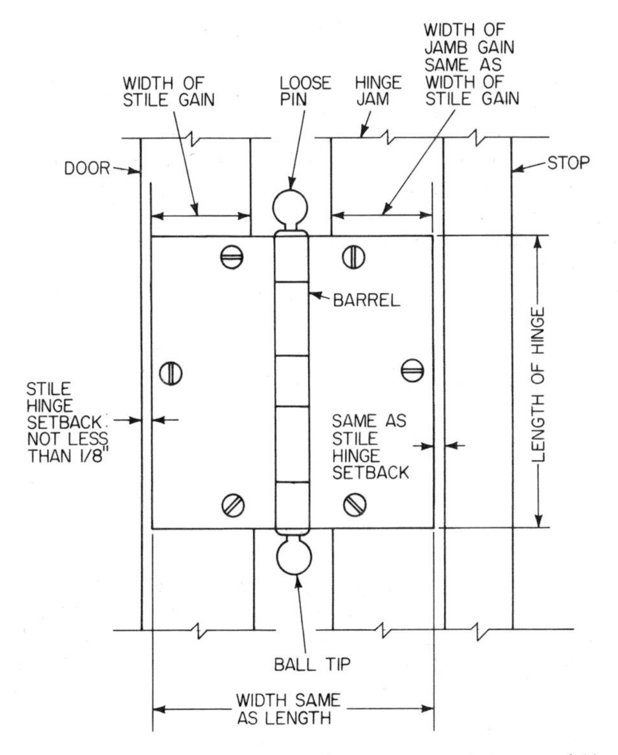

Fig. 5. A loose-pin type of mortise butt hinge *(U.S. Bureau of Naval Personnel).*

Butts may be either square-cornered or round-cornered. Round-cornered butts are installed by carpenters who have a special door-hanging template that is used with an electrically driven router. Round-cornered butts are also an indication of a prefabricated door. The square-cornered butts must be mortised in place using a hammer and chisel; they are usually an indication of better-class workmanship.

The top butt is installed first. After it has been screwed securely in place, the door is hung on it and its clearances are checked. If these are satisfactory, the door is removed from the doorframe and the bottom butts are installed. The door is then hung once again to recheck the clearances and the manner in which it swings out into the room and closes into the doorframe.

If the door clearances aren't satisfactory, the carpenter can adjust the position of the door slightly by inserting thin strips of cardboard under one of the leaves (which shifts the door away from the butt jamb), or he can deepen the mortise slightly (which shifts the door toward the butt jamb).

If the door binds in a hinge as it swings back and forth, the carpenter will have to adjust the position (or positions) of one (or both) of the butts until the door does swing freely.

Finally, the bottom of the door is checked for clearance and for parallelism with the floor.

Doorstop

The two main kinds of doorstop are: (1) a strip of wood that is nailed completely around the jambs and (2) a rabbeted stop that is ploughed into the edges of the jambs. The first type is used for interior doors, the second for exterior doors (see Fig. 1). In both kinds, the stop is so located that, when the door is closed, its surface is flush with one edge of the doorframe. In both kinds of doorstop also, the stop prevents light or air passing through the edges of the door.

A variation of the nailed-on strip is a stop that rests partially within a groove ploughed into the jambs. This increases the strength of the stop, which can absorb more punishment, and it also makes the door more burglarproof, if the door is to be kept locked for any reason. With the usual nailed-on stop, all a burglar need do to open the door is force a thin metal blade between the stop and the jamb. He can in this way push the latch back until the door opens. When the stop is inserted into a groove, this is impossible.

Casing Installation

It is the casing that gives distinction to a doorway. Modern houses tend to have very simple, narrow casings, for both stylistic and economic reasons. Typically, the wood will be somewhere between 2¼ and 3½ in. wide and from ½ to ¾ in. thick. Casings intended for traditional houses are much more decorative, although there is nothing today like the heavy, ornately carved casings that surround the doors and windows in a Victorian mansion, which sometimes have spiral or fluted pilasters or half-columns set into the main moulding. Nevertheless, even today a homeowner can build up an elaborate casing around the doors and windows of his home if he wants to, limited only by the amount of money he is willing to spend, his taste, and his common sense.

Basically, a casing consists of three pieces of wood—two side pieces and a top piece. Large casings are usually hollowed out at the backside of the wood. This lightens the wood, prevents warping, and it also keeps most of the wood away from the construction, especially at the gap between jamb and plaster. If the plastering should have been done unevenly, high spots or roughness in the wall that might otherwise hold the casings away from the wall are prevented from doing so, and the casing will be able to rest flush against the wall.

Casings are always set back slightly from the edges of the jambs, leaving these edges slightly exposed. Usually 3/16 to 5/16 in. of the jamb edges will show, and this dimension must be the same all around the doorframe. An uneven-looking installation is an indication of a poorly done job.

Drywall Construction

Drywall construction may perhaps be best defined by what it is not—it is any method of finishing the interior walls and ceilings of a house that is *not* traditional plastering.

A plastered wall hs a great many virtues (*see* PLASTERING). It provides a smooth, even surface that is hard and strong, that can breathe, that takes paint and wallpaper extremely well, that is sound-resistant and very fire-resistant, and that can be patched easily if it should crack or be damaged. But, for a builder, plastering has many disadvantages. It makes for a wet, messy job that takes a considerable amount of time between application and final drying, which slows down the overall pace of the construction; the general dampness produced by the wet plaster may cause studs and joists to swell, which may result in the dried plaster's cracking when the wood finally shrinks back down again; plastering is especially difficult and time-consuming in winter because drying times are longer and the house must be fully enclosed and heated before the plastering can begin—freezing plaster is weak plaster; and, finally, plastering is expensive, a cost that is, of course, borne ultimately by the homeowner.

So drywall construction was invented. Drywall construction may be thought of as a kind of portable gypsum-plaster wall that is manufactured elsewhere in the form of large, paper-covered panels. These panels are carried to the jobsite where they are nailed or screwed (or nailed *and* screwed), and sometimes glued to the studs and joists as well, without any of the messiness or bother attendant upon a wet plastering job, though drywall does have faults that will be described in the course of this article.

By a natural extension of meaning, *drywall* has come to mean the installation of materials other than gypsum wallboards to finish a room, including plywood, wood paneling, fiberboard or insulation board, hardboard, particleboard, asbestos-cement panels, and so on. For the purposes of this article, however, drywall construction—which is also known as *plasterboard* construction—will mean a wall constructed of gypsum wallboard. For a description of the manufacture of the panels themselves as well as the range of sizes and types that are available, *see* GYPSUM PANELS.

The gypsum wallboards used in drywall construction are usually 4 × 8 ft in size, though panels from 6 to 16 ft long are available. The panels are made in thicknesses that range from ¼ in. to ⅝ in., with thicker panels available for use in special construction.

The ¼-in.-thick panels are used mainly for remodeling over an existing wall or ceiling, when for some reason the wall or ceiling no longer provides a satisfactory surface. The ¼-in.-thick panels may also be used to form a curved surface—to form an arched doorway, for example—for which the thicker panels are too rigid. Other than this, however, ¼-in. panels are never used for an original wall construction.

The ⅜-in.-thick panels are also used for repair and remodeling jobs and for double-ply construction, as described below. The ½-in.-thick panels are the size most often used in new construction, and the ⅝-in.-thick panels are used where a more solid wall is needed or where a more fire-resistant construction is desired. The most important determinants of which thickness panel should be used are, in fact, the stud and joist spacing in wood-framed houses and whether the panels are attached parallel or crosswise to the studs and joists.

The panels have tapered edges on their long sides that allow them to be joined together without the joints showing using the technique described below.

The gypsum core is protected against moisture and minor injuries by a layer of hard, smooth-surfaced paper, which also provides a base to which paint and wallpaper can be applied. (There are also available gypsum wallboards that are used as lathing, to which plaster is applied; these wallboards are covered by a porous, rough-textured paper that offers a good grip to the plaster. These wallboards should never be used for ordinary drywall construction in which the surfaces of the panels remained exposed.)

Before the panels are installed, certain preliminary matters must be taken care of. For one thing, the builder must make certain that the studs and joists provide a flat plane to which the panels can be attached. In a plastered wall, by contrast, the plasterer need not be as concerned how wavy or uneven the supporting surface is, within limits, of course. It is the plasterer's job to make the surface of the plaster flat no matter what the support is like. In drywall construction, however, the panels are relatively stiff and inflexible. If the supporting studs and joists are not aligned with each other, the panels will not be aligned with each other, and in a gypsum-wallboard wall nothing looks worse than the wall's weaving in and out every four feet—unless it is having the joints between panels clearly visible.

Unevenly installed studs and joists may have to be reinstalled. Warped or bowed studs and joists may have to be replaced, or forced into alignment by the installation of special bracing cut into the lumber. Another technique that is used to achieve a flat supporting surface is to nail long strips of wood,

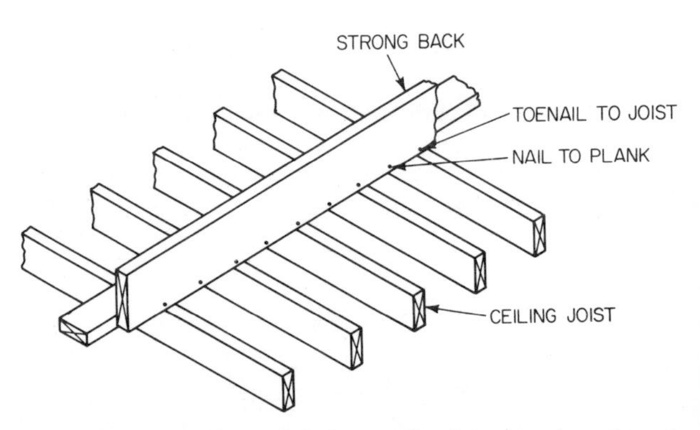

Fig. 1. A strongback may be nailed over ceiling joists in order to force the bottom edges of the joists into alignment *(U.S. Forest Service)*.

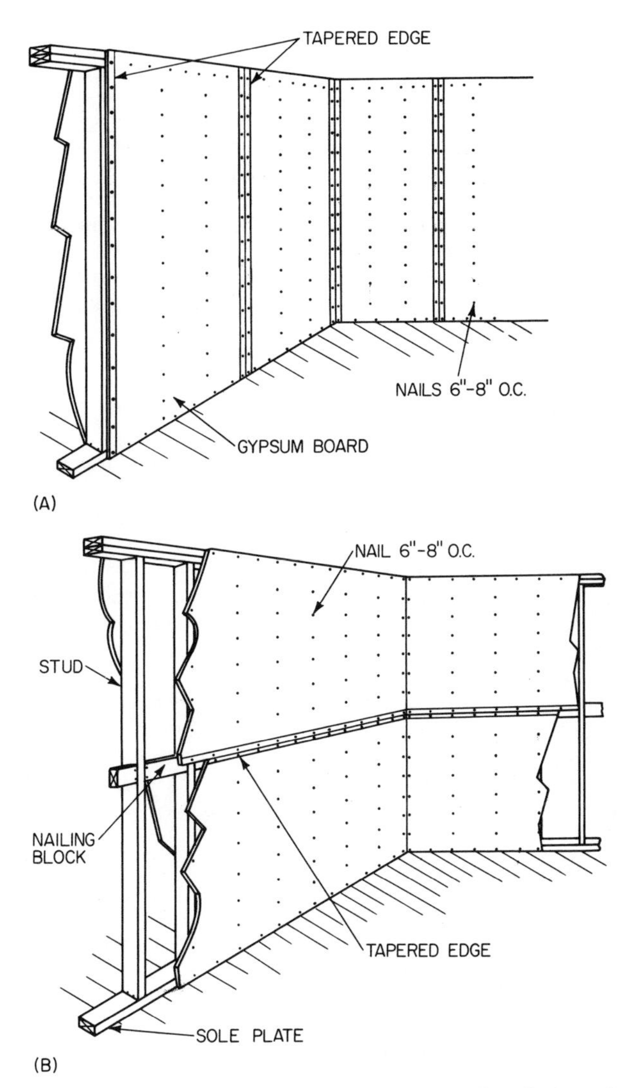

Fig. 2. Gypsum wallboards may be installed either vertically or horizontally *(U.S. Forest Service).*

suitably wedged to align and level them, to the edges of the studs and bottoms of the joists.

Joists are less likely to bow or warp because of their larger size, but when they are out of shape they are more difficult to straighten. Furring the joists with 1 × 3 in. lumber may be necessary. Sometimes, when the tops of the joists form the floor of an unfinished attic, out-of-line joists can be forced into alignment by the installation of a *strongback,* as shown in Fig. 1.

Another thing the builder must do beforehand, even before the panels have been ordered, is make a careful to-scale layout of the walls and ceilings, including the positions of the doors and windows in the walls. The purpose of this layout is to discover how best to install the panels to minimize both the number of visible joints and the amount of cutting and fitting that will be required. Making an advance layout will probably reduce the overall cost of the installation as well.

The panels can be installed on the walls either vertically or horizontally (refer to Fig. 2). When only one ply of wallboard is being installed, the preferred direction of installation is with the long sides of the panels horizontal. This construction is stronger, since a horizontal panel ties more studs together and, in addition, as much as 25 percent less jointing is necessary between panels. Horizontal joints are also less obvious to the eye than vertical joints.

These factors also apply to the ceiling panels. The preferred direction of panel installation is crosswise to the direction of the joists. If this also happens to be the long dimension of the ceiling, good. If not, the builder will have to choose between installing the panels crosswise or parallel to the direction of the joists. On the whole, it is preferred that the long sides of the panels run crosswise to the joists for the sake of the stronger construction.

As much as possible, single panels should run from one corner of a room to another. Since panels 16 ft long are available, this should be possible for most of the walls in a house. If not, the builder will have to calculate where best to make the breaks or whether to install the panels vertically. The least conspicuous breaks are over the centers of doors and windows.

A layout will also enable the builder to locate the places in the walls and ceilings where he will have to install 2 × 4 in. blocking, since all the edges of all the panels must have a solid backing into which the fastenings (i.e., nails, screws, or staples) can be driven and/or adhesive applied.

There are two general methods of installing gypsum wallboards. One method, called *single-ply* construction, consists of attaching a single layer of wallboard to the walls and ceilings. The other method, called *two-ply* or *laminated* drywall construction, consists of attaching two layers of wallboard to the walls and ceilings. Two-ply construction results in a wall that is much stronger and more resistant to any structural stresses that might crack or crush the panels of a single-ply installation. It is also a more fire- and sound-resistant construction than single-ply, and it enables the installers to lay a flatter wall surface, which gives the finishers a better chance to disguise the joints between panels. On the other hand, two-ply construction is twice as expensive to install as single-ply construction.

SINGLE-PLY CONSTRUCTION

In single-ply construction the panels are first installed on the ceiling. The panels should butt loosely against each other and against the adjacent wall studs. The panels should never be jammed up tight against each other or against the studs as they tend to swell slightly when they become damp, which may cause the drywall to crack or buckle.

Once the ceiling panels are in place, the wall panels can be installed. If these panels are being installed horizontally, the row of panels adjacent to the ceiling panels should be installed first. All the panels can either be nailed or screwed into place, or a combination of these fastenings plus adhesive can be used.

Nailing and Nail Popping

Nailing is the usual method of attaching panels; the spacing between nails on both ceilings and walls is shown in Fig. 3. One should always nail from the centers of the panels outward, leaving the edges that abut the wall and ceiling junctures unnailed. This is thought to provide a slight amount of give to the construction, in case the framing should settle or shrink, that will help prevent stressing the panels.

The nailing-on of gypsum wallboards has had an unforeseen and unfortunate consequence—the phenomenon of *nail popping.* Assume a gypsum panel has been nailed in place. As the wood to which the panel is attached dries out, it shrinks, which is normal enough. But as the wood shrinks down, the nails stay where they are in relation to the central axis of the wood. Consequently, the nails will appear to have popped out of the wood by 1/16 to 1/8 in. (If the panel should follow the popped

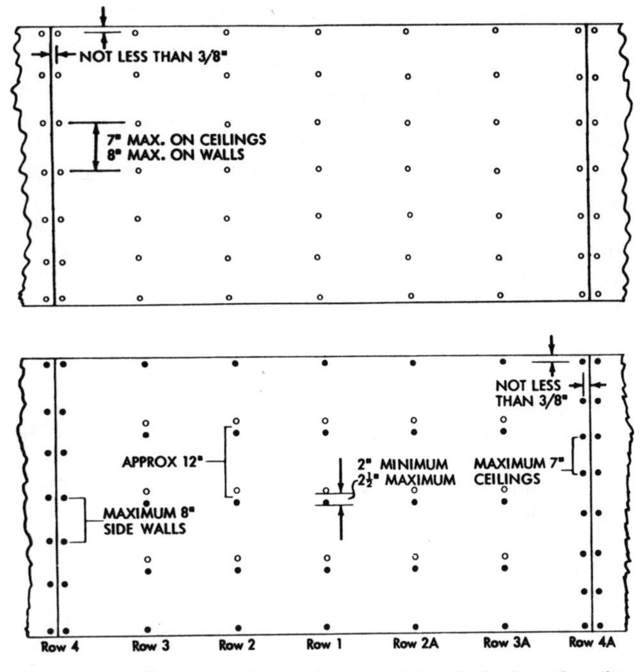

Fig. 3. The nailing pattern for single-ply installation, for both single-nailing and double-nailing techniques. In the lower figure, the solid dots represent the first nailing, the open dots represent the second nailing *(Callender, Time-Saver Standards, McGraw-Hill).*

nails away from the wood, this will result in bulges appearing here and there instead of nail popping.)

Nail popping is likely to occur whenever the wood framing has an inordinate amount of moisture in it, which is, roughly speaking, any moisture content in excess of 19 percent. To reduce the probability of nail pops, therefore, the builder must either order kiln-dried lumber, which will increase the cost of the construction, or he must order the lumber long enough in advance of construction to enable him to store the lumber under suitably dry conditions until it does dry out to a moisture content of 19 percent or less. Or he can put up the wood framing and then go away until the framing has had a chance to dry out. Or he can install the panels using one of the adhesive techniques described below, which does away with the need for most of the nails altogether.

In this connection, it should also be remembered that, in most sections of the country, the structural framework of a house will stabilize at a moisture content of around 10 percent after the house has been occupied for a while, so even the 19 percent moisture content that is considered a minimum to prevent nail popping means the wood will still shrink down a bit after the house has been completed.

Manufacturers of gypsum products have spent considerable time and money investigating the phenomenon of nail popping and one of the conclusions that has been reached is that the less distance a nail is driven into wood, the less it will protrude when the wood does shrink down. It follows, therefore, that the shorter the nail, the better. As a consequence, special ringshank nails ⅛ in. shorter than ordinary finishing nails but having an equal holding ability are recommended for use with gypsum wallboards.

The nails should always be driven slightly below the surface of the panels. This requires the use of a hammer having a crown head, so that when a nail is struck that one last blow, it will be driven below the surface of the panel and a small indentation, or dimple, about 1/32 in. deep will be left in the surface. These dimples will be covered later with joint compound to hide the nails.

Other Kinds of Fastenings

A contractor specializing in the installation of gypsum wallboards will use power screwdrivers to drive special Phillips-head screws that secure the panels in place. These screws are double-threaded and have a diamond point. They are reputed to hold the panels more firmly against the support than nails do. The labor output is also higher and the screws do reduce the amount of nail popping.

Staples are not recommended for single-ply construction, as the panels are not as securely fastened as when they are nailed or screwed in place.

Special adhesives compounded by manufacturers of gypsum wallboards can also be used to attach the panels to wood framing. These adhesives are of the rubber-latex type, and they are applied to the studs and joints with a caulking gun as a continuous ribbon. Use of an adhesive does not eliminate the necessity for nails or screws; it merely reduces the total quality of these fastenings required. The idea behind using an adhesive is that the fewer the fastenings, the less chance of nail pops appearing. When an adhesive is used, therefore, the nail spacing need be only 12 in. on ceilings and 16 in. on walls, while screw spacing may be reduced to 16 in. on ceilings and 24 in. on walls.

Double nailing is also practiced. The spacing between nails is also shown in Fig. 3. The first set of nails must be completely installed before the second set is installed. Double nailing does not stop or reduce nail popping. What it does is secure the panels more firmly to the wood, which of course makes the construction that much stronger.

Joint Finishing

All joints, including those in the corners of a room, are covered by *joint compound,* which is spread over the tapered edges of the panels with a 5 to 6-in.-wide flexible broadknife (see Fig. 4). Joint compound is a gypsum product formulated by wallboard manufacturers for this purpose, and it can be purchased either ready-mixed and ready for application or as a dry powder.

A strip of paper tape is used to reinforce and strengthen the joint. The tape is 2 to 2¼ in. wide and it is usually purchased in 250-ft-long rolls. A thin layer of joint compound (approximately 1/32 in. thick) is then spread over the tape, the edges of the cement being feathered out smoothly. The joint compound is then allowed to dry for 24 hr.

Ceiling joints are taped first, then the wall corners, then the joints between wall panels, and finally, the corners between ceiling and wall panels. The corners are taped using the same

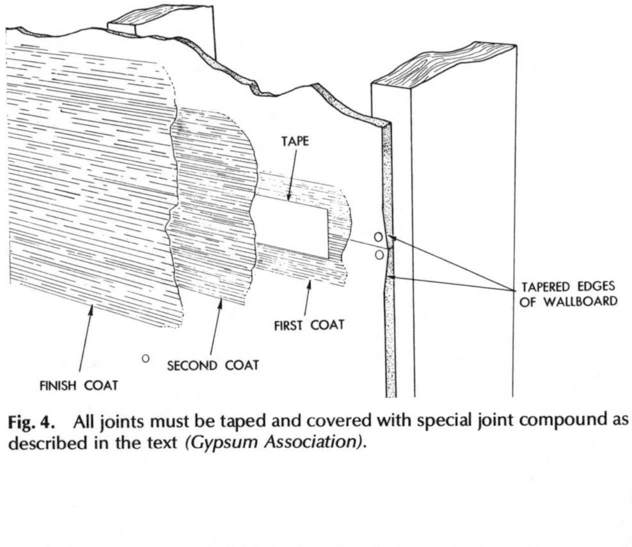

TAPE
TAPERED EDGES OF WALLBOARD
FIRST COAT
SECOND COAT
FINISH COAT

Fig. 4. All joints must be taped and covered with special joint compound as described in the text *(Gypsum Association).*

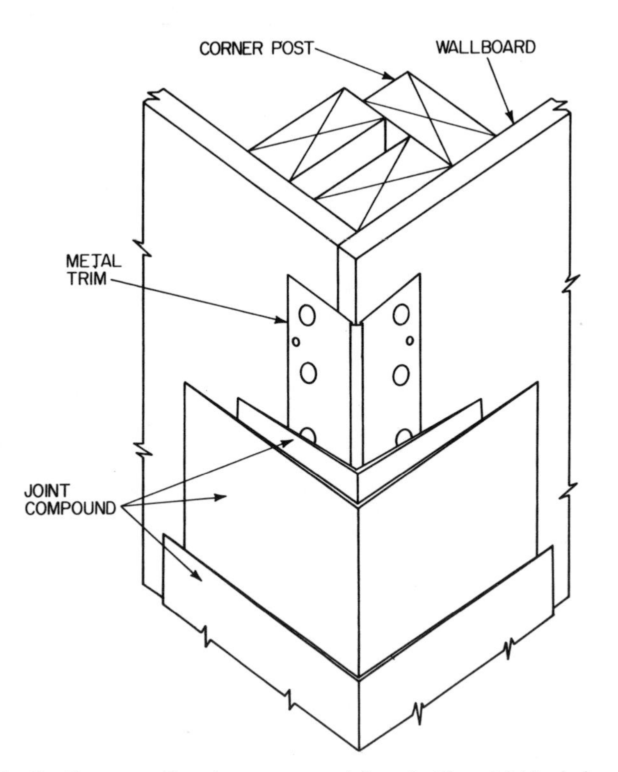

Fig. 5. Gypsum wallboard corners are reinforced with metal trim before being covered with joint compound.

general technique described above for ordinary joints; however, with interior corners, the tape must first be folded and creased before it is pressed into the compound. Special V-shaped broadknives are available that enable the plasterer to spread the compound evenly and firmly into corners.

Exterior corners, and all other cut or otherwise unprotected edges of the panels (such as at the edges of door and window openings) are first covered by metal cornerbead trim especially manufactured for the purpose (see Fig. 5). Once the trim has been installed, the exterior corners can be treated as shown in Fig. 4.

As the joint compound dries, it will shrink slightly. The joint is then sanded smooth and a second thin layer of the compound is applied over the first. This second layer should extend beyond the edges of the first layer by about 1½ in., with the edges again being feathered out smoothly. Wallboard manufacturers recommend the use of a premixed *finish compound* (or *topping compound*) for this second layer. The finish compound has a more buttery consistency than joint compound, and it spreads very smoothly. After it has dried, it has excellent sanding properties. The finish compound should be allowed to dry for 24 hr before it is sanded.

If the walls are to receive a paint finish that has a texture to it, the joints can be considered completed. But if the wall is to be papered, or if a glossy coat of paint is to be applied, which will show up the slightest unevenness in the joints, another thin layer of finish compound should be applied to the joints. This third layer should extend beyond the edges of the second layer by another 1½ in. and it, too, should be feathered out smoothly. The third layer should also be allowed to dry for 24 hr before it is lightly sanded. The joints may now be considered finished.

At the same time that the first layer or finish joint compound is being applied over the joints, the nail indentations mentioned above should also be filled with finish compound. Two coats of compound should be applied to all the indentations, with each of the coats being allowed to dry for a 24-hr period and each coat being sanded smooth after it has dried.

An alternative to cementing and taping the joints at the ceiling lines is the installation of wood cornices that will hide the juncture, in the same way that a baseboard hides the juncture between wall and floor. A variety of moldings is available, and cornices do enhance the appearance of a room just as a baseboard does.

TWO-PLY INSTALLATION

In all essential points, the installation of two plies of wallboard is identical to the installation of only a single ply. The first ply of a two-ply installation is installed in exactly the same way as already described for single-ply construction, except that it is not necessary to treat the joints, and staples may be used to attach the panels. When laying out the panels for the second ply, however, the edges must be offset at least 10 in. from the edges of the first ply. The plies should also run crosswise to each other, if possible, since this makes the construction much stronger. Thus, if for appearance's sake the second ply is to be installed horizontally, the first ply should be installed vertically—though there is no law that says so. If convenience dictates that both plies must run vertically, or horizontally, so be it.

Somewhat greater freedom is allowed in the placement of the second-ply panels if the second layer is laminated to the first since the installer does not have to be concerned about ending the panels along studs and joists. The second ply, of course, will be adequately supported by the first ply.

Most domestic two-ply installations use ⅜-in.-thick wallboards for both plies, but any combination of ⅜-in. and ½-in.-thick panels is satisfactory. The thicker the panels, the stronger the walls will be, the greater the fire resistance, and the better the sound resistance.

The plies may be attached to the studs and joists and to each other in a number of different ways. Both plies can be nailed or screwed in place, with both plies having the same nail or screw spacing specified above for single-ply installation. The only difference between the two is that the fasteners used for the second ply must be long enough to penetrate into the wood the same distance as the fastenings of the first ply. Or a combination of fastener and adhesive can be used for the second ply as already described for the first ply.

But the most common method of fastening the two plies in place is to nail the first ply (as already described) and then fasten the second ply to the first ply with an adhesive. The adhesive most commonly used is joint compound, and it is applied to the back of the second ply with a special trowel that has ¼ × ¼ in. notches located 2 in. apart. After the adhesive has been applied to a panel, the panel is placed against the first ply and held there temporarily by nails until the adhesive sets. The nails are then removed and the nail holes filled with finish compound.

The joints between the panels of the second ply are treated as already described for a single-ply installation.

DRYWALL FOR MASONRY WALLS

Gypsum wallboards can also be used to cover masonry, that is, concrete, brick, or cement-block walls. The simplest, though not the preferred, technique is simply to nail the wallboards directly to the masonry surface, assuming the wall is above ground and the climate is temperate. This can be done most easily against a concrete wall because of the smooth surface usually offered by concrete, but even with concrete it will probably be necessary to remove projections with a hammer and chisel before the panels are installed. On any exterior masonry wall, nails driven directly into the masonry very often provide what are called *cold bridges* that will transmit cold outside temperatures into the building. The nails, being colder than the interior face of the wall, will enable moisture within the building to condense on the nail heads, with the result that the wallboards will become discolored, as will any paint or other finish that is applied over them.

For these reasons, it is more sensible to apply furring to the wall and attach the wallboards to the furring. Furring consists of 1 × 2 in. or 1 × 3 in. strips of wood nailed to the masonry. The wallboards are then nailed, screwed, or glued to these furring strips in the same way that has already been described for single-ply construction. This method of providing a suitable support for the wallboards has the advantage that the furring strips can be carefully plumbed and aligned with each other by placing wedges under the strips before the wallboards are installed, which means the wallboards will present a flat and even appearance.

Another advantage in using furring is that an air space is created between the masonry and the wallboards that acts as insulation and prevents dampness reaching the wallboards. If, in addition, aluminum-foil-backed wallboards are installed, the insulating value of the air space will be further increased (*see* INSULATION).

Electrical Conductors and Cables

This article describes the types of electrical wire (or *cables* or *conductors*) used and the methods of installing the wire in residential electrical systems. For a description of how electrical switches, receptacles, and fixtures are wired together into complete, functional circuits, *see* ELECTRICAL WIRING CIRCUITS. *See also* ELECTRICAL SYSTEM; GROUND, ELECTRICAL.

An electrical *wire* consists of a single *conductor* enclosed in a suitable protective wrapping of some kind, the purpose of which is to isolate the conductor electrically from its surroundings. A *cable* consists of two or more *conductors* enclosed in a common wrapping.

ELECTRICAL CONDUCTORS

A conductor can consist of either a single solid wire or of several strands of fine wire twisted together. As far as residential wiring is concerned, stranded conductors are limited to extension cords, appliance cords, and similar applications. Otherwise, the cables making up the basic electrical system of a house contain conductors consisting of solid copper or aluminum wires. Although aluminum cables are less expensive than copper cables, aluminum cables are more difficult to work up into complete circuits, as it is almost impossible to solder two aluminum wires together. To get around this difficulty, manufacturers now supply aluminum cables in which the conductors are coated with a thin film of copper, which makes it possible to solder the conductors together as easily as solid copper conductors. In addition, there are many different mechanical connectors or splicing devices available for use with aluminum and/or copper wires that do not require soldering at all.

Table 1. Allowable Ampacities of Insulated Copper Conductors
Not more than three conductors in raceway or cable or direct burial (based on ambient temperature of 30°C 86°F)

	Temperature rating of conductor. See table 310.13 in the NEC		
	60°C (140°F)	75°C (167°F)	90°C (194°F)
Size AWG MCM	Types RUW (14-2), T, TW, UF	Types RH, RHW, RUH (14-2), THW, THWN, XHHW, USE	Types TA, TBS, SA, AVB, SIS, FEP, FEPB, RHH, THHN, XHHW*
18	. . .	. . .	21
16	. . .	. . .	22
14	15	15	25†
12	20	20	30†
10	30	30	40†
8	40	45	50
6	55	65	70
4	70	85	90
3	80	100	105
2	95	115	120
1	110	130	140
1/0	125	150	155
2/0	145	175	185
3/0	165	200	210
4/0	195	230	235

*For dry locations only. See Table 310-13 in the NEC.
These ampacities relate only to conductors described in Table 310-13 in the NEC.
†The ampacities for Types FEP, FEPB, RHN, THHN, and XHHW conductors for sizes 14, 12, and 10 shall be the same as designated for 75°C conductors in this table.
Source: The National Electrical Code, NFPA No. 70-1975 (ANSI C-1-1975), Table 310-16, p. 68.

Table 2. Allowable Ampacities of Insulated Aluminum and Copper-clad Aluminum Conductors
Not more than three conductors in raceway or cable or direct burial (based on ambient temperature of 30°C 86°F)

	Temperature rating of conductor. See table 310-13 in the NEC.		
	40°C (140°F)	75°C (167°F)	90°C (194°F)
Size AWG MCM	TYPES RUW (12-2), T, TW, UF	TYPES RH, RHW, RUH (12-2), THW THWN XHHW, USE	TYPES TA, TBS, SA, AVB, SIS, RHH THHN XHHW*
12	15	15	25†
10	25	25†	30†
8	30†	40	40
6	40	50	55
4	55	65	70
3	65	75	80
2	75	90	95
1	85	100	110
1/0	100	120	125
2/0	115	135	145
3/0	130	155	165
4/0	155	180	185

These ampacities relate only to conductors described in Table 310-13 in the NEC.
*For dry locations only. See Table 310-13 in the NEC.
†The ampacities for Types RUH, THHN, and XHHW conductors for sizes 12 and 10 shall be the same as designated for 75°C conductors in this table.
Source: The National Electrical Code, NFPA No. 70-1975 (ANSI C1-1975), Table 310-18, p. 69.

Conductor Sizes

For residential electrical installations, conductors are available in diameters that range in size from 0.0641 to 0.528 in.—or, in equivalent American Wire Gauge (AWG) sizes, from No. 14 to No. 0000. AWG sizes Nos. 0, 00, 000, and 0000 are usually

Table 3. Properties of Conductors

Size AWG MCM	Area cir. mils	Concentric lay stranded conductors: No. wires	Concentric lay stranded conductors: Diam. each wire in.	Bare conductors: Diam. in.	Bare conductors: Area* sq in.	DC resistance ohms/M Ft at 25°C (77°F), Copper: Bare Cond.	DC resistance ohms/M Ft at 25°C (77°F), Copper: Tin'd. Cond.	DC resistance ohms/M Ft at 25°C (77°F): Aluminum
18	1620	Solid	.0403	.0403	.0013	6.51	6.79	10.7
16	2580	Solid	.0508	.0508	.0020	4.10	4.26	6.72
14	4110	Solid	.0641	.0641	.0032	2.57	2.68	4.22
12	6530	Solid	.0808	.0808	.0051	1.62	1.68	2.66
10	10380	Solid	.1019	.1019	.0081	1.018	1.06	1.67
8	16510	Solid	.1285	.1285	.0130	.6404	.659	1.05
6	26240	7	.0612	.184	.027	.410	.427	.674
4	41740	7	.0772	.232	.042	.259	.269	.424
3	52620	7	.0867	.260	.053	.205	.213	.336
2	66360	7	.0974	.292	.067	.162	.169	.266
1	83690	19	.0664	.332	.087	.129	.134	.211
0	105600	19	.0745	.372	.109	.102	.106	.168
00	133100	19	.0837	.418	.137	.0811	.0843	.133
000	167800	19	.0940	.470	.173	.0642	.0668	.105
0000	211600	19	.1055	.528	.219	.0509	.0525	.0836
250	250000	37	.0822	.575	.260	.0431	.0449	.0708
300	300000	37	.0900	.630	.312	.0360	.0374	.0590

*Area given is that of a circle having a diameter equal to the overall diameter of a stranded conductor.

The values given in the table are those given in Handbook 100 of the National Bureau of Standards except that those shown in the 8th column are those given in Specification B33 of the American Society for Testing and Materials, and those shown in the 9th column are those given in Standard No. S-19-81 of the Insulated Power Cable Engineers Association and Standard No. WC3-1964 of the National Electrical Manufacturers Association.

The resistance values given in the last three columns are applicable only to direct current. When conductors larger than No. 4/0 are used with alternating current, the multiplying factors in Table 9 in the NEC compensate for skin effect.

Source: The National Electrical Code, NFPA No. 70-1975 (ANSI C1-1975), Table 8, p. 144.

written 1/0, 2/0, 3/0, and 4/0. Note in Tables 1 and 2, the purpose of which is described below, that the diameters of aluminum conductors tend to run two AWG sizes larger than copper conductors carrying the same amperage. This is because aluminum transmits an electrical current with less efficiency than copper. Note also that the diameter of the conductors increases as the gauge size number decreases (refer to Table 3).

Tables 1 and 2, which are reproduced from the National Electrical Code (NEC), spell out the *minimum* size conductors that can be used in circuits intended to carry a given *maximum ampacity,* as it is called in Tables 1 and 2—ampacity will be defined below. An electrician has no discretion in the smallest size conductors he must install, though he can, if he wishes, always install conductors larger than the minimum sizes specified in Tables 1 and 2. And, in fact, in the 15-ampere branch circuits installed in the typical dwelling (see ELECTRICAL SYSTEM), an electrician will usually install No. 12 copper wiring (or No. 10 aluminum wiring) rather then the No. 14 copper wiring (or No. 12 aluminum wiring) permitted by the NEC. The reason is safety—over a period of years most homeowners tend to increase gradually the number of appliances and lamps they plug into their electrical outlets, which tends gradually to overload the circuits.

Ampacity

Ampacity is the ability of a conductor to carry continuously an electrical current of a given amperage without the conductor becoming dangerously overheated. It is apparent that the circuits in any electrical system must carry their rated loads without becoming excessively hot. If the conductors do become excessively hot, they might burn off their insulation, which might result in the bare conductors shorting out. The purpose of sizing conductors by ampacity, therefore, is to ensure that the conductor sizes and the types of insulation used in different circuits are appropriate and safe for the expected service conditions.

Different types of insulation vary in their ability to withstand heat. Therefore, the ability of a conductor of a given size to carry a given amperage is limited by the type of insulation surrounding it. Several conductors located within the same cable or conduit will generate a greater amount of heat than will a single conductor. Thus, the number of conductors in the cable or conduit also affects the ampacity of the cable. All these factors are implicitly taken into account in Tables 1 and 2.

Tables 1 and 2 also implicitly assume that a conductor will carry its maximum rated amperage continuously, and that a maximum of three conductors will be contained within any given cable or conduit. If there are more than three conductors within a cable or conduit, the amperage values shown in Tables 1 and 2 must be reduced accordingly. Thus, if the number of conductors in a cable or conduit is between four and six, the maximum amperage is reduced to 80 percent of the values shown in the tables. If the number of conductors is between 7 and 24, the allowable amperage is reduced to 70 percent of the values shown.

As for the ambient temperatures shown in the tables, all conductors are classified according to the maximum temperatures they can safely withstand during normal service. For example, in the living quarters of the typical dwelling, the ambient temperatures are assumed not to exceed 86°F (or 30 °C). Under these conditions, the maximum operating temperatures of the conductors will be unlikely to exceed 140°F (or 60°C), and the types of cable insulations shown in the 60°C columns can be used for these service conditions.

For such poorly ventilated spaces as an attic, or when the conductors are exposed to the direct rays of the sun, the ambient temperatures are assumed not to exceed 113°F (or 45°C). Under these conditions, the maximum operating temperatures of the conductors will be unlikely to exceed 167°F (or 75°C), and the types of cable insulations shown in the 75°C columns can be used for these service conditions.

If the anticipated ambient temperatures are likely to be higher than 113°F (or 45°C), the types of cable insulations shown in the 90°C columns must be used. These insulations are rated for service conditions that may become as hot as 194°F (or 90°C).

CONDUCTOR INSULATION

Once a conductor of the required AWG size has been selected for a particular installation, it is the type of insulation surrounding the conductor that will determine which of the many different kinds of cable manufactured will be most appropriate for that use.

The primary function of insulation has already been mentioned. It is to isolate electrically the conductors within a cable from each other and from external objects as well. Insulation serves other purposes as well. It protects the conductors from mechanical damage and from the corrosive effects of moisture and chemicals in the atmosphere or the soil. How well a particular insulation performs these functions will determine how suitable the cable is for a given location; that is, whether the cable must be used only in dry, protected locations or whether it can also be used in damp or wet locations, underground or outdoors.

There is a distinction between damp locations and wet locations. There is, as a matter of fact, a distinction between dryness, dampness, and wetness, and an insulation material that is suitable for a dry or damp location may not be suitable for a wet location. A *damp* location is partially protected by canopies, overhangs, and so on, and is subject to a moderate degree of moisture, such as may exist in a basement or barn. A *wet* location may be saturated with water, either underground or directly exposed to the weather.

Insulation Materials

The basic difference among insulation materials is between those made of a rubber compound and those made of a thermoplastic compound. At one time all insulation was made from natural or synthetic rubber. And, although cables are still manufactured with rubber insulation, its use declines more and more every year. The trend today is toward thermoplastic insulation made of polyvinyl chloride, polyethylene, and similar compounds. These thermoplastic compounds have outstanding advantages as electrical insulation. They are (1) relatively inexpensive, (2) capable of extrustion onto a length of wire at high rates of speed, which reduces manufacturing costs considerably, (3) excellent dielectrics, which means they strongly resist the passage of an electric current, (4) physically tough and resistant to mechanical abrasion, and (5) chemically inert to most chemicals and to oxygen and ozone as well, which rubber compounds are not.

If one has ever had occasion to replace or resplice an old rubber-and-cotton-insulated cable that has been buried within the walls of a house for a great many years, one will remember how brittle the rubber had become. It cracks and crumbles away at a touch. This embrittlement is due to the long-term effects of oxygen upon the rubber. Therefore, as far as cable insulations are concerned, the wave of the future is definitely polyvinyl chloride, polyethylene, and similar thermoplastic compounds.

The symbols T, RH, THW, XHHW, and so on, at the tops of the columns in Tables 1 and 2 identify the kinds of insulating materials that are satisfactory for the temperature conditions indicated. The symbols also indicate whether a particular kind of insulating material is suitable for damp and wet locations, or whether it must be used only in dry locations. These symbols are printed at 24-in. intervals on the surface of the cable so the user can tell at a glance whether a particular cable is suitable for the intended service.

TYPES OF CABLE

Having said all this, we must now add that cables intended for installation in dwellings may also be identified according to the intended function of the cable in the electrical system. Cables can be used either above ground or below ground for (1) service-entrance equipment, (2) feeder lines, or (3) branch circuits (*see* ELECTRICAL SYSTEM). Therefore, instead of the designations that we have just described, cables are often identified with a symbol such as NM, NMC, AC, ACL, ACT, SE, USE, or UF, all of which describe the *function* of that cable in an electrical installation. As long as the electrician uses an appropriately identified cable, he need not concern himself with the insulation material used to make the cable.

Nonmetallic-sheathed Cable, Type NM and NMC

Nonmetallic-sheathed cable is a low-cost cable that is widely used for the interior of dwellings. It is better known under a variety of trade names—Romex, Loomwire, and Cresflex, for example. Type NM cable is used in locations that are normally dry. Type NMC cable—the C stands for *corrosion-resistant*—is used in potentially damp and wet locations or in locations where plain NM cables might corrode or be attacked by fungi.

Both NM and NMC cables may use either rubber or plastic insulation. At one time, all NM cable was built up of several layers of rubber, paper, and cotton. Each conductor in the cable was first surrounded by a layer of rubber, the thickness of which depended on the diameter of the conductors. The rubber was enclosed in a layer of cotton or silk braid, over which was wound a layer of paper tape. All the conductors were then laid side by side, with lengths of jute string packed between the conductors to close up the gaps. The conductors were then wrapped together in heavy paper, over which was applied a final jacket made of heavy cotton braid. This jacket was impregnated with an asphalt compound that increased the cable's resistance to moisture and flame.

Almost all the NM and NMC cable used for dwellings is 1- or 3-wire, plus an additional *grounding conductor* which is used to ground metal outlet boxes, receptacles, and the frames of electrical equipment.

In 2-wire cables, the current-carrying conductor is black and the neutral conductor is white. In 3-wire cables, the current-carrying conductors are red and black, and the neutral conductor is white. The grounding conductor may either be a bare wire or an insulated one; if it is covered with insulation, the insulation is green, or green with yellow stripes.

Service-Entrance Cable, Types SE and USE

The service-entrance cable, which carries power from the public utility to the main distribution panel in the house should carry a minimum of 100 amperes and may carry as much as 200 amperes. *See* ELECTRICAL SYSTEM. Service-entrance cables, therefore, are large, heavy-duty cables, the insulation of which must resist the temperatures and climatic conditions usually found out of doors. The insulation may be either rubber or thermoplastic. In either case, the outer sheathing must be moisture-resistant and flame-retardant, though not necessarily resistant to mechanical abuse.

Type USE cable is intended for underground service, and therefore, the outer sheathing must not only be completely waterproof but be corrosion-resistant and fungi-resistant as well, though not necessarily flame-resistant.

Underground Feeder and Branch-Circuit Cable, Type UF

This type of cable has an outer covering that is moisture-resistant, fungi-resistant, corrosion-resistant, and flame-retardant. The conductors are color-coded as already described.

The NEC requires that UF cable be buried a minimum of 24 in. in the earth if the cable is not otherwise protected, or a minimum of 18 in. if there is a 2-in concrete pad over the cable, or a minimum of 6 in. if the cable is contained within a rigid metal conduit. If UF cable is buried directly in the earth, some sort of additional protection is usually required to protect the cable against possible damage by earth movements (especially the heaving that takes place during the winter as the earth alternately freezes and thaws). This protection usually takes the form of placing the cable on a bed of fine sand and then shoveling additional fine sand over the cable before the trench is refilled.

Metal-clad Cable, Type AC, ACT, and ACL

Metal-clad cable is more commonly known as *armored* cable (see Fig. 1). It consists of insulated conductors installed inside a flexible steel conduit.

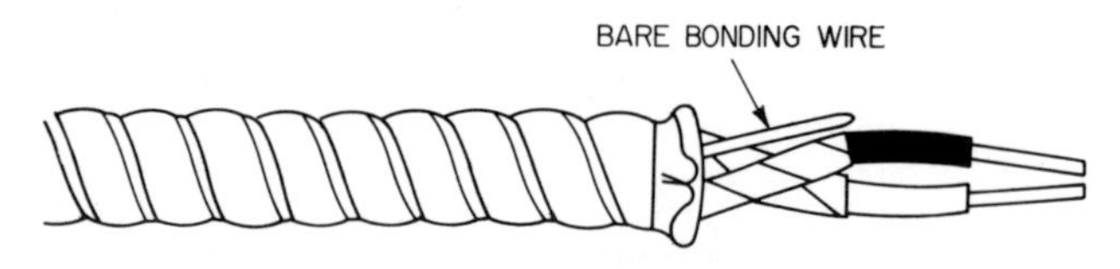

Fig. 1. Metal-clad cable. The current-carrying conductors are color-coded as are the conductors of all cables, and, in addition, a bare bonding wire runs through the metal wrapping to ensure the entire cable can be properly grounded.

Types AC and ACT metal-clad cable are used for branch and feeder circuits just as Type NM cable is. The conductors in Type AC cable are wrapped in rubber insulation. When the conductors are wrapped in a plastic insulation, the cable is designated ACT. In all types of metal-clad cable, the insulation surrounding each conductor is covered with a paper wrapping that protects the insulation against abrasion by the outer steel wrapping. Both AC and ACT cables are for use only in dry locations. If a metal-clad type of cable must be installed in a damp or wet location, or if it is to be used below grade or embedded in concrete, Type ACL cable must be used. Type ACL cable is metal-clad cable in which the individual conductors are sheathed in a film of lead.

CABLE INSTALLATION

The wiring is installed in a new house after the wood framework has been completed and the interior has been enclosed against the weather. The wiring plan provided by the architect or builder merely shows the locations of the switches, receptacles, and fixtures in each room. The plan reveals little to a layman, but it provides sufficient information to the electrician to enable him to wire the house from cellar to attic. It will be noted that the architect's or builder's wiring plan says nothing about the number and sizes of the conductors that must run between the switches and the fixtures or outlets they control. The electrician is expected to know this sort of detail by heart.

The first step in the wiring installation is called *boxing out.* The electrician goes through the entire house with a crayon or piece of chalk marking on the studs and joists to which they will be attached the location and type of switches, receptacle outlets, and fixtures that are to be installed. The locations of ceiling fixtures are marked on the subflooring directly below their place of installation.

Once the boxing out has been completed (and checked by the builder or architect), the outlet boxes in which the switches, receptacles, and fixtures will be mounted are nailed to the studs and ceiling joists. All cable runs must begin and end at an outlet box, which must be large enough to accommodate not only whatever switches, receptacles, or fixtures will later be attached to the box but also the cables that will enter and leave the box.

Switches are located 36 to 48 in. above floor level, and duplex receptacles about 12 in. above the floor. There is no law about this, however, and, for appearance's sake or for the convenience of the homeowner, switches can be placed as high or as low as desired. The same is true of the duplex receptacles to which lamps will later be connected. The usual 12-in height allows lamp cords to dangle visibly from the wall, and a homeowner may well prefer to have these outlets located as low as possible to reduce the visibility of the cords. In fact, in a large room, the receptacle outlets can be located in the floor itself, near the anticipated locations of floor and table lamps.

Once the outlet boxes have been nailed in place, the electrician will begin drilling innumerable holes through the studs and joists through which he will run the cables that connect the distribution panels to the switch outlet boxes and the switches to the fixtures they will control. The NEC requires that these holes be bored not less than 1½ in. from the nearest edge of all structural lumber. If this is impossible, then a steel plate several inches long and at least ¹⁄₁₆-in. thick must be fastened against the edge of the lumber to protect the cable against any nails or screws that would otherwise be driven into the lumber.

When holes are drilled through floor joists, the holes should be located at the midpoint of each joist, this being the place at which the strength of the joist is least affected by holes. Alternatively, when cables run exposed under first-floor joists, a wooden board called a *runner* can be nailed to the underside of the joists, and the cables are then stapled to this runner.

Each run of cable from one outlet box to another must consist of one unspliced length of cable. Splicing lengths of cable together between outlet boxes is not permitted. About 6 in. of cable should be allowed to hang free at each outlet box to give the electrician sufficient material to work with when he later makes the final connections to the switches and receptacles.

The cables are secured in place at the outlet boxes by special clamps or connections. If splices are required between two runs of cable, the splice must be made at an outlet box. A splice having been made, the cable is folded neatly into the outlet box to await the installation of the switch, receptacle, or fixture. The installation of this equipment (which is called *trimming out a house*) is accomplished after the interior has been painted and wallpapered.

Electrical System

This article describes in general terms the overall design and installation of a complete residential electrical system. An elementary knowledge of electrical fundamentals is assumed. For a description of how wiring, switches, and receptacles (i.e., convenience outlets) are connected together into complete systems, *see* ELECTRICAL WIRING CIRCUITS. *See also* GROUND, ELECTRICAL.

In principle, residential electrical systems are quite simple, as may be seen by Fig. 1. There are only two basic parts to a typical residential system—the *service-entrance equipment,* which consists of that part of the electrical installation between the utility's power lines and the main fusebox in the house, and the *branch circuits,* which run from the main fusebox and connect to the individual fixtures, switches, and receptacles installed around the house. In a two-story house, a large rambling house, or a heavily electrified house, there may also be *feeder circuits* that run from the main fusebox to subsidiary fuseboxes located closer to the actual electrical loads. The branch circuits will then begin at these subsidiary fuseboxes, not at the main fusebox.

It might be noted that *fusebox* is a layman's term that electricians use when speaking to anyone who is ignorant of house wiring. In the trade, a fusebox is known as a distribution panel, service panel, panelboard, load center, or control center—or just plain panel. We will use the term *distribution panel.*

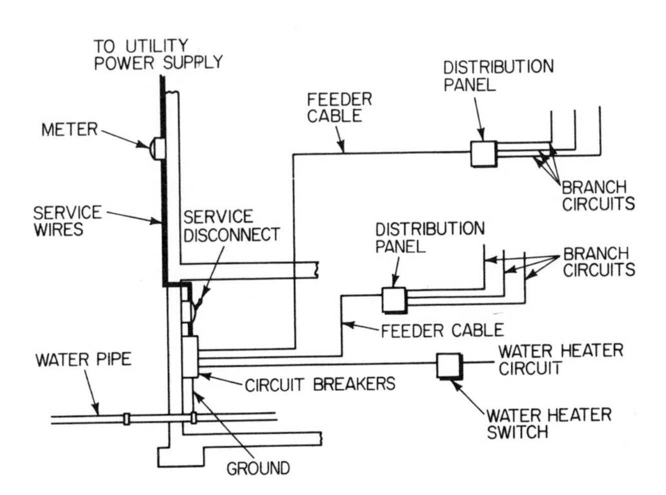

Fig. 1. A residential electrical installation.

THE NATIONAL ELECTRICAL CODE

Before describing residential electrical systems, mention must be made of the National Electrical Code (NEC), which is published by the National Fire Protection Association (NFPA). The NFPA is a private, nonprofit organization founded in 1897 that publishes standards and codes on fire prevention, protection, and suppression. The Association also carries on an extensive public education program in fire prevention. In its present form the NEC has been published since 1911. In 1951 NFPA began publishing the NEC and continues to do so with triannual revisions.

The NEC is concerned primarily with establishing safe methods of installing electrical wiring and equipment, not only in dwellings but in any type of structure. The influence of this code has been profound. Innumerable communities have adopted the NEC in toto as their basic electrical code, although some large cities or smaller communities with special problems have modified the NEC as necessary to suit local conditions. In most states a *licensed* electrician is a person who has, among other things, mastered the NEC sufficiently to pass a written examination based on its provisions, since a competent electrician must know how to install an electrical system in conformity with this code.

It should be clearly understood that the NEC contains *minimum* requirements for a safe electrical installation. Anyone is perfectly free at any time to install wiring or equipment that is larger or heavier than the minimum specified by the NEC. Many communities do, in fact, insist on particular standards that are more rigorous than those required by the NEC. This should be kept in mind because the electrical code adopted by a community is a legal document that must be adhered to, and the NEC is merely advisory in nature.

Nor does the NEC attempt to establish standards for the electrical generating and distribution equipment owned and operated by local public utilities, since it must be assumed that these utilities are competent enough technically to know their own business. Therefore, in addition to the provisions of the NEC, which most communities have adopted, each local public utility will have its own rules and regulations on how the power supplied by it is to be connected to the houses and businesses of its customers, and these rules and regulations also have the force of law.

In sum, there are the provisions of the NEC (as modified here and there by local communities) and there are the regulations of the local public utility. Both must be adhered to strictly by builders, electricians, and electrical contractors.

The provisions of the basic NEC, which make up quite a large volume, apply mainly to commercial and industrial establishments, though residential installations are included as part of the overall code. For the convenience of builders, electricians, and inspection officials who are concerned mainly with residential electrical systems, the National Fire Protection Association published in 1975 a 163-page book in which are excerpted all the code requirements that pertain to one- and two-family dwellings. A copy of this abridgement (*1975 Electrical Code for One and Two-family Dwellings*—NFPA No. 70A) can be purchased from the Association.

ELECTRIC SERVICE

The electric power used in a house is supplied by the local utility, which, in the lingo, provides "electric service" to its customers. This service consists usually of 115/230-volt, 60-cycle, single-phase, 3-wire service. *Three-wire service* means that the electricity is transmitted from the utility's lines to the customer's house via three separate conductors. Two of these conductors will carry 115 volts each (or 230 volts between them), and the third conductor will be a neutral that leads ultimately to ground. Depending on the way in which the neutral conductor and the power conductors are connected across an appliance, a switch, a receptacle, or a motor, the consumer has available to him through these three conductors electric power that can take the following forms:

1. *Two-wire, 115-volt service,* which is obtained by connecting only one of the 115-volt conductors to the appliance, motor, or whatever, with the neutral conductor connected to the frame of the appliance, motor, etc.
2. *Two-wire, 230-volt service,* which is obtained by connecting both 115-volt conductors to the appliance, etc. The neutral is connected to the frame of the appliance, etc.
3. *Three-wire, 115/230-volt service,* which is obtained by connecting both 115-volt conductors and the neutral cable to three appropriate connections on an appliance or motor. For example, electric ranges are often rated for 115/230-volt service, which means that their internal circuits are so designed that some of these circuits use 115-volt current and others use 230-volt current.

Some utilities offer their customers 120/240-volt service instead of 115/230-volt service. The difference in voltage is due to the system of electrical generation and distribution adopted by the utility rather than to any inherent difference in the quantity or quality of the power delivered to the customers. As far as the customers are concerned, the end result is, if not identical, inconsequential.

Other utilities offer 2-wire, 115-volt (or 120-volt) service. This means mainly that 230-volt (or 240-volt) equipment, such as large 220-volt room air-conditioning units, for example, cannot be used by their customers. The most obvious sign that only 2-wire, 115-volt service is available in an area (to a prospective home buyer who has a sharp eye for such things) is the fact that only two conductors will be tapped off the utility's lines. These will consist of a 115-volt conductor and a neutral conductor. The use of such 2-wire, 115-volt systems has been declining and will no doubt continue to decline.

Electrical service may be supplied by the utility either via pole-mounted lines or via underground ducts. Each will be described in turn.

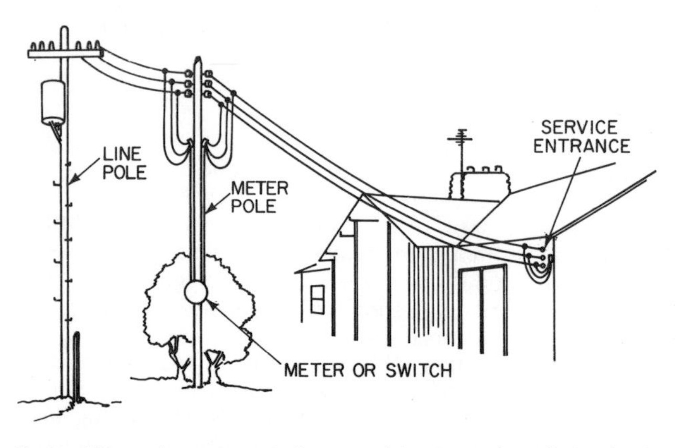

Fig. 2. The *service entrance* is that part of the electrical installation that is located between the utility's lines and the dwelling.

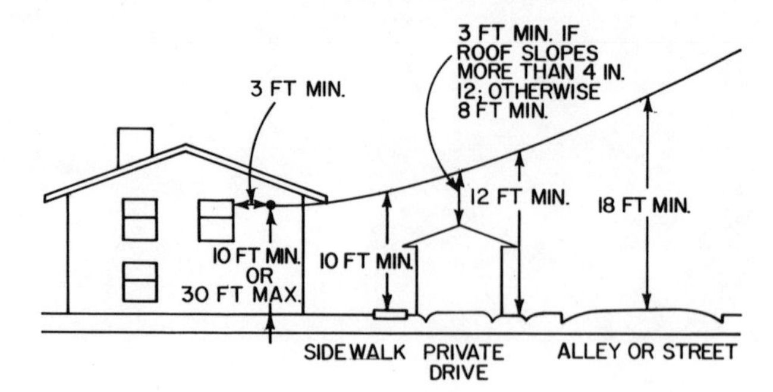

Fig. 3. The clearances between the service entrance and any buildings or walks and driveways is closely regulated by the public utility.

Overhead Service

Figure 2 shows a typical overhead service installation and the way in which the utility's lead-in conductors are attached to a house. These three conductors are known collectively as the *service drop* and they run from the nearest (or most convenient) pole to special insulated fittings bolted or screwed securely to the side of the house. As shown in Fig. 2, the three conductors form deep loops just before they enter the building, the purpose of which is to prevent rain running down alongside the conductors and into the electrical equipment installed in the house.

The installation of the service drop is the responsibility of the utility. It is installed by utility linemen and its height above obstructions, the direction in which it runs, and the way in which it terminates at the house are all closely regulated by the utility. The main concern of these requirements is with the overall safety of the service-drop installation.

For example (see Fig. 3), the NEC requires that the service drop be at least 18 ft above any public alley or street, 12 ft above any private driveway, and 10 ft above any sidewalk. The service drop must also clear the roof of any building it passes over by at least 8 ft (or by at least 3 ft if the building has a roof that slopes more than 4 in 12). In addition, the service drop must be attached to the side of the house somewhere between 10 to 30 ft above the ground and without coming closer than 3 ft to any window or building exit.

It is easy to understand why the utility must have a look at the plot plan and building elevations before deciding on the installation of any particular service drop, and why the builder and electrical contractor must be familiar with the utility's regulations before deciding how and where to install the service-drop equipment. In fact, the "how" and "why" of the installation will be determined in large measure by the service-drop and electrical installation requirements of both the utility and the local electrical code.

One aspect of the service-drop installation that the builder must keep in mind is the heavy pull that will be exerted by the service-drop cables when they have been attached to the house. It is not unheard of for a house to be pulled (slightly) lopsided by this load, or for framing members to be pulled loose from the building.

Structurally, the strongest parts of a wall are its corners and, in a wood-frame house, the service-drop fittings are usually bolted into a corner post, with the post itself being securely fastened to the rest of the house framework by metal strapping. Figure 4 shows a typical construction. Sometimes in the utility's opinion a house may not be capable of sustaining the heavy pulling load exerted by a long service drop, or it may be inconvenient to attach the service drop directly to the house. In such cases the utility will require that a pole, sometimes called a *meter pole,* be installed alongside the house to which the service drop can be connected. This is shown in Fig. 2.

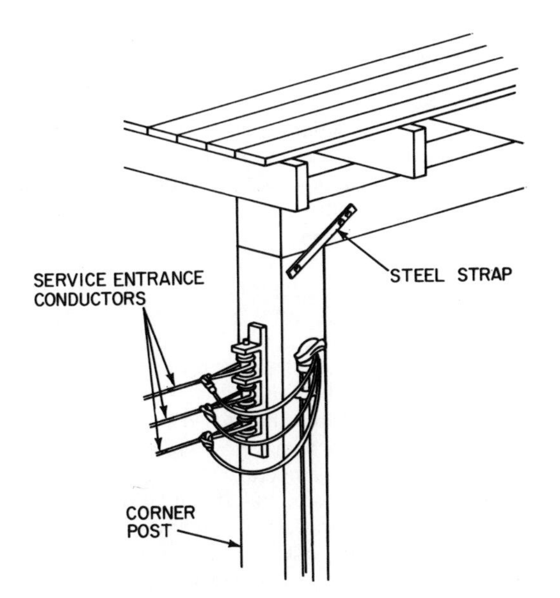

Fig. 4. The service-entrance fitting must be securely bolted to the corner post of a house, and the post itself must be secured firmly to the rest of the building frame in order to resist the pull of the service-entrance conductors.

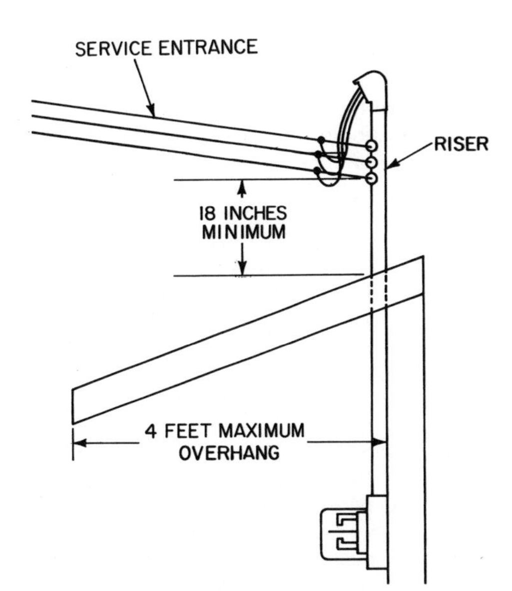

Fig. 6. The steel riser that encloses the service-entrance conductors is often extended through a roof so that the service entrance can be connected to it. When this is done, the riser must be adequately braced by guy wires.

If a house is built too low to meet the minimum height requirements required by the service-drop installation, a 4 × 4 in. length of lumber can be bolted to the side of the house, as shown in Fig. 5, to act as a mast. Alternatively, the steel conduit (the *riser*) that encloses the service-drop cables between the service drop and the service-entrace equipment can be extended up the side of the house as high as is necessary. The top of the riser may require support by guy wires against the pull of the service drop. Another possibility, especially if a house has an overhanging roof, is to extend the riser through the roof, as shown in Fig. 6. When this is done, however, the riser must extend at least 18 in. above the roof line at the point of attachment. The amount of roof overhang must not exceed 4 ft.

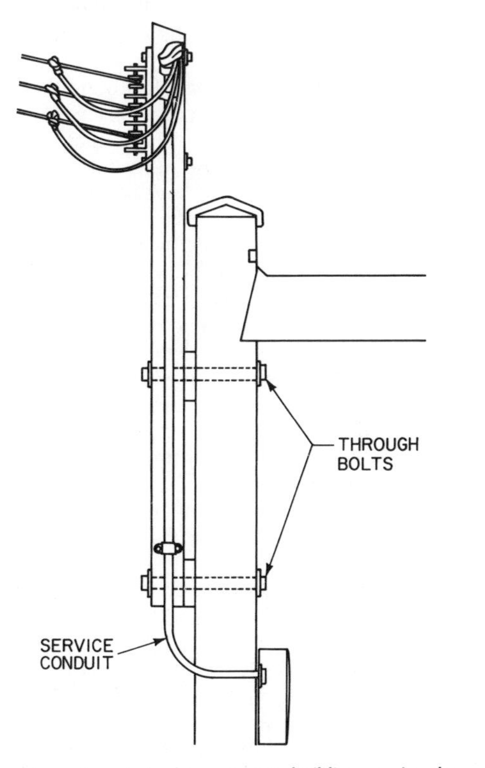

Fig. 5. A mast can be attached to a 1-story building to give the service entrance conductors the necessary clearances. This mast must be securely bolted to the building structure.

Underground Service

It is far more common nowadays, for aesthetic reasons as well as for safety's sake, to connect the utility service to a house through underground conductors, regardless of whether the utility's conductors are pole-mounted or run underground. Figure 7 shows a typical underground service installation.

The underground portion of the installation is called the *service lateral,* and digging the trench in which the conductors are to be buried and covering up the trench afterward is the responsibility of the electrical contractor. The conductors must be buried at least 24 in. underground.

The electrical contractor must also run the service-entrance

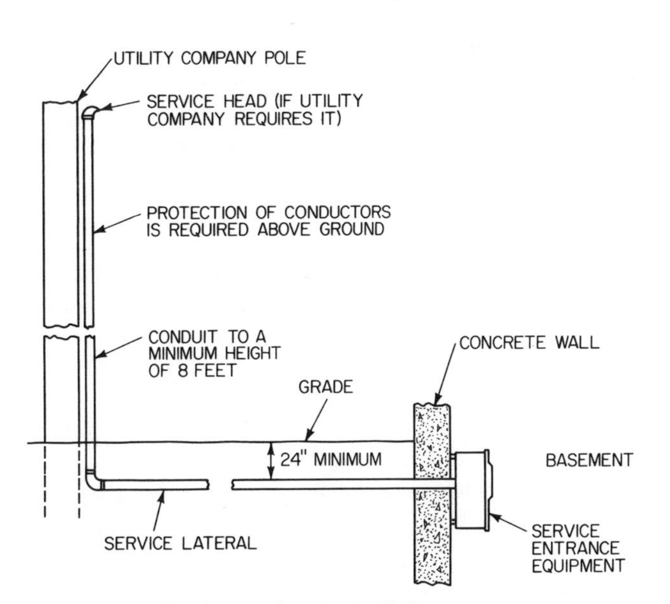

Fig. 7. A typical underground service installation.

conductors up the utility pole (enclosed in a suitable conduit that ends at least 8 ft above the ground) so that the utility linemen can make the necessary connections to them. Or, if the utility's lines should run underground also, the electrical contractor must extend the cables into the nearest utility manhole or handhole so that the linemen can make the connections there.

The other end of the service lateral usually extends through the foundation wall of the house so that the conductors can be connected to the service-entrance equipment located within the house. It is preferred, therefore, that the conduit be installed after the foundations have been excavated and while the formwork is being built for the placing of the concrete. The service-lateral conduit will then pierce this formwork. After the concrete has been placed, the conduit will be sealed tightly within the foundation wall. If the service-lateral trench cannot be dug at the same time as the foundations, a metal sleeve can be inserted through the formwork and the conduit can be installed through this sleeve at a later time.

Once, any electrical conductors that ran underground had to be protected against corrosion by a lead coating, regardless of whether they were installed within a conduit or were protected by a steel wrapping. But now, since the introduction of plastics that are completely unaffected by any substances that might be in the soil (*see* ELECTRICAL CONDUCTORS AND CABLES), an electrical contractor is permitted to run plastic-covered conductors directly through soil. But when plastic-covered conductors are to run underground, special precautions must be taken in those sections of the United States where the soil is likely to freeze during the winter. Local building codes usually require that the conductors rest on a bed of fine sand and be covered by a layer of fine sand before the trench is filled to prevent the conductors being stressed by heaving soil as the earth freezes and thaws.

SERVICE-ENTRANCE EQUIPMENT

The service-entrance equipment has three functions: (1) it measures the amount of electrical power consumed within the building; (2) it protects the electrical circuits against overloads or short circuits; and (3) it distributes the electricity to the different feeder and/or branch circuits installed within the building.

Watt-Hour Meter

The amount of power consumed within a building is measured by a *watt-hour meter,* which is connected directly across the incoming service-entrance conductors. This meter is invariably the property of the public utility, and it is installed by utility employees. The meter fits into a socket that is usually the property of the utility also, but it is installed by the electrical contractor.

Most utilities require that the meter socket be installed in such a way that the meter can be read from outside the house. This requirement is intended to reduce the expense and nuisance of having to make repeat visits if the occupants should be away. The utility also requires that the meter be located 5 to 6 ft above the ground, and that the soil directly under the meter be covered over with concrete or a wood platform. If it is inconvenient to install the meter on the side of the house, it may be located on a *meter pole,* or on the side of a garage or gate.

In most large cities, however, or in older communities that were in existence before the promulgation of these rules and regulations, the meter is usually installed indoors, the most likely location being in the basement, where it is safe from the weather.

Disconnect Switch

Next in line is the main disconnect switch, the size of which is determined by the maximum amount of power likely to be consumed within the house. In new construction, this switch will have a minimum rating of 100 amperes or a maximum rating of perhaps 200 amperes. The disconnect switch must meet a considerable number of code requirements if it is to be approved as electrically safe. It must be fully enclosed in a metal box that meets the requirements of the Underwriters' Laboratories, and the switch must be operable from outside the box. There must be some plain indication on the outside of the box as to whether the switch is open or closed. The switch must be so constructed that when it is in the open position it cannot accidentally fall into the closed position. The switch must be protected against damage by a fuse or circuit breaker that usually has the same amperage rating as the switch. And the switch must be so located that it can be reached easily—especially in an emergency. Some utilities require that the switch be mounted on the outside of the house, the idea being that the most usual reason for disconnecting the switch is a fire inside the house, and no one wants to enter a burning building to disconnect the electricity.

If the disconnect switch is located inside the house, it must not be hidden inside a closet or cupboard, or be barricaded by a washing machine, a furnace, or some other piece of equipment, nor should it be located so high off the floor that it can only be reached by standing on a ladder or a chair.

Distribution Panel

From the disconnect switch, the service-entrance conductors enter the main distribution panel, where a number of heavy copper bars distribute the incoming power to the different branch and/or feeder circuits. The distribution panel also contains fuses or circuit breakers, one for each branch or feeder circuit in the house. If any of these branch or feeder circuits should have an overload or short out, the appropriate fuse will blow or circuit breaker pop and the branch or feeder circuit will thus be protected against damage.

Main distribution panels are manufactured in a wide variety of sizes, shapes, and electrical capacities. Small, subsidiary distribution panels are also manufactured that have only two, four, or six branch circuits for use in feeder circuits.

A newly built house is likely to have an all-in-one enclosure installed that contains not only the distribution busses and fuses (or circuit breakers), but also the main disconnect switch, main circuit breaker, and a socket to which the meter can be attached. The virtue of having such an all-in-one cabinet is that it *does* include all the electric-service equipment in one simple low-cost enclosure. The disadvantage of having such a cabinet is that it limits the possible ways in which the service-entrance equipment can be installed. For example, because many utilities require that the meter face outward, the simplest method of meeting this requirement is to install the enclosure on the outside of the house. The enclosure is raintight, of course, but, even if it is, the circuit breakers may not stand up to the rigors of an outdoor installation and they may have to be replaced at intervals. An exterior installation is also rather unattractive, especially if the service entrance or service lateral is located toward the front of the house. And if additional electrical capacity should ever be required at a later date (because, say, the builder has installed only the minimum amount of electrical capacity necessary to meet the minimum requirements of the local electrical code; see below), one will have *two* plain-looking boxes hanging on the outside of one's house.

A better way of installing an all-purpose enclosure is to recess it within the wall, with a weathertight cover protecting the meter so that the meter reader must first open the cover to do his job. An even better way of installing an all-purpose enclosure is to purchase an enclosure in which the meter socket faces outward and the rest of the box opens to the inside of the house. Best of all, if the meter must face outward, forget the savings of an all-in-one box, mount the meter by itself in some convenient location and install the rest of the service-entrance equipment indoors where it will remain dry and be out of the way; not forgetting, however, that the service-entrance conductors are heavy and expensive and the shorter the distance between the meter and the rest of the service-entrance equipment, the better. Besides, service-entrance conductors should not run within the cavities of a wood-frame wall unless they are suitably protected.

Finally, it might be noted that an extremely important part of the service-entrance equipment that hasn't been mentioned is the conductor that grounds the entire electrical system; *see* GROUND, ELECTRICAL.

BRANCH CIRCUITS

It was not too many years ago that a house with a 60-ampere main circuit breaker and four 15-ampere branch circuits was considered to be adequately wired for any appliances or fixtures the homeowner might care to purchase and install. Such houses still exist by the millions, of course, but the continually increasing number of larger and heavier electrical appliances that are being developed and used has made a 60-ampere system obsolescent.

Nowadays, a new house is required by the NEC to have a minimum service capacity of 100 amperes, and the NEC has divided the branch circuits into four groups, each with its own minimum requirements. These groups are (1) the general-purpose convenience circuits, (2) small-appliance circuits, (3) a laundry circuit, and (4) fixed-appliance circuits and circuits supplying the heating and cooling equipment.

General-Purpose Circuits

The general-purpose circuits supply power for room lighting and for the duplex receptacles (or convenience outlets) that are spaced at intervals around the walls and supply power for such things as TV sets, vacuum cleaners, small fans, hi-fi equipment, and the like. A duplex receptacle is required for every 12 ft of wall length or portion thereof. Thus, a room that has 25 linear ft of wall will require a minimum of three duplex receptacles.

The total *minimum* number of general-purpose circuits required can be calculated according to the total habitable floor area of a house, which includes the basement and an unfinished attic that is potentially habitable. One branch circuit is usually recommended for every 500 sq ft of floor area. This is a minimum requirement. In addition, a branch circuit is, or should be, installed for every three or four outlets, the general idea being that although installing extra branch circuits adds very little to the original cost of the overall electrical installation, not having sufficient circuits in the first place may cost a substantial amount if one has to install them at a later date.

Another method of calculating the total number of general-purpose circuits, and the method that is also used to estimate the total electrical load in a house, is to allow 3 watts for every square foot of habitable floor area. The total wattage thus obtained is divided by 1500 watts (this being considered a reasonable load for the branch circuits to carry) to obtain the total number of circuits required.

Small-Appliance Circuits

Small appliance means portable kitchen applicances mainly; such things as electric roasters and rotisseries, deep friers, hotplates, blenders, and so on, though a 115-volt refrigerator or refrigerator-freezer unit is also included under this heading (see Table 1). The NEC requires, in addition to the general-purpose circuits, a minimum of two 20-ampere small appliance circuits in a dwelling, which are expected to handle a total load of 3000 watts (that is, the two circuits will each have 1500-watt capacity). These small-appliance circuits are installed not only in the kitchen of a house but also in the family room, breakfast nook, dining room, or wherever portable appliances are likely to be used.

Table 1. Typical Power Requirements, in Watts, for Household Appliances

Type	Watts
Air conditioner	850–1200
Attic fan	500–1500
Clothes drier	Up to 4500
Dishwasher	530–1000
Disposer	380–530
Electric fan	50–300
Furnace blower	380–670
Grill	1000
Hair drier	250–1000
Heater	1000–1650
Home freezer	300–670
Hot plate	600–1000
Infrared lamp	500
Iron, hand	660–1000
Ironer, home	1275–1620
Mixer	125–150
Oil burner	300–550
Percolator	400–600
Power tools	Up to 1000
Radio	50–200
Range	7000–14000
Refrigerator	200–670
Roaster	1150–1650
Sewing machine	75
Sunlamp	250
Television	200–400
Toaster	600–1350
Vacuum cleaner	300
Waffle iron	660–1000
Washing machine	350–900
Water heater	750–3000

The duplex receptacles to which these appliances are connected should also be installed above every counter top where an appliance is likely to be used. Over a long counter, 4 ft between receptacles is considered a reasonable spacing, and a total of three or four duplex receptacles is considered a reasonable number of outlets to install for one small-appliance circuit.

An additional 20-ampere branch circuit may be required if the homeowner has a home workshop that includes such things as a power saw, drill press, and/or lathe. This circuit is considered a special kind of small-appliance circuit.

Laundry Circuit

A separate 20-ampere (i.e., 1500-watt) branch circuit is required for laundry appliances, such as a 115- or 120-volt washer or ironer (see Table 1). Note that this laundry circuit is distinct from the circuits required for large 230- or 240-volt washers and driers, which require individual circuits as described below.

Fixed-Appliance Circuits

Electrically operated kitchen ranges and ovens, central air-conditioning systems, electric water heaters, and space and baseboard heaters consume so much power (see Table 1) that each must be separately connected to its own branch circuit. (Furnaces and boilers also have separate circuits, not because they consume very much power but to isolate them electrically from the rest of the system in case any of the other circuits should short out.) The fuses or circuit breakers protecting these fixed-appliance circuits usually have a 20-ampere capacity, though a kitchen range, a central air-conditioning system, or an electrical heating system may require as much as a 50-ampere circuit. In fact, a heavily electrified house may require a separate distribution panel altogether to handle the circuits of the large fixed appliances that are installed in the house.

LOAD CALCULATIONS

Before a house is built, it is necessary to calculate the maximum loads that the electrical system may carry. These calculations are necessary in order that (1) the minimum size of the incoming service cables and the sizes of the branch-circuit cables can be determined; (2) to ensure that the service-entrance equipment is correctly sized; and (3) to make sure that a sufficient number of general-purpose and small-appliance circuits are installed.

The most convenient method of making these load calculations is as follows.

First, divide all the branch circuits into the four groups described above. Then calculate the total wattage that will be required for the general-purpose circuits, using the 3-watt-per-sq-ft figure mentioned above.

As an example of how load calculations are made, let us assume a house having a total habitable floor area of 3000 sq ft. Therefore,

$$3 \text{ watts/sq ft} \times 3000 \text{ sq ft} = 9000 \text{ watts}$$

which is the power required for the general-purpose circuits.

Next, let us calculate the number of small-appliance circuits that will be required. At least two circuits are required by the NEC, and each circuit must carry no more than 1500 watts maximum. Therefore, we multiply the number of circuits by 1500 watts, and

$$2 \text{ circuits} \times 1500 \text{ watts} = 3000 \text{ watts}$$

There is also the laundry circuit, which will also have a maximum load of 1500 watts, and

$$1 \text{ laundry circuit} \times 1500 \text{ watts} = 1500 \text{ watts}$$

Common sense tells us that not every light, not every piece of electrical equipment that is connected to the duplex receptacles, and not every small appliance in a house will be operated at the same time. Studies by the NFPA and other organizations have shown that different kinds of circuits and different types of electrical equipment have different maximum *demand factors*. For example, a laundry drier or a central air-conditioning system will have a demand factor of 100 percent because once they have been energized, their heating elements or motors will operate continuously at 100 percent capacity. A kitchen range, on the other hand, is never (or very rarely) operated with all its burners going full blast at the same time. It will usually operate at only 60, 70, or 80 percent of its rated capacity. Its demand factor, therefore, will be only 60, 70, or 80 percent, depending on its size and its nameplate capacity.

The same considerations apply also to the general-purpose and small-appliance circuits. Studies have shown that in the typical dwelling only 3000 watts of the total installed capacity will operate 100 percent of the time, with the balance of the electrical equipment being operated only 35 percent of the time. Therefore, since the total load we have calculated above is 13,500 watts (that is, 9000 watts + 3000 watts + 1500 watts = 13,500 watts), we can calculate the total power required for these circuits in the following manner:

Total load, watts	*Demand factor, %*	*Power required, watts*
3,000	100	3000
10,500	35	3675
	Total	6675

As for the fixed appliances, we will assume that the house will have the following equipment installed, the equipment having the following nameplate ratings:

Appliance	*Rating, watts*
Range	8,000
Freezer	500
Clothes washer	600
Clothes drier	4,500
Water heater	2,000
Total	15,600

We will further assume that the clothes drier has a demand factor of 100 percent, that the range has a demand factor of 80 percent, and that the remaining equipment have demand factors of 75 percent each. These factors are taken from the NEC. Therefore, the total amount of power required for the fixed appliances will be as follows:

Appliance	*Rating, watts*	*Demand factor, %*	*Power required, watts*
Range	8000	80	6,400
Freezer	500	75	375
Clothes washer	600	75	450
Clothes drier	4,500	100	4,500
Water heater	2,000	75	1,500
		Total	13,225

We have so far ignored the heating and cooling loads, as they are calculated separately. Assume now that the house will have a central air-conditioning system (which will consume 6000 watts) as well as several electric room heaters (which together will consume a total of 3000 watts). The NEC requires that only the greater of the heating and cooling loads be considered in the overall load calculations, since it is obvious that the heating and cooling equipment will never operate at the same time and that it is the higher of the two loads that will determine the overall load. In this case the air-conditioning load is greater and we can, therefore, ignore the heating load (though, of course, it will show up on the electric bill).

Air-conditioning equipment is assumed to operate at a demand factor of 100 percent. Therefore, the total estimated load that will be carried by the entire electrical system will be

Estimated load	*Power required, watts*
General purpose plus small appliance circuits	6,675
Fixed appliance circuits	13,225
Heating/cooling circuits	6,000
Total	25,900

Now then, to find the required capacity of the service-entrance equipment, we must first convert watts into amperes, using the formula

$$\text{watts/volts} = \text{amperes}$$

Inserting the total wattage obtained above into this formula, and the incoming voltage also, we obtain

$$25{,}900 \text{ watts} \div 230 \text{ volts} = 112.6 \text{ amperes}$$

for which a 125-ampere service will be necessary, since for our requirements this is the closest circuit-breaker size manufactured. However, any house having a calculated load that is this close to the circuit-breaker size of 125 amperes would be well advised to have 150- or 200-ampere equipment installed, as suggested above, in case future additions should ever be made to the electrical system.

Note that our calculations say nothing about the size of the conductors that are required for any particular circuit; for that discussion, *see* ELECTRICAL CONDUCTORS AND CABLES. What our calculations do show is how one determines the *minimum* capacity of the service-entrance equipment, although one can also get some idea of the minimum number of general-purpose circuits that would be required in a house similar in size to the one in our example.

Our figures show, for example, based on 3 watts per sq ft in a 3000-sq-ft house, that 9000 watts is the general-purpose load. Since these will be 2-wire, 115-volt circuits, in order to find the amperage of these circuits, we again divide watts by volts to get

$$9000 \text{ watts} \div 115 \text{ volts} = 78.2 \text{ amperes}$$

Since each general-purpose circuit is a 20-ampere circuit, the house will require a minimum of four 20-ampere, or six 15-ampere, branch circuits to handle the anticipated lighting and miscellaneous electrical equipment.

We also mentioned above that an average-sized house should have a branch circuit for every 500 sq ft of floor area, and 3000 sq ft ÷ 500 = 6, which is the number of 20-ampere branch circuits required by this calculation.

Optional Calculations

The NEC permits the use of an optional, and simpler, method of calculating the total estimated electrical load for a house. One takes the first 10,000 watts at a demand factor of 100 percent and the remainder of the load at a demand factor of 40 percent. Any heating and/or cooling loads are then added to the result obtained. As an example, we will take the same wattage figures we used in our previous example, the wattage of the fixed appliances being taken at their full nameplate ratings.

Load type	*Watts*
3000 sq ft at 3 watts per sq ft	9,000
Two 20-amp small-appliance circuits	3,000
One 20-amp laundry circuit	1,500
Range	8,000
Freezer	500
Clothes washer	600
Clothes drier	4,500
Water heater	2,000
Total	29,100

The first 10,000 watts at 100 percent (10,000 watts) plus the remaining 19,100 watts at 40 percent (7640 watts) gives a total of 17,640 watts. We now add to this total the air-conditioning load of 6000 watts at a 100 percent demand factor, which gives us

$$17{,}640 \text{ watts} + 6000 \text{ watts} = 23{,}640 \text{ watts}$$

To find the amperage, we again divide the wattage by the voltage to obtain

$$24{,}840 \text{ watts} \div 230 \text{ volts} = 102.7 \text{ amperes}$$

a figure that is somewhat lower than the 112.6 amperes we obtained using the first method of calculation. Either result can be used; both are valid, although, again, it should be emphasized that a builder is always prudent if he installs excess capacity in the electrical system to take care of as-yet unanticipated loads.

Electrical Wiring Circuits

This article describes the ways in which switches, duplex receptacles, and lamp sockets—the principle operating devices used in residential electrical circuits—are wired together to produce functional circuits. For an overall description of residential electrical systems, *see* ELECTRICAL SYSTEM. For a description of the cables and conductors used for electrical wiring and their manner of installation, *see* ELECTRICAL CONDUCTORS AND CABLES. *See also* GROUND, ELECTRICAL.

Anyone who examines an electrical diagram for a house, such as the one shown in Fig. 1, can see how inadequate such

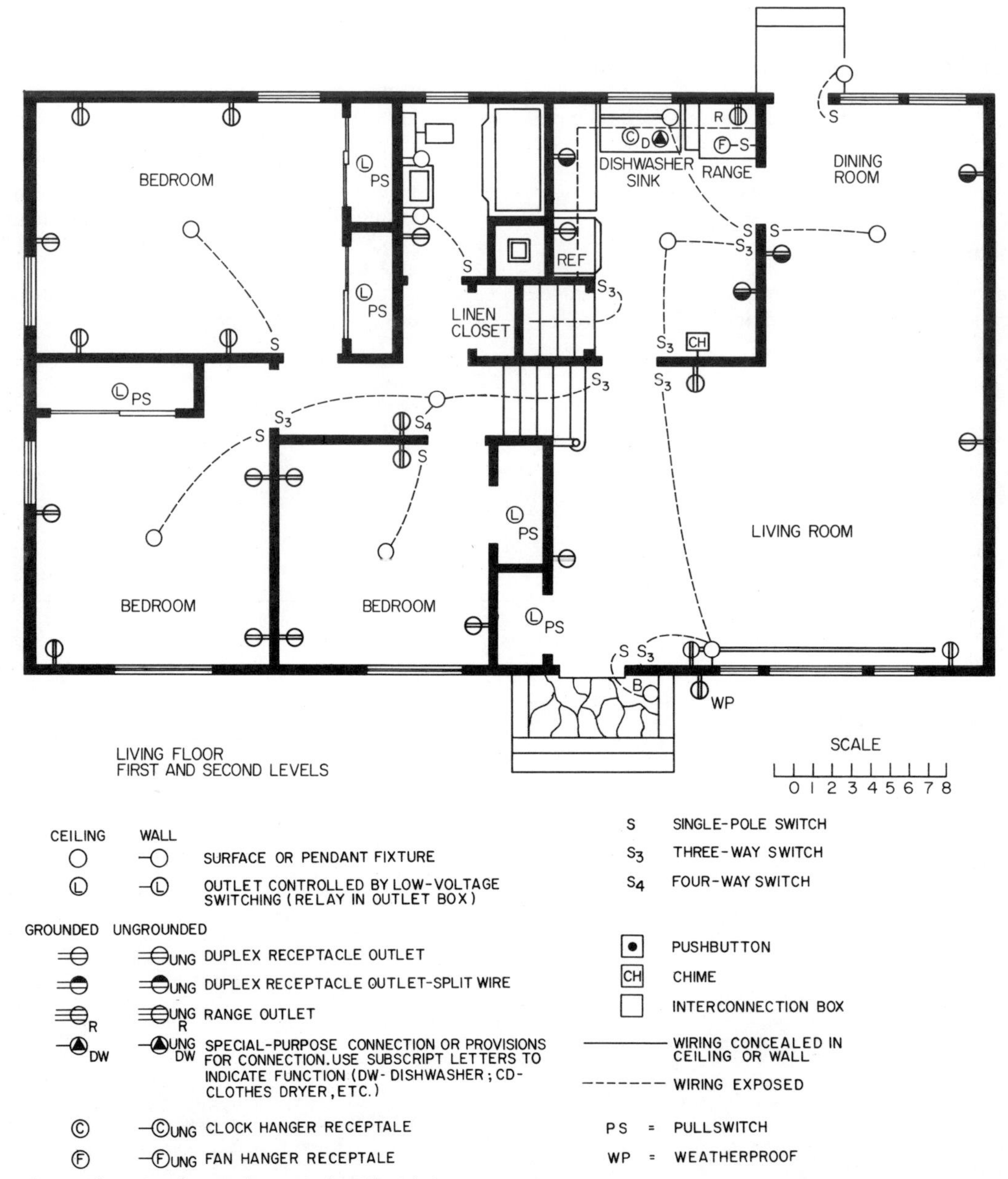

Fig. 1. The wiring diagram for a typical residential electrical installation.

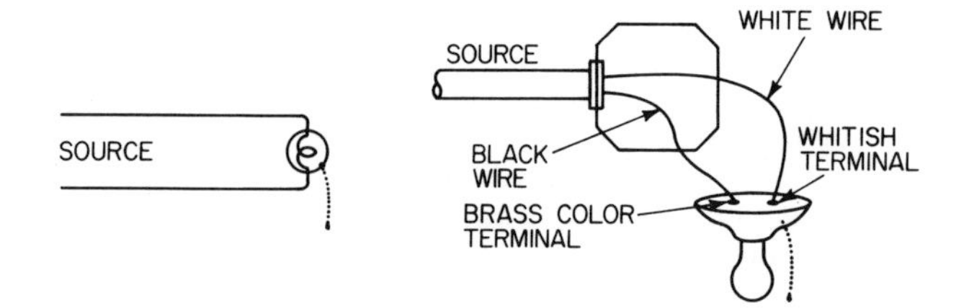

Fig. 2. Electrical installation for a pull-chain lamp socket. The electrical schematic is also shown *(Richter, Practical Electrical Wiring, McGraw-Hill).*

a diagram is as a guide to anyone who is ignorant of electrical systems and who wants actually to install wiring. The architect or engineer who prepares an electrical diagram depends on a body of understood convention and the expertise of the electrician who will actually convert these symbols into a working system.

As a matter of fact, the actual circuits are quite simple, as the illustrations in this article demonstrate, and are easily understood by anyone having a minimal knowledge of electricity. But what an electrician knows that an ordinary person doesn't know are the practical methods by which the wiring is installed. The electrician also has an intimate knowledge of the electrical code that prevails in his area, which is probably based on the National Electrical Code (NEC). This local code describes in great detail the materials that must be used and the procedures followed to ensure that a complete electrical system is both safe and reliable. *See* ELECTRICAL SYSTEM.

The basic elements of a simple residential electrical circuit are shown in Fig. 2. The illustration shows two wires (or conductors), one white and the other black, leading to and from a pull-chain-operated lamp socket. Lamp sockets of this type are usually installed nowadays in attics, basements, closets, garages, and similar out-of-the-way places.

This type of circuit is called a *2-wire* circuit. The black and white colors in the illustration are the actual colors of the insulation that surround the wires in the actual circuit. The black wire is always the *hot* wire. The convention is that it conducts the current of AC electricity from its source (which is usually a *distribution panel,* or *fusebox,* located in the house) to the base of the socket and thence through the lamp filaments.

The white wire is the *neutral* wire. It is connected to the shell of the lamp socket and conducts electricity away from the socket. Ultimately, the neutral wire conducts the electricity to *ground,* or *earth,* quite literally in fact because the main grounding connection in most domestic electrical installations is to an underground water pipe; *see* GROUND, ELECTRICAL. As far as the circuit shown in Fig. 2 is concerned, the neutral wire *grounds* the socket shell so that anyone who touches the shell will not receive an electric shock.

Switches, receptacles, and lamp fixtures are so constructed nowadays that the neutral side of these devices is always identified in some unmistakable way; usually the neutral terminal is a lighter color than the hot terminal. Thus, for example, the head of the screw to which the neutral wire should be attached will have a whitish color, and the other screws on the device will be brass colored. The box in which the device is packed will usually indicate clearly enough how the manufacturer has identified the neutral terminal on the device.

The reason for these precautions is to preserve the *polarity* of the electrical system, that is, to keep the hot side and the neutral side of the system completely separate at all times. The reason for preserving system polarity is safety. If the function of the terminals on the equipment used in a system is uncertain, it is possible that the wires will be connected together in such a way as to overload the neutral circuit. The possibility would then exist that the wires will overheat, the insulation will burn away, and a fire will result. (On the other hand, the actual danger of such an excessive overload occurring in the typical residential installation is quite remote.)

Since for reasons of safety it is more important always to know the identity of the neutral circuit, the white wire is called the *identified* wire, and the colors white or light gray are permitted to be used only for the insulation on the neutral wires of a circuit.

In Fig. 2, the pull chain on the lamp socket controls the operation of this circuit. Pulling the chain opens and closes a switch built into the lamp housing. When the switch is open, the circuit is open, and electricity cannot flow through the lamp filament. When the switch is closed, the circuit is closed, and electricity can and will flow through the filament, which will now glow brightly.

In essence this is all there is to most of the electrical circuits in a house, with the difference that most of the circuits in a house will supply power to several fixtures or devices at the same time. As shown in Fig. 3, for example, a circuit may supply power to several lamp fixtures. The lamps in the illustration are connected together in what is called a *parallel* circuit, which means that the hot wire branches off at each socket to supply that socket individually with electricity. Similarly, the neutral wire branches off at each socket to individually ground out each socket.

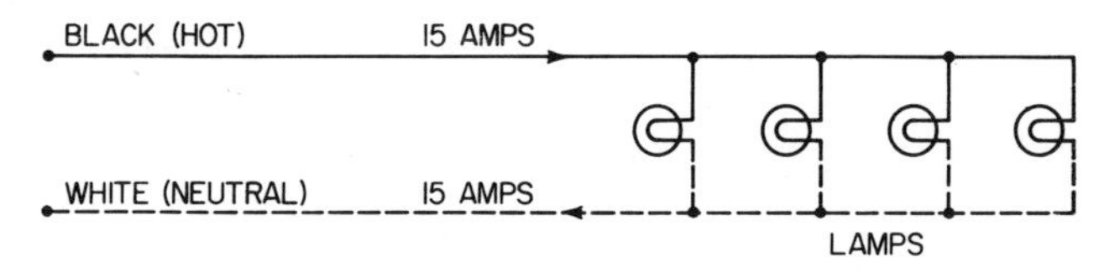

Fig. 3. A number of lamps connected in parallel in an electrical circuit.

The advantages of connecting devices to a circuit in parallel are two. First, the current of electricity in the hot wire is not dissipated by having to pass through several lamps successively. If the hot wire were connected to a number of lamps in series, each lamp in the circuit would consume a portion of the available electrical power, leaving less for the lamps that followed after. In such a circuit each lamp would glow less brightly than its predecessor instead of all the lamps glowing with equal brightness. Second, the failure of one lamp in a parallel circuit cannot possibly affect the operation of the other lamps in the same circuit because the hot wire never passes exclusively through any of the lamps. If it did, the failure of any one lamp would act like an open switch in the circuit, and all the lamps in the circuit would go out.

SWITCHED CIRCUITS

Ceiling and wall-mounted lights in the living quarters of a house are usually operated by single-pole ON-OFF switches mounted in the wall next to a doorway. A typical lighting circuit is shown in Fig. 4. Note that the hot side of the circuit passes through the switch. This fact illustrates another basic principle of all wiring systems—the continuity of the neutral wire is *never* interrupted in any way whatsoever. In a distribution panel, for example, the fuses or circuit breakers are always installed on the hot side of the circuits, never on the neutral side. The integrity of the neutral circuit is maintained to ensure that, in the event of a short circuit, the surge of electricity is transmitted immediately and harmlessly to ground.

In Fig. 4, that portion of the hot wire that descends to the

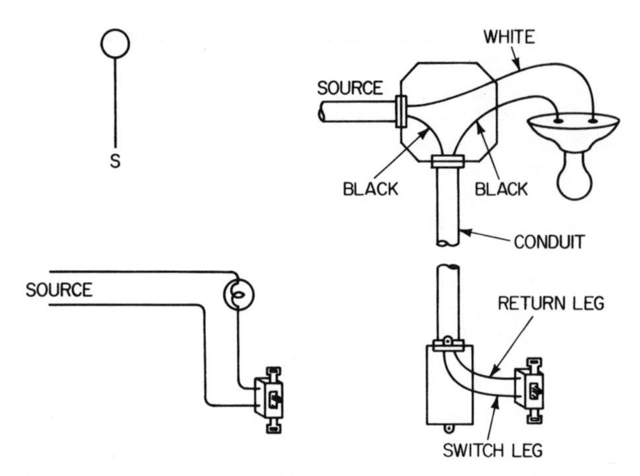

Fig. 4. A typical lamp circuit controlled by a wall switch *(Richter, Practical Electrical Wiring, McGraw-Hill).*

switch and then returns to the ceiling outlet box is called the *switch loop.* The descending wire is the *switch leg* of the loop and the ascending wire is the *return leg.*

When wiring this or any other circuit in a new house, the builder can use one of two methods, conduit or cable. With a conduit system, a network of steel or aluminum pipes (i.e., the conduits) is first installed throughout the house and the wires are inserted into these conduits. With a cable system, lengths of cable containing as many individual wires (or conductors) as are required between any two outlet boxes are attached to the framework of the house. For a description of a cable system, *see* ELECTRICAL CONDUCTORS AND CABLES.

With a conduit installation, an electrician can easily insert into any particular conduit as many wires as are required for any given circuit. More to the point here, the electrican can also select the colors of the insulation on the wires. Thus, when he installs the wiring for a switch loop, he can insert two lengths of black-insulated wire in the conduit for the switch loop. This is shown in Fig. 4.

When cable is used, however, difficulties arise because a cable consisting of two black-insulated conductors is not manufactured; one of the conductors in a 2-wire cable will always be white. But electrical codes forbid the use of a conductor with white insulation in a hot circuit. To repeat what was said above, it is a basic principle of the NEC that the neutral circuit always be clearly identified; that is, the neutral wires must always be white or light gray.

A dilemma, but a dilemma easily resolved since the NEC bows to reality and makes an exception in this particular case. When a 2-wire *cable* is used for a switch loop, the black conductor in the cable will be the return leg of the loop and it will connect to the hot terminal on the lamp socket. The white conductor, therefore, will be the switch leg of the loop. This means that within the ceiling outlet box, this white conductor will be connected to the source of electricity. This is shown in Fig. 5. Anyone who has occasion hereafter to look into an outlet box and notes that a black wire is connected to a white wire, should immediately realize that he is dealing here with an *un*identified, or hot, wire.

He should further realize that this connection is the one exception to the general rule regarding the colors of electrical wires and, therefore, of the two wires traveling down to the wall switch, the black wire must be the return leg and the white wire the switch leg of the circuit.

(We might note in passing that the NEC allows electricians to paint the ends of a white conductor black. Many electricians do this as a way of showing a white- or gray-covered conductor is the hot wire.)

Figure 6 shows the same basic switching circuit as in Fig. 5, but this time the hot wire travels first to the switch and then to the ceiling fixture. There can be no confusion in this case regarding which wire should be installed where, regardless of whether conduit or cable is used, since, as the illustration shows, black is connected to black and white to white.

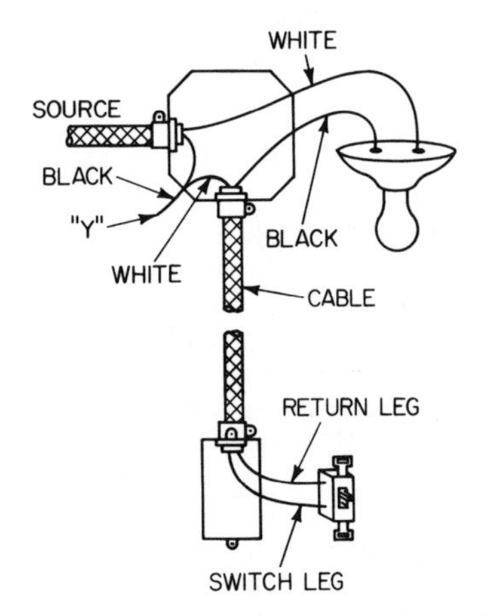

Fig. 5. The same circuit shown in Fig. 4 but with cable used for the installation *(Richter, Practical Electrical Wiring, McGraw-Hill).*

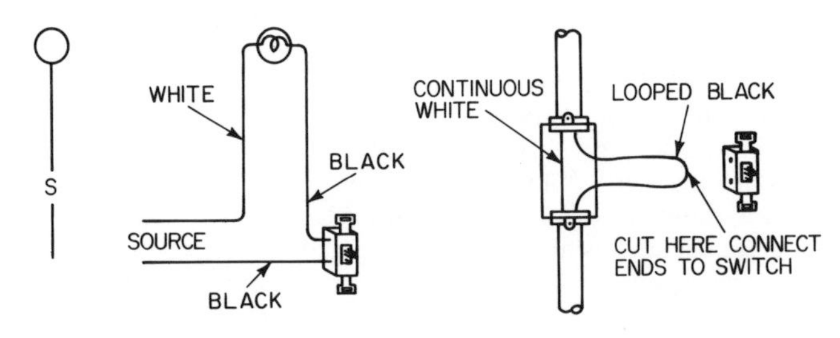

Fig. 6. The same circuit as in Figs. 4 and 5 except that the wires in the circuit are connected to the switch first *(Richter, Practical Electrical Wiring, McGraw-Hill).*

Three-Way Switch Circuit

It is often convenient if a ceiling light can be turned on or off from either of two switches. It should be possible, for example, to turn off a hallway light from either end of the hallway; and, similarly, one should be able to turn off a stairway light from either the top or bottom of a flight of stairs. In situations of this kind, three-way switches are used to open and close the circuit. A typical installation is shown in Fig. 7.

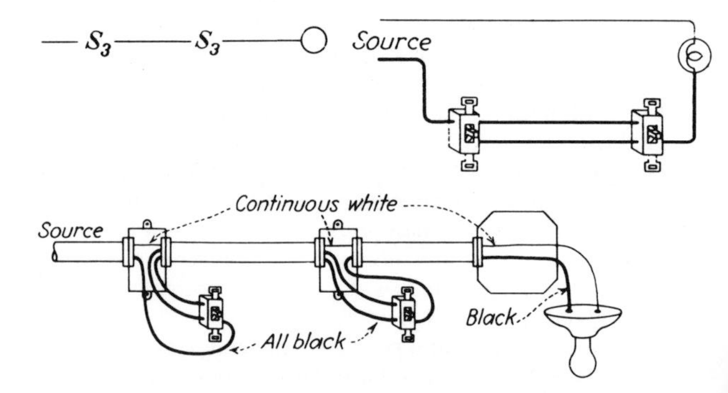

Fig. 7. A lamp circuit controlled by two switches *(Richter, Practical Electrical Wiring, McGraw-Hill).*

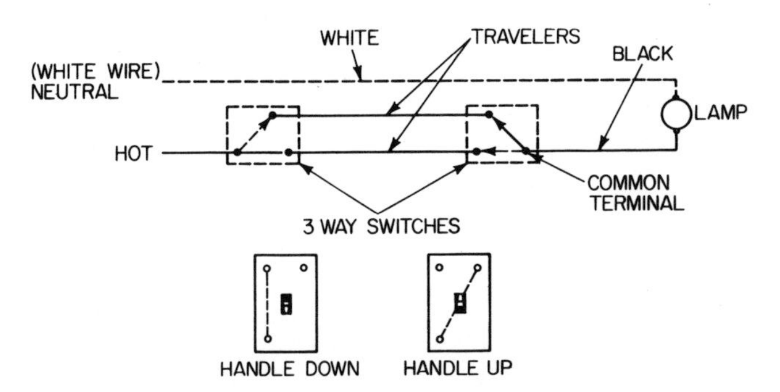

Fig. 8. The operation of an electrical circuit containing 3-way switches.

Three-way switches do not have an OFF or an ON position. Instead, the switch may be said always to be ON. What the switch does is choose between two ON positions, as shown in Fig. 8. When the switch is in one position, the electrical current takes one path through the switch; when the switch is in the other position, the current takes the other path.

The common connection in the switch circuit is called either the *common* terminal or the *hinge point.* In an actual switch, the common terminal need not necessarily be opposite the other two terminals as is shown in Fig. 8. Whatever the actual design of the switch, the box in which the switch is packed will identify each terminal and its function in the circuit. Usually, the common terminal will have a different color than the traveler terminals.

A comparison of Fig. 7 with Fig. 4 will show that the two circuits are quite similar to each other. The basic switching circuit of Fig. 4 remains but with the two switches connected together by two new wires called *travelers.*

Figure 8 shows the circuit open and the lamp off. If either of the two switches should now be turned to its opposite position, the circuit would be closed and the lamp will light. Thereafter, turning either switch to its opposite position would open the circuit and the light will go off again. And so on. A study of the diagram should make this clear.

The actual wiring of a three-way switch circuit is simple enough as long as the wires run through conduits. As shown in Fig. 7, the electrician simply inserts the required number of black and white wires into the conduits, connects the appropriate wires to the switches and the lamp socket, and the circuit will operate as required.

But if cable is used, the situation can become quite complicated because in some parts of the circuit the electrician will have to install 2-wire cable and in other parts of the circuit he will have to install 3-wire cable; which cable is used where will depend on the positions of the switches and lamp in the circuit. Furthermore, the three conductors in a 3-wire cable are always colored black, white, and red. All this is shown in Fig. 9. It is situations like this that separate the experienced electrician from the do-it-yourselfer who is updating the wiring in his home a bit. But, despite the apparent complexity of the wiring installations shown in Fig. 9, the basic circuit shown in Fig. 7

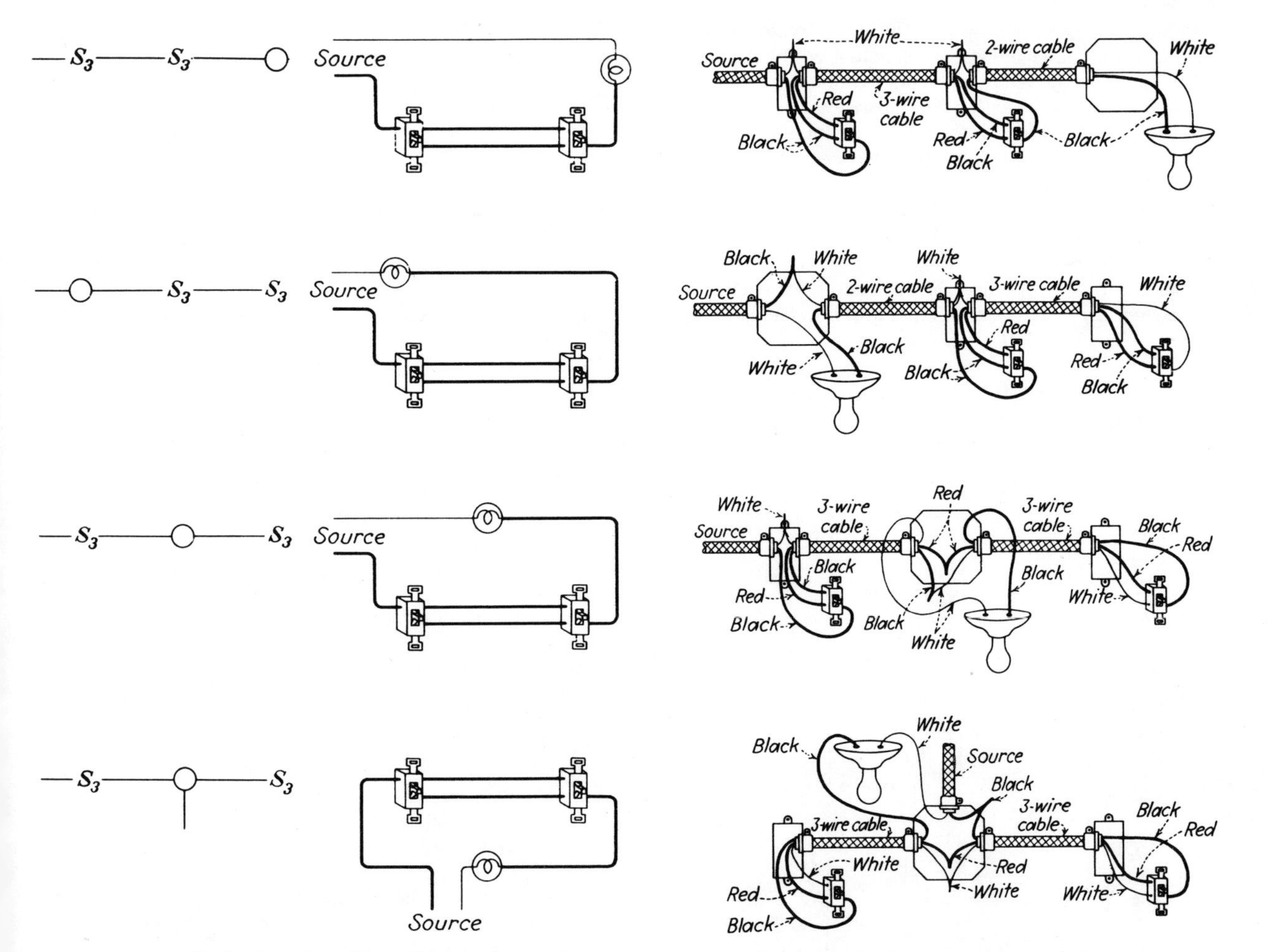

Fig. 9. Several possible methods can be used to wire up a circuit containing 3-way switches when cable is used. It depends on the sequence in which the lamp and switches are connected *(Richter, Practical Electrical Wiring, McGraw-Hill).*

lies hidden within, and anyone interested can trace out the wiring connections for himself and see that this is so, regardless of the particular sequence of switches and lamps in any particular installation.

One might wonder how an electrician keeps track of all the wires going into and coming out of an outlet box. The convention is to wrap the end of the switch leg around the ends of the two traveler wires.

Four-Way Switch Circuit

It may happen that a lamp must be switched on or off from *three* different locations. For example, a large living room may have three entrances, or it may be desirable to control a driveway light from a garage and two house entrances.

In a situation such as this, the circuit must consist of two three-way switches and one *four-way* switch, as shown in Fig. 10. Note that the only difference between this circuit and the three-way switch circuit shown in Fig. 8 is the connection of the four-way switch across the traveler wires. Otherwise, the circuits are identical.

A four-way switch, like a three-way switch, has no OFF position. What a four-way switch does is cross and uncross the traveler wires as the switch is moved from one position to the other. This is shown in Fig. 10, in which the circuit is shown closed and the lamp lit. Examination of this diagram will show that turning any one of the three switches to its opposite position will cause the light to go out, and that turning any one of the three switches to its opposite position subsequently will cause the light to go on again.

When conduit is used, wiring this circuit is straightforward enough. But when cables are used, the actual wiring connections will depend on the relative positions of the switches and lamp in the circuit, just as with a three-way switch circuit.

An electrician keeps track of the wires in a four-way switch box by twisting together the two traveler wires coming from the same three-way switch.

And what if one wants to wire a circuit having four, five, six, or even more switches controlling lamps or other devices in a circuit? The solution is simple enough in principle. One need only add to the traveler portion of the circuit, between the three-way switches, as many additional four-way switches as are required.

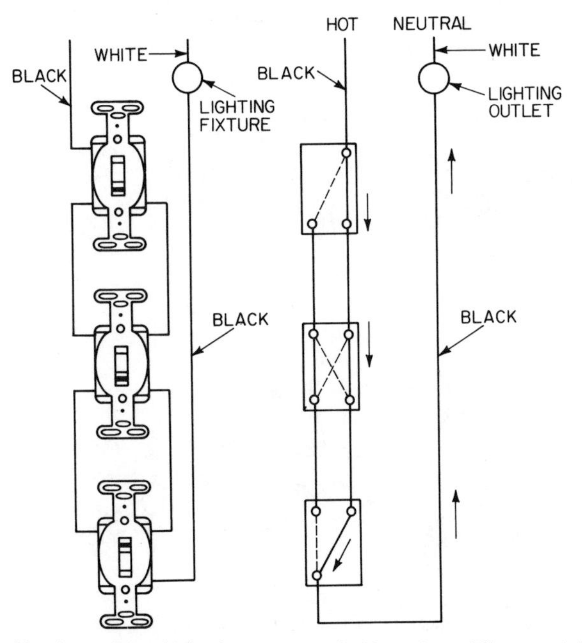

Fig. 10. A circuit in which a lamp is controlled from three different switch locations.

RECEPTACLE CIRCUITS

The branch circuits that supply power to the duplex receptacles mounted along the bottoms of the walls and above kitchen counters are always hot. One need only plug into such an outlet a vacuum cleaner, a floor lamp, a television set, a toaster, or whatever, and the device will function immediately. Receptacle circuits are wired in parallel, as shown in Fig. 11, for the reasons already given for lighting circuits.

Receptacles also have a hot terminal and a neutral terminal, which are distinguishable in the same way that the terminals on switches are. It doesn't make any difference to the operation of the receptacle which line is connected to which terminal but it is important to keep the hot and neutral circuits separate from each other, or polarized. It is always the neutral side of a receptacle that is distinguished in some special way; usually the heads of the terminal screws are a lighter color than the terminal screws on the hot side. The box in which the receptacle is packed will usually describe exactly how the neutral terminal is identified.

Increasingly, duplex receptacles are being installed that have a separate ground connection that connects to a special grounding prong on the plug, which is in turn connected to the metal framework of the device that is connected to the receptacle. (For an illustration, *see* GROUND, ELECTRICAL, Fig. 11.) This ground connection has nothing whatsoever to do with the electrical circuit per se or with the supply of electrical power. It is a safety measure. The purpose of this connection is to ground the metal frame of whatever equipment is being plugged into the receptacle outlet, in order to prevent anyone's

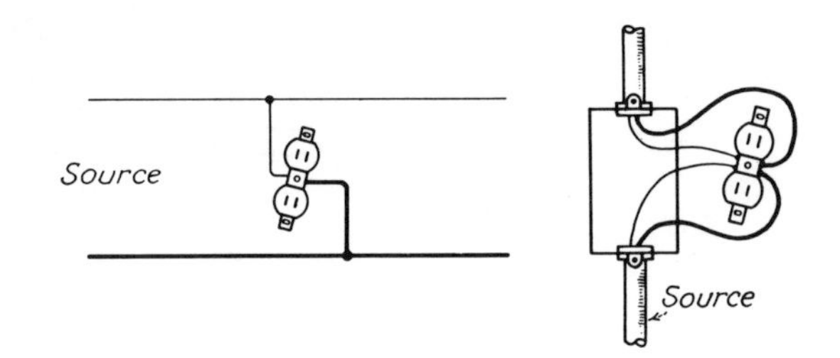

Fig. 11. Duplex receptacles are wired in parallel in an electrical circuit *(Richter, Practical Electrical Wiring, McGraw-Hill).*

receiving a shock if there should be a short within the equipment that has made the frame hot.

This grounding circuit connects to earth either via a separate system of ground wires that is installed along with the hot and neutral conductors within the conduits or cables making up the electrical circuits or, in an old house, by connecting the grounding connections to nearby conduits or water pipes via special clips or screws. For a description *see* GROUND, ELECTRICAL. This ground wire may be distinguished either because it has green insulation around it or because it is completely bare of insulation. The wire is connected to a special grounding terminal on each receptacle, which can be identified either by its green tint or by its having a hex head.

Installing three-pronged duplex receptacles without taking the trouble to ground out the receptacles properly is an extremely deceptive, not to say potentially dangerous, practice that gives the illusion of safety when there is no safety. When these receptacles are used, they should be grounded. *See* GROUND, ELECTRICAL.

Electrical codes forbid the attachment of two or more wires to any one terminal. However, as Fig. 11 shows, there will usually be at least two connections that have to be made to the hot and neutral terminals of the receptacle. Receptacles, therefore, are manufactured with two pairs of terminal screws, one pair for each half of the receptacle.

Double-wired Circuits

In the duplex-receptacle circuit just described, if a fuse should blow or circuit breaker trip on the branch circuit for any reason, all the receptacle outlets on that branch circuit would go dead. In the typical living-room installation, this would mean that all the floor and table lamps in the room would become inoperative. In a kitchen, all the small-appliance circuits would become inoperative. The latter event would be especially inconvenient since these receptacles usually supply power to the refrigerator and freezer also.

In receptacle circuits, therefore, it is possible to *double-wire* the receptacles; that is, to install two complete and independent branch circuits to the receptacles by connecting one half of each duplex receptacle to one branch circuit and the other half to another branch circuit. This is shown in Fig. 12. The two halves of the receptacles are made independent of the other merely by snapping off with a pair of pliers a section of a brass plate that connects the two terminal screws together. Thereafter, if the outlets connected to one branch circuit should go dead, the outlets connected to the other branch circuit will still be capable of supplying electricity until the fault has been found and corrected.

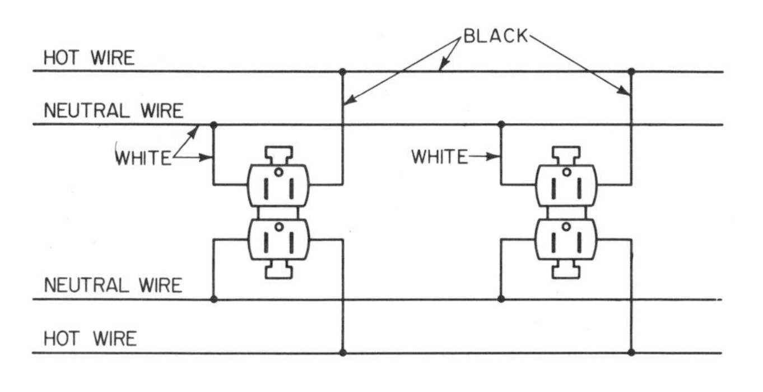

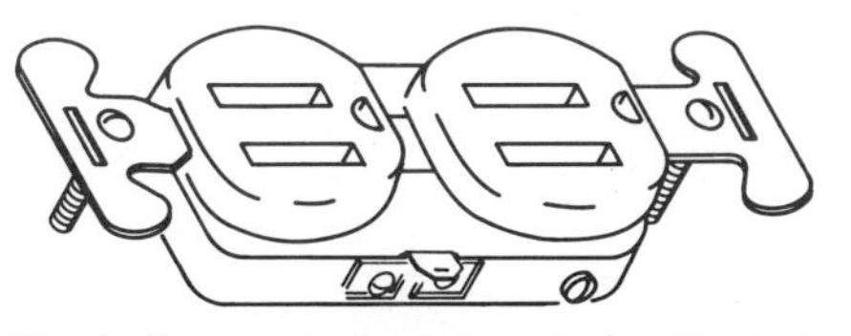

Fig. 12. A double-wired receptacle circuit. In order for this circuit to function correctly, a brass tab that joins the two outlets of the receptacle together must be snapped off.

Switched Receptacle Circuits

Many homes today are built without any ceiling outlets at all. Instead, all the room lighting is supplied by floor or table lamps connected to duplex receptacles. When this sort of lighting system is installed in a house, it may be inconvenient to go around a room switching on the lights one by one, or switching off the lights one by one at the end of the day. There may even be some danger in entering a dark room and searching for a lamp to switch on.

To do away with this inconvenience, duplex receptacle circuits are wired through a wall switch located near the doorway of each room. The lamps are connected to the duplex receptacles with their pull chains or ON-OFF switches left in the ON position. Thereafter, the lamps are switched either on or off through the wall switch (see Fig. 13).

This type of switched receptacle circuit is wired together in exactly the same way as has been described for switched ceiling-lamp circuits. Everything that has been said regarding the installation of the hot and neutral wires of lamp circuits, the

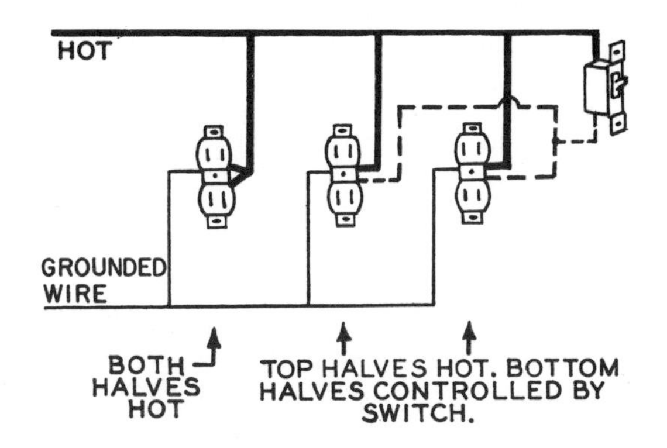

Fig. 13. Duplex wall receptacles wired through a wall switch. Two alternatives are shown, one in which power to the bottom halves of the receptacles is controlled by the switch and one in which both halves of the receptacles are continuously hot *(Richter, Practical Electrical Wiring, McGraw-Hill).*

connection of their wires, and their color-coding, applies also to switched receptacle circuits.

Usually, only one of the two halves of a duplex receptacle is wired through a wall switch. The brass plates on the sides of the receptacle are converted into two independent units by snapping off the brass tab connecting the plates, and the receptacle is then double-wired as shown in Fig. 13. One outlet on the receptacle will continue to supply electricity directly from a branch circuit, and the other outlet will supply electricity via the wall switch. Thus, one has the best of both worlds—one of the receptacle outlets will supply light and the other will supply power. And, although we have described this switched circuit as if only one receptacle outlet were connected to each switch, there is no reason why several recepta-

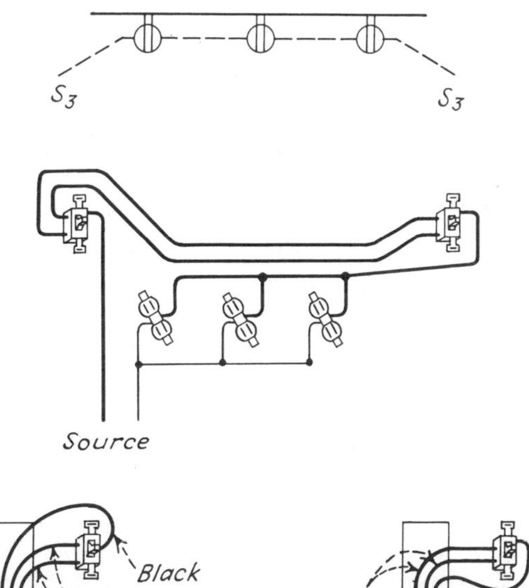

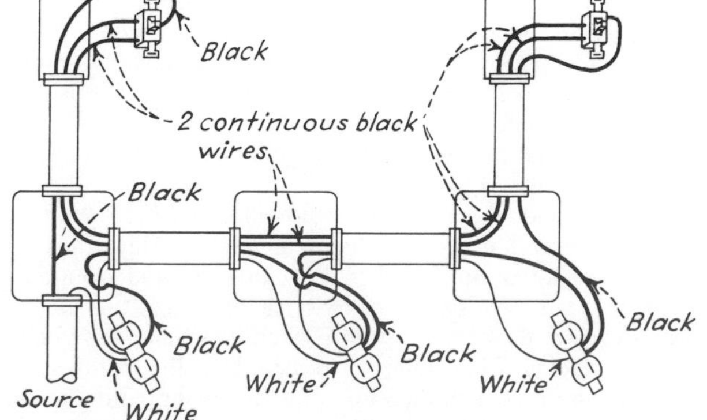

Fig. 14. Duplex receptacles controlled by wall switches. *(Richter, Practical Electrical Wiring, McGraw-Hill).*

cles, each connected to a separate lamp, couldn't be connected to the same wall switch so that all the lamps in a room operate together (see Fig. 14); or why a double-wired lamp circuit of this type couldn't be operated through several three-way and four-way switches, just as the lamp circuits described above.

Pilot Lights in Switched Circuits

In many switched circuits it is convenient to be able to tell whether the circuits are energized or not. For example, a basement light may be controlled by a switch from the top of a flight of stairs, from where it is impossible to tell whether the lamp is actually on or not, or whether the lamp has burned out or not. Or a switch controlling driveway lights or a sump-pump motor may be located inside the house, from where it is impossible to tell whether the driveway lamps are on or not or whether the sump pump is energized or not. In cases such as these, a pilot light located in the same outlet box as the switch will give a visual indication of the state of the circuit. Fig. 15 shows a typical circuit. All it consists of basically is two lamps wired in parallel, one of which is the pilot light itself, both lamps being controlled by the same switch. The circuit is identical to that shown in Fig. 4.

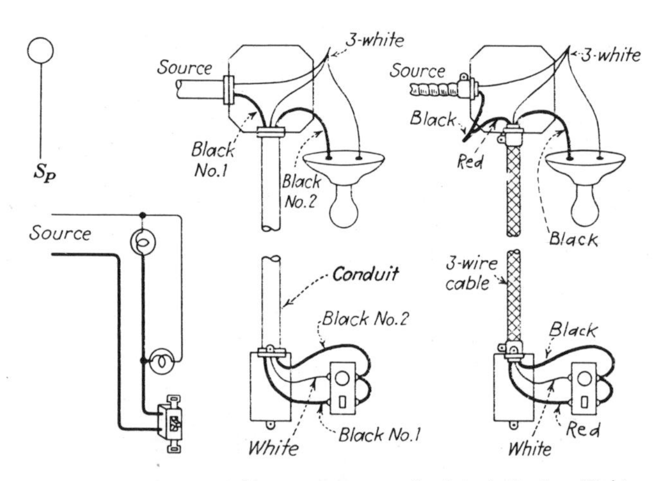

Fig. 15. Installation of a switch having a pilot light indication (*Richter, Practical Electrical Wiring, McGraw-Hill*).

WIRING CONNECTIONS

The wires in an electrical circuit are always connected together at outlet boxes by connections called *pigtail* splices. The splices are quite simple to make and the technique should be obvious from Fig. 16.

As many wires as necessary can be twisted together into a pigtail splice. Such a multi-wire splice is called a *bunch* splice. The only precaution that need be taken when making a bunch splice is to be sure that *all* the wires are twisted about each other equally, and that several wires are not twisted about a central wire that has remained straight; such a splice can pull apart very easily.

We mentioned that the NEC forbids attaching more than one wire to a terminal of any electrical device. This rules out making a pigtail or bunch splice that can be twisted around the terminal screw of a switch or receptacle; yet, as the illustrations show, multi-wire splices are often necessary at switches and receptacles. The solution is to include in the pigtail splice an extra wire that is about 6 in. long. This is shown in Fig. 16. This is called a *jumper;* it connects the wires making up the splice to the terminal screw, with the splice itself then being

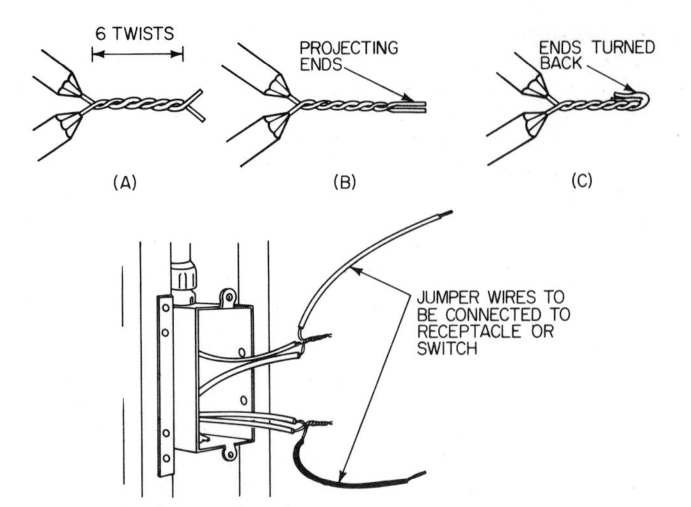

Fig. 16. Making a simple pigtail splice. When pigtail-spliced wires must be connected to a switch or receptacle, a jumper wire is made part of the splice and the connection is actually made via the jumper.

tucked into a corner of the outlet box where it is out of the way.

Lighting fixtures are usually manufactured with short lengths of black and white wire extending from the fixture, and the splices that connect these fixtures to the circuit wiring are made to these wires. These fixture wires are made of stranded wire, not solid wire, which makes it a bit difficult to twist these wires and the solid circuit wires together. In such cases a pigtail splice is made as shown in Fig. 17, with the solid wire being bent back against the stranded wire to hold the strands in place. Once a splice with stranded wire has been completed, the NEC requires that the wires be soldered together.

After a splice has been completed, electrician's tape must be wound about the spliced wires until the splice is as well insulated as the original wires were. Before the introduction of plastic tape, unvulcanized rubber tape was used exclusively to insulate a splice. This tape, which is also known as *splicing compound,* is supplied in rolls, with a strip of glazed fabric between the layers of rubber to prevent the layers adhering to each other, which they will do on the slightest contact.

The tape is unwound about 4 in., the glazed fabric is pulled away from the rubber, and the rubber is then wound around the splice helically, each turn of rubber covering the preceding turn by about half the width of the tape. The rubber should be pulled taut—it will stretch quite a bit—and, as it presses against the turn of rubber that has already been applied, the pressure will cause the two layers of rubber to vulcanize together. In effect, the two layers will become one solid layer. Once the tape has been wound completely over the splice, it is pulled around the splice squarely and is then wound back down over the first layer of rubber until the starting point has been reached.

Since the rubber is very soft and abrades easily, it must be covered with electrician's *friction tape,* which is made of

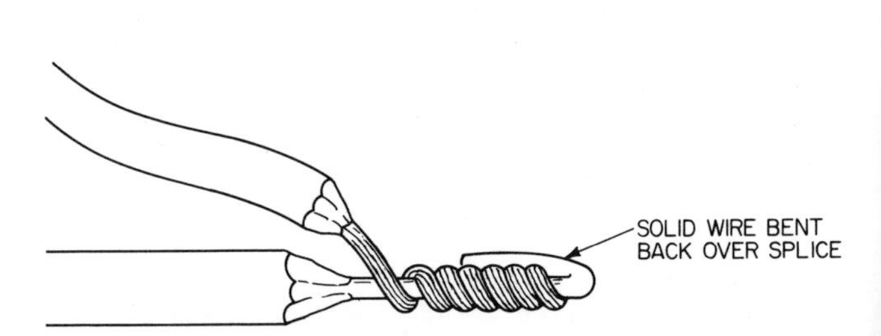

Fig. 17. A pigtail splice joining together solid and stranded wire. Such a splice must be soldered if it is to be mechanically secure.

fabric and has no electrical insulating properties whatsoever. Friction tape is used simply to protect the rubber. It should never be used by itself to cover bare wire, not only because it lacks insulating properties but also because it tends to dry out over a period of time, and it may then simply fall away from the splice. Very often the tape is coated with shellac to hold it together and in place.

Plastic electrician's tape has largely replaced rubber tape. Electrically, it has a higher insulating value than rubber, and thus it need not be wrapped about the splice as thickly; nor does it have to be covered by friction tape afterward. The method by which plastic tape is wound about a splice is exactly the same as when rubber tape is used.

An electrician may be tempted sometimes to splice two wires together between two outlet boxes when one wire runs a bit short. Splicing between outlet boxes is strictly forbidden by the NEC for any reason whatsoever. If a wire isn't long enough to reach from one outlet box to another, it must be removed and a longer wire installed in its place. The reason, of course, is the difficulty that would be encountered if the splice proved defective, but this defect weren't discovered until the house had been completed. There is also the danger that a defective splice might short out and be the cause of a fire.

For the same reasons, the NEC forbids the splicing of extension cords and appliance cords, though any number of how-to-do-it books give detailed, fully illustrated instructions on how to make such splices. Defective appliance cords must be discarded. Extension cords that are too short must be replaced by longer cords.

Wire Nuts

Wire nuts, or *solderless connectors,* or *splicing* devices, as they are also called, Fig. 18, are intended for use only on pigtail splices. When they are used, they do away with the requirement that the splice be soldered and taped, or even that the wires be twisted together. This labor-saving feature of wire nuts has increased their use greatly.

The most widely used type of wire nut consists of a coil of wire fastened inside a plastic shell, which acts as insulation. The straight ends of the wires to be joined are pushed into the coil, which is screwed down over them. The coil is so fashioned that when it is turned in a clockwise direction (which is the normal direction in which all threaded fittings are turned to tighten them), the coil untwists a bit. Once the wire nut has been completely screwed down and the pressure released, the coil clamps down on the wires, holding them securely together.

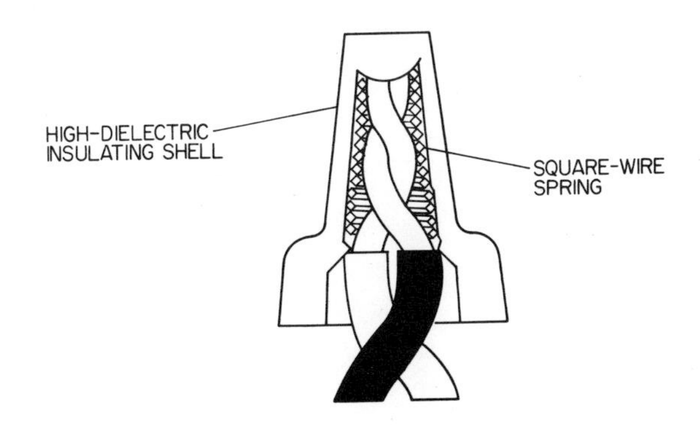

Fig. 18. Cross section of a wire nut.

TESTING ELECTRICAL CIRCUITS

Once the wiring has been installed in a house, the circuits must be checked to make certain that the wiring has been correctly installed and, of equal importance, that shorts have not inadvertently been introduced into the system.

The tests are performed after all the wiring has been completely installed but before any connections have been made to any switches, duplex receptacles, or lamp fixtures. All the required pigtail splices should have been completed, although the splices should not yet have been soldered and taped together or wire nuts installed over them. The ends of those wires that will later be connected to switches, receptacles, and fixtures must be exposed and hanging freely from the outlet boxes.

During all this preliminary splicing of wires, the electrician will of course make certain that the bare wires do not touch either each other or the outlet boxes. It is also important that the electrical system be disconnected from the main source of power, and that any operating equipment such as motors, the control circuits of boilers and water heaters, and so on, be disconnected from the electrical system. Finally, in what follows we shall assume that all the outlet boxes are made of metal, not plastic.

As his test equipment, an electrician uses two dry cells connected in series with each other, and with test leads connected to the negative and positive terminals of the dry cells. He will also have a bell that he can connect in series with the dry cells.

The first check is made at the main distribution panel. This check is for possible shorts between the neutral and hot sides of all the branch circuits. One test lead is connected to the neutral buss and the other test lead is touched, in turn, to each of the branch circuits leading from the distribution panel. The bell should not ring, of course, since no devices have as yet been connected to the branch circuits to complete any circuit. If the bell does ring when one of the black wires is touched with the test lead, it means there must be a short somewhere in that branch circuit, and the electrician must go around to all the outlet boxes in the circuit and make sure that none of the wires is touching an outlet box or another wire. If this check doesn't clear up the short, the electrician will have to investigate further.

If this first check is negative, the fuses are now installed in the distribution panel (or the circuit breakers are all closed). The bell is removed from the tester. One lead of the tester is again connected to the neutral buss in the distribution panel, and the other lead is now connected to the distribution panel busses on the load side of the main switch. The main switch is open, of course.

What this hookup does is supply a temporary source of electricity from the two dry cells to all the branch circuits, thus energizing them.

Taking his bell, the electrician now goes to each outlet box in the dwelling. He touches each pair of black and white wires in each outlet box to the bell terminals. The bell should ring every time. If the bell does ring, it means of course that electrical energy is reaching the bell from the distribution panel. If the bell doesn't ring, it means there is an open connection somewhere in the circuit and the electrician must find the fault and correct it.

After having touched the pair of black and white wires to the bell terminals, the electrician then touches each black wire to one of the bell terminals and touches the other bell terminal to the side of the outlet box. The bell should ring again, to indicate that the outlet box is properly grounded. In this test, the bell will probably ring less loudly than before because of the higher electrical resistance offered by the outlet box and the conduits (if installed).

This continuity test to ground cannot be performed, of course, if plastic outlet boxes have been installed in the system. A system using plastic outlet boxes will usually have separate grounding wires installed in the cable runs along with the hot and neutral wires, and the grounding wires will be connected to the receptacles and fixture frames as already described. In this case, the electrician connects each hot wire and each green (or bare) ground wire together through the bell to test for continuity to ground. The bell should ring every time this test is made.

The entire test procedure will probably have to be performed again in the presence of a building or electrical inspector, whose responsibility it is to make sure that the electrical system has been installed in conformity with the local electrical code. The inspector will also check the overall quality of the installation, making sure that all the wire sizes, splices, and connections have been made in conformity with the local code. Once the inspector has approved the installation, and after the electrician has installed the switches, receptacles, and fixtures, made his final connections, installed all the wallplates and coverplates, and the building has been tested and certified for occupancy, the electrician can send in his bill.

Fireplace

A fireplace is not a very efficient method of heating a room. About 90 percent of the heat produced goes up the chimney. The only utilitarian justification for building a fireplace in these days of central heating might be for use on those relatively few evenings during the year when starting up the boiler or furnace just to take a chill out of the air would be too wasteful of fuel. But even for this limited purpose, a good stove is three times more efficient a heat producer than a fireplace.

On the other hand, nothing seems homier, more soothing, relaxing, and comfortable than sitting before the fireplace in one's home, watching the flames dance and the sparks fly up the chimney. The feeling is primordial and not to be denied. It is this feeling that is the real justification for having a fireplace, and for most people it is justification enough.

FIREPLACE DESIGN

General Considerations

The fire in a fireplace radiates heat both directly and by reflection off the back and sides of the combustion chamber, or *firebox,* as it is also called. This radiant heat has exactly the same characteristics as all other forms of radiant energy, including radio waves, light, and the ultraviolet radiations given off by the sun. It suffices here to note two important characteristics of radiant heat—it warms only those objects it strikes, and it travels only in straight lines.

An obvious example of these characteristics is our experience with the heat energy given off by the sun. As we know, on a sunny but chilly day, the only parts of our body that are warmed by the sun are those parts that face the sun; the rest of our body remains chilled. And if we move to a shady spot, we do not feel any warmth at all.

Exactly the same thing happens when we sit before an open fire. Those parts of our body that are facing the fire are as warm as toast while the rest of our body remains unwarmed. And if we move out of the fire's line-of-sight, we are not warmed at all.

It is another characteristic of radiant heat that the amount of heat radiated by an open fire falls off drastically as our distance from the fire increases. As those familiar with physics will remember, the amount of heat energy radiated by a fire is inversely proportional to the square of its distance from us. That is, if we were to sit twice as far from a fire as we had been sitting originally, the amount of heat that would reach us would decrease to one-quarter the original amount. If we were to sit three times as far from the fire as we had been sitting originally, the amount of heat that would reach us would decrease to one-ninth the original amount, and so on. The closer we are to a fire, therefore, the warmer we will feel.

Another reason for the poor heat output of a fireplace is that the air within a room is not warmed by an open fire at all. Radiant heat has almost no warming effect at all upon air. Therefore, air currents in the room cannot pick up and distribute the heat that is given off by the fire, as happens with a radiator or convector unit.

In fact, even if the air in a room were warmed a significant amount by an open fireplace (because of direct contact with surfaces that *have* been warmed by the fire), it is unlikely that this warmed air would remain in the room at all. This is because a fire requires a large amount of air if it is to continue burning satisfactorily. It has been estimated that an open fire in a room consumes in one hour an amount of air equal to three to four times the volume of the room. This air must come from somewhere, and this "somewhere" is obviously from outside the room. If the doors and windows are closed, and if there is not a special vent for the fireplace, air for combustion will enter through cracks and openings around the doors, windows, and floorboards. Having a fire going, therefore, means there will be a continual draft, or drafts, of air blowing through the room, which is hardly conducive to comfort in cold weather (and it is one reason, also, why old houses with open fireplaces are so often referred to as "drafty" when cold weather comes). On the other hand, if one is living in a house that has been very thoroughly weatherstripped and insulated, so there are no cracks in the construction at all through which air can enter the house (and assuming again, there is no special vent for the fireplace), it may be impossible to keep a fire going at all because of insufficient air.

Practical Design Requirements

The most basic requirement in the design of any fireplace is not really practical at all but aesthetic—the size of the fireplace must be in keeping with the overall size of the room, not too big, not too small (refer to Fig. 1). If this point is not considered carefully while the house is being planned, the appearance of both the fireplace and the room in which it is located will be spoiled. This is entirely a question of taste and judgment since there are no practical rules or formulas that will help one decide on the correct size of a fireplace for any given room. For example, one writer (Plummer) says that a fireplace opening 36 in. wide by 30 in. high (that is, 7½ sq ft in size) is just about right for a room having 300 sq ft of floor area (a room 15 × 20 ft in size, say). Another writer (Dietz) says that in a room this size a fireplace opening 42 in. wide by 36 in. high (that is, 10½ sq ft in size) is just about right. Who is to say which of them is right?

In fact, there cannot be such a thing as an optimum size for a fireplace unless the dimensions of the room are specified, the height of the flue is known, and the location of the fireplace in the room has been determined. All that one can really aim for is an attractive, well-proportioned appearance, a flue that draws properly at all times, and firesafe construction. Whatever the final size of the fireplace opening, once the proportions of this opening have been decided upon, the other dimensions and proportions of both fireplace and flue will follow after, as described below.

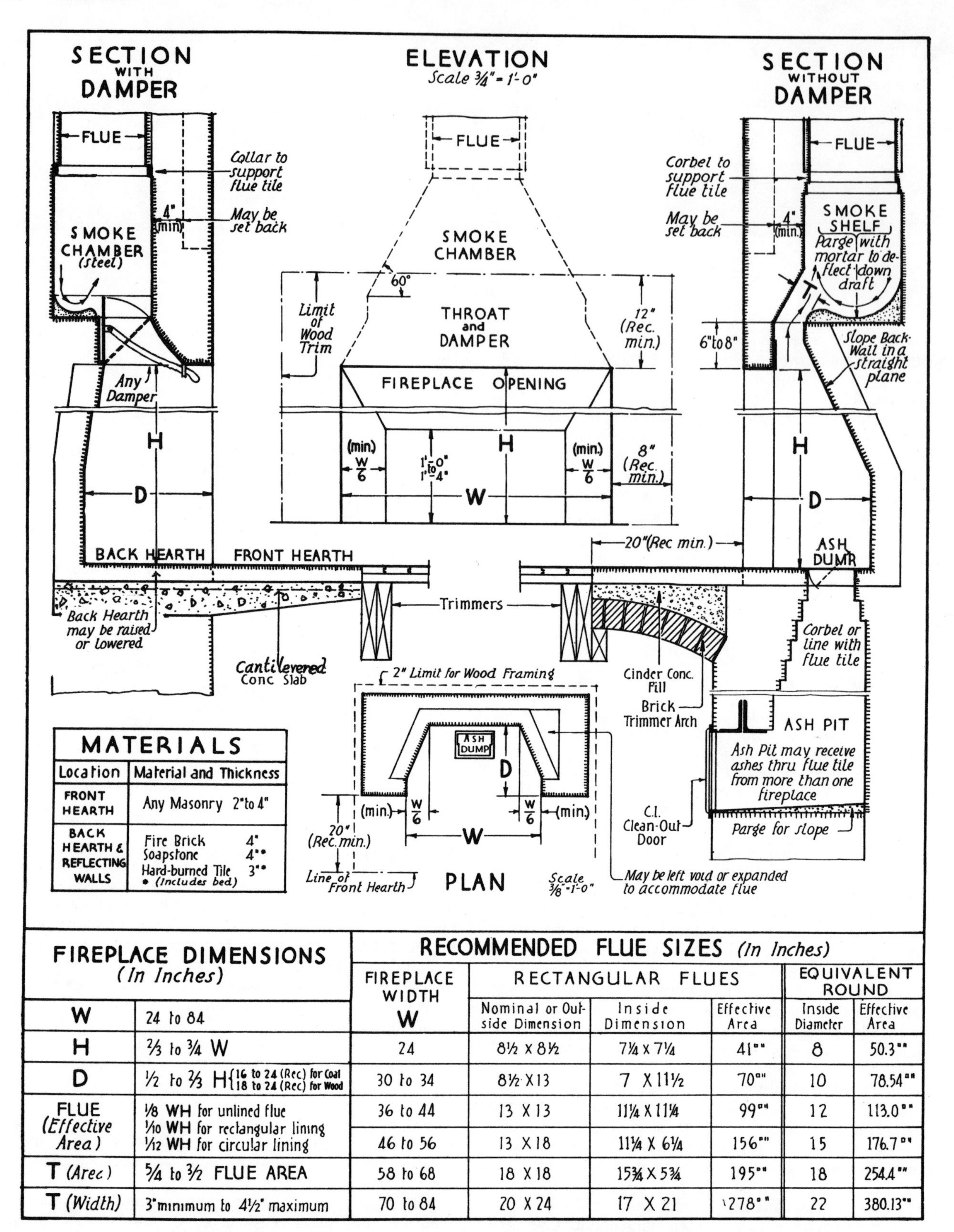

MATERIALS	
Location	Material and Thickness
FRONT HEARTH	Any Masonry 2" to 4"
BACK HEARTH & REFLECTING WALLS	Fire Brick 4" Soapstone 4"• Hard-burned Tile 3"• • (Includes bed)

FIREPLACE DIMENSIONS (In Inches)	
W	24 to 84
H	⅔ to ¾ W
D	½ to ⅔ H {16 to 24 (Rec) for Coal; 18 to 24 (Rec) for Wood}
FLUE (Effective Area)	⅛ WH for unlined flue 1/10 WH for rectangular lining 1/12 WH for circular lining
T (Area)	5/4 to 3/2 FLUE AREA
T (Width)	3" minimum to 4½" maximum

RECOMMENDED FLUE SIZES (In Inches)					
FIREPLACE WIDTH	RECTANGULAR FLUES			EQUIVALENT ROUND	
W	Nominal or Outside Dimension	Inside Dimension	Effective Area	Inside Diameter	Effective Area
24	8½ X 8½	7¼ X 7¼	41""	8	50.3""
30 to 34	8½ X 13	7 X 11½	70""	10	78.54""
36 to 44	13 X 13	11¼ X 11¼	99""	12	113.0""
46 to 56	13 X 18	11¼ X 6¼	156""	15	176.7""
58 to 68	18 X 18	15¾ X 5¾	195""	18	254.4""
70 to 84	20 X 24	17 X 21	278""	22	380.13""

Fig. 1. The proportions of an efficiently operating fireplace. The overall size of the fireplace can change as long as its proportions remain those shown in the illustration *(Callender, Time-Saver Standards, McGraw-Hill)*.

In addition to its size, the position of a fireplace within a room is important, and not only for aesthetic reasons. If one has a choice, the fireplace should preferably be located against an interior wall, not against an exterior wall, as is usually done. If the back of a fireplace is part of an exterior wall, in cold weather a large portion of the hot combustion gases passing up the chimney flue will become chilled, and this chilling will adversely affect the strength of the draft. If the fireplace has not been correctly proportioned in the first place, the end result will probably be downdrafts and a fluttery, smoky fire. (For a discussion of the entire subject of chimney design and flue efficiency, *see* CHIMNEY.)

But when the fireplace is built as part of an interior wall, the efficiency of the draft within the flue will be unaffected by cold outside temperatures, and in addition, a portion of the heat passing up the flue will be transferred through the chimney brickwork to the interior of the house, which will help make the fireplace that much less wasteful of heat.

In any case, a fireplace should not be located where it is likely to be affected by drafts that pass across the front of the fireplace opening. This can happen, for example, if there are doors on both sides of the fireplace. A draft entering through one door and exiting out the other may adversely affect the draft passing up the flue, which may result in a smoky fire. The same thing might happen if there is a warm-air or air-conditioning outlet located on one side of the fireplace and a door located on the opposite side. As a general rule, therefore, one can say that a fireplace located next to a doorway is always badly placed for efficient operation.

Combustion Chamber

The width of the combustion chamber opening is usually between 30 and 42 in., with the most common width being 36 in. The width is often determined by the type of fuel that will be burned. Wood, for example, is sold by the cord, the wood being supplied in 4-ft lengths. For use in a fireplace, the wood must be cut in half, and 2-ft-long logs seem to fit most appropriately in a fireplace that is 36 in. wide. When coal is to be burned, a narrower fireplace is satisfactory since the heat output of coal is much greater than that of wood; and, besides, the coal can be held in a grate.

The height of the combustion chamber opening is usually two-thirds to three-quarters the width. If the opening is made any higher than this there is a good chance that smoke will seep into the room from the top of the opening.

The depth of the combustion chamber is usually one-half to three-quarters the height. The shallower the chamber, the greater the proportion of heat that is thrown out into the room, so the shallower the better. On the other hand, for a wood-burning fireplace, the deeper the combustion chamber, the more wood it can hold and the larger the fire can be. Thus, the depth of the combustion chamber also depends in part of the type of fuel that will be burned. A coal- or gas-burning fireplace can be made much shallower than a wood-burning fireplace having the same size opening. In addition, for safety's sake, to keep burning brands from falling onto the floor, a wood-burning fireplace is usually required to have a minimum depth of 16 to 18 in. This danger is minimized in a coal-burning fireplace by holding the coal in a grate, and it is nonexistent in a gas fireplace.

The back wall of the combustion chamber is usually built vertical for about one-third the height of the opening, and it then begins to slope forward. The slope throws the heat forward into the room, and it also helps guide the combustion gases up toward the flue. The amount of slope will depend on the size and position of the fireplace's *throat,* which is described below.

The side walls of the combustion chamber are angled toward the room slightly to further increase the amount of heat that is thrown into the room. This angle is somewhere between 2 to 5 in. per ft of depth. The steeper the angle, the more effectively is heat radiated into the room. Angling the sides of the combustion chamber serves another purpose also. If the corners were at 90-degree angles, eddies and smoke pockets would form in the corners and interfere with the efficiency of combustion.

Throat

As the sides, back, and top front of the combustion chamber rise toward the flue, the slope of the back wall gradually reduces the cross-sectional area of the combustion chamber to a certain minimum area. This minimum area is the *throat* of the fireplace (refer to Fig. 2), and it is not too much to say that sizing and positioning the throat correctly does more for the overall efficiency of a fireplace than any other single factor in its design. A well-made, accurately sized throat will result in a trouble-free draft. On the other hand, a badly made or incorrectly sized throat will ruin the draft, no matter how well proportioned the remainder of the fireplace may be.

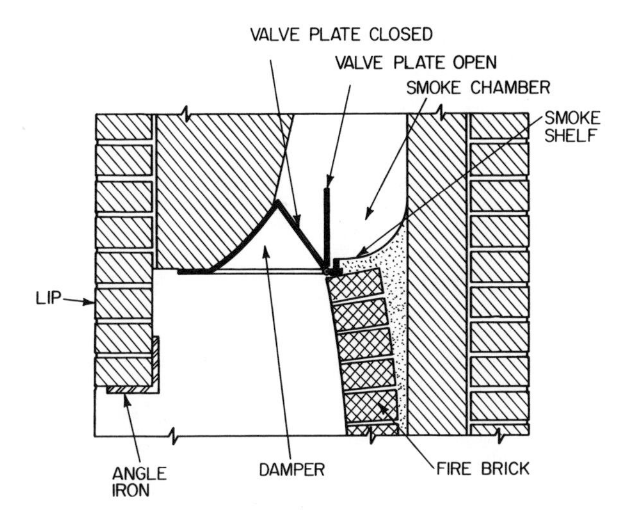

Fig. 2. The throat of a fireplace, showing the relationship between the damper and the smokeshelf.

The throat serves the same function in a fireplace as a venturi (which is a streamlined constriction) does in the test section of a wind tunnel. Its smoothly decreasing cross section induces a faster, smoother flow of combustion gases out of the combustion chamber and into the flue. For maximum combustion efficiency, therefore, the walls of the throat must be smooth and free of obstructions.

The throat should be located from 6 to 8 in. above the top of the combustion chamber opening, no more, no less. In addition, the cross-sectional area of the throat should be somewhere between 1¼ and 1½ the cross-sectional area of the flue. This area will depend in part on the height of the flue, since the shorter a flue, the less well it draws. And the less well it draws, the smaller the throat must be to stimulate a good draft. But whatever the actual size of the throat, it must never be smaller than the flue, otherwise the result will be a constricted flue opening and, therefore, a poor draft and a smoky fire.

The length of the throat is determined by the width of the combustion chamber, the side walls of which usually rise straight up from the floor (or hearth) of the fireplace. The width of the throat is determined by the angle at which the back of the combustion chamber slopes upward. This slope should be such that the width of the throat is between 4 and 8 in., depending on the overall size of the fireplace and the kind of fuel that is going to be burned. Wood-burning fireplaces require larger throats than coal- or gas-burning fireplaces. In any case, it is not the exact size in inches that is important but the relationship in area between the throat and the flue.

Damper

A *damper* (see Fig. 2) is not an absolute necessity for the proper functioning of a fireplace, but the advantages of installing one are so great that all fireplaces should have one. The damper is located at the fireplace's throat, and in fact, it functions as a kind of variable throat. By varying the size of the throat opening, the damper controls the draft of the fireplace according to the kind of fuel that is being burned. For example, pine logs with their high resin and volatile-oil content require a very strong draft if they are to burn properly. A hardwood fire, on the other hand, will burn efficiently with a less powerful draft of air. But whatever the fuel, sometimes one wants a roaring fire and at other times a banked fire. A damper allows one this choice.

A damper has other functions as well. In the winter, when the fireplace is not being used, closing the damper prevents the loss of heat up the chimney. In the summer, closing the damper keeps insects from flying down the chimney into the house.

Most dampers are made with cast-iron frames. A valve sits on top of the frame and is secured to it by a hinge. This valve is opened and closed by means of a handle that projects from the front of the fireplace. The entire assembly should be fitted into the throat in such a way that, when the valve is opened, it will tip backward toward the *smoke shelf,* which is described below.

The overall size of the damper opening should be the same as the throat's. In fact, dampers are made that act as combination throats and dampers. That is, the damper frame is made about 8 in. deep and it is so shaped that it guides the smoke and gases leaving the combustion chamber smoothly toward and through the damper. All the bricklayer need do is set the damper assembly in position and then build the masonry around it.

Smoke Shelf

The smoke shelf (see Fig. 2) is merely a ledge at the same height as the throat. Its function is to keep cold air that may be flowing down the flue from entering the combustion chamber.

Combustion gases rise within a flue in a circular column that spirals upward through the center of the flue. At the same time, a downward-flowing current of cold air hugs the walls of the flue (see Fig. 3). If it weren't for the smoke shelf, this column of cold air would slip past the throat and into the combustion chamber. Even if it didn't succeed in doing so, it would still interfere with the combustion gases rising into the flue, and this would result in a poor draft and a smoky fire. By deflecting the cold air back up again, the smoke shelf prevents this interference.

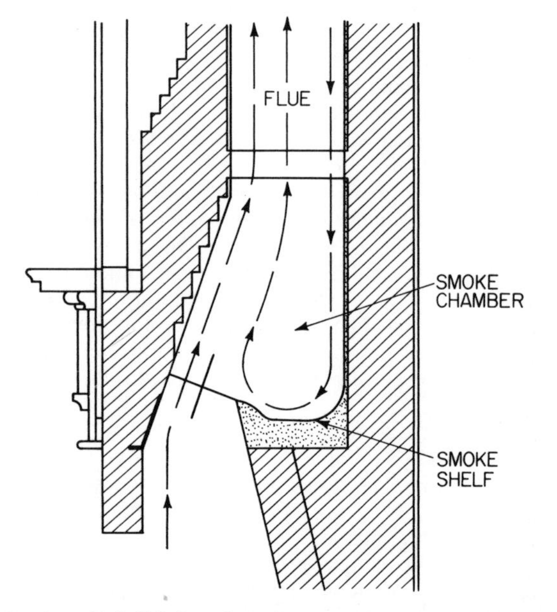

Fig. 3. A smokeshelf deflects down-flowing cold air away from the fireplace.

Smoke Chamber

That part of the fireplace between the smoke shelf and the bottom of the flue is the *smoke chamber.* The smoke chamber is merely a large funnel-shaped space that joins together the throat and the bottom of the flue. It must be correctly proportioned if the smooth flow of combustion gases is to be maintained. The back of the smoke chamber rises straight up from the smoke shelf to the backside of the flue. The sides of the smoke chamber slope toward each other, at the rate of about 7 in. per ft, until they meet the bottom of the flue. The front of the smoke chamber also slopes upward at whatever angle is necessary to have it meet the flue at the same place that the side walls do.

The smoke chamber must be built symmetrically along the centerline of the fireplace. That is, the flue must not be located off-center in relation to the fireplace opening. If it is, the smooth flow of air into the fireplace and the smooth flow of combustion gases out of the fireplace may be upset, which will cause the fire to burn unevenly. If the flue must be offset for any reason, the base of the flue must still be installed on the same centerline as the fireplace and allowed to rise vertically for about 2 ft before it is permitted to angle off to one side.

Flue

For a description of the materials the flue is constructed of, and for the actual construction itself, *see* CHIMNEY. All that need be emphasized here is that the cross-sectional area of the flue will depend on the cross-sectional area of the fireplace opening and on whether the flue is to be *lined* or *unlined.* (For a description, *see* CHIMNEY.)

If the flue is to be *unlined,* its area must not be less than one-eighth the area of the fireplace opening, and in any case not less than 100 sq in.

If the flue is *lined* and either *square* or *rectangular,* its area must not be less than one-tenth the area of the fireplace opening, and in any case not less than 64 sq in.

If the flue is *lined* and *round,* its area must not be less than one-twelfth the area of the fireplace opening, and in any case not less than 50 sq in.

FIREPLACE CONSTRUCTION

A fireplace and its chimney can be built either at the same time as the rest of the house is being built or it can be built any time after the rest of the house has been completed. In a masonry house it is obviously more sensible to build the fireplace/chimney at the same time as the walls are being built. But in a wood-frame house, it doesn't make as much difference because the fireplace and its chimney are separated completely from the rest of the construction (quite literally) and two different sets of craftsmen—carpenters and bricklayers—work on each. In a wood-frame house, the wall, floor, and roof framing can be completely finished if necessary, or desired, before construction of the fireplace/chimney is begun. It would still be necessary, of course, to construct the foundations for the walls and the fireplace/chimney at the same time.

It is also possible for one to build the fireplace and the chimney at completely different times, the chimney being built at the same time as the rest of the house, with the fireplace being built at some convenient time in the future. The builder need only leave a gap in the front of the chimney construction in which the fireplace can be built, the weight of the chimney above the fireplace opening being supported temporarily by a lintel.

The reason that a fireplace is often left for construction later is that if it is to be built of expensive (and fragile) tile or terra cotta, the builder will not want to take the chance of having this material dirtied or damaged accidentially by the workmen. The big problem with leaving a fireplace for later construction is that a gap is almost sure to open up at a later date where the fireplace and chimney meet because of differences in the rates at which the fireplace and chimney settle. If such a gap should open up it might become a fire hazard.

Fireplaces are traditionally constructed of brick, although any solid masonry material is satisfactory. The exception is hollow concrete blocks which do not have the strength to support the weight of both the fireplace and the chimney. Concrete blocks can be used, however, if their hollow cores are packed solidly with concrete or mortar as construction of the fireplace progresses, and if the concrete blocks are reinforced with steel bars. (*See* CONCRETE-BLOCK CONSTRUCTION.)

Fireplaces can also be made of reinforced concrete quite easily as long as a suitable formwork is prepared beforehand. Prefabricated *modified* fireplaces are available (see below) that simplify considerably the construction of a reinforced-concrete fireplace. The modified fireplace is set in position, and a suitable form is constructed around it, a space 4 to 8 in. wide being left between the fireplace and the formwork. The concrete is poured into this space and allowed to set. Afterward, a decorative brick or tile veneer can be built over the concrete to disguise it and provide a decorative finish.

If a fireplace is being built of brick in the traditional manner, then if at all possible a bricklayer who is experienced in fireplace construction should be hired to do the work. There are a great many niceties about brick-fireplace construction that take time to learn and skill to do properly. The bricklayer must also be a first-class mechanic because sloppily done brickwork shows clearly in a fireplace. If an experienced bricklayer cannot be found, or if a do-it-your-selfer wants to try his hand at constructing a fireplace, then installing a prefabricated fireplace shell beforehand will simplify both the amount of work that has to be done and the amount of skill required to do it.

Clearances and Firestopping

The structural framework of a wood-frame house must not be allowed to come into direct contact with any part of a fireplace or chimney. If the brickwork is 4 in. thick, the studs and joists must not be closer than 6 in. to the chimney, closer than 4 in. to the back of the fireplace, or closer than 2 in. to any other outside surface of the fireplace. If the brickwork is 8 in. thick, these gaps can be as little as ½ in. A floor or subfloor made of wood, or any other combustible material, should not come closer than ¾ in. to the fireplace.

Furthermore, the gaps between the floor construction and the surface of the fireplace and chimney must be *firestopped;* that is, the gaps must be closed by a noncombustible material to prevent a fire in the house traveling to the upper floors by way of these gaps. The gaps are usually closed, or firestopped, by nailing strips of sheet metal to the bottoms of the joists adjacent to the fireplace/chimney. About 1 in. of a noncombustible material such as mineral wool is then laid on top of the metal strips to act as insulation. *See also* CHIMNEY.

Hearth

There are two parts to a hearth—the *back* hearth, which is actually the floor of the combustion chamber, and the *front* hearth (or *finished* hearth), which extends out into the room. We are interested here in the front, or finished, hearth.

Building codes specify strictly the kinds of materials that a hearth can be made of and the overall size of the hearth, as there is always the danger that a burning brand may tumble out of the fireplace or that a spark may fly out of the fire. Either may set fire to any inflammable materials that happen to be near the fireplace. The purpose of the front hearth is to provide a safety zone between the fireplace opening and the room proper that will keep accidents from happening.

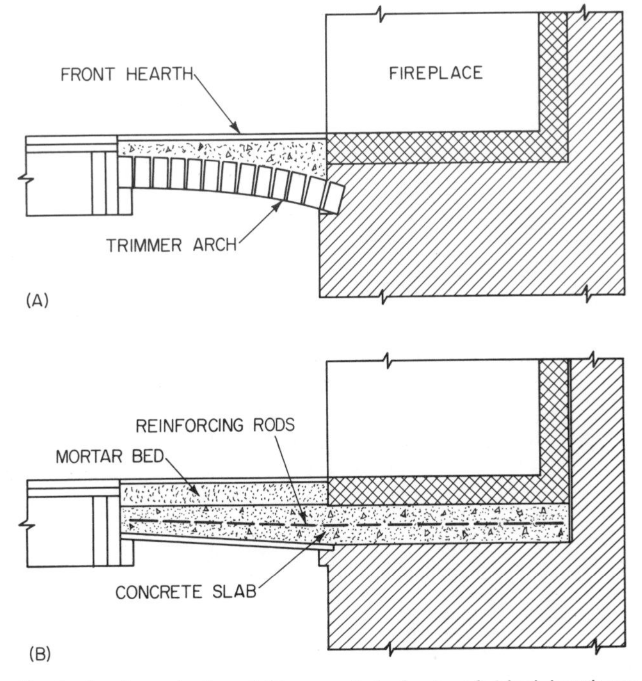

Fig. 4. In a house having a full basement, the front, or finished, hearth can be supported by either a trimmer arch or a reinforced-concrete slab.

If the overall size of the fireplace opening is less than 6 sq ft, the hearth must extend at least 16 in. beyond the opening and at least 8 in. beyond each side of the opening. If the size of the fireplace opening is 6 sq ft or more, the hearth must extend at least 20 in. beyond the opening and at least 12 in. beyond each side of the opening.

The hearth must be made of a noncombustible material such as tile, brick, cut stone, or concrete. This material must be at least 4 in. thick. It must also be adequately supported. In a house that has either a full basement or a crawl space under the hearth, this support is provided either by a reinforced-concrete slab or a brick arch called a *trimmer arch.* The constructions are shown in Fig. 4.

In a house built on a concrete slab, any kind of additional support for the hearth is unnecessary (*see* SLAB-ON-GRADE).

MODIFIED FIREPLACE

A *modified* fireplace (refer to Fig. 5) is a complete, prefabricated unit that consists of a combustion chamber, throat, damper, smoke shelf, and smoke chamber. This unit is made of sheet steel, with the metal making up the combustion chamber being at least ¼ in. thick. The unit can be set into position as-is with the brickwork then being built up around it.

There are many advantages to the purchase of a modified fireplace for the amateur fireplace builder, and for the skilled bricklayer as well. Most important, the homeowner is assured of a correctly proportioned fireplace that has been designed to achieve the maximum amount of heat output and a trouble-free draft, although it should be added that a prefabricated fireplace of this type has no inherent advantages over an all-masonry fireplace constructed by a skilled bricklayer. And the

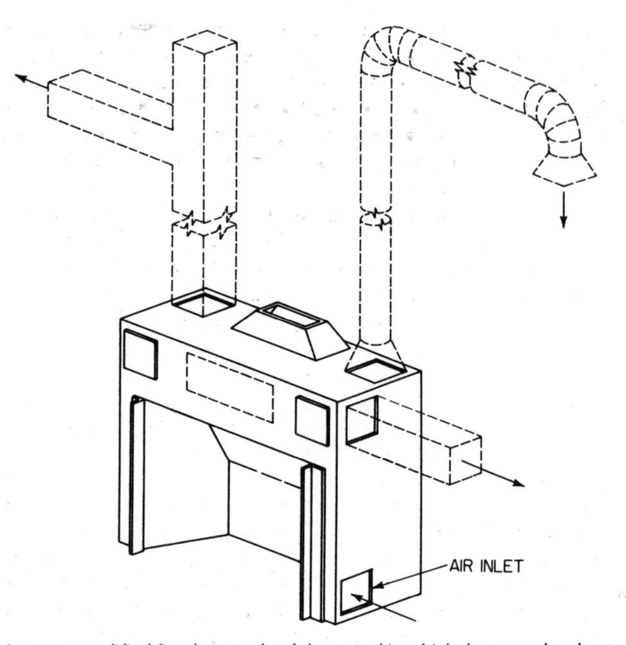

Fig. 5. A modified fireplace made of sheet steel in which ducts conduct heat to other parts of the roof.

operation of a prefabricated fireplace, like all fireplaces, will also depend in large part on having a correctly sized and installed flue within the chimney.

The uniqueness of a modified fireplace, however, lies in the fact that the combustion chamber is double-walled, which allows air to circulate behind the combustion chamber. As shown in Fig. 5, cold air can enter the interior of the fireplace through two cold-air ducts. The air is heated by direct contact with the hot surfaces of the combustion chamber and it is then discharged into the room through two warm-air ducts. This modification to the fireplace design increases enormously the heat output of the fireplace, making it the equal of a conventional stove in heat output, and also making it possible for the fireplace to heat the room in which it is installed adequately and without any increase in fuel consumption. In temperate climates in which the nights are often chilly but never cold, a modified fireplace of this type may very well prove more convenient and cheaper to install and operate than a central heating system.

All modified fireplaces are tested and, if approved, listed by Underwriters' Laboratories. The UL listing will specify exactly how the fireplace must be installed.

PREFABRICATED FIREPLACES

There is another type of prefabricated fireplace design that is quite popular in homes of contemporary design that consists of nothing more than an insulated shell. These prefabricated fireplaces are available in a wide variety of free-standing designs. They can be purchased complete with hearths and insulated metal chimneys. Unlike the traditional fireplaces that have been described in this article, these prefabricated fireplaces are lightweight, ingenious in design, and they can be installed almost anywhere in a room without the need for a reinforced foundation, additional masonry work, or fireproofing. Their light weight enables them to be placed on a wood-frame floor without exceeding the designed dead-load limit of the usual wood-frame floor construction.

Whatever prefabricated fireplace a homeowner selects, it should have been tested and listed by Underwriters' Laboratories. They must be installed in accordance with the manufacturer's instructions.

Flashing

Flashing is both a construction material and a construction technique. As a material, flashing consists of a flat sheet of waterproof material, made either of metal, plastic, or asphalt-impregnated fabric. As a construction technique, flashing is a method of installing this waterproof material in such a way that it will divert away from the interior of the house any rain that may otherwise find its way through a gap in the construction.

All houses—all buildings, for that matter—are inherently leaky, especially at junctures exposed to the weather. These locations include the places where walls meet roof, where roof and chimneys intersect, where two sloping roof surfaces meet, at door and window openings, and wherever two different kinds of material abut each other. Leaks occur because of the nature of the materials out of which houses are built. The wood in a newly constructed house will shrink over a period of time, sometimes a considerable amount. In all wood-frame houses, there is a natural tendency for the wood to swell and shrink in response to changing humidity conditions. In a wood-frame house, therefore, wherever two flat wood surfaces meet, there is a natural tendency for these surfaces to separate from each other, opening up a gap through which rain may enter.

In a masonry house, the masonry tends to expand and contract with large changes in temperature. This movement, which is ordinarily imperceptible to the eye, may in time open up cracks in the mortar joints, especially alongside doors and windows. Cracks also tend to open up where masonry and wood constructions adjoin each other; for example, where a stuccoed wall meets a wood siding.

There are many other places in the structures of both wood-frame and masonry houses through which rain may enter a house; they are described below. The point we wish to make here is that this leakage cannot be prevented by better workmanship or the use of better materials or different construction techniques. The leakage occurs because of the inherent tendency of both wood and masonry to expand and contract. It occurs because it is impossible to seal completely a juncture where two different types of material abut each other. In addition, in even the best-made houses, gaps may open up in the structure because the foundation settles slightly over a period of time or because strong winds may rack and twist the house frame.

Leakages cannot be prevented by brute force. But we can outwit the rain, so to speak, by installing suitable flashing materials over all gaps or potential gaps in the structure where rain may otherwise enter; then the water is directed away from the gaps, and the building is free to expand and contract, to twist or settle according to the nature of the materials used in the construction.

FLASHING MATERIALS

Flashings are made of a wide variety of metals, bituminous-saturated fabrics, plastics, plus combinations of these materials. Whatever the material, it must have certain properties if it is to make a satisfactory flashing. Above all, the material must be waterproof. Second, it must be long-lasting, which means highly resistant to damage by corrosion and to the effects of weathering and of extreme temperature changes. The material must be flexible enough so that it will accommodate itself to slight shifts in the position of the underlying structure without tearing or cracking. It must be pliant enough so that it can be folded or worked into any desired shape without cracking.

There are few places in the structure of a house where the quality of the materials makes such a long-term difference to the life of the house as does the quality of the flashing, particularly when one considers the cost of the flashing compared to the overall cost of the house. An initially expensive but excellent flashing material like copper, for example, that will last the lifetime of a house, will prove far less expensive in the long run than a cheaper flashing material that may crack or corrode through in a few years. The cost of replacing this cheaper flashing (including the cost of disassembling and replacing the roofing and/or siding and/or masonry surrounding it) may be many times the cost of having installed the more expensive flashing material in the first place. But it is the first cost that is important to a builder, not the long-term costs. Therefore, anyone who is considering buying a particular house should examine the exposed flashings carefully, particularly the flashing around the chimney, because the quality of the flashings will prove an excellent indication of the overall quality of construction.

The selection of a flashing material for a particular location depends on whether the flashing will be concealed entirely within the house structure (and thus protected from the weather and from extreme temperature changes) or whether it will be exposed, in whole or in part. Exposed locations include almost all flashings installed on the roof and along roof-wall and roof-chimney junctures. Flashing that will be buried within a wall but with a portion extending out of the wall a short distance so that it may act as a *water drip* is also considered to be exposed. Metals are far and away the preferred materials for use in all exposed locations.

Concealed locations include mainly the walls abutting door and window openings, as well as those roof locations where the flashing will be completely buried within brickwork or covered by the roofing material. For these concealed locations, materials other than metals can also be used as flashing.

Copper

Copper is the preferred material for all flashing. It is ductile and easily shaped; it is highly resistant to atmospheric corrosion as well as to most other types of corrosion; it is strong yet flexible enough to accommodate itself without cracking to shifts in the position of the underlying structure; and it is manufactured in a variety of special shapes especially designed for specific applications.

The copper used for flashing is cold rolled. Annealed (soft or *roofing-temper*) copper should not be used since it may prove less durable in service. For all exposed flashing, the copper should be 16-oz material (i.e., 0.020 in. thick), which means that 1 sq ft weighs 16 oz. For concealed locations, 10 oz (i.e., 0.012 in. thick) copper is satisfactory.

Combination flashing materials are also available for use in concealed locations that consist of thin sheets of copper or lead-coated copper attached to a backing made of kraft paper or an asphalt-impregnated cloth. The copper may also be sandwiched between two layers of this backing material. The copper used in these combination flashings may weigh 1, 2, 3, 5, or 7 oz per sq ft. The copper is made in different weights in order to meet the minimum requirements of different building codes, which have different standards. In general, of course, the thicker the metal, the better.

Copper can be used anywhere in a wood-frame house. It is equally satisfactory in masonry walls, because it is unaffected by the alkaline substances that are normally present in mortars. Thus, it can be embedded in fresh mortar without any danger of its corroding.

One must be careful, however, never to install another metal (except stainless steel or lead) in direct contact with copper any place in the construction, as the other metal will be preferentially corroded. This means that only copper-alloy nails and copper-alloy fastenings should be used to attach copper flashings in place.

Copper develops a green patina, a tarnish, that is considered attractive but which may cause problems on a masonry wall if this tarnish washes down the wall and stains the masonry. The effect is particularly unwelcome on light-colored masonry. To prevent staining, copper sheets coated with lead are available. They can be used in place of plain copper flashings whenever the flashings are exposed above a masonry wall. Or one can install generous ¾ to 1-in.-wide drip edges that will throw water clear of the masonry.

Lead

Lead is as satisfactory a flashing material as copper but because of its cost it is little used today. Lead's most outstanding virtue as a flashing material is its extreme ductility, which allows it to be cut or shaped so that it fits closely against an underlying support. Like copper, lead is flexible enough that it can adjust itself easily to shifts in the underlying structure without cracking. It is also highly resistant to atmospheric corrosion. Unlike copper, lead can be attacked by fresh mortar, or by old mortar if it is damp. Lead, therefore, must first be protected by a coating of asphalt if it is to be installed in or adjacent to masonry.

For exposed locations, sheet lead weighing at least 2½ lb per sq ft (i.e., approximately 0.039 in. thick) should be used. For concealed locations, lead weighing at least 1½ lb per sq ft (i.e., approximately 0.023 in. thick) is satisfactory. Lead flashings should be held in place with fastenings made either of copper or galvanized steel.

Aluminum

Aluminum is another satisfactory flashing material. Like lead, however, it is attacked by the alkalies in mortar and, therefore, cannot be embedded in masonry or placed against a masonry surface without first being coated with asphalt. Another potential difficulty with exposed aluminum flashings is that it is likely to corrode if exposed to salt air. Aluminum should not, therefore, be used in a house located near the ocean.

For exposed locations, aluminum having a minimum thickness of 0.019 in. and a minimum tensile strength of 16,000 lb per sq in. should be used. For concealed locations, aluminum having a minimum thickness of 0.015 in. and a minimum tensile strength of 14,000 lb per sq in. should be used.

There is also available a combination flashing material that consists of a thin sheet of aluminum sandwiched between two layers of asphalt-saturated cloth. The aluminum is either 0.004, 0.005, or 0.016 in. thick. This combination flashing can be used in all concealed locations and can even be embedded in mortar.

The fastenings used with aluminum flashings must be made either of an aluminum alloy or stainless steel; fastenings made of other metals are likely to corrode the aluminum. And because of aluminum's high coefficient of expansion, it should never be nailed tightly in place but fastened so that it can expand and contract with changes in temperature.

Stainless Steel

Stainless steel has the reputation of being a stiff metal, difficult to work, but in the thicknesses in which it is used for flashing it is pliable enough to be worked easily into corners and against irregular surfaces. Different varieties of stainless steel differ in their degree of resistance to corrosion. The preferred material for flashings is AISI type 304 stainless steel having a No. 2B finish.

This material makes extremely effective flashing. It has none of the limitations regarding usage that have been mentioned for copper, lead, and aluminum. That is, it is completely unaffected by the alkalies in mortar and by salt air; nor will it stain adjacent surfaces.

For exposed locations, the stainless steel should be at least 30 gauge. For concealed locations, it should be at least 28 gauge.

When used in contact with another metal (except aluminum), the contacting surfaces should be coated with asphalt to prevent the preferential corrosion of the other metal. All fastenings used with stainless steel should be made of stainless steel.

Galvanized Steel

Galvanized (i.e., zinc-coated) steel is perhaps the most widely used of the metal flashing materials because it is the least expensive. Galvanized steel also makes the least satisfactory flashing because it must be inspected at periodic intervals to make sure it has not started to corrode. Exposed galvanized steel flashings must also be painted at periodic intervals to maintain the integrity of the zinc coating.

Galvanized steel cannot be embedded in mortar or placed against a masonry wall, as the metal will be corroded by the alkalies in the mortar. Not that the zinc itself is affected by the alkalies, but if there should be pinholes or small cracks in the zinc coating (and there probably will be), the base metal will be attacked. As with lead and aluminum flashings, therefore, galvanized-steel flashings must first be coated with asphalt if they are to be embedded in or placed against masonry. In addition, if galvanized steel is to be placed in contact with another metal, the contacting surfaces must be protected against corrosion by painting them with an asphalt coating.

For use in exposed locations, the base metal must be at least 26 gauge. For use in concealed locations, the base metal must be at least 28 gauge. There are two thicknesses of zinc coating that may be applied on the base metal, one of which weighs 1¼ oz per sq ft (coated both sides) and the other 1½ oz per sq ft (coated both sides). The heavier coating is preferred for both exposed and concealed locations; the thinner coating can be used satisfactorily for concealed locations only. The fastenings used with galvanized-steel flashings must also be galvanized.

Terneplate

Terneplate is sheet steel that has been coated on both sides with a lead-tin alloy. The metal thus has the strength and mechanical qualities of sheet steel combined with the corrosion-resistant properties of the lead-tin mixture. For use as flashing, the base metal must be at least 30 gauge, with the minimum weight of the coating being 1.45 oz per sq ft, coated both sides. The fastenings used with terneplate should be made either of terneplate or galvanized steel.

Bituminous Flashing Materials

There are no especially manufactured bituminous flashing materials, per se, apart from the combination flashing materials described above. The bituminous products that are used as flashings are the standard asphalt-saturated cloths and felts that are used to cover flat roofs (for a description of these products *see* BUILT-UP ROOFING). These same materials are usually installed as flashings along roof valleys and the eaves of roofs, especially when the roofing material also consists of asphalt shingles or asphalt roll roofing. Bituminous flashings are also used over door and window frames, under window sills, and as a base flashing (or damp check) in masonry-wall construction. In general, bituminous materials are used as flashings mainly in houses in which keeping the costs down is the major consideration.

The main problem with using bituminous flashing materials is they they become hard and brittle over a period of time, especially in climates that have very cold winters. When cold, they crack quite easily if they are stressed. These flashings are also soft and weak (especially in hot weather) and a workman can puncture or tear a sheet of the material quite easily while he is installing it, without realizing he has done so. These holes or tears, of course, reduce the effectiveness of the flashing considerably. Great care should always be taken when handling and cutting this material.

Metal flashing materials are always preferred to bituminuous flashing materials. The rationale for using bituminous flashings at all, apart from their low cost, goes something like this. They are convenient to install when the roofing material itself is also made of a bituminous material. Since this roofing material will last for only 10 to 20 years before it must be replaced, the flashing can be replaced at the same time at very little additional cost. There is no point, therefore, in spending money on an expensive metal flashing, no matter how long-lasting it may be. But the homeowner should appreciate that it is the builder who saves money in the long run by installing bituminous flashings, not the person he has sold the house to.

Plastic Flashings

Plastic sheet materials, particularly thin films of polyethylene and polyvinyl chloride, are being used more and more as flashing materials. The polyvinyl chloride film is usually 60 mils (i.e., 0.060 in.) thick, and the polyethylene is usually 6 mils (i.e., 0.006 in.) thick. These and other plastic flashing materials have in common all the virtues of plastic building materials. That is, they are completely waterproof, chemically inert, noncorrosive, flexible, and strips can be joined together very easily using special cements.

Whether or not plastics can or should be used as flashing depends at the present time on the confidence the builder has in these materials, since there are not as yet any industrywide or government standards regarding strength, minimum thick ness, method of installation, the composition of the plastics, or their expected life under service conditions. All one has to go on really is the public reputation of plastics. Caveat emptor!

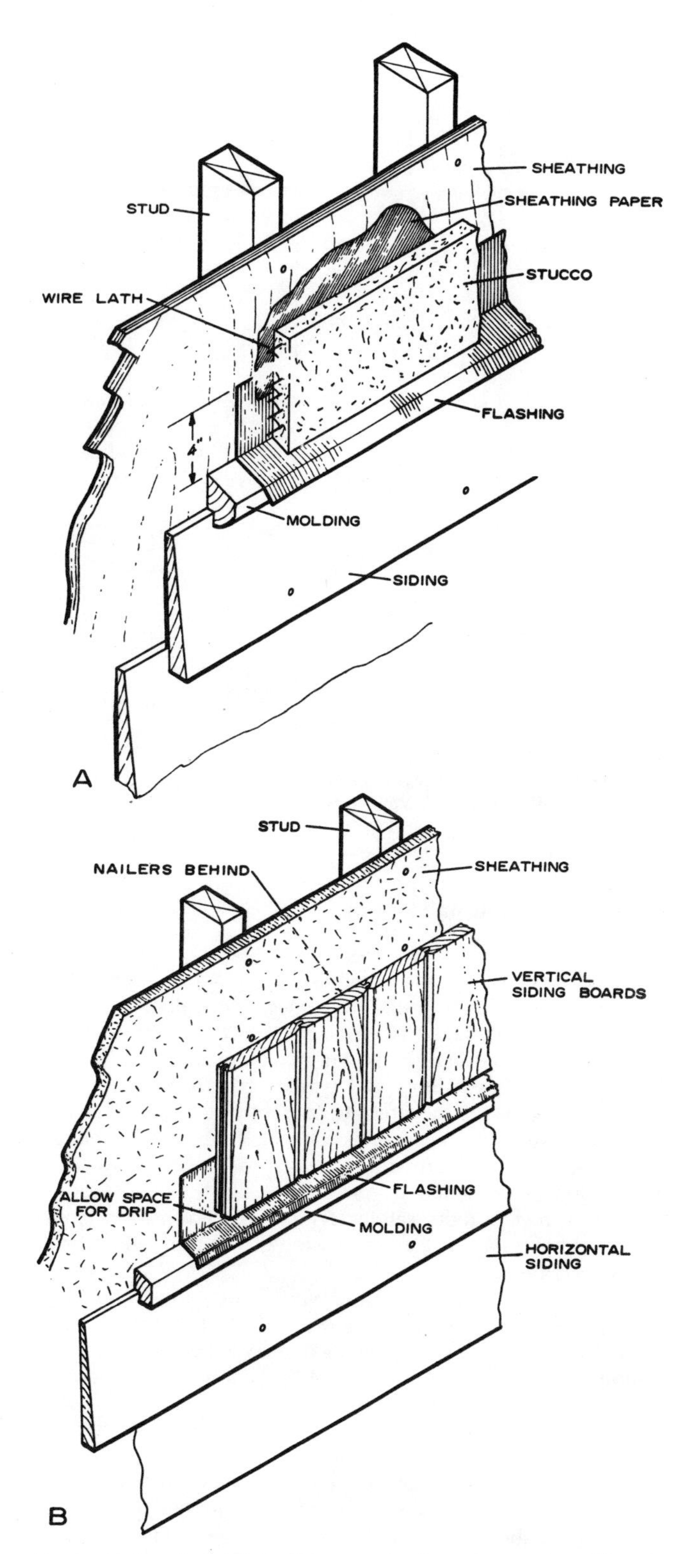

Fig. 1. Flashing is installed within a construction wherever two different materials meet or wherever there is a gap in the construction *(U.S. Forest Service).*

WALL FLASHINGS

Wall flashings (see Fig. 1) must be installed in homes located in those sections of the United States where the rainfall is either quite heavy and the winds strong (that is, along the Gulf of Mexico, the Atlantic coast, the Great Lakes region, and the Pacific Northwest) or where rainfall and winds are only moderately heavy and strong (that is, throughout the Midwest). In those sections of the United States where the rainfall tends to be light or practically nonexistent (that is, in the Mountain states and in the Southwest), wall flashings are optional and dependent on local conditions.

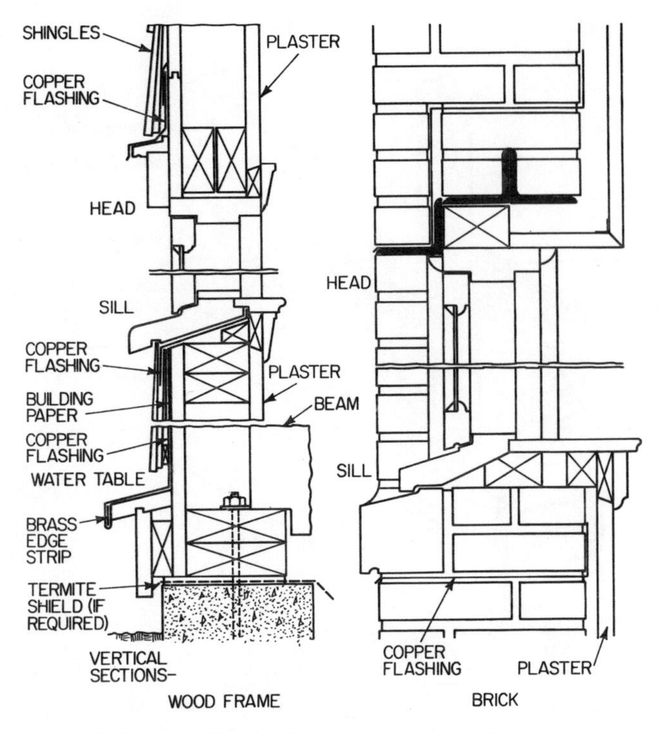

Fig. 2. Flashings must be placed above and below all wall openings and at the bases of all walls as well. Wood-frame and brick shown *(Copper Development Association).*

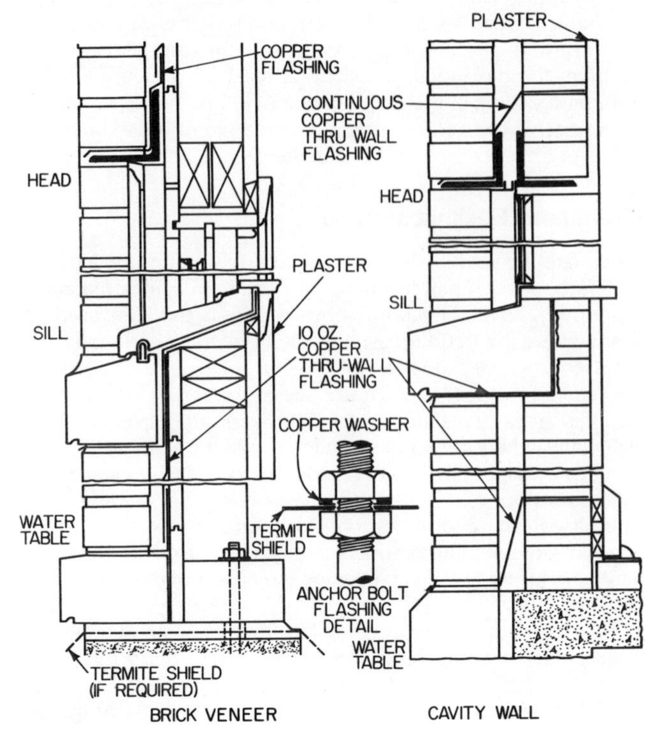

Fig. 3. Flashings in brick-veneer and cavity-wall construction must also be placed in all the locations given in Figure 2.

Door and Window Flashing

Figures 2 and 3 show how flashing is placed above and below all doors, windows, and other wall openings in both wood-frame and masonry walls. In a masonry wall the flashing should always be laid on top of a thin bed of fresh mortar with another thin layer of fresh mortar applied over the flashing.

Flashings above doors and windows must always extend beyond the ends of the jambs to prevent water making its way into the house around the ends of the flashing. In addition, the flashings should be turned up at least 1 in. The 2-in. turned-up edge under the sheathing shown in the illustrations is a minimum; 4 in. is a more satisfactory construction as the longer overlap makes it less likely that wind-driven rain will force its way past the flashing.

The flashings above door and window openings should also extend out from the wall by about ¼ in. and be turned down slightly to form drips. The slight gap between wall and flashing should not be caulked, as is often recommended, because caulking only prevents any water that may have found its way into the wall from making its way out again. The sides of door and window openings are not flashed, however, and these joints in the construction should be caulked, in both wood-frame and masonry walls, to prevent the entrance of moisture.

Flashing above windows and doors can be omitted when the amount that the roof overhangs the wall is at least four times greater than the distance between the tops of the doors and windows and the bottom of the eaves.

Base Flashing

In any wall, moisture tends to migrate up from the foundation into the wall. A *base flashing* (or *damp check*) is required at the top of the foundation walls, therefore, to block this diffusion of moisture into the wall construction. The methods of installing a base flashing are shown in Figs. 4, 5, and 6.

Any of the noncorrodible metals mentioned above can be used as base flashing, which is one of the concealed flashings in a house. A bituminous flashing made of 30-lb, asphalt-saturated felt is also satisfactory, despite some slight tendency for the flashing to be squeezed out of the wall.

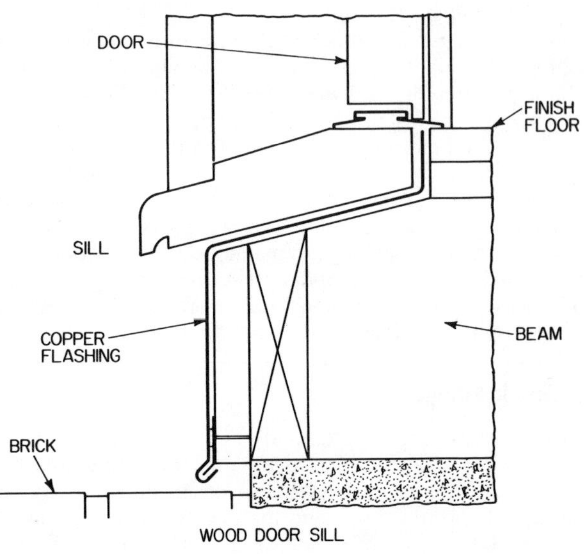

Fig. 4. The installation of a base flashing under a wood-frame wall *(Copper Development Association).*

ROOF-WALL JUNCTURES

Pitched-Roof Flashing

Where the edge of a pitched roof abuts a wall (see Fig. 7), the juncture is made watertight by installing pieces of flashing

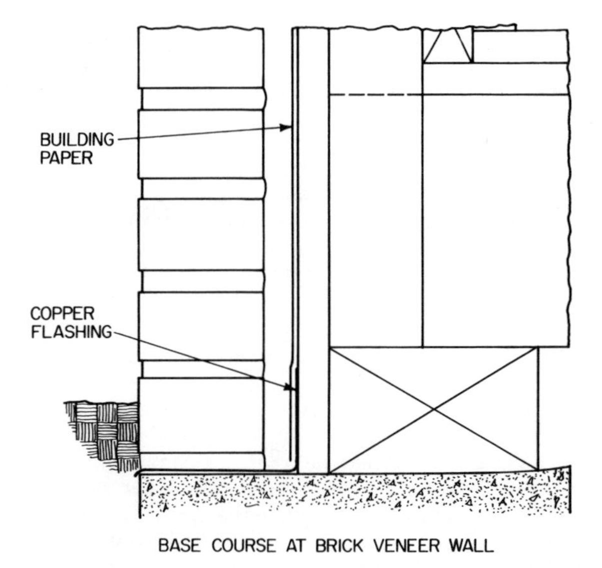

Fig. 5. The installation of a base flashing under a brick-veneer wall *(Copper Development Association).*

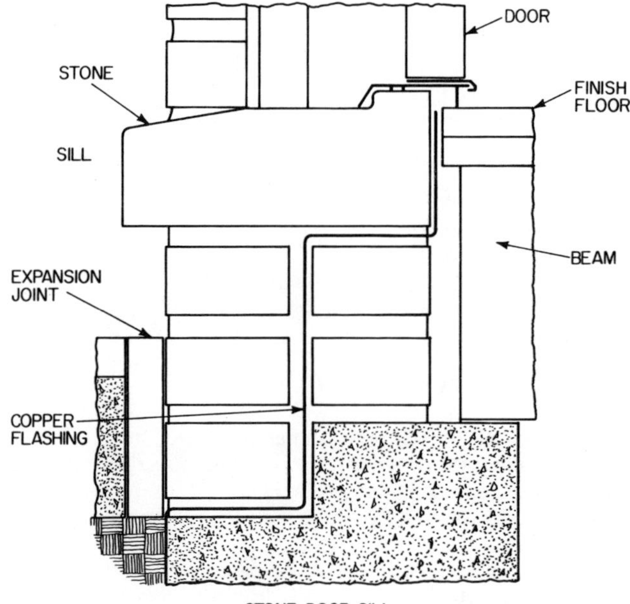

Fig. 6. The installation of a base flashing under a brick-masonry wall *(Copper Development Association).*

material cut to the approximate size of the roof shingles. Each flashing is made about 7 in. wide and 3 in. longer than the shingles. Each flashing is then bent lengthwise so that about 4 in. of its width will lie flat upon the roof.

The method of installing the flashing is called *step flashing.* The first piece of flashing along the bottom of the roof is set in position so that it extends slightly beyond the eave. A shingle is nailed in place on top of the flashing and the rest of the first course of shingles is then installed (*see* SHINGLE, ASPHALT; SHINGLE, WOOD).

A second piece of flashing is then set in position against the wall and the second course of shingles is then installed, and so on, until the top of the roof has been reached. The result of this step method of installing the flashings is that the flashings are sandwiched between adjacent courses of shingles, and they also overlap each other in such a way that they redirect to the surface of the roof any water that may have worked its way under the shingles.

It is still possible for rain washing down a wood-framed wall to work its way under the flashings. If the wall is covered either by shingles or siding, leakage is prevented by having the shingles or sidings overlap the flashing when they are installed, as shown in Fig. 7. Thus, water running down the wall is directed over the flashings.

If the wall is made of masonry, then another row of flashing, called *cap* flashing, or *counterflashing,* must be installed in the masonry that will overlap the flashings installed under the roof shingles. Strips of counterflashing are inserted into the mortar joints as shown in Fig. 7. In new construction, each strip of counterflashing is inserted at least 1 in. into the mortar joint. As shown in the illustration, these strips of flashing are inserted into the brickwork in step fashion to conform to the slope of

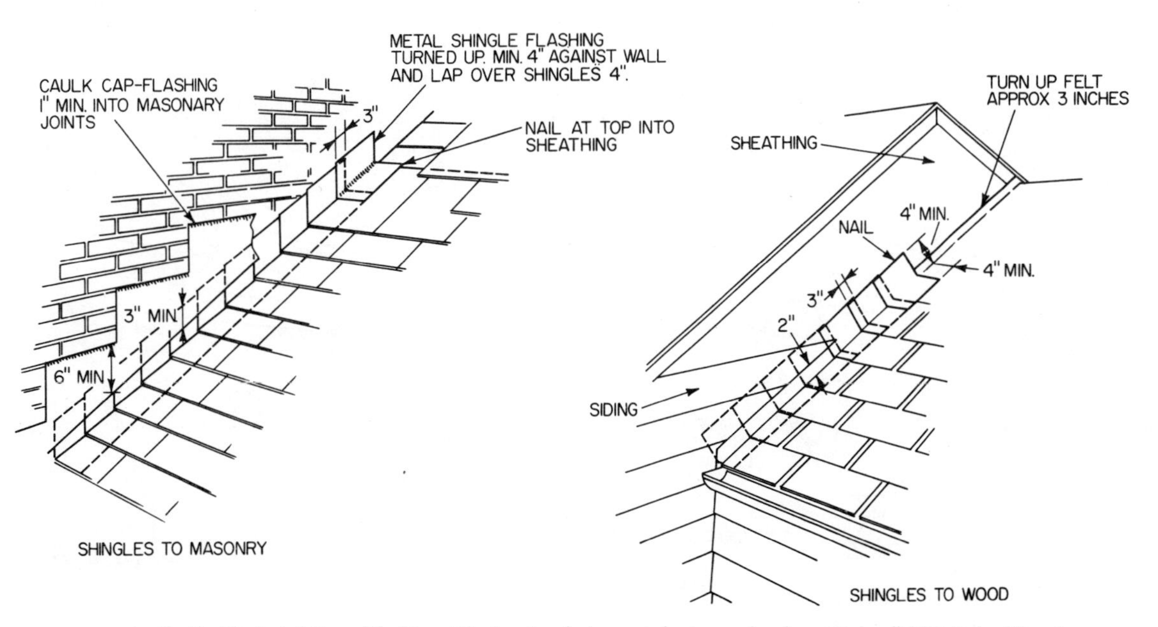

Fig. 7. The installation of flashing at the juncture between a sloping roof and a vertical wall *(U.S. Federal Housing Administration).*

the roof. If the mortar joints must be chiseled or sawed out to install the counterflashing, the cut must be at least 1½ in. deep. After the counterflashing has been installed, they are held in place by lead wedges driven into the joints. The joints are then filled solidly with either mortar, asphalt, or caulking compound.

Finally, the portion of the counterflashing extending out from the wall is bent down over the base flashing. The base flashing and counterflashing should overlap by at least 4 in. In addition, the bottom edges of the counterflashing should be cut at an angle to conform to the slope of the roof.

ROOF-VALLEY FLASHING

Where the two surfaces of a pitched roof meet, they form a V-shaped *valley*. When it rains, the water running down the roof tends to concentrate in this valley. Flashing must be installed in the valley, therefore, to prevent any leakage through the roof construction.

How this flashing is installed will depend on (1) the slope of the roof, (2) whether the runoffs from the two surfaces are equal or unequal, (3) whether the roofing material will be installed in the valley in an open or closed manner, and (4) the kind of roofing material that will be installed. For the sake of simplifying the discussion, we will assume that either asphalt or wood shingles are to be installed on the roof.

Sheet-Metal Flashing (See Fig. 8)

The width of the flashing will depend on the slope of the roof since, naturally enough, the steeper the roof, the faster the water will run off and the less the likelihood that it will run in a broad stream through the valley. Therefore, the width of the flashing will depend on the slope of the roof as follows:

1. For slopes less than 4 in 12, the flashing should be at least 24 in. wide.
2. For slopes between 4 in 12 and 7 in 12, the flashing should be at least 18 in. wide.
3. For slopes greater than 7 in 12, the flashing should be at least 12 in. wide.

The flashing should consist of one continuous strip from the top of the roof to the bottom (except for copper, the strips of which should be not more than 8 ft long; these strips should then be overlapped by at least 6 in.). If necessary, short lengths of metal can be soldered or brazed together, or the ends of short lengths can be overlapped and crimped together to form watertight joints. The sides of the flashing should be bent up and over about ½ in. to form barriers that will catch and turn away any water that may have found its way under the roofing.

Although the flashing (except for copper and aluminum) may be nailed loosely in place using a minimum number of nails, the preferred method of attaching metal flashing to the roof is with *nailing cleats* that engage the turned-over edges of the flashing as shown in Fig. 8. This method of attachment not only holds the flashing in place, it also allows the flashing to expand and contract.

If one of the roof surfaces has a larger area than the other, or if one surface has a steeper slope than the other, the larger or steeper surface will discharge a larger quantity of water into the valley. During a heavy rainstorm, this water may overrun the flashing covering the smaller, or shallower, roof surface. To prevent this possibility, the center of the flashing should have a 1-in.-high, inverted V crimp that will form a barrier between the two sides of the valley.

The builder has the option of laying the shingles with an *open* or *closed* valley. In open-valley construction, the shingles stop on either side of the valley, leaving an open lane that gives rainwater a clear run to the eaves. In closed-valley construction, the shingles are laid right through the valley. The open-valley method of laying shingles is simpler, cheaper, and less likely to cause problems than the closed-valley method, which, although it results in a more attractive-looking roof, is more likely to lead to leakage problems. In any case, it is unwise to install a closed valley unless the roof slope is at least 10 in 12.

To lay an open valley, a 4-in. gap is measured off through the center of the valley at the top of the roof. This gap widens by ⅛ in. for every foot down to the eaves. The sides of the gap are marked off by snapping chalk lines the length of the valley. The shingles are then laid to these lines.

Bituminous Flashing (See Fig. 9)

If bituminous flashing is installed, a 36-in.-wide strip of 15-lb, asphalt-saturated felt is first laid down through the center of the valley as an *underlayment*. This underlayment is held in place with nails. If the valley is to remain open, two layers of

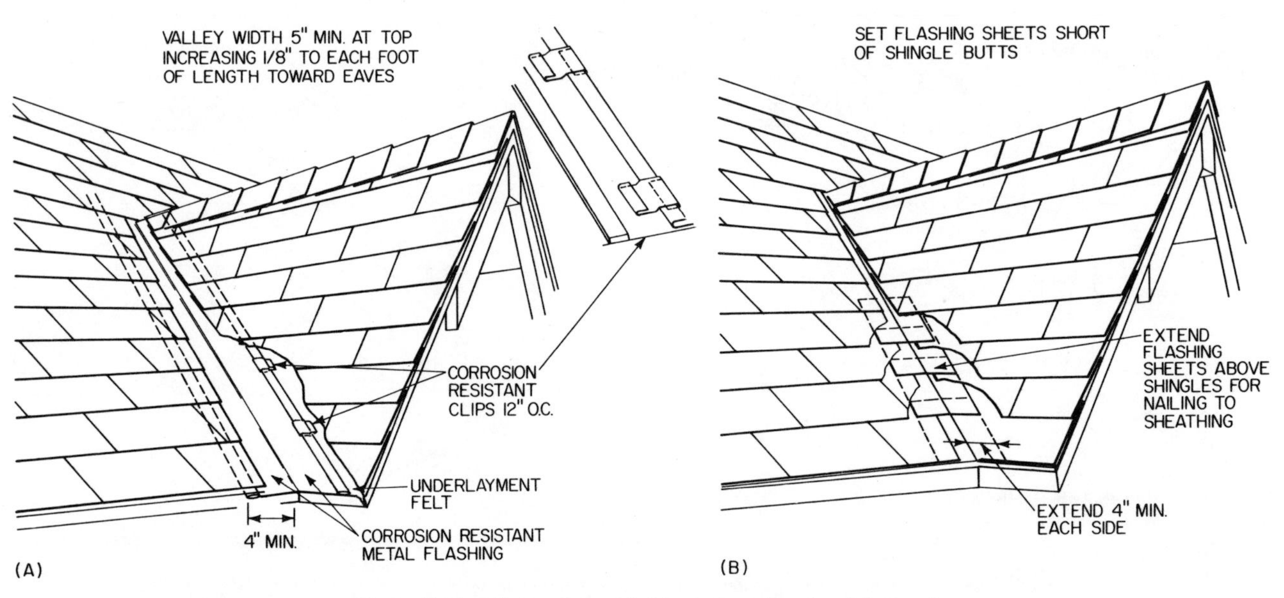

Fig. 8. The installation of metal flashing in the valley of a pitched roof.

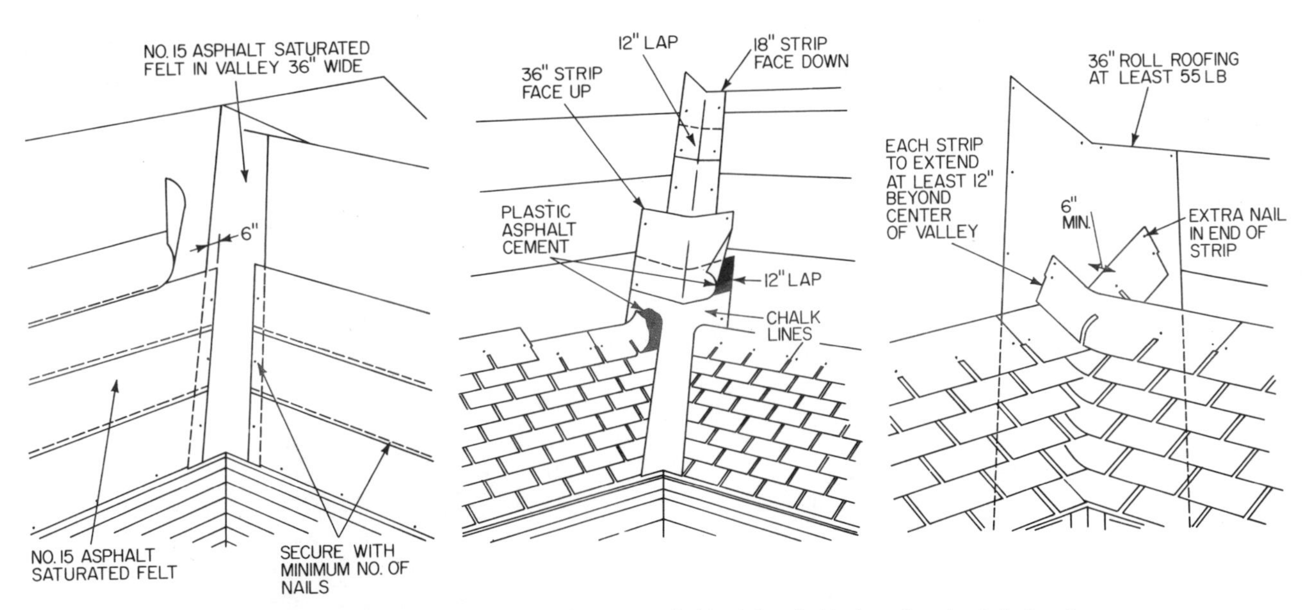

Fig. 9. These illustrations show how bituminous flashing is installed in the valley of a pitched roof, for both open and closed valley installation *(Asphalt Roofing Manufacturers Association).*

bituminous flashing are laid down the length of the valley; if the valley is to be closed, only one length of bituminous flashing is laid down.

For open-valley construction, a layer of 18-in.-wide, 90-lb, mineral-surfaced roll roofing is laid down through the middle of the valley over the underlayment, mineral side down. If it is necessary to join two short pieces of flashing material together, the top piece should overlap the bottom piece by at least 12 in., and the two pieces should then be cemented together with asphalt cement. Only sufficient nails should be used to hold the flashing in place. The nails should be located 1 in. from the edges of the flashing.

Over this, a 36-in.-wide strip of the same material is laid down, mineral side up. This strip is also secured to the roof using a minimum number of nails, which are also located 1 in. from the edges of the flashing.

For closed-valley construction, a layer of 15-lb, asphalt-saturated felt is first laid down through the center of the valley as an underlayment, just as with open-valley construction. A 36-in.-wide strip of 55-lb (or heavier) roll roofing material is then laid down over the underlayment. It is nailed in place as described above.

A closed valley can have a bituminous flashing installed only when the roofing material will consist of asphalt *strip* shingles (see Fig. 9), since installing individual shingles will require so many nails through the flashing material that water will almost certainly leak through the flashing. If wood shingles are to be installed and if the flashing is to be made of a bituminous material, then a closed valley should not be considered at all.

CHIMNEY FLASHING

A chimney is often built on a foundation that is separate from the house foundation to permit the chimney and house to settle into the soil at different rates, if settlement should occur. In addition, the structure of a wood-frame house is purposely kept a short distance away from the chimney to allow (1) for this differential movement, (2) for the slight movement of the chimney that occurs in high winds, and (3) as a fire-prevention measure.

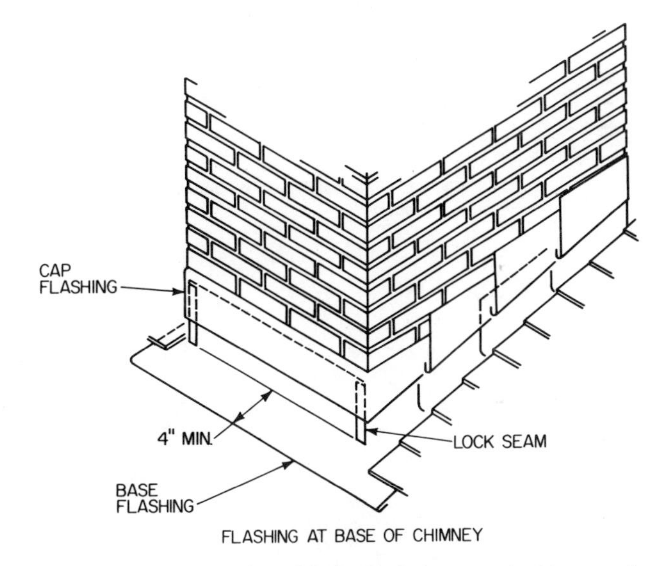

Fig. 10. The installation of metal flashing at the juncture of a chimney and a pitched roof closely resembles the installation of flashing at a wall/roof juncture *(Copper Development Association).*

The gap between the roof and the chimney, in particular, is sealed by the two-part base flashing and counterflashing technique that has been described for roof-wall junctures. The preferred flashing material is, as always, a metal. In low-cost houses, however, a combination bituminous and metal flashing is often installed, especially when the roofing material is also made of a bituminous material. The installation for both types of flashing are shown in Figs. 10 and 11.

Chimney Saddle

On a sloped roof, whenever the chimney is at least 30 in. wide and there is a likelihood that snow or heavy rains may collect in the pocket between the chimney and the roof, a chimney *saddle* (or *cricket*) is installed against the high side of the chimney as shown in Fig. 12. The purpose of the saddle is to divert away from the chimney any snow or rain that would otherwise collect against the chimney.

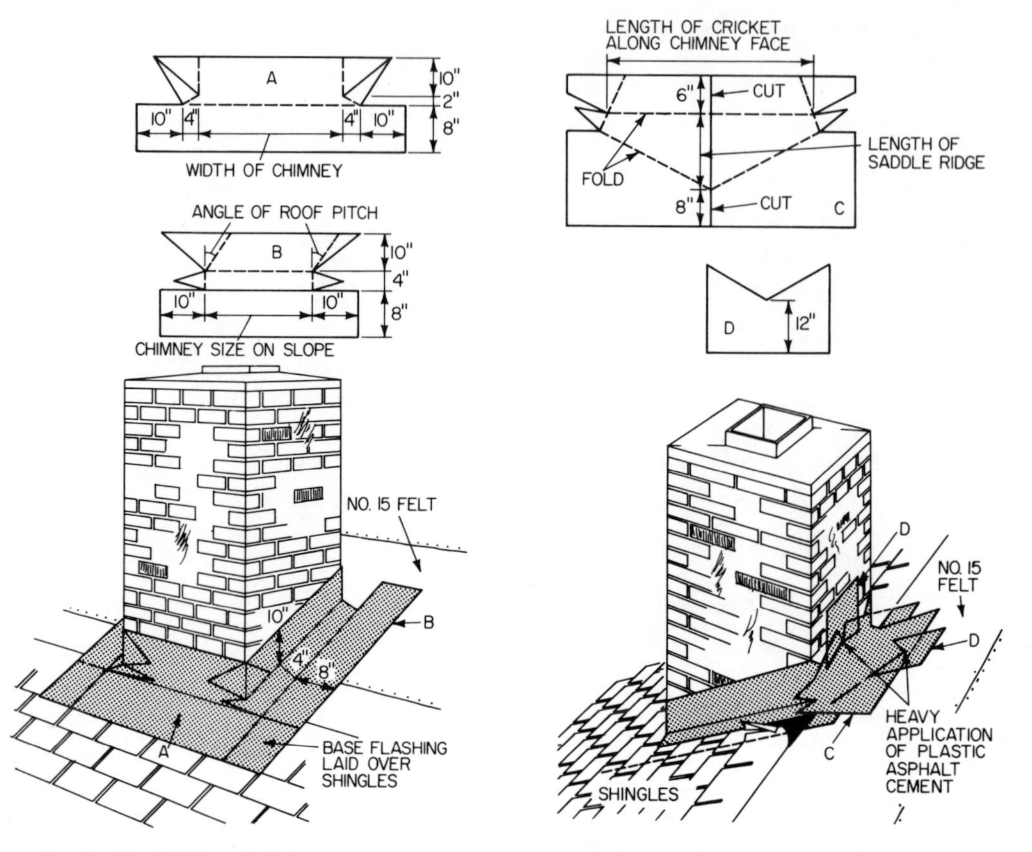

Fig. 11. The installation of bituminous flashing at the chimney/roof juncture *(Asphalt Roofing Manufacturers Association).*

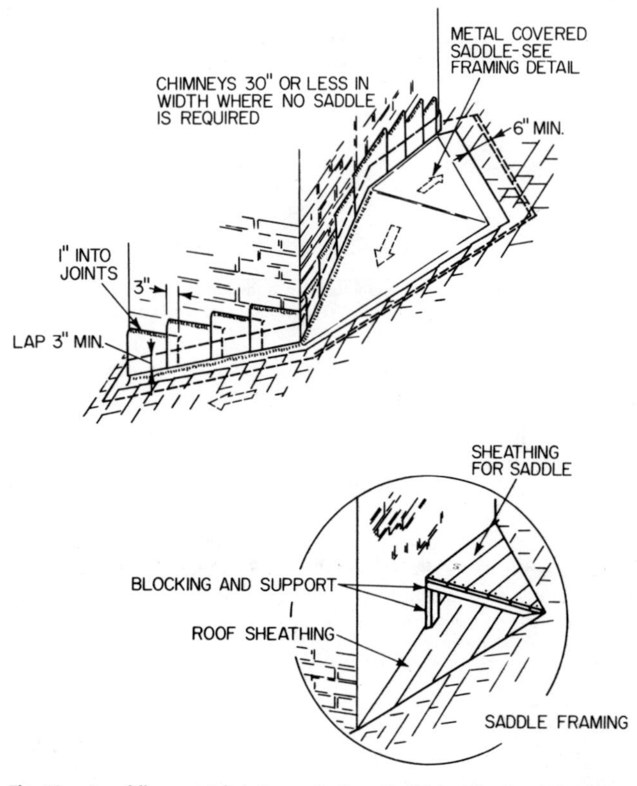

Fig. 12. A saddle, or cricket, is required on the high side of a chimney to prevent snow or rain collecting in the pocket and perhaps soaking into the house.

A wood framework is first constructed as shown in Fig. 12. This framework is then covered over completely with metal flashing or with the same bituminous material that is used for the base flashing around the chimney. This flashing should extend at least 6 in. away from the framework so that when the shingles are laid over the flashing, rain or melted snow will be unable to work its way under the saddle. The saddle flashing should also extend up the side of the chimney by about 6 in.

Counterflashing is then installed in the chimney brickwork to cover the saddle flashing as has already been described. Because the saddle is sloped, this counterflashing must be installed in the brickwork step-fashion.

VENT-PIPE FLASHING

There is always at least one vent pipe from the plumbing system that passes through the roof. The gap between this pipe and the roof is usually sealed by a prefabricated, two-piece metal flashing (see Fig. 13). One part of the flashing consists of a round sleeve that slips over and covers the vent pipe. The top of this sleeve is turned down inside the end of the vent pipe to keep water from running down alongside the pipe. The other part of the flashing consists of a square flange, or collar, about 16 in. on a side, that rests on the roof. This flashing is so made that the angle between the sleeve and its flange can be adjusted to conform to the slope of the roof.

If a suitable prefabricated flashing assembly isn't available, it is easy enough for a sheet-metal worker to make the necessary parts out of flat stock. By soldering or brazing the two pieces of metal together at the desired angle, he can make a one-piece

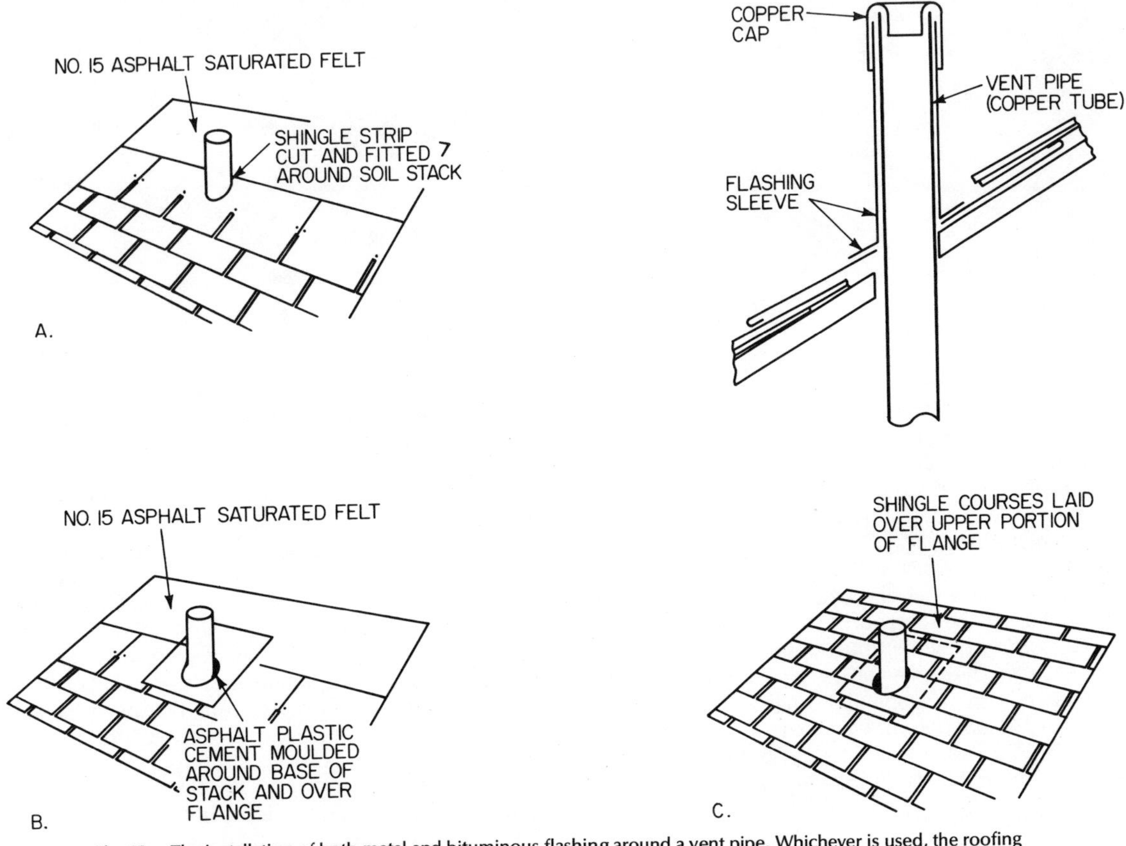

Fig. 13. The installation of both metal and bituminous flashing around a vent pipe. Whichever is used, the roofing is installed under and over the flashing as shown in order to direct water away from under the roofing.

unit be that matches the slope of the roof.

Flashings made of bituminous material can also be used. This is usually the case when asphalt shingles are installed on a sloped roof and also when built-up roofing is installed on a flat roof.

A collar from 12 to 16 in. sq is cut out of 55-lb (or heavier) roll roofing material. A hole the size of the vent pipe is then cut in the center of this collar.

On a sloped roof, the shingles are first laid up to the base of the vent pipe. The collar is then slipped over the vent pipe and laid flat on the roof. The gap between the collar and the pipe is filled by troweling asphalt *flashing compound* for a distance of about 2 in. completely around both pipe and flashing. The remainder of the shingles are then laid down. Wherever a shingle touches the vent pipe it, too, is bedded solidly in asphalt flashing compound to seal the juncture.

Floor Framing

This article describes the traditional method of building a wood-frame floor, as shown in Fig. 1. In most wood-frame houses, the basement or crawl space under the house is spanned by *joists* that run in parallel from one side of the house to the other, their ends resting on the foundation walls. If the span is greater than, say, 15 ft, two sets of joists are installed, one end of each joist resting on the foundation wall and the other end being supported by a large *girder* that runs from wall to wall down the center of the house crosswise to the direction of the joists. The girder thus supports about one-half the total floor loads. The center of the girder is usually supported by one or more *posts,* depending on the span. The joists are covered over by *subflooring,* which consists of boards or plywood panels. The subfloor ties the joists together into a single structural unit and it also provides a base for the finish flooring.

GENERAL CONSIDERATIONS

Before a builder can obtain a building permit, he must submit plans, or written specifications attached to the plans, that show the joist size and spacing, and the species and grade of the lumber to be used. What sorts of things must the builder take into account when laying out the joists?

The most obvious factors are the overall size of the house and the applied floor loads, which will determine the sizes and lengths of both the girder and the joists.

The next most obvious factor is the cost of the lumber, which increases with the cross-sectional area of the joists and especially with their length. The builder, therefore, tries to lay out the floor framing to minimize the lengths of the joists. The reasoning is as follows. The shorter the joists, the less weight each joist will have to support. The less weight each joist has to support, the smaller it need be. And the smaller it need be, the less it will cost. The cost of the joists will decrease even more as their length decreases.

But still other factors must be taken into account that may complicate the joist layout. If the house is to be more than one story high, the builder must take into account the location (or locations) of the second-floor bathroom (or bathrooms), since space must be found underneath the flooring in which to run the plumbing that will lead to and from the bathroom fixtures. If it is at all possible, this plumbing should run parallel with the joists; otherwise the installation of the joists may become quite complicated.

If the house is to be heated by a warm-air heating system

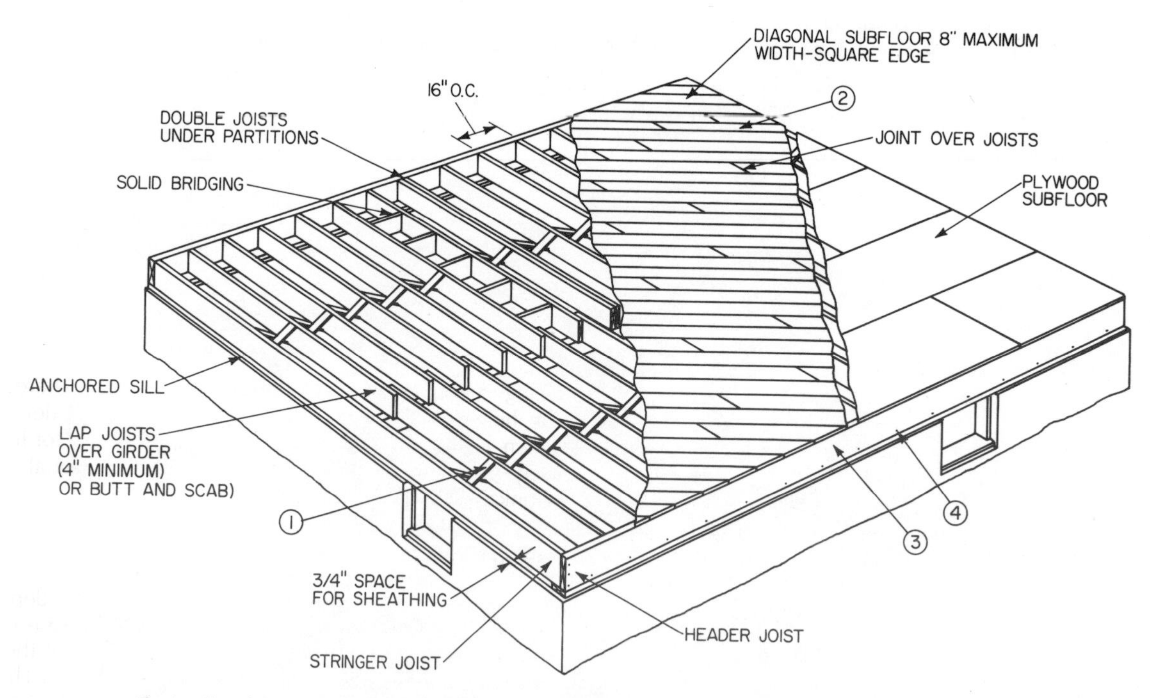

Fig. 1. Floor framing: (1) Nailing bridging to joists, (2) nailing board subfloor to joists, (3) nailing header to joists, (4) toenailing header to sill *(U.S. Forest Service).*

and/or cooled by a central air-conditioning system, the builder must also take into account the location of the ductwork and its outlets, since the cheapest and most efficient duct installation is one in which the ducts run the shortest possible distances and have the fewest possible bends. If there seems to be an insoluble conflict between the locations of the plumbing, the ducts, and the joists, it must be resolved by the architect, builder, and plumber before construction begins.

The builder must also be sure the lumber is dry. If the lumber is green, it will shrink as it dries out, sometimes by a considerable amount. The shrinkage lengthwise is trivial and may be ignored. Green lumber does, however, shrink considerably across its grain. A 12-in.-deep joist may shrink as much as ½ in., for example. When one length of green lumber rests upon another, as a joist may rest upon a girder, the total amount of shrinkage will double, assuming both pieces of lumber are the same size and equally green, which will cause the floors to sag perhaps an inch or more as the lumber dries out.

POST

Most girders are supported at their centers by a post (refer to Figs. 2 and 3). Posts made of wood or steel are usually installed in a house having a full basement. Masonry or concrete *piers* are usually built when there is a crawl space under the house, although a masonry pier can also be built in a full basement, of course.

Whatever the material, the post must be supported by a large footing if it is to be prevented from settling into the soil. For one- and two-story houses, most building codes require that footings be at least 12 in. sq and 6 in. deep, and for three-story houses, the footings must be at least 16 in. sq and 6 in. deep.

Wood Post

Wood posts (Fig. 2) are usually made from a solid timber. The timber supporting first-floor girders is usually 6 in. sq. The footing on which the end of the post rests should be raised 2 to 3 in. above the basement floor to prevent any dampness in the floor from rising through the footing and into the wood, which would cause the wood to rot. In addition, the top of the footing should be made waterproof by pouring a layer of hot asphalt over it before the post is placed in position.

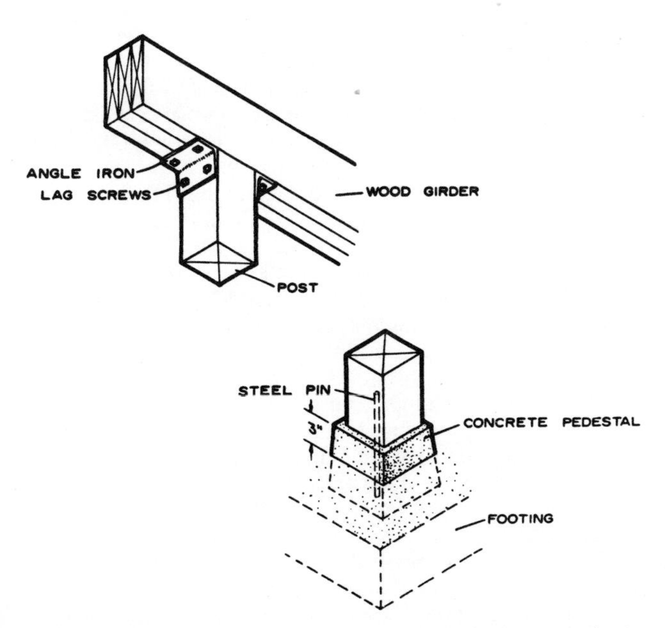

Fig. 2. Installation of a wood post to support a girder: (top) connection to girder; (bottom) installation of base *(U.S. Forest Service)*.

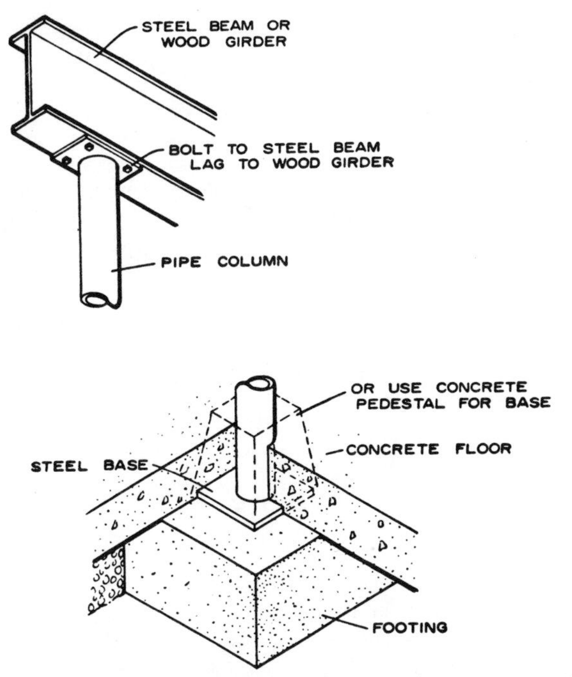

Fig. 3. Installation of a steel post to support a steel girder: (top) connection to beam; (bottom) base plate supported by a footing and embedded in a concrete floor. The base plate may also be mounted on and anchored to a concrete pedestal as shown in Fig. 2. *(U.S. Forest Service)*.

Figure 2 shows how the post is held securely to the footing by a pin. The top of the post is sometimes capped by a steel plate to prevent crushing the wood fibers of the girder where the girder rests upon the post.

Steel Post

For most houses, a steel post need be only 3 or 4 in. in diam. and it may look rather spindly under a girder but, rest assured, it can safely support all the loads likely to be imposed on it. Steel plates are welded to both ends of the post. The top plate is secured to the girder by bolts (if the girder is made of steel) or lag screws (if the girder is made of wood).

The bottom plate is secured to the footing either by bolts or by being embedded within the footing, assuming the footing is made of concrete.

Steel posts having a screw thread in one end are available. After the post has been set in position and the girder has been installed on top of it, the post can be lengthened by turning the screw until the post is bearing firmly against the girder. Thus, an exact fit can be obtained without metal or wood wedges being required between the girder and the post. Thereafter, if the house should begin to settle because the girder shrinks down slightly, the settlement can be compensated for immediately by lengthening the post an amount equal to the shrinkage.

Masonry Piers

Piers made from brick are usually 12 to 16 in. sq, depending on the loads they are expected to carry. Piers made of solid concrete blocks are usually 16 in. sq regardless of the loads they will carry, as this is the size of concrete blocks. The piers are built on concrete or brick footings that extend 4 to 8 in. away from each side of the piers. The actual size of the footings will depend on the type of soil under the house. A clay soil, for example, will require a larger bearing surface than a gravel soil.

The top of the pier can consist either of a steel plate set into mortar or of a solid masonry cap 4 in. thick, which is also set in mortar. The height of the piers must be such as to maintain a minimum distance of 12 in. between the soil and the bottom of the girder to minimize the possibility of termite infestation.

Concrete Post

A pier made of reinforced concrete is usually 10 in. sq. Both the pier and its footing are poured at the same time and in the same form. The dimensions of the footing should be the same as for masonry piers.

GIRDER

Wood Girder

A wood girder may consist of one solid timber, be built up out of nominal 2- or 3-in.-thick planks, or be made of laminated construction. Although a solid timber is usually stronger than a built-up girder, depending on the species and grade of the lumber, a timber is more likely to shrink after it has been installed because it will have a much higher initial moisture content. It is, therefore, also much more likely to develop splits. In addition, a solid timber that is long enough to span the basement of a house is quite expensive, and it would also probably require a crane to set it in place. Most girders made of solid timbers usually consist, therefore, of several short timbers that abut each other at center posts.

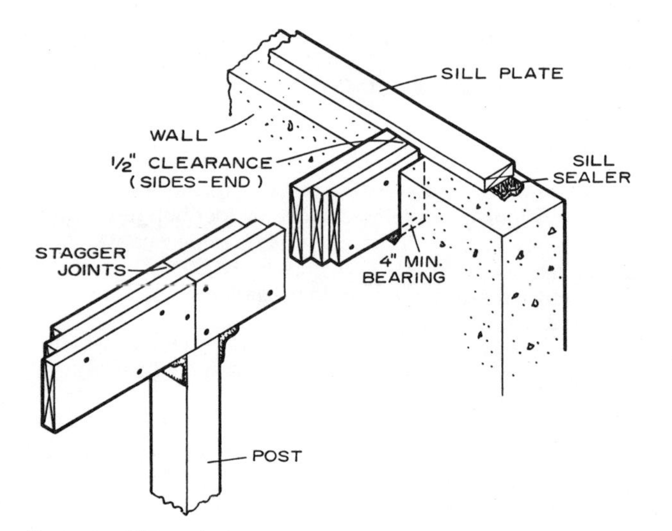

Fig. 4. Installation of a built-up wood girder *(U.S. Forest Service)*.

Built-up girders (see Fig. 4) are usually cheaper than solid timbers because the lumber out of which they are made usually costs less than solid timbers. Lumber that is only 2 to 3 in. thick is also more likely to be thoroughly seasoned than a solid timber, which means the wood will shrink much less. By spiking together a number of short lengths of lumber, a built-up girder of almost any size and length can be constructed, within limitations, of course. One cannot have an excessive number of joints, for example, and the joints must be located as much as possible over posts.

Steel Girder

In a comparison of wood and steel girders, if they have the same strength and stiffness, the steel girder will be both smaller and heavier than the wood girder. Steel girders are used because they can span wider spaces with less trouble than wood girders and because, whatever else may happen, the girder won't shrink down with the passing of time. Steel girders and steel posts are usually, though not necessarily, installed together and fastened by bolts as shown in Fig. 3.

Girder Installation

The ends of both wood and steel girders may be supported in either of two ways on the foundation walls. The girders can rest in niches formed in the walls during their construction, or they can rest on top of the foundation walls. When a girder rests in niches, as shown in Fig. 4, the niches must be at least 4 in. deep if they are to provide sufficient bearing surface. In addition, a steel plate is usually set in the niche to give the girder a solid support, with a ½-in. space left all around the girder to allow air to circulate. It is poor practice to seal the ends of a girder in a wall by pouring concrete or mortar around it under the belief that the construction is thereby made more secure. It is far more likely that the ends of a wood girder will rot under such circumstances, in which case it may suddenly collapse one day.

JOIST

Joists are usually made of 2-in.-thick lumber and are anywhere from 6 to 14 in, deep, depending on the expected floor loads, the spacing between joists, the total span, and the kind and grade of lumber used. These factors are summarized in joist tables, such as Table 1, which are used by the builder as a guide to the selection of the joists. It should be remembered, however, that joist tables always give the *minimum* required joist sizes; it is always possible, if thought necessary, to increase the size of the joists to carry heavier than usual loads.

Joists with Wood Girder

There are three ways in which wood joists can meet at and be connected to a central girder: (1) the joists can rest on top of the girder; (2) they may be installed level with the top or bottom of the girder and rest on *ledgers* that are nailed to the bottom of the girder; or (3) they may be level with the top or bottom of the girder and supported by *joist hangers* or *framing anchors* resting on or attached to the girder. All three methods are shown in Fig. 5. The illustrations should make these construction methods clear.

The simplest method of installing joists is to place them on top of the girder, allowing abutting joists to overlap so they can be spiked together. Simple as it is, this method of installation may also lead to the greatest problem with wood shrinkage.

The total amount of shrinkage at the girder will be the shrinkage through the girder itself plus the shrinkage through the joists it is supporting. If green lumber is used, this total shrinkage can amount to as much as 1 in. Furthermore, since the total depth of the lumber at the foundation wall will be about half the depth of the lumber at the girder, the amount of shrinkage at the foundation walls will be about half of what it is at the girder. As the wood dries out over a period of months, the center of the house will settle by as much as ½ in., which is sufficient to cause the plaster to crack and the doors to jam as the house frame twists.

Table 1. Floor and Ceiling Joists
(Tables such as this are used to select joists having the required strength and stiffness to support floor loads without deflecting.)

FLOOR JOISTS
For all rooms except sleeping areas and attic floors
(40 lb per sq ft live load)

Design criteria: Deflection: for 40 lb per sq ft live load. Limited to span in inches divided by 360. Strength: Live load of 40 lb per sq ft plus dead load of 10 lb per sq ft determines required fiber stress value. *Note:* the required extreme fiber stress in bending F_b in lb per sq in. is shown below each span.

Joist Size, in.	Joist Spacing, in.	Modulus of elasticity E in 1,000,000 psi																		
		0.4	0.5	0.6	0.7	0.8	0.9	1.0	1.1	1.2	1.3	1.4	1.5	1.6	1.7	1.8	1.9	2.0	2.2	2.4
2 × 6	12.0	6-9	7-3	7-9	8-2	8-6	8-10	9-2	9-6	9-9	10-0	10-3	10-6	10-9	10-11	11-2	11-4	11-7	11-11	12-3
		450	520	590	660	720	780	830	890	940	990	1040	1090	1140	1190	1230	1280	1320	1410	1490
	13.7	6-6	7-0	7-5	7-9	8-2	8-6	8-9	9-1	9-4	9-7	9-10	10-0	10-3	10-6	10-8	10-10	11-1	11-5	11-9
		470	550	620	690	750	810	870	930	980	1040	1090	1140	1190	1240	1290	1340	1380	1470	1560
	16.0	6-2	6-7	7-0	7-5	7-9	8-0	8-4	8-7	8-10	9-1	9-4	9-6	9-9	9-11	10-2	10-4	10-6	10-10	11-2
		500	580	650	720	790	860	920	980	1040	1090	1150	1200	1250	1310	1360	1410	1460	1550	1640
	19.2	5-9	6-3	6-7	7-0	7-3	7-7	7-10	8-1	8-4	8-7	8-9	9-0	9-2	9-4	9-6	9-8	9-10	10-2	10-6
		530	610	690	770	840	910	970	1040	1100	1160	1220	1280	1330	1390	1440	1500	1550	1650	1750
	24.0	5-4	5-9	6-2	6-6	6-9	7-0	7-3	7-6	7-9	7-11	8-2	8-4	8-6	8-8	8-10	9-0	9-2	9-6	9-9
		570	660	750	830	900	980	1050	1120	1190	1250	1310	1380	1440	1500	1550	1610	1670	1780	1880
	32.0					6-2	6-5	6-7	6-10	7-0	7-3	7-5	7-7	7-9	7-11	8-0	8-2	8-4	8-7	8-10
						1010	1090	1150	1230	1300	1390	1450	1520	1590	1660	1690	1760	1840	1950	2060
2 × 8	12.0	8-11	9-7	10-2	10-9	11-3	11-8	12-1	12-6	12-10	13-2	13-6	13-10	14-2	14-5	14-8	15-0	15-3	15-9	16-2
		450	520	590	660	720	780	830	890	940	990	1040	1090	1140	1190	1230	1280	1320	1410	1490
	13.7	8-6	9-2	9-9	10-3	10-9	11-2	11-7	11-11	12-3	12-7	12-11	13-3	13-6	13-10	14-1	14-4	14-7	15-0	15-6
		470	550	620	690	750	810	870	930	980	1040	1090	1140	1190	1240	1290	1340	1380	1470	1560
	16.0	8-1	8-9	9-3	9-9	10-2	10-7	11-0	11-4	11-8	12-0	12-3	12-7	12-10	13-1	13-4	13-7	13-10	14-3	14-8
		500	580	650	720	790	850	920	980	1040	1090	1150	1200	1250	1310	1360	1410	1460	1550	1640
	19.2	7-7	8-2	8-9	9-2	9-7	10-0	10-4	10-8	11-0	11-3	11-7	11-10	12-1	12-4	12-7	12-10	13-0	13-5	13-10
		530	610	690	770	840	910	970	1040	1100	1160	1220	1280	1330	1390	1440	1500	1550	1650	1750
	24.0	7-1	7-7	8-1	8-6	8-11	9-3	9-7	9-11	10-2	10-6	10-9	11-0	11-3	11-5	11-8	11-11	12-1	12-6	12-10
		570	660	750	830	900	980	1050	1120	1190	1250	1310	1380	1440	1500	1550	1610	1670	1780	1880
	32.0					8-1	8-5	8-9	9-0	9-3	9-6	9-9	10-0	10-2	10-5	10-7	10-10	11-0	11-4	11-8
						990	1080	1170	1230	1300	1370	1450	1520	1570	1650	1700	1790	1840	1950	2070
2 × 10	12.0	11-4	12-3	13-0	13-8	14-4	14-11	15-5	15-11	16-5	16-10	17-3	178	18-0	18-5	18-9	19-1	19-5	20-1	20-8
		450	520	590	660	720	780	830	890	940	990	1040	1090	1140	1190	1230	1280	1320	1410	1490
	13.7	10-10	11-8	12-5	13-1	13-8	14-3	14-9	15-3	15-8	16-1	16-6	16-11	17-3	17-7	17-11	18-3	18-7	19-2	19-9
		470	550	620	690	750	810	870	930	980	1040	1090	1140	1190	1240	1290	1340	1380	1470	1560
	16.0	10-4	11-1	11-10	12-5	13-0	13-6	14-0	14-6	14-11	15-3	15-8	16-0	16-5	16-9	17-0	17-4	17-8	18-3	18-9
		500	580	650	720	790	850	920	980	1040	1090	1150	1200	1250	1310	1360	1410	1460	1550	1640
	19.2	9-9	10-6	11-1	11-8	12-3	12-9	13-2	13-7	14-0	14-5	14-9	15-1	15-5	15-9	16-0	16-4	16-7	17-2	17-8
		530	610	690	770	840	910	870	1040	1100	1160	1220	1280	1330	1390	1440	1500	1550	1650	1750
	24.0	9-0	9-9	10-4	10-10	11-4	11-10	12-3	12-8	13-0	13-4	13-8	14-0	14-4	14-4	14-11	15-2	15-5	15-11	16-5
		570	660	750	830	900	980	1050	1120	1190	1250	1310	1380	1440	1500	1550	1610	1670	1780	1880
	32.0					10-4	10-9	11-1	11-6	11-10	12-2	12-5	12-9	13-0	13-3	13-6	13-9	14-0	14-6	14-11
						1000	1080	1150	1240	1310	1380	1440	1520	1580	1640	1700	1770	1830	1970	2080
2 × 12	12.0	13-10	14-11	15-10	16-8	17-5	18-1	18-9	19-4	19-11	20-6	21-0	21-6	21-11	22-5	22-10	23-3	23-7	24-5	25-1
		450	520	590	660	720	780	830	890	940	990	1040	1090	1140	1190	1230	1280	1320	1410	1490
	13.7	13-3	14-3	15-2	15-11	16-8	17-4	17-11	18-6	19-1	19-7	20-1	20-6	21-0	21-5	21-10	22-3	22-7	23-4	24-0
		470	550	620	690	750	810	870	930	980	1040	1090	1140	1190	1240	1290	1340	1380	1470	1560
	16.0	12-7	13-6	14-4	15-2	15-10	16-5	17-0	17-7	18-1	18-7	19-1	19-6	19-11	20-4	20-9	21-1	21-6	22-2	22-10
		500	580	650	720	790	860	920	980	1040	1090	1150	1200	1250	1310	1360	1410	1460	1550	1640
	19.2	11-10	12-9	13-6	14-3	14-11	15-6	16-0	16-7	17-0	17-6	17-11	18-4	18-9	19-2	19-6	19-10	20-2	20-10	21-6
		530	610	690	770	840	910	970	1040	1100	1160	1220	1280	1330	1390	1440	1500	1550	1650	1750
	24.0	11-0	11-10	12-7	13-3	13-10	14-4	14-11	15-4	15-10	16-3	16-8	17-0	17-5	17-9	18-1	18-5	18-9	19-4	19-11
		570	660	750	830	900	980	1050	1120	1190	1250	1310	1380	1440	1500	1550	1610	1670	1780	1880
	32.0					12-7	13-1	13-6	13-11	14-4	14-9	15-2	15-6	15-10	16-2	16-5	16-9	17-0	17-7	18-1
						1000	1080	1150	1220	1300	1380	1450	1520	1580	1650	1700	1770	1830	1950	2070

(continued)

Table 1. Floor and Ceiling Joists (Continued)
(Tables such as this are used to select joists having the required strength and stiffness to support floor loads without deflecting.)

CEILING JOISTS
Limited attic storage where development of future rooms is not possible
(20 lb per sq ft live load)
Design criteria: Deflection: For 10 lb per sq ft live load. Limited to span in inches divided by 360. Strength: Live load of 20 lb per sq ft plus dead load of 10 lb per sq ft determines required fiber stress value. *Note:* The required extreme fiber stress in bending F_b in lb per sq in. is shown below each span.

Joist Size, in.	Joist Spacing, in.	Modulus of elasticity E in 1,000,000 psi																		
		0.4	0.5	0.6	0.7	0.8	0.9	1.0	1.1	1.2	1.3	1.4	1.5	1.6	1.7	1.8	1.9	2.0	2.2	2.4
2 × 4	12.0	5-5 430	5-10 500	6-2 560	6-6 630	6-10 680	7-1 740	7-4 790	7-7 850	7-10 900	8-0 950	8-3 990	8-5 1040	8-7 1090	8-9 1130	8-11 1170	9-1 1220	9-3 1260	9-7 1340	9-10 1420
	13.7	5-2 450	5-7 520	5-11 590	6-3 650	6-6 720	6-9 770	7-0 830	7-3 880	7-6 940	7-8 990	7-10 1040	8-1 1090	8-3 1140	8-5 1180	8-7 1230	8-8 1270	8-10 1320	9-2 1400	9-5 1490
	16.0	4-11 470	5-4 550	5-8 620	5-11 690	6-2 750	6-5 810	6-8 870	6-11 930	7-1 990	7-3 1040	7-6 1090	7-8 1140	7-10 1200	8-0 1240	8-1 1290	8-3 1340	8-5 1390	8-8 1480	8-11 1570
	19.2	4-8 500	5-0 580	5-4 660	5-7 730	5-10 800	6-1 870	6-3 930	6-6 990	6-8 1050	6-10 1110	7-0 1160	7-2 1220	7-4 1270	7-6 1320	7-8 1370	7-9 1420	7-11 1470	8-2 1570	8-5 1660
	24.0	4-4 540	4-8 630	4-11 710	5-2 790	5-5 860	5-8 930	5-10 1000	6-0 1070	6-2 1130	6-4 1190	6-6 1250	6-8 1310	6-10 1370	7-0 1420	7-1 1480	7-3 1530	7-4 1590	7-7 1690	7-10 1790
2 × 6	12.0	8-6 430	9-2 500	9-9 560	10-3 630	10-9 680	11-2 740	11-7 790	11-11 850	12-3 900	12-7 950	12-11 990	13-3 1040	13-6 1090	13-9 1130	14-1 1170	14-4 1220	14-7 1260	15-0 1340	15-6 1420
	13.7	8-2 450	8-9 520	9-4 590	9-10 650	10-3 720	10-8 770	11-1 830	11-5 880	11-9 940	12-1 990	12-4 1040	12-8 1090	12-11 1140	13-2 1180	13-5 1230	13-8 1270	13-11 1320	14-4 1400	14-9 1490
	16.0	7-9 470	8-4 550	8-10 620	9-4 690	9-9 750	10-2 810	10-6 870	10-10 930	11-2 990	11-5 1040	11-9 1090	12-0 1140	12-3 1200	12-6 1240	12-9 1290	13-0 1340	13-3 1390	13-8 1480	14-1 1570
	19.2	7-3 500	7-10 580	8-4 660	8-9 730	9-2 800	9-6 870	9-10 930	10-2 990	10-6 1050	10-9 1110	11-1 1160	11-4 1220	11-7 1270	11-9 1320	12-0 1370	12-3 1420	12-5 1470	12-10 1570	13-3 1660
	24.0	6-9 540	7-3 630	7-9 710	8-2 790	8-6 860	8-10 930	9-2 1000	9-6 1070	9-9 1130	10-0 1190	10-3 1250	10-6 1310	10-9 1370	10-11 1420	11-2 1480	11-4 1530	11-7 1590	11-11 1690	12-3 1790
2 × 8	12.0	11-3 430	12-1 500	12-10 560	13-6 630	14-2 680	14-8 740	15-3 790	15-9 850	16-2 900	16-7 950	17-0 990	17-5 1040	17-10 1090	18-2 1130	18-6 1170	18-10 1220	19-2 1260	19-10 1340	20-5 1420
	13.7	10-9 450	11-7 520	12-3 590	12-11 650	13-6 720	14-1 770	14-7 830	15-0 880	15-6 940	15-11 990	16-3 1040	16-8 1090	17-0 1140	17-5 1180	17-9 1230	18-0 1270	18-4 1320	18-11 1400	19-6 1490
	16.0	10-2 470	11-0 550	11-8 620	12-3 690	12-10 750	13-4 810	13-10 870	14-3 930	14-8 990	15-1 1040	15-6 1090	15-10 1140	16-2 1200	16-6 1240	16-10 1290	17-2 1340	17-5 1390	18-0 1480	18-6 1570
	19.2	9-7 500	10-4 580	11-0 660	11-7 730	12-1 800	12-7 870	13-0 930	13-5 990	13-10 1050	14-2 1110	14-7 1160	14-11 1220	15-3 1270	15-6 1320	15-10 1370	16-1 1420	16-5 1470	16-11 1570	17-5 1660
	24.0	8-11 540	9-7 630	10-2 710	10-9 790	11-3 860	11-8 930	12-1 1000	12-6 1070	12-10 1130	13-2 1190	13-6 1250	13-10 1310	14-2 1370	14-5 1420	14-8 1480	15-0 1530	15-3 1590	15-9 1690	16-2 1790
2 × 10	12.0	14-4 430	15-5 500	16-5 560	17-3 630	18-0 680	18-9 740	19-5 790	20-1 850	20-8 900	21-2 950	21-9 990	22-3 1040	22-9 1090	23-2 1130	23-8 1170	24-1 1220	24-6 1260	25-3 1340	26-0 1420
	13.7	13-8 450	14-9 520	15-8 590	16-6 650	17-3 720	17-11 770	18-7 830	19-2 880	19-9 940	20-3 990	20-9 1040	21-3 1090	21-9 1140	22-2 1180	22-7 1230	23-0 1270	23-5 1320	24-2 1400	24-10 1490
	16.0	13-0 470	14-0 550	14-11 620	15-8 690	16-5 750	17-0 810	17-8 870	18-3 930	18-9 990	19-3 1040	19.9 1090	20-2 1140	20-8 1200	21-1 1240	21-6 1290	21-10 1340	22-3 1390	22-11 1480	23-8 1570
	19.2	12-3 500	13-2 580	14-0 660	14-9 730	15-5 800	16-0 870	16-7 930	17-2 990	17-8 1050	18-1 1110	18-7 1160	19-0 1220	19-5 1270	19-10 1320	20-2 1370	20-7 1420	20-11 1470	21-7 1570	22-3 1660
	24.0	11-4 540	12-3 630	13-0 710	13-8 790	14-4 860	14-11 930	15-5 1000	15-11 1070	16-5 1130	16-10 1190	17-3 1250	17-8 1310	18-0 1370	18-5 1420	18-9 1480	19-1 1530	19-5 1590	20-1 1690	20-8 1790

Source: National Forest Products Association.

For this reason, it is preferable that the joists and girder be placed at the same height. The choice of construction methods to accomplish this will then be among ledgers, joist hangers, or framing anchors.

A ledger is nothing more than a 2 × 2 or 2 × 4 in. length of wood that is securely nailed to the bottom edge of the girder. The joists rest on the ledger and are also spiked to the girder or to each other for additional support, as shown in Fig. 5. A girder is usually deeper than the joists, which means that even with a ledger strip nailed to the bottom of the girder the tops of the joists will usually be level with the top of the girder. When this method of installing joists is used, the amount of wood shrinkage at both the girder and foundation walls will be approximately equal.

It is possible, however, for the joists either to be deeper than the girder or for the joists to be the same depth of the girder. When this is the case, and when ledger strips are used, the joists can be fastened to the girder by notching them as shown

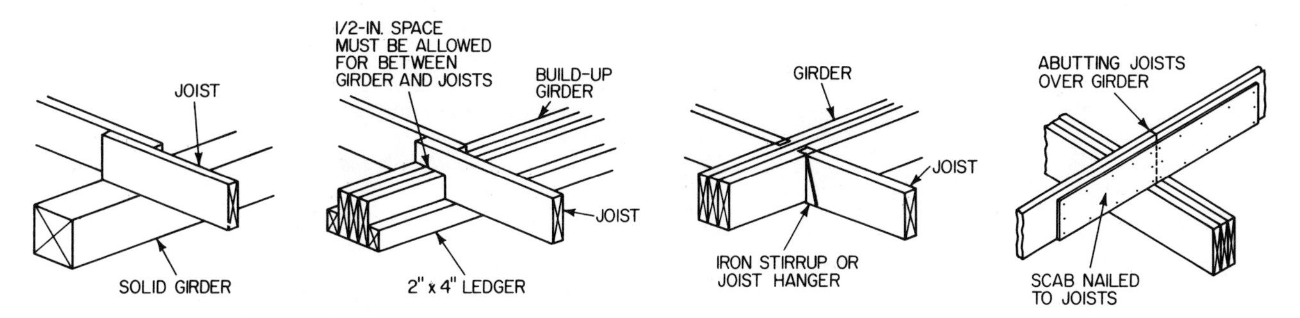

Fig. 5. Joists may either rest on a girder, they may rest on ledgers nailed to the girder, or they may be supported by metal joist hangers.

in Fig. 5 so that abutting joists will abut or overlap each other on top of the girder. When joists are notched above a girder, however, a gap of at least ½ in. must be allowed for between the girder and the joists; otherwise, as the joists dry out, the notched ends may shrink down upon the girder and the stresses that result will cause the joists to split at the notch cutouts.

Another alternative is to leave the ends of the joists square and to nail *scabs* across the top of the girder to join the joists, as shown in Fig. 5.

The use of joist hangers is also illustrated in Fig. 5; the illustration should be self-expanatory. If joist hangers *are* used, one must be careful that they are spiked into the sides of the girder and not merely hung over the top of the girder. If a joist hanger rests on top of a girder, it will allow the same total amount of shrinkage through green wood as if the joists were laid on top of the girder. This may not be obvious at first but if one remembers that all the wood in a house shrinks *down* toward the foundations, one can see that the joist hanger will drop down ½ in., say, as the girder shrinks, and that the joist will shrink down another ½ in.

Joists with Steel Girder

Everything that has been said above about the installation of wood girders will apply also to steel girders, except that the builder won't have to worry about shrinkage in a steel girder (refer to Fig. 6). There may still be a small amount of differential shrinkage between the two ends of each joist when a steel girder is used, and the builder must take suitable precautions to prevent such differential shrinkage.

Very often in wood-frame construction, for example, the ends of the joists at the foundation walls will rest upon a 2 to 4-in.-thick wood *sill* that is bolted to the top of the foundation walls (see Fig. 7) with the other ends of the joists resting on a steel girder installed in the center of the house. The result may be a slight amount of shrinkage at the foundation walls. The solution is simple. As shown in Fig. 6, bolt to the top or bottom flange of the girder a length of wood that is as thick as the sill and rest the joists on this length of wood. A wood support is usually necessary, in any case, to give the joists a material to which they can be nailed for added support.

If the steel girder and the joists are to be at the same height, ledger boards can be bolted to the bottom flange of the girder as shown in Fig. 6. A ½-in. gap must be allowed for between the notched ends of the joists and the top of the girder to avoid splitting the wood.

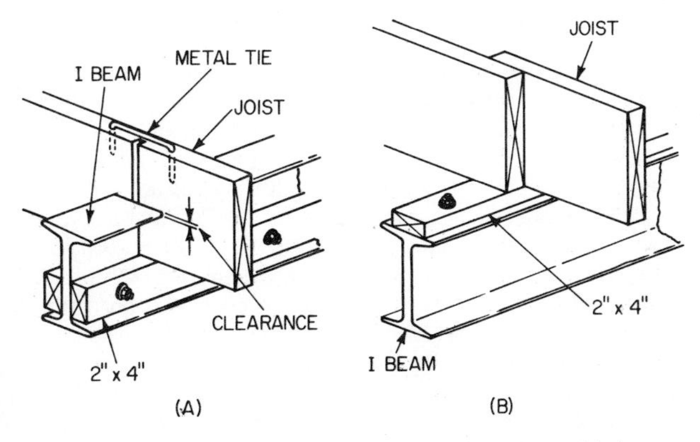

Fig. 6. Joists supported by steel girders usually rest on wood ledgers to equalize the amount of wood shrinkage at both ends of the joists.

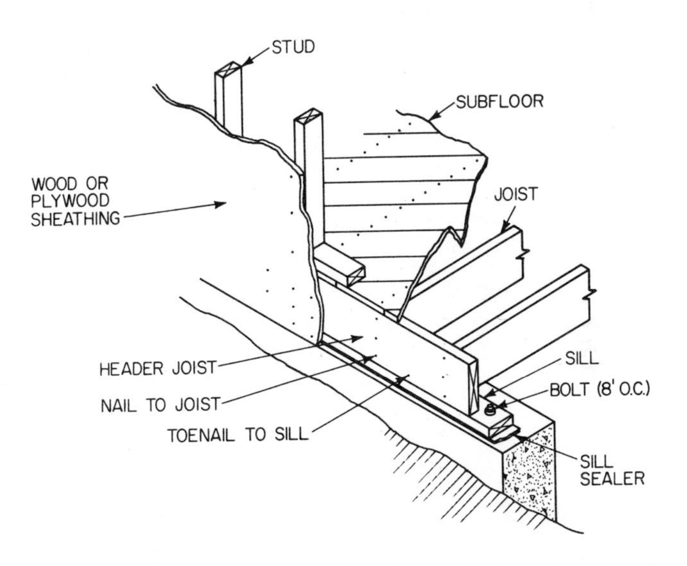

Fig. 7. The assembly of joists to the foundation wall of a platform-framed house *(U.S. Forest Service).*

JOIST INSTALLATION

The way in which the joists are secured at the foundation walls will of course depend on the type of wood framing that is used in the original construction. (For details *see* WOOD-FRAME CONSTRUCTION.)

In *platform-frame* construction the joists rest on top of a sill that runs around the perimeter of the building, the sill being fastened directly to the top of the foundation walls (see Fig. 7). A *header joist* is spiked into this sill (the two pieces of lumber thus forming what is called a *box sill*) and the joists are then attached to the header joist by spikes driven through the header joist and into the ends of the joists.

In *balloon-frame* construction, the joists also rest upon a sill but they are spiked into the wall studs rather than to a header joist (see Fig. 8).

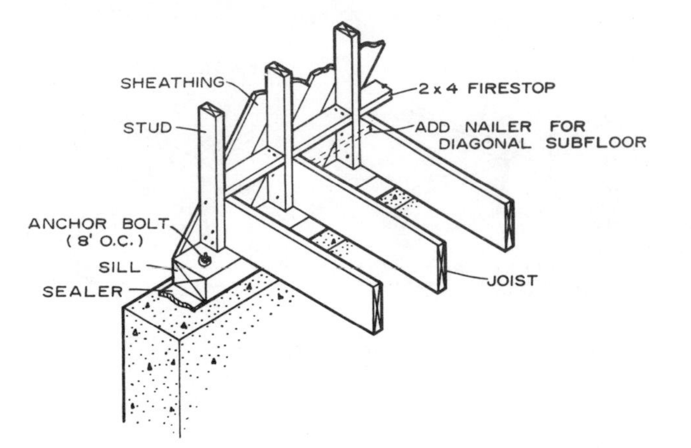

Fig. 8. The assembly of joists at the foundation wall of a balloon-framed house *(U.S. Forest Service).*

Masonry Walls

Neither of the above methods of supporting the ends of the joists will work in a foundation wall made of brick or concrete blocks. In a masonry wall, the builder may either set the ends of the joists in niches, as is done for the girder, or he may rest the joists in joist hangers built into the masonry. If the joists are supported in niches, they should rest on bearing surfaces that are 4 in. deep.

In a masonry house, the second-floor joists that rest in niches have one construction peculiarity. The ends of the joists must be cut at a "self-releasing" angle, as shown in Fig. 9. The angle allows the joists to fall clear of the walls if they should ever collapse during a fire. If the ends of the joists were left square, there is a chance that they would catch against the masonry as they fell, thus causing the entire wall to collapse.

It is also usual in brick-masonry construction, and especially in parts of the country where earthquakes or hurricane-force winds occur, to tie the walls and joists together at each floor level with steel straps or anchors embedded in the wall. For a description of this construction *see* CONCRETE-BLOCK CONSTRUCTION.

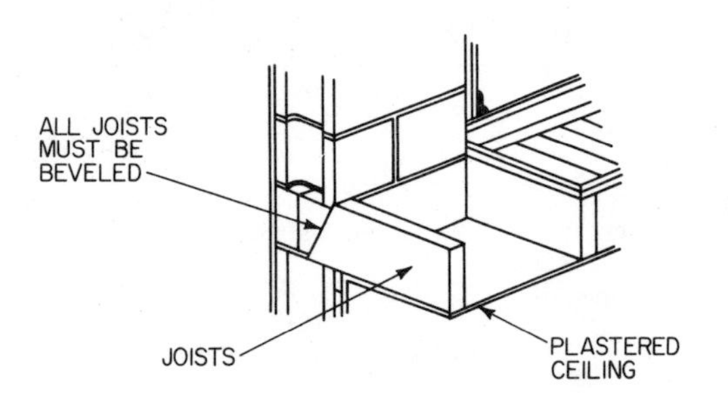

Fig. 9. The ends of second-floor joists in a masonry wall are beveled to enable them to fall clear of the wall in case they collapse during a fire.

Floor Openings

In two-story dwellings, or in a one-story dwelling with a full basement, stairs must obviously lead from one level to another. And it is equally obvious that, if the joists are cut through to make a stairwell opening, they will collapse. The construction, therefore, must be reinforced at the stairwell in such a way as to maintain its integrity.

When a floor opening is less than 4 ft wide, short lengths of joist called *headers* span the opening and tie the cut joists together (refer to Fig. 10). When the floor opening is more than 4 ft wide, these headers must be doubled. When the headers are doubled, the joists running parallel to the sides of this opening must be doubled also. These added parallel joists are called *trimmers.*

When the floor opening is 6 ft or less in width, it is sufficient if the doubled headers are attached to the trimmers by spikes that are driven through the trimmers into the ends of the headers. But when the opening is more than 6 ft in width, then the headers must, in addition, be attached to the trimmers by joist hangers or framing anchors.

The joists that have been cut through in order to make the opening are called *tail joists, tail beams,* or *header joists*—different sections of the country have their own nomenclature. These tail joists (as we shall call them) are spiked into the headers as shown in Fig. 10. If the tail joists are longer than 12 ft, they must also be supported by framing anchors or rest on 2 × 2 in. ledgers nailed to the bottoms of the headers.

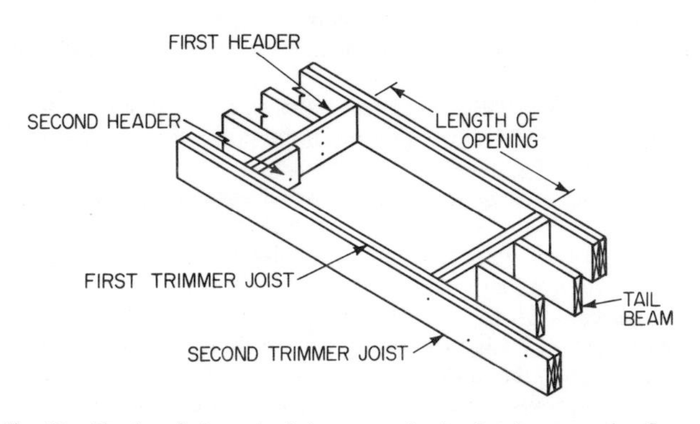

Fig. 10. The installation of reinforcement in the framing around a floor opening.

Chimney Openings

The method of framing a floor opening through which a chimney passes is the same as for a stairwell opening. The only difference between the two is that the gap between the brick-work and the floor framing at the chimney opening must be closed off to prevent smoke or hot gases from traveling from floor to floor, if a fire should break out in the house. The gap is there in the first place to prevent any part of the brickwork, which can get very hot, from touching and possibly igniting the wood. No part of the wood framing should, therefore, be closer than 2 in. to the chimney, if the walls of the chimney are at least 8 in. thick.

Bathroom Joists

Bathroom joists must support much greater loads than the joists in other parts of a house. For one thing, the bathroom floor may consist of ceramic tiles bedded in from 1 to 3 in. of concrete; this floor can weigh as much as 30 lb per sq ft.

For another thing, bathroom fixtures are usually made of porcelain-coated cast iron and they are quite heavy. The fixtures may add perhaps another 10 to 20 lb per sq ft to the floor. The joists must also support the weight of a bathtub filled with water. Water weighs 62.4 lb per cu ft and an ordinary sized bathtub filled with water can easily weigh 1250 lb. The weight of this water, spread over the floor area covered by the bathtub (10 sq ft, say) represents an additional, if temporary, load of 125 lb per sq ft. Therefore, it should be obvious that the floor framing must be reinforced if it is to support these loads.

This reinforcement can be provided in different ways. If the bathroom floor is otherwise unsupported, doubled headers located between doubled joists can be installed to help support these loads. The principles of construction are pretty much the same as described above for the construction of floor openings.

It is often possible to locate a second-floor bathroom over a first-floor partition. When this is the case, the partition can help support the weight of the floor, which will reduce the need for doubled joists. Better still, if the second-floor bathroom can be located above a first-floor hallway so that the partitions on either side of the hallway will help support the weight on the joists, no doubled framing may be required at all.

There is, however, another problem with second-floor bathroom floors that additional framing cannot cope with, and this is the problem of having to cut into the joists to make room for the plumbing.

The *closet bend* located under the water closet, which transports solid and liquid wastes from the water closet to the *soil stack,* can prove an especially difficult problem (for a description of plumbing systems *see* PLUMBING). The closet bend can be as much as 4 in. in diam., and it requires space under the flooring to make the sweeping turns that take it from under the water closet to the soil stack. The plumber and carpenter must decide in advance on the best location for the water closet and on the best way of running the closet bend. The carpenter can then allow for the weight of this fixture and also for the space required for the closet bend when he frames out the floor. What he usually does is treat the space required for the closet bend as another floor opening and he frames this space in the same way as he would a stairwell opening.

Partition Loads

There are two kinds of partitions: loadbearing and nonloadbearing. Loadbearing partitions support a load in addition to their own weight. Nonloadbearing partitions support only their own weight.

Joists supporting nonloadbearing partitions need not be especially strengthened in order to support the partitions, if the partitions run crosswise to the joists. The bottom plate of the partition is simply nailed to the rough flooring (*see* WALL FRAMING). If, however, the wall should run in a direction parallel to the joists, care should be taken that the partition is located over a joist.

For loadbearing partitions, if a partition runs parallel with the joists, the joists under the partition may either be doubled or spaced apart under the partition as shown in Fig. 11. If the joists are doubled, the carpenter must make sure that the partition is located directly above the joists. The joists must, in turn, be supported all the way to the footings by other partitions, beams, or walls.

When a loadbearing partition runs across the joists, every other joist under the partition is usually doubled. If the load being carried by the partition is unusually heavy, as when it is supporting the weight of a bathroom floor, for example, *every* joist can be doubled to help carry the load. Again, the joists must, in turn, be supported all the way to the footings in some way.

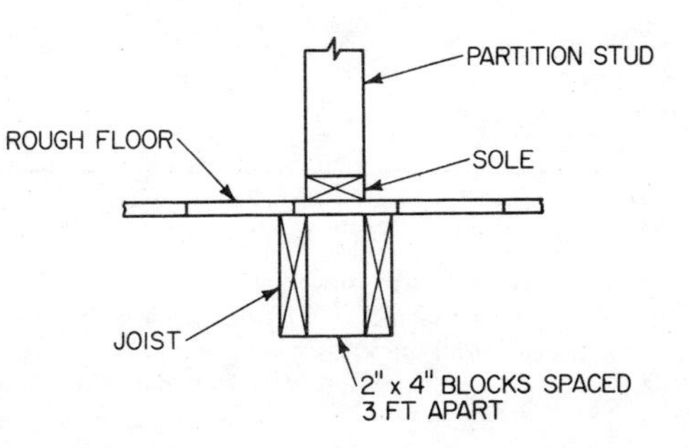

Fig. 11. A loadbearing partition running parallel to the joists must be properly supported by doubled joists. Sometimes, 2 × 4 in. blocks are nailed to the joists under the partition to give added support.

Installation Techniques

A few additional remarks on joist installation before we leave the subject. Joists should be installed carefully so that their top surfaces provide a level surface for the finish flooring. For the best construction, the carpenter should sight along the edges of each joist before installing it to see if it has warped in such a way that one edge has a crown to it. If it has, the carpenter must install that joist crown-side-up. After the subfloor and finished flooring have been nailed in place and the floor has been in service for a while, the joist will straighten itself out.

Since there may be slight differences in the depth of the joists, the carpenter must also bring the joists up to the same level surface before nailing them to their supporting members. He does this by placing shims under the low joists.

BRIDGING

Once the joists have been installed, *bridging* is nailed between them. Bridging consists of wood or metal bracing that runs from joist to joist in a crisscross pattern, as shown in Fig. 12; or it can consist of solid blocking.

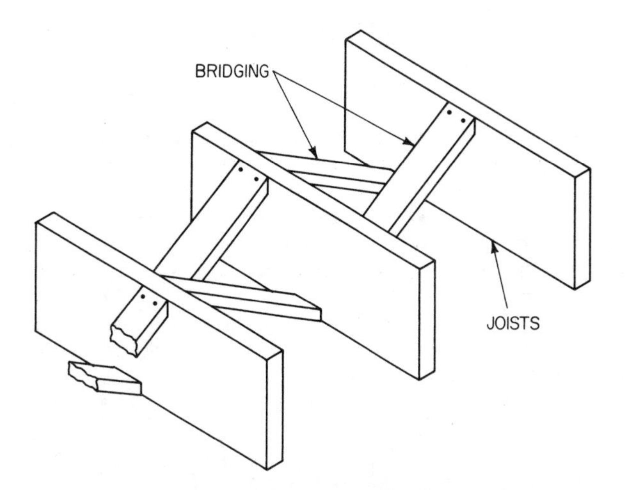

Fig. 12. Bridging installed between joists to stiffen the construction.

The main function of bridging is to stiffen long spans of unsupported joists since the joists would otherwise tend to bow sideways under a load. If a plastered ceiling were to be installed under such an unsupported floor, the plaster, not being able to resist this sideways pressure, would crack. A lack of bridging is also responsible sometimes for floor squeaks that develop because the nails holding the subflooring to the joists work loose as the joists flex back and forth.

It had long been believed that bridging helped to transfer a load concentrated upon one joist to the rest of the floor construction. The ability of bridging to spread loads in this way has been put in doubt as the result of experiments conducted by several building research organizations. It is now the general opinion that having a subfloor that is solidly nailed to the joists is of far greater value in helping to spread floor loads evenly.

Nevertheless, bridging is still required by some local building codes. The consensus at present seems to be that a line of bridging *may* be installed for every 8 ft of unsupported span—it depends on how solid the builder wants the overall floor construction to be; but a line of bridging *should* be installed if the joists are unusually deep and if the unsupported span is greater than 8 ft. An unusually deep joist means a joist in which the nominal depth is more than six times the nominal thickness. Thus, for example, a joist that is 2 × 14 in. in size requires bridging to give it an additional amount of stiffness, assuming the joist is also more than 8 ft long.

The most common type of bridging consists of strips of wood nailed crisscross between joists. For joists 2 × 10 in. or less in size, the wood is 1 × 4 in. in size. For joists larger than 2 × 10 in., the bridging may be either 2 × 2, 2 × 3, or 2 × 4 in. in size, depending on the depth of the joists.

The ends of wood bridging must always be cut accurately at such an angle as to provide solid bearing against the joists. When the carpenter first installs the bridging, he will nail only the upper ends of the bridging to the joists. The bottom ends are allowed to hang free until after the subfloor and finish floor have been completely installed, at which time it can be assumed that the joists will have settled into their final positions relative to each other. The bottom ends of the bridging are then nailed in place.

Metal bridging can also be used. If it is, the bridging must have a V-shaped cross section to give it stiffness. Bridging made of flat metal strapping is useless and should never be used.

SUBFLOORING

The subfloor is the final portion of the floor framing to be built. A subfloor has several functions: (1) it provides a working platform for the workmen; (2) it gives a base for the finish flooring; and most important, (3) it transforms the joists from a collection of individual beams into a single, rigid structure. This is especially true when the boards are laid diagonally to the joists rather than at right angles to them. A diagonally laid subfloor not only makes a single structural entity of the floor framing, but the boards that tie the corners of the framing together also help to stiffen and strengthen the entire structure of the house; subflooring that is laid at right angles to the joists cannot do this as well. In addition, a diagonally laid subfloor allows the finish flooring to be laid parallel to or across the joists in any room to suit the taste of the owner.

The lumber used for subflooring consists either of *matched* boards (which are boards that have their long edges, and sometimes their ends also, tongue-and-grooved or shiplapped) or *common* boards (which are boards having straight edges); or the subfloor can be made of plywood.

Board Subflooring

In the best-quality construction, nominal 1-in.-thick lumber, 6 to 8 in. wide, and with tongue-and-grooved or shiplapped edges is used. For ordinary construction, the boards are nominally ¾ in. thick.

To lay the boards diagonally, the carpenter starts in one corner of the house with a board that has been cut into an equilateral triangle. This is called the *starter board.* The carpenter then gradually works his way across the joists toward the opposite corner of the house. Each board must end over a joist, unless end-matched boards are used. Often, however, a gap of about ⅛ in. is left between the ends of abutting boards to allow for any expansion caused by wetness (and a surprising amount of water is used in a house during its construction). A gap of ⅛ in. must also be left wherever a board meets a partition or wall. If the boards are 6 or 8 in. wide, two nails are required at each joist and at both ends of each board. If the boards are more than 8 in. wide, three nails are required at each joist and at each end of the boards.

Plywood Subflooring

Plywood subflooring has the great advantage over wood-board subflooring in that it takes much less time to install—the cost of installing plywood can be as much as 50 percent less. One carpenter can usually install the plywood subflooring in a house in one day.

Plywood panels can be used in two different ways when they are used as subflooring. First, the panels can be installed merely as a substitute for wood boards, with the builder still intending to install a finish wood floor over the subfloor (refer to Fig. 13). For a description of this type of installation *see* FLOORING, WOOD. Wood flooring has considerable structural value of its own, which adds to the overall strength of the

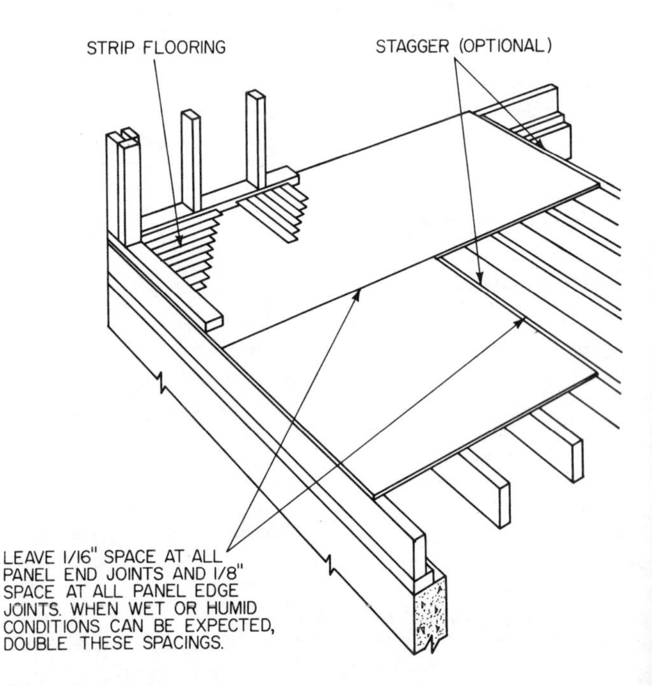

Plywood Subflooring

Panel indentification index	Plywood thickness, in.	Maximum span, in.
30/12	⅝	12*
32/16	½, ⅝	16†
36/16	¾	16†
42/20	⅝, ¾, ⅞	20†
48/24	¾, ⅞	24
1⅛" Groups 1 & 2	1⅛	48
1¼" Groups 3 & 4	1¼	48

*May be 16" if 25/32" wood strip flooring is installed at right angles to joists.
†May be 24" if 25/32" wood strip flooring is installed at right angles to joists.

Use 6d common nails for ½-in. plywood, 8d for thicknesses from ⅝ to ⅞ in., and 10d for 1⅛- and 1¼-in. thicknesses. Space nails at 6 in. along panel edges for all thicknesses. Along intermediate supports, space nails at 10 in., except when plywood spans 48 in., space nails at 6 in.

Fig. 13. The installation of plywood subflooring *(American Plywood Association).*

construction. The subflooring, therefore, need not be too thick, whether it consists of wood boards or plywood panels.

Second, instead of wood flooring the builder may intend to install a resilient flooring material, such as asphalt tiles, vinyl tiles, or linoleum, or he may intend to install a carpeting material over the subfloor. Apart from their obvious physical

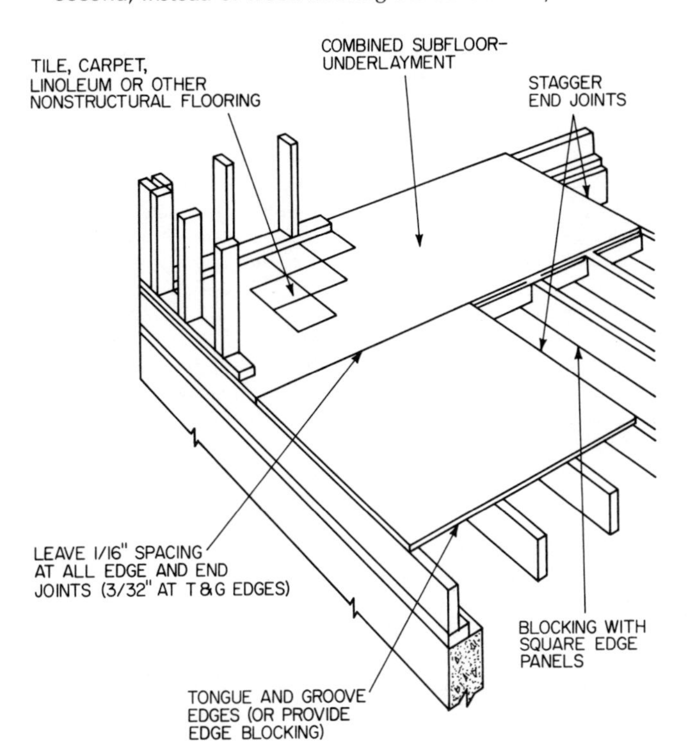

Combined Subfloor-Underlayment

Plywood grade	Plywood species group	Maximum support spacing 16" o.c. Min. panel thickness, in.	20" o.c. Min. panel thickness, in.	24" o.c. Min. panel thickness in.
UNDERLAYMENT INT-APA (with interior, intermediate or exterior glue)	1	½	19/32	23/32
or UNDERLAYMENT EXT-APA	2 & 3	19/32	23/32	⅞
(C-C Plugged)	4	23/32	⅞	1

To minimize the effects of framing shrinkage, ring-shank or spiral-thread nails should be used. Use 6d deformed-shank nails for thicknesses through ¾ in. Use 8d for panels ⅞ in. and thicker. Space nails at 6 in. along panel edges and at 10 in. along intermediate supports. Unless joists are of thoroughly seasoned material and have remained dry during construction, countersink nail heads 1/16 in. below surface of the underlayment just prior to laying finish floors to avoid nail popping. Do not fill holes. If resilient flooring is to be applied, thoroughly sand joints.

The T&G joint is designed so that the upper plies of the panel will be spaced to avert ridging if the panel picks up moisture and expands. Joints should not be tightly butted, but left open slightly. Normally a space of about 3/32 in. (the width of a 6d box nail) between the upper plies of the panels will be enough for the T&G joint. A space of 1/16 in. is recommended for panel end butt joints.

If wet conditions are anticipated, additional spacing of up to 1/16 in. at both sides and ends is advisable.

Fig. 14. Installation of a combined subfloor and underlayment when the finish flooring will consist of tiles or carpeting *(American Plywood Association)*.

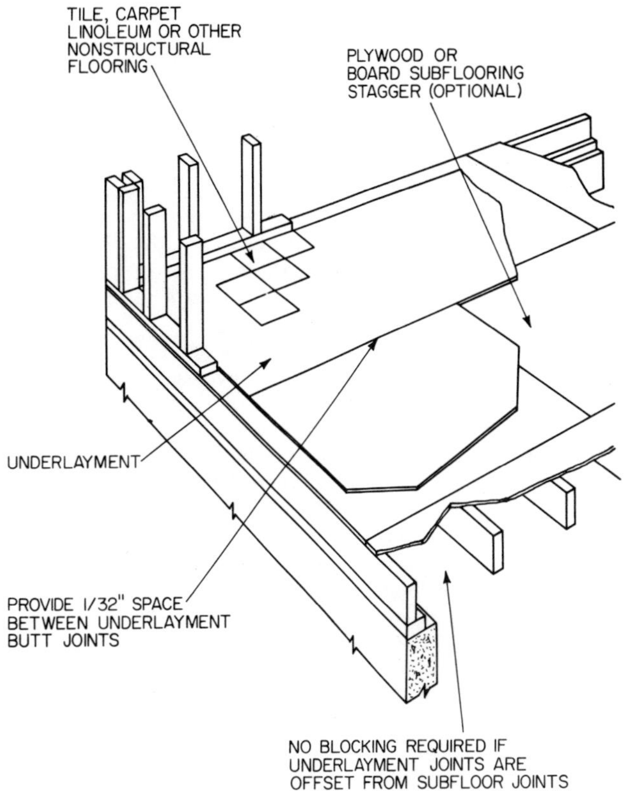

Plywood Underlayment

Plywood grades and species group	Application	Minimum plywood thickness, in.
Groups 1, 2, 3, 4, 5 UNDERLAYMENT INT-APA (with interior, intermediate or exterior glue)	Over plywood subfloor	¼
or UNDERLAYMENT EXT-APA (C-C Plugged)	Over lumber subfloor or other uneven surfaces	⅜
Same grades as above, but Group 1 only	Over lumber floor up to 4" wide; face grain must be perpendicular to boards	¼

Use 3d ring-shank nails for underlayment up to ½ in. thickness, 4d for ⅝ in. and thicker; 16 gauge staples may be used, except 18 gauge may be used with ¼-in. thick underlayment. Crown width should be ⅜ in. for 16 gauge staples, 3/16 in. for 18 gauge. Length should be sufficient to penetrate subflooring at least ⅝ in. or extend completely through.

Space fasteners at 3 in. along panel edges and at 6 in. each way along intermediate supports except for ⅜-in. or thicker underlayment applied with ring-shank nails. In this case, use 6-in. spacing along edges and 8-in. spacing each way along intermediate supports.

Fig. 15. The installation of underlayment plywood over a wood board or plywood subfloor, when the finish floor will consist of tiles or carpeting *(American Plywood Association)*.

differences, all these materials have in common the fact they have no structural value whatsoever. The subflooring, therefore, must provide the strength that would otherwise have been provided by the wood flooring. In this case, therefore, the plywood subfloor must be much stronger and stiffer than if a wood floor were to be laid down over it (refer to Fig. 14).

The plywoods used for both types of installation are manufactured in different qualities and thicknesses to suit different requirements. If, for example, the subflooring is going to be exposed to the weather for a considerable period of time, and the climate is a rainy one, then the plywood must be manufactured with a waterproof glue that will resist separation of the plies. That is, the plywood must be an Exterior plywood. If the climate is dry, or the subfloor will be exposed to the weather for only a brief period of time, then an Interior panel made either with an interior or an exterior glue can be used (*see* PLYWOOD).

The thickness of the panels will depend in part on the spacing between the joists. Obviously, the wider the joist spacing, the thicker (and, thus, stiffer) the panels must be. The stiffness of any particular panel will also be dependent in part on the wood used in its manufacture, some woods being stronger and more rigid than others. This is reflected in the tables of Figs. 13 and 14, which show that plywoods in Group 1 can be thinner than plywoods in Group 4 used for the same service, the reason being that the plywoods in Group 1 are made from woods that are inherently stronger than the woods used for Group 4 plywoods. (For a description of the grading of plywood panels *see* PLYWOOD.)

In sum, the builder has a choice among plywoods, and the particular plywood he selects will depend on the costs of the different panels balanced against the purpose to which he wants to put them.

Plywood panels are usually installed with their long sides laid crosswise to the direction of the joists. This construction is stronger since more joists are tied together by a single panel. The long sides of the panels should also be staggered so that adjacent panels do not begin or end on the same joist. Nor should the panels be butted tightly against each other, whether they are square-edged or tongue-and-grooved. Instead, a gap of about 1/16 in. should be left between panels along their 4-ft edges, and an 1/8-in. gap should be left along their 8-ft edges. The gaps are necessary because, despite popular belief to the contrary, plywood panels do expand slightly when they become wet.

Sometimes a builder will install a wood-board subfloor over which he intends to install a resilient flooring material or perhaps a carpeting material. When this is the case, he must then install lightweight plywood panels over the entire subfloor, as subflooring made of boards is neither flat enough nor smooth enough to prevent irregularities and gaps between the boards from showing through the resilient flooring or carpeting.

These lightweight plywood panels are called *underlayment* (refer to Fig. 15). Like other types of plywood paneling, underlayment is made in Exterior and Interior grades, from different woods, and with exterior and interior glues, to suit the particular application.

The underlayment should be installed just before the finish flooring, which means the house will have been enclosed by this time and there is no danger of rain soaking into or damaging the plywood.

Flooring, Wood

Wood is the preferred material for residential flooring because of its attractive appearance, durability, and warmth (warm to the touch, that is); because it has an elasticity to it that makes it a comfortable material to walk and stand on, and because it *is* the traditional flooring material.

Wood flooring consisted originally of long softwood planks, 3½ in. to 10 in. wide, that were fastened to the underlying joists with wood pegs. This *plank* flooring is still available and in a variety of hardwoods as well as softwoods, as described below.

Planks gradually evolved into long, narrow (3½ in. or less in width) *strip* flooring that is manufactured from both hardwoods and softwoods to very close tolerances and in standardized patterns. Each strip of wood has a tongue cut into one side and a groove into the opposite side. These tongues and grooves (t & g) mate with matching tongues and grooves machined into adjoining strips of flooring as shown in Fig. 1; flooring with t & g sides is said to be *side-matched.* The ends of strip flooring are often cut t & g also, and the ends of the strips are then said to be *end-matched.* In fact, the term *matched* flooring is often used synonymously with *strip* flooring to distinguish t & g flooring from flooring on which the sides and/or ends are left square.

The third basic type of flooring is *parquet* flooring, of which a subclassification is *block* flooring. Parquet flooring consists of narrow, relatively short pieces of wood (18 in. or less in length) that are laid in a wide variety of decorative patterns collectively called *parquetry,* some of which can be extremely elaborate.

One of the simpler methods of forming a parquet pattern is simply to lay the strips of wood so they form squares, with the long sides of the strips in adjoining squares running in opposite directions. This gives a checkerboard effect to the flooring. This style of laying a parquet pattern has evolved into block flooring, in which strips of wood are preassembled into square or rectangular blocks at a mill and then installed as complete units by a carpenter to form a variety of block patterns. All four sides of these assembled blocks are very often cut t & g to simplify their installation.

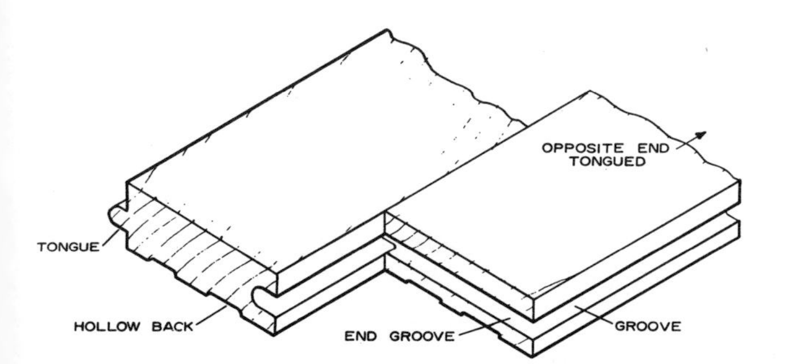

Fig. 1. Side- and end-matched wood flooring *(U.S. Forest Service).*

HARDWOODS VERSUS SOFTWOODS

About 80 percent of all wood flooring is made from one species or another of hardwood. Hardwoods are stronger and harder than softwoods, they wear better, and the grain doesn't splinter under prolonged use as happens with softwoods. Hardwoods are also much less likely to be abraded, or marked, than softwoods. Finally, hardwoods have more interesting grain patterns, and they can take a transparent wood finish that brings out these patterns with excellent effect.

When a softwood is selected as flooring by a reputable builder it is usually for use in an attic or other out-of-the-way place, or because the builder intends using the wood flooring as a base upon which he will lay carpeting or a resilient flooring material. Softwoods are also used as the finish flooring in the bedrooms and closets of low-cost houses because the traffic is light and the wood is unlikely to receive much wear. But when a softwood is installed as the finish flooring in the main living quarters of a house (aside from a house intended as low-cost dwelling), any potential purchaser should take it as a sign that the builder has cut costs in the construction as much as possible.

Varieties of oak account for about 92 percent of all hardwood flooring installed, with maple following far behind with only 6 percent of the total. Straggling behind maple with the remaining 2 percent are several other hardwoods such as beech, birch, and pecan.

Yellow pine accounts for about 50 percent, and Douglas fir about 40 percent, of all softwood flooring installed, mainly because these two woods are the principal softwoods logged for home construction lumber. The remaining 10 percent is divided among eastern and western hemlock, white pine, ponderosa pine, larch, spruce, cypress, and redwood. There is little to choose from among the softwoods, structurally speaking. The selection of any particular species of softwood for flooring depends more on price and local availability than on any other factor.

HARDWOOD FLOORING

Manufacture

Hardwood flooring is more expensive than softwood flooring for three reasons: (1) there are fewer suitable hardwood trees available for lumbering, (2) only a relatively small amount of usable lumber can be obtained from a hardwood tree, and (3) the manufacturing process is more complex.

The log is first rough-sawn into the approximate sizes of the strips. The lumber is then left to air-dry in open sheds for several months until the moisture content of the wood has equalized throughout, which is necessary to give the lumber time to work out the stresses that invariably develop during the drying-out process. Defects such as warping, twisting, cupping, etc., are much less likely to occur thereafter. The lumber is then placed in a kiln and further dried until the moisture content has been reduced to between 6 and 9 percent.

Manufacture can now begin. The rough-cut strips of lumber are first planed smooth on all sides. Defects are then cut out of the strips, which results in a considerable number of short and odd-length pieces. The strips then pass through a number of machines that trim them to their final sizes and patterns, including the cutting of tongues and grooves.

Grading

Grading of the strips is accomplished according to rules promulgated by a trade association to which the mill belongs. Thus, manufacturers of oak flooring belong to the National Oak Flooring Manufacturers' Association (NOFMA), and manufacturers of maple, beech, and birch flooring belong either to the NOFMA or the Maple Flooring Manufacturers' Association (MFMA).

Grading consists of separating the strips according to color, the number and size of visible defects, the direction of the grain, and length. Résumés of the hardwood associations' grading rules are shown in the tables later in this article. Every piece of flooring that has been manufactured and graded according to an association's rules has that association's trademark stamped on the bottom of the flooring. This stamp will also indicate the grade of the flooring and a number that identifies the mill where the flooring was manufactured. Flooring that does not meet the minimal grading rules of the association is not stamped.

The grading rules also differentiate between flooring in which the grain of the wood runs parallel with the floor or at right angles to the floor. (The nomenclature describing the grain directions in hardwoods and softwoods is complicated; for details, *see* LUMBER).

The distinction is important because in hardwoods the more attractive grain patterns are visible when the grain runs at right angles to the floor (that is, vertically through the flooring; see Fig. 2). Vertical-grained flooring is also less affected by changes in humidity than horizontal-grained flooring, and vertical-grained flooring also wears longer. Financially, the distinction is also important because vertical-grained flooring costs more.

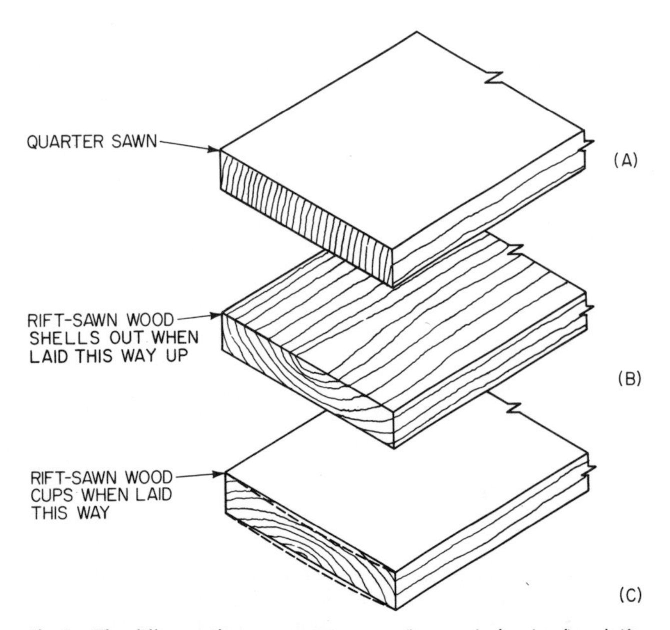

Fig. 2. The difference between quarter-sawn (i.e., vertical grained) and rift-sawn (i.e., horizontal grained) wood flooring.

Flooring Sizes

Table 1 shows the sizes in which hardwood flooring is made. The flooring installed in dwellings is usually $\frac{3}{4}$ in. thick and $2\frac{1}{4}$ in. wide. The pattern of the flooring is shown in Fig. 3. Note particularly the wide channel, or hollow, that is cut along the bottom of the strip, the off-center positions of the tongue and groove, and the fact that the bottom edges of the tongue and groove are undercut by about $\frac{1}{32}$ in.

Table 1. Standard Sizes, Counts, and Weights Manufactured under the Grading Rules of the NOFMA

Nominal is the size designation used by the trade, but it is not always the actual size. Sometimes the actual thickness of hardwood flooring is $\frac{1}{32}$ in. less than the so-called nominal size. *Actual* is the mill size for thickness and face width, excluding tongue width. *Counted* size determines the board feet in a shipment. Pieces less than 1 in. in thickness are considered to be 1 in.

Oak			
Nominal, in.	Actual, in.	Counted, in.	Weights m ft, lb
Tongued and grooved—end matched			
$\frac{3}{4} \times 3\frac{1}{4}$	$\frac{3}{4} \times 3\frac{1}{4}$	1×4	2210
$\frac{3}{4} \times 2\frac{1}{4}$	$\frac{3}{4} \times 2\frac{1}{4}$	1×3	2020
$\frac{3}{4} \times 2$	$\frac{3}{4} \times 2$	$1 \times 2\frac{3}{4}$	1920
$\frac{3}{4} \times 1\frac{1}{2}$	$\frac{3}{4} \times 1\frac{1}{2}$	$1 \times 2\frac{1}{4}$	1820
$\frac{3}{8} \times 2$	$\frac{11}{32} \times 2$	$1 \times 2\frac{1}{2}$	1000
$\frac{3}{8} \times 1\frac{1}{2}$	$\frac{11}{32} \times 1\frac{1}{2}$	1×2	1000
$\frac{1}{2} \times 2$	$\frac{15}{32} \times 2$	$1 \times 2\frac{1}{2}$	1350
$\frac{1}{2} \times 1\frac{1}{2}$	$\frac{15}{32} \times 1\frac{1}{2}$	1×2	1300
Square edge			
$\frac{5}{16} \times 2$	$\frac{5}{16} \times 2$	face count	1200
$\frac{5}{16} \times 1\frac{1}{2}$	$\frac{5}{16} \times 1\frac{1}{2}$	face count	1200
Beech, birch, hard maple, and pecan			
Nominal, in.	Actual, in.	Counted, in.	Weights m ft, lb
Tongued and grooved—end matched			
$\frac{3}{4} \times 3\frac{1}{4}$	$\frac{3}{4} \times 3\frac{1}{4}$	1×4	2210
$\frac{3}{4} \times 2\frac{1}{4}$	$\frac{3}{4} \times 2\frac{1}{4}$	1×3	2020
$\frac{3}{4} \times 2$	$\frac{3}{4} \times 2$	$1 \times 2\frac{3}{4}$	1920
$\frac{3}{4} \times 1\frac{1}{2}$	$\frac{3}{4} \times 1\frac{1}{2}$	$1 \times 2\frac{1}{4}$	1820
$\frac{3}{8} \times 2$	$\frac{11}{32} \times 2$	$1 \times 2\frac{1}{2}$	1000
$\frac{3}{8} \times 1\frac{1}{2}$	$\frac{11}{32} \times 1\frac{1}{2}$	1×2	1000
$\frac{1}{2} \times 2$	$\frac{15}{32} \times 2$	$1 \times 2\frac{1}{2}$	1350
$\frac{1}{2} \times 1\frac{1}{2}$	$\frac{15}{32} \times 1\frac{1}{2}$	1×2	1300
Special thicknesses (t & g, end matched)			
$\frac{17}{16} \times 3\frac{1}{4}$	$\frac{33}{32} \times 3\frac{1}{4}$	$\frac{5}{4} \times 4$	2400
$\frac{17}{16} \times 2\frac{1}{4}$	$\frac{33}{32} \times 2\frac{1}{4}$	$\frac{5}{4} \times 3$	2250
$\frac{17}{16} \times 2$	$\frac{33}{32} \times 2$	$\frac{5}{4} \times 2\frac{3}{4}$	2250
Jointed flooring—i.e., square edge			
$\frac{3}{4} \times 2\frac{1}{2}$	$\frac{3}{4} \times 2\frac{1}{2}$	$1 \times 3\frac{1}{4}$	2160
$\frac{3}{4} \times 3\frac{1}{4}$	$\frac{3}{4} \times 3\frac{1}{4}$	1×4	2300
$\frac{3}{4} \times 3\frac{1}{2}$	$\frac{3}{4} \times 3\frac{1}{2}$	$1 \times 4\frac{1}{4}$	2400
$\frac{17}{16} \times 2\frac{1}{2}$	$\frac{33}{32} \times 2\frac{1}{2}$	$\frac{5}{4} \times 3\frac{1}{4}$	2500
$\frac{17}{16} \times 3\frac{1}{2}$	$\frac{33}{32} \times 3\frac{1}{2}$	$\frac{5}{4} \times 4\frac{1}{4}$	2600

Source: National Oak Flooring Manufacturers Association.

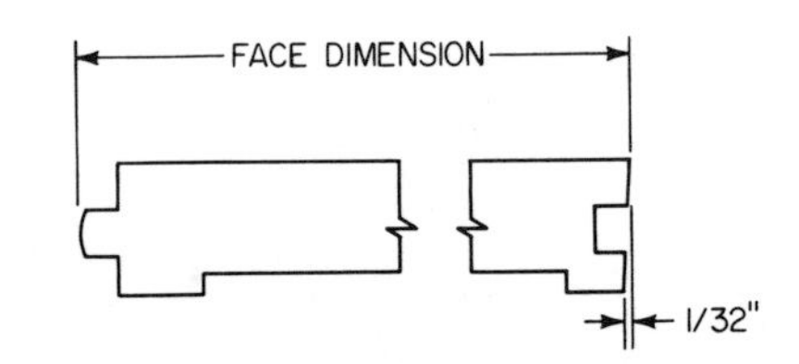

Fig. 3. Standard pattern of hardwood flooring.

The channel (some flooring has two narrow channels instead of one wide channel) has several functions: (1) it allows a more level floor to be laid since the channel makes it less likely that imperfections in the subflooring will affect the level installation of the flooring, (2) it makes it less likely that any moisture in the subflooring will be transferred to the finish flooring, (3) it helps to give a bit more springiness to the floor, and (4) it counteracts any tendency the flooring may have to *cup* (that is, for the two long edges of the wood to warp upward at the same time) in case moisture should ever get into the wood.

The off-center position of the tongue and groove allows a greater thickness of wood to be scraped or sanded away as the floor is refinished with the passing of years.

The undercuts along the bottom edges of the flooring enable the top edges of the strips to butt tightly against each other, which makes for a more attractive and watertight construction.

Flooring that is only 11/32 in. (⅜ in. nominal) and 15/32 in. (½ in. nominal) is also manufactured. They are intended for use on top of an existing wood floor or for use as the finish flooring in bedrooms and other less-used rooms in a house. The use of flooring thinner than ¾ in. is not recommended, however, because there is an increased likelihood that nails will work loose, which will result in squeaking floors; it is also more difficult to nail thin flooring down in the first place because the wood has a greater tendency to split, especially at the ends of the strips; the strips are also more likely to cup if the wood becomes damp; and, finally, there is less wood present for later refinishing. The use of thinner flooring is not, therefore, as economical a proposition as may at first appear, and certainly not in the long run.

Flooring thicker than ¾ in. is also made, but it is intended for use in institutions and commercial establishments in which the floors receive very hard use. There is no reason to consider this flooring for use in the home at all.

Flooring wider than 2¼ in. is also manufactured. However, strip flooring that is wider than 2¼ in. has a greater tendency to cup, and the wider strips require closer nail spacing to counteract this tendency.

Bundling

Because of the wide range of lengths that are produced in the manufacturing process, hardwood strip flooring has traditionally been sold in bundles that contain strips of approximately the same length, give or take 6 in. Thus, a bundle supposedly containing 4-ft-long strip flooring will in fact contain strips that range from 3½ to 4½ ft in length. The longer the average length of strip flooring in a bundle, the more expensive the bundle will be.

Bundles are also sold to contain so many board feet of flooring per bundle (see Table 2). That is, the size of the strips is

Table 2. Calculating the Quantity of Strip Flooring to Use in Board Feet

Strip Dimensions, in.		Pieces per	Bundle Quantity, bd ft§		Required Quantity, bd ft¶	
Actual*	Counted†	Bundle‡				
¾ × 3¼	1 × 4	12	4		1.29	
¾ × 2¼	1 × 3	12	3	times	1.383	times
¾ × 2	1 × 2¾	12	2.75	the	1.425	the
¾ × 1½	1 × 2¼	12	2.25	average	1.55	area to
15/32 × 2	1 × 2½	18	3.75	bundle	1.30	be
15/25 × 1½	1 × 2	18	3	length	1.383	covered
11/32 × 2	1 × 2½	24	5	(ft)	1.30	(ft)
11/32 × 1½	1 × 2	24	4		1.383	

*Nominal thickness for 15/32 and 11/32 is ½ and ⅜ respectively; ¾ is the same.
†Equivalent dimensions used for calculating bd ft quantity.
‡1 × 4 and 1 × 3 pine flooring commonly is packaged 6 pieces to the bundle.
§Mulitiplier is equal to counted size times number of pieces per bundle divided by 12.
¶Multiplier makes allowance for side-matching, end-matching and normal waste.

Table 3. Summary of the Grading Rules for Oak Flooring According to NOFMA*

Type of wood	Grade	Description	Length of strips, ft†	Thickness & width, in.
Plain-sawed or Quarter-sawed	Clear	Face practically clear, admitting an average of ⅜ in. bright sap	1¼ and up; average 3¾	
	Select	Face may contain burls, small streaks, pinworm holes, slight imperfections in working and small tight knots that do not average more than 1 to every 3 ft	1¼ ft and up; average 3¼	¾ × 3¼ × 2¼
Plain-sawed	No. 1 common	Lays good residential floor and may contain varying wood characteristics as flags, heavy streaks, and checks, wormholes, knots, and minor imperfections in working	1½ ft and up; average 2¾	× 2 × 1½ ½ × 2 × 1½
	No. 2 common	May contain sound natural variations of the forest product and manufacturing imperfections. Furnishes an economical floor suitable for homes, general utility use or where character marks and contrasting appearance are desired.	1¼ and up; average 2¼ lengths	⅜ × 2 × 1½

*Data are condensed from grading rules of the National Oak Flooring Manufacturers Association (NOFMA) and are for t & g flooring. All graded flooring is kiln-dried grade-marked and trade-marked. For square-edged flooring see current NOFMA grading rules.
†Also available in nested flooring.

rounded off to the next highest full inch in both width and thickness and the next highest full foot in length, and the result is a certain measure in board feet. This method of measuring lumber seems tricky to the uninitiated (who always feels the lumber yard is short-changing him), but it is the traditional method of calculating a quantity of lumber and everyone associated with the lumber trade in any shape, manner, or form knows this convention and takes it into account when estimating, buying, or selling lumber.

The NOFMA also packages oak, maple, beech, and birch flooring in bundles containing mixed lengths. An individual bundle will contain twelve courses of strip flooring, each course being between 7½ and 8½ ft long. The individual lengths of flooring in each course will range from 9 in. to 8½ ft in length. The total number of short pieces, that is, those pieces between 9 and 18 in. in length, and the average length of all the pieces in a mixed bundle are specified by the grading rules of the NOFMA, which permits a larger proportion of short pieces and a shorter average length the lower the grade. This is shown in Table 3. In the highest grade of oak flooring, for example, the average length of the pieces must be at least 3¾ ft. In the lowest grade of oak flooring, on the other hand, the pieces can average as little as 2¼ ft in length. The price per bundle varies accordingly, of course.

Oak Flooring

Oak is classified as being either white oak or red oak. The difference in color is fairly inconspicuous, but if samples of red and white oak are compared directly, it can be seen that red oak has a faint pinkish tinge to it. The grading rules, however, make no distinction at all regarding color, but the mills will usually pack either all-red-oak or all-white-oak bundles anyway. The white oak does have a more prominent and attractive "ray" pattern to it when the flooring is cut to show the vertical grain; on the other hand, red oak has a more even-looking appearance. As far as structural properties are concerned, there is no difference at all between red oak and white oak.

Maple Flooring

The maple used for flooring is obtained from the sugar maple, which is also known as hard maple and rock maple. The wood is exceptionally strong and hard, which makes it suitable for industrial flooring as well as residential flooring. Gymnasium floors are almost invariably made of maple. In the "old days," kitchen floors were invariably laid in maple also.

Maple is graded under the rules of the NOFMA (see Table 3) as well as the MFMA, since some members of the NOFMA log maple as well as oak. Although there are some differences in the nomenclature of the grading rules, for all practical purposes the rules are the same.

Since maple is very strong regardless of the way the grain runs, the grading rules make no distinction as to grain direction in the flooring. Maple does have a special color grade for First Grade flooring, however, this being First Grade White Hard Maple, for those who want uniformity in the color of their floor, as the strips in an ordinary bundle will have a range of tones.

Birch and Beech Flooring

Almost everything said above regarding maple flooring applies also to birch and beech, except that these two woods are not quite as strong as maple and they are, therefore, somewhat less wear-resistant and less expensive. Both woods are graded under the rules of both the NOFMA and the MFMA, just as maple is.

SOFTWOOD FLOORING

The manufacture of softwood strip flooring is much simpler than that of hardwood strip flooring. This follows from the less dense and more straight-grained nature of most softwoods. The wood is first rough-cut to the approximate size of the strips and then placed in a kiln for about 96 hr until the moisture content of the wood has been reduced to about 9 or 10 percent. The wood is then finish-planed to size in one operation—tongue, groove, channels, and all—with the ends of the strip being trimmed in a subsequent operation. Flooring sizes and patterns are very similar to those of hardwood flooring.

Softwood flooring can be cut into equal-sized lengths without any difficulty at all. For ease of installation, however, the flooring is bundled in packages that contain lengths that vary between 1 ft and 8 ft.

Each species of softwood flooring is graded in accordance with the rules of the lumber association that regulates the manufacture of that species. The principal grading associations are the Southern Pine Inspection Bureau for yellow pine, and the West Coast Lumber Inspection Bureau for Douglas fir. Neither of these associations requires specific trade or grade marking for strip flooring. The grade markings that are used are merely those of the usual lumber grades.

INSTALLATION OF STRIP FLOORING

Preliminary Considerations

Strip flooring can be laid satisfactorily over a wood subfloor (boards or plywood), directly upon joists (i.e., without a subfloor), or on a concrete slab. Whatever method is followed, it is essential that before the flooring is installed the rest of the house be completed as much as possible. Laying the finish flooring should be one of the last jobs started, and it should not be started until all the plastering has been completed, all the wallpaper has been hung, and all the paint has been applied. All the water absorbed by these materials must have evaporated and the interior of the house must be bone-dry before the laying of the floor begins.

There are two reasons for these precautions. One is that the flooring is a finished material manufactured to very close tolerances. All the workmen should have finished their jobs and be out of the house to make sure that there is no one about who might accidentally nick, scratch, or otherwise damage the flooring.

But the more important reason is the sensitivity of the wood to any moisture in the air (refer to Fig. 4). When hardwood flooring is shipped from the mill it will have a moisture content of somewhere between 6 and 9 percent. This is equal to or below the maximum permissible required moisture content for wood flooring installed in any given part of the United States. In the South, for example, wood flooring should not have a moisture content greater than 10 percent. In the rest of the United States, except for the northern states, the maximum permissible moisture content is about 8 percent, and in the northern states it should be 6 or 7 percent.

If, before it has been installed, this flooring is allowed to absorb moisture from wet plaster, wet wallpaper, or a wet water-based paint, the flooring will swell. After it has been installed, as the flooring dries out again, it will shrink down, and gaps will open up between the strips, or the strips may warp or cup. If these defects are serious enough, the flooring will have to be relaid.

It is necessary, therefore, to take certain precautions to keep the flooring dry. For one thing, the builder must make sure that the lumber yard from which he buys the flooring has stored it

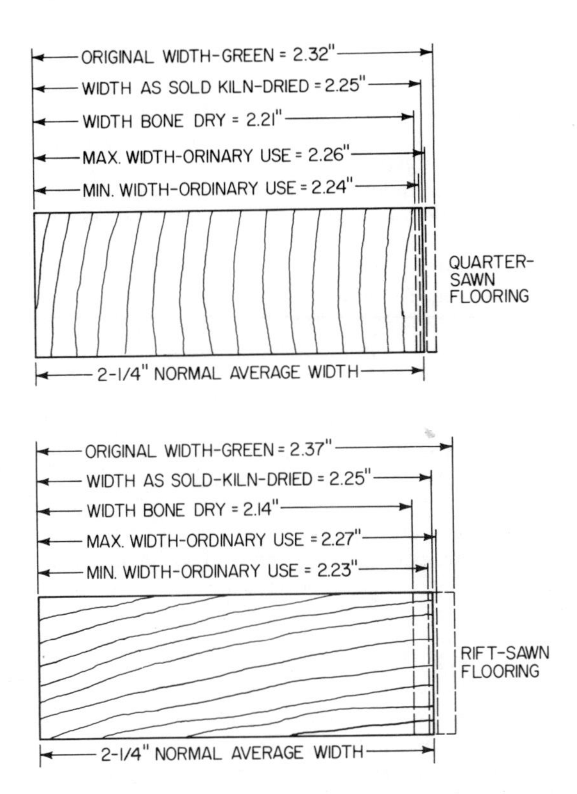

Fig. 4. These two illustrations show how much green lumber will shrink as it is dried to the kiln-dried and bone-dry states. Note also how much more the rift-sawn wood will shrink compared to quarter-sawn wood.

indoors under suitably dry conditions. Once the flooring has been delivered to the jobsite, it must continue to be stored under dry conditions and in a location where it is unlikely to be accidentally damaged. It should not be delivered, certainly, as long as the interior of the house is damp for any reason.

The best time of year to install strip flooring is during the winter months, when the humidity tends naturally to be low and the house is entirely closed from the weather. The heating system should also be in operation. A few days before the flooring is due to be installed, it can be unpacked and the strips laid out in loose piles to allow the wood to reach a moisture content that is in equilibrium with the moisture content of the wood in the rest of the house. When these precautions have been taken, it is unlikely that any problems will develop with the flooring after it has been installed.

Conditions are, however, entirely different during the summer months, particularly in the South. To maintain its dryness, the flooring should not be unpacked until just before it is to be installed. To reduce the amount of humidity within the house to an acceptable level, the heating system should be turned on during the night for a week or two before the floor-laying is to begin. The thermostat should be set about 15°F higher than the expected nighttime temperatures and turned down again during the day. The house should be kept dry in this way until the flooring has been completely laid. During the fall, winter, and spring, a constant minimum temperature of 60°F should be maintained in the house.

An alternative method of drying out a house, if air conditioning has been installed throughout it, is to turn on the air-conditioning system and allow it to operate continuously until the humidity level within the house has been reduced to a low equilibrium condition. The flooring should be stored inside the house during this period to make sure it reaches the same equilibrium condition.

There is a considerable advantage in installing vertical-grain flooring in humid sections of the United States since this type of flooring tends to expand and contract much less with changes in humidity than horizontal-grain flooring (see Fig. 4). By using vertical-grain lumber, therefore, one is much less likely to encounter problems either with the laying of the floor or with later changes in humidity.

Installation Over A Wood Subfloor

The subfloor is first covered with a layer of building paper (see Fig. 5), preferably 15-lb, asphalt-saturated felt, the edges of which overlap by at least 4 in. Sometimes 30-lb felt is used to increase the thermal insulation of the floor. In any case, the building paper will prevent drafts passing up through the flooring, which might bring dust up from the basement or under the house. The building paper will also help to exclude moisture from the finish flooring, it will help to muffle the sounds of footsteps a bit; and it will add a bit of springiness to the floor.

The direction in which the flooring should be laid must be decided. The subflooring, if it consists of wood boards, should have been laid diagonally to the joists. This will allow the finish flooring to be laid in either direction in the room (*See* FLOOR FRAMING.)

The preferred direction of installation is with the finish flooring running parallel with the length of the room. Less cutting and fitting are necessary, the flooring will swell and shrink less with seasonal changes in humidity, and the flooring presents a more attractive appearance. If the joists happen to run across the short side of the room so much the better, as this will help strengthen the construction.

Flooring should always run the length of a hallway. When two rooms merge into each other, as when a living room and dining room are separated only by a wide archway, the flooring is usually run parallel to the combined lengths of the living and dining rooms, regardless of what the individual dimensions of these rooms might be. Again, the total amount of dimensional change that is likely to take place because of humidity changes will be less, and the overall effect will be more attractive. If it is impossible for some reason to run the flooring in the same direction through adjoining rooms, then

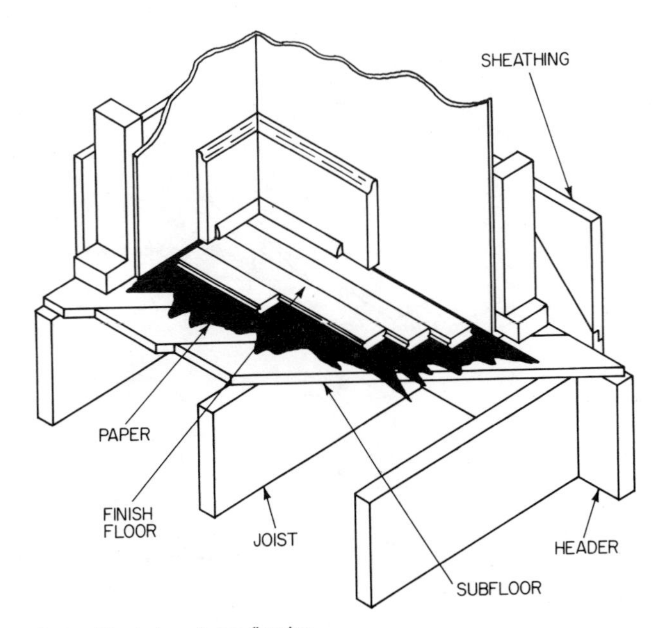

Fig. 5. The laying of strip flooring.

the change in direction should occur at the door, where it will look most natural.

To start the actual floor-laying, a long, straight strip of flooring is selected and nailed in place about ½ to ⅝ in. away from the wall surface, as shown in Fig. 6. This gap in the flooring, the purpose of which is to allow space for the flooring to expand in humid weather, will be covered later by a base molding and/or base shoe. The nails holding the flooring in place will be hidden by the same moldings. The tongue side of the strip is then toe-nailed to the subfloor as shown in Fig. 6.

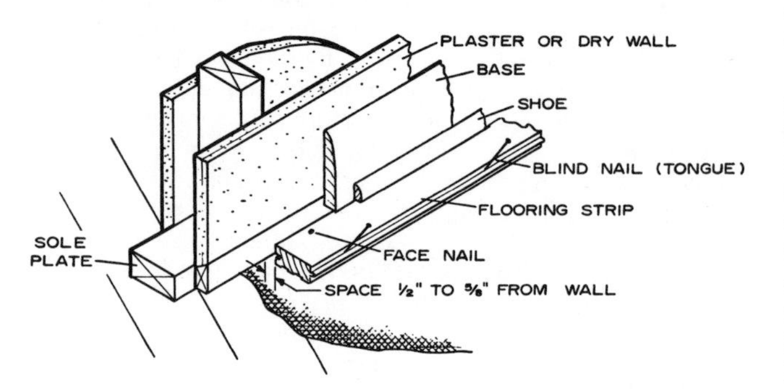

Fig. 6. Installing the starter strip along the wall.

A two- or three-man crew usually works together to lay the flooring, one or two men cutting and fitting the lengths of flooring together into courses while the third man does the actual nailing. The nailer works bent over like a football lineman, facing the direction in which the flooring is being laid. He sets a strip in place by driving it with a hammer against the flooring that has already been placed. To prevent damaging the strip, he uses a scrap piece of flooring as a buffer. The nailer than stands on the strip to hold it in position while he drives the nails. A nail is driven into the wood as far as it will go without the head of the hammer actually striking the wood. The nail is then driven flush with the surface of the wood with a nail set (see Fig. 7).

As the opposite wall is approached, any lack in parallelism between the walls will become apparent. If the difference is slight, the strips making up the final course can be sawed or planed at a slight angle until their edges are aligned with the wall surface. They can then be nailed in place. As with the starter strip, a ½ to ⅝ in. gap is left between the flooring and the wall to allow for expansion.

If there is a considerable difference in parallelism, the floor layer will have to follow another procedure. First, to backtrack a bit, before the floorlaying had begun, the floor layer should have measured the distance between the walls to check on their parallelism. Therefore, any lack of parallelism should not come as a surprise to him, and he should, in fact, have planned on an appropriate course of action beforehand. There are two things he can do.

Either he can measure the difference in the width of the room at opposite ends of the walls and then split this difference between the two opposite sides of the room, or he can plan on absorbing this lack of parallelism entirely on one side of the room. In the first case, this means cutting both the starter course and the final course at the same offset angles, so that the remaining courses of flooring can be laid parallel with each other. In the second case, he lays the starter course parallel with the wall as if the room were perfectly square, and then as he approaches within 2 or 3 ft of the opposite wall, he can begin to plane the strips at slight angles so that when the final course is laid down, it will be parallel with the wall.

The final course of flooring cannot be toe-nailed in place as there is no room in which to swing a hammer. The course must be face-nailed. The flooring is pressed tight against the preceding course by driving wedges between it and the wall. The strips are then nailed down in the same way as described above for the starter course. These nails will also be hidden by the base and/or shoe molding.

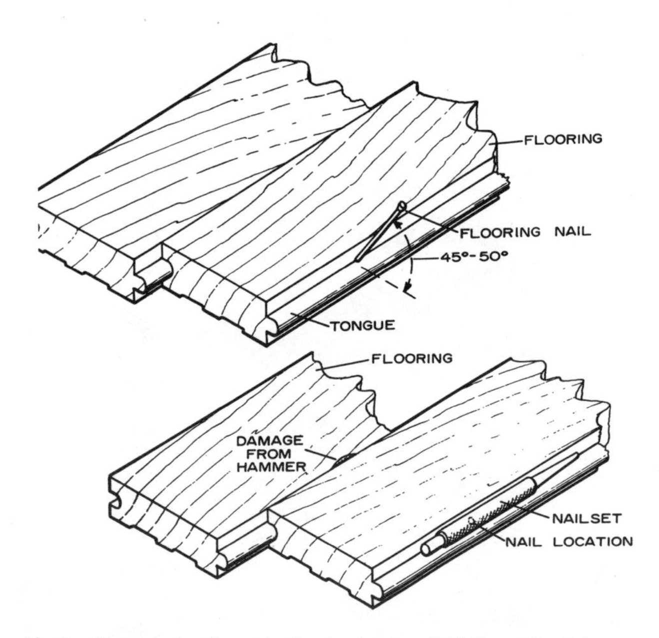

Fig. 7. Method of nailing strip flooring in place *(U.S. Forest Service)*.

Installation Over Joists

The subflooring is often omitted in low-cost construction as a way of reducing the overall building costs. Strip flooring can certainly be installed satisfactorily directly over joists and the construction will be strong enough to support all the loads that are normally imposed on a floor. In one series of tests, for example, side- and end-matched strips of ¾ in.-thick flooring were laid directly over joists spaced 16 in. apart on centers (see Fig. 8). One short length of flooring was installed in such a way

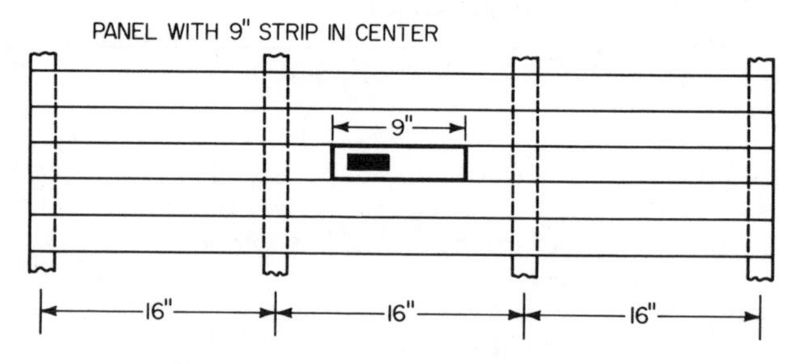

Fig. 8. The short piece of flooring shown, supported only by the matched strips of flooring that surround it, was able to support a load of 845 lb.

that it was not supported by a joist at all but was held in place by the tongues and grooves of the adjoining flooring strips. This piece of flooring was able to support a concentrated load of 845 lb before giving way. The leg of a grand piano will exert a concentrated load of about 400 lb, so this type of construction is certainly satisfactory as far as strength is concerned.

Nevertheless, this method of installing finish flooring does introduce certain problems. If we assume the flooring is to be installed after the interior of the house has been completed, planks or a temporary floor of some kind will have to be laid down to enable the workmen to go about their business. Since the workmen will not be able to work as efficiently as when a solid subfloor is in place, one will probably end up spending as much money in additional labor costs as is saved by omitting the subflooring.

On the other hand, if we assume that the finish flooring is installed before the interior is completed, the flooring will be splattered with plaster, wallpaper paste, and paint, and scored and gouged by the workmen's tools and equipment. The flooring will probably be swollen with moisture before it is installed, and after the house has been finished, the workmen gone, and the heating system turned on for the first time, large gaps will probably open up between the strips as the wood dries again, not to mention the stains and marks that will be left on the surface of the flooring.

Installation Over A Concrete Slab

Wood flooring should not be installed on a concrete slab that is below grade because of the difficulty in preventing moisture from penetrating the concrete and soaking into the wood. Even when the concrete slab is at grade level, an effective vapor barrier must be installed under the slab (*see* SLAB-ON-GRADE) to prevent moisture in the soil from making its way through the concrete. In addition, a framework of wood strips called *sleepers,* or *screeds,* must be installed on top of the concrete. The sleepers are necessary to keep the finish flooring away from the concrete, thus further reducing the possibility of moisture making its way into the flooring. The sleepers also add springiness to the construction, and they provide a base into which flooring nails can be driven.

The preferred method of installing the flooring is shown in Fig. 9. The surface of the concrete is first coated with a thin layer of asphalt to seal the surface. A layer of asphalt mastic is then poured or spread over the entire slab to a depth of about 3/32 in. The mastic acts as a combination adhesive and cushion for the sleepers. Alternatively, "rivers" of asphalt mastic 1/4 in. thick can be spread or poured over the slab at 12 to 16 in. intervals. The sleepers will later be placed on top of these rivers of mastic.

The sleepers consist of short lengths of 2 × 4's that have been pressure-treated with a wood preservative (other than creosote) to increase their resistance to moisture. (The trouble with treating the wood with creosote is that the creosote tends to seep out of the sleepers and into the flooring, via the flooring nail holes, which will stain the flooring.)

The sleepers are cut in random 2 to 4 ft lengths. They are placed, also somewhat randomly, at 12-in. intervals (if there will *not* be a subfloor under the finish flooring) or at 16 in. intervals (if there *will* be such subflooring installed), in rows that run crosswise to the intended direction of the finish flooring. Figure 10 shows how the sleepers are set in position. Note

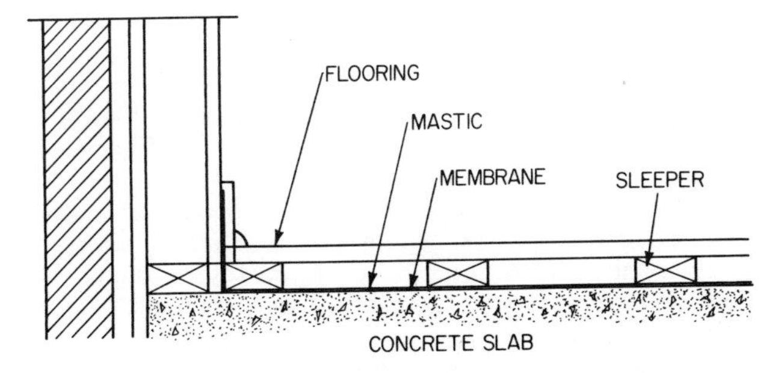

Fig. 9. The installation of strip wood flooring on a concrete slab.

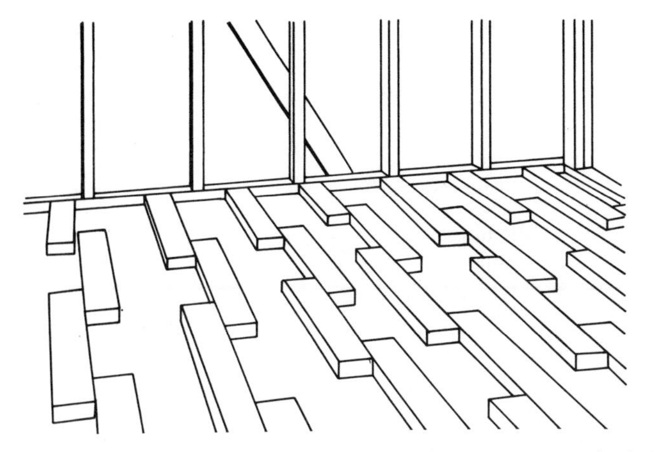

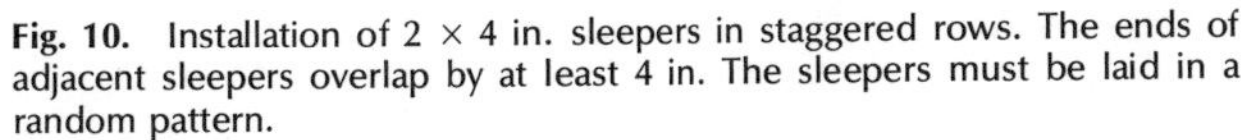

Fig. 10. Installation of 2 × 4 in. sleepers in staggered rows. The ends of adjacent sleepers overlap by at least 4 in. The sleepers must be laid in a random pattern.

especially that sleepers in line with each other overlap each other by about 4 in. and that there is also a slight space between overlapping sleepers. There is also a 1-in. gap left between the sleepers and the walls. The idea is to allow air to circulate completely underneath the flooring and thus prevent the buildup of excessive moisture.

Another method of installing the sleepers must be used when the builder or flooring contractor is not sure whether or not a vapor barrier is located under the concrete slab. In this case, a vapor barrier must be laid down as shown in Fig. 11. (*See* SLAB-ON-GRADE.)

The surface of the slab is first coated with a layer of asphalt paint. Strips of 1 × 4 in. wood, which have been pressure-treated with a wood preservative (other than creosote), are placed 12 in. apart and securely fastened to the slab with anchor bolts. Shims are placed under the sleepers where necessary to ensure that the floor will be be level. Sheets of polyethylene film 4 mils thick (i.e., 0.004 in. thick) are then laid over the strips. The overlapping edges of these sheets must be sealed together using a suitable adhesive.

Other strips of 1 × 4 in. wood, which have a moisture content similar to that of the flooring (that is, somewhere between 6 and 9 percent), are now nailed to the anchored sleepers using 1½-in.-long nails spaced 12 to 16 in. apart. The flooring is then installed as already described.

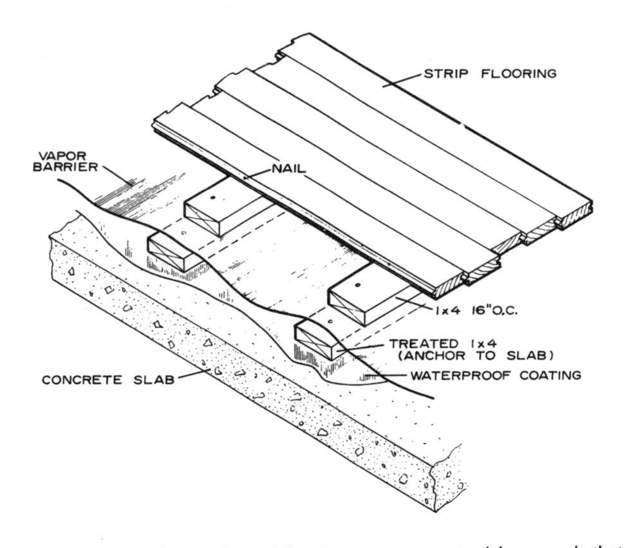

Fig. 11. The installation of wood flooring over a concrete slab-on-grade that doesn't have a vapor barrier underneath it *(U.S. Forest Service).*

INSTALLATION OF WOOD-BLOCK FLOORING

There are two basic kinds of wood-block flooring, *unit-block* flooring and *laminated-block* flooring. There is also a pseudo-parquet type of block flooring which is discussed under Installation of Parquet Flooring later in this article.

Unit-Block Flooring

A unit block (see Fig. 12) consists of several parallel strips of standard ¾-in.-thick flooring fastened together to form a large square- or rectangular-shaped block. Not only are oak and maple used to make these unit blocks but also such attractive woods as cherry, teak, and walnut. The overall size of the unit is determined by the width and length of the individual flooring strips. For example, if three strips of 2¼-in.-wide flooring cut 6¾ in. long are fastened together, they form a square block that is 6¾ in. on a side. If four strips of 2¼-in.-wide flooring cut 9 in. long are fastened together, they form a square block that is 9 in. on a side. And so on.

The sides of the assembled block are usually trimmed t & g, although grooves can also be cut into all four sides of the block to enable several blocks to be fastened together with splines. This latter technique is often used when the blocks are to be laid in asphalt mastic. But blocks in which the sides are merely cut square can also be laid in asphalt mastic.

Laminated-Block Flooring

A laminated block is a plywood product consisting of either three or five plies of wood assembled together using a waterproof adhesive. The top ply consists of a decorative hardwood of some kind. The laminated nature of the block prevents moisture having any expansive effect upon the block. That is, the block will remain dimensionally stable under varying conditions of humidity. It is, therefore, especially suitable for installation on potential damp surfaces, such as concrete slabs.

The most commonly manufactured laminated block is 9 in. on a side, althrough blocks 8 in. and 8½ in. on a side are also manufactured. Although laminated blocks ¾ in. thick with t & g sides are made, the usual thicknesses are ⅜ and ½ in., for laying in asphalt mastic. The sides of these thinner blocks are usually cut square.

Both unit blocks and laminated blocks can be purchased prefinished, which means the blocks have been fine-sanded and coats of a penetrating-oil sealer have been applied to it; see the section below on floor finishing. All the floor layer need do is wax the floor after it has been installed, if desired.

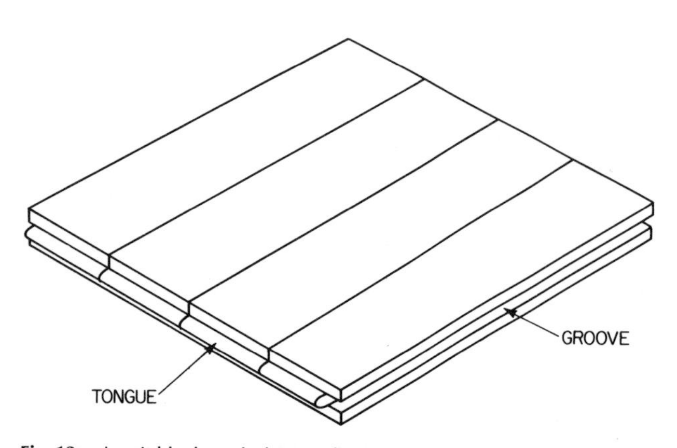

Fig. 12. A unit block made from ordinary strips of flooring that are splined together to form a solid unit.

Installation Over a Wood Subfloor

If unit blocks are to be installed on top of either a wood subfloor or a concrete slab, the same precautions regarding the moisture content of the blocks and the house in which they are installed must be observed as already described for strip flooring. When the blocks are delivered, they will have a moisture content of between 6 and 9 percent, and this dry state must be maintained unitl the blocks are fastened in place. If the blocks should be installed while they are swollen with moisture, large gaps will open up between the blocks when they dry out again. On the other hand, if the blocks are allowed to absorb moisture *after* they have been installed, usually because they become soaked with water while being washed, they will swell, and the pressure between the blocks may cause several blocks to lift up from the floor; this is especially likely to happen if the blocks are laid in mastic, as described below. None of these remarks applies to laminated blocks, of course.

Both unit blocks and laminated blocks ¾ in. thick are usually installed over a wood subfloor, the blocks being toe-nailed to the subfloor in much the same way as described above for strip flooring. A building paper, preferably 30-lb, asphalt-saturated felt, must first be laid down, and a ½ to ⅜-in. gap must be left between the blocks and the walls, as already described for the installation of strip flooring. At least two nails must be driven through the two tongue sides of each block. If the subfloor consists of plywood panels, or if laminated blocks are being nailed in place, threaded nails must be used to increase the nails' holding ability.

If thin wood blocks are to be laid in mastic over a wood subfloor, the floor layer must first install an underlayment as a base. Underlayment consists either of ⅜-in.-thick plywood panels, 7/32-in.-thick hardboard panels, or ⅜-in.-thick particleboard panels. These panels will provide the smooth, flat base necessary for the blocks, and they also act as a barrier that prevents moisture rising up through any cracks in the subflooring. Of course, if the subfloor already consists of plywood paneling, the underlayment panels can be omitted (*see* FLOOR FRAMING).

INSTALLATION OF PARQUET FLOORING

"Real" parquet flooring consists of individual *slats*, or *fingers*, of wood trimmed to very close tolerances. These slats are laid in place one at a time to form a more or less elaborate pattern (see Fig. 13). The slats are available in thicknesses that range from 5/16 to ¾ in., in lengths up to 18 in., and in widths between 1 in. and 2¼ in. They are made from a wide variety of hardwoods that include a number of fruitwoods and exotic tropical woods as well as the more usual oak, maple, etc.

Before the invention of the flooring-set-in-mastic technique, slats of this type were selected for grain and color and fitted accurately together on a work table according to the artisan's preconceived plan. The entire pattern was then transferred to the floor piece by piece, each piece being nailed in position individually. The nail heads perforce remained visible and, in an old parquet floor, they do not detract from the appearance of the floor at all. In fact, they add authenticity to the flooring.

Nowadays, a parquet floor made by hand and installed piece by piece would undoubtedly be embedded in mastic, following the general procedure described above for block flooring, and why not? But, in fact, parquet floor patterns are made up today in preassembled blocks by firms specializing in this type of work. The slats are selected for color and grain and assembled into a complete block on a paper coated with a suitable adhesive.

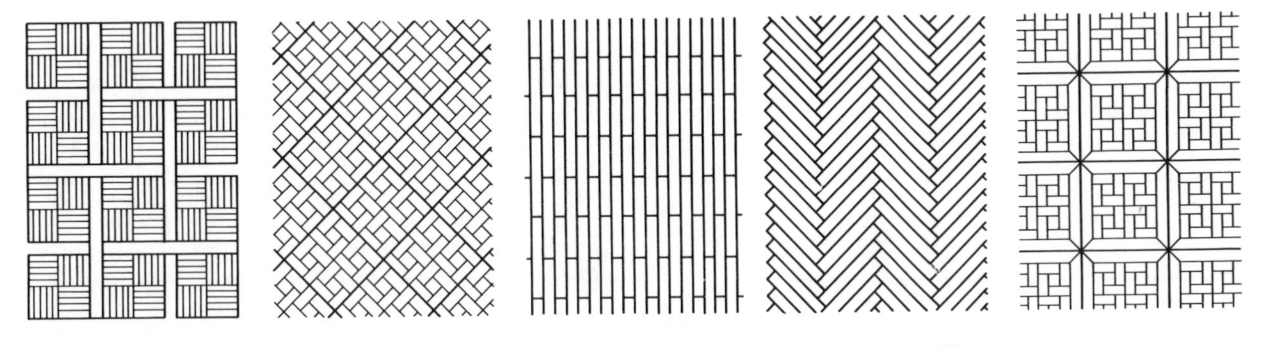

Fig. 13. A few of the many parquet-flooring patterns that can be made using slats of different sizes.

To install the flooring, the entire block, paper backing uppermost, is lowered upon the mastic, which has been troweled beforehand over the surface of the floor. The block is pressed firmly in place, the mastic is given time to set, and the paper backing is then peeled away. Repetitive premade blocks of this kind are made up into sizes as large as 30 in. sq, although the usual sizes are 9½, 18, and 19 in. sq.

In a variation of this technique, the slabs can be assembled face-side-up on a suitable backing, which may consist of a cloth or felt saturated with asphalt. The face-up position of the preassembled slats allows them to be prefinished in the shop, if desired, and once they have been embedded as a unit in the mastic—backing included—the slats need only be waxed.

There are also available short lengths of standard strip flooring machined very accurately to size and side- and end-matched. These slats ¾ in. thick, either 1, 1½ or 2¼ in. wide, and from 6 to 18 in. long. The slats are available in all the standard hardwoods, and they can be nailed in place or embedded in mastic piece by piece in simple repetitive patterns such as herringbone, basketweave, or one of the other patterns shown in Fig. 13. Thus, anyone who has the inclination and the time can assemble his own parquet floor to his own design using these slats, just as was done by artisans in times past.

FINISHING A WOOD FLOOR

The manufacturers of wood flooring will supply prefinished flooring in all the most commonly used sizes. That is, the individual strips, blocks, slats, or planks will have been sanded smooth and a penetrating sealer applied to the wood. The wood may also have been stained.

The use of prefinished flooring can be of considerable advantage to a flooring contractor since it enables him to deliver a complete job with much less time and labor. All he needs to do to the floor after he has installed it is to apply a coating of wax, and the waxing is optional.

To make the most effective use of prefinished flooring, it is important that the subfloor or concrete slab supporting the flooring be as level as possible; otherwise any irregularities will be transferred to the flooring. If this should happen, the flooring may have to be sanded to remove the irregularities, which, of course, obviates the point of installing prefinished flooring in the first place. The floor layer must be extremely careful when handling and laying the flooring, since any scratches or nicks will remain visible on the surface of the wood. The alternative is to throw away all the damaged pieces of wood, which can be very wasteful.

In what follows, we will assume that the flooring has been installed unfinished. It must now be sanded and a protective finish applied.

Sanding Wood Floors

The purpose of sanding a newly laid wood floor is to make the surface of the floor as plane and smooth as possible, and to remove any superficial marks or scratches on its surface. If the floor is old and is being refinished, the old finish is first removed by sanding it off with a very coarse sandpaper.

Before the invention of electrically operated sanding machines, wood floors were scraped flat by hand. Many people still speak of having their floors "scraped," when what they mean is that their floors are being sanded. Scraping does bring out the grain of the wood to better advantage than sanding because the pores of the wood are not filled with wood dust; but the cost of having floors hand-scraped is prohibitively expensive today.

The quality of a sanding job depends primarily on the number of sandings the floor receives. A job of ordinary quality usually is sanded twice, once with a coarse sandpaper, then with a fine sandpaper. In a first-class job, the floor will be sanded four times, each time with a successively finer grade of sandpaper. The first sanding removes all irregularities and waviness in the floor. The subsequent sandings will remove the marks made by the previous sandings and will give the floor a progressively finer finish that brings out the grain to best advantage.

The general procedure for doing a first-class job on a newly installed hardwood strip floor is as follows.

A drum-type floor-sanding machine is used. The first sanding is with a No. 1½ or No. 2 sandpaper. The machine operator must continually replace the sandpaper as the grits wear or else the sandpaper will slide over the surface of the wood instead of cutting into it. The floor is always sanded in the direction of the grain.

For the second sanding, No. 1 or No. ½ sandpaper is used. After this sanding has been completed, a disc-type hand-held sander, sometimes called a *spinner* and sometimes an *edger,* is used to sand the edges and corners of the room. A No. 1 or No. ½ sandpaper is used. Along those sides of the room where the sandpaper may cut across the grain of the wood, the flooring must be hand-scraped and so must the edges and corners of the room where the edger cannot reach.

This completes what may be considered the coarse sanding. If the machine has a built-in dust collector, the sandpaper need only be changed and the sanding continued. Otherwise, the floor must be swept clean of all wood dust and sandpaper grits before sanding is continued.

For the last two sandings, No. 0 and No. 00 sandpapers are used, respectively. After the machine has completed each sanding, the edges and corners of the room must be sanded with the edger again, using the same grades of sandpaper that were used with the machine. To complete the job, the room is swept clean again—vacuum-cleaned, preferably—and the protective finish can then be applied.

Very often the final sanding with the No. 00 sandpaper is omitted to reduce the cost of the job, but the minimum number of sandings should never be less than three.

The above procedure applies mainly to hardwood strip or plank flooring. If softwood flooring has been installed, a very fine sandpaper cannot be used, as the resinous nature of the softwood will only cause the sandpaper to clog up with gum. Softwood floors are usually sanded only twice—once with a No. 3 or No. 2½ sandpaper and then with a No. 1½ or No. 1 sandpaper.

To remove the old finish on a floor, it will be necessary to sand the floor with a No. 4 or No. 3½ sandpaper. This applies to all floors, hardwood and softwood, strip, plank, block or parquet. Once the old finish has been removed, the flooring should be sanded as described above. If the floor is badly worn, additional coarse sandings may be necessary to bring it to a plane surface.

Block or parquet floors will require a different sanding technique as the grain of the wood runs in at least two different directions. The general procedure is as follows.

A drum-type floor-sanding machine is used first. The first sanding is done with a No. 1½ or No. 1 sandpaper. The sanding machine is operated up and down the length of the room.

The sandpaper is changed to a No. 1 or No. ½ grit, and the machine is then operated crosswise to the direction of the first sanding. The edges and corners of the room are then sanded or scraped, using an edger and/or hand scraper as already described. The entire floor is then swept clean to remove all wood dust and coarse grits.

A large rotary-type floor-sanding machine is now used. The machine is operated up and down the length of the room as before, using a No. ½ sandpaper. The floor is swept clean for the second time, and the sandpaper is changed to a No. 0 or No. 00 grit. The machine is now operated first in a direction crosswise to the first sanding and then up and down the length of the room again. The floor is swept clean for the third and final time. It is now ready for the application of the protective finish.

Applying Protective Finish

Once, the standard method of protecting and enhancing the appearance of a wood floor was to apply two or three coats of a good-quality shellac or varnish. With the appearance on the market of new, extremely hard varnishes which are based on polyurethane, epoxy, phenolic, and alkyd resins, the trend among homeowners has been toward the use of one of these new, super-modern products.

But all varnishes and shellacs have one fatal flaw when they are used as floor finishes, no matter how hard they may be. These coatings lie on the surface of the wood; they do not penetrate into the pores of the wood. Sooner or later, unless the floor is kept continually protected by fresh layers of varnish or wax, the finish will be worn down to the bare wood by foot traffic, especially in doorways and along hallways, where foot traffic is always the heaviest. When this happens, dirt and dust will work their way into the pores of the wood and the wood will gradually turn black. This dirtiness can be seen in the heavily traveled sections of floor in any number of old houses. There is no way of repairing the damage except by resanding the entire floor and applying an entirely new finish.

Flooring manufacturers, therefore, no longer recommend the use of varnishes or shellacs at all, except as a surface coating to give a high polish to the floor *after* the flooring has been properly protected in the manner described below.

The recommended method for finishing a wood floor is simply to apply one or two coats of a penetrating wood sealer, which is a varnish thinned to a watery consistency. The sealer also contains compounds that help the varnish soak deep into the pores of the wood. The sealer can be either brushed on or flowed on.

If brushed on, the first coat should be heavily applied and well brushed against the grain of the wood. A gallon should cover about 500 sq ft. This coat is allowed to dry for about 24 hr and a second coat, thinned down a bit with mineral spirits or turpentine, is brushed on. This second coat should be thin enough to cover 800 to 1000 sq ft per gal. It should be allowed to dry for about 48 hr.

In the flow method of application, the sealer is poured over the floor and spread about with a cloth pad or lamb's-wool applicator. As the sealer soaks into the wood, dull spots will appear here and there. Additional sealer should be worked into these dull-looking spots (which are more porous than the other parts of the floor) until the entire floor has an even sheen. The floor should be left as-is for perhaps a half hour to give the sealer time to soak well into the wood. Any excess sealer remaining on the surface of the wood can then be squeegeed off. The floor is finally wiped with a clean rag to remove any footprints or superficial marks. The floor should be left to dry for about 48 hr.

Whatever method of application is used, after the sealer has dried completely, the floor should be burnished with a floor-polishing machine and No. 000 steel wool to remove any highlights and smooth down any raised grain.

And that's all. Nothing more need be done to the floor to protect it. Thereafter, as the wood wears down under foot traffic, the sealer will wear down along with the wood. Dirt cannot possibly enter the pores of the wood, and the appearance of the floor will, therefore, remain always the same. If necessary, fresh sealer can be applied to the floor at any time without altering its appearance at all. This certainly cannot be said when fresh varnish or shellac is applied to a floor that already has been varnished or shellacked.

So much for the basic procedure, which applies to all wood flooring.

To obtain a high polish, the floor may be waxed, if desired. Two coats of a paste wax should be applied, each coat being buffed with a machine. The wax also acts as part of the protective coating, of course, helping to absorb any wear the floor would otherwise receive. If the floor is rewaxed at frequent intervals, it should remain as new-looking as the day it was first sanded. Paste waxes are especially recommended because they are more durable than liquid waxes, which require applications at very frequent intervals if the wax coating is to be maintained. Waxes suspended in a water emulsion are not recommended at all as they tend to raise the grain of the wood.

Foundation

By *foundation* one can mean the soil or rock under the house, the base upon which the house rests, or one can mean those parts of the house that are below grade, that is, the *substructure* of the house. It is the second meaning that we will use in this article. This article will discuss the design of substructures (that is, the design of footings and foundation walls) insofar as their design is dependent on soil conditions. For the actual construction of footings and foundation walls *see* BRICK-MASONRY CONSTRUCTION; CONCRETE; CONCRETE-BLOCK CONSTRUCTION.

The structural integrity of any house depends in large part on the adequacy of the soil on which it is built; yet there is no part of a house that builders and homeowners take so much for granted. Unless a house is resting on bedrock or on a very stable, permeable soil, the possibility always exists (however remote for most dwellings) that one day the soil will become less capable of supporting the house than it had been in the past.

If a house has a full basement, the substructure consists of the *foundation walls,* which support those parts of the house that are above grade (i.e., its superstructure), and the *footings,* upon which the foundation walls rest. The footings carry the entire weight of the house and transmit this weight to the soil below. The footings are usually thought of as being merely the base of the foundation walls, and in a sense they are, of course, but they have an entirely different structural function than foundation walls and they are designed on different principles.

If a house lacks a full basement, it may still have foundation walls and footings, but these will be built much shallower. Instead of foundation walls, a basementless house is very often supported by *piers,* which are short columns that rest, in turn, on individual footings. Or a basementless house may rest on *pilings,* which are long columns made either from tree trunks, precast or poured-in-place concrete, or steel beams. Pilings, whatever material is used, are driven deep into a stable stratum of soil.

A basementless house may also be supported by *grade beams,* which are beams made of poured-in-place or precast reinforced concrete that appear to rest on the soil but are actually supported by piers or posts. Grade beams support the weight of the superstructure just as foundation walls do but they transmit this weight to their supporting piers or posts instead of to footings.

Whatever the method by which a house is supported on its foundations, it is the soil beneath that counts. A house rarely fails structurally because of a defect or weakness in its substructure. Most apparent defects in a substructure can be traced back to an inadequate supporting foundation. That is, for one reason or another the soil is, or has become, incapable of supporting the loads placed upon it, a fact that is, or was, unknown to the architect, engineer, or builder at the time the house was designed and built.

SOILS

Soil Characteristics

For the purposes of this article, we are concerned mainly with the particle size, or grain size, of soils, since it is the size of the particles that to a large extent determines a soil's loadbearing characteristics. To simplify the discussion, since there exists an enormous variety of soils, ranging from very coarse gravel types to extremely fine clay types, and every mixture in-between (refer to Table 1), we will concentrate on two very opposite kinds of soil that have very different characteristics—that is, soils that are either very sandy or very clayey.

It should be intuitively obvious that the particles of matter making up a sandy soil do not make very close contact with each other. There are comparatively large voids, or spaces, between the particles that give the soil a very porous, or permeable, quality. These voids may contain air or they may be more or less filled with water. If the voids are filled with water, this water tends to drain away fairly quickly, assuming there is some place this water can drain to.

What happens when we place a heavy concentrated load upon a sandy soil? The load will compact the particles of sand together more closely, but not much more closely, and the load will settle into the soil a slight amount. Because of the porosity of the soil, this settlement takes place as quickly as the load is applied and, once this settlement has occurred, it is unlikely that any further settlement will take place.

If any water has collected in the voids, this water will be squeezed out of the soil by the load until the particles are packed together as closely as if the soil were completely dry. But if there should happen to be a considerable hydrostatic pressure behind the water, it may happen that the load will settle into the soil just to that point where it and the hydrostatic pressure balance each other. If this should happen, the load will, to some extent, be floating in the soil. (Many lightweight summer cottages located close to the ocean will rise and fall very slightly with the changes of the tides.)

As for clay soils, it is much less obvious that voids also exist between the extremely small particles of matter that make up the soil, especially since in some clays the particles are too small to be seen even with a low-powered microscope. Nevertheless, voids do exist between the particles. In fact, these voids occupy a much greater proportion of the total volume of the soil than they occupy in an equivalent volume of sandy soil. Because of this, a clay soil can absorb a much greater volume of water than a sandy soil, but because of the small size of the voids, the soil is very impermeable. That is, the flow of water into and out of the soil takes place very, very slowly. Clay soils tend to hold tenaciously onto their water content.

If we were to place the same heavy load on a clay soil that we placed on the sandy soil, the rate of settlement would be much, much slower. It would also continue over a few months

Table 1. Types of soils.

DIVISION	SYMBOLS LETTER	SYMBOLS HATCHING	SOIL DESCRIPTION	VALUE AS A FOUNDATION MATERIAL
GRAVEL AND GRAVELLY SOILS	GW		Well graded gravel, or gravel-sand mixture, little or no fines	Excellent
	GP		Poorly graded gravel, gravel-sand mixtures, little or no fines	Good
	GM		Silty gravels, gravel-sand-silt mixtures	Good
	GC		Clayey-gravels, gravel-clay-sand mixtures	Good
SAND AND SANDY SOILS	SW		Well-graded sands, or gravelly sands, little or no fines	Good
	SP		Poorly graded sands, or gravelly sands, little or no fines	Fair
	SM		Silty sands, sand-silt mixtures	Fair
	SC		Clayey sands, sand-clay mixtures	Fair
SILTS AND CLAYS LL <50 †	ML		Inorganic silts & very fine sands, rock flour, silty or clayey fine sands, or clayey silts with slight plasticity	Fair
	CL		Inorganic silts of low to medium plasticity, gravelly sands, silty clays, lean clays	Fair
	OL		Organic silt-clays of low plasticity	Poor
SILTS AND CLAYS LL >50 †	MH		Inorganic silts, micaceousor diatomaceous fine sandy or silty soils, elastic silts	Poor
	CH		Inorganic clays of high plasticity, fat clays	Very poor
	OH		Organic clays of medium to high plasticity, organic soils	Very poor
HIGHLY ORGANIC SOILS	Pt		Peat & other highly organic soils	Not suitable

† LL indicates liquid limit

certainly, and perhaps for as long as a few years. As long as the load was applied upon the soil, it would continue to force water out of the voids, settling deeper and deeper into the soil with the passing of time until either all the water had been removed or some sort of balance had been established between the load and any hydrostatic pressure acting on the water. In the long run, also, the total amount of settlement would probably be greater than in a sandy soil.

But once a balance had been established between the load exerted on a clay soil and this hydrostatic pressure, the soil would thereafter be as stable as a sandy soil. If the hydrostatic pressure increased for any reason, water would again be forced into the soil. Indeed, clay soils are noted for their ability to absorb large quantities of water, even under load, and to expand in volume as their content of water increases. During a very rainy season, for example, if a clay soil should become soaked with water to a depth of 8 ft or more, the surface of the clay may expand upward by as much as 1 in. The deeper into the soil one goes, the less the expansion will be, of course. Even though the 1-in. expansion may not be obvious to the eye or appear to be of great consequence, it would certainly be sufficient to have potentially serious consequences on a house having a shallow substructure, if the expansion were uneven. And as the soil dried out, it would slowly shrink back down to its original volume, again with potentially serious consequences for the substructure of the house.

The Bearing Capacity of Soils

Table 2 shows the number of tons per sq ft that different types of soil can support. (Looking through the table, one might wonder why anyone would worry about the ability of a foundation to support a load at all; we will take up this point later.) As Table 2 shows, different soils have different *bearing capacities*. That is, they vary in their ability to resist the pressures exerted on them by an exterior load. If a soil has no bearing capacity at all, any load applied to the soil would simply sink into it until it reached a stratum of soil that could support it. The outstanding example of a soil with an exceptionally low bearing capacity is quicksand.

Table 2. Presumptive Surface Bearing Values of Foundation Materials

	Class of material	Ton per sq ft
1	Massive crystalline bedrock including granite, diorite, gneiss, trap rock, hard limestone, and dolomite	100
2	Foliated rock including bedded limestone, schist, and slate in sound condition	40
3	Sedimentary rock including hard shales, sandstones, and thoroughly cemented conglomerates	25
4	Soft or broken bedrock (excluding shale), and soft limestone	10
5	Compacted, partially cemented gravels, and sand and hardpan overlying rock	10
6	Gravel and sand-gravel mixtures	6
7	Loose gravel, hard dry clay, compact coarse sand, and soft shales	4
8	Loose, coarse sand and sand-gravel mixtures and compact fine sand (confined)	3
9	Loose medium sand (confined), stiff clay	2
10	Soft broken shale, soft clay	1.5

Source: Building officials and Code Administrators International.

Subsoil Investigation

The actual bearing capacity of any particular soil can be determined only by suitable tests at the site itself and by laboratory examination of soil samples taken from the site, procedures that are both expensive and unnecessary for the great majority of houses.

As far as most communities are concerned, there is a background of experience that local builders can fall back on insofar as the kinds of soil in the neighborhood and their bearing capacities are concerned. The local buildings department will probably have a file of records regarding subsoil investigations for particular building lots, information that can be extrapolated for nearby building sites. In addition, most local buildings departments also publish as part of the local building code conservative rule-of-thumb tables of permissible footing sizes and foundation-wall thicknesses that are based on local experience and that will almost positively guarantee that a house built in that community will be safely supported on its foundation. Local builders must, of course, follow the recommendations of these tables if they wish to get their building plans approved.

Why, then, undertake any subsoil investigation at all? For several reasons. (1) Because the possibility always exists, given the great variability of most soil strata, that any given building site may have an underlying stratum of soft clay or silt, or the land may at one time have been filled in; if so, the substructure of the house will have to be carried down to more solid strata since neither soft clay, silt, nor landfill has much bearing capacity; (2) because the possibility also exists that groundwater may run through the site, which will not only greatly increase the cost of excavating the soil and building the substructure but may also make the site unusable; and (3) because there may be large boulders or an outcrop of rock just below the surface that will make the digging of the basement or foundation walls prohibitively expensive. If no problem shows up, the builder can go ahead with a deep sense of relief that construction costs will not exceed original estimates.

Subsoil investigation usually means drilling a hole into the ground to a depth of about 1½ times the depth of the substructure in order to bring up samples of soil for examination. The hole is drilled this deep to check whether there is groundwater present at this depth, since if there is any water it might travel upward through the soil and flood the excavation.

The drilling is done by a firm that specializes in this work. Usually an on-the-spot examination of the soil samples will be sufficient to determine whether or not the soil has the necessary bearing capacity. If the soil samples look suspiciously weak or are ambiguous, a test pit may be dug into the soil as deep as the bottom of the substructure so that the soil can be examined in situ.

The Settlement of Houses

All houses settle, except when they are built on bedrock. The amount of settlement is, for the great majority of houses, very slight, amounting at most to perhaps ¼ to ½ in. If the settlement is even throughout, no harm is done and the homeowner will probably not even be aware that any settlement at all has occurred. Wood-frame houses are, after all, very lightweight structures, and very few will exert a pressure exceeding 2000 lb per sq ft on their foundation. A two-story house with masonry walls may exert a pressure in excess of 3000 lb per sq ft, but even this load is not considered especially heavy for a foundation, as Table 2 shows.

If all this is true, the reader may be wondering by now what all the fuss is about, since most houses will settle only a fraction of an inch at most, even in what seem to be the most unsatisfactory of soils.

The problem is not one of settlement, since all houses settle, but *uneven* settlement, which may place very great stresses on the structural framework of any house, especially if the settlement occurs across the corner walls of a house. The amount of differential settlement we are talking about is quite small—½ to ¾ in. A differential settlement as little as this may result in plaster cracks and jammed doors and windows. Any differential settlement greater than ¾ in. may open up large cracks in the foundation walls of the house.

Furthermore, the problem of differential settlement is usually of concern only in certain types of soil—clays, silts, and landfill, although any soil completely soaked with water is potentially unstable if for any reason its water content should drain away one day (but why build on a wet soil in the first place?).

The fact is, large numbers of houses *are* built on clays, silts, or landfill today because more suitable soils are either economically more useful for agriculture or for large commercial structures; because houses have already been built upon these more suitable soils in most communities; or because clay is the predominant soil in an area. In any expanding community, the last parcels of land that are considered for home construction (by commercial builders) are those in which the soil has a low bearing capacity, since the cost of excavating the soil and building a strong enough substructure to support the house adequately are greater in a poor soil than in a soil having a high bearing capacity. In short, in many parts of the country today, a builder has no choice except to build on a poor or inadequate soil.

FOOTINGS

Most foundation walls and piers rest on footings (refer to Fig. 1), the bases of which are in direct contact with the soil. The footings, therefore, transmit all the building loads to the foundation. The type of footing most often used in dwellings is the

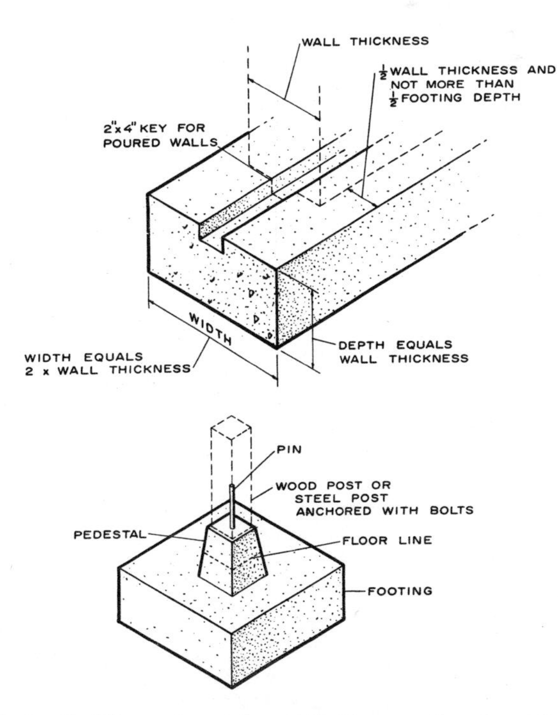

Fig. 1. (Top) The usual proportions of a footing supporting a foundation wall for a dwelling. (Bottom) A footing that supports a post in a dwelling. Masonry piers are supported by similar footings *(U.S. Forest Service)*.

spread footing, which simply means that the footing is wider than the foundation wall or pier it is supporting.

The principle behind the use of spread footings is quite simple. Assume a building has a total weight that is equal to 4000 lb per linear ft of foundation wall. If the foundation walls are in fact 12 in. wide, then each linear foot of wall will bear down upon the soil with a force of 4000 lb per sq ft. But when spread footings 24 in. wide are placed under the foundation walls, then each linear foot of foundation wall will bear down on the foundation with a force of only 2000 lb per sq ft. The weight of any house can, therefore, within limits, always be accommodated to the bearing capacity of the soil.

This principle cannot be extended indefinitely (with spread footings, anyway) because the footing itself will crack if it is made too wide. The weight of the foundation walls or piers will shear or punch through the footing, unless the footing is heavily reinforced, which will increase its cost unduly.

To give the footings adequate rigidity, they are always made of concrete, even when the foundation walls or piers they support are made of masonry. The footings are usually made as deep as the foundation walls are thick, while the total width of the footings is twice the thickness of the foundation walls (refer to Fig. 1). Thus, for example, if a foundation wall is 8 in. wide, the footing will usually be 8 in. deep and 16 in. wide. If the footings are made much wider than twice the thickness of the foundation walls, they may crack because the load exerted upon them by the foundation wall will be concentrated too much at the center of the footing.

A footing should be reinforced with steel rods placed lengthwise within it wherever a tunnel is dug under it to lay a drain pipe or an electrical conduit. In addition, this excavation should not be refilled with soil (which will consolidate too much and thus fail to support the footing) but with gravel tamped firmly into place.

For most houses, the footings are usually the same width all around the perimeter of the house. This may be satisfactory for box-like houses, under which the bearing pressures are the same everywhere, but if a house is built so that the weight upon the foundation is uneven, the house may settle into the soil unevenly. Assume, for example, that one half of a masonry house is two stories high and the other half is only one story high. If the footings are the same width under the entire foundation wall, the result will be that one half of the footing will be supporting a weight that is twice as great as that being supported by the other half of the footing. The result might very well be a differential settling of the house.

What is in fact done under these circumstances is to calculate the size of the footing according to the heaviest anticipated loads and to construct the entire footing to this load, even under the one-story section of the house. This is cheaper than attempting to calculate different sized footings and constructing different sized forms for the concrete.

The footing must be dug deeply enough that it is unlikely to be affected by any soil heaving caused by freezing soil or swelling clay. Whatever the depth at which the soil freezes in any particular locale, which may be as much as 8 ft below the surface in some northern sections of the United States, the footing must be located 1½ ft below this level. Even in southern states where frost is extremely unlikely, it may still be necessary to place the footing 5 ft below the surface to avoid any soil heave caused by swelling clay. In short, footings must always rest upon solid, undisturbed soil, even when—or *especially* when—the soil happens to be a clay.

FOUNDATION WALLS

Like footings, foundation walls are usually sized according to some rule-of-thumb guide based on local experience. These wall thicknesses are very conservative and can be considered more than strong enough to resist any compressive loads that may ever be placed upon them.

If a house has a full basement, the foundation walls enclosing the basement must also be strong enough to resist the lateral loads imposed on them by the weight of the soil lying against the house. Included among the lateral loads are earthquake loads, as well as such above-ground loads as wind pressures, insofar as these wind pressures may be transmitted to the foundation walls.

PIERS

There is very little point in constructing footings and foundation walls under a basementless house. Piers spaced at appropriate intervals along the perimeter of the house and under its center (see Fig. 2) are much cheaper to construct and equally capable of supporting the superstructure via the beams that run from pier to pier.

Building codes usually specify the minimum allowable size of a pier since, of course, the thicker the pier the more stable it is and the more capable it is of supporting a heavy load. Whatever the size of a pier, its height is usually required to be less than 10 times its minimum thickness. In the case of concrete-block piers, the piers cannot be more than four times their minimum thickness.

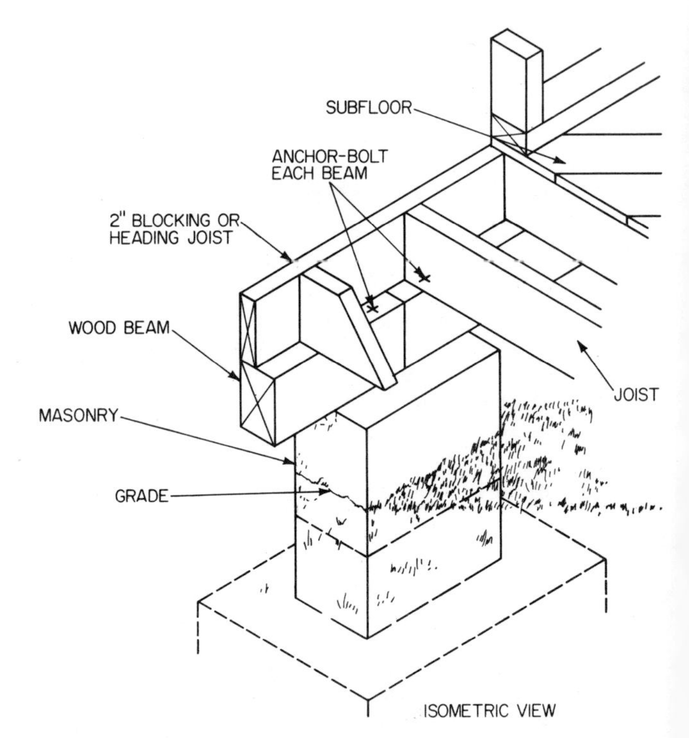

Fig. 2. A dwelling supported by free-standing piers.

Ground, Electrical

Just as all the water in the world tends to fall to a common level, which is sea level, so does all generated electrical energy tend to fall to a common zero level, which we call *ground potential.* And just as gravity is the driving force that compels water to flow to the sea, so is the difference in electrical potential between a generated voltage and the ground potential of the earth the driving force that compels the flow of the generated voltage through an electrical circuit. In any electrical circuit, therefore, if there isn't a connection to ground, there cannot be a flow of electricity through that circuit. Indeed, we can say that in any circuit the basic function of a switch or a fuse is to interrupt the flow of electrical energy to ground.

Occasionally, defects in the wiring or insulation of a circuit—defects that are called *faults*—allow the electrical current to bypass its normal route, to take a short cut to ground, so to speak. This short cut is called a *short circuit,* and it can occur (1) between two hot wires, (2) between a hot wire and a neutral wire, or (3) between a hot wire and some unintentional ground. What we are going to discuss in this article is the third case, a short circuit (or fault) to ground. Any such short circuit is potentially very dangerous. If the flow of current is great enough, the short circuit can kill a person, or it may so overheat the wires through which the current is flowing that the insulation melts and combustible materials touching the bare wires are set afire.

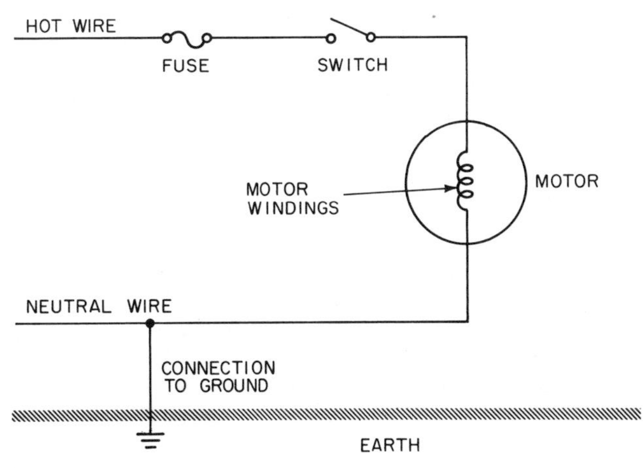

Fig. 1. A typical electrical circuit.

> NOTE: Residential electrical systems are described in the article ELECTRICAL SYSTEM; particular electrical circuits are described in the article ELECTRICAL WIRING CIRCUITS; it is suggested that these two articles be read first.

Figure 1 shows a typical electrical circuit containing a fuse, a switch, and a motor. Electricity is assumed to flow through the hot wire to the motor windings and thence through the neutral wire to ground. If either the fuse or switch is opened, the circuit will remain hot up to the point of disconnection. The circuit beyond that point will be at ground potential. Once the fuse or switch has been opened, we can disconnect the motor or do any kind of repair or maintenance work we want without receiving a shock.

Note that both the fuse and the switch are on the *hot* side of the circuit. If the switch, for example, were to be located on the neutral side of the circuit, opening the switch would mean that the circuit would be hot right through the motor, even though current could not flow through the circuit to ground, and, consequently, the motor would not operate. But if we were to try to work on the motor we would, the moment we touched a hot wire, complete the circuit to ground and would, therefore, receive a severe shock. For this reason, installing a switch, fuse, or circuit breaker on the neutral side of any electrical circuit can be highly dangerous and is strictly forbidden by all electrical codes.

It should be emphasized that in a 115- or 120-volt residential electrical system it is primarily the rate of current flow through a circuit, that is, the amperage, that makes any short circuit potentially dangerous. And the amount of current it takes to kill you in a 115- or 120-volt system can be quite small. For example, a *milliampere* (ma) is one-thousandth of an ampere. If you were to touch a 115- or 120-volt wire through which 5 ma or less of current were flowing, you would feel a slight tingle in your hand. As the current flow increased to about 15 ma, the shock would become increasingly unpleasant, and most people would jerk their hand violently away from the wire. If the current flow were to exceed 15 ma, and if you were actually grasping the wire, "muscle freeze" might occur in which you could not let go of the wire at all. If, at the same time, you happened to be standing in a puddle of water and therefore were part of an extremely low resistance path to ground, the electricity would very likely kill you.

Whether or not any given flow of current will prove fatal depends on many things—one's body size, the amount of voltage driving the current through the circuit, how well one is grounded or isolated from ground—so that a current flow that would kill one person would leave another with nothing more than a tingling sensation. In a 115- or 120-volt system, it is only when a current exceeds 75 ma that it is almost invariably fatal, and even 75 ma is a very small amount of electricity.

The techniques by which an electrical system is safely grounded in a domestic installation are spelled out in very complex detail in the National Electrical Code (NEC). To understand the reasoning behind the requirements of the NEC, one has to make a basic distinction between the main *system* ground that protects the entire electrical installation against short circuits and an *equipment* ground in a branch circuit that protects an individual piece of electrical equipment. The main system ground is installed close to the point at which the public utility conductors enter the house, and the equipment ground connections are located in the individual branch circuits.

SYSTEM GROUND

Electricity is supplied to most dwellings through three large conductors (or wires) that are connected to the public utility's distribution system. These conductors are known collectively as the *service entrance,* and it will suffice here to note that two of these conductors carry 115 or 120 volts each (depending on the utility's practice); the third conductor is the neutral cable. For a detailed description of the service entrance *see* ELECTRICAL SYSTEM.

The electricity supplied to most dwellings is generated and transmitted by the utility at 2300 or more volts. This voltage is reduced to the 115/230 or 120/240 volts used in most dwellings by nearby transformers.

The neutral conductor of the service entrance is connected to ground at both the transformer and at the dwelling, the latter connection being described below. Thus, the neutral conductor of the service entrance is at ground potential over its entire length, and, in addition to its basic function of maximizing the efficiency with which voltage can push current through the system, the neutral conductor also protects the entire electrical system against such major calamities as lightning striking the service entrance or storm damage that causes the 2300-volt power line to short across the incoming 115- or 120-volt service-entrance conductors. If the neutral conductor weren't present and properly grounded, either of these events could result in a surge of high-voltage electricity through the wiring system of the house. The entire wiring system might very well burn out. In addition, any appliances that happened to be connected to the system at the time of the short would have their circuits burned out also. But if the neutral conductor is adequately grounded, as it most probably would be, this surge of high-voltage electricity will be conducted immediately to ground and harmlessly dissipated.

Figure 2 shows a typical system ground installation including the connections between the incoming neutral conductor and a *ground wire* and between the ground wire and the main ground itself; what the NEC refers to as the *grounding electrode.*

In most dwellings the service entrance (and the utility's responsibility) ends at a service disconnect switch that is located within a metal enclosure (i.e., a distribution panel often called a *panel box*) mounted either on the outside or just inside the dwelling. This panel box usually contains the main and branch-circuit fuses or circuit breakers as well.

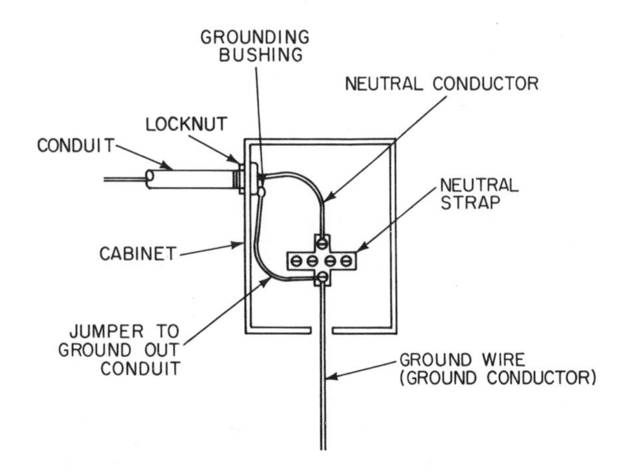

Fig. 2. Typical grounding connections at a main panel box.

The incoming neutral conductor, or wire, which is easily recognized since its insulation *invariably* has a white or light gray color, unless it is completely bare (the hot wires have either black or red insulation), ends at a heavy *buss* or *neutral strap* made of copper or a copper alloy to which it is securely fastened by a large bolt. (There may also be terminals on the neutral strap to which are connected the neutral conductors of such major household appliances as an electrical range or freezer, as well as smaller terminals to which are attached the neutral conductors that are part of the house's branch circuits.) Since the neutral strap is at the same time securely bolted to the panel box housing, it, the panel box, the neutral conductor, and the ground wire are all at ground potential.

The ground wire is usually made of copper because this metal is both an excellent conductor of electricity and noncorrodible. It can be in the form of a solid bar or rod, or stranded or solid wire, and it may or may not be covered with insulation. The main thing is that the ground wire be of a certain minimum size; this size depends on several factors, the main one being the capacity in amperes of the incoming electrical service. The ground wire must certainly be large enough that it does not introduce any electrical resistance of its own into the circuit. If the ground wire is No. 6 American Wire Gauge (AWG) in size or larger, it can be installed as is. If it is No. 8 AWG or smaller, it must be located within a conduit for protection, and the conduit, if it is made of metal, must then be grounded itself to both the panel box and the main ground.

Main Ground

In all communities that have an underground water-supply system in which the water pipes are made of metal, these pipes serve as the main electrical grounds. In each house, the end of the ground wire is connected to the cold-water supply pipe just where the pipe enters the house. The water pipe is used not because water happens to be flowing through it, but because the exterior of the metal pipe is in solid contact with damp soil, which is of course at zero potential.

The ground wire is connected to the water pipe using special clamps that are specifically designed for this purpose. When the water pipe is made of iron or steel, the clamp must be made of iron or steel; when the water pipe is made of copper, the clamp must be made of copper or brass—the similarity of materials being to prevent the electrolytic corrosion of the clamp, which might occur if two dissimilar metals were joined together. If corrosion should occur, it would increase the electrical resistance of the ground circuit and thus decrease its effectiveness. Under no circumstances can any connection be made by soldering the ground wire to the main ground.

If a water meter should be installed in the water-supply line, the ground connection must be made on the supply side of the meter. Then, if the meter should ever be removed for repairs, the electrical system won't inadvertently be disconnected from its ground. Besides, the pipe joints connecting the meter to the water-supply pipes often contain rubber or cork gaskets that will also prevent the transmission of ground currents; again, the system would in effect be disconnected from its ground.

If there should be a main shutoff valve installed in the water-supply line, the ground connection must also be on the supply side of this valve, in case the valve is ever removed for repairs.

Sometimes the layout of the house makes it difficult to install the ground wire on the supply side of a water meter and/or shutoff valve. In this case jumper lines are connected to the pipe around both meter and valve. In many communities in which the water is hard, a homeowner may install a water-softener unit in the main water-supply line. Very often this equipment is connected to the water-supply lines by rubber hoses. When this is the case, a jumper line must also be connected to the water-supply lines around the water-softening unit.

As for the water pipe itself, it should not be painted, enameled, or coated with an asphalt compound, all of which tend to reduce the ability of the pipe to transfer an electrical current into the soil. Before he connects the ground wire to the pipe, the electrician must also make sure that the pipe actually extends at least 10 ft into the soil without interruption. This may sound like a curious precaution to take, but many new water-supply installations have plastic gaskets installed between pipe sections to secure good seals at these joints. These gaskets will, of course, electrically isolate the sections of the pipe from each other. Plastic pipe is also being used more and more for cold-water supply systems. The electrician, therefore, must also make sure that he is not connecting the ground wire to a plastic pipe, which of course would result in no ground at all.

If the water-supply line cannot be used as the main ground, it may be necessary to connect the ground wire to an incoming gas line. But before a gas line can be used as a system ground, the electrician must obtain the permission of both the local gas utility and the local building inspectors. The electrician must also make certain that nonmetallic gaskets haven't been installed between the sections of pipe. Lacking this alternative, it will be necessary for the electrician to install special *made electrodes* into the soil to which he can connect the ground wire.

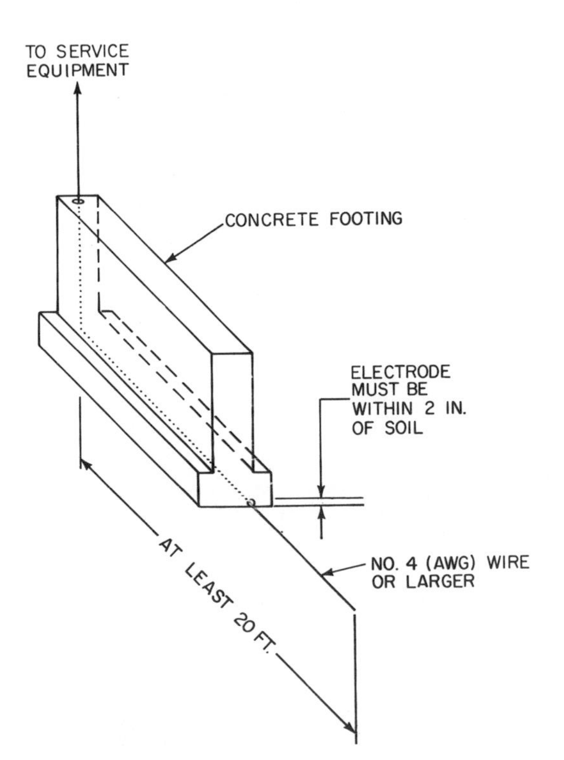

Fig. 4. A UFer grounding electrode, buried in the bottom of a concrete footing, can be used to ground a residential electrical system.

Made Electrode

A made electrode is usually necessary for an isolated house. The electrode consists either of a length of pipe or rod (see Fig. 3), a flat metal plate, or perhaps the metal casing of a well. The preferred materials are either copper or galvanized steel. Whichever of these two metals is used, it should not be painted, varnished, or otherwise covered with a finish that will reduce its ability to conduct a ground current into the soil.

Rods and pipes must be driven at least 8 ft into the soil, leaving the end of the electrode exposed a few inches above the ground to enable the electrician to clamp the ground wire to it. The fact that the connection is above ground also allows it to be inspected from time to time.

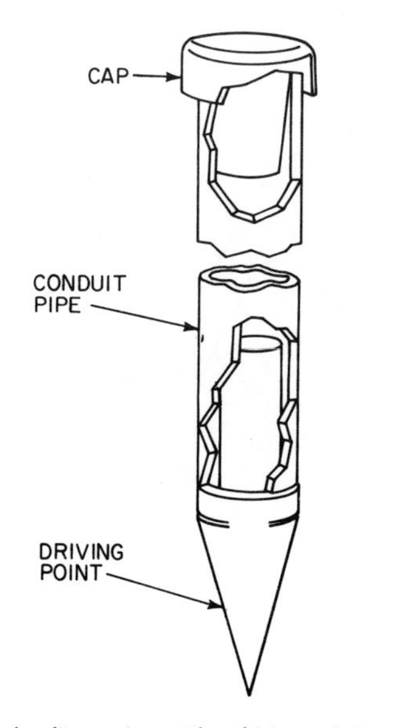

Fig. 3. A length of ordinary pipe, with a driving point on one end and a cap on the other, can be used as a made electrode to ground a residential electrical system.

Whatever the type of electrode used, the soil into which it is driven or buried must be damp; otherwise it is very likely that ground currents will not be adequately dissipated in the soil. This would most likely be a problem in arid regions. It may be necessary, therefore, to drive the electrode a considerable distance into the earth before permanently damp soil is reached, or, alternatively, several electrodes can be driven into the earth and the ground wire connected to them in parallel. But when more than one electrode is installed, the electrodes should be at least 6 ft apart.

Since 1968, another type of electrode has been approved by the NEC. It is called the Ufer electrode, after its inventor, and it consists of a bare copper wire having the same diameter as the ground wire (but at least No. 4 AWG in diam.) and at least 20 ft long that is buried within the concrete footing of the house (see Fig. 4). The copper wire must not actually touch the soil, but it must be located at the bottom of the footing, within 2 in. of the soil. One end of the Ufer electrode is clamped to the ground wire at the point where it emerges from the concrete. The Ufer electrode works, and it works well, because the soil underneath most foundation walls is damp, because concrete is permeable to water, and because the weight of the house pressing down upon the damp soil assures a solid electrical contact between the wire and the soil below.

An equipment ground has an entirely different purpose from a system ground. The purpose of an equipment ground is to ground the housing of an electrical appliance in case an internal short should occur between the appliance's electrical system and its housing. The situation is shown in Fig. 5, which is the same circuit as Fig. 1 but with an internal short between the motor windings and the motor housing.

This internal short circuit will not prevent the motor from operating, nor will it result in excess current flow that will cause the branch-circuit fuse to blow. But anyone who touches

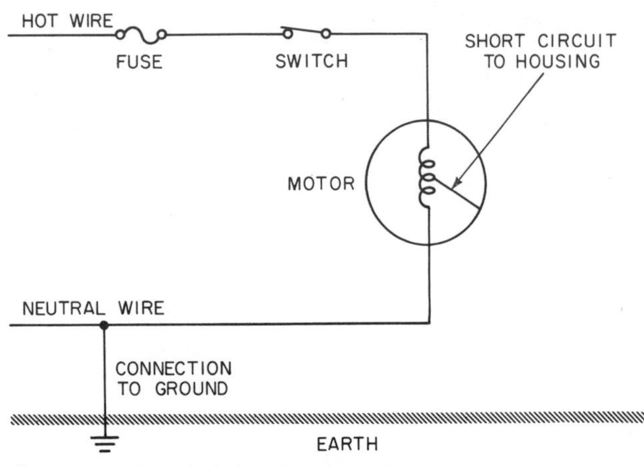

Fig. 5. A circuit in which there is an internal short from the motor windings to the motor housing. Anyone touching the housing would complete a short circuit to ground and receive a shock.

the motor housing will complete a low-resistance path to ground and receive a shock. The severity of the shock will depend on the extent of the fault and on how well the person is insulated from ground.

For safety's sake, therefore, the housing must be grounded. This is accomplished by connecting a special *grounding wire* to the housing. What the other end of this grounding wire is connected to depends on the type of wiring installation in the dwelling, of which there are two basic types.

In one type of wiring installation, the wires making up the electrical system are contained within metal conduits that run from one outlet box to another throughout the house. These conduits are securely fastened to each outlet box by metal bushings and locknuts. The conduits are also securely connected to the main panel box by special bushings to which short jumper wires can be fastened. These jumper wires are then fastened to the neutral strap located within the panel box. Thus, the entire system of conduits is grounded.

In the second basic type of wiring installation, the wiring consists of what is called *nonmetallic sheathed cable,* which means that the wires are surrounded only by rubber or plastic insulation. (*Armored* cable is also used for dwellings, but the principle of installation remains the same.) In older houses, those built before the 1960s, say, electrical cables contained only two insulated wires—the hot wire and the neutral wire. But beginning in the 1960s cable that included a separate grounding wire within the outer cable wrapping has come to be used more and more. This grounding wire is easily identified because it is completely bare of insulation.

In a house that is wired with this type of cable, each outlet box is grounded by attaching the bare grounding wire to the outlet box with a screw or special clamp. In this way, the entire

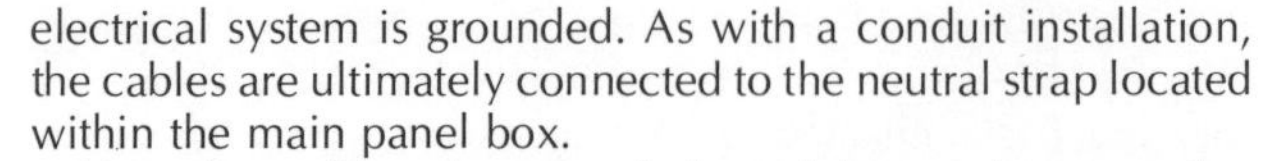
electrical system is grounded. As with a conduit installation, the cables are ultimately connected to the neutral strap located within the main panel box.

Note that neither the grounded conduits nor the grounding wires of the cable system have anything at all to do with the neutral side of the electrical wiring circuits, even though everything eventually ends up at the same neutral strap. For safe and efficient operation of the electrical circuits, it is important that the neutral wires and grounding circuits be kept completely separate from each other.

To return to our discussion of electrical-equipment grounds, in almost all motor-driven appliances manufactured today, the appliance housing is grounded through a wire enclosed in the cord that supplies electricity to the motor. This grounding wire is identified by the green insulation that covers it. One end of this grounding wire is connected to the appliance housing. The other end of the wire is connected to a blade on the connecting plug. There are three such blades (see Fig. 6). Two serve the usual function of connecting the hot and neutral wires to the electrical circuit via the duplex receptacle. The third blade, which may have either a round or U shape, leads to the system's grounding circuit.

To accommodate a three-bladed plug, the receptacle into which it is inserted must have three mating slots—two for the hot and neutral wires and another for the grounding connection. Such a three-slot receptacle, which is called a *grounding receptacle,* is shown in Fig. 7. The grounding connection from the appliance housing emerges from the side of the receptacle at a green-tinted hex-head screw. A grounding wire from one of the cables entering the outlet box is connected to this screw, or, in a conduit system, a jumper wire connected to the side of the outlet box grounds the appliance housing. In this way the safe operation of the appliance is assured.

Appliances manufactured before the 1960s are connected to receptacle outlets via two-bladed plugs. These plugs can be inserted into the three-slot grounding receptacles shown in Fig. 7, of course. The opposite is not true, however. An appliance having a three-bladed plug cannot be connected to a two-slotted receptacle. The preferred solution to this problem is to replace the old receptacle with a new grounded receptacle. But when this is done, it is important that the newly installed receptacle be grounded to the outlet box with a clip—after one has made certain that, in fact, the outlet box is grounded out. If it isn't, all that one has gained is a false sense of security for which one may pay dearly some day.

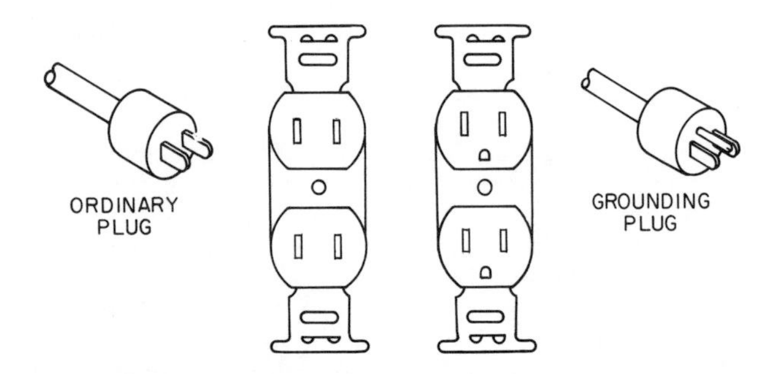

Fig. 6. An ordinary plug and the duplex receptacle into which it is inserted compared with a grounding plug and its duplex receptacle. Either type of plug will fit into the grounding type of receptacle, but the opposite is not true.

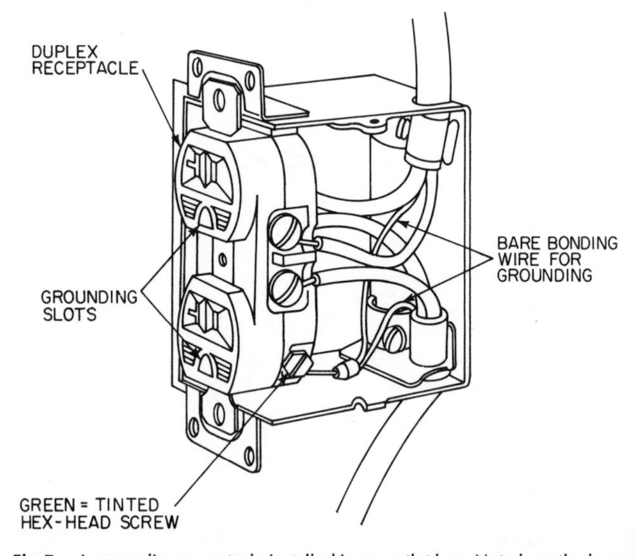

Fig. 7. A grounding receptacle installed in an outlet box. Note how the bare grounding wires from the cables are connected to the receptacle grounding connection through the hex-head screw.

Gypsum

Gypsum is a white, crystalline mineral that is usually found in nature in the form of large rock deposits, though it also exists as an earthy substance known as *gypsite* and as a sand. Pure gypsum is a *hydrous* calcium sulfate, which means that each molecule of the calcium sulfate, $CaSO_4$, is chemically combined with two molecules of water, as expressed in the formula, $CaSO_4 + 2H_2O$. About 20.9 percent of the weight of pure gypsum consists of this bound water.

Gypsum is rarely found pure in nature, however. Usually mixed with it are small amounts of silica, alumina, iron oxide, calcium carbonate, and/or magnesium carbonate. These impurities give the raw gypsum a gray, brown, or reddish-brown color.

Gypsum has the interesting property that if it is crushed into small pieces and heated, the heat will drive off most of the combined water. The gypsum is then ground into a powder. When water is added to the powder, a plastic mass is formed that remains plastic until the water and gypsum recombine, which causes the gypsum to revert to its original rock-like state. During the short period of time that the gypsum is plastic, it can easily be molded into any desired shape, a shape it will retain as it sets. It is this temporary plasticity that gives gypsum its value as a construction material (*see* PLASTERING.)

This property of gypsum was known to the ancient Egyptians 5000 years ago. They used gypsum as a plaster, as did the Greeks and Romans after them. Its use was limited, however, because pure gypsum sets very quickly. In fact, it was not until the nineteenth century that retarders were discovered that were capable of delaying the setting time of gypsum for up to two hours and more. Once these methods of controlling the setting time were known, the use of gypsum increased enormously.

The fact that gypsum contains about 21 percent water gives the products made from it another very interesting property—they cannot burn. In fact, they won't even get very hot. Gypsum is an inherently fire-resistant material. If, for example, a gypsum wallboard should be subjected to an intense flame, the water in the gypsum will be converted into steam. Because of this process, the maximum temperature of the wallboard is limited to 212°F, the boiling point of water. If the fire is prolonged, and as the water content of the gypsum on the side of the wallboard facing the fire is gradually consumed, the calcium sulfate that remains turns into a fine powder. The powder acts as an insulator while the water content of the gypsum behind it continues to restrict the temperature of the wallboard to 212°F. The wallboard will not exceed this temperature until all the water has boiled out of the gypsum. For effective fireproofing, therefore, all one need do is line the surface of a wall with a sufficiently thick layer of gypsum plaster or wallboard. This does not mean, of course, that smoke damage cannot occur, nor can there be gaps in the construction (such as doors and windows or duct openings) through which flames, gases, and smoke may pass.

The essential step in the manufacture of gypsum is the *calcination* of the raw material, which means the subjection of the gypsum to a high enough temperature that most of its water content is driven off. The rock is first crushed into 2-in. pieces. These pieces are further reduced in size according to the method of manufacture used.

There are two such methods. One is a batch-type process in which a large kettle is loaded with a charge of gypsum, the gypsum is calcined, and the kettle is then unloaded. The raw gypsum must be reduced to a fine powder before this method can be used.

The kettle used in the batch process is a large steel vessel, 8 to 10 ft in diam. and from 6 to 9 ft high. Seven to ten tons of gypsum can be calcined at one time. A fire is lit under the kettle and the charge of gypsum is loaded into the kettle. The gypsum is turned continuously by paddles attached to a rotating shaft that extends into the kettle. As the temperature increases, the gypsum begins to boil violently as the water that is merely soaked into the particles of gypsum is driven off. The boiling subsides as the temperature continues to increase, only to increase again once the chemically bound water begins to boil away. The temperature within the kettle is held to some temperature below 374°F until the boiling subsides. When it does, it means that about three-quarters of the bound water has been driven off. The gypsum is then discharged from the kettle through a door located near its base and allowed to cool.

The second method is a continuous process in which a large, cylindrically shaped kiln is loaded at one end. The kiln is tilted a bit so that the loading end is higher than the discharge end. The kiln is heated over most of its length, the temperature increasing as the discharge end of the kiln is approached. The kiln also rotates, and, as the gypsum passes the length of the kiln, it is heated more and more until its water content has been driven off. When this method of manufacture is used, the pieces of raw gypsum can be as large as ½ in. in size.

The cylindrical kiln is from 12 to 15 ft in diameter and about 15 ft long. The interior is lined with firebrick. What happens inside the kiln is the same as inside a kettle, except that higher temperatures are required to drive off the bound water. The hot gypsum is discharged into bins where the water content of the gypsum continues to boil away for perhaps another 36 hr. The gypsum is then rapidly cooled.

What emerges from the kettle or kiln depends on the kind of raw gypsum that went in. If the original charge was pure (or almost pure) gypsum, the final product is a *half-hydrated* calcium sulfate ($2CaSO_4 + H_2O$) known as *plaster of paris,* so-called because of the huge deposits of gypsum located under the city of Paris that were first mined toward the end of the nineteenth century.

If the original charge contained impurities, either as part of the raw gypsum or because additives were purposely mixed with the charge to give the final product certain desired characteristics, the final product is called *gypsum plaster* or *hard-wall plaster.* A gypsum plaster, for example, usually has a retarder added to it to extend the setting time of the gypsum. On the other hand, impurities in the raw gypsum may retard the setting time too much, in which case accelerators will be added to speed up the setting time. The amount of additives required is usually quite small, amounting to perhaps 0.2 to 0.4 percent of the quantity of gypsum.

What happens if the calcining process is allowed to continue until all the bound water has been driven off? This will happen if the temperature within the kettle or kiln is increased to somewhere between 750°F and 930°F. The temperature should not exceed 930°F, however, nor should the gypsum be held at too high a temperature for too long a time as the final product will fail to set at all when water is added to it again.

If the original charge consisted of pure gypsum, and all the bound water is driven from it, the final product is *anhydrous* calcium sulfate, sometimes known as *flooring plaster.* If the original charge contained impurities, either as part of the original raw gypsum or because additives were mixed with the charge, the final product is called *hard-finish plaster.* A particularly well-known type of hard-finish plaster is called *Keene's cement,* after the Englishman who developed it. Keene's cement is made by quenching red-hot calcined gypsum in a 10 percent solution of alum, and by then recalcining the gypsum.

Gypsum Panels

Panels consisting of sheets of gypsum plaster sandwiched between two layers of paper are used variously as wallboards, sheathing boards, lath boards, formboards, backing boards, and roofing planks. Although the uses to which these panels are put makes it convenient that they be made in different sizes and thicknesses (see Table 1), there is not really much difference between them. The materials used are much the same for all the panels, and all the panels are manufactured in much the same way.

Manufacture of the panels begins with clacined gypsum powder (for a description of the manufacturing process *see* GYPSUM). The powder is mixed with water to form a slurry, which is then extruded between two sheets of paper to form a continuous 4-ft-wide sheet. As this endless sheet moves along a conveyor belt, its edges may be left square or they may be pressed to form either tapered edges, beveled edges, V-shaped tongues-and-grooves, or rounded edges (see Fig. 1). All these shapes have different purposes, as described below.

Three to four minutes after it has been extruded, the plaster has set hard enough that the sheet can be cut into panels. The finished panels are then sent to a drying oven where they are heated to hasten the evaporation of excess moisture as much

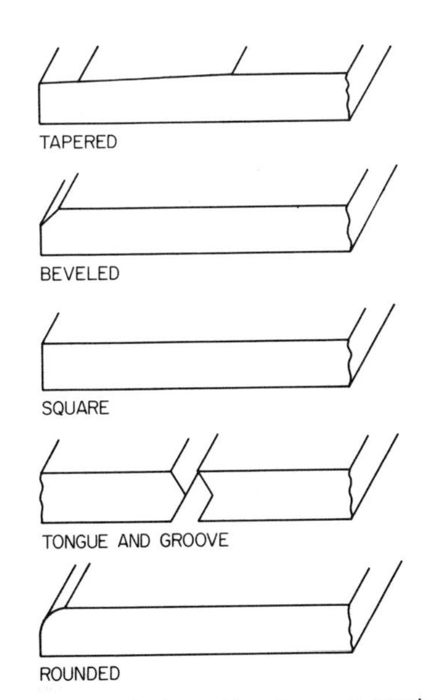

Fig. 1. Types of edges on the long sides of gypsum panels.

Table 1. Different Products Made from Gypsum Panels

Type	Thickness, in.	Width, in.	Length, ft	Edge detail
Wallboard				
Regular	¼	48	8, 10, 12,	
	⅜	48	7, 8, 9, 10, 12	
	½	48	7, 8, 9, 10, 12, 14	Tapered
Insulating	⅜	48	8, 10, 12	
	½	48	8, 9, 10, 12	
Regular Type X	½	48	8, 9, 10, 12,	
	⅝	48	8, 9, 10, 12	
Lath				
Plain	⅜	16	4, 8	
	½	16	4	
Insulating	⅜	16	4, 8	Round
	½	16	4	
Plain Type X	⅜	16	4, 8	
Backing board				
Regular	⅜	48	8	Square
	½	24	8	Tongue and Grooved
Insulating	⅜	48	8	Square
Regular Type X	⅝	24	7	Tongue and Grooved
Sheathing		24	8	Tongue and Grooved
	½	48	8, 9	Square
Formboard	½	32	8, 10	Square

as possible. The panels are then ready for packing and shipping, unless they are to receive one of the special treatments described below.

WALLBOARDS

Regular Wallboards

Regular wallboards are made in the sizes, thicknesses, and with the edge shapes shown in Table 1. One side of the panel is covered with a rough, gray-colored lining paper, and the other side is covered with a smooth, hard, ivory-colored manila paper. The manila-paper side of the panel is considered the front or exposed side, to be painted or wallpapered over as desired.

The surfaces of regular wallboards taper gradually near the long edges of the panels so that abutting panels can be taped and covered with joint compound to make the joint as inconspicuous as possible (*see* DRYWALL CONSTRUCTION).

Insulating Wallboards

Insulating wallboards have a sheet of aluminum foil lamined to the backsides of the panels. The foil acts as both a vapor barrier and reflective insulation (*see* CONDENSATION; INSULATION) when the wallboards are installed on the exterior-wall surfaces of a house. In other respects, the wallboards are identical to regular wallboards, and they are made in the same sizes and thicknesses.

Fire-Resistant (Type X) Wallboards

Gypsum wallboards are inherently fire-resistant (*see* GYPSUM). To further increase the length of time that the thin wallboards used in ordinary house construction can resist the heat of a fire, the plaster cores of these panels may have mixed with the gypsum an aggregate consisting of either vermiculite or perlite, plus a binder of fiber glass. These panels are known as *Type X* panels. These Type X wallboards are made in the same sizes and thicknesses as regular wallboards. They are also available with the aluminum-foil backing described above.

Water-Resistant Wallboards

Gypsum plasters deteriorate badly if they should ever absorb moisture. This failing pretty much limits the use of gypsum panels to interior locations where they are very unlikely to come into contact with an excessive amount of moisture. To extend the use of gypsum panels to rooms where they are likely to absorb moisture—bathrooms, kitchens, laundry rooms, and the like—one type of gypsum panel is manufactured in which emulsified asphalt is mixed with the gypsum powder. The asphalt is supposed to coat the gypsum and thus prevent moisture from getting at it. In addition, the paper covering the panel is especially treated with water-repellent chemicals. Otherwise, these wallboards are made in the same sizes and thicknesses as regular wallboards.

BACKING BOARD

Backing board is merely regular wallboard having both sides of the panel covered with rough, gray paper. Backing boards are intended as a base upon which any of the wallboards described above can be attached in a two-ply type of construction (*see* DRYWALL CONSTRUCTION). They are also used as a base to which acoustical ceiling tiles can be attached.

In addition, a special type of water-resistant backing board is made for use as a base in exceptionally damp locations, such as shower stalls, for example, over which ceramic tile is to be applied. To maintain the moisture-resistance of the backing board wherever it is cut, the cut edges must be covered with special vinyl tapes that are provided by the manufacturer for this purpose.

There is no need for decorative finishes or a variety of edge treatments on backing boards because they will be covered with some sort of finish material, and the range of available sizes and thickness is limited to those shown in Table 1. The long edges of the panels are usually made with V-shaped tongues-and-grooves that help to speed up installation of the boards and ensure that the mating edges are flush with each other.

Backing boards are also available with aluminum-foil backing and with fire-resistant (Type X) cores, as described above.

GYPSUM SHEATHING

Special gypsum panels 2 × 8 ft in size are made for use as wall sheathing over wood-framed structures (*see* SHEATHING). Emulsified asphalt is sometimes added to the gypsum mix and the paper covering the panels is treated with asphalt to further increase the water resistance of the panels. Each panel so manufactured has the words *Water Resistant* printed on it. It is unnecessary, therefore, to nail sheathing paper over the panels to increase the water resistance of the construction. In addition, the long edges of the panels have V-shaped tongues-and-grooves that help to speed up installation of the sheathing and increase (a bit) the moisture- and air-infiltration resistance of the construction.

The great advantage of using gypsum sheathing is the improved fire resistance of the construction. But gypsum panels have no great nail-holding ability and, therefore, they do not add anything to the rigidity of the construction, as lumber and plywood sheathing do because the nails will evenually work loose. Local building codes, therefore, usually require that the wood framing to which gypsum sheathing is nailed must be strengthed with additional bracing.

Gypsum sheathing panels 4 × 8 ft and 4 × 9 ft in size are also manufactured. Their purpose is to add the same rigidity to the construction that plywood panels do. *See also* SHEATHING.

GYPSUM LATH

Gypsum lath (which is also known as *rock lath*) is manufactured in 16- and 24-in. widths and in 4- and 8-ft lengths. Gypsum lath is used as a base over which gypsum plasters are applied (*see* PLASTERING). The paper covering the panels has a rough, fibrous surface that allows the plaster to bond exceptionally well to the lath. The edges of the panels are usually rounded to increase the thickness of the plaster at the joints, thus increasing the strength of the plaster at these weak points and helping to prevent plaster cracks. Gypsum lath is available with the special Type X fire-resistant core described above. It is also available with an aluminum-foil backing.

Heating Systems and Human Comfort

This article is a general introduction to domestic heating systems and the factors that are (or should be) taken into account in their design in a new house. Anyone considering the purchase of an existing house should take these same factors into account when judging the adequacy of the existing heating system.

The size of a heating plant (that is, its output in Btu per hr) will depend on (1) the average outside winter temperatures, (2) the size and shape of the house and the locations of the rooms relative to the sun, and (3) the amount of heat that is lost through the walls, roof, and floor of a house and the amount of cold air that leaks into the house.

The basic point that must be decided on before any heating system can be designed is the difference between the design *outdoor* temperature and the design *indoor* temperature, the latter usually being assumed to be 70°F. In fact, the most comfortable indoor temperature will depend not only on the temperature of the air but also on the humidity, the amount of physical activity being performed, the temperatures of the walls, ceilings, and floors, which will differ from the air temperature, as well as on psychological factors that are both extremely important and extremely difficult to pin down in a formula. Women, for example, usually feel most comfortable at room temperatures that men consider too high. How is this subjective fact to be measured?

A number of experiments have shown that a temperature of 68°F—not 72°F—is probably the optimum room temperature for most people who are working. In experiments with students, for example, it was found that the students did their best work when the room temperature was 68°F. When the room temperature was increased above 68°F, the students become increasingly restless and dull-witted; at 74°F they were doing 15 percent less work than they had been doing at 68°F. When the temperature was decreased below 68°F, the amount of work accomplished fell off again.

These experiments involved sedentary labor. When the average person is sitting quietly in a room, the metabolic processes of his body produce about 6⅔ Btu of heat per minute. When this heat is dissipated from the body at the same rate at which it is produced, the skin temperature will be somewhere between 82°F and 89°F and the individual will feel "comfortable." If his skin temperature should fall below 82°F, he will feel chilly. On the other hand, if his skin temperature should rise above 89°F, he will feel warm. Thus, it is not the temperature of the air within a room that is of basic importance for human comfort but the balance that exists between body temperature and the physiologically perceived room temperature.

Whenever the skin temperature is within the normal comfort range of 82 to 89°F, about 92 percent of the body's excess heat is dissipated from the skin by a combination of convection and radiation. Only about 8 percent of this excess body heat is dissipated by respiration and the evaporation of moisture from the skin. As the body temperature increases, however, the body begins to perspire and the proportion of body heat dissipated by the evaporation of this perspiration increases until, when the skin temperature is much above 89°F, it amounts to about 30 percent.

Several conclusions follow. If the air within a room is too dry, as is usually the case during the winter, the average individual will feel that the room is a bit on the cool side, even though the air temperature is within his comfort range, because the amount of moisture evaporating from his skin will be greater than if the air were more humid. A draft of air through a room has the same cooling effect, since the draft will increase the rate of evaporation as it strikes the skin, which is why a drafty room always feels a bit cooler in hot weather and a bit chilly in cold weather.

On the other hand, maintaining the relative humidity at a fairly high level, 50 percent and more, increases the apparent air temperature because the humidity inhibits the evaporation of moisture from the skin. The amount of humidity cannot increase indefinitely, however, because the excess moisture will begin to condense on cold surfaces, with potentially harmful consequences. (For details of this process, *see* CONDENSATION.)

Of the 92 percent of body heat that is lost through convection and radiation at the low end of the comfort range, about half is lost through convection and half through radiation. The amount of body heat lost through convection depends on the air temperature in the room, and this factor can be controlled by the heating plant through the setting of a room thermostat.

The amount of body heat lost through radiation, however, depends entirely on the temperature of the surrounding walls, floor, and ceiling—the room air temperature has nothing whatever to do with this radiation loss. The reason is that heat radiates from a warm object (like the human body) to nearby cold objects, and never vice-versa. The greater the difference in temperature between the two objects, the greater the rate at which heat is transferred from the warmer object to the colder object. It follows that in a room in which the walls, floor, and ceiling are cold, the rate at which the body gives up its heat will be high. In the ordinary, uninsulated wood-frame house, for example, the rooms on the north side of the house are much colder during the winter than the rooms on the south side of the house. The rooms are cold, not because "cold" is being radiated into the house but because the heat being radiated by the occupants and any other warm objects in the room is being transferred at a high rate to the walls. The walls, in turn, are transferring this heat to the outdoors with equal rapidity.

Sizing a Heating Plant

When a heating contractor designs the heating system for a new house, the main objective is to select a heating plant having a certain optimum heat output. That is, the heat output (in Btu per hr) must be sufficient to maintain the design indoor temperature within the house during cold weather while at the same time keeping the fuel consumption as low as possible.

Basically, the size of any heating plant is determined by the difference between two assumed temperatures—the design indoor temperature and the design outdoor temperature. The latter temperature is *not* the lowest temperature likely to be encountered during any given winter but a statistic. It is obtained by examining the weather records for past years and then striking a reasonable balance between the coldest temperature that *may* occur during any given year and the average temperatures that usually *do* prevail during the winter. Typically, about 97½ percent of the temperatures that do occur during any given year will be at or above the design outdoor temperature and about 2½ percent will be below. If the heating plant has been correctly sized, therefore, the heating plant will be unable to maintain the design indoor temperature for about 2½ percent of the time. If, for example, the design indoor and outdoor temperatures are 70°F and 10°F, respectively, and the outdoor temperature drops to, say, 5°F for a few hours, the heating plant will be capable of heating the house only to 65°F during that period.

This may seem unreasonable to the reader, but it is plain wasteful to install an oversize heating plant merely because once every few years the outside temperature may fall as low as 5°F for a few hours. Besides, the lowest temperatures occur at night, when the thermostat is usually turned down anyway. Of course, if money is no object, the heating contractor can be told to go ahead and design the heating system around a lower assumed outdoor temperature.

The design indoor and outdoor temperatures having been determined, the heating contractor can proceed to calculate the total amount of heat that will be lost from the house per hr. The basic fact that must be taken into account here is that different types of wall construction lose heat to the outdoors at different rates. It should be obvious, for example, that in any given time period a lesser amount of heat will be lost through a 2-ft-thick masonry wall than will be lost through the thin sheets of paper that are pasted to the exterior wall screens of the traditional Japanese house.

A considerable amount of research has been conducted at universities and by different trade and engineering associations into the heat-transmission characteristics of different kinds of material and different types of construction. This research has been summarized in the form of *heat-transmission coefficients* that allow one to compare these materials and constructions directly with each other. Formulas have been developed into which these coefficients can be "plugged" when the heat loss is being calculated in square feet of wall area per hr. By examining the plans of a new house or the actual construction in an existing house, the heating contractor can determine what coefficients to use in order to determine the heat loss for that house. Separate calculations are required to determine the heat loss for the exposed wall areas and for the doors and windows in each room.

If the ceiling of a room is also the underside of a roof surface or an attic, another calculation is necessary. The heat loss through a roof is treated the same as the heat loss through a wall. As for the heat loss through an attic floor, the calculation must take into account whether the attic is heated or whether it is unheated and vented to the outdoors. The heat loss through the walls and floor of a basement, or through a slab-on-grade floor, must also calculated separately, the calculations being based on an assumed soil temperature, which will vary according to the latitude.

Another cause of heat loss is the crevices that always exist between doors and windows and their frames. In an old house these crevices can be large enough that a complete change of air occurs three or four times every hour. Even in a new house they may allow one or two air changes every hour. This air-infiltration loss through the crevices must also be calculated separately.

Finally, the individual calculations completed, the heating contractor adds them all together and comes up with a number that represents the total heat loss in Btu per hr. The required size of heating plant is determined by this number. A correctly sized heating plant will have a heat output that is sufficient to maintain the house at the desired indoor temperature without its either operating continuously or switching on and off at frequent intervals. If a heating plant operates continuously, it means it is undersized. If it switches on and off continually, it means it is oversize. In either case the fuel consumption will be greater than necessary and, in the games that homeowners play, one of the games is called minimizing the fuel bill.

In addition to maintaining a desired air temperature, the heating plant has another function, which is to quickly heat a house that has gone completely cold. This quick pick-up is needed when the family returns from a winter vacation, or if a malfunction of some kind makes it necessary to shut down the heating system for a day or two. It is also needed on a regular basis in most households when, to conserve fuel, the thermostat is turned down to about 62°F every night. Every morning, therefore, there is a start-up period during which the heating plant must operate continuously until the design indoor temperature is reached. If the capacity of the heating plant were just sufficient to replace only the heat lost as calculated above, then, from a cold start, the heating plant would not be able to achieve the desired indoor temperature. It would run continuously until a spell of warm weather allowed it to catch up with a decreasing air temperature. For this reason boilers and furnaces are rated at both net and gross output. The net-output rating should be slightly above the calculated heat loss, and the gross-output rating should provide the additional heat required for morning pick-up.

Hot-Water Supply System

There are two main kinds of hot-water supply system, storage and demand. In a *storage* system, the water is heated and then stored in a tank until needed. When the storage tank and heating unit are combined into a single assembly, with the heating unit located directly under the tank, the system is a *direct-fired* storage system. Most domestic hot-water supply systems are of this type (see Fig. 1). If the heating unit and storage tank are completely separate, the system is called an *indirect-fired* storage system.

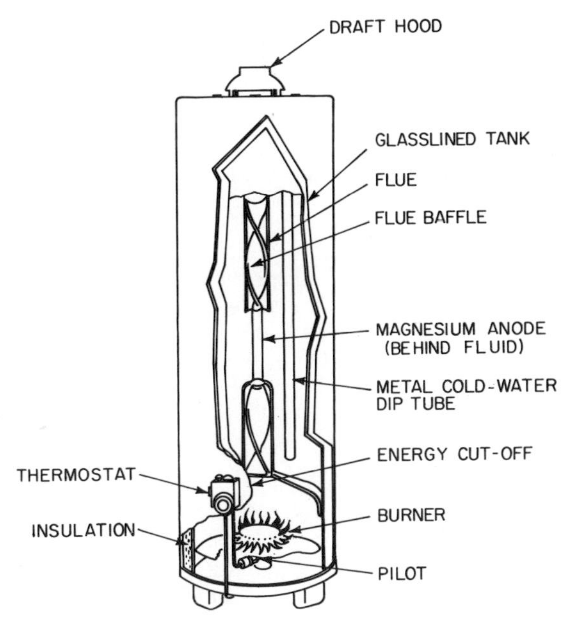

Fig. 1. A typical gas-fired water heater.

In the *demand* type of hot-water storage system, the water is heated and supplied on demand—there is no storage tank at all. This system is also known as an *instantaneous,* or *tankless,* system. In a demand system a copper coil—usually called a *tankless coil*—is surrounded by hot water or steam obtained from a boiler. Cold water circulates through the coil and is heated by the surrounding hot water or steam. The domestic hot water then flows either directly into a distribution system for immediate delivery upon demand or into a storage tank, from which it is withdrawn upon demand. Demand systems are common in dwellings when a boiler has been installed as part of the heating system, except in parts of the country where the water is excessively "hard." Tankless-coil installations (for those dwellings that already have hot-water or steam boilers installed) are quite practical and can save about 30 percent on the cost of heating water during the heating season.

STORAGE TYPE HOT-WATER SYSTEMS

Storage Tank

The storage tanks used in direct-fired systems are usually made of galvanized steel, although copper-lined tanks are also manufactured. The interior of galvanized-steel tanks may be coated with a layer of vitreous porcelain enamel, known as *glass* in the trade to further increase the life of the tank. *Glass-lined* tanks are usually guaranteed against corrosion for 7 to 10 years; unlined galvanized tanks are guaranteed for only 2 to 5 years, depending on the thickness of the zinc coating.

Galvanized-steel tanks will also have a magnesium rod inserted into the tank that runs the length of the tank and acts as the anode of a galvanic couple. That is, the rod is slowly consumed by an electrolytic process and, by being consumed, helps to protect the steel against corrosion. Since the magnesium rod does waste away in time, one of the important duties of a conscientious homeowner should be to remove the rod from time to time, inspect its condition, and replace it when necessary. Otherwise, once the rod has been destroyed, the tank will sooner or later corrode through.

In copper-lined tanks the usual thickness of the copper is 3 lb per sq ft. Copper-lined tanks are, of course, immune from the corrosion that attacks steel.

Tank sizes vary according to the method used to heat the water. Storage tanks used with gas- or oil-fired heating units usually have 30-, 40-, and 50-gallon capacities. Electric units heat more slowly, so they require tanks with a greater storage capacity, usually 40-, 52-, 66-, and 80-gallons.

All the tanks are surrounded by insulation at least 1 in. thick to maintain the temperature of the water stored within the tanks. Mineral wool and fiber glass are the usual insulating materials used. The entire tank is then covered by sheet metal having a baked-on enamel finish.

Tank Safety Precautions

The tanks used with direct-fired systems must be factory-tested at 300 lb per sq in. and be capable of withstanding a normal operating pressure of 150 lb per sq in. Any tank meeting these requirements, as well as the design and construction standards of the American Society of Mechanical Engineers, will have an ASME stamp located somewhere on the tank.

It is always possible—if not very probable—that a system malfunction of some kind may cause pressure to build up in the storage tank or the temperature of the water to reach the boiling point. In either case, the result is a potentially dangerous condition.

An excessive buildup in tank pressure (that is not the result of an excessive temperature increase) is not particularly dangerous, though it can result in water damage to the house. The result of excessive pressure is that the tank ruptures rather than explodes. The water bursts from the tank at the instant of rupture, but the amount of pressure will not have been exceptionally high in the first place, and, in the second place, the water pressure drops abruptly once the rupture has occurred. The result will be a flood of hot water on the floor and a ruined tank but no particular danger to the house or its occupants.

If, however, the pressure buildup is the consequence of a temperature buildup, the result might very well be an explosive release of energy that will completely demolish the house.

When water is heated, the molecules of which it consists move more and more rapidly. This increased molecular motion causes the molecules to knock against each other with greater force and frequency, which gives each molecule a little more elbow room, so to speak. One of the effects of this increased elbow room is a decrease in the density of the water, which is why hot water tends always to rise above cold water.

Another effect of the increased molecular motion in a closed hot-water storage tank is an increase in the pressure exerted by the hot water against the walls of the tank. As long as the temperature of the water remains below the boiling point, the pressure within the tank will be normal system pressure plus the weight of the water pressing against the walls of the tank. But once the temperature of the water increases above 212°F, it begins to exert an excessive pressure against the tank walls. At some point the pressure within the tank will exceed the strength of the metal, and the tank will rupture. As the superheated water escapes into the atmosphere, it expands into steam with explosive force.

The table below shows how the pressure within a 30-gallon hot-water tank increases as the temperature increases. It should be understood that it is not the increase in pressure that is responsible for the explosion. The explosion is the result of the almost instantaneous expansion in the volume of the escaping superheated water as it flashes into steam. The total expansion of water into steam is about 1700 times the original volume of the water. The energy contained in this expansion, assuming a water temperature of about 300°F, is equal to the force of 1 lb of nitroglycerine

Temperature of water in closed tank, °F	*Pressure of water above atmospheric, psi,*	*Energy contained within tank, ft·lb*
212	0	
240	10	480,000
274	30	1,305,000
300	50	2,022,000
316	70	3,640,000
330	90	4,140,000

Pressure and Temperature Relief Valves

The need for installing protective devices on a hot-water tank should be evident. These devices usually take the form of pressure and temperature relief valves (see Fig. 2).

Separate pressure and temperature relief valves can be installed on a hot-water tank—if the tank has been tapped for two relief valves; but most direct-fired storage tanks contain a combination pressure-temperature relief valve.

The pressure setting on either an independent or a combination valve is usually 25 lb per sq in. above the normal water supply pressure. If, for example, the pressure of the incoming water is 40 lb per sq in., the valve setting would be 65 lb per sq in. But, whatever the incoming water pressure, the valve setting should never exceed the tank's working pressure, of course. Some pressure valves have a fusible plug installed in them. The plug melts at about 210°F. Thus, if the valve should stick in the closed position, the plug will melt and relieve any excess pressures that would otherwise result from a temperature increase.

The temperature setting on both an independent and a combination relief valve is also 210°F. The temperature relief

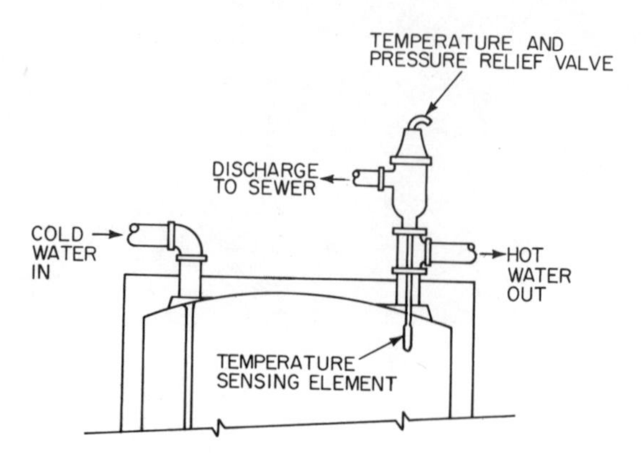

Fig. 2. A 'combination temperature-and-pressure-relief valve mounted on a hot-water supply tank *(U.S. Federal Housing Administration).*

valve operates through a sensing bulb immersed in the tank. The sensing bulb must always be located within 6 in. of the top of the tank, where the water is hottest.

All relief valves include a safety test lever. When this lever is pulled up, the relief valve will open and water will be discharged from the tank. Thus, the valve can always be checked for correct operation. When the lever is released, the valve must reseat itself firmly to prevent any further discharge of water.

The outlet of a relief valve should always be piped to a floor drain or into a laundry tub. Failure to do this may cause the valve to spit water onto the floor, or even flood the floor if it should ever stick open for any reason, or be set too low. This drain line should never be connected directly into a floor drain as it will then be impossible to check the operation of the valve visually.

A tank may have a *high-temperature energy shutoff device* installed instead of a temperature relief valve (see Fig. 3). The energy shutoff device is part of the thermostat circuit that controls the operation of the heating unit. If the temperature within the tank should ever exceed 210°F, the energy shutoff device will immediately shut off the fuel flowing to the heating unit.

Some manufacturers supply only an energy shutoff device on their tanks, with the pressure or pressure-temperature relief valve being an optional extra. A hot-water tank should never be installed with only an energy shutoff device as protection against excessive tank temperatures. If the device should become inoperative for any reason, there would not be any

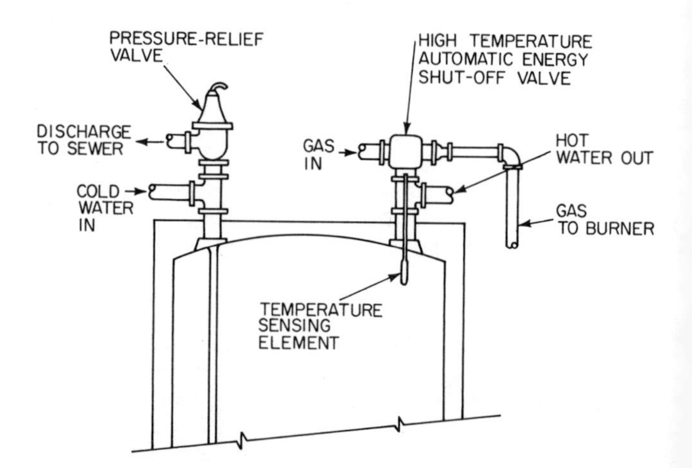

Fig. 3. A hot-water supply tank with a seperate high-temperature energy-shutoff device installed *(U.S. Federal Housing Administration).*

other way available of preventing an excessive temperature rise. Either a pressure relief valve or a combination pressure-temperature valve should, therefore, be installed at the same time as the energy shutoff device.

Gas- and Oil-Fired Heating Units

Both gas- and oil-fired heaters operate in much the same way. The fuel passes through piping into a burner. If the burner is not already in operation, the fuel is ignited by a pilot flame (for gas) or a spark (for oil), and the hot combustion gases pass up a flue that extends through the center of the tank, thus heating the water. The gases emerge from the top of the tank into an external flue or vent that leads either to a chimney or directly to the outside of the house.

In both gas- and oil-fired systems, the water temperature is sensed by a thermostat that maintains the water temperature at the desired value by regulating the flow of fuel to the burner. The thermostat can be adjusted to provide hot water between 120°F and 180°F, with the usual setting being 140°F. If the pilot flame should go out for any reason, or if a spark fails to ignite vaporized fuel oil, the fuel flow is automatically shut off. The burner must then be restarted manually.

Gas-fired installations have the advantage over oil-fired installations in that a fuel storage tank is not required with gas. Thus, anyone considering the purchase of an oil-fired heater must already have an oil tank installed, usually because he has an oil-burning boiler or furnace installed; otherwise, he must install an oil storage tank plus the necessary piping, which adds considerably to the cost of the installation.

An oil-fired burner will cost two to three times as much as a gas-fired burner, assuming the two burners have the same water-heating capacity. The basic reason is that oil requires a more complex burner mechanism than gas. Gas comes into the house under pressure and can be burned as-is. The oil must first be vaporized and blown into the combustion chamber before it is ignited. The complexity of the pump, fan, and burner that are required thus adds to the cost of the installation. Anyone having a choice between gas and oil, therefore, should take into account the comparative costs of the fuel-burning equipment as well as the cost of the fuel.

Electric Hot-Water Heating Units

The great advantage of an electric hot-water system over gas or oil is the simplicity and cleanliness of the installation. Heat is supplied by two immersion heating elements installed within the tank (see Fig. 4). These heating elements operate either independently or together to produce the required amount of hot water. Hot-water tanks having a relatively small capacity, 40 gal, say, need only one or two 110-volt, 1500-watt heating elements installed in the tank. Larger tanks require 208- or 220-volt circuits and heating elements having an output of up to 6000 watts.

The great disadvantage of using electricity to heat water is the cost of the electricity itself. Assume, for example, that we have two hot-water heaters, one gas-fired and the other electrically operated. Both units have a 30-gal capacity and a *recovery rate* of 33.6 gal per hr. (The recovery rate of a hot-water system is the number of gallons of cold water per hour that the burner is capable of heating to the desired temperature.) In this example, a 33.6 gal per hr recovery rate means the burners can completely replenish the hot water supply in the tanks in one hour. To provide this recovery rate, the gas-fired burner must have a heat output of 34,000 Btu per hr, and the electric heater must have a heat output of 8200 watts per hr, which is equivalent to about 27,900 Btu per hr.

What these figures show is that even though fewer Btu per hr are required for the electrically operated heater because of the greater efficiency with which electrical energy is converted into heat energy, nevertheless, the heater must operate longer; and, because of the higher cost of electricity per Btu, the dollar cost of obtaining hot water from electricity is much higher.

The cost of the electricity can be reduced in one of two ways. One method is to reduce the size of the hot-water tank. The smaller the tank, the wattage input of the heating elements remaining the same, the faster the water will heat up. One can in this way avoid the waste involved in having to heat a large quantity of water and in keeping the water hot. In effect, what one is doing is converting a storage-tank system into a modified demand system, and experience shows that demand systems operate with about 30 percent less fuel consumption than direct-fired storage-tank systems.

The other method of reducing electrical costs is to use a very large tank and heat the water in the tank only during certain off-peak periods when electrical rates are much lower. Many utilities offer lower off-peak rates to encourage the installation of electric hot-water heaters.

The utility will install a separate line that supplies the electricity going to the storage-tank heating elements. A clock installed in this circuit by the utility closes the circuit only during the off-peak hours, at which times the heating elements can be energized. To make the most efficient use of this cheaper power, the capacity of the tank must be large enough to meet all the hot-water requirements of the family during times when the hot-water electrical circuit is deenergized.

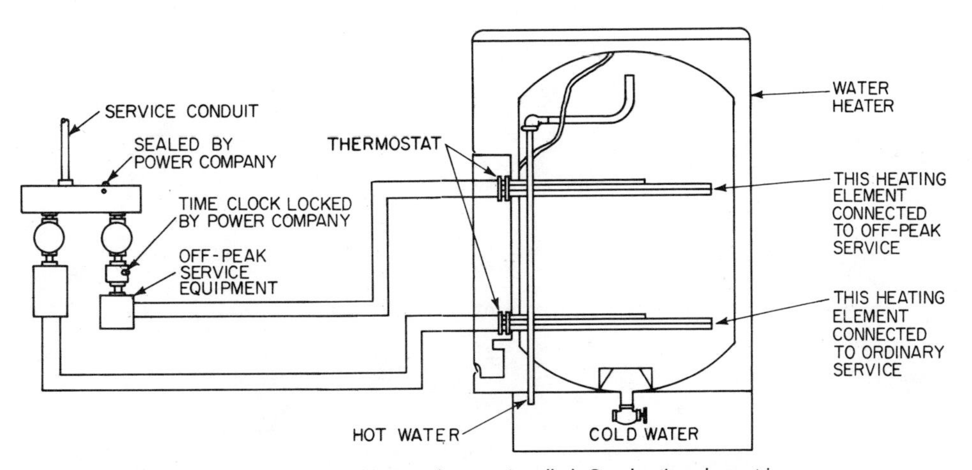

Fig. 4. An electric hot-water heating unit has two immersion heating elements installed. One heating element is connected to an off-peak service line to reduce the cost of heating the water.

DEMAND HOT-WATER SYSTEMS

In a demand hot-water system (see Fig. 5), the incoming cold water passes through a copper coil, which may be located either inside the boiler or outside. If inside, the coil is submerged within the boiler water. If outside, the coil is located inside a metal housing through which hot water or steam from the boiler circulates. This assembly of housing and coil is a simple version of what is better known as a shell-and-tube heat exchanger. The heated water flows from the coil directly into the distribution system whenever a faucet is opened.

There is no way in which the temperature of the water within the coil can be controlled directly in a demand system, since the hot-water temperature will depend on the temperature of the fluid within the boiler, the temperature of the cold water, and the rate at which the cold water flows through the coil. Manufacturers, therefore, rate the heating capacity of their coils on the basis of assumed temperature values. The usual assumptions are that the boiler water is either at 180° or 200°F, and that the coil has sufficient surface area so that x gallons of cold water flowing through the coil per minute will have its temperature raised 100°F. The larger the coil, the greater the heating capacity of the coil in gallons per minute. For domestic use, a tankless coil should have a minimum capacity of 2¾ to 3¾ gal of 140°F water per minute, depending on the size of the family and the number of bathrooms in the house (see Table 1).

When there is no demand for hot water, there is no hot-water flow. Whenever there is a demand, the hot water flows through a thermostatically controlled mixing valve and into the hot-water distribution system. Since the water may have been recirculating through the coil for a considerable period of time, its temperature may be somewhere between 180 and 200°F.

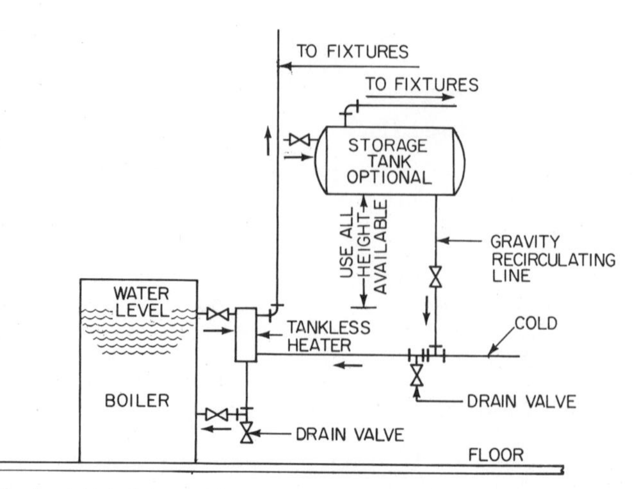

Fig. 5. A typical demand hot-water system *(Emerick, Heating Handbook, McGraw-Hill).*

The mixing valve is necessary, therefore, to cool, or temper, the hot water to 140°F. The valve may, of course, be set to a lower or higher temperature, if desired. The mixing valve must also be of the *fail-safe* type. That is, if it should develop a defect, it will go automatically to its full-cold setting, thus preventing the accidental scalding of anyone who might be using hot water at the time.

A pressure relief valve should be installed in the line between the coil and the mixing valve. The valve is usually set at the operating pressure of the boiler, that is, 15 lb per sq in. (gauge) for steam and 30 lb per sq in. (gauge) for water.

Table 1. Guide to the Selection of Hot-Water Heaters (Based on the number of bathrooms and bedrooms in a dwelling)
Storage and tankless type domestic water heaters

Number of bathrooms		1–1½			2–2½			3–3½		
Number of bedrooms	2	3	4	3	4	5	3	4	5	6
Storage type, gas and oil fired										
Storage gal	30	30	40	40	40	50	40	50	50	50
1000 Btu input	30	30	30	33	33	35	33	35	35	35
Storage type, electric										
Storage gallons	52	52	66	66	66	80	66	80	80	80
Input in kilowatts: Upper elements	1.5	1.5	2.0	2.0	2.0	2.5	2.0	2.5	2.5	2.5
Input in kilowatts: Lower element	1.0	1.0	1.25	1.25	1.25	1.5	1.25	1.5	1.5	1.5
Input in kilowatts: Single elements	2.5	2.5	3.0	3.0	3.0	4.0	3.0	4.0	4.0	4.0
Boiler connected indirect water heaters Internal or external Type with tank, boiler water 180°F										
IWH rated; gal in 3 hr; 100°F, rise	40	40	66	66	66	66	66	66	66	66
Manufacturer rated; gal in 3 hr; 100° F, rise	49	49	75	75	75	75	75	75	75	75
Tank capacity in gal	66	66	66	66	66	82	66	82	82	82
Internal or external type, tankless, boiler water 200° F										
IWH rated; gpm; 100°F. rise	2.75	2.75	3.25	3.25	3.25	3.75	3.25	3.75	3.75	3.75
Manufacturer rated; draw in 5 min; 100°F rise	15	15	25	25	25	35	25	35	35	35

Note: (a) FHA field offices may require proof that manufacturer rated indirect water heaters produce the required amount of hot water at the indicated temperature rise.
(b) The vertical columns in the table show the minimum permissible sizes of heaters for the listed bath and bedroom count when heaters utilize the fuel indicated. It is not to be assumed that heaters of the sizes shown in any single vertical column are comparable in their ability to produce hot water.
Source: U.S. Federal Housing Administration.

Storage Tanks Used with Tankless Coils

The minimal-sized tankless coils installed in most dwellings may be said to have a steady but small output of hot water. In most of these dwellings the coil cannot be depended upon to supply sufficient hot water if the demand should ever be excessive; when, for example, in a two-bathroom house, both bathtubs are run at the same time. It is sensible, therefore, to install a storage tank that will increase the capacity of the system to meet any unusually heavy demand. The tank should have a minimum capacity of 30 gal, and it should be covered with at least 1 in. of insulation, which will reduce the fuel costs by about 30 percent.

Such a tank is installed in the system as shown in Fig. 5. As may be seen, whenever there is no demand, the hot water recirculates between the coil and the tank, which keeps the tank filled with hot water. The system is a *gravity* system, in which the water recirculates because of the difference in its density between the two units. The greater the temperature differential of the water, the greater the difference in its density, and the faster the water will recirculate. In installations where the temperature difference isn't all that great, the difference in height between the tank and the coil becomes important, since height emphasizes slight differences in density. For most efficient circulation, therefore, the tank must be as high as possible above the outlet of the coil. A difference of 5 ft should be considered a minimum. If this cannot be managed because the ceiling is too low, it might be necessary to install a pump in the circuit to recirculate the water.

Sometimes a tankless coil is used to boost the temperature of the incoming cold water when a direct-fired storage tank is actually connected to the domestic hot-water system. This is called a *modified* tankless-coil installation. Combining the two systems in this way can reduce substantially the cost of supplying hot water to the fixtures during the winter months.

The storage tanks used in modified tankless-coil installations are made nowadays either of galvanized steel, monel metal, or stainless steel, but in an old installation the tank may be made of copper. Tanks made of steel should be glass-lined; if not, they should at least have a magnesium rod installed to prevent corrosion. All tanks should have either separate pressure and temperature relief valves installed on them or a combination pressure-temperature valve, for the reasons already described. The tanks should also be insulated to help maintain the temperature of the water within.

HOT-WATER DISTRIBUTION SYSTEM

The Cold-Water Supply Line

In a hot-water storage tank, the incoming cold water enters the top of the tank through a pipe that extends about 6 in. from the bottom of the tank. This extension prevents dilution of the hot water at the top of the tank with cold water; and the fact that the pipe ends about 6 in. from the bottom of the tank prevents water currents from stirring up any sediment that may have collected at the bottom.

A check valve is sometimes installed in the water-supply lines to prevent backflow from the tank. Backflow may occur because of an overtemperature condition in the tank. The increased pressure that is a consequent of this overtemperature forces the hot water out of the tank and into the cold-water line. And, in fact, this is a very convenient way of relieving a potentially dangerous condition. But if other equipment should be installed in the cold-water line—a water meter or water-softening equipment, say—a check valve must be installed in the line to protect this equipment against any such backflow. And if a check valve is installed in the line for this reason, a relief valve must then also be installed between the tank and the check valve to relieve any overpressure that might develop in the line.

Noncirculating versus Circulating Distribution Systems

A *noncirculating* distribution system transmits hot water from the tank, via one or more risers, to the branch lines that supply hot water to the plumbing fixtures. This is a one-way system, and, though extremely common because it is the least expensive method of installing a hot-water distribution system, it has several drawbacks. Hot water will flow through the system only when a faucet is opened. Between times, the water cools in the lines. The next time a hot-water faucet is opened, the water must run for a while before hot water finally emerges from the faucet. When the faucet is turned off again, all the hot water now filling the line will cool off again. Apart from the slight inconvenience in having to wait for hot water, a noncirculating system is very wasteful of hot water and of the energy it takes to heat the water. (To which it might be added that insulating the lines with at least ¾-in.-thick insulation will reduce the heat loss by 75 to 80 percent.)

A *circulating* system, on the other hand, though more expensive to install because of the additional piping required to return hot water to the storage tank again, has the advantage that hot water is always immediately available at every faucet, since the hot water is circulating continuously between the plumbing fixtures and the storage tank. Circulating systems are, however, very uncommon in dwellings.

SIZING HOT-WATER HEATING EQUIPMENT

Storage Tank Capacity

When a storage tank is part of a hot-water installation, the demand for hot water is usually calculated on a gallon-per-hour basis. Most households use most of their hot water during two peak periods every day—in the morning and at night. Between times, hot-water usage is very light, except on clothes-washing days. To ensure there is always sufficient hot water available for the family's needs, the capacity of the storage tank must be calculated on the basis of these periods of heaviest demand.

The capacity of the storage tank can be calculated in several different ways, all of which give different results. The simplest method is to consult a table, such as Table 1, which is published by the Federal Housing Administration. It shows the minimum tank and heater sizes considered satisfactory for most dwellings.

Another widely used method for detemining the capacity of the hot-water storage tank is based on the cold-water consumption of the dwelling. Experience shows, for example, that in large private dwellings and in most apartment buildings the cold-water demand will average around 100 gal per person per day. Experience also shows that the hot-water demand averages about one-third the cold-water demand. Thus, for example, in a house having five occupants, the cold-water demand will be about 500 gal per day. The hot-water demand, therefore, will be one-third this figure, or 167 gal per day.

It has also been found that the maximum hourly demand for hot water will average about one-tenth the total daily demand, or, using the above figure, 167 ÷ 10 = 16.7 gal per hr.

Finally, since some of the water in the tank is bound to cool, only about three-quarters of a tank is considered to be filled with hot water. To take this cooled water into account, the maximum hourly demand is multiplied by a factor of 1.25. In our example, therefore, the maximum hourly demand—16.7 gal—is multiplied by 1.25 = 20.875 gal as the tank size. In

Table 2. Calculations for Determining Hot-Water Heater Size

No. of occupants	Max. hr H.W. demand	Av. hr H.W. demand	Btu input rating	Recov. gal per hr*	Nominal size in gal
2	6.7	2.8	16,000	20.0	15
3	10.05	4.2	20,000	20	20
4	13.40	5.6	25,000	26	30
5	16.75	7.0	30,000	30	40
6	20.10	8.4	40,000	32	50
7	23.45	9.8	50,000	52	60
8	26.80	11.2	75,000	79	75
9	30.15	12.6	75,000	79	75
10	33.50	14.0	75,000	79	100
11	36.85	15.4	100,000	105	100
12	40.20	16.8	100,000	105	100

*80 degree rise.

practice, a 30- to 40-gal storage tank is universally recommended for a family of five. Whether in fact this tank capacity is excessive or not would be difficult to say, since the amount of hot water required will also depend on the pattern of hot-water usage and the recovery rate of the heater.

In fact, the size of the storage tank depends on a family's pattern of water usage and on the recovery rate of the heater. As we mentioned above, if a family concentrates its demand for hot water in one or two periods of the day, then a large tank is necessary. If the hot-water demand is more uniform throughout the day, a smaller tank will be satisfactory. On the other hand, the greater the recovery rate of the heating unit, the faster will additional hot water be produced and the smaller the tank capacity need be.

Another difficulty in selecting the correct size of hot-water equipment depends on how much the water must be heated. For example, the temperature of the groundwater in the northernmost states may be as low as 40°F, but in the southernmost states the groundwater temperature may be as high as 75°F. The optimum temperature of domestic hot water is usually assumed to be 140°F. Water hotter than this increases the rate at which carbonate deposits build up in the hot-water tank, in those parts of the country where the water is hard. In addition, a steel tank corrodes at a faster rate as the water temperature increases.

Therefore, if the water is to be heated to 140°F, in the northern parts of the United States the incoming water must be heated 100°F (i.e., 40°F + 100°F = 140°F), while in the southern parts of the United States the cold water need by heated only 65°F (i.e., 75°F + 65°F = 140°F). This, of course, affects the selection of the hot-water heating equipment. In a southern state, the burner need not have the Btu output that is necessary in a northern state. Table 2 can be used as a guide in the selection of a suitable hot-water system. The table shows the input in both Btu and kilowatts for storage tanks of different capacities. Tables 3 and 4 show the amount of heat input required to heat cold water to achieve different recovery rates. Since the size of most hot-water equipment is calculated on a per-hour basis, the recovery rate per gallon can be used as a guide in selecting a tank size; that is, tank capacity per gallon will equal the recovery rate in gallons per hour.

Table 3. Gas Water Heater–Recovery Capacity (gal per hr) for Different Btu Inputs and Water Temperature Rises

Input Btu	130°F	120°F	110°F	100°F	80°F	60°F
5,000	3.2	3.5	3.8	4.2	5.3	7.0
21,000	13.5	14.6	16.0	17.6	22.1	29.4
23,000	14.8	16.0	17.5	19.3	24.2	32.2
25,000	16.1	17.5	19.0	21.0	26.3	35.0
27,000	17.4	18.9	20.6	22.7	28.4	37.8
28,000	18.0	19.5	21.3	23.5	29.4	39.2
30,000	19.3	21.0	22.9	25.2	31.5	42.0
31,500	20.3	22.0	24.0	26.5	33.1	44.1
33,000	21.3	23.0	25.1	27.7	34.7	46.2
33,500	21.6	23.5	25.6	28.2	35.2	55.8
35,700	23.0	25.0	27.2	30.0	37.5	50.0
36,000	23.2	25.1	27.4	30.2	37.7	50.1
40,000	25.8	28.0	30.5	33.6	42.0	56.0
49,980	32.3	35.0	38.1	42.0	52.5	70.0
50,000	32.3	35.0	38.1	42.0	52.5	70.0
55,000	35.5	38.5	42.0	46.2	57.7	77.0
60,000	38.7	42.0	45.8	50.4	63.0	84.0
65,000	42.0	45.5	49.6	54.6	68.2	91.0
100,000	64.6	70.0	76.3	84.0	105.0	140.0

Source: W. L. Jackson Manufacturing Co.

Table 4. Electric Water Heater–Recovery Capacity (gal per hr) for Different Btu Outputs and Water Temperature Rises

Wattages	130°F	120°F	110°F	100°F	80°F	60°F
600	1.9	2.0	2.2	2.5	3.1	4.1
750	2.3	2.5	2.8	3.1	3.8	5.1
1,000	3.1	3.4	3.7	4.1	5.1	6.8
1,250	3.9	4.2	4.6	5.1	6.4	8.5
1,500	4.6	5.0	5.5	6.1	7.6	10.1
2,000	6.3	6.8	7.4	8.2	10.2	13.6
2,500	7.8	8.5	9.2	10.2	12.8	17.1
3,000	9.4	10.2	11.1	12.3	15.4	20.5
3,500	11.0	11.9	13.0	14.3	17.9	23.9
4,000	12.6	13.6	14.9	16.4	20.5	27.3
4,500	14.1	15.3	16.7	18.4	23.1	30.7
5,000	15.7	17.0	18.6	20.5	25.6	34.2
5,500	17.3	18.8	20.5	22.5	28.2	37.6
6,000	18.9	20.5	22.3	24.6	30.7	40.8
7,200	22.6	24.5	26.8	29.5	36.9	49.1
9,000	28.3	30.7	33.5	36.9	46.1	61.5
10,000	31.5	34.1	37.2	41.0	51.2	68.3

Source: W. L. Jackson Manufacturing Co.

Tankless-Coil Capacity

The size of a tankless coil is calculated on the basis of maximum hot-water consumption per *minute*. A hot-water consumption table should be used as a guide in making up an estimate. For example, the typical shower will consume about 30 gal of hot water in about 5 min, or 6 gal per min. A clothes washer will complete a wash cycle in about 10 min, which means its per-minute consumption of hot water will be about 7.5 gal, and so on. By adding up the maximum probable demands in this way, based on the habits of the particular family, one can arrive at a reasonable estimate of the required hot-water output of a tankless coil.

Illumination

This article describes those factors that should be taken into account when the lighting of a house is being planned, if the lighting is to be considered adequate. Adequate lighting is *not* synonymous with bright lighting but with *sufficient* lighting that is so placed as to avoid glare or distracting shadows.

The importance of good lighting cannot be overestimated. About one in four individuals under 20 years of age has defective vision of some kind. One-half of those between 30 and 40 years of age have defective vision. Of those between 50 and 60, the proportion has increased to four out of five. Of those over 60 years of age, 19 out of 20 have defective vision.

It has not been shown to everyone's satisfaction that poor lighting *causes* visual defects, but it is a fact that individuals who must use their eyes intensively are more likely to have visual defects than those who do not. For example, over 80 percent of all draftsmen have eye problems of some kind and less than 20 percent of all laborers do. Whether or not these problems are caused by poor lighting, it is obvious that those who do have visual defects need good lighting if they are to see adequately, and in the long run this will include 95 percent of those reaching old age.

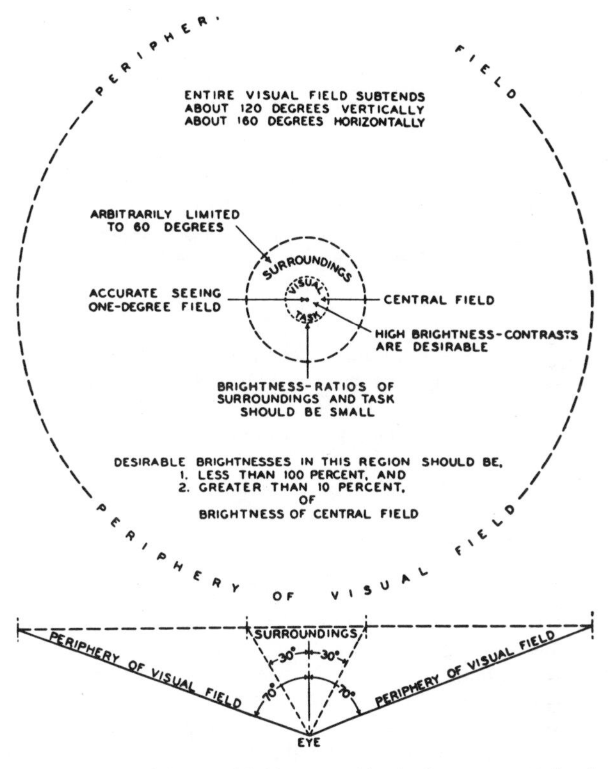

Fig. 1. Diagram of the visual field perceived by the human eye *(Callender, Time-Saver Standards, McGraw-Hill).*

ILLUMINATION AND THE EYE

How We See

When we concentrate our attention intently upon some object, we can see that object with great distinctness. This quality of *acuity,* or keeness of vision, is, however, limited to a relatively narrow cone, or *central field,* that extends outward from the eye at an included angle of about 1 degree (see Fig. 1). Within this central cone we can see whatever we are looking at in all its color and detail.

Beyond this central field, our acuity begins gradually to fall off. Although we are still aware of the color and shape of objects, we do not really look at objects outside of the central field, and they make a less vivid impression on our mind. This area of less-acute vision is called the *surround.* It extends about 30 degrees beyond the central field, forming a cone that has an included angle of about 60 degrees.

Beyond the surround is the area of *peripheral vision.* We don't see the color or shape of objects within this area of vision at all. All we really notice are movements. The area of peripheral vision extends outward from the surround in an oval-shaped cone that has an included angle of about 160 degrees horizontally and about 120 degrees vertically.

Light Intensities and the Eye

The eye can respond to an enormous range of light intensities. A bright summer sky, for example, shines with an intensity as great as 10,000 footcandles, and at the other extreme the light from a star may have an intensity of only 0.00003 footcandle. The eye accommodates itself to this range of intensities mainly by altering the diameter of the pupil and, when the light is extremely bright, by squinting, which helps to reduce the amount of light that enters the eye. In addition, in extremely dim light, the retina will adapt itself over a period of time to the darkness and become sensitive to what light there is.

The eye cannot take in too great a range of light intensities at the same time. If the range is too great, the pupil will contract until the eye can look at the brightest objects without discomfort, but this will leave the less-bright objects to shift for themselves. These objects will appear washed out and without detail. Light from objects that are too poorly lit may fail to register on the retina at all. This sort of thing happens, for example, on a very sunny day when objects that are directly in the sun appear brilliant and sharp while objects in the shade may disappear from view entirely.

Glare

Unwanted bright light that shines directly into the eyes or that is reflected from a bright source into the eyes interferes greatly with the acuity of vision. This interfering light is called *glare.*

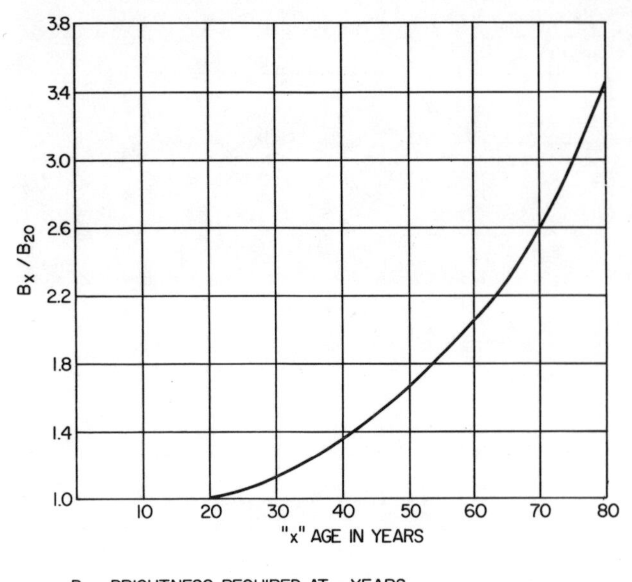

Fig. 2. As a person ages, objects must be more brightly illuminated if they are to continue to be clearly seen *(Callender, Time-Saver Standards, McGraw-Hill).*

Perhaps the most common example of extreme glare occurs when we are driving into the sun in the late afternoon, with the sun low in the sky and directly in our line of sight. The glare can make driving extremely hazardous for we can see nothing on the road ahead of us.

This sort of interference occurs indoors as well, if not in as extreme a form. The eyes will adjust themselves automatically to this glare. But when they do so, the object upon which our attention is fixed will appear less vivid to our sight than it should, and the eyes must strain to see this object accurately and in detail. If the strain continues for a period of time, the eye muscles tire. The entire body may even become fatigued, all because a light is shining in one's eyes. Because of these physical consequences of glare, it is not too much to say that, in addition to having adequate illumination in the first place, the chief object in planning the lighting in a house is to prevent glare as much as possible.

Having insufficient light will also strain the eye muscles, especially when one is engaged in a task that requires concentration, such as reading, sewing, drawing, model building, watch repairing, and so on. The eyes of a young person—20 years old and under—adapt themselves very readily to poor light conditions, but this adaptability gradually disappears as one grows older. The pupils lose their facility for opening as wide as they once did, or for altering their diameter as rapidly as they once could. The muscles that control the shape of the eyeball, and thus allow the eye to focus on near and far objects, lose their strength, and the eyeball itself becomes less flexible with increasing age. This sort of physical degeneration is inevitable in all of us, but it can be compensated for in part by increasing the brightness of the object we are looking at. Figure 2 shows how much brighter an object must be lighted as we grow older in order to compensate for the increasing deterioration of the eyes.

THE MEASUREMENT OF LIGHT

Candlepower

The intensity of the light emitted from a source is measured in *candlepower* (or *candela* in the metric system of light measurement). Once there were actual candles made of wax, the residue of refined whale oil, and 1 candlepower was defined as the amount of light emitted from such a candle when it was consuming wax at the rate of 120 grains per hour. But times have changed and nowadays there are several different kinds of laboratory light sources, all more stable and reproducible than a wax candle, the outputs of which are defined as being equal to 1 candlepower.

The word *candlepower* by itself, strictly defined, means the intensity of the light emitted *in a particular direction,* not in every direction from the light source.

Candlepower by itself is thus a rather useless measure since what is usually wanted for practical purposes is a measure of the *total* light emitted from a source, not the amount of light emitted in one direction. And the total quantity of light emitted from a source is measured in *lumens.*

Lumen

And what is a lumen? Imagine that there is a point source of light that is emitting 1 candlepower of light in every direction. Imagine that this light source is located in the center of a transparent sphere that is 2 ft in diameter (see Fig. 3). Therefore, every point on the inner surface of this sphere will be 1 ft from the light source, and all these points will be receiving an equal quantity of light. If we were now to mark off on the surface of this sphere an area of exactly 1 sq ft, then the total quantity of light falling on this 1-sq-ft area is defined as being equal to 1 lumen.

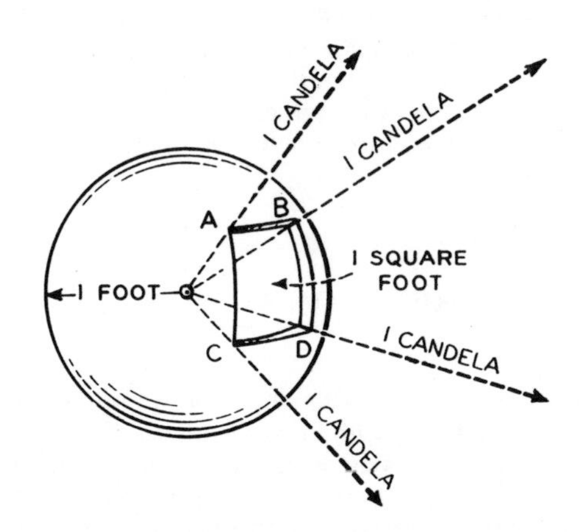

Fig. 3. The relationship between candelas, lumens, and footcandles. A uniform point source (luminous intensity or candlepower = 1 candela) is shown at the center of a sphere of 1 ft radius. It is assumed that the sphere is perfectly transparent. The illumination at any point on the sphere is 1 ft-c (1 lumen per sq ft). The solid angle subtended by the area A, B, C, D is 1 steradian. The flux density is, therefore, 1 lumen per steradian, which corresponds to a luminous intensity of 1 candela, as originally assumed. The sphere has a total area of 12.57 (4 π) sq ft, and there is a luminous flux of 1 lumen falling on each square foot. Thus, the source provides a total of 12.57 lumens (*Callender, Time-Saver Standards, McGraw-Hill*).

Footcandle

Lumens are used by the manufacturers of incandescent and fluorescent lamps as a measure of the total amount of light emitted from their products. Another very common measure of light is the *footcandle,* which is defined as the amount of light that is falling on a surface 1 ft square when that surface is 1 ft distant from a light source that is emitting 1 candlepower.

Note that footcandle and lumen measure light from different points of view. Lumen is a measure of light intensity from the point of view of the source. Footcandle is a measure of light intensity from the point of view of the object being illuminated. If, for example, a light source were to emit light with an intensity such that 100 lumens were to fall on a surface 1 sq ft in area, no matter how distant that surface is from the source, that surface would be illuminated by 100 footcandles of light.

The Law of Inverse Squares

Light, like all other forms of electromagnetic radiation, follows the law of inverse squares. That is, the intensity of the emitted light falls off sharply as the distance from the source increases. The relationship between footcandles, light intensity, and distance is defined by the following equation

$$\text{Footcandle} = \frac{\text{candlepower}}{\text{distance}^2}$$

Going back to our definition of footcandle for a moment, we can see that 1 footcandle is equal to 1 candlepower falling on a surface 1 ft from the source, or

$$1 \text{ Footcandle} = \frac{1 \text{ candlepower}}{1 \text{ ft}^2}$$

Assume now that we have a source of light that is emitting 100 candlepower and that this light is 1 ft distant from a surface. The amount of light falling on that surface is, therefore

$$\text{Footcandle} = \frac{100 \text{ candlepower}}{1 \text{ ft}^2} = 100$$

If we were to increase this distance to 10 ft, the intensity of the light falling on the surface would become

$$\text{Footcandle} = \frac{100 \text{ candlepower}}{10 \text{ ft}^2} = 1$$

That is, the amount of light striking the surface has decreased by 100 times even though the distance has increased by only 10 ft. Obviously, the closer a light is to a surface, the greater the illumination falling on that surface.

Brightness

Interesting as these facts no doubt are to the reader, the difficulty with footcandles and lumens is that we do not see an object because a certain amount of light happens to fall upon it but because a portion of that light is reflected from that object into our eyes. That is, a certain minimum amount of light must always be *reflected* from an object before we can see it. The object must have a certain amount of *brightness,* or what is more accurately called *luminance.*

To give an extreme example, if we were to enter a room illuminated by a 1000-lumen lamp, the walls and ceilings being painted a light color, we would see every object in the room, including themselves, with great distinctness. If, on the other hand, we were to enter the same room illuminated by the same 1000-lumen lamp, but with the walls and ceiling painted a dull, flat black, the furnishings in the room also being black, we would not see anything at all, or very little, because none, or almost none, of the light given off by the lamp would be reflected from the walls and furnishings into our eyes. All that we would really see would be the glare of the lamp itself.

Footlamberts

Brightness (or luminance) is measured in *footlamberts,* which is defined as the illumination falling on an object times the *reflectance* of the surface expressed as a percent. That is,

$$\text{Footlambert} = (\text{footcandle} \times \text{reflectance})/100$$

All objects have a certain amount of reflectance; that is, they will reflect a certain percentage of the light striking them and absorb the remainder. No object ever reflects all the light striking it, which is to say that no object ever has 100 percent reflectance. The percentage of light reflected by any particular surface will depend on the color and surface texture of that surface. Dark colors, for example, reflect much less light than light colors (see Table 1), and rough surfaces reflect much less light than smooth surfaces.

Table 1. Brightness with which Different Colors Reflect Light Striking Them, Expressed as a Percent

Color	Approximate percent reflection
Whites	
Dull or flat white	75–90
Light tints	
Cream or eggshell	79
Ivory	75
Pale pink and pale yellow	75–80
Light green, light blue, light orchid	70–75
Soft pink and light peach	69
Light beige or pale gray	70
Medium tones	
Apricot	56–62
Pink	64
Tan, yellow-gold	55
Light grays	35–50
Medium turquoise	44
Medium light blue	42
Yellow-green	45
Old gold and pumpkin	34
Rose	29
Deep tones	
Cocoa brown and mauve	24
Medium green and medium blue	21
Medium gray	20
Unsuitably dark colors	
Dark brown and dark gray	10–15
Olive green	12
Dark blue, blue-green	5–10
Forest green	7
Natural wood tones	
Birch and beech	35–50
Light maple	25–35
Light oak	25–35
Dark oak and cherry	10–15
Black walnut and mahogany	5–15

Source: U.S. Dept. of Agriculture.

To return to our formula, assume that 70 footcandles of light are falling on a wall painted a light beige. Table 1 shows that this color reflects 70 percent (i.e., 0.70) of the light falling on it. Therefore,

$$\text{Footlambert} = 70 \text{ footcandles} \times 0.70 = 49$$

The actual brightness of the wall is 49 footlamberts, not the 70 footcandles that we might assume if we were to take into account only the intensity of the light and its distance from the wall. Furthermore, if the wall were dirty or dusty, as is usually the case, even less light would be reflected from it. The wall, therefore, will appear even less bright than 49 footlamberts.

It should also be noted that if we were to measure the amount of light falling upon the wall surface using a lightmeter calibrated in footcandles, the meter would indicate 70 footcandles when we pointed it toward the light source. But if we were then to point the meter directly at the wall, it would indicate only 49 footcandles—which is the amount of light striking the wall less the reflectance. As a practical matter, a lightmeter can be used to measure both footcandles and footlamberts; it depends on where you point it.

INCANDESCENT LAMPS

When 115-volt electricity passes through the tungsten filament of the ordinary household incandescent lamp, the resistance offered by the filament to the current of electricity causes it to heat up until it is glowing white-hot. Visible light is then emitted by the filament. The filament can glow as hot as it does because the bulb has been evacuated of air (which contains oxygen, which will cause the filament to burn up). The bulb is instead filled with an inert gas such as argon, or it may be completely evacuated. The filament will give off a considerable amount of heat, an indication of an incandescent lamp's basic inefficiency.

The electricity enters the lamp through a contact at the bottom of the screw-type base. The circuit is completed through the side of the base, which is insulated from the contact.

Lamp Efficiency

The amount of electrical energy consumed by the lamp is measured in *watts*. The efficiency of the lamp is measured by comparing the amount of light emitted (in lumens) to the amount of electricity consumed (in watts). That is,

$$\frac{\text{Light output of lamp (lumens)}}{\text{Electricity consumed (watts)}} = \text{lumens per watt}$$

If, for example, we have a 75-watt lamp with an output of 1150 lumens (the lumen rating of a lamp is marked on the box it is packed in), then

$$\frac{1170 \text{ lumens}}{75 \text{ watts}} = 15.3 \text{ lumens per watt}$$

Table 2. Output in Lumens per Watt and Average Life of Standard Incandescent Lamps

Size of lamp, watts	Average life, hours	Total lumens	Lumens per watt
25	2500	235	9.5
40	1500	455	11.5
60	1000	870	14.5
75	750	1,190	15.9
100	750	1,750	17.5*
150	750	2,780	18.5
200	750	4,010	20.0
300	750	6,360	21.2
500	1000	10,850	21.7

*The first lamps made by Edison in 1888 produced less than 2 lumens per watt, in the 100-watt size.

Source: Richter, *Practical Electrical Wiring*, McGraw-Hill.

Table 2 shows the lumen-per-watt output of standard incandescent lamps of different sizes. Note that the efficiency of these lamps increases with wattage. A 25-watt lamp, for example, has an output of 9.5 lumens per watt, but a 150-watt lamp has an output of 18.5 lumens per watt. One could save on one's electric bill by installing one 150-watt lamp instead of six 25-watt lamps, assuming one has a choice. The 150-watt lamp has nearly twice the light output of the six 25-watt lamps, but it uses only one-half as much electricity.

Table 2 also shows the variability in the lifetimes of lamps having different wattages. These figures are average lifetimes. Any one lamp may have a lifetime that varies ±20 percent from these figures, as it is impractical to manufacture lamps to any closer tolerances.

Extended-Life Lamps

Several lamp manufacturers have *extended-life* lamps that are advertised to last two to three times as long as ordinary lamps. These lamps are made by increasing the diameter and length of the filaments. As long as the wattage remains the same, the temperature of the filaments is thereby reduced and they will last longer. One pays a higher price for the lamp, but this additional cost is more than made up for by not having to purchase one or two standard lamps as replacements.

There may be a hidden cost, however. Since the filaments operate at a reduced temperature, their lumen-per-watt output is 5 to 20 percent less than the output of an ordinary lamp having the same rated wattage. That is, the extended-life lamps are somewhat dimmer. Whether this is important or not depends on the user. But if, in order to secure adequate illumination, it becomes necessary to purchase an extended-life lamp having a wattage greater than that of an ordinary bulb, then obviously the operating cost of the lamp will increase. The cost of the additional electricity consumed by the extended-life lamp over its lifetime may well exceed the money saved originally by not having purchased ordinary lamps.

FLUORESCENT LAMPS

Lamp Operation

Fluorescent lamps are much more complicated electrically than incandescent lamps. The basic construction and circuit of a *preheat* type of fluorescent lamp (which is the type of lamp most commonly used in dwellings) are shown in Fig. 4. A tungsten filament is located at each end of the lamp. Each filament is connected to the circuit through two pins that project from the lamp. The filaments are coated with a material that emits a stream of electrons whenever the filaments are heated. The inner surface of the glass tube is coated with a white phosphorescent powder that glows with light wherever it is struck by ultraviolet rays. The source of the rays is a small amount of mercury that is injected into the lamp during its manufacture, together with a much larger quantity of argon gas.

When the circuit is first energized, electricity flows to both filaments via a closed switch located in the circuit between the filaments. This switch, which is called a *starter,* operates automatically and is normally open when the lamp is inoperative. As the filaments heat up, the substance with which they are coated emits electrons, which travel at extremely high speeds through the lamp. The motion of these electrons excites the argon gas, which in turn excites the mercury.

Thereafter, whenever one of the electrons collides with an atom of the mercury, it causes the atom to emit a ray of ultraviolet light. Wherever this ray strikes the phosphor coating on the surface of the lamp, the phosphor glows with light for a moment.

By this time the ends of the fluorescent lamp will be glowing with light while the main portion of the glass tube is still dark. The starter now opens automatically, and it will remain open for as long as the circuit remains energized. The fact that the starter has opened forces the electrical energy to flow through the tube, which it can do since the mercury vapor within the tube is capable of conducting an electrical current. Once such an arc has been established between the two filaments, the entire tube begins to glow with light, and it will continue to do so as long as the circuit remains closed.

To complicate this description a bit, a pulse of high-voltage electricity is required to initiate the arc in the first place, but this pulse of high voltage must be very brief, otherwise the lamp will be irreparably damaged. Furthermore, even if this high-voltage pulse is brought back quickly to its original 115-volt value, the current flowing through the lamp thereafter must be closely controlled; otherwise the filaments will burn out.

The solution to both these difficulties is to insert a *choke*

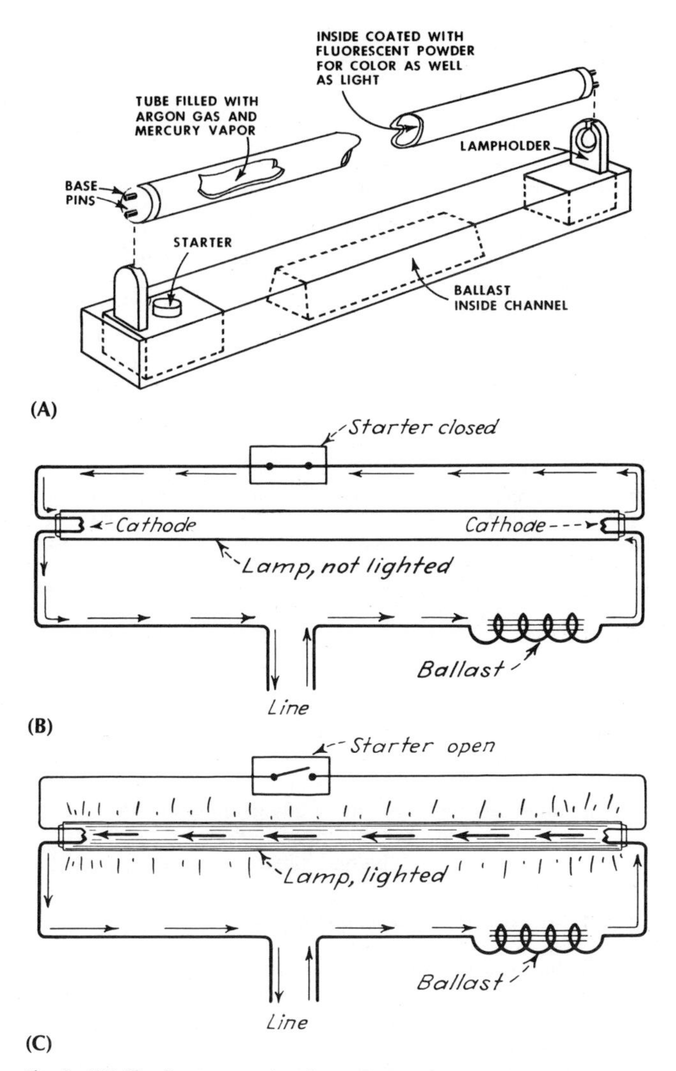

Fig. 4. (*A*) The fluorescent phosphor coating on the inside of the tube is activated by electrical energy passing through the tube. Light is thereupon given off. The starter in standard starter-type fixtures permits preheating of the electrodes in the ends of the tube to make it easier to start. The ballast limits the current to keep the tube functioning properly. The channel holds ballast and wiring and spaces the lampholders. (*B*) shows the flow of current in a fluorescent lamp circuit at the instant the lamp is turned on. (*C*) shows the flow of current after the lamp has started. The starter has opened, and current cannot flow through it now (*U.S. Dept. of Agriculture; Richter, Practical Electrical Wiring, McGraw-Hill*).

coil, or *reactor coil,* in the circuit, as shown in Fig. 4. A choke coil is widely used in AC electrical and electronic circuits. In a fluorescent-lamp circuit it is called a *ballast,* presumably because it acts as a kind of weight, a resistance, that holds down the amount of current that can flow through the circuit. The ballast has two properties that make fluorescent lighting practical: (1) it resists any change in the direction of the current flowing through its windings, and (2) whenever the current flowing through it is cut off, a momentary surge of high-voltage electricity is induced in its windings, which is, of course, transmitted throughout the circuit.

Now then, when the starter opens, such a surge of high-voltage electricity is induced in the ballast windings. The ballast is so designed that the amount of voltage induced is sufficient to force an arc between the filaments but not so high as to damage the lamp. And once the arc has been established, the ballast thereafter acts as a resistance in the circuit, limiting the amount of current that passes back and forth within the lamp.

Other Types of Fluorescent Lamp

Since the introduction of the preheat type of fluorescent lamp in 1938, two other methods of initiating the arc have been commercially developed. Both eliminate the need for a separate starter. Both are also used much more in commercial and industrial installations than in domestic installations.

Instant-start lamps do, as a matter of fact, start instantly. The circuit and ballast are so designed that the moment the circuit is closed there is a momentary surge of 450- to 600-volt electricity through the circuit, which produces the arc within the lamp. Thereafter, the ballast will maintain the current at the proper level for continuous operation.

Rapid-start lamps will light up in a fraction of a second, not far behind instant-start lamps. In a typical rapid-start lamp the filaments are heated by a special 3½-volt winding in the ballast, in addition to the heat provided by the arc. This reduces the voltage required for starting to 250 to 400 volts. Otherwise, both instant-start and rapid-start lamps operate continuously in the same way that preheat lamps do.

Fluorescent versus Incandescent Lighting

The advantages of fluorescent lighting over incandescent lighting are considerable. First, fluorescent lamps are much more efficient than incandescent lamps (see Table 3). The preheat type of lamp, for example, produces from 58 to 75 lumens per watt compared with the 14 to 23 lumens-per-watt output of the ordinary household incandescent lamp. The preheat type of lamp has a lifetime from five to ten times greater than that of an incandescent lamp. A lifetime of 18,000 hours is not unusual; it depends on the way the lamp is operated. The greater efficiency of fluorescent lamps also shows up in lower operating temperatures compared to incandescent lamps. Finally, their color apart, fluorescent lamps produce a visually more comfortable light than incandescent lamps because the light is emitted along the length of the tube, which results in a more diffused, shadowless illumination. Fluorescent lamps also have much less surface brightness—they produce much less glare—since the light output is distributed over a very large area compared to the small surface area of an incandescent lamp.

The disadvantages of fluorescent lighting are: (1) the initial installation is much more expensive than for incandescent lighting; (2) a fluorescent lamp cannot operate below 50°F, unless it is used with a special circuit designed to operate at low temperatures; (3) the ballast may produce a low hum that can be distracting in a very quiet room; and (4) fluorescent lamps will go out if there is a pronounced drop in voltage while incandescent lamps will continue to operate, merely dimming until the voltage picks up again.

The Color of Fluorescent Lamps

But the main disadvantage of fluorescent lighting for most people has to do with the color of the light emitted. The phosphors used in most fluorescent lamps radiate energy chiefly in the blue and green portions of the color spectrum. That is, the illumination tends to emphasize the cooler-looking blues, greens, and light-yellows in a room, and the warmer browns, reds, and oranges appear dull and faded. Fluorescent lighting is especially unflattering to the complexion; it tends to make people look pale, if not sallow. Incandescent lamps, on the other hand, radiate more in the red end of the color spectrum. Incandescent lighting tends to emphasize the browns, reds, and oranges at the expense of the blues and greens. The complexion has a healthier, more ruddy look under incandescent lighting.

There are seven different *whites* available in fluorescent lamps now, each of which is different from the others. This

Table 3. Characteristics of Typical Fluorescent Lamps

Variety	Watts	Bulb type	Diameter, in.	Length, in.	Type of starting	Lumens* Total	Lumens* Per watt	Type of base
Ordinary	15	T-8	1	18	Preheat	870	59	Medium bi-pin
	30	T-8	1	36	Preheat	2,200	73	Medium bi-pin
	15	T-12	1½	18	Preheat	800	53	Medium bi-pin
	20	T-12	1½	24	Preheat	1,300	65	Medium bi-pin
	40	T-12	1½	48	Preheat/ Rapid	3,150	79	Medium bi-pin
Slimline	40	T-12	1½	48	Instant	3,000	75	Single-pin
	55	T-12	1½	72	Instant	4,500	82	Single-pin
	75	T-12	1½	96	Instant	6,300	84	Single-pin
High output	60	T-12	1½	48	Rapid	4,300	72	Recessed double contact
	85	T-12	1½	72	Rapid	6,650	78	
	110	T-12	1½	96	Rapid	9,200	84	
Extra-high output	110	PG-17	2⅛	48	Rapid	7,000	63	Recessed double contact
	165	PG-17	2⅛	72	Rapid	11,500	70	
	215	PG-17	2⅛	96	Rapid	16,000	74	

*Total lumens and lumens per watt are for cool-white lamps after 100 hr of use. The average during the useful life of the lamps will be about 15% less. The figures are based on watts consumed by the lamps, not including power consumed by the ballasts.

Source: Richter, *Practical Electrical Wiring,* McGraw-Hill.

may not be obvious when the lamps are viewed at separate times, but if the same multicolored object is observed under these different whites in a showroom, one after the other, the color changes can be striking.

The practical consequence is that one can be very deceived about the colors one is selecting for one's home. Paint, fabric, and wallpaper colors for a home that is to be illuminated by incandescent lamps are usually chosen in a shop illuminated by fluorescent lamps. But the greens and blues that looked so brilliant in the shop turn dull and quiet in the home, and apparently dull-looking reds and browns brighten up considerably. A brown selected in a shop for its look of quiet richness may prove to be rather garish-looking in the home. The obvious rule to follow is always to select or match colors under the same kind of lighting that exists in the home.

The Efficiency and Output of Fluorescent Lamps

In principle, the voltage level is irrelevant for the efficient operation of a fluorescent lamp. The lamp will work as well in a 230-volt circuit as in a 115-volt circuit. However, the ballast used with the lamp can be affected by the voltage level, and, since ballasts and lamps must be very closely matched, lamps should in fact be operated only at their designated voltages.

The wattage rating marked on fluorescent lamps (see Table 3) is a bit deceptive in the sense that the electricity consumed by the ballast is not included in the wattage rating of the lamp, even though the lamp will not operate at all without its ballast. In the smaller size lamps, the ballasts require an additional 30 percent over the wattage consumed by the lamps, whatever that wattage is. In larger size lamps, the ballasts require an additional 10 percent of the wattage consumed by the lamps. On the average, then, the power required by the ballast will increase the total power consumption of the circuit by 10 to 15 percent.

Although the lumens-per-watt ratings of incandescent lamps are arrived at in a straightforward way, the lumens-per-watt ratings of fluorescent lamps are a more arbitrary figure. For one thing, the light output depends on the color of the lamp.

Furthermore, the light output of a fluorescent lamp is not the same over its entire lifetime. A fluorescent lamp does not produce a steady output of light until it has been in use for about 100 hr. Thereafter the light output of the lamp begins to drop slowly and steadily over its lifetime until by the end of its life its lumens-per-watt output may be only 70 percent of its output after its first 100 hr of operation. As a practical matter, therefore, over the entire lifetime of a lamp, its average lumens-per-watt output will be only about 85 percent of its published rating.

The lumens-per-watt output of any given lamp also depends on its size and type. As with incandescent lamps, larger fluorescent lamps have a higher lumens-per-watt output than smaller lamps.

The Lifetimes of Fluorescent Lamps

The lifetimes of fluorescent lamps are also highly variable. A lamp's lifetime will depend directly on the number of times the lamp is turned on and off. The more times this happens within a given period, the shorter the lamp's lifetime. For example, if a lamp is turned on and off every five minutes, its lifetime may be no more than 500 hr. If the lamp is turned on and left on, its lifetime may be as much as 30,000 hr. The reason for this enormous variability is the chemical coating on the filaments. A small but definite amount of this material is consumed every time the lamp is started; this is one reason that the initial voltage surge must be controlled. Once this material has been entirely consumed, the lamp will no longer operate and must be replaced.

Given these circumstances, it is difficult for manufacturers to establish typical lifetime for their products, since they have no idea how any given lamp will be used. If, however, they arbitrarily assume that the typical lamp operates for 3 hr every time it is turned on, then it is possible to assume what the probable lifetime of the lamp will be. Under these conditions it is possible to say that the average lifetime of the typical 20-watt lamp is about 6000 hr, and the lifetime of the typical 40-watt lamp is about 12,000 hr.

Types and Sizes of Fluorescent Lamps

Table 3 shows the range of fluorescent lamps available, their wattage ratings, sizes, and lumen outputs. In general, the wattage rating and size of a fluorescent lamp go together. Since each type and size of fluorescent lamp requires a ballast especially designed for it (as well as a specific starter for preheat types of lamps), the fixture-plus-lamp must be considered the basic functional unit. One cannot interchange fluorescent lamps in fixtures in order to increase or decrease the light output of the lamp, as one can do with incandescent lamps, because the lamps are designed *not* to be interchangeable; nor can one interchange preheat lamps with either instant-start or rapid-start lamps (except for a 40-watt 48-in.-long lamp called a *preheat/rapid-start*), nor, indeed, can one interchange instant-start with rapid-start lamps. When fixtures for different lamps are the same length, the fixtures are designed with bases that make it impossible to install the wrong lamp in them.

DESIGNING LIGHTING SYSTEMS

Table 4 shows the *general* illumination levels recommended for the different rooms in a house. In working areas such as kitchens, laundries, workshops, and, sometimes, garages, this kind of general room illumination is necessary for safety and efficiency as well as to prevent eye strain.

Table 4. Footcandles Recommended for General Domestic Use (Also shows lighting levels in lux)

Activity or area	Footcandles (dekalux) minimum at any time
Conversation, relaxation, and entertainment	10* (11)
Passage areas, for safety	10* (11)
Areas involving visual tasks, other than kitchen	30 (32)
Kitchen	50 (54)

*General lighting in these areas need not be uniform.
Source: Illumination Engineering Society.

But a general room illumination is less important in the living quarters of a house where illumination is usually necessary only for conversation, watching television, eating, and getting ready for bed. In the living quarters, the decorative aspects of home illumination take precedence and, often enough, only enough light is present to keep from tripping over the cat. More intense illumination must, however, also be available for reading, sewing, writing letters, doing homework, and performing similar sight-intensive tasks. Providing adequate illumination for these tasks is perhaps the most important part of home-lighting design, and it should be done before the general room illumination. One can always fill in the dark areas of a room with whatever illumination is thought necessary.

When planning the illumination for light-intensive activities, there are three visual zones, or areas, that must be taken into account (see Fig. 5). These are (1) the area of the task itself—the book or magazine one is reading, the piece of cloth one is sewing, and so on, (2) the area immediately surrounding the task, and (3) the overall room area.

Table 5 shows the amount of light required, in footcandles,

Table 5. Illumination Levels Recommended by the Illumination Engineering Society for Different Seeing Tasks

Seeing task	Footcandles* (dekalux) Primary task plane	Secondary task plane
Dining	15 (16)	
Grooming, shaving, make-up	50 (54)	
Handcraft		
Ordinary seeing tasks	70 (75)	
Difficult seeing tasks	100 (110)	
Very difficult seeing tasks	150 (160)	
Critical seeing tasks	200 (220)	
Ironing (hand and machine)	50 (54)	
Kitchen duties		
Food preparation and cleaning (at sink, range, and counter) involving difficult seeing tasks	150 (16)	
Serving and other noncritical tasks	50 (54)	
Laundry tasks		
Preparation, sorting, inspection	50 (54)	30 (32)
Tub area—soaking, tinting hand wash	50 (54)	30 (32)
Washer and dryer areas	30 (32)	
Reading and writing		
Handwriting, reproductions, and poor copies	70 (75)	
Books, magazines, and newspapers	30 (32)	
Reading piano or organ scores		
Advanced (substandard size)	150 (160)	50 (54)
Advanced	70 (75)	30 (32)
Simple	30 (32)	10 (11)
Sewing		
Hand (dark fabrics)	200 (220)	
Hand (medium fabrics)	100 (110)	
Hand (light fabrics	50 (54)	
Hand (occasional, high contrast)	30 (32)	
Sewing		
Machine (dark fabrics)	200 (220)	70 (75)
Machine (medium fabrics)	100 (110)	50 (54)
Machine (light fabrics)	50 (54)	20 (22)
Machine (occasional, high contrast)	30 (32)	10 (11)
Study	70 (75)	30 (32)
Table Games	30 (32)	
Table Tennis	20 (22)	

*Minimum on the task plane at any time. These levels are based on young eyes with 20–20 vision. Older eyes, even when properly corrected by glasses, have reduced visual acuity, a longer period of adaptation and decreased resistance to glare. To state it simply, older persons need more light and special precautions against glare.
Source: Illumination Engineering Society.

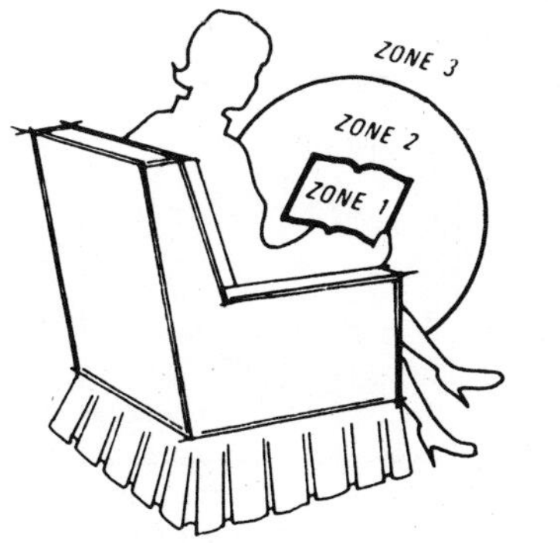

ZONE 2 THE IMMEDIATE SURROUNDINGS (AREA ADJACENT TO THE VISUAL TASK)

DESIRABLE RATIO . . . . 1/3 TO EQUAL TO TASK*

MINIMUM ACCEPTABLE RATIO . . . . 1/5 TO EQUAL TO TASK*

ZONE 3 THE GENERAL SURROUNDINGS (NOT IMMEDIATELY ADJACENT TO TASK)

DESIRABLE RATIO . . . . 1/5 TO 5 TIMES TASK*

MINIMUM ACCEPTABLE RATIO . . . . 1/10 TO 10 TIMES TASK*

*TYPICAL TASK LUMINANCE RANGE IS 12 TO 35 FOOTLAMBERTS (SELDOM EXCEEDS 60 FOOTLAMBERTS)

Fig. 5. Seeing zones and brightness (i.e., luminance) ratios for visual tasks *(Illumination Engineering Society).*

to perform different kinds of tasks. The table shows the *minimum* footcandle requirements, that is, the amount of light that should actually fall on the task *after* the factors that we have discussed in the previous section of this article have been taken into account. This illumination is usually provided by floor and table lamps or, with modern decor, perhaps by pole-mounted spotlights or recessed ceiling fixtures.

The area immediately surrounding the task (the top of a desk, say) must also be adequately illuminated; otherwise the person performing the task will be unpleasantly aware of any excessive difference in brightness every time he glances up or is distracted from his work. This difference in brightness will eventually result in eye fatigue. For optimum visual comfort, therefore, the brightness of the surfaces immediately surrounding the task should be at least as bright as the task area itself or, at a minimum, one-third as bright.

For the same reason, the brightness of the general area within the person's view should never be more than five times brighter than the task area nor less than one-fifth as bright. The greater the concentration required, and the longer the period of concentration, the more closely should both the area immediately surrounding the task and the general room area approach the task area in brightness.

Illuminating the Task Area

As we mentioned, the task area is usually illuminated by a table or floor lamp or by some kind of specialized fixture. The manufacturers of these lamps and fixtures usually publish a graph or table of some kind that shows how light is distributed from their product. Figure 6, for example, is a *candlepower distribution curve,* which is typical of the kind of information that is available from manufacturers. The graph shows the amount of light, in candlepower, that is emitted in various directions from a particular lamp-fixture combination. In order to use the graph, it is necessary to know the distance of the task from the light source and to convert candlepower into footcandles. This can be done using the formula already described

$$\text{Footcandles} = \frac{\text{candlepower}}{\text{distance}^2}$$

Actually, the best and simplest method of determining the level of illumination received from any fixture is to use a lightmeter calibrated in footcandles. (A photographic lightmeter calibrated in *f*-stops is satisfactory if the meter measures incident light and if the manufacturer can provide a conversion table.) One need then only place the meter on the task area, pointing the photoelectric cell toward the light source to obtain the footcandles directly. If there isn't sufficient light, one can then change the wattage of the lamp or change the position of the fixture, if this is possible, until the desired level of illumination has been obtained. One must, of course, still take into account the probable depreciation in the light output of the lamp-and-fixture combination.

Lamp Placement and Glare

Just as important as the brightness level of the task area is the direction from which the light is shining. When one is working at a desk or table, especially, it is very common to place the lamp directly in front of or directly above the task in order to maximize the illumination. The amount of light falling on the task will certainly be at a maximum, but so is the probability that some of this light will bounce up from the task area directly into one's eyes. That is, this placement of the lamp will result in reflected glare. This is particularly true if one is reading a book or magazine printed on smooth, white paper, or if the top of the desk is polished.

A light source should always be positioned in relation to the task so that the light falling on the task will not be reflected into one's eyes. An incandescent desk lamp, for example, should be placed to one side of the table, and the shade should be low enough that the incandescent lamp itself cannot possibly shine directly into the eyes. A fluorescent lamp fixture mounted on a flexible support should be positioned rather close to the reader's head so that the light emitted from the fixture will bounce off the task and away from the eyes. At a worktable illuminated by an overhead ceiling fixture, the best position for the fixture is above the front edge of the worktable. This position provides maximum illumination, and it also reduces the possibility that light will be reflected back into the eyes.

General Room Illumination

Having determined what concentrated sources of light are necessary in a room, one can then plan the general lighting. This planning consists of tying the concentrated sources together with fixtures that provide general illumination. The basic idea is to avoid or prevent having too great a contrast in brightness between the different parts of the room. As already mentioned, the general lighting level should not be more than five times as bright or less than one-fifth as bright as the light provided for the performance of particular tasks. The same principle should be followed when lighting adjacent rooms, as between a living room and a hallway, say.

Stairways in particular should be well lighted, which means that the stairs should be at least one-fifth as bright as the adjacent hallways or rooms. The walls along a stairway should be painted a light color to increase their apparent brightness. Under no circumstances should a light be placed at the bottom of a flight of stairs in such a way that the bulb is directly visible to anyone descending the stairs.

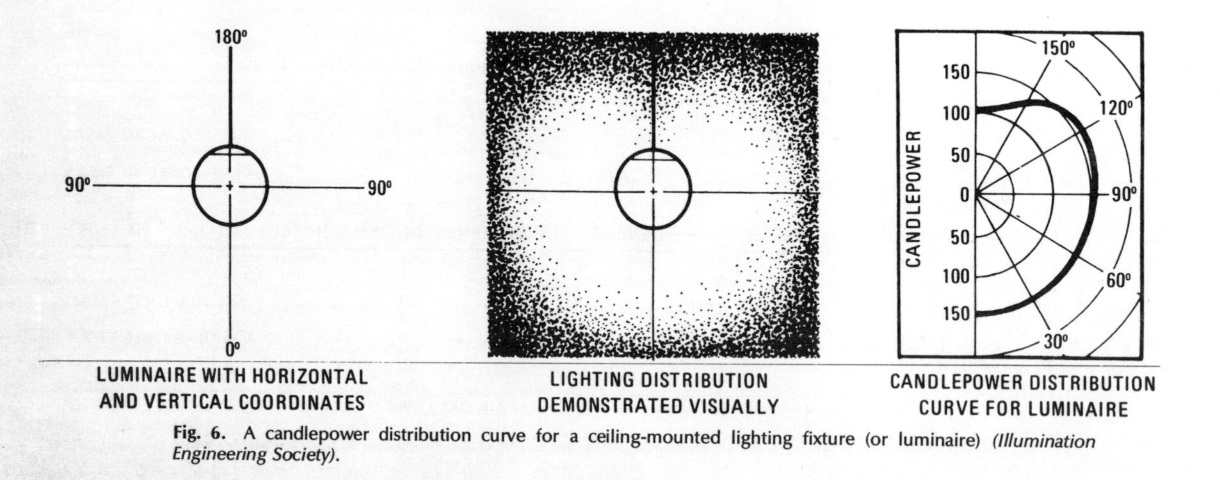

Fig. 6. A candlepower distribution curve for a ceiling-mounted lighting fixture (or luminaire) *(Illumination Engineering Society).*

Insulation

All materials conduct heat, some at a relatively fast rate, others at a relatively slow rate. A material that conducts heat at a *very* slow rate is considered a *thermal insulator.* If a sufficiently thick layer of this material is installed in the outer shell of a house, it will reduce considerably the passage of heat into and out of the house.

The primary justification for insulating a house is economic. During the winter, thermal insulation prevents the loss of heat from within the house, which reduces the amount of fuel that must be burned to maintain the desired indoor temperature. This, of course, saves on fuel costs.

During the summer, thermal insulation prevents the transfer of solar heat into the house, which makes the house more comfortable to live in during hot weather. If the house, or part of the house, should also be air-conditioned, the insulation, by preventing the entrance of heat, increases the effectiveness of the air-conditioning system. This, in turn, reduces the amount of electricity required to operate the air-conditioning system.

One should realize, however, that the economic justification for installing insulation decreases as the thickness of the insulation increases. There is a limit to how much insulation can be installed. Consider, for example, a house in which the walls and roof are insulated with 1-in.-thick material. It is possible to calculate very closely the effect this 1 in. of insulation will have on the total amount of heat that is transmitted through any particular construction. This heat transmission is calculated as U, the coefficient of heat transmission. Furthermore, once the U value is known, one can calculate the amount of fuel that is saved because of this 1 in. of insulation. It is then a simple enough matter to put a dollar value on the fuel savings.

However, as the amount of insulation is increased, the amount of money saved decreases proportionately. Say, for example, that 1 in. of insulation results in an annual savings of $100 in fuel costs. If we add another inch of insulation to the house, the amount of heat transmitted from the house will be halved. Therefore, we will save an additional $50 in fuel costs, for a total savings of $150. Suppose we add still another inch of insulation. The amount of heat lost from the house will now be one-third the original amount. Therefore, we will save an additional $33 in fuel costs, for a total savings of $183. Suppose we add a fourth inch of insulation. The amount of heat lost from the building will be one-quarter the original amount, for an additional savings in fuel costs of $25. Our total savings are now $208. But note how, as we continued to increase the thickness of the insulation by 1-in. increments, our additional savings amounted to only $50, $33, and $25. At the same time, the cost of the insulation will have been increasing proportionately. A point will be reached where it will cost more to install that one last additional inch of insulation than will be saved in fuel.

Where that point will be is difficult to say. It depends on the price of fuel and how cold and how long the winters are. Four inches of insulation may be economically justifiable in Minnesota and North Dakota, but in Georgia 1 in. of insulation may be excessive. In Georgia, however, it may be economically justifiable to install 2 in. of insulation because of the savings that will result in the cost of operating an air-conditioning system during the summer, whereas in Minnesota or North Dakota these air-conditioning savings may be ignored, since the house will be more than adequately insulated already.

Furthermore, the amount of insulation required depends on the shape of the house. A boxy, 2-story house with an attic is easier to heat and keep warm, with or without insulation, than a long, low, rambling 1-story house, especially when the ceiling is left exposed. The reason is that the boxy house, being more compact, has a smaller proportion of its volume exposed to the weather, which means that less of its heat will be transmitted to the outdoors, assuming the two houses have the same total interior volume. Eskimo are short and stocky for the same reason; their body shape helps them conserve body heat. In the tropics, however, a thin, lanky build is preferable, because such a body shape has a much larger surface area, which helps to dissipate body heat more efficiently.

In addition, in a 2-story house, the effect of having an uninsulated roof is felt less in the first-floor living quarters because the ceiling that separates the second floor from the attic has some insulating value. In a 1-story house, on the other hand, the insulating effect of an upper floor is absent, and, if the roof insulation should be inadequate, the entire house will feel colder and require more fuel to heat than a 2-story house.

An insulated house has other advantages over an uninsulated house, if we assume the design and construction of the two are the same. In an insulated house, the temperature throughout tends to be more equitable, especially the temperatures of the walls, which makes the entire house more comfortable to live in. The importance of having warm walls (and ceiling and floor) lies in the fact that most of one's body heat is lost by radiation from the skin. The greater the difference in temperature between the skin and the surrounding walls, the greater the amount of heat that is radiated from the body and the colder one feels. When the walls are warm, therefore, there is less heat lost from the body by radiation and the more comfortable one feels—even though the air temperature in the room may not have been increased. For a more complete discussion *see* HEATING SYSTEMS AND HUMAN COMFORT. In an uninsulated house, however, not only are the walls colder but there also tend to be differences in the temperatures of the different rooms, and even in the temperature of different parts of the same room, which makes for discomfort. During the winter, because there is a lower rate of heat loss through the exterior walls, an insulated house also heats up much more quickly whenever the heating plant is started up, as on a cold winter morning after the thermostat changes from its nighttime setting of, say, 55°F to its daytime setting of 68°F or 70°F; this also helps to conserve fuel. During the summer, an air-conditioning unit installed in an insulated house cools the house down more quickly than in an uninsulated house, and this reduces the costs of operating the air-conditioning unit.

In a very cold climate, it may even pay to install insulation between some of the room partitions if certain of the rooms are unused most of the time or if fuel costs are exceptionally high. Guest rooms, for example, can thus be isolated from the rest of the house. It is not difficult to calculate the probable fuel saving if this were done. For the same reason, in a very hot climate it may pay to isolate certain rooms from the air-conditioned parts of the house.

HOW INSULATION INSULATES

The three ways heat is transferred through the walls, roof, and floor of a house are by *conduction, convection,* and *radiation.*

Little can be done about reducing the amount of heat that is lost by conduction through the solid structure of a house. The house necessarily consists of solid materials fastened securely to each other, and it is only natural that these materials will conduct heat. Reducing the amount of heat lost from a house, therefore, means reducing the amount of heat radiated from the house or transferred from the house by convection currents.

A very large proportion of the heat lost from a building is lost through the doors and windows. It was found in one study, for example, that in the typical uninsulated, 2-story wood-frame house approximately 55 percent of the heat lost through the shell of the house was lost through the doors and windows. This heat loss can be reduced considerably by double-glazing the windows and by installing weather stripping and storm windows and doors.

This leaves about 45 percent of the total heat loss that can be reduced by installing insulation materials in the outer shell. In a wood-frame house, this heat is lost by a combination of convection currents and heat radiation. If this heat loss is to be prevented, or reduced significantly, the insulation materials installed within the wall cavities must be able either to trap the air—that is, prevent convection currents—or be able to reflect into the house the heat that is being radiated across the cavity.

A layer of perfectly still air is considered to be the best thermal insulation. Theoretically, such a layer of air 1 in. thick has an *r* value of 5.95, which indicates just how good an insulating material air is. (The insulating ability of a material is measured in *r* or R values, as described below. The higher a material's *r* or R value, the better an insulator it is.) As may be seen in Table 1, only polyurethane-foam insulation, which has

Table 1. Insulating Value of Common Insulating Materials and of Air Spaces

Insulating Material & Description	Temperature °F	Density, lb per cu ft	Resistance, R: Per in. thickness, 1/k	Resistance, R: For thickness listed, 1/C
Blanket and batt				
Mineral wool, fibrous form processed from rock, slag, or glass		0.5	3.12	
		1.5–4.0	3.70	
Wood fiber		3.2–3.6	4.00	
Boards and slabs				
Cellular glass	90	9	2.44	
	60		2.56	
	30		2.70	
	0		2.86	
	−30		3.00	
Corkboard	90	6.5–8.0	3.57	
	60		3.70	
	30		3.85	
	0		4.00	
	90	12	3.22	
	60		3.33	
	30		3.45	
	0		3.57	
Glass fiber	90	4–9	3.85	
	60		4.17	
	30		4.55	
	0		4.76	
	−30		5.26	
Expanded rubber (rigid)	75	4.5	4.55	
Expanded polyurethane (R 11 blown) (Thickness 1 in. & greater)	100	1.5–2.5	5.56	
	75		5.88	
	50		6.25	
	25		5.88	
	0		5.88	
Expanded polystyrene, extruded	75	1.9	3.85	
	60		4.00	
	30		4.17	
	0		4.55	
	−60		5.26	
Expanded polystyrene, molded beads	75	1.0	3.57	
	30		3.85	
	0		4.17	
Mineral wool with resin binder	90	15	3.45	
	60		3.57	
	30		3.70	
	0		400	

(continued)

an *r* value of 6.25 at 50°F, exceeds air in thermal insulating ability. All the other materials listed are well behind air in their ability to prevent the transfer of heat.

In practice, it is extremely difficult, if not impossible, to provide a layer of absolutely still air in the outer shell of a house. The next best thing is to install a material that can trap air—and this defines the unique characteristic of most insulating materials. They may trap air within closed cells (as in the foamed plastics mentioned above, and in cork), or they may trap air because the material consists of a dense network of fibers through which air finds it very difficult to move. Mineral wool is an example of this latter type of insulating material. In addition to preventing air currents, a microscopically thin film of air tends to cling to each fiber of a fibrous material, which further increases the material's insulating ability.

The other basic type of thermal insulation—reflective insulation—insulates because it presents a barrier to radiant heat. The radiant heat is reflected off the surface of the insulation like light off a mirror. Reflective insulating materials include polished metal foils made of aluminum and copper, metallic paints such as aluminum paint, and anodized metallic finishes that are coated onto a backing such as paper or cardboard. The importance of reflective insulation within the walls and roofs of most buildings may be gauged by the fact that between 65 and 85 percent of the heat that passes through these structures is radiated heat.

To sum up, therefore, the basic function of both fibrous and cellular insulation materials is to provide a layer of still air that enhances the ability of these cavity spaces to act as insulation in the frame of a house. In addition, the bulkiness of the insulation prevents the establishment of convection currents within the cavities that would otherwise transfer heat across them.

The purpose of reflective insulation is to prevent the transmission of radiant heat across these air cavities. The maximum degree of insulation is usually achieved when fibrous (or cellular) insulation material is used in combination with a reflective insulation material.

THERMAL RESISTANCE AND *r*, R, AND R(T) VALUES

Thermal conductivity, *k*, is defined as the amount of heat (in Btu per hr) that is transferred from one side of a material to the other when that material has an area of 1 sq ft, is *1 in. thick,*

Table 1. Insulating Value of Common Insulating Materials and of Air Spaces (continued)

			Resistance, R	
Insulating Material & Description	Temperature °F	Density, lb per cu ft	Per in. thickness, 1/k	For thickness listed, 1/C
Mineral fiberboard, wet felted				
Core or roof insulation		16–17	2.94	
Acoustical tile		18	2.86	
Acoustical tile		21	2.73	
Mineral fiberboard, wet molded				
Acoustical tile		23	2.38	
Wood or can fiberboard				
Acoustical tile ½ in.				1.19
Acoustical tile ¾ in.				1.78
Interior finish (plank, tile)		15	2.86	
Insulating roof deck				
Approximately 1 ½ in.				4.17
Approximately 2 in.				5.56
Approximately 3 in.				8.33
Wood shredded (cemented, preformed slabs)		22	1.67	
Loose Fill				
Mineral wool	90	2.0–5.0	3.33	
(glass, slag, or rock)	60		3.70	
	30		4.00	
	0		4.35	
Perlite (expanded)	90	5.0–8.0	2.63	
	60		2.78	
	30		2.94	
	0		3.12	
Vermiculite (expanded)	90	7.0–8.2	2.08	
	60		2.18	
	30		2.27	
	0		2.38	
	90	4.0–6.0	2.22	
	60		2.33	
	30		2.50	
	0		2.63	
Roof insulation				
Preformed, for use above deck				
Approximately ½ in.				1.39
Approximately 1 in				2.78
Approximately 1½ in.				4.17
Approximately 2 in.				5.26
Approximately 2½ in.				6.67
Approximately 3 in.				8.33
Cellular glass			2.56	

(continued)

Table 1. Insulating Value of Common-Insulating Materials and of Air Spaces (continued)

Air spaces

Position of air space and thickness, in.		Heat flow dir.	Season	Types of surfaces on opposite sides: Both surfaces nonreflective materials — Resistance, R	Aluminum coated paper/ Nonreflective materials — Resistance, R	Foil/ nonreflective materials — Resistance, R
Horizontal	¾	Up	W	0.87	1.71	2.23
	¾		S	0.76	1.63	2.26
	4		W	0.94	1.99	2.73
	4		S	0.80	1.87	2.75
45° slope	¾	Up	W	0.94	2.02	2.78
	¾		S	0.81	1.90	2.81
	4		W	0.96	2.13	3.00
	4		S	0.82	1.98	3.00
Vertical	¾	Down	W	1.01	2.36	3.48
	¾		S	0.84	2.10	3.28
	4		W	1.01	2.34	3.45
	4		S	0.91	2.16	3.44
45° slope	¾	Down	W	1.02	2.40	3.57
	¾		S	0.84	2.09	3.24
	4		W	1.08	2.75	4.41
	4		S	0.90	2.50	4.36
Horizontal	¾	Down	W	1.02	2.39	3.55
	1½		W	1.14	3.21	5.74
	4		W	1.23	4.02	8.94
	¾		S	0.84	2.08	3.25
	1½		S	0.93	2.76	5.24
	4		S	0.99	3.38	8.08

and the temperature differential across the material is 1°F.

When, however, one is considering the ability of a material to *resist* the transfer of heat, it is more natural to think of the reciprocal, or inverse, of k, which is $1/k$. The inverse of thermal conductivity is called *thermal resistivity, r;* and $r = 1/k$. If, for example, the k value of a material is 1.25, then its r value is 1/1.25, or 0.80.

Also, when a material is *other* than 1 in. thick, its thermal conductance, C, is defined as the amount of heat (in Btu per hr) that is transferred from one side of a material to the other when that material has an area of 1 sq ft and the temperature differential across the material is 1°F. As in the previous paragraph, when one is discussing the ability of a material to *resist* the transfer of heat, it is more natural to think of the reciprocal, or inverse, of C, which is 1/C. This reciprocal of thermal conductance is *thermal resistance,* R, and R = 1/C.

Note that both r and R may refer to exactly the same material but that r measures the thermal resistance of that material *when it is 1 in. thick,* and R measures the thermal resistance of that material when it is some thickness *other* than 1 in. To convert from r to R, one need only multiply the r value by the actual thickness of the material. For example, in Table 1, 1-in.-thick mineral wool at a temperature of 30°F has an r value of 3.70. If one is considering installing a layer of mineral wool that is 3 in. thick, then its R value at 30°F would be 3.70 × 3 in. = 11.10, and R = 11.10 would be the value inserted in one's calculations when calculating what effect this insulation will have on the total heat transmitted through a wall.

Insulation R Values

The manufacturers of insulating materials assign R values to their products, rounding out the R value to the nearest full odd number. These R values are commonly used when their effect on the heat loss of a construction is being calculated.

Glass-fiber insulation, for example, will have printed on its packing, or envelope, an R value that depends on its thickness, thus,

Thickness, in.	R value
2½	7
3½	11
6	19
6½	22

If we examine Table 1, we see that the $1/k$, or r, value of glass-fiber and other fibrous insulations depends on the outside air temperature, the r value decreasing as the air temperature increases. The reason for this decrease in r value is the increased thermal motion of the air molecules trapped within the fibers when the weather is hot. This means, of course, that the insulation is less effective in the summer when one wants to keep hot air outside the house than in winter when one wants to keep warm air in. We might also add that there is an optimum density to a fibrous insulating material (not only glass fiber) that maximizes its ability to insulate. Densely compressed insulation is not as effective as loosely compressed insulation—but only up to a certain point. The insulation cannot be too loosely packed.

To return to R values and Table 1, glass-fiber insulation, for example, actually has not one but a range of R values that depends not only on its thickness but also on the outside air temperature. The range of these R values between 90°F and 30°F is as follows

Thickness, in.	Range of R values
2½	9.6–13.2
3½	13.5–18.4
6	23.1–31.6
6½	25.0–34.2

Comparing this table with the one above, one will note how conservative the manufacturers' ratings are.

In any actual building, the construction consists of a variety of materials, each of which has its own r or R value. The total amount of heat transmitted through a wall or roof is summarized by the construction's overall heat transmission coefficient, U. Thus, U equals the total amount of heat (in Btu per hr) that will be transmitted through an entire wall or roof section having an area of 1 sq ft when the temperature differential across that section is 1°F. The reciprocal of U is the *total thermal resistance,* R(T), and R(T) = 1/U. In practice, the R(T) value of any building construction is obtained by adding all the R values in that construction or, if the U value is known, simply by finding the reciprocal of U.

TYPES OF INSULATION

As we said, the basic division between insulating materials is between those that trap air and those that reflect radiant heat. Practically, however, insulating materials are separated into groups according to the way in which they are manufactured and used in building construction. Thus, the insulations used in most houses are classified as *blankets, batts, fills,* and *reflective foils.* There are, in addition, various *slab* and *foam* insulations that have as yet only a limited, though important, use in dwellings.

Blankets and Batts

Fibrous Insulating Materials

Blankets and batts are made of fibrous materials, of which mineral wools and wood fibers are the most common.

Mineral wool is a generic term that includes fibrous insulations made from limestones and shales (rock wools), blast-furnace slag (slag wools), and silica (glass wools).

To manufacture a rock wool, the raw material is heated with coke in a blast-furnace cupola until it melts. The melting temperature is somewhere between 2300° and 3400°F, depending on the raw material used. As the molten rock is discharged in a thin stream from the bottom of the cupola, a blast of steam blows it into long fibers that are from 0.0002 to 0.0004 in. in diam.

Slag wools are manufactured in the same way, the major difference being that slags melt between 2000° and 2850°F.

The silica from which glass wools are manufactured is melted and then forced through fine orifices into the path of a jet of air or steam that blows the molten glass into fibers. All three of these mineral wools can also be made into a granular form by altering the thickness of the molten stream and the velocity of the impinging jet of air or steam.

Mineral wools are inherently fungus-proof, vermin-proof, moisture-proof, and fire-resistant. They are usually treated with a chemical that enables them to shed surface water quickly, to prevent the buildup of moisture within a closed wall or roof construction.

Wood fibers and the other natural fibers have none of these inherent virtues, and they must be especially treated if they are to be resistant to damage from fungi, vermin, fire, and moisture.

Blankets

Blankets (which are sometimes called *quilts*) consist of long rolls of insulation that are usually 1, 1½, 2, and 3 in. thick (although they are available as thin as ½ in. and as thick as 6 in.), and in widths of 15, 19, or 23 in. These widths allow the blankets to be installed between studs, rafters, or joists that are spaced 16, 20, or 24 in. apart. The blankets are manufactured in 40- to 100-ft lengths, the actual length depending on the thickness of the material.

Although blankets are manufactured in which the fibers are simply matted together, most blankets are enclosed in a paper or vinyl envelope, or they may have a paper, cardboard, or wire-mesh backing cemented to one side. This backing, which may be reinforced for extra strength, extends ¾ to 2 in. away from the sides of the blanket, forming two flanges that allows the blanket to be stapled or nailed to the adjacent framing members. The envelope enclosing a blanket has similar flanges made by folding and cementing the edges of the paper together.

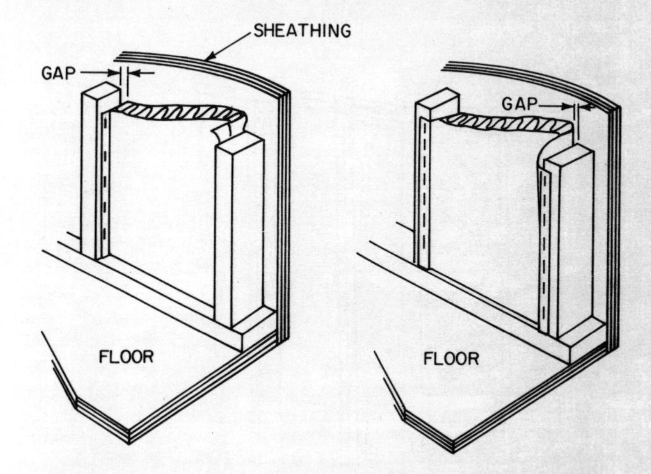

Fig. 1. The flanges on blanket insulation can be attached either to the sides or ends of the studs within a wall cavity.

To install a blanket, these flanges are stapled or nailed either to the faces or sides of the framing members, as shown in Fig. 1. If the flanges are attached to the sides of the framing, this will automatically cause the blanket to extend into the cavity by ¾ to 2 in., which provides an air space between the blanket and the sheathing on the opposite side of the cavity. When the flanges are attached to the ends of the framing members, the blankets will form an effective vapor seal (see below). Care must be taken that the staples or nails used to hold the blankets in place are driven flush with the flanges and that the flanges themselves lie smoothly against the framing members; otherwise there is a chance that the flanges will interfere with the installation of gypsum wallboard, if gypsum wallboard is to be used to finish the interior surface of the wall (*see* DRYWALL CONSTRUCTION).

Very often one side of a paper envelope (usually the flange side) is coated with plastic or asphalt to make the paper impermeable to water vapor. When the flanges are correctly attached to the framing, this side of the envelope will form a *vapor barrier* that prevents the passage of water vapor into the wall or roof construction. Insulation having a backing on one side can be treated in the same way to make the backing vapor-resistant.

Although a combined blanket and vapor barrier appears to have advantages, especially to the builder, who is always intent on reducing costs, these advantages will prove illusionary unless the flanges are secured tightly against the framing. If a vapor barrier is to prevent the passage of water vapor, it must not have any gaps or openings through which water vapor can pass; for if water vapor does pass into the air cavity and condense against the construction, the consequences can be very serious (*see* CONDENSATION).

The warmer the climate, the less necessary a vapor barrier is in any case, since there is less likelihood that any vapor will condense within the air cavity. The colder the climate, however, the greater the probability that water vapor will make its way past an inadequately installed barrier and condense within the air cavity. When a vapor barrier is necessary, a sheet of polyethylene film installed against the inner side of the framing will make a far more effective vapor barrier than vapor-proofed blanket insulation (see Fig. 2), and the use of such a separate vapor barrier is always to be preferred in these cases.

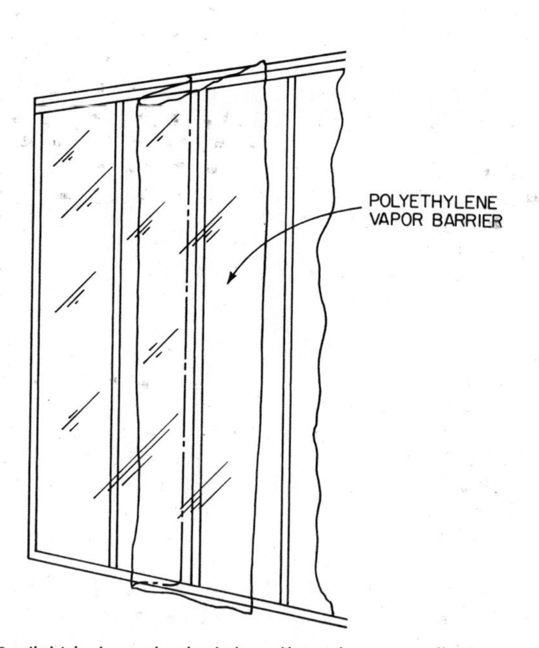

Fig. 2. A 2-mil-thick sheet of polyethylene film makes an excellent vapor barrier. The film is unrolled and installed as a single sheet over the entire wall. After it has been nailed or stapled to the wall framing, any door and window openings are cut out of the film.

Batts

Batts are similar to blankets in every way in both their construction and use. They may be enclosed within envelopes or have backings attached to one side, and they may have a vapor barrier and/or reflective foil added, just as blankets have. The major differences between batts and blankets are that batts are usually thicker—3, 4, 5, or 6 in.—and they are only 24 to 48 in. long, although some manufacturers of 8-ft-long blankets refer to their products as *batts*.

Batts are usually manufactured to be slightly wider than the usual 16, 20, and 24 in. spacing that exists between framing members, which enables them to fit snugly between these structural members. They need only be pushed into place and there they remain. If, however, they are installed horizontally between the rafters of an unfinished roof, or between the first-floor joists over a basement or crawl space, and the builder is concerned lest they slip out one day, they can be held in place by paper flanges that are provided as part of the batts or they can be held in place by lengths of wire that are pressed into place under the batts. If batts are to be installed inside a finished ceiling, they can be pushed into place and kept from falling out until the finish ceiling has been installed by a polyethylene vapor barrier that is fastened in place under them.

There is a question whether blankets and batts should be installed in a wall in such a way that they press against the wall sheathing or whether an air gap should be left between the insulation and the sheathing. Certainly, whether the insulation is placed flat against the sheathing or whether a space is left between them will not affect the insulating value of the insulation at all. But insulation does lose its effectiveness when it gets wet, and one never knows whether a wall that is sound and watertight on the day it is finished will remain sound and watertight as the years pass, even when a sheet of waterproof building paper has been nailed against the sheathing to keep moisture out. It would seem more prudent, therefore, to allow an air gap between the insulation and the sheathing.

When a mineral-wool insulation is to be installed within or against a masonry wall, then under no circumstances should the insulation touch the exterior wythe of brick or concrete blocks since moisture will almost positively make its way through the wall during extremely wet weather, either by soaking through the masonry or by following cracks in the mortar joints (*see* BRICK-MASONRY CONSTRUCTION). The only way that blankets or batts should be installed against a masonry wall is to nail 1 × 2 in. or 2 × 2 in. furring strips to the masonry and then attach the insulation to the furring. The thickness of the insulation must be watched (or the size of the furring strips increased) to make sure that a gap remains between the masonry and the insulation after the flanges have been nailed or stapled to the furring.

Loose-Fill Insulation

Both mineral and vegetable wools are available as *loose fill*. The insulation material is packed loosely into bags that weigh 40 lb each and contain 4 cu ft of the material. The insulation is used in spaces too small for blankets or batts; or it may be blown into a cavity by machine.

Loose granules, or pellets, of insulation made of vermiculite and perlite are also available in 40-lb 4-cu-ft capacity bags.

Vermiculite

Vermiculite is a mica-like mineral known chemically as hydrated magnesium-aluminum-iron silicate. Like all micas, it consists of very thin, flat layers of material pressed tightly together. Molecules of water are trapped between these layers. When the mineral is crushed into small pieces and the pieces are then heated to 2200° to 2400°F, the trapped water expands into steam, which forces the layers of mica apart. When the mica cools, the volume of each granule will have increased about 12 times, the spaces between the layers now being filled with trapped air.

Perlite

Perlite is a silica-like volcanic rock, the molecules of which hold combined water. When the rock is crushed into small pieces and the pieces are heated to about 1500°F, the rock softens. At the same time, the water content is liberated and converted into steam. As a result, the entire granule expands. When it cools again, it contains a considerable volume of trapped air.

Loose-Fill Installation

Both of these materials are usually poured or blown into wall cavities or between the floor joists of an unfinished attic. For use in a wall space, the exterior wall is, as usual, built first. But before the interior wall surface is constructed, a polyethylene vapor barrier is nailed or stapled to the bottom of the wall framing. The wall cavity is then gradually filled with the granules from the floor up as the interior wall is constructed.

The big problem with using loose granular insulation in a wall is that over a period of time the granules may pack down because of the small but continual vibrations any house is subjected to. Eventually an air gap may open up at the top of the wall across which heat may be transmitted by convection or radiation.

The walls of an old, uninsulated wood-frame house are often filled with insulation by blowing these granules between the wall studs. Holes are drilled in the top of the wall between each pair of studs (or the top strip of siding may be removed from the wall) and the insulation is then blown into the wall cavities through a hose, like a giant vacuum cleaner operating in reverse.

The same technique is used to blow the granules between the floor joists of an unfinished attic. Blowing the granules into

NAME OF MANUFACTURER
(Example)

Pneumatic Insulation — **Weight of insulation per bag: 30 lb**

The manufacturer recommends these maximum coverages at these minimum thicknesses to provide the levels of installed insulation resistance (R) values shown:

R value	Minimum thickness	Maximum net coverage	Minimum weight per sq ft
To obtain an insulation resistance R of:	Installed insulation should be not less than:	Contents of this bag should not cover more than:	The weight per square foot of installed insulation should be not less than:
R-24	8¼ in.	28 sq ft	1.07 lb per sq ft
R-19	6½ in.	35 sq ft	0.86 lb per sq ft
R-13	4½ in.	50 sq ft	0.60 lb per sq ft
R-9	3¼ in	75 sq ft	0.40 lb per sq ft

Builder's statement:

This insulation has been installed in conformance with the above recommendations to provide a value of R-19, using 37 bags of this insulation to cover 1290 square feet of area.

Edward P. Frank	Webildit, Corp.	May 27, 1977
(Builder's signature)	(Company name)	(Date)

Fig. 3. A label cut from a bag of loose insulation must be placed in the attic where it can be inspected. The insulation must be installed in conformity with the label recommendations.

place is quicker and cheaper than emptying bags by hand, and the insulation packs into place more evenly and more densely than if the work were done by hand.

The effectiveness of blown insulation, that is, its R value, depends on the density with which the granules are packed together as well as the overall depth of the material. Most manufacturers base the R values of their insulation on an assumed density of from 6 to 8 lb per cu ft. For a homeowner to specify only the depth to which attic insulation is to be laid is not enough. It is possible, for example, for a builder or contractor to skimp on the amount of insulation laid down by failing to pack it in densely enough.

To ensure that the correct density is laid down, Federal Specification HH-I-1030 requires that each bag of loose-fill insulation have a label that shows (1) the minimum thickness, (2) the maximum net coverage, and (3) the minimum weight of the insulation required per unit area to achieve a specified R value. Figure 3 is an example of such a label. The FHA requires that a label cut from one of the bags (or a card duplicating the label information) be signed by the builder and nailed somewhere in the attic as a guarantee that the insulation has been correctly installed.

Reflective Insulation

Bright metal surfaces reflect heat because of a property called *emissivity*. The emissivity of a material is a measure of how well it absorbs radiant heat energy. An ideal black object that absorbed all the radiant heat energy falling on it would have an emissivity of 1.0. The emissivities of ordinary construction materials such as wood, plaster, and concrete are around 0.80. Metals with brightly polished surfaces have extremely poor emissivities. Aluminum foil, for example, has an emissivity of only 0.05. That is, it will absorb only 5 percent of the radiant heat energy striking it. The rest is reflected away. Aluminum foil, therefore, reflects 95 percent of the radiant heat that strikes it.

However, in order that a bright metal surface be capable of acting as reflective insulation, it must face an air gap that is at least ¾ in. wide. The effectiveness of the reflective insulation decreases as the width of the air gap decreases from ¾ in. If there is no air gap at all, the material loses its reflective-insulation ability altogether and becomes merely a conductor of heat, like any other metal. On the other hand, an air gap wider than ¾ in. does not add significantly to the effectiveness of reflective insulation.

Aluminum foil, the most widely used reflective insulation, is too fragile a material to be used by itself. The foil is usually cemented to a paper or cardboard support. Two-sided reflective foils are made by cementing two of these single-sided sheets to both sides of a sheet of asbestos.

Aluminum foil is often cemented to one side of gypsum wallboards and insulation-board panels. These panels and wallboards must then be installed so that the foil side is facing the wall cavity. Foil cemented to a panel in this way has two functions: it serves as reflective insulation, and it also acts as a vapor barrier that prevents the passage of water vapor into the cavity from the interior of the house.

Paper- or cardboard-backed foil is made in rolls of standard widths to fit between framing members. The edges of the rolls can be bent to form flanges that are nailed or stapled to the framing members.

Double-sided foil is intended to be placed in the center of a wall cavity, thus dividing the cavity into equal halves. The inward-facing foil reflects heat into the interior of the house during the winter; the outward-facing foil reflects solar heat away from the interior of the house during the summer. Reducing the width of the air cavity by half in this way also helps to reduce the strength of convection currents within the wall space and thus helps to improve the insulating properties of the air itself.

The same general technique can be used with blanket insulation having foil on one side. If the blanket is installed in the center of a wall cavity, two air spaces are created where there had been only one. At the same time, the foil backing, which should be installed so that it is facing toward the interior of the house, reflects heat into the house. The fact that the foil is facing the interior of the house also enables it to act as a vapor barrier. In fact, if a foil is cemented to one side of a blanket or batt, the foil *must* face toward the warm side of the wall to prevent the condensation of water vapor on the insulation material.

Reflective insulation is particularly effective in preventing the transfer of heat through a roof during the summer, since so much of the solar heat that enters a house is radiated through the roof. Reflective insulation within a roof is much less important during the winter; instead, blanket insulation is required in the winter to prevent the setting up of convection currents within the roof construction that will transfer heat out of the house. The most effective insulation for roofs, therefore, is a combination of blanket or batt insulation to which is attached a layer of reflective foil facing upward, with an air gap between the roofing and the foil. But when this method of installing the insulation is followed, a vapor barrier *must also* be installed on the ceiling side of the insulation to prevent the condensation of water vapor in the insulation during the winter months.

Blanket or batt insulation placed between roof rafters must in any case be separated from the roofing itself to allow air to circulate between the insulation and the roofing (see Fig. 4). This will enable air currents to carry away a large proportion of the heated air that would otherwise be trapped within the roof space, which will allow the insulation to perform its function more effectively.

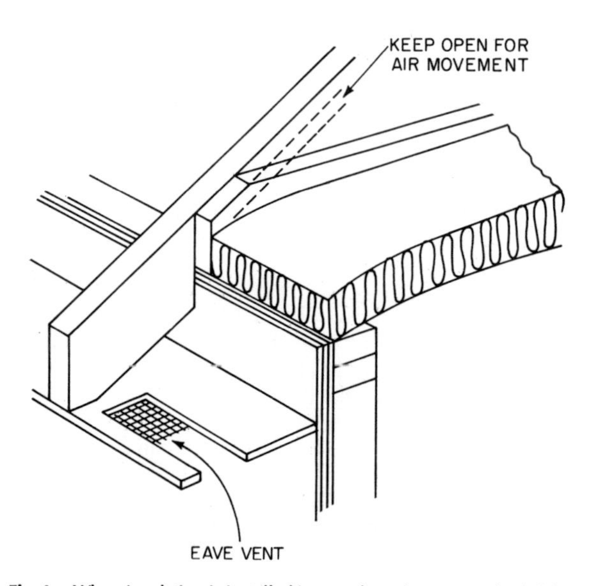

Fig. 4. When insulation is installed in a roof, an air gap must be left between the insulation and the bottom of the roof to allow air to circulate next to the roofing.

Slab Insulation

There is one place in a house where the flexible insulating materials described so far are unsuitable. This is around the perimeter of a poured-concrete slab-on-grade floor (see Fig. 5; and *see also* SLAB-ON-GRADE). A considerable proportion of the heat contained within such a floor is transmitted to the perimeter of the slab and thence into the surrounding soil. Insulation is required to prevent this loss of heat. But a flexible insulating material such as a mineral wool blanket has not the strength to support the weight of the slab nor is it resistant to moisture in the soil—and wet insulation loses almost all its insulating properties.

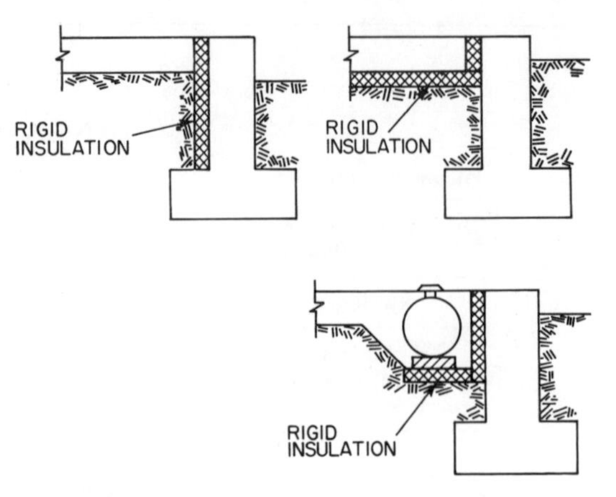

Fig. 5. The ways in which rigid insulation can be placed around the perimeter of a slab floor to prevent the loss of heat from the slab.

In order to be effective, any insulating material installed around the perimeter of a floor slab must be impermeable to water, or immune to water damage, and it must be strong enough to resist the crushing loads imposed by the slab. It must also be unaffected by fungi or termites. Materials having these qualities, more or less, include cellular glass, mineral wool that has been impregnated with a binder that stiffens the wool into a rigid mass, and foamed plastics, such as polystyrene and polyurethane.

Slab insulation materials are available as rigid panels that are anywhere from 12 × 32 to 20 × 96 in. in size, though the most common size is 24 × 48 in. The usual panel thicknesses are 1, 2, 3, and 4 in.

Cellular Glass

Cellular glass insulation is made in slabs 2, 3, 4, and 5 in. thick. This material has a crushing strength of about 150 lb per sq in. and can thus be used under a floor slab. It is possible, however, for moisture to soak into the insulation, and the slabs should, therefore, be dipped in hot coal tar or be completely coated with asphalt before they are installed. Otherwise, if the insulation should be subjected to wetting and freezing cycles during the winter, the surface of the insulation may spall away, exposing the interior to moisture. The insulation should also be surrounded by a vapor barrier to further protect it against moisture.

Rigid Mineral Wool

Rigid panels of *mineral wool* are available in thicknesses of ¾, 1, 1½, and 2 in. This insulation has a crushing strength of only 12 lb per sq in. and, therefore, is satisfactory for use as perimeter insulation only if it is installed vertically, so that it is not subjected to any crushing loads. The slabs of insulation should also be completely coated with asphalt to render them waterproof; otherwise if the binder should be exposed continuously to moisture, the slabs will disintegrate. The slabs should also be covered by a vapor barrier to further protect them from moisture.

Foamed Plastic Insulation

There are two foamed plastics in common use, polystyrene and polyurethane. Both these materials are widely used in the solid form. The only difference between these plastics as solid materials and as foams is that a gas is introduced into them during the manufacturing process that causes the plastic to expand about 40 times its original volume. Both of these foamed plastics are strong, lightweight, and easy to cut and shape with simple hand tools such as knives and saws. They are immune to attack from fungi or vermin and they are completely waterproof.

Although polystyrene insulation must be bought in slab form because the manufacturing process requires a source of steam heat as well as molds, the polyurethane insulation can be purchased either as solid slabs or in the form of two liquids that, when they are mixed together, foam in place.

Of the two, polystyrene costs about half as much as polyurethane, but the insulating value of polyurethane is about twice that of polystyrene, so that the price per R value is approximately the same.

Neither of these materials is widely used for home construction, however, since both cost much more than the more traditional insulating materials described in this article. Where they are widely used is around and under floor slabs. Here their insulating properties as well as their resistance to moisture and their ability to resist crushing loads makes them the preferred material, despite their high cost.

Lime

Lime has been used in building construction, mainly as a plaster and mortar, for over 5000 years. The Egyptians used lime plaster as a wall finish, as did the Greeks and Romans. The Romans also made a mortar from lime and a volcanic ash called *pozzolana* that has proved to be exceptionally strong and long-lasting, the mortar in many Roman ruins being as sound today as the day it was made; sounder, in fact. The Chinese also made a lime mortar that was used for building the Great Wall.

Up to the beginning of the twentieth century, lime plasters and mortars continued to be used almost exclusively in building construction, but, with the introduction of gypsum plasters and cement mortars, the use of pure lime plasters and mortars has declined precipitously, the virtues of lime as a building material being unsuited to present-day methods of construction in which speed and minimal labor costs are of primary concern to a builder. Today, lime is used mainly as an additive in gypsum plasters and mortars to impart more desirable characteristics to them.

The principal virtues of lime plasters and mortars are their resilience under stress and their resistance to moisture—once the plasters and mortars have set. It is the slowness with which lime sets and gradually attains its full strength that is its principal drawback in present-day construction. For a comparison between lime and gypsum plasters *see* PLASTER; PLASTERING. For a similar comparison between lime and cement mortars *see* MORTAR.

THE MANUFACTURE OF LIME

Limestone

Lime is obtained from limestone, a common sedimentary rock found in large deposits in most parts of the world. Some deposits are made up of the fossil remains of sea animals such as diatoms, shellfish, and molluscs. Other deposits are made up of precipitated chemicals. It is the latter type of limestone that is used primarily for the manufacture of lime.

Chemically, pure limestone consists of calcium carbonate, $CaCO_3$, which is formed by the chemical combination of calcium oxide, CaO, and the gas, carbon dioxide, CO_2, thus,

$$CaO + CO_2 \rightarrow CaCO_3$$

When pure, or almost pure, limestone such as this is subjected to great pressure, it is metamorphosed into marble.

Pure limestone rarely exists in nature, however, and most deposits consist mainly of calcium carbonate and magnesium carbonate, $MgCO_3$, plus small amounts of other minerals such as silica, alumina, and iron oxide.

The principal type of limestone used for the manufacture of building lime in the United States is called *dolomitic* limestone, which is a mixture of 54 percent calcium carbonate and 46 percent magnesium carbonate. Ninety-five percent of all building lime is dolomitic. There are also *high-calcium* limestones, which consist of from 90 to 98 percent calcium carbonate, and *magnesian* limestones, which consist of from 75 to 90 percent calcium carbonate, the remainder being mainly magnesium carbonate. These latter two types of limestone are of comparatively minor importance in the building trades.

Calcining of Limestone

Lime is prepared from limestone by *calcining,* or heating, the limestone until the calcium and magnesium carbonates dissassociate themselves from the carbon dioxide. The carbon dioxide is driven off as a gas, and all that remains are the oxides, of which the calcium oxide is the more important, chemically speaking.

The calcining takes place in a kiln that is heated to about 2500°F. There are two main types of kiln in use, a *vertical* kiln, which resembles a squat chimney, and a *rotary* kiln, which consists of a long metal cylinder laid horizontally with one end tipped higher than the other. About 95 percent of the lime manufactured is produced in rotary kilns.

Rotary kilns are 100 to 400 ft long and from 8 to 13 ft in diameter. They revolve continuously. Before the limestone is dumped into the higher end of the kiln, it is crushed into pieces from ½ to 1½ in. in size. As the limestone tumbles its way through the kiln, the temperature increases and it is gradually heated. The carbon dioxide is driven off at a much faster rate in a rotary kiln than in a vertical kiln, mainly because of the smaller size of the limestone particles.

The final product is cooled and packed into double-walled, waterproof bags—80 lb to the bag. The product of a vertical kiln is usually shipped in the form of lumps, that of a rotary kiln as pebbles. This final product, whether it is mainly calcium oxide or a mixture of calcium and magnesium oxides, is better known as *quicklime,* a highly reactive, very caustic, very alkaline substance.

QUICKLIME

Quicklime has a very intense affinity for water. When quicklime with a high proportion of calcium oxide comes into contact with water, there is a violent reaction in which the amount of heat that is evolved is great enough to cause the water to boil violently. Anyone standing near this reaction is likely to suffer skin burns if he is splattered with the lime or the hot water.

Before the quicklime can be used, therefore, it must be *slaked,* or *hydrated.* That is, the quicklime's thirst for water must be quenched. This is accomplished by adding water to the quicklime under controlled conditions so that the generation of excessive heat is prevented, for if the quicklime should overheat, it will become *burned,* and as a result it will be useless as either plaster or mortar. On the other hand, the temperature of the reaction must not be reduced too much because the hotter the reaction, the better the resultant *lime putty*, which is the end product of slaking and the material that is actually used for plasters and mortars.

With pure calcium oxide, about 25 gal of water per 100 lb of quicklime are required to slake the quicklime completely. The heat of the reaction causes the lumps or pieces of quicklime to break apart and crumble into a fine powder. The powder then combines chemically with the water to form calcium hydroxide, thus,

$$CaO + H_2O \rightarrow Ca(OH)_2$$

During this reaction, the volume of the limestone increases by 2½ to 3 times. Lime putty, therefore, consists of calcium hydroxide plus enough additional water to form a putty of the desired consistency.

As a practical matter, since very few quicklimes consist of pure calcium oxide, the actual amount of water required and the total amount of expansion that occurs will be less than this, as any impurities in the quicklime do not take part in the reaction. In particular, the magnesium oxide content of dolomitic quicklimes (which is not actually an impurity) slakes very, very slowly. For dolomitic quicklimes, therefore, in which about 46 percent of the quicklime consists of magnesium oxide, only about 15 gal of water are required to slake the quicklime completely. Not only do dolomitic quicklimes slake more slowly than high-calcium quicklimes, but much less heat is evolved during the reaction and the final product expands much less in volume.

The Slaking of Quicklime

At one time all quicklimes were slaked on the job by a man experienced in this work. Slaking on the job is rarely done today as quicklimes that have been preslaked (that is, *hydrated*) at the mill are used almost universally. Nevertheless, the procedure is interesting for itself, and a description of the procedure will help to introduce hydrated limes.

The quicklime is first tested. A sample of the quicklime, about two cups or so, is thrown into a pail, and sufficient water is then added to cover the quicklime. After a while the lumps or pieces of quicklime will begin to break apart. This signals the beginning of the slaking reaction. The time it takes for the slaking to begin after the water has been added is then noted. If the slaking begins within 5 minutes, the quicklime is considered a fast-slaking high-calcium type. If slaking begins in 5 to 30 minutes, the quicklime is considered a medium-slaking magnesian type. And if it takes more than 30 minutes for slaking to begin, then the quicklime is considered a slow-slaking dolomitic type. The test completed, the workman now knows how to proceed with the slaking of a large batch.

For a *fast-slaking* quicklime, a large mortar box is filled with an amount of water sufficient to cover the quicklime completely. The quicklime is then added so that it is distributed evenly throughout the mortar box. The slaking is allowed to proceed at a slow boil. That is, the water must bubble, but no steam must arise from the water. The reaction is watched carefully for signs of excessive heat, which is signaled by the appearance of steam. If steam does appear, additional water must be added immediately, and the mixture stirred with a hoe to dissipate the excess heat.

As long as no steam appears, the reaction is allowed to proceed as rapidly as possible, since rapid slaking produces a better quality lime putty, one in which the particles of calcium hydroxide will be very fine. This type of lime putty is actually a colloidal suspension of calcium hydroxide in water, in the same way that thick cream is a colloidal suspension of fat, casein, and other particles in water. Such a lime putty sets rapidly (for a lime, that is); it has a great amount of plasticity; that is, it will have a buttery quality that makes it very easy to spread and work; the yield of putty is maximized, since the largest possible quantities of water and quicklime will have combined to form the hydroxide; and the lime putty can be mixed with the largest possible quantity of sand, which means the mortar or plaster made from this mixture will shrink the least possible amount as it sets.

For a *medium-slaking* quicklime, the quicklime is first distributed evenly throughout the mortar box and the water is then added to it, but only enough water to half submerge the quicklime. As soon as slaking has begun and steam appears, additional small amounts of water are added to stop the steaming, the mixture being stirred with a hoe all the while. As the water is absorbed by the quicklime, additional water is added to prevent the quicklime's becoming dry and crumbly, but the reaction should not be slowed down.

Note that with a fast-slaking quicklime, the workman does what he can to slow down the reaction, but with a medium-slaking quicklime, he does what he can to speed up the reaction, mainly by being miserly with the water.

For a *slow-slaking* quicklime, the quicklime is also distributed throughout the mortar box first and the water is then added, but only enough water to wet the quicklime. The object with a slow-slaking quicklime is to speed up the reaction as much as possible. Therefore, only enough water is added to enable the reaction to proceed as rapidly as it can, care being taken not to add too much water at any one time, which might cool the mixture. Sometimes warm water is added to increase the rate of slaking. To further conserve heat, the mixture is hoed until the slaking has just about run its course.

After the reaction has stopped, the lime putty is run through a sieve to screen out any particles of unslaked quicklime and the putty is allowed to cool.

Lime putty is never used immediately. Apart from its having to cool, lime putty improves with age, becoming smoother, creamier, and more workable the longer it stands. This is because it continues to slake slowly with the passing of time, with the particles of hydroxide becoming finer and finer and the suspension more and more colloidal.

As we mentioned, almost all the quicklimes used in the building trades in the United States are dolomitic quicklimes, in which there is about 46 percent magnesium oxide. During the slaking, very little of this magnesium oxide is converted into magnesium hydroxide. Another reason that the lime putty has to age, therefore, is to give the magnesium oxide a chance to slake. Failure to slake a dolomitic quicklime properly in the first place, or to age it for a long enough period afterward, may result in serious construction defects.

For example, if a dolomitic lime putty is being used as a plaster, any unslaked particles of magnesium oxide in the plaster will continue slowly to absorb moisture from the air long after the plaster as a whole has set and hardened. This slaking will sometimes continue for years. As the particles absorb moisture, they swell and exert a pressure against the surrounding hardened plaster. In time, bits of plaster may be forced off the wall. The result may be innumerable tiny pockmarks covering the surface of the wall.

To give the magnesium oxide in a dolomitic quicklime time to slake, therefore, the lime putty must be allowed to age for at least 2 weeks, protected all the while from the air by a film of water or sand. In fact, as long as it continues to be kept from contact with air, lime putty can be stored almost indefinitely and used months after it has been slaked.

THE HARDENING OF LIME

Lime hardens, or sets, by a process of *recarbonation*. That is, after the excess water in the lime has evaporated, the hydroxide is gradually replaced by carbon dioxide to form the original calcium or magnesium carbonate again, thus,

$$Ca(OH)_2 + CO_2 \rightarrow CaCO_3 + H_2O$$

or

$$Mg(OH)_2 + CO_2 \rightarrow MgCO_3 + H_2O$$

which completes the cycle.

This recarbonation of lime proceeds very, very slowly. It continues long after the lime has set hard. It may, in fact, take from 6 months to several years for all the hydroxide to be replaced by carbon dioxide, depending on the location and thickness of the construction; this is all right, as the slower the recarbonation proceeds, the better the final plaster or mortar. Lime that has set too quickly will never achieve its full potential strength.

The rate at which recarbonation proceeds depends on the availability of carbon dioxide, which means the lime must be exposed to the atmosphere. As long as part of the lime remains a hydroxide and as long as air can reach this hydroxide, the lime will continue to revert to its original carbonate form and the harder and stronger it will become. Indeed, it has been discovered that in some Roman walls 2000 years old the lime deep within the mortar joints still hasn't reverted to its original carbonate form because the joints were made so tightly that air just hasn't been able to make its way that deeply into the wall.

Since the addition of water during slaking has caused the quicklime to expand, the evaporation of this water and the replacement of the hydroxide by carbon dioxide will cause the lime to shrink back down again. This shrinkage will cause large cracks to open up in a plaster coat or mortar joint. Sand is added to the lime putty, therefore, to reduce the shrinkage as much as possible. The more sand the lime putty can carry without losing any of its strength, the less the total shrinkage. Different types of lime can hold different quantities of sand. These quantities have been determined by tests. Usually the lime manufacturer will specify the amount of sand that is to be used with his product under different circumstances. As a general rule, high-calcium limes can carry more sand than dolomitic limes. On the other hand, dolomitic limes tend to shrink less than high-calcium limes in the first place and, therefore, less sand is needed with them.

HYDRATED LIME

Hydrated lime, or quicklime that has been slaked at the mill, is the manufacturer's response to all the problems that have ever been experienced with quicklimes slaked on the job. The use of hydrated lime does away entirely with the need to slake quicklime, and it makes the use of lime as simple and direct as the use of gypsum and cement.

The quicklime is slaked at the mill by a process that is similar to that already described for job-slaked quicklime, but the process is accomplished automatically and under the complete control of the manufacturer. The quicklime is first crushed into particles ½ to 1 in. in size (if it is not already a coarse powder). The particles are then mixed thoroughly with just enough water so that what results at the conclusion of the hydration is a dry powder. This powder is screened to remove any unslaked particles, it is ground very fine, and it is then packed in 50-lb bags for shipment.

The result is a lime having uniform properties and several advantages over quicklime that has been slaked on the job. Putty made from hydrated lime is quicker-setting than job-slaked lime putty, it is stronger after it has set, and it shrinks less while it is setting. It need not be aged after it has been mixed with water but can be used immediately. And it is a far more convenient material to handle and store than quicklime is.

Putty made from hydrated lime is not as good as job-slaked lime putty in several respects, however. The putty is stiffer and more difficult to work than job-slaked lime putty, it cannot carry as much sand, and the amount of putty obtained from a given quantity of hydrated lime is less than that obtained from quicklime slaked on the job.

Nevertheless, when hydrated and job-slaked dolomitic lime putties are compared, the advantages are all with the hydrated lime putty, mainly because of a modification that has been introduced into the slaking process by manufacturers. After the quicklime has been hydrated and while it is still hot, it is slaked again under pressure. This "cooking" of slaked dolomitic quicklimes effectively converts all but about 8 percent of the magnesium oxide into magnesium hydroxide. As a result, hydrated dolomitic limes have none of the defects of job-slaked dolomitic limes, and the inherent virtues of dolomitic limes can be made full use of. These virtues are that they make stronger mortars and plasters than high-calcium limes do, they are more workable and more plastic, and they shrink less as they set. Dolomitic limes that have been slaked under pressure are sold as *double-hydrated, pressure-hydrated,* or *autoclaved* limes.

Types of Hydrated Limes

There are two main types of hydrated lime—*finishing limes,* which are used mainly for the final coat of plastered walls, and *building,* or *mason's, limes,* which are used for mortars and for the base coats of plastered walls. Each type has its own characteristics, and, although finishing limes may be used as building limes, the converse is not true. Within each type there are two subtypes, called *type N* (normal) and *type S* (special).

Building or Mason's Lime

Type N mason's lime is the standard product used for mortars, for the base coat of plastered walls, and in concrete. It has only a fair amount of plasticity, much less than job-slaked lime putties have. On the other hand, since it sold as a dry product, it can be mixed more thoroughly and more easily with sand than lime putty can, and the necessary amount of water can always be added afterward to make a homogenous paste. In addition, this paste can be used immediately, without having to wait for it to age.

Type S mason's lime is the double-hydrated variety. It has a much greater degree of plasticity than type N mason's lime, a plasticity that is achieved immediately upon the addition of water. Type S mason's lime is almost as workable and plastic a product as job-slaked lime putties are.

Finishing Lime

Type N finishing lime is a very white, very plastic product that must be soaked in water from 12 to 24 hr before it is used if it is to achieve its full amount of plasticity. A mortar box or a tank is filled with the required amount of water, as given in the manufacturer's instructions, and the lime is then sifted through a screen into the water. The proportion of water to lime should be such that the lime comes just to the surface of the water after the full amount has been sifted into the mortar box. The putty is allowed to stand undisturbed for the required period of time, which is usually overnight. As it slakes, the putty will turn stiff. Just before it is to be used, it is shoveled through a ¼-in. screen to break up any lumps that may have formed.

Type S finishing lime, which is the double-hydrated variety, does not require soaking beforehand but can be used immediately after it has been mixed with water. Type S finishing lime does not have the plasticity of type N finishing lime, and plasters made from it are not of equal quality.

Lumber

Although the term *lumber* applies to both hardwoods and softwoods, it should be understood that in this article when we say "lumber," we mean the *softwoods* that are used in house construction. In the United States, the wood-frame structures of houses consist of various species of softwood lumber, hardwoods being used mainly for interior trim, cabinets, and flooring. For a description of the characteristics of hardwoods and softwoods and a comparison between the two types *see* WOOD.

LUMBER MANUFACTURE

There are two basic methods by which a log can be sawed into lumber: either parallel to the annual rings or across the annual rings (see Fig. 1). The first method is called *slash-cutting* and it produces (in softwoods) *flat-grained* lumber. The second method is called *rift-cutting,* and it produces (in softwoods) *edge-grained* or *vertical-grained* lumber.

At the sawmill, a log first passes through a *barker,* in which jets of high-pressure water strip the bark from the log. The log then passes back and forth through either a large bandsaw or circular saw, which saws the log into rough lumber. The log may be sawed into large slabs, which are then sent on to other saws called *trimmers, edgers,* and *gang saws* that roughly saw the wood into smaller sized pieces, or the log may be roughly sawed directly into boards and planks of various sizes.

After the rough-sawed lumber has been sorted out according to size and grade, it is stored outdoors for several months until it has seasoned. Most of the lumber is then passed through a *planer,* which smooths the surfaces of the lumber and reduces it to its final, finished size or pattern. The lumber may then be dried further in a kiln or shipped as-is to wholesale or retail lumber yards for sale.

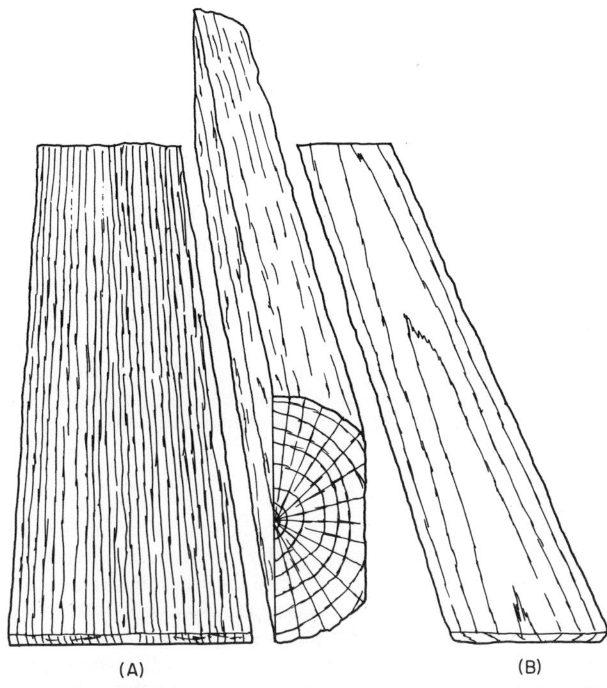

Fig. 1. (*A*) Rift cutting results in vertical-grained lumber. (*B*) Slash cutting results in flat-grained lumber (*U.S. Forest Service*).

Flat-Grained and Edge-Grained Lumber

The simplest method of reducing a log to lumber is by slash-cutting, since there is very little waste. As may be seen in Fig. 1, slash-cutting brings out the pattern of the grain, which is attractive in some species and much desired for paneling. As may also be seen, flat-grained pieces of lumber gradually merge into pieces of edge-grained lumber. The dividing line between flat-grained and edge-grained lumber is 45 degrees. All lumber in which the annual rings make an angle of 45 degrees or less with the width of the lumber is flat-grained lumber. All lumber in which this angle is greater than 45 degrees is edge-grained lumber.

Rift-cutting logs into edge-grained lumber is more complicated than slash-cutting logs into flat-grained lumber. A log must first be cut into halves or quarters, or in some more complicated way, before it is sawed into finished sizes. The advantages of edge-grained lumber over flat-grained lumber are: (1) the lumber is dimensionally more stable; (2) edge-grained lumber swells and shrinks less with changes in humidity; and (3) it tends to warp, twist and check less as it is being seasoned. Edge-grained lumber is also longer-wearing than flat-grained lumber and is for this reason preferred for flooring.

LUMBER DEFECTS

The most casual look at a piece of finished lumber immediately reveals the obvious—lumber is not a homogenous material like steel or concrete. Most pieces of lumber contain defects of one kind or another that will affect its strength, durability, or utility. Very, very few pieces of lumber are clear and without a defect of some kind. Such lumber commands a high price and is used where the appearance or natural finish is very important.

Most lumber does contain knots, splits, and other defects that affect its strength, durability, or utility more or less. After lumber has been sawed and planed to size, therefore, it must be *graded* to separate the excellent, very good, ordinary, and poor pieces from each other. The grading is accomplished by men experienced in this work, and it is limited to a visual inspection of each piece of lumber. Grading requires skill and judgment. Althougb no two graders will always place a particular piece of lumber in the same grade, the variation of opinion among graders is believed to amount to no more than 5 percent.

Knots

The most obvious kinds of defects are knots. They may also be the most serious defects in a piece of lumber. Knots are the remains of branches that have been surrounded completely by the trunk as the trunk has increased in girth during the life of the tree. Except in species such as Douglas fir and red cedar, which continue to grow from their tops, all the main branches of most trees originate in the sapling and knots are present, therefore, throughout the trunk, increasing in size as they approach the outer diameter of the trunk.

As long as a branch remains alive, the annual rings of the trunk will grow outward with the branch. Knots of this type are called *intergrown* knots (see Fig. 2). They may be recognized in a piece of lumber by the large amount of grain distortion around the knot. The grain resembles the widely spaced contour lines on a map that indicate a mountain with gentle slopes.

If the branch should die, which happens very often in growing softwood trees, the annual rings of the trunk will no longer extend outward along the branch. The grain surrounds the knot more tightly, and there will be less distortion of the grain. The grain pattern will more nearly resemble the closely spaced contour lines on a map that indicate a cliff or a steep slope. This kind of knot is called an *encased* knot (see Fig. 2). An encased knot tends to be more loosely held by the surrounding wood than an intergrown knot, and the knot may also be decayed and fall out.

A knot of any kind, intergrown or encased, reduces the strength of the lumber because its presence distorts the direction of the wood fibers; it is these fibers that give wood what strength and stiffness it possesses. A piece of wood is obviously strongest and stiffest when its fibers all run lengthwise in the lumber, parallel to each other, and without distortion.

How serious a defect a knot may be depends mainly on its size and its location in the piece of lumber. A small, round knot less than ¼ in. in diam. (i.e., a *pin* knot) is obviously less of a defect than a knot that is 1½ in. in diam. The seriousness of the defect also depends on the relative size of the knot in a

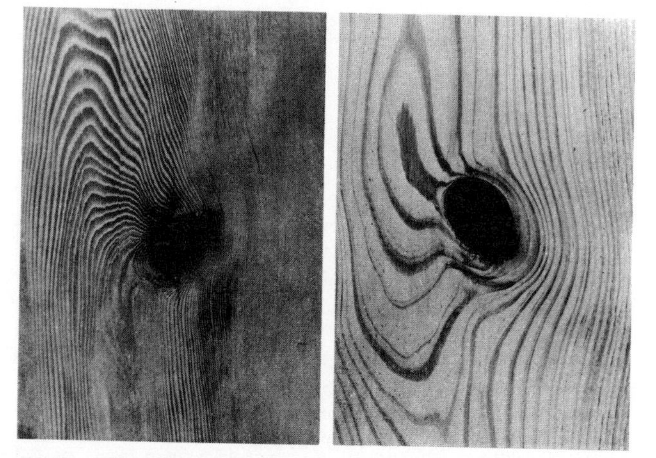

Fig. 2. (Left) An encased knot; (right) an intergrown knot (*U.S. Forest Service*).

given piece of lumber, a 1-in. knot in a board 12 in. wide being less of a defect, proportionately, than a ½-in. knot in a 2 x 4.

The seriousness of the defect also depends on the use to which the piece of lumber is to be put. Knots have almost no effect at all on the compressive strength of a piece of lumber such as a stud; they have more of an effect on the stiffness and bending strength of a joist or rafter.

Whether or not any particular knot or group of knots is acceptable in a piece of lumber that is to be used as a joist or rafter is a complicated problem. It depends not only on the number and size of the knots themselves, but also on their location in the lumber. Knots above a certain size are not permitted along the top and bottom edges where the compressive and tensile stresses will be greatest, nor are knots above a certain size acceptable in the middle third of a length of lumber because the bending moments are greatest here.

Knots are a much less reprehensible defect in boards intended as sheathing or underflooring, since the main function of these boards is that of a covering that ties the structural members together.

Cross Grain

When the direction of the wood fibers is not parallel with the edge of the piece, the lumber is said to be *cross-grained*. Cross-graining occurs when the wood fibers in the trunk happen to grow with a twist, or spiral, around the central axis of the trunk. When the wood from such a tree is cut into lumber, if the lumber is not properly seasoned, it will tend to twist, which makes the lumber difficult to fasten securely to another piece of lumber or to bear solidly against another piece of lumber.

The true direction of the wood fibers can be discovered by splitting a sample of the wood across the annual rings and observing the direction in which the fibers lie. Or one can look for checks or pitch pockets (which are described below) and note whether or not these defects run parallel to the edge of the piece because they do tend to follow the direction of the fibers.

Akin to spiral-grained wood is wood in which the grain runs diagonally the length of the piece of lumber, not parallel, as it should. This kind of defect results in lumber that is not as strong as lumber in which the grain does run straight.

Shake

A shake is the separation of the wood fibers. It occurs *between* the annual rings (its distinguishing characteristic) and runs parallel with them. A shake can run for a considerable distance in the wood. The cause is unknown. They are found in freshly felled logs, and the most probable cause is believed to be the stresses that result when high winds bend the trunk of a growing tree. Shakes have little effect upon any compressive loads that may be supported by the lumber, and they are, therefore, usually acceptable in lumber intended for posts and studs. In joists and other beams, however, the shakes reduce the wood's resistance to shearing stresses. Therefore, they are unacceptable when they are larger than a certain size.

Check

A check is also a separation of the wood fibers that runs parallel to the annual rings, but it is a separation that occurs *across* the annual rings. Checks are the result of stresses set up in the lumber as the lumber is being seasoned. As with shakes, checks have little effect on the strength of a piece of lumber that is carrying a compressive load, but, as with shakes, checks reduce the wood's ability to resist shearing stresses when the lumber is used as a beam. For a fuller description, *see* WOOD.

Wane

A wane is the absence of wood along an edge of a piece of lumber. A wane occurs when there is not enough wood in the log for the required size of the lumber. Wane may also consist of a strip of bark that runs along one edge of a piece of lumber. Since wane means the cross section of the lumber has been reduced, the piece of lumber is weaker than it should be. As a result, wane is a defect that limits the usefulness of the lumber for structural purposes.

Pitch Pocket

Pitch pockets are lengthwise separations of the wood fibers that have filled up with solid or liquid resin. Though unsightly, pitch pockets have no or very little effect on the strength of the lumber and are, therefore, usually disregarded when the lumber is graded for strength, unless the pitch pockets are unusually numerous or large, which may be an indication of shake. In addition, pitch pockets are difficult to paint over or stain, and the lumber is thus unsuitable as siding.

Decay

Decay is another defect the full extent of which it is difficult to estimate by eye. Incipient decay is difficult to detect. Advanced decay is easily recognized because the wood turns soft and spongy, or crumbly—dead looking—and it offers no resistance when it is stabbed with the point of a knife.

There are two kinds of decay—decay that merely affects the appearance or color of the wood and decay that destroys the substance of the wood. Both forms are caused by fungi that live off the wood. Staining caused by fungi, the so-called "blue stain," for example, is, by itself, merely a blemish that does not in any way affect the strength of the wood. But conditions that are suitable for the growth of the staining fungi are equally suitable for the growth of the wood-destroying fungi. For this reason, evidence of any kind of decay is considered sufficient cause for lowering the grade of lumber that is to be used structurally.

Warp

Warp is a defect that results from the uneven shrinkage of wood during its seasoning. Some warpage is inevitable—all wood warps as it shrinks (*see* WOOD)—but excessive warpage is a defect caused by improper seasoning. Any lumber that has warped an excessive amount is rejected or downgraded because it is difficult to attach the lumber to other pieces of lumber, nor can the lumber be made to bear fully on other pieces of lumber.

Manufacturing Defects

Manufacturing defects are defects caused by improper handling at the mill: the grain may be chipped, loosened, raised, or torn as the wood passes through the planer. The planer may skip spots on the wood, leaving a roughened surface, or the planer may leave gouge or burn marks. Saw marks may be left in the wood which are visible even after planing, or the wood may be incorrectly trimmed to size.

Other possible manufacturing defects are that the tongues and grooves of *matched* boards (see below) are not aligned with each other or that the grooves may not be cut to their full depth. These and many other imperfections may occur. Since they all restrict or limit the use of the lumber to some extent, they are all classified as defects or blemishes, and the lumber is downgraded accordingly.

THE GRADING OF LUMBER

Grading is a system of classifying lumber according to certain rules agreed upon jointly by the lumber industry and the U.S. Department of Commerce. The lumber industry is divided into associations according to the location and/or species of lumber each association manufactures. For softwoods the most important associations are by far the Southern Forest Products Association (yellow pine), the West Coast Lumber Association (Douglas fir, western hemlock, western red cedar, white fir, and Sitka spruce), and the Western Wood Products Association (Douglas fir south, western larch, Engelmann spruce, western hemlock, white pine, and ponderosa pine). They, and all other lumber associations, have established *inspection bureaus,* the personnel of which are responsible for inspecting and grading the lumber cut by members of each association (and others) before the lumber is shipped from the mills.

Each inspection bureau publishes *grading rules* that describe in great detail the sizes, kinds, and grades of lumber manufactured by the members of its association. Most important, these grading rules describe in detail the criteria by which the grade of each piece of lumber is determined, based on the visible defects or blemishes the lumber contains. Once it has been graded, each piece of lumber is stamped (see Fig. 3) to indicate the inspecting authority, the grade, species, and the identity of the lumber mill.

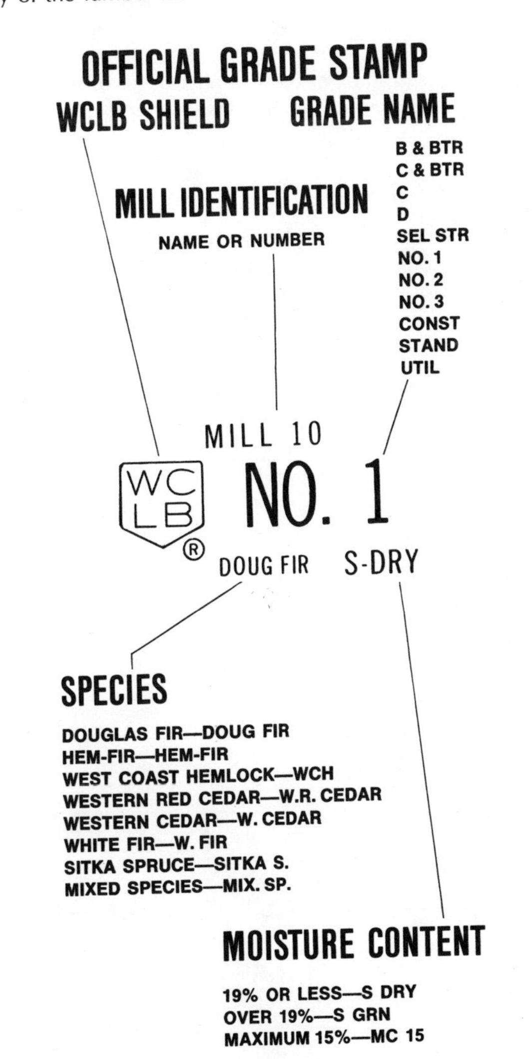

Fig. 3. The grade stamp used by the WCLIB for Douglas fir *(West Coast Lumber Inspection Bureau).*

As far as the buyer is concerned, these grading stamps assure him that the lumber he is buying is satisfactory for the purposes he intends using the lumber for. The grading stamps also guarantee him (to a large extent) that he is getting what he is paying for. Lumber is one commodity for which the prices are determined very much by supply and demand, and a buyer who shops around will find that within each grade prices vary up and down in response to market conditions.

This preamble over, lumber can be classified in any of three different ways: (1) by intended use (the most important classification), (2) by method of manufacture, and (3) by size. We will take the last two methods of classification first.

Grading by Method of Manufacture

As far as grading by method of manufacture is concerned, lumber is sold either as rough lumber, dressed lumber, or worked lumber.

Rough Lumber

Rough lumber is lumber as it comes from the saw; it is rarely used in home construction.

Dressed Lumber

Dressed lumber is lumber that has passed through the planer and has been smoothed to a uniformly sized piece of known dimensions. Any combination of sides and edges of the lumber can be dressed, though most lumber used for home construction is dressed on all four sides and referred to as surfaced 4 sides (S4S).

Worked Lumber

Worked lumber, in addition to being dressed, has its surface planed to some particular shape. *Matched* lumber, for example, has a tongue cut along one edge and a groove cut along the opposite edge, as shown in Fig. 4. When two similarly matched boards are placed edge-to-edge so that tongue and groove mate, the tongue and groove form a close-fitting joint. Flooring, for example, is almost always matched tongue-and-groove; siding very often is. *Shiplapped* lumber has a *rabbet* worked into each edge of the board, as shown in Fig. 4. The rabbets of two mating boards form an overlapping, weather-tight joint. Siding, for example, is frequently shiplapped. *Patterned* lumber is lumber that has a more or less elaborate curved surface planed along one edge or side. Moldings, for example, also shown in Fig. 4, are examples of patterned lumber.

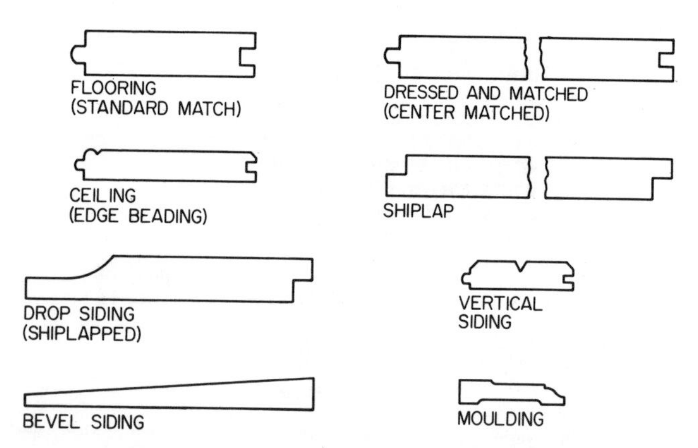

Fig. 4. Typical patterns of worked lumber *(U.S. Forest Service).*

Grading According to Size

As far as grading by size is concerned, there are two ways of describing the size of a piece of lumber—its dimensions may be either nominal or actual (refer to Table 1).

Nominal dimensions are the dimensions a piece of lumber has after it has been rough cut but before it has been planed.

In dressed lumber, however, the nominal dimensions are reduced by the amount of wood that is removed from the surface as it passes through the planer. What remains is the *actual* dimensions of the lumber. A dressed 2 × 4 is not, for example, actually 2 × 4 in. in size. It is actually 1½ × 3½ in. in size, after it has been seasoned. The same thing is true of all the other standard lumber sizes that are, by custom, referred to by their nominal dimensions.

Many people feel they are being cheated when they discover a difference exists between the nominal and dressed size of a piece of lumber. These people should understand that lumber of a given size is sold on the basis of the job it is expected to do. Whether a stud is 1½ × 3½ in. or 2 × 4 in. doesn't really matter as long as it does the job. Once this fact has been accepted, it will be found that the actual dimensions of dressed lumber are adhered to very closely. If, for example, one orders a piece of seasoned 2 × 4-in. lumber, dressed S4S, one can be quite certain that one is going to receive a piece of lumber that measures 1½ by 3½ in. to within very close tolerances.

As for the lengths of lumber, all lengths are actual and to the even foot—6, 8, 10, 12, and so on.

Grading According to Use

As far as use is concerned, lumber is divided into three categories: structural timber, factory and shop lumber, and yard lumber (see Fig. 5).

Structural Timber

Structural timbers are lumber that is 5 in. or more in both thickness and width. It is used for such things as railroad trestles, mine timbers, ship decking, and the like. In the house, structural timbers are used occasionally as posts and girders.

Factory and Shop Lumber

The builder can ignore *factory and shop lumber,* which is lumber having too many defects in it for it to be used structurally in home construction. Factory and shop lumber is not graded according to the overall quality of the entire piece of lumber but according to the number and sizes of clear cuttings that can be taken from it. This lumber is sold to manufacturers of doors, window sash, cabinets, and similar finish products, who will find enough clear pieces of wood in the lumber to manufacture their products.

Yard Lumber

Yard lumber is so-called because it is the lumber that is usually found in your neighborhood lumber yard. It includes all lumber less than 5 in. thick that is too high in quality to be cut up as factory or shop lumber.

The situation regarding the classification and grading of yard lumber is complicated. First, the lumber is divided into two major groups—*finish* lumber and *dimension* lumber—depending on the thickness and intended use of the lumber. Finish lumber (which also includes ordinary *boards;* see below) is lumber that is (1) nominally 1 in. thick and (2) used either decoratively or to enclose the underlying framework of a house. Thus, finish lumber includes lumber categorized as *casing* and *base, paneling, flooring, siding, ceiling* and *parti-*

Table 1. Differences between the Nominal and Actual Dimensions of Dressed Lumber*
(The thicknesses apply to all widths and all widths to all thicknesses.)

	Thicknesses, in.			Face widths, in.		
		Minimum dressed			Minimum dressed	
Item	Nominal	Dry†	Green†	Nominal	Dry†	Green†
Boards	1	¾	25/32	2	1½	1 9/16
	1¼	1	1 1/32	3	2½	2 9/16
	1½	1¼	1 9/32	4	3½	3 9/16
				5	4½	4⅝
				6	5½	5⅝
				7	6½	6⅝
				8	7¼	7½
				9	8¼	8½
				10	9¼	9½
				11	10¼	10½
				12	11¼	11½
				14	13¼	13½
				16	15¼	15½
Dimension	2	1½	1 9/16	2	1½	1 9/16
	2½	2	2 1/16	3	2½	2 9/16
	3	2½	2 9/16	4	3½	3 9/16
	3½	3	3 1/16	5	4½	4⅝
				6	5½	5⅝
				8	7¼	7½
				10	9¼	9½
				12	11¼	11½
				14	13¼	13½
				16	15¼	15½
Dimension	4	3½	3 9/16	2	1½	1 9/16
	4½	4	4 1/16	3	2½	2 9/16
				4	3½	3 9/16
				5	4½	4⅝
				6	5½	5⅝
				8	7¼	7½
				10	9¼	9½
				12	11¼	11½
				14		13½
				16		15½
Timbers	5 and thicker		½ off	5 and wider		½ off

*Two-in.-thick dimension lumber, for example, is actually 1½ x 3½ in. in size when dry.
†Dry lumber is defined as lumber which has been seasoned to a moisture content of 19 percent or less. Green lumber is defined as lumber having a moisture content in excess of 19 percent.
Source: National Forest Products Association.

tion, stepping, and so on, as shown in Fig. 5. *Boards* include all nominally 1-in.-thick lumber that is used for sheathing, roofing, subflooring, and concrete formwork.

Dimension lumber includes all lumber that is (1) nominally at least 2 in. thick and (2) used structurally to build the frame of the house; that is, it consists of framing lumber, such as studs, rafters, joists, and decking.

Grading of Finish Lumber

Finish lumber is in general graded as either *Select* or *Common.* Select grades are the lettered grades—A, B, C, D, and E—and the Common grades are the numbered grades—Nos. 1, 2, 3, 4, and 5. Select grades are used where the appearance of the wood is of some importance; Common grades are used where the appearance of the wood is unimportant.

Grade A lumber, sometimes called *Clear* lumber, is rarely found in a lumber yard. Most lumber associations do not even manufacture it except on special order. It is lumber having only the most superficial blemishes and is intended for use as moldings, paneling, and flooring in those relatively few homes that are built to the highest standards of quality. Grade A lumber is meant to be left with a natural finish.

Grade B lumber has a few small, barely noticeable imperfections. It, too, is intended to be left with a natural finish. This grade is usually combined with Grade A lumber to make up a grade called *B & Better.* B & Better is used most often for the trim, paneling, and cabinet work of high-quality homes. Most softwood flooring is also made from Grade B & Better lumber.

Grade C lumber has more and larger imperfections than Grade B lumber, but the defects are still trivial enough to be completely hidden by a coat of paint. Grade C, therefore, is the best grade to use when the wood is to be painted. It is used chiefly for cornices, exterior trim, siding, porches, and painted kitchen cabinets in high-quality homes. Grade C is often combined with higher grades of finish lumber to form a grade called C & Better.

Grade D lumber has more and larger imperfections than Grade C has, but it is still possible to hide these defects under a coat of paint. The grade is used for the same purposes as Grade C lumber but in less expensively built homes.

Grading of Common Lumber

The Common grades are used mainly for boards in which the appearance of the wood is unimportant, such as the boards used for sheathing, subflooring, and similar purposes, because they will be covered up when the house is completed.

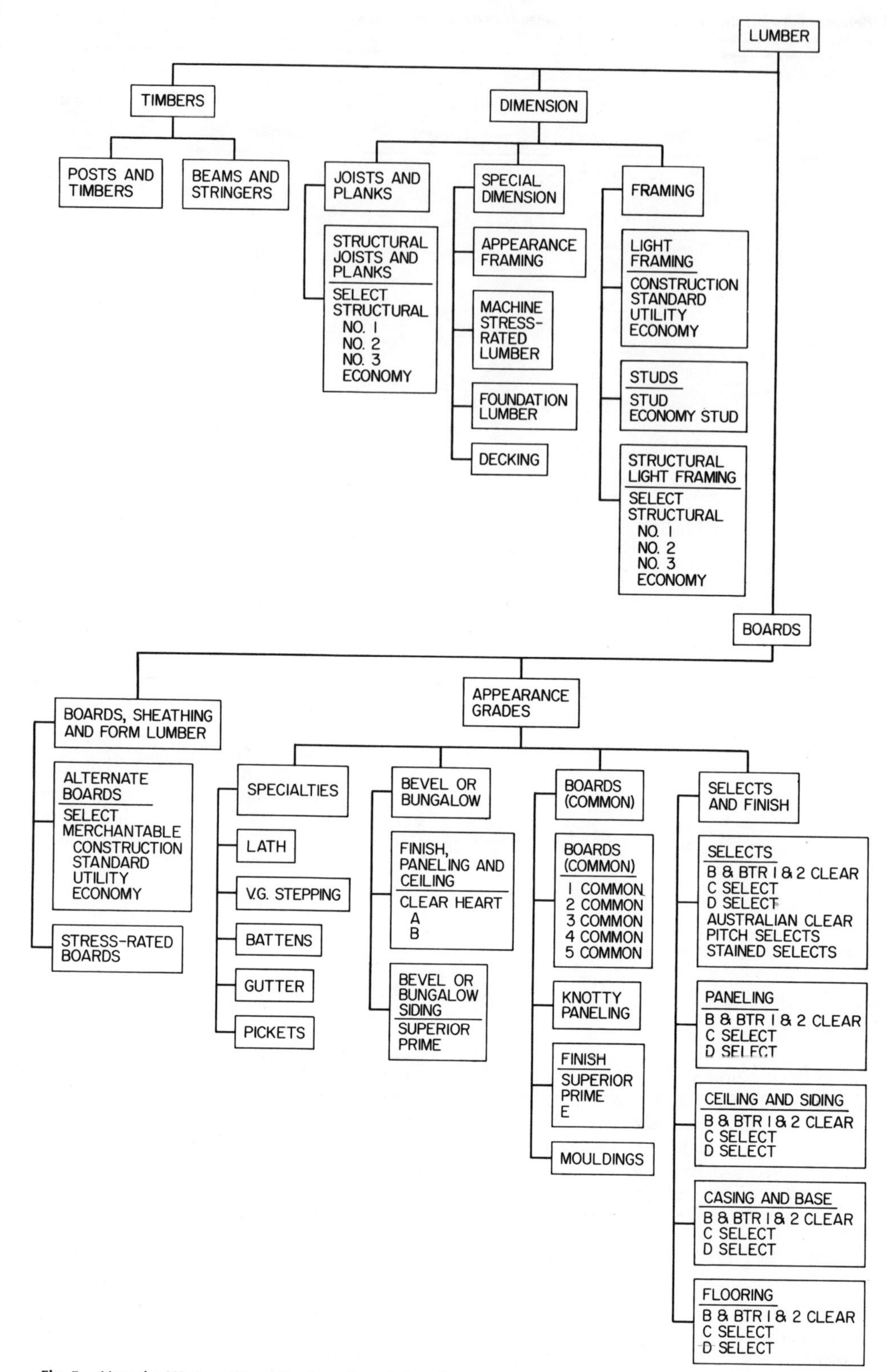

Fig. 5. How the Western Wood Products Association classifies Douglas fir lumber manufactured by its members according to use, size, and grade *(Western Wood Products Association).*

No. 1 boards (see Fig. 6) are used as sheathing and subflooring in expensively built homes and as siding, cornices, shelving, and paneling in medium- and low-cost homes. The knots in No. 1 lumber are required to be sound and watertight.

Nos. 2 and *3* boards serve the same structural purposes as No. 1 boards, but they are used in houses of increasingly cheaper construction.

No. 4 boards are used for temporary bracing, or they are purchased to be cut up for the short, usable pieces that are always required when a house is being built; for blocking, for example.

As for *No. 5* boards, the best that can be said of them is that somewhere, in some way, they are of some possible use to someone.

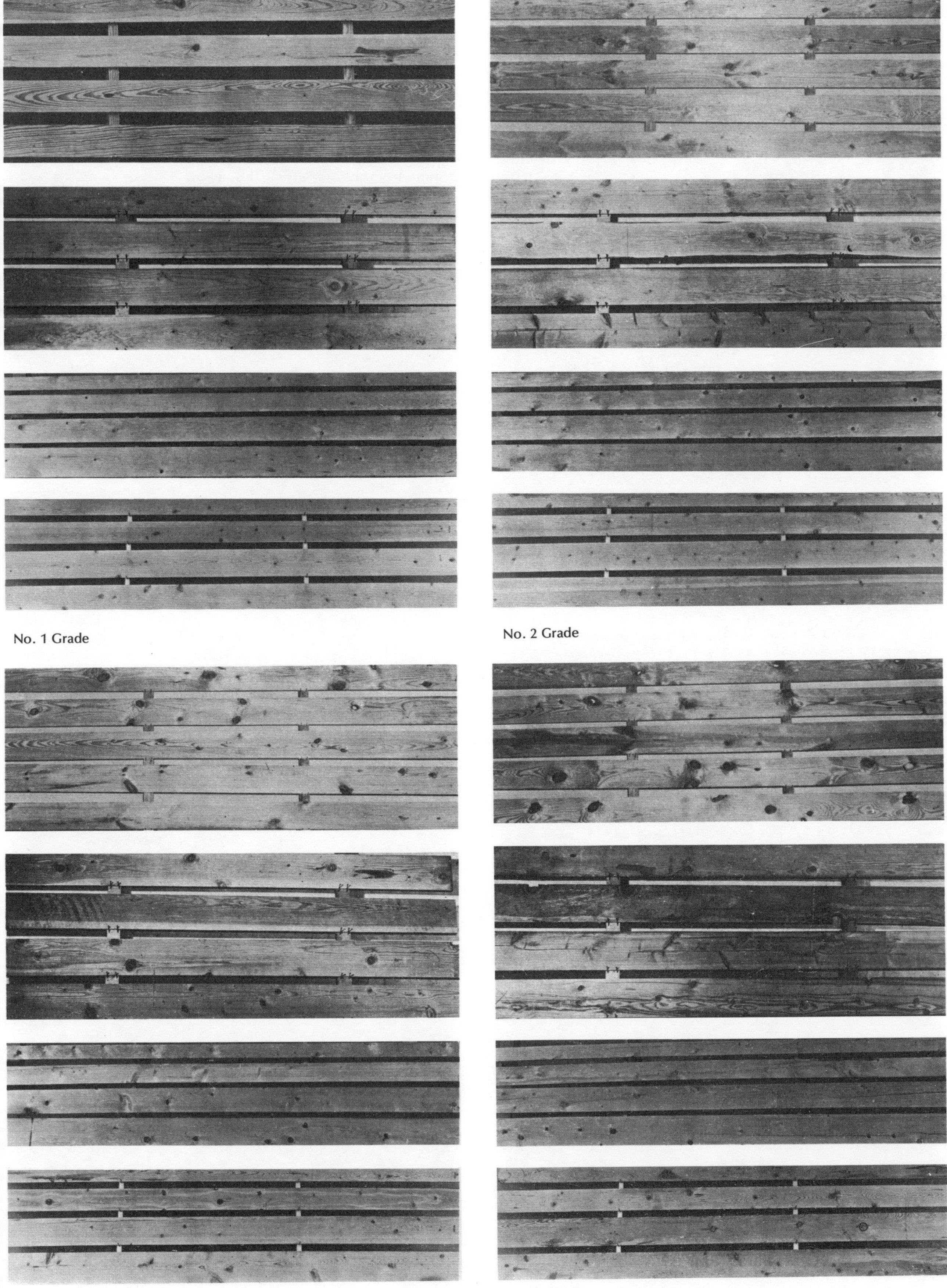

Fig. 6. The typical appearance of Common lumber graded as No. 1, No. 2, No. 3 and No. 4 *(U.S. Forest Service).*

Differences in Grading Systems

The grading of finish and common lumber is complicated by the fact that the different lumber associations use different grading systems to grade the same quality lumber. As may be seen in the following table, for example, the Southern Forest Products Association (SFPA) offers a greater variety of Select grades than does the West Coast Lumber Association (WCLA), and their systems of classifying Common lumber are different:

SFPA	*WCLA*
B & Better	....
C	....
C & Better	C & Better
D	D
....	E
No. 1	Select Merchantable
No. 2	Construction
No. 3	Standard
No. 4	Utility
....	Economy

One should never try to compare finish with dimension lumber. The only thing they have in common is that they are both considered yard lumber.

Softwood Lumber Standard PS 20–70

In 1970, the U.S. Department of Commerce, together with the softwood lumber associations, published a new, simplified guide to lumber grading rules, *Softwood Lumber Standard* 20–70 (the *70* is the year of publication), that replaced a standard that had been in effect since the 1920s. PS 20–70 applies *only* to dimension lumber, which includes, as we have mentioned, the lumber that is used to construct the structural framework of a house; that is, where strength and stiffness are the important considerations.

The standards in PS 20–70 do several things. They standardize the *actual* dressed lumber sizes at new and smaller dimensions (which reduces the finished widths and thicknesses of *seasoned* dressed dimension lumber by about ⅛ in. under their former dressed sizes; *green* lumber, which is expected to continue shrinking after it has been dressed, is 1/16-in. larger than seasoned dressed dimension lumber); they standardize the moisture content of "dry" lumber at 19 percent or less; they standardize the moisture content of *kiln-dried* lumber at 15 percent or less; they classify lumber according to intended use, that is, as *framing lumber, joists, planks, studs,* and *appearance* framing lumber; and, last, they relate the lumber grading system to strength tests that have been conducted through the years by the Forest Products Laboratory of the U.S. Department of Agriculture.

Stress-graded Lumber

In these tests, the Forest Products Laboratory has bent, crushed, and twisted representative samples of every species of wood that grows in the United States in order to derive the basic structural properties of these woods. In the new grading system, each piece of dimension lumber is still graded according to its visible defects and blemishes, but the new grading system now relates these grades to the basic strength properties of clear samples of the wood, specifically to the ability of the wood to resist bending stresses (see Table 2).

Table 2. Allowable Unit Stresses for Douglas Fir and Southern Pine
(Allowable unit stresses listed are for normal loading conditions. The lumber is graded visually.)

		Allowable unit stresses, lb per sq in.							
		Extreme fiber in bending F_b							
Commercial grade	Size classification, in.	Single-member uses	Repetitive-member uses	Tension parallel to grain F_t	Horizontal shear F_v	Compression perpendicular to grain F_{ct}	Compression parallel to grain F_c	Modulus of elasticity E	Grading rules agency
		Douglas Fir-Larch (Surfaced dry or surfaced green. Used at 19% max. m.c.)							
Dense select structural	2 to 4 thick 2 to 4 wide	2450	2800	1400	95	455	1850	1,900,000	West Coast Lumber Inspection Bureau and Western Wood Products Association
Select structural		2100	2400	1200	95	385	1600	1,800,000	
Dense No. 1		2050	2400	1200	95	455	1405	1,900,000	
No. 1		1750	2050	1050	95	385	1250	1,800,000	
Dense No. 2		1700	1950	1000	95	455	1150	1,700,000	
No. 2		1450	1650	850	95	385	1000	1,700,000	
No. 3		800	925	475	95	385	600	1,500,000	
Construction		1050	1200	625	95	385	1150	1,500,000	
Standard		600	675	350	95	385	925	1,500,000	
Utility		275	325	175	95	385	600	1,500,000	
Stud		800	925	475	95	385	600	1,500,000	
Dense select structural	2 to 4 thick 6 and wider	2100	2400	1400	95	455	1650	1,900,000	
Select structural		1800	2050	1200	95	385	1400	1,800,000	
Dense No. 1		1800	2050	1200	95	455	1450	1,900,000	
No. 1		1500	1750	1000	95	385	1250	1,800,000	
Dense No. 2		1450	1700	950	95	455	1250	1,700,000	
No. 2		1250	1450	825	95	385	1050	1,700,000	
No. 3		725	850	475	95	385	675	1,500,000	
Appearance	2 to 4 thick 2 to 4 wide	1750	2050	1050	95	385	1500	1,800,000	
Appearance	2 to 4 thick 6 and wider	1500	1750	1000	95	385	1500	1,800,000	
Dense select structural	Beams and Stringers	1900		1100	85	455	1300	1,700,000	West Coast Lumber Inspection Bureau
Select structural		1600		950	85	385	1100	1,600,000	
Dense No. 1		1550		775	85	455	1100	1,700,000	
No. 1		1300		675	85	385	925	1,600,000	
Dense select structural	Posts and Timbers	1750		1150	85	455	1400	1,700,000	
Select structural		1500		1000	85	385	1200	1,600,000	
Dense No. 1		1400		950	85	455	1200	1,700,000	
No. 1		1200		825	85	385	1000	1,600,000	

(continued)

Table 2. Allowable Unit Stresses for Douglas Fir and Southern Pine (continued)
Allowable unit stresses listed are for normal loading conditions. The lumber is graded visually.)

		Allowable unit stresses, lb per sq in.							
		Extreme fiber in bending F_b							
Commercial grade	Size classification, in.	Single-member uses	Repetitive-member uses	Tension parallel to grain F_t	Horizontal shear F_v	Compression perpendicular to grain $F_{c\perp}$	Compression parallel to grain F_c	Modulus of elasticity E	Grading rules agency
Select dex	Decking	1750	2000			385		1,800,000	
Commercial dex		1450	1650			385		1,800,000	
Dense select structural		1900		1250	85	455	1300	1,700,000	
Select structural	Beams and	1600		1050	85	385	1100	1,600,000	
Dense No. 1	stringers	1550		1050	85	455	1100	1,700,000	
No. 1		1350		900	85	385	925	1,600,000	
Dense select structural		1750		1150	85	455	1350	1,700,000	Western
Dense structural	Posts and	1500		1000	85	385	1150	1,600,000	Wood
Dense No. 1	timbers	1400		950	85	455	1200	1,700,000	Products
No. 1		1200		825	85	385	1000	1,600,000	Association
Selected decking	Decking		2000					1,800,000	
Commercial decking			1650					1,700,000	
Selected decking	Decking		2150	(Surfaced at 15% max. m.c. and				1,900,000	
Commercial decking			1800	used at 15% max. m.x.)				1,700,000	
		Southern Pine (Surfaced dry. Used at 19% max. m.c.)							
Select structural		2100	2400	1250	90	405	1600	1,800,000	
Dense select structural		2450	2800	1450	90	475	1850	1,900,000	
No. 1		1750	2000	1000	90	405	1250	1,800,000	
No. 1 Dense	2 to 4	2050	2350	1200	90	475	1450	1,900,000	
No. 2	thick	1250	1450	725	75	345	850	1,400,000	
No. 2 Medium grain	2 to 4	1450	1650	850	90	405	1000	1,600,000	
No. 2 Dense	wide	1700	1950	1000	90	475	1150	1,700,000	
No. 3		825	950	475	75	345	600	1,400,000	
No. 3 Dense		950	1100	550	90	475	700	1,500,000	
Construction		1050	1200	625	75	345	1150	1,400,000	
Standard		600	700	350	75	345	925	1,400,000	
Utility		275	325	175	75	345	600	1,400,000	
Stud		825	950	475	75	345	600	1,400,000	
Select structural		1800	2050	1200	90	405	1400	1,800,000	
Dense select structural		2100	2400	1400	90	475	1650	1,900,000	
No. 1		1500	1750	1000	90	405	1250	1,800,000	Southern
No. 1 Dense	2 to 4	1800	2050	1200	90	475	1450	1,900,000	Pine
No. 2	thick	1050	1200	700	75	345	700	1,400,000	Inspection
No. 2 Medium grain	6 and	1250	1450	825	90	405	1050	1,600,000	Bureau
No. 2 Dense	wider	1450	1650	975	90	475	1250	1,700,000	
No. 3		725	825	475	75	345	650	1,400,000	
No. 3 Dense		850	975	575	90	475	750	1,500,000	
Dense std. factory	2 to 4	2000	2300	1200	90	475	1450	1,900,000	
No. 1 factory	thick	1400	1600	825	90	405	1000	1,600,000	
No. 1 Dense factory	2 to 4	1650	1900	975	90	475	1150	1,700,000	
No. 2 factory	wide	1400	1600	825	90	405	1000	1,600,000	
No. 2 Dense factory		1650	1900	975	90	475	1150	1,700,000	
Dense std. factory	2 to 4	1750	2000	1200	90	475	1450	1,900,000	
No. 1 Factory	thick	1250	1450	825	90	405	1050	1,600,000	
No. 1 Dense factory	6 and	1450	1650	975	90	475	1250	1,700,000	
No. 2 Factory	wider	1250	1450	825	90	405	1050	1,600,000	
No. 2 Dense factory		1450	1650	975	90	475	1250	1,700,000	
Dense structural 86	2 to 4	2750	3150	1850	150	475	2050	1,900,000	
Dense structural 72	thick	2300	2650	1550	125	475	1700	1,900,000	

Source: National Forest Products Association.

Bending stresses have been selected as the grading criterion because the most important stresses that structural lumber must resist are the bending stresses imposed on a beam when the beam is supporting a load. The ability of a particular wood to resist bending stresses has, therefore, been taken as the basic criterion of the inherent strength properties of that wood.

Any actual sample of wood will, of course, contain defects of one kind or another. The tests it has conducted have enabled the Forest Products Laboratory to estimate how much the strength of any given species of wood is reduced by different types of defects. If, say, knots of a certain size and location reduce the ability of the wood to resist bending stresses by one-third, then an actual piece of lumber having knots of that size and in that location will be only 67 percent as strong as an idealized clear piece of that wood, which will have an assumed stress value of 100 percent. This reduction in bending strength is expressed as a ratio, in percent, of the difference between the strength of a clear sample of that wood compared with the reduced strength of a piece of that wood with defects.

For framing lumber, for example, the minimum bending loads the lumber can sustain, compared with the bending loads that could be sustained by a clear sample of that wood are as follows:

Grade	*Bending strength ratio, %*
Select Structural	67
No. 1	55
No. 2	45
No. 3	26
Construction	34
Standard	19
Utility	9

Once graded, each piece of lumber receives a stamp that identifies the inspection bureau, the grade of the lumber as indicated above, the mill the lumber comes from, the *use grade* (see below), and a number that indicates the strength of the lumber. This number is shown in Table 2 under the heading "Extreme Fiber in Bending."

Machine Stress-Rated Lumber

It is obvious that although there is a correlation between the visible defects in a piece of lumber and the amount of stress that piece of lumber can sustain, the correlation must of necessity be somewhat inexact when the lumber is graded visually. The lumber may in fact be either weaker or stronger than its stress rating will indicate. If it is weaker, the builder may unwittingly use it in such a way that the lumber will break. If it is stronger, its strength will be wasted.

For these reasons, some mills now have machines installed that can accurately and automatically stress-rate 2-in.-thick dimension lumber. After the lumber has been finished to size, it passes flatwise through a machine in which a standardized bending load is exerted against the wood. The resistance of the wood to this stress is converted by the machine into a numerical rating, and this rating is stamped on the wood. Table 3 shows machine stress ratings used by two lumber associations. The F_b is the extreme fiber in bending, and is an indication of the bending strength of the wood. The E is the modulus of elasticity, and it is an indication of the stiffness of the wood.

Dense Grades

In additional to all this, another lumber classification has been established for stress-rated lumber that is intended for use where the structural stresses are much greater than usual. Lumber intended for use in these locations is classified as: Dense Select Structural, No. 1 Dense, No. 2 Dense, and No. 3 Dense.

And what makes a piece of lumber dense? Density is directly related to two qualities of any piece of wood—the number of annual rings and the amount of summerwood it has per inch.

It has been known for a long time that the denser a piece of lumber, the stronger it is. What makes a piece of wood dense in the first place is the proportion of summerwood in each annual ring, since the wood fibers of summerwood are denser

Table 3. Allowable Unit Stresses for Structural Lumber that is Machine Stress-Graded
(The allowable unit stresses listed are for normal loading conditions.)

		Allowable unit stresses in pounds per square inch									
		Extreme fiber in bending F_b§				Compression perpendicular to grain F_c (DRY)*					
Grade designation	Size classification, in.	Single-member uses	Repetitive-member uses	Tension parallel to grain F_t	Compression parallel to grain F_c	Douglas fir-larch	Hem-fir	Pine†	Engelmann spruce	Cedar‡	Modulus of elasticity E
					Western wood products association						
1200f-1.2E	Machine	1200	1400	600	950	385	245	240	195	295	1,200,000
1500f-1.4E	rated lum-	1500	1750	900	1200	385	245	240	195	295	1,400,000
1650f-1.5E	ber 2	1650	1900	1020	1320	385	245	240	195	295	1,500,000
1800f-1.6E	thick or	1800	2050	1175	1450	385	245	240	195	295	1,600,000
2100f-1.8E	less	2100	2400	1575	1700	385	245	240	195	295	1,800,000
2400f-2.0E	all widths	2400	2750	1925	1925	385	245	240	195	295	2,000,000
2700f-2.2E		2700	3100	2150	2150	385	245	240	195	295	2,200,000
3000f-2.4E		3000	3450	2400	2400	385	245	240	195	295	2,400,000
3300f-2.6E		3300	3800	2650	2650	385	245	240	195	295	2,600,000
900f-1.0E	Machine	900	1050	350	725	385	245	240	195	295	1,000,000
900f-1.2E	rated joists	900	1050	350	725	385	245	240	195	295	1,200,000
1200f-1.5E	2 thick	1200	1400	600	950	385	245	240	195	295	1,500,000
1350f-1.8E	or less	1350	1550	750	1075	385	245	240	195	295	1,800,000
1800f-2.1E	all widths	1800	2050	1175	1450	385	245	240	195	295	2,100,000
					West coast lumber inspection bureau						
900f-1.0E		900	1050	350	725	385	245				1,000,000
1200f-1.2E	Machine	1200	1400	600	950	385	245				1,200,000
1500f-1.4E	rated	1500	1750	900	1200	385	245				1,400,000
1650f-1.5E	lumber	1650	1900	1020	1320	385	245				1,500,000
1800f-1.6E	2 thick	1800	2050	1175	1450	385	245				1,600,000
2100f-1.8E	or less	2100	2400	1575	1700	385	245				1,800,000
2400f-2.0E	all widths	2400	2750	1925	1925	385	245				2,000,000
2700f-2.2E		2700	3100	2150	2150	385	245				2,200,000
900f-1.0E	Machine	900	1050	350	725	385	245				1,000,000
900f-1.2E	rated joists	900	1050	350	725	385	245				1,200,000
1200f-1.5E	2 thick	1200	1400	600	950	385	245				1,500,000
1500f-1.8E	or less	1500	1750	900	1200	385	245				1,800,000
1800f-2.1E	6 and wider	1800	2050	1175	1450	385	245				2,100,000

*Allowable unit stresses for horizontal shear "F_v" (dry) for all grade designations are as follows:

Douglas fir-larch	Hem-fir	Pine	Englemann spruce	Cedar
95	75	65	70	75

†Pine includes Idaho White, Lodgepole, Ponderosa, or Sugar Pine.
‡Cedar includes Incense or Western Red Cedar.

§Tabulated extreme fiber in bending values "F_b" are applicable to lumber loaded on edge. When loaded flatwise, these values should be multiplied by the following factors:

Nominal width (in.)	4	6	8	10	12	14
Factor	1.10	1.15	1.19	1.22	1.25	1.28

Source: National Forest Products Association.

and have thicker walls than the wood fibers of springwood. It follows that the greater the number of annual rings per inch and the higher the proportion of summerwood in each ring, the denser the wood and, therefore, the greater its strength.

These dense grades apply mainly to yellow pine and Douglas fir, these being the two species of wood that are most used as structural lumber. To be considered dense, a piece of yellow pine or Douglas fir lumber must average at least six annual rings per inch in a representative cross section of the lumber and, in addition, one-third or more of the wood must consist of summerwood. Or, if the number of annual rings is between five and six per inch, one-half of the wood must consist of summerwood. When these conditions are met, the lumber is considered dense and the following minimum bending strength ratios will apply:

Grade	*Bending strength ratio, %*
Dense Select Structural	86
No. 2 Dense	72
No. 2 Dense	65
No. 3 Dense	55

To bring this complicated subject to a conclusion, there is one more grading classification called *Medium Grade* that is between the ordinary stress-grade rating and a dense-grade rating. Medium-grade lumber is lumber that has at least four annual rings per inch in a representative cross section of the wood and that otherwise complies with the grading rules for No. 2 lumber.

Use Grades

One last word. In looking over Table 2, you will note there are two use classifications, *Stud* and *Appearance*. The Stud classification has been established primarily for 2 x 4 in. lumber that is to be used for studs. In a stud, the structural loads borne by the lumber are almost entirely compressive and parallel to the grain of the wood. Since knots have almost no effect on the compressive loads sustained by a stud, the Stud classification enables lumber with an excessive number of knots to be used where their presence would otherwise be unacceptable. Stud lumber has a bending strength ratio of only 26 percent.

As for the Appearance classification, it is intended for lumber that is to be used for exposed beams and planks, in houses that are built according to plank-and-beam principles of house construction (*see* PLANK-AND-BEAM CONSTRUCTION). Appearance lumber is graded to have the strength characteristics of No. 1 lumber and the appearance grade of Select grade C & Better lumber.

Modular Construction

Modular construction is a method of designing and building a house in which the component parts are sized according to some agreed-upon basic unit of length, or *module*. In a house designed and built according to modular principles, all the lumber, masonry materials, and prefabricated factory-built assemblies, such as doors and windows, are manufactured to some multiple or submultiple of 4 in., 4 in. being the basic modular unit of length in house construction. When the house is put together, all the parts will then fit together with a minimum amount of cutting and reworking being required on the job.

The advantages of having modular-sized components should certainly be obvious enough. Manufacturers need make their products in only a relatively few standard lengths and widths, bricks and other masonry products need be molded in a relatively few standard sizes, lumber need be milled in a relatively few standard sizes and lengths, and the same thing is true of the thousand other items that go into a house. Manufacturers would save on their manufacturing, inventory, and storage costs, architects would save on design and drafting costs, builders would save on estimating and labor costs, and it is hoped, homeowners would save on the price in buying their homes.

However obvious the idea of modular construction may be, that the idea is coming to fruition at all is the result of a great many years' effort on the part of the American Institute of Architects, the American National Standards Institute, the U.S. Department of Commerce, and various building industry groups that, together, formed the Modular Building Standards Association to represent their joint interests.

That houses be built to some modular standard was first proposed in 1936. The difficulty in achieving any kind of industrywide agreement on a standard lay in the fact that house construction is an extremely tradition-minded craft industry in which things are a certain size because they have always been that size (for example, the spacing between studs, joists, and rafters is usually 16 in. on centers), and because manufacturers have an enormous investment in equipment designed to manufacture and assemble components based on these traditional sizes. Before any industrywide agreement could be reached, therefore, it was necessary that the standard selected satisfy the requirements of manufacturers, builders, and craftsmen and result in a minimum of change and expense.

The agreed-upon module is 4 in. Ideally, all the components that make up a house should be sized to some multiple or submultiple of 4 in. Practically, this is impossible, given the enormous range of materials used in house construction. In practice, therefore, only the basic framework of a modular house is built to the modular measure, and everything that is

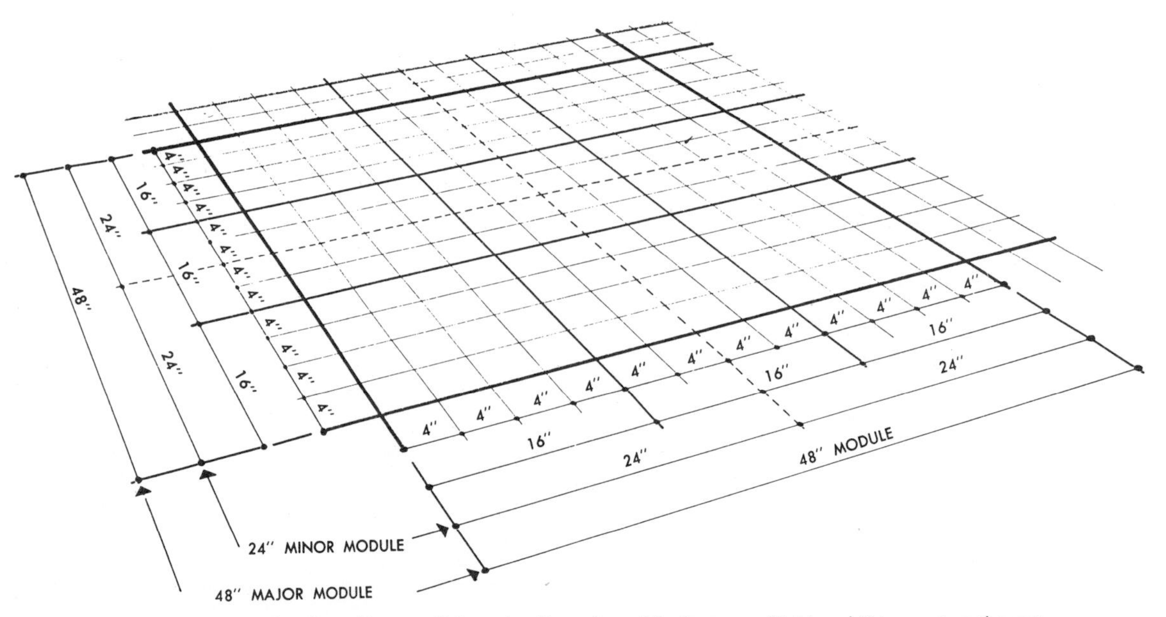

Fig. 1. In a modular-planned house, all the major dimensions of the house are 16, 24, and 48 in. apart, or they are multiples of these dimensions *(National Forest Products Association)*.

attached to this basic framework—finishes, roofing materials, flooring, and so on, is manufactured in the most convenient sizes.

A modular house must be well-planned beforehand. The overall dimensions of the house, including the locations of the room partitions, are laid out on a grid on which the grid lines represent a 4-in. spacing or some multiple of 4-in. spacing, depending on the scale of the drawing.

In a wood-frame house, it is sensible to plan the major dimensions in multiples of 16, 24, and 48 in., as shown in Fig. 1. The studs and joists are then centered on these grid lines, which allows all the other materials that are to be installed in the house to be attached to the framing in the most expeditious manner. Framing lumber, for example, is manufactured in 6-, 8-, 10-, 12-, and 14-ft lengths, and so on, and this lumber can be assembled so that the spacing of adjacent pieces is 16 or 24 in. on centers. In addition, the most widely used wall panels are 4 × 8 ft in size.

On building plans that are prepared according to modular conventions, only those dimensions that are modular have arrowheads. These are the dimensions that begin and end on a grid line. All the other dimensions shown on the plans will begin and end in small dots. The arrowheads and dots immediately indicate which dimensions are modular and which are not. By glancing over the plans, one can quickly see just how modular the construction is and thus be able to gauge in a rough sort of way just how much cutting and fitting will be necessary.

Wood-Frame Modular Construction

The National Forest Products Association has developed what it calls the Unicom method of house construction that seeks to fit traditional lumber sizes and building practices into the 4-in. module. The object is to maximize the use of modular components and minimize the cost of construction.

Although in the Unicom method the basic modular dimentions remain 4 in., in practice house plans are drawn on the basis of 24- and 48-in. modules, as shown in Fig. 1; and, though the framing is nominally spaced 16 or 24 in. on centers, in practice the framing consists not of individual studs or joists or rafters but studs, joists, and rafters assembled into large, modular-sized components. The plans are so drawn that, for example, the wall sections consist of complete units that are either 32, 48, 64, 96, or 144 in. wide, complete with doors and windows, which are also sized on a modular basis. This is shown in Fig. 2. The object has been to develop large, standardized "building blocks" that can be manufactured as units in a shop or on the job and then installed with a minimum of cutting and fitting.

In the Unicom method, the first-floor room heights have been planned on the basis of an 8 ft 1½ in. distance between the floor and ceiling framing. The fact that the wall framing is 8 ft 1½ in. high enables standard 4 × 8 ft wall panels to be installed, with space left over in which one can install the finish flooring and ceiling. Second-floor room heights have been planned either on the basis of the 8 ft 1½ in. wall height or, for 1½-story houses, on the basis of a 7 ft 7½ in. wall height.

All this may seem unduly restrictive from the architect's point of view, but a considerable variety in room layout and door and window placement is possible, and the interior and exterior finishes are, as always, left completely to the architect's or builder's taste and imagination. It is doubtful that one person in 10,000 would notice anything unusual in a Unicom-designed house.

The Unicom method has been thought out as a complete system that takes into account 1-story, 2-story, 1½-story, split-level, and bilevel houses, which may have either gable, hip, or flat roofs constructed with conventional rafters or with a system of framed trusses, and with or without overhangs. About the only major restriction on the use of the Unicom method—if it is a restriction—is the fact that one must design the house on the basis of 24-in. and 48-in. modules. And, in fact, if one wishes to depart from the modular dimensions in some respect, one is always free to do so.

Masonry Modular Construction

Modular masonry construction is also based on the 4-in. module, although the individual bricks, tiles, and concrete blocks

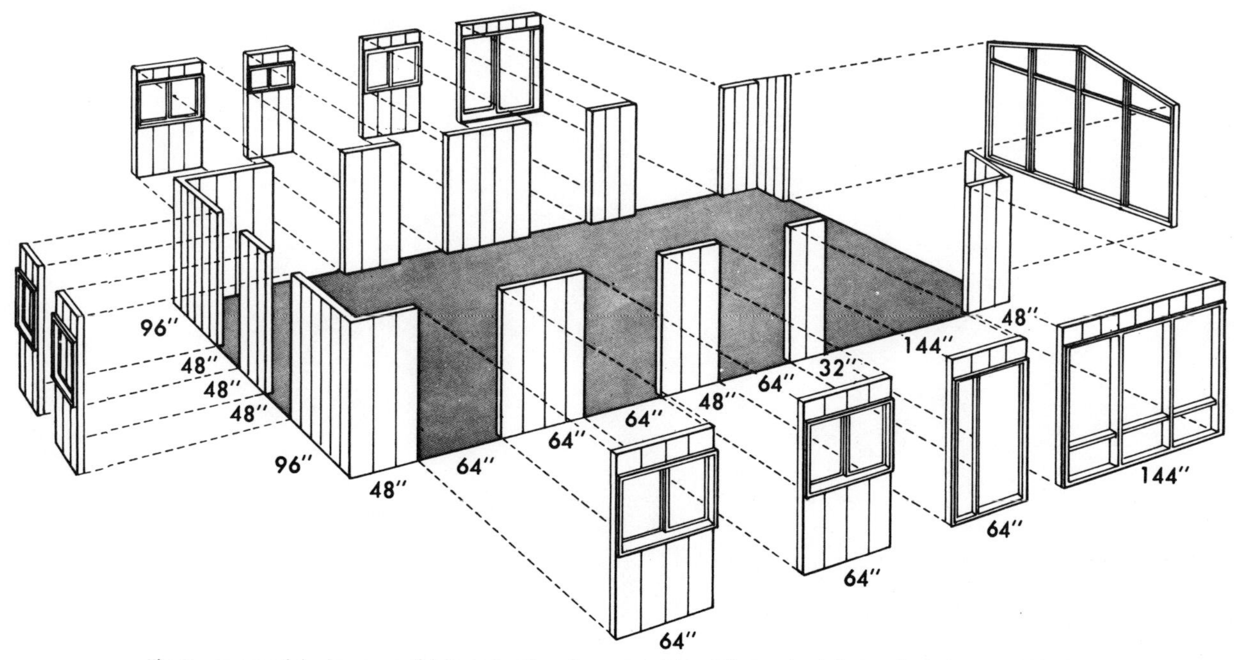

Fig. 2. In a modular house, prefabricated wall sections containing different sized door and windows, and blank wall sections, can be assembled into a complete wall *(National Forest Products Association).*

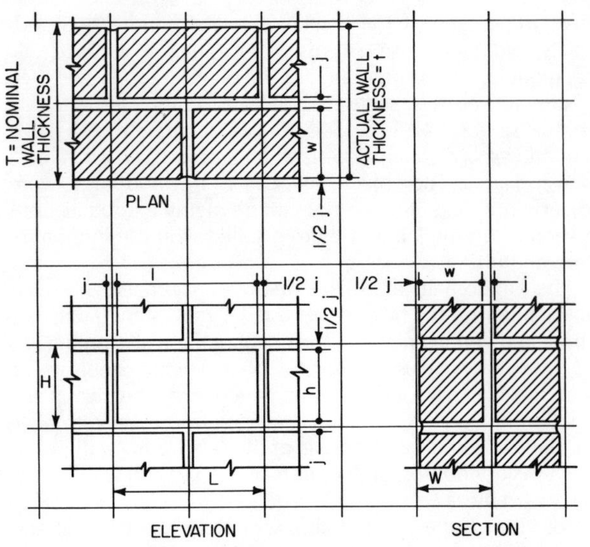

Fig. 3. Bricks molded to size and laid in place according to modular principles. Note the use of both arrowheads and dots in the dimensioning. In addition, capital letters signify nominal dimensions. Lower-case letters signify standard or specified dimensions. The lower-case *j* indicates a standard-thickness mortar joint (*J. H. Callender,* Ed., *Time-Saver Standards,* 5th ed. *McGraw-Hill,* 1974.)

are not. A masonry wall is, of course, constructed with mortar joints between the masonry units, and in modular construction the thickness of these mortar joints is considered part of the modular dimension (see Fig. 3). Modular-sized bricks, for example, are $2\frac{1}{6} \times 3\frac{1}{2} \times 7\frac{1}{2}$ in. in size, which makes for a rather odd-sized brick. But when one includes the ½ in. mortar joints, three courses of brick will be exactly 8 in. in height, and the length of the bricks in each course will equal 8 in. also. It should be noted that the grid lines split all the mortar joints, and, therefore, only half the thickness of the mortar joint at each grid line is considered part of the basic 8-in. dimension.

When a house with brick walls is being designed, the overall dimensions of the house are calculated from the exterior faces of the masonry units, which should be aligned with the grid lines. The grid lines should split the mortar joints, as already described. Because all the masonry units are made to the same modular standard, an architect or builder has considerable freedom to lay out an attractive design in the brickwork if he should so desire (for brickwork patterns *see* BRICKLAYING) as well as in planning the overall size and proportions of the house, assuming the walls are of all-masonry construction. When the house is to have a brick-veneer exterior, the architect or builder is still bound to the 16 in., 24 in., and 48 in. modules required for the wood framing, assuming the entire house is being built according to modular principles.

Mortar

Mortar is a mixture of a *cementitious material* (which may be portland cement or lime, or both) and *sand.* When water is added to these ingredients, the result is a plastic substance that is used to bind together brick, tiles, concrete blocks, and other kinds of masonry units. After the mortar has set, the masonry units are bound together by the mortar in such a way that they form a single structural unit.

Mortar is closely related to other cementitious materials such as concrete, plaster, and stucco, but it would be a mistake to confuse mortar with these other materials or attempt to use them as a substitute for mortar; the properties required of each are distinctive and differ from the others.

PROPERTIES

By a mistaken analogy with a chain and its weakest link, it is a common belief that for any masonry construction to be strong, the mortar must be strong also. Very often, for example, a person who is familiar with concrete will infer that mortar, being a cementitious material like concrete, should have properties similar to those of concrete and be mixed and used in much the same way. Since, for example, concrete has, or should have, a high compressive strength, mortar should have a high compressive strength also. But the primary function of mortar is to *bind the masonry units together,* not to resist compressive loads or add to the strength of the masonry units.

A great many tests have been made of brick walls built with mortars having a wide range of strength characteristics. These tests show uniformly that a brick wall is strongest when the mortar used to bind the brick is weaker than the brick. Indeed, the mortar can be substantially weaker than the brick without much affecting the overall strength of the construction. As long as the mortar is strong enough to resist the erosive effects of the weather and of freezing water, it is strong enough for use in the ordinary exterior wall.

But suppose for the sake of argument that a brick wall has been built using a mortar that does have a compressive strength greater than that of the brick. Any stress this wall may be subjected to—the result of the settlement of the soil under the wall, say—will cause the brick to fracture along the line of greatest stress. This fracture will run in a single jagged crack right through the brick, from the top of the wall to the bottom.

But when the mortar is weaker than the brick, as it should be, any stresses in the construction will be absorbed entirely by the mortar. The mortar will absorb these stresses in the form of a multitude of minute cracks invisible to the eye that leaves the basic strength of the construction unimpaired. The overall appearance of the wall and its structural integrity will be unchanged.

Masonry constructions can, however, suffer from another type of failure. Sometimes stresses are relieved by a separation of the mortar from the brick. The result is a zig-zag crack through the mortar that follows the brick pattern. What has happened here is that the bond between the brick and the mortar was too weak, a consequence either of ignorance or poor workmanship, or both, since the last thing that should happen in a well-made masonry wall is for there to be a poor bond between the masonry units and the mortar. Either the mortar was incorrectly proportioned and mixed or it was improperly applied to the brick, or both.

Workability

Freshly prepared mortar in which the cement, lime, and sand are accurately proportioned and mixed with the required amount of water has a quality called *workability* or, sometimes, *plasticity*. Workability is as difficult to describe in words as the consistency of pancake batter or soft butter, but fresh mortar that doesn't have this quality will be incapable of bonding masonry units together as tightly as they should be. A workable mortar can be spread with a trowel smoothly, evenly, and without effort. The mortar has a cohesive quality that enables it to hold its shape and keeps it from falling of its own weight when it is troweled onto the side of a brick. A workable mortar has a give to it that enables a bricklayer to bed masonry units solidly into place. One can say that on a microscopic scale a workable mortar makes such intimate contact with the surface of a masonry unit that the bond between the mortar and the masonry unit is as strong as possible.

Consistency

A closely related aspect of workability is *consistency*. By consistency we mean the relative softness or stiffness of a batch of fresh mortar. Consistency depends on the water content of the mortar. By changing the amount of water added to the dry ingredients, we can change the consistency of the mortar. If too much water is added, the mortar merely becomes sloppy. Up to this point of sloppiness, however, the consistency of a mortar is always improved by the addition of water, but not necessarily its workability. Workability depends on the proportions and quality of the ingredients making up the mortar. Two mortars having the same consistency need not, therefore, have the same workability.

Although workability cannot as yet be measured in a laboratory, consistency can, on a device called a *flow table*. The mortar to be tested is packed into a small, truncated cone that is 4 in. in diam. at the bottom. The mortar is emptied onto a metal plate mounted on the flow table. The plate is then rapidly vibrated up and down 25 times, the total up-and-down movement being ½ in. This vibration causes the mortar to spread out on the plate like pancake batter poured on a griddle. The amount of spread is a measure of the mortar's consistency. If, for example, the mortar spreads out 8 in. from its original 4-in. diam., then the original diameter has doubled, and the mortar is said to have a consistency of 100 percent. In practice, most workable mortars will have a consistency somewhere between 130 and 150 percent.

Water Retentivity

The ability to retain its water content is another quality of freshly mixed mortar that is closely related to workability. A mortar that has good water retentivity is not only workable, it will also retain its workability for a considerable period of time. The water does not evaporate too quickly from the mortar if it should be exposed to the air, nor is the water sucked too easily from the mortar by dry, porous masonry units.

What happens if mortar having poor water retentivity is used to build a brick wall? The bricklayer first lays down a line of mortar on which he beds the first course of brick (for a description of the technique *see* BRICKLAYING). After the first course is in place, he spreads mortar over the brick preparatory to laying the second course. If the brick are porous and the mortar has poor water retentivity, the brick in the first course will have sucked the water from the mortar, and the mortar will have stiffened considerably before the second course of brick can be laid in place. As a result, the second course of brick cannot be bedded securely in place and the bond between the mortar and the brick will not be strong.

In addition, as the wall is built higher and higher, the brick in the lower courses will not be able to adjust themselves to the stresses imposed upon them by the weight of the brick above. The entire wall, therefore, although it certainly won't collapse, will be poorly bonded together. In time, large cracks will undoubtedly open up in the mortar joints, resulting in the zigzag appearance described above. In addition, in rainy weather the entire construction will leak, and if the weather is cold any moisture that makes its way into the cracks will freeze and gradually cause the mortar to erode away.

This sort of problem can be prevented by using a mortar that has good water retentivity. And how is good water retentivity obtained? By making the mortar with well-graded sand and by using lime in the mortar. Water retentivity can also be increased by adding an air-entrainment chemical to the mortar, as described below. Finally, if the brick are too porous, their porosity can be reduced by soaking them with water beforehand.

The water retentivity of a batch of mortar can be tested on a flow table similar to that used to test the consistency of mortar. The difference is that when testing for water retentivity the plate has small holes drilled through its surface, these holes being connected to a vacuum pump. The cone-shaped pile of mortar is placed on the table as already described and allowed to spread under vibration. A consistency measurement is then taken. The mortar is then subjected to a controlled amount of suction to draw out some of its water content. The mortar is then repacked into the truncated cone and the consistency test is repeated. The second time around the mortar will not spread as readily as it did the first time because of the loss of some of its water. The difference between the two consistency measurements, expressed as a percent, is an indication of how well the mortar was able to retain its water. If, for example, the mortar had a consistency of 100 percent before being subjected to suction and a consistency of 80 percent afterward, then its water retentivity would be $80/100 \times 100$ percent $= 80$ percent.

In practice, it has been found that a mortar having a water retentivity of 85 percent is the most workable, although mortars with water retentivities as low as 70 percent are acceptable. The greater the amount of water retained, the better, as long as the mortar does not have so much water in it that it turns loose and sloppy.

INGREDIENTS

Mortars are classified as either *cement* mortars, *cement-lime* mortars, or *lime* mortars. All three types are mixed with sand, the primary purpose of which is to reduce to a minimum the shrinkage that occurs in cement and lime as they set.

Both portland cement and lime bind masonry units together as follows. When the cement or lime hardens, it does so in the form of a dense network of microscopic-sized interlocking crystals. It is this interlocking, crystalline structure of cement and lime that gives them their strength in the first place. When these crystals interlock with the minute crevices, protuberances, and voids that lie exposed on the surface of a brick, the result is a strong bond between the two.

Cement Mortar

The ordinary Types I, II, and III portland cements that are used to make concrete are also used to make mortar. The only difference between a portland-cement mortar and a portland-cement concrete is that the concrete contains a coarse aggregate consisting of large stones and gravel in addition to sand, while the mortar contains only sand.

The great defect of a mortar that is made entirely from portland cement is that when the mortar is mixed to achieve the maximum possible amount of strength, it will be too stiff to be workable or bond tightly to the masonry units. On the other hand, when a portland-cement mortar is mixed so that it is workable, it will contain so much water that it will be unable to develop its full strength.

An all-portland-cement mortar does have its place in construction, however, and this is where exceptionally strong brick are being used to construct a foundation wall, or when a foundation wall will be subjected to a hydrostatic pressure. Under these conditions, the strength of the mortar and its imperviousness to water penetration are valuable attributes. If, however, an all-portland-cement mortar is used with brick or concrete blocks having ordinary strengths, the mortar may be so strong that the masonry units will crack instead of the mortar if the construction is stressed excessively.

Lime Mortar

All-lime mortars are of great antiquity and were used by both the Greeks and Romans in exactly the same way as they were until quite recent times. There are many Roman-built brick and stone structures that are still standing in which the mortar joints are as solid as ever, 2000 years after these structures were completed.

Lime is made from calcium carbonate or a combination of calcium and magnesium carbonates (*see* LIME). The carbonates are crushed and heated in a kiln to drive off their chemically bound water. The resultant product is called *quicklime*. In order that quicklime can be used as a mortar, it must first be *slaked,* which means water is added to it to form the compound calcium hydroxide, which, when it has an excess of water, forms a slurry called *lime putty*. This lime putty can rest for a considerable period of time before it is used. Indeed, the longer it rests the more plastic and workable it will become, and the better a mortar it will make. In times past it was not unusual to rest lime putty for 10 weeks and more before using it. This is the way that the Greeks and Romans prepared their lime mortars. That their structures have remained standing for so long a time is a tribute to the thoroughness with which they slaked and mixed their lime.

If, during the slaking of the quicklime, just enough water is added to form the calcium hydroxide without there being any surplus water left over, the result is a dry powder called *hydrated lime.* Hydrated lime can be mixed with sand and

Table 1. Strength of Mortar Expressed as the Ratio of Portland Cement to Lime Used in the Mortar

Proportion of cement and lime to sand, by volume	Ratio of lime and cement, by volume						
	All cement	50:50	60:40	70:30	80:20	90:10	All lime
1:1		72	70	66	58	47	
1:1½		87	84	77	68	56	
1:2	96	94	90	84	74	60	
1:3	100	96	92	87	79	65	48
1:4		92	87	81	71	59	

water and used immediately, but the mortar will be greatly improved if the hydrated lime is mixed with water first and this mixture is allowed to rest for about 24 hr before it is used.

The great disadvantage of lime mortars is that they take a considerable time to set hard, since their setting depends on a recombination of the lime with the carbon dioxide in the atmosphere to form the original carbonate again.

Cement-Lime Mortar

Nowadays, a mixture of portland cement and hydrated lime is used in most of the mortars made in the United States. In this way the virtues of each are made use of, and their defects are minimized. The portland cement gives strength, durability, and impermeability to the mortar, as well as greatly increasing the speed with which the mortar sets. The lime gives workability and water retentivity to the mortar. That is, the lime acts mainly as a plasticizer.

When a very strong, durable, and impermeable mortar is desired, the proportion of portland cement is increased relative to the amount of lime. On the other hand, where strength is less important than workability and water retentivity, the proportion of lime is increased relative to the amount of cement. This is shown in Table 1 where the compressive strengths of the mortars increase with an increase in the proportion of cement used.

Sand

The sand used in mortars must be clean and completely free of any organic materials, salts, or clay. The particles must be sharp-edged and well-graded as to size, the sizes ranging from very fine up to ¼ in. in diam. The importance of having a wide range of sizes cannot be overemphasized. A sand in which a high proportion of the particles is too large will result in a stiff, unworkable mortar. On the other hand, a sand in which a high proportion of the particles is too small will result in a very workable mortar, but the water retentivity will be poor.

Admixtures

It is possible, though not really necessary, to add certain chemicals to mortar that will improve its characteristics. A small amount of an air-entrainment compound may be added, for example. This compound forms millions of extremely small bubbles in every cu in. of the mortar, and these bubbles greatly improve the mortar's workability and retentivity. In an all-portland-cement mortar, for example, the presence of these bubbles dramatically improves the workability and water retentivity of the mortar, at the expense of some loss in compressive strength and bonding ability. The air-entrainment compound also improves the long-term durability of the mortar since the bubbles will absorb the expansion that occurs when damp mortar freezes. Air-entrained mortars can withstand hundreds of freezing-thawing cycles without suffering any deterioration due to weathering.

Nevertheless, all this having been said, it must be added that the introduction of an air-entrainment compound in a mortar will reduce the bonding strength of the mortar, the amount of reduction being proportional to the amount of compound added. Since the primary function of the mortar is to bond the masonry units together, this loss in bonding strength may be a serious deficiency in the mortar.

PROPORTIONING AND MIXING

It is impossible to proportion the ingredients of a mortar in such a way as to achieve all the desirable properties required of a mortar at the same time. For example, the strength of a mortar can be increased by increasing the proportion of portland cement used, but if the proportion of cement to lime is more than 2 parts cement to 1 part lime, the strength of the bond will be reduced. The strength of the mortar can also be increased if a smaller quantity of water is used, but reducing the amount of water also reduces the workability and the water retentivity of the mortar, which again reduces the strength of the bond.

Both workability and water retentivity are increased, however, if the proportion of lime is increased, but increasing the amount of lime reduces the strength and durability of the mortar. The workability and water retentivity can also be increased if the amount of entrained air in the mortar is increased, but this also reduces the strength of the bond.

The best thing that one can do, therefore, is mix the portland cement and lime together according to the intended use of the mortar. For example, in a mortar intended to bond high-strength brick together, a larger proportion of portland cement should be used; as the strength of the bricks decreases, the proportion of portland cement in the mortar should be reduced accordingly.

Types of Mortars

ASTM Spec C270 recognizes four basic types of mortar, which are identified in Table 2a as types M, S, N, and O. (Table 2b shows the types of masonry with which each may be used.)

Type M mortar is a high-strength mortar in which the proportion of portland cement to lime varies from 1:0 to 1:¼. Type M mortar is used for reinforced-masonry foundations and retaining walls, and wherever the masonry is in contact with soil, as in brick walks, sewers, and manholes. Type M mortar is especially recommended for use whenever the compressive strength and impermeability of the construction are important.

Type S mortar is a medium-high-strength mortar that is used where the tensile bond between the mortar and the masonry units is important, as in a wall subjected to high lateral (i.e., sideways) stresses. The proportion of portland cement to lime ranges from 1:¼ to 1:½. Type S mortar is also recommended where maximum adhesion is desirable, as when ceramic tiles are cemented to a wall.

Table 2a. Four Basic Types of Mortar and the Proportions of Cement to Lime in Each, by Volume

Mortar type	Portland cement	Hydrated lime	Sand, measured in a damp, loose condition
M	1	¼	Not less than 2¼ and
S	1	½	not more than 3 times
N	1	1	the sum of the volumes
O	1	2	of cement and lime used

Table 2b. Types of Mortar Used with Various Masonry Constructions

Type of masonry	Types of mortar permitted
Masonry in contact with earth	M or S
Grouted and filled cell masonry	M or S
Masonry above grade or interior masonry:	
Piers of solid units	M, S or N
Piers of hollow units	M or S
Walls of solid units	M, S, N or O
Walls of hollow units	M, S or N
Cavity walls and masonry bonded hollow walls:	
Design wind pressure exceeds 20 psf	M or S
Design wind pressure 20 psf or less	M, S or N
Glass block masonry	S or N
Nonloadbearing partitions and fireproofing	M, S, N, O (or Gypsum)
Linings of existing masonry, above or below grade	M or S
Masonry other than above	M, S or N

Type N mortar is a medium-low-strength mortar that is used for general abovegrade construction where the masonry is exposed to severe weather, that is, for parapets, chimneys, and exterior walls generally. The proportion of portland cement to lime ranges from 1:½ to 1:1¼.

Type O mortar is a low-strength mortar that is used for nonloadbearing interior walls and for exterior walls that are neither exposed to the weather nor subject to freezing temperatures. The proportion of portland cement to lime ranges from 1:1¼ to 1:2½.

The amount of sand in all these mixes will vary between 1 part cementitious material to 2¼ to 3 parts sand, with 3 parts sand being the preferred proportion. If the sand is poorly graded, a larger proportion of cementitious material will be required to obtain a workable mortar. As we have mentioned, a well-graded sand helps improve the workability of a mortar. If the sand is well-graded, therefore, the proportion of sand can be increased to the desired 1:3 ratio. The maximum possible amount of sand should always be used, as the more sand there is in a mortar the less chance there will be that shrinkage cracks will occur as the mortar sets. On the other hand, too much sand (that is, a ratio greater than 1:3) results in a stiff, "harsh" mortar in which it is difficult to bed the masonry units properly. In addition, a too-sandy mortar will probably prove porous and leaky, and the mortar is also quite likely to erode away if it is exposed to severe weather conditions.

Proportioning Mortar

The ingredients of a mortar are proportioned by volume. The unit of measurement is usually cu ft, for the very good reason that portland cement and hydrated lime are sold by the bag, which always contain 1 cu ft of material. Thus, when a batch of mortar is being made up, the cement and lime are usually proportioned by the bag as follows:

Mortar type	Cement-lime ratio, bag *Cement* : *Lime*
M	From 1:0 to 4:1
S	From 4:1 to 2:1
N	From 2:1 to 1:1¼
O	From 1:1¼ to 1:5

For odd jobs in which full bags of material would result in an excessive amount of mortar, one can take a small container of some kind, an empty coffee can, say, and measure out the proportions that way.

Whatever the quantity of cementitious material required, the volume of sand will be approximately 3 times the volume of the cementitious material. The preferred method of measuring out sand is to obtain a container that holds 1 cu ft exactly and see how many shovelfuls of damp sand are required to fill the container. One can then multiply the number of shovelfuls by the number of cu ft required to obtain the total number of shovelfuls. If, for example, one were preparing a batch of mortar consisting of 2 bags cement to 1 bag hydrated lime, then 9 cu ft of sand would be required. If it takes 5 shovelfuls of sand to fill a 1-cu-ft measuring container, then it will take $5 \times 9 = 45$ shovelfuls of sand to obtain the correct amount of mortar.

The problem with calculating sand by the shovelful is that the dampness of the sand affects its cohesiveness. One shovelful of dry sand has not the same volume of sand as one shovelful of very damp sand. Furthermore, the inside of a sandpile is likely to be damper than the outside, and unless one takes this difference into account, the proportion of sand in successive batches of mortar will be different. The preferred method of measuring out sand, therefore, is not to shovel the sand directly into a mortar box or mixer but to fill a wheelbarrow with a carefully measured quantity of sand and thereafter continue to fill the wheelbarrow the same amount.

Mixing Mortar

If a small amount of mortar is to be mixed by hand in a mortar box, the preferred method of mixing the ingredients is to fill the mortar box with sand first and then add the cementitious material. The ingredients are mixed thoroughly with a hoe until the color of the mixture is even throughout. Approximately three-quarters of the required water is then added, the ingredients being mixed all the while with the hoe until the mixture is equally damp throughout. The remainder of the water is then added in small amounts, the mortar being mixed continuously until the desired consistency has been obtained. The mortar is allowed to rest for about 5 min; it is then thoroughly remixed once again, and it is ready to be used.

Whenever bagfuls of cementitious material are being mixed, machine mixing is the best and most efficient method of doing so. In machine mixing, about three-quarters of the required amount of water is poured into the mixer first, followed by one-half the required sand, then all the cementitious material, and finally the remainder of the sand. The mortar is allowed to mix thoroughly for a minute or two. Small amounts of additional water are then added until the consistency of the mortar is satisfactory. The mortar should continue to be mixed for at least another 5 min. All the mortar should be discharged from the mixer before another batch is started.

The Use of Mortar

The use of mortar is described elsewhere (*see* BRICKLAYING; CONCRETE-BLOCK CONSTRUCTION). The only point we will make here is that one should never try to reset a masonry unit once it has been placed in position in a bed of mortar, since moving the masonry unit will break the bond that has been established between the unit and the mortar. Once a masonry unit has been tapped, leveled, and otherwise adjusted into position, it should be left alone. A bond once broken will never re-adhere as well as it adhered originally.

Also, mortar can be used for up to 2½ hr after it has been mixed. If the mortar stiffens because its water content evaporates away, additional water can always be added to the mortar until it has regained its original plasticity.

MASONRY CEMENT

All manufacturers of ordinary portland cements also manufacture proprietary mortar mixes that go under the generic name of *masonry cements*. Each manufacturer has a secret formula, but, generally speaking, most of these products consist of equal parts of portland cement and powdered limestone, the limestone taking the place of lime insofar as improving the workability and water retentivity of the mortar are concerned. Some formulas may, however, contain, in place of the limestone, one or another of such ingredients as hydrated lime, "natural" cement, which is a clayey limestone found naturally in the soil that is burned to produce a cementitious material, and pozzolana-like materials such as ground blast-furnace slag, fly ash, or volcanic ash.

In addition, each manufacturer adds his own combination of additives such as resins, tall oils, water-repellent fats, fatty acids, stearates, and wetting agents to the mortar to improve its workability and water retentivity and increase the mortar's resistance to water penetration. Most masonry cements will also include an air-entrainment agent to improve both workability and resistance to weathering.

The result is a compound that makes an excellent mortar when mixed and used as directed on the package. However, there is nothing these proprietary products can do that straight portland-cement-and-lime mortars can't do also. Their principal virtue, to a homeowner especially, is that they provide a convenient, all-in-one package to which only water and sand need be added—and some manufacturers even sell masonry cements in which the correct proportion of sand has already been added. One just adds water and stirs.

ASTM Spec C91 defines two types of masonry cement, Types I and II, of which only Type II is used as a mortar. Table 3 shows the proportions of Type II masonry cement and portland cement that are mixed to produce mortars that are equivalent to the Types M, S, N, and O mortars described above. In all cases, the proportion of cementitious material to sand is 1:3. These masonry cements are mixed and used in exactly the same way that regular mortars are.

Table 3. Four Basic Types of Mortar and the Proportions of Portland Cement and Masonry Cement in Each, by Volume

Mortar type	Portland cement	Masonry cement	Sand, measured in a damp, loose condition
M	1	1 (Type II)	Not less than 2¼
S	½	1 (Type II)	and not more than
N		1 (Type II)	3 times the sum of
O		1 (Types I or II)	the volumes of cements
PM	1	1	used.

Note: PM mortar is used as a grout.

Paneling, Wood

This article describes interior wall paneling, which is made of wood. This paneling may be (1) an open framework of wood strips into which are inserted thin wood panels that give the entire paneling a solid-looking front, (2) large sheets of plywood, which have a decorative or hardwood-veneer surface, or (3) wood boards that may also be decoratively finished in some way or have a hardwood-veneered surface. There also exist large panels made of hardboard, insulation board, or some similar material, that are covered with a simulated wood-grain pattern or a decorative finish; these panels are not discussed in this article, though their installation is similar to that of plywood.

RAIL-AND-STILE PANELING

Nothing about the interior of a house better evokes an atmosphere of luxurious comfort than a traditional wood-paneled library or study, which most people associate with Elizabethan England. The original purpose of the paneling was to cover, in a manner more elegant than the usual interior plastering, the rough brick or stone construction, or rough wood framing, that made up the exterior and interior walls of the house.

This paneling (see Fig. 1) consisted of an open framework of narrow wood strips—horizontal *rails* and vertical *stiles,* with thin wood panels inserted between the rails and stiles. The paneling was constructed in this way not primarily because of its aesthetic effect, though it was attractive, of course, but because it was the only practical way (before the invention of plywood) of building a large decorative wall surface out of ordinary pieces of wood.

Wood shrinks as it ages. It shrinks most of all across its grain. If, in order to form one large surface, the paneling had been made of wide boards fastened tightly to an underlying support, gaps would have opened up between the boards as they shrank; either that or the boards would have eventually split or warped.

By constructing the paneling in the form of an open framework made up of relatively narrow rails and stiles, not only was a nice decorative effect achieved, but the shrinkage of the wood across its grain was minimized. The center panels could then be set loosely in place between the rails and stiles (see Fig. 2). This allowed the panels to shrink or expand with changes in their moisture content and to shift their position as necessary without stressing either themselves or the rails and stiles. Wood-panel doors are constructed according to exactly the same principles (*see* DOOR, WOOD). The advantages of this type of construction are such that wood paneling of this type continues to be built today in exactly the same way as it was built 400 years ago; indeed, there are many examples of rail-and-stile paneling in existence that are as sound and as attractive today as the day they were built, 400 years ago.

Wood paneling may cover the walls of a room up to ceiling height; this is called *room height* paneling. The paneling may cover the walls up to a height of only 6 or 7 ft; this is called *partition-height* paneling. Or the paneling may be only 3 ft or so in height; this is called *dado* paneling or *wainscotting.* Partition-height paneling is often built as screening or as a room divider. In this case it stands away from a wall, it is finished on both sides, and it is called *screen* paneling.

Wainscotting is the most common type of paneling installed, as it is the least expensive to construct and the least difficult to install. Wainscotting also has something of a utilitarian function, and it is often installed along the walls of a hallway or entrance way, or along a stairway, to take the knocks and bumps that might damage a plaster wall.

The best-quality rail-and-stile wood paneling is manufactured in a cabinetmaker's shop and then transported as a complete unit or in several large sections to the room in which it is to be installed, just like a piece of fine furniture. In this type of construction, the inset panels between the rails and stiles are inserted from the rear of the construction and held in place by wood strips. Thus, the manner of construction is completely invisible.

PLYWOOD PANELING

The invention of plywood has made possible an entirely new type of wood paneling, one in which it is possible to cover the walls of a room with 4-ft wide sheets of decoratively finished wood that reach from floor to ceiling without a break.

Hardwood veneers of walnut, oak, cherry, hickory, birch, mahogany, and ash, among others, are available in these panels, as are exotic foreign woods that have very ornate, not to say theatrical, grain patterns. These grain patterns can be installed on a wall in a number of different ways, as shown in Fig. 3, to achieve different effects.

Softwood panels are also available in which the face of the panel has been textured by striating the wood, by cutting fine grooves in it, by rough-sawing or rough-sanding the wood, or by sandblasting or wire-brushing it to give an embossed effect to the grain pattern. Softwood panels are also available in which the face veneer simulates random-width planking, with the edges of the planks also cut sometimes in V grooves or beading to make joining the panels easier.

Installation

The panels are usually attached to furring strips spaced 16 in. apart, as shown in Fig. 4. They can also be attached to large backing sheets made of plywood, gypsum wallboards, or fiberboards that have been nailed to the studs. Whichever method is used to support the plywood panels—furring strips or back-

Fig. 1. The construction of traditional rail-and-stile wood paneling *(Architectural Graphic Standards, John Wiley & Sons).*

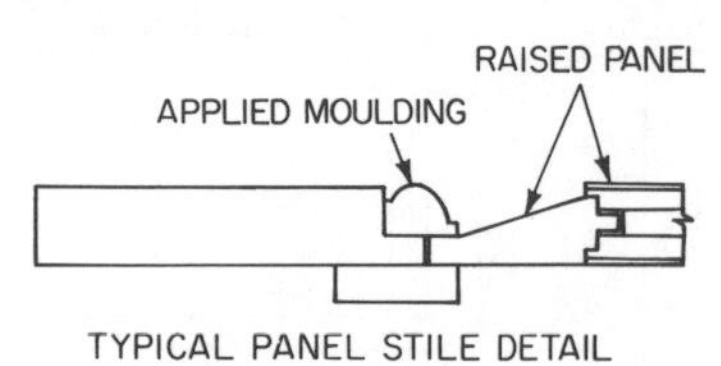

Fig. 2. Center panels must be loosely set in place to allow the wood a chance to swell and shrink in response to humidity changes.

ing—it must provide a completely flat surface for the panels. Nothing spoils the appearance of paneling more than a wavy look to the walls, especially when lights are reflected from its surface, unless it is gaps between the panels. Backing sheets are particularly desirable if the paneling is made from ¼-in.-thick plywood, which lacks the rigidity to support itself stiffly.

Furring strips are the preferred method of supporting the paneling if it is to be installed against an exterior masonry wall or against a masonry wall located belowgrade, for, even if the wall has been waterproofed, it is still likely that moisture will condense on the inside surface of the wall during cold weather. To keep this moisture from soaking into the paneling, it must be kept at least 1 in. away from the wall, which is one of the functions of the furring, of course.

The panels are usually attached to the furring strips with a contact-type adhesive supplied by the manufacturer of the panels. When the panels are to be attached to furring strips, each panel must first be located accurately in position and the location of the furring strips then marked on the back of the panel. The panel is then removed and a ribbon of adhesive is applied with a caulking gun to both the furring strips and the back of the panel. The adhesive is allowed to set partially, the panel is set in position again, and it is pressed carefully and firmly against the furring strips. Small brads are then nailed along the top and bottom edges of the panel to hold it in place until the adhesive has set completely. These brads will be hidden later by the baseboard and ceiling moldings that finish off the paneling.

BOARD PANELING

Wood boards are often used to panel the sides of a hallway or stairway. This board paneling is attractive, and it offers an alternative to either the traditional rail-and-stile paneling or to plywood paneling. Lumber mills sell nominally ½-in. and ¾-in.-thick lumber that is especially finished for this purpose, the lumber being kiln-dried and then either left unfinished, primed with a sealer, or lacquered. The ¾-in.-thick stock is used for first-class work and the ½-in.-thick stock for less expensive installations. Boards are also available in which the surface of the wood is rough sawed, striated, sandblasted, or wire-

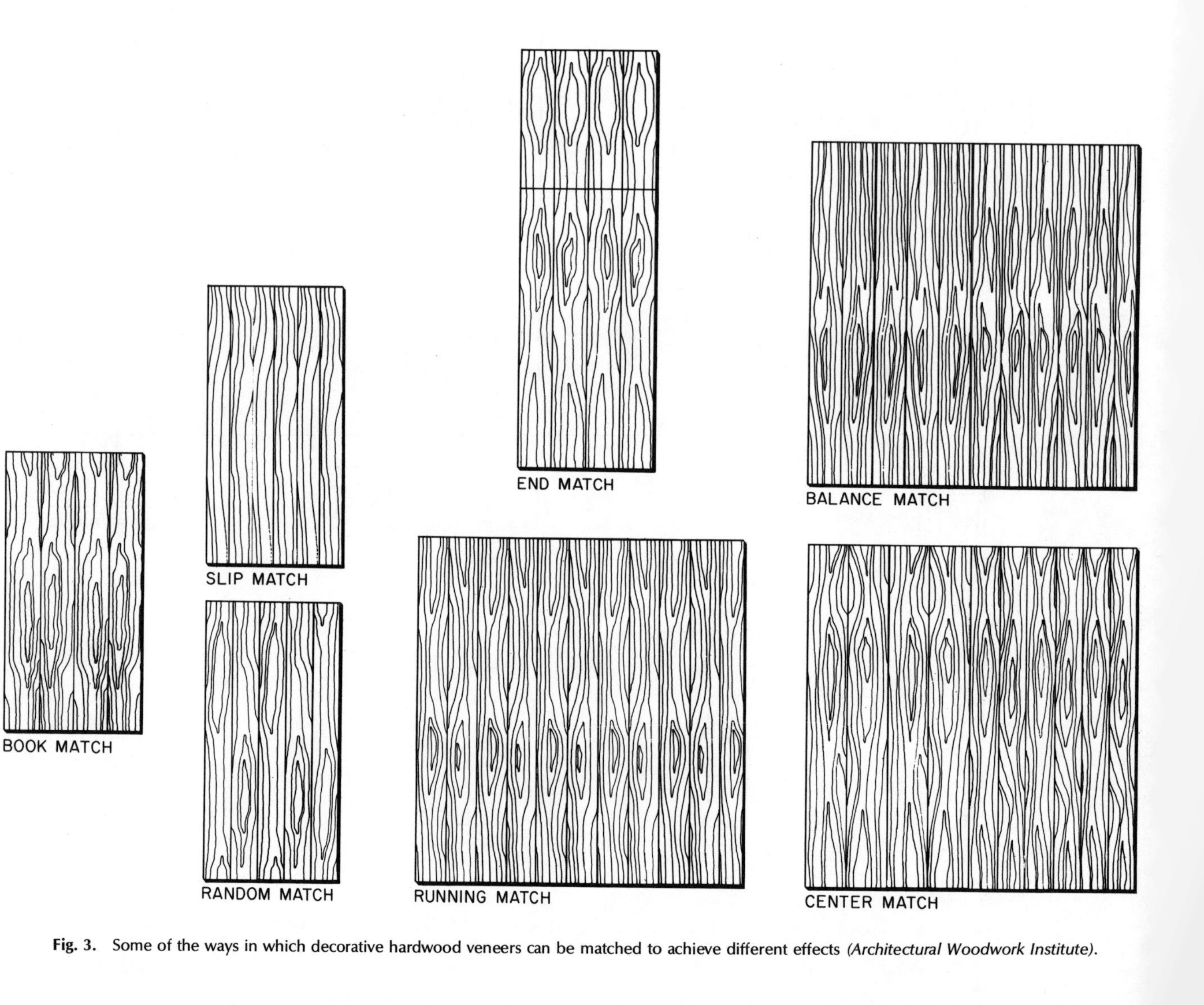

Fig. 3. Some of the ways in which decorative hardwood veneers can be matched to achieve different effects *(Architectural Woodwork Institute)*.

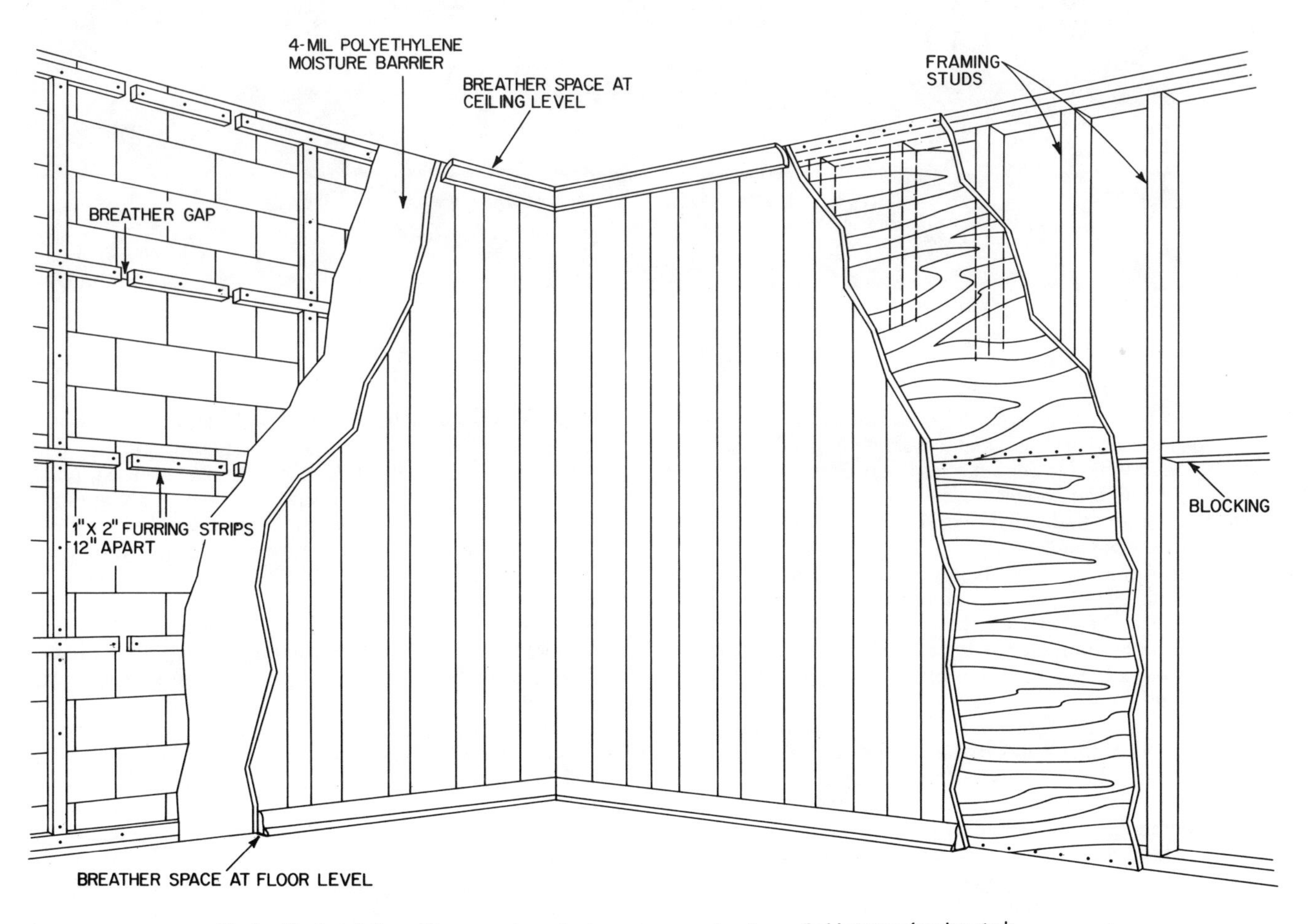

Fig. 4. The installation of large panels on furring strips or on headers nailed between framing studs.

brushed. Boards to which a hardwood veneer has been glued are also available.

The edges of these boards may be tongue-and-grooved, and/or have a V-shaped bevel and/or a bead to set off each board. The boards are usually blind-nailed to furring strips or blocking. These furring strips or blocking are spaced about 24 in. on centers, if the boards are to be installed vertically, as they usually are. The installation is given a finished appearance by cap and base moldings installed at the top and bottom of the paneling, respectively.

Plank-and-Beam Construction

Plank and beam construction (also known as *post-and-beam* construction; see Fig. 1) is a method of building the framework of a house in which fewer but much heavier timbers are used instead of the lighter and more closely spaced joists, studs, and rafters of more traditional wood-frame construction (*see* WOOD-FRAME CONSTRUCTION).

Instead of the floor loads being carried by 2-in.-thick joists spaced 16 to 24 in. apart, the floor loads are carried by 4-in.-thick beams spaced 6 to 8 ft apart.

Instead of the roof loads being carried by 2-in.-thick rafters spaced 16 to 24 in. apart, the roof loads are carried by 4-in.-thick wood beams that are also spaced 6 to 8 ft apart.

And instead of the traditional 2 × 4 in. wall studs spaced 16 to 24 in. apart that support the weight of the entire roof construction and the weight of the interior and exterior wall finishes as well, there is no such wall construction. Instead, the weight of the roof is supported by large wood posts or by large girders, which are in turn supported by posts. This construction enables lightweight *curtain* walls to be built; it also allows for the use of glass walls, or, indeed, for the walls to be eliminated altogether.

In addition, a plank-and-beam house does not have the traditional subfloor and finished-floor construction that is attached to floor joists, or the traditional roof sheathing that is attached to the roof rafters. Instead, both the floor and roof consist of 2-in.-thick planks fastened directly to the floor and roof beams, respectively. The floor and roof are tied together by these planks into a single rigid and solid structure.

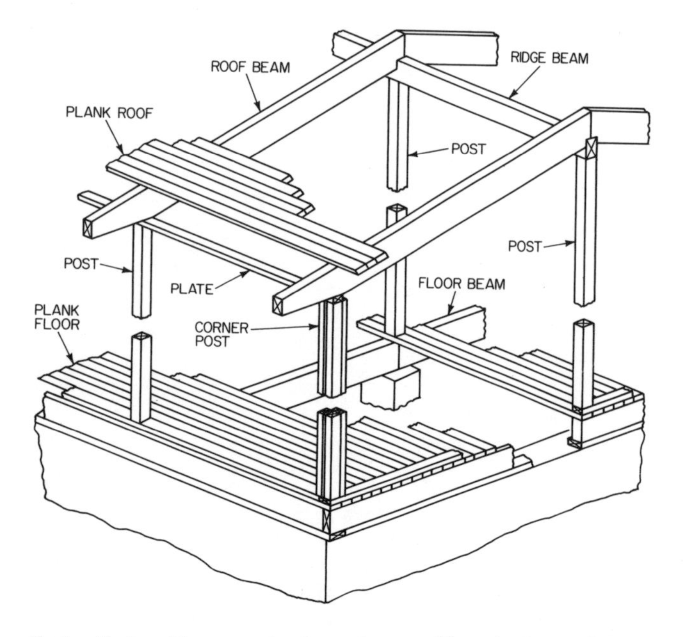

Fig. 1. Plank-and-beam construction makes use of fewer but heavier beams and posts to construct the framework of a house.

Figure 2 shows a typical dwelling. As may be seen, plank-and-beam construction can result in an extremely attractive appearance. The open, spacious effect that is achieved is very similar to that of traditional Japanese architecture, a similarity enhanced by the grain and coloring of the exposed wood flooring, roofing, and roof beams.

Because each beam and each post carries such a large proportion of the total loads, the entire building structure must be carefully designed in a way that is unnecessary for conventional wood-frame construction, in which traditional lumber sizes and spacings are depended upon to support the house. The loads carried by the posts, beams, and planking must be calculated according to traditional engineering analysis and the sizes of their structural members determined accordingly.

One inherent structural weakness of plank-and-beam construction that must be reckoned with is that the construction is not capable of resisting strong lateral loads, such as the sideways loads imposed by high winds and earthquake shocks. It is important that the posts be supported in such a way as to resist these loads. This is usually accomplished by attaching stiff wall panels to the corners of the building to increase the strength and rigidity of the walls.

The overall roof and floor construction is further strengthened by using floor and roof planks that are as long as possible. If, for example, the floor beams are spaced 6 ft apart, then the floor planking should be 12 ft long and bear on at least three beams. If the floor beams are spaced 8 ft apart, then the floor planks must be at least 16 ft long and bear on at least three beams, and so on. A plank that extends across and is fastened to three beams in this way results in a construction that is 2½ times stronger than when a plank extends across, and is fastened to, only two beams.

The labor cost of constructing a plank-and-beam dwelling can be considerably less than the labor cost of constructing a conventional wood-frame house, mainly because there are fewer individual pieces of lumber that must be fitted and fastened together. And if the beams and posts are cut to exact size at the mill, the erection cost can be reduced even further.

Another reason for careful advance planning is that there is no place in plank-and-beam construction to hide the electrical wiring and the plumbing lines. If solid timbers are being used, one has to plan beforehand which beams and posts must be routed out, where, and by how much. Since such routing out will weaken the beams and posts substantially, the strength of these structural members must be carefully calculated in advance, and steps taken either to use larger beams and posts or to reinforce them in some inconspicuous way. It is also possible to install spaced beams and posts, as shown in Fig. 3, or use laminated timbers within which a space has been left.

One cannot plan the layout of the rooms in a plank-and-beam dwelling as freely as one might wish, either. For one

Fig. 2. The attractive, spacious effect that can be obtained in a house constructed according to plank-and-beam principles *(Weyerhaeuser Corp.).*

thing the locations of the beams and posts will determine the runs of the plumbing and the electrical wiring. This will in turn determine the locations of the bathroom and kitchen fixtures.

Furthermore, interior partitions are best located where they are directly in line with posts. This not only avoids the awkward appearance that would result if the posts were located at odd positions within the rooms, the floor beams or piers supporting the posts can then support without difficulty the weight of the interior partitions. If, for example, a partition should run parallel to and between two floor beams, it is doubtful that the floor planks by themselves can support the weight of the partition, especially if the partition is supporting the weight of plumbing fixtures. In this case, the planks could not support the weight of the partition. It would then be necessary to install supplementary 2 × 4 in. beams directly under the partition, these beams being connected to, and supported by, the main beams.

The floor planks are not intended to support heavy concentrated loads, such as that of bathtubs, bookcases, and grand pianos, in the first place. Either supplementary beams will be required to help carry these concentrated loads and transmit their weight to the main beams or a finish floor of some type must be installed on the planks to help distribute the floor loads.

Finally, plank-and-beam construction is better suited to mild climates in which one doesn't have to worry about heat loss and fuel bills. In cold climates, lightweight curtain-wall construction and 2-in.-thick roof planking can make the house prohibitively expensive to heat. If roof insulation is required on the underside of the roof planking, then the appearance of the ceiling may be spoiled (or, at least, it will no longer look like wood planking); and having large expanses of thick, well-insulated walls will not only increase the construction costs, they may also detract from the attractiveness of this type of construction unless thought is given to the appearance of the insulation.

When insulation is necessary, however, it may be possible to insulate the roof by installing rigid-foam panels on top of the roof planking. These panels are embedded in mastic that is troweled over the roof planking, and they are then covered over and made watertight by installing built-up roofing on top of them (*see* BUILT-UP ROOFING). If one wants to keep the open appearance of glass walls in a northern climate, one might be able to reduce the loss of heat through the glass by installing sealed, double-paned glass.

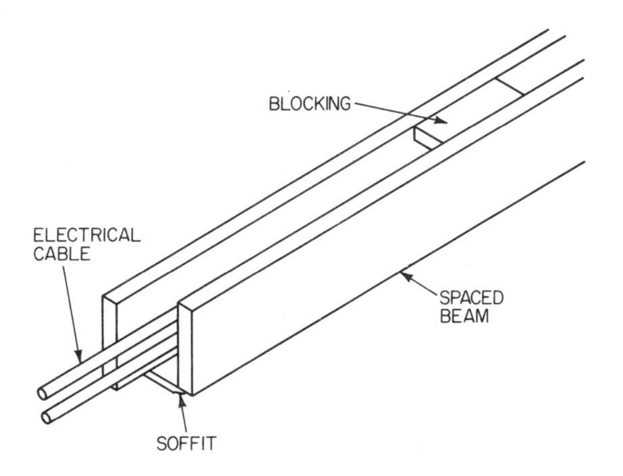

Fig. 3. Plumbing lines and electrical cables can be buried within hollow beams.

Plaster

By *plaster* we mean *gypsum* plaster, since in the United States plasters made from gypsum are used almost exclusively. There also exist plasters made from *lime,* which were widely used up to the beginning of the twentieth century but have since been supplanted almost completely by gypsum plasters, mainly because lime plasters are so slow-setting. They are, however, still widely used in Europe. Lime is used in the United States nowadays mainly as an admixture to gypsum plaster or portland-cement mortar to make the gypsum and mortar more workable than this would otherwise be.

Gypsum plasters may be divided into two main groups: (1) *hemihydrated* plasters, in which the gypsum has been heated somewhat above the boiling point of water (but below 374°F) during manufacture of the gypsum, and (2) *anhydrous* plasters, in which the gypsum is heated above 374°F. In plasters belonging to the first group, about three-quarters of the water that is chemically bound to the gypsum (the chemical formula of which is $CaSO_4 + 2H_2O$) is driven off. This results in the *hemihydrated,* or half-hydrated plaster, as it is also called, the formula of which is $CaSO_4 + \frac{1}{2}H_2O$. In the second group, the remainder of the chemically bound water is driven off, leaving an anhydrous (i.e., without water) plaster that has the formula $CaSO_4$.

Included in the group of half-hydrated plasters are *plaster of paris* and *hardwall* plaster. Included in the anhydrous group are *hard-finish* plaster and *Keene's cement.* Within each group are varieties of plaster that differ among themselves according to the purity of the gypsum used and whether other substances have been mixed with the gypsum.

HALF-HYDRATED PLASTERS

Plaster of Paris

Plaster of paris is made from pure gypsum (for a description of manufacture, *see* GYPSUM). When water is added to plaster of paris, a paste, or putty, results that sets hard in 5 to 20 minutes, depending on the temperature of the water and the purity of the gypsum. The plaster sets because the water and the particles of half-hydrated gypsum plaster quickly combine with each other to form the original gypsum compound, $CaSO_4 + 2H_2O$. The plaster sets hard because the particles form a dense, interlocking network of needle-like crystals. It is this crystallization of the plaster that gives plaster of paris what hardness and strength it possesses.

The identical sort of process takes place in all gypsum plasters as they set, but it takes place so rapidly in plaster of paris that it makes this plaster completely unsuitable for general plastering use, or even for the repair of large areas. The only practical use of plaster of paris in the home is to fill relatively small cracks and holes, jobs that can be done quickly and without fuss. For some reason, however, plaster of paris is the only plaster product found in most local hardware stores, and a great many homeowners think it is the only kind of plaster that exists.

Hardwall Plaster

In fact, the standard plastering material that is used to plaster the interior walls and ceilings of buildings is *regular or gypsum* plaster, which is also known as *neat* or *hardwall* plaster. Regular plaster differs from plaster of paris in that (1) it is not made from pure gypsum but from gypsum to which clay has been added, or in which clay exists naturally, or (2) it is actually plaster of paris to which clay or chemicals have been added to retard the time of set. Depending on the amount of clay present, which can range up to 15 percent of the amount of plaster by volume, or depending on the amount and kind of retardant added (as described below), regular plasters take from 2 to 4 hr to set.

Regular plaster is used mainly for preparing the scratch and brown coats of a plaster wall or ceiling (*see* PLASTERING). Although it is much slower setting than plaster of paris, it is much harder once it has set. The plaster is manufactured neat—that is, as-is—or fibers, sand, or lime, or combinations of these materials, can be added to the plaster at the mill in order to impart certain desirable properties to it.

Sisal fibers are used to bind the freshly made plaster together until it has had time to set; they make the wet plaster more cohesive. Fibers also make it more difficult for the plasterer to force the plaster too far through the openings in the lathing, though, in fact, gypsum plasters are pretty much self-supporting; it is the lime plasters that require binding together with sisal fibers.

Wood fibers are sometimes added to regular plaster at the mill. These fibers are obtained by shredding a nonstaining wood. Wood-fibered plaster is stronger and lighter in weight than the sanded plaster with which it competes, and nails can be driven into it without its crumbling apart; wood-fibered plaster is, however, much more expensive than sanded plaster. It finds its chief use in those sections of the country where sand of the requisite quality is difficult to obtain.

Sand is, of course, the aggregate that is usually added to a plaster (*see* PLASTERING). The sand is usually added on the job, though it may be added to the plaster at the mill, the manufacturer then referring to his product as *ready-sanded* plaster.

Ready-sanded plasters are convenient for small repair jobs, since they eliminate the bother of obtaining and transporting a small quantity of sand. Since the sand and plaster will have been premixed in the correct proportions, one has the assurance that the plaster will not be oversanded and thus weaker than desired.

Ready-sanded plasters may also be necessary in those sections of the country where suitable sand is not readily available, though there is a limit to how far one can ship ready-sanded plasters because of the expense. It may be more sensible to use a wood-fibered plaster instead.

Gauging Plaster

Gauging plaster is another variety of regular plaster. It is used as the finish coat on a plastered wall or ceiling. Sold by the bag, gauging plaster consists of 1 part gypsum plaster to 2 parts dry hydrated lime, by weight. The lime is added to increase the plasticity and workability of the fresh plaster, which enables the plasterer to apply the plaster smoothly and easily. Retarders are usually added to gauging plaster so that it will set either 2, 4, or 6 hr after water has been added to it.

Gauging plaster may also be mixed on the job by the plasterer from regular plaster and hydrated lime, in which case a batch of *lime putty* must first be prepared (*see* LIME).

Lightweight Plaster

Lightweight plaster is regular plaster that contains either perlite or vermiculite as the aggregate in place of sand or shredded wood fibers. It is sold ready-mixed by the bag. The advantages of using a lightweight plaster are (1) it makes plastering easier for the plasterer, (2) it imposes less deadweight on the supporting structure, (3) it has some insulating value (*see* INSULATION), (4) it absorbs sounds somewhat (*see* ACOUSTICAL MATERIALS), and (5) it increases the fire resistance of the construction. As an added virtue, lightweight plasters do not expand as they set, and, therefore, they can be applied directly to a concrete surface if this should be desired. Its principal disadvantage is that it is much more expensive than the other kinds of plaster that have been described.

ANHYDROUS PLASTERS

Hard-Finish Plaster

Hard-finish plaster is an anhydrous plaster that is prepared by heating pure gypsum to a temperature between 750 and 925°F until all the chemically combined water in the gypsum has been driven off. After it has cooled, the *calcined* gypsum is ground into a fine powder and it is ready for use.

The plaster thus prepared has several properties that are quite different from those of half-hydrated plasters. If this plaster is mixed with water, the resultant paste is very slow-setting. Accelerators must be mixed with it in order to speed up the rate at which the plaster sets so that it will set in 4 to 6 hr. Hard-finish plasters can also be *retempered,* which means that additional water can be added to the paste to maintain its workability if its water content should evaporate away, which will make the paste difficult to work. If the same sort of thing were done to regular plaster while it was still plastic, the plaster would be weakened considerably. Hard-finish plaster can also be troweled to a high polish and a very dense surface. This last property enables hard-finish plaster to be used decoratively as imitation floor tile or marble.

One must be careful, however, not to apply hard-finish plaster to a too-absorbent support. If its water content is sucked from the plaster prematurely, the hydration of the setting plaster (that is, its reabsorption of moisture) will be incomplete, and it will not develop its full strength. If the plaster should become wet at a later time, hydration will begin again, the plaster will absorb an additional quantity of water, and it may then expand enough for it to force itself away from its support. Either the plaster coating will then bulge away from its support or large pieces will spall (i.e., flake) off the wall.

Keene's Cement

Keene's cement is a variety of anhydrous plaster that is prepared by a double calcination process. It is prepared from a very pure gypsum that is first heated above the boiling point of water, as described above for half-hydrated plasters. It is then quenched in a 10 percent solution of alum or in borax, both substances being accelerators that hasten the setting time of the plaster. After the plaster has dried, it is reheated to 750°F until the remainder of the combined water has been driven off. It is then ground into a very fine powder.

The result is a plaster having exceptional hardness, strength, and resistance to moisture. Its hardness makes it the preferred plaster to use where walls are likely to receive hard knocks, that is, along corridors and hallways, in playrooms, and in workshops. Its resistance to moisture makes it the preferred gypsum plaster to use in those parts of the house subject to dampness, such as bathrooms, kitchens, and laundry rooms. It should be understood, however, that Keene's cement is merely water-resistant. It is not waterproof in the way that concrete is waterproof, and, if subjected to continual moisture, Keene's cement will eventually disintegrate just as other gypsum plasters disintegrate. It just takes longer to do so.

And it should also be noted that Keene's cement is also much more expensive than other kinds of gypsum plaster, and for this reason it is not used as often as it should be.

Once it has been mixed with water and applied, Keene's cement sets very, very slowly. Accelerators can be added to the cement, however, to increase the setting time to 2, 4, or 6 hr. Once the plaster has set, it will continue slowly to increase in strength.

ACCELERATORS AND RETARDERS

Retarders

All half-hydrated gypsum plasters set rather too quickly to suit a plasterer. Manufacturers, therefore, add retarders to the plaster to slow down the rate at which the plaster sets.

Most of these retarders are organic materials of some kind. They include gelatin, glues, dextrine, gums, starches, and vegetable and animal oils, among other substances. Some inorganic substances such as ammonia and zinc sulfate are also used. All these substances work because they interfere with the ability of the plaster particles to absorb water. They thus delay the rate at which the particles crystallize.

The amount of retarder required depends on the kind used and how much retardation is wanted. In any case, the amount usually added to the plaster is very small, something on the order of 0.2 percent of the plaster, by volume.

The retarder is usually incorporated into the plaster at the mill, where the amount added can be controlled very accurately. It is always possible, however, for a plasterer to add a retarder to a batch of plaster himself if he wishes, but he must do so with delicacy, as a little retarder goes a long way. The packets of retarder sold by manufacturers for use with their products should be used as recommended. The retarder is usually dissolved in the water that is to be used to mix a batch of the plaster.

Accelerators

Only anhydrous plasters require accelerators to hasten their setting times. Most accelerators are crystalline substances, and they act as nuclei around which the particles of plaster will crystallize. Hardened gypsum plaster that has been reground into a powder makes an excellent accelerator because the particles already have a needle-like crystalline structure that will quickly induce crystallization in freshly made plaster. Other crystalline substances that are used include common table salt and potassium sulfate. As with retarders, a little goes a long way. Follow the manufacturer's recommendations.

Plastering

Plaster consists of any of several powdered *cementitious materials* (that is, gypsum and/or lime, or portland cement, or portland cement and lime) to which sand, fibers, or other substances are added to give the plaster specific characteristics. When water is mixed with any of these cementitious materials, the result is a plastic, more or less easily workable *mortar* that may be applied to a suitable base by a plasterer. When the mortar sets, it will retain the shape imposed upon it by the plasterer.

Plaster has a great many advantages when used as an interior finish. It can be troweled into a smooth, flat coat that covers the entire wall and ceiling area of a room without break or interruption and hides the roughness and irregularities of the underlying structure. If necessary, once the plaster has hardened, it can be drilled or cut or nailed into, although it is not an especially strong material and will not support a heavy load by itself. If plaster should crack or be damaged in any way, the damage can be repaired easily and in such a way that the repaired area is indistinguishable from the rest of the plastered area. Plaster will neither rot nor burn; indeed, it is fire-resistant and will prevent the spread of fire. It is vermin-proof. It makes an excellent base for paints, wallpapers, fabrics, wood paneling, or any other kind of decorative finish. Plaster can also be molded into highly elaborate and intricate shapes and patterns, but, sad to say, it is not the fashion to decorate an interior in this way any more, even in a modest way.

The disadvantages of plaster follow from the fact that large quantities of water are necessary for mixing and applying it. It has been estimated, for example, that, in a moderate-size six-room house in which all the walls and ceilings have been plastered, about 1000 gallons of water will evaporate from the plaster as it sets and hardens. As the plasterer completes each coat of plaster (there will be two or three such coats, as described below), he must wait for each to dry out before he can apply the next coat. Depending on the weather, the complexity of the job, and the kind of plaster used, it will take from one to several days for each coat to dry. And once the plastering has been completed, it will take additional time for the plaster, and the rest of the house structure as well, to dry out completely. In the summer, during a spell of warm, dry weather, this complete drying out may take only a week or so; during a cold, damp winter it may take several weeks. These waiting times translate into expense to a builder, since nothing else can be done to the interior of the house until it has dried out. (In the old days, when the walls of a house were painted with an oil-base paint, the paint could not be applied to fresh plaster for two full months after the plastering had been completed because small but significant amounts of moisture remained in the plaster that would cause the paint to blister.) Even leaving oil-base paints out of account, the delays that are inherent in plastering have been chiefly responsible for the widespread switch by most builders to gypsum wallboards (*see* DRYWALL CONSTRUCTION). It has been estimated that, today, only 1 in 5000 homes is plastered.

In addition, plastering requires a dimensionally stable support if cracks are to be avoided in the completed job. If, for example, green lumber has been used to construct the framework of the house, this lumber must be given time to dry out and shrink down to its final size before the plastering is begun. Otherwise, if the plastering has been completed and the house shrinks down subsequently, large cracks will open up in the walls, especially at the corners of doors and windows and along the tops of the walls. Cracks may also appear in both the walls and ceilings if the house should settle after construction has been completed.

For these reasons it is unwise to paint or wallpaper a plastered interior for at least 6 months after a house has been completed. By that time, any plaster cracking that will occur because of wood shrinkage or house settlement will have occurred and the walls and/or ceilings can then be patched. When the house is finally painted or wallpapered after any necessary patching is done, the end result will be that much more attractive and long-lasting.

PLASTERS AND PLASTERING MATERIALS

Cementitious Materials

The three basic cementitious materials used as interior plasters are *gypsum, lime,* and *portland cement.* Each is manufactured in several different grades or types. It will suffice here to note the chief differences among them.

Gypsum plasters predominate in the United States because they set hard within a day, weather permitting, which allows three coats of gypsum plaster to be applied on three successive days. The principal defect of gypsum plasters is their susceptibility to moisture. If hardened gypsum plaster ever gets wet, it will disintegrate.

Up to the beginning of the twentieth century, lime plasters were used almost exclusively, with the (then) more expensive gypsum plasters being used only for ornamental plastering, where the value of their quick-setting characteristic outweighed their cost in dollars. The great virtues of lime plasters are their plasticity and workability. A lime plaster, properly aged, will spread as smoothly and easily as soft butter (*see* LIME). Also lime plasters are not affected by moisture once they have set hard. They may stain if they become wet, but they will retain their hardness and strength. The great defect of lime plasters is the fact that, though they may set quickly and stiffen within a few hours after application, it takes something like two weeks for them to dry hard and begin developing their full strength. For this reason, lime plasters are used today mainly as an admixture to both gypsum and portland-cement plasters and mortars. The lime improves the plasticity and workability of both gypsum and portland cement, both of which would otherwise be too stiff and hard-working to apply easily.

Portland-cement plasters are the strongest and most water-resistant of the three. They are not used much in the interior of dwellings, though they are preferred wherever there is likely to be a considerable amount of water present, as in a shower stall, a laundry room, or wherever the plaster is likely to receive hard knocks, as in a garage or workshop. They are also preferred in basements, where a gypsum plaster could be badly damaged by any moisture that penetrates the walls. Portland-cement plasters will not be discussed in this article, however, because their use indoors is so limited.

Sand

Neat plaster, that is, plaster without anything else mixed in it except water, shrinks badly as it dries out. As it shrinks, it cracks. This is true of all plasters—gypsum, lime, and portland cement. This shrinkage can be reduced to acceptable limits by adding sand to the plaster. Sand is nonshrinking. It is normally added to dry plaster in proportions that range from 2 to 3 parts sand to 1 part plaster, by weight. When the sand and plaster are wetted and mixed together, the plaster paste coats the particles of sand and, after the plaster has set, it binds everything together into one solid mass. In return, the particles of sand prevent the plaster from shrinking. It should be recognized that the mere addition of sand to a plaster prevents the plaster from achieving its full strength. As a practical matter, however, the loss in strength is more than compensated for by the fact that both shrinkage and cracking have been eliminated.

Sand is so much cheaper than plaster that adding sand to a plaster reduces considerably the overall cost of the plastering material. Therefore a builder may be tempted to oversand a plaster in order to reduce the plastering costs as much as possible. But oversanding a plaster weakens it considerably. Whenever the hardened plaster in a completed house falls spontaneously from a wall or ceiling, the most probable cause is excess sand in the plaster.

The sand used must be clean, which means free of any alkaline or salty substances, clays, loams, or any organic matter. Any of these impurities in the plaster mix will reduce its strength. An alkaline substance will attack the cementitious material directly. Salts accelerate considerably the time it takes the plaster to set, which may make it impossible for the plasterer to apply and smooth a batch of freshly made plaster before it hardens. Organic substances may, on the other hand, retard the setting of the plaster or even inhibit its setting altogether. If, therefore, the sand to be used is not clean, it must be washed until it is clean.

The sand should also consist of a full range of grain sizes from about 1/8 in. down to less than 1/64 in., and these grain sizes must be present in approximately equal proportions. The point in having a full range of grain sizes is that it enables the grains to pack together closely in the plaster. The plaster must be as dense as possible if it is to be as strong as possible, and this is best achieved by having a wide range of grain sizes in the sand. On the one hand, if there should be too great a proportion of large grains, the plaster will be porous and weak. If, on the other hand, there should be too great a proportion of small grains, the plaster will be difficult to work, and when it dries it will very likely shrink and crack excessively.

The sand should also be sharp-edged, not rounded. Rounded edges will reduce the strength of the plaster because the plaster paste cannot grip rounded grains as securely as it can grip sharp-edged grains.

Water

The water used for mixing plaster must be clean. The general rule is that if the water is good enough to drink, it is good enough for mixing plaster. In particular, it is a mistake to use any water that the plasterer has used to clean his tools with. If this water should then be used to make up a fresh batch of plaster, the particles of hardened plaster that are transferred into the water when the tools are washed will initiate the setting process prematurely, which may make it impossible to use all the fresh plaster before it hardens.

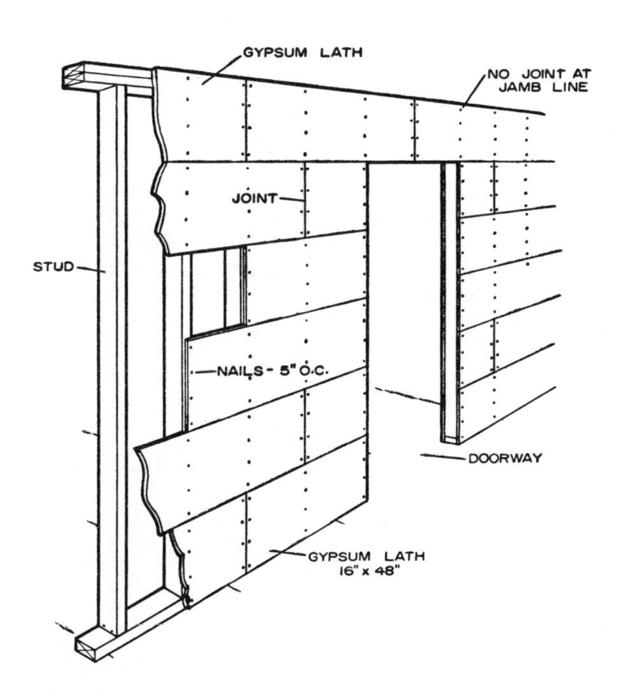

Fig. 1. The application of gypsum lath to a stud wall *(U.S. Forest Service)*.

PLASTER BASES

2-Coat and 3-Coat Plastering

Plaster may be applied to a wall or ceiling in either two or three individual coats. The number of coats applied and their total thickness will depend primarily on the type of base that is to support the plaster. If the base consists of wood or metal lath, then three coats of plaster will be required, which are called, in order of application, the *scratch* coat, the *brown* coat, and the *finish* coat.

If the base consists of gypsum lath (see Fig. 1), insulation-board lath, masonry of any kind (that is, brick, concrete blocks, gypsum tiles, or clay tiles), or concrete, then only two coats of plaster will be required, which are the brown and finish coats. In this *2-coat* work, the brown coat may be thought of as a combination scratch-and-brown coat.

Whether a separate scratch coat is required will depend primarily on the ability of the base to provide, by itself, adequate support for the brown coat, the brown coat being the principal structural layer of plaster. If the base *will* provide adequate support, then a scratch coat is unnecessary; if it won't, then a preliminary scratch coat is necessary. The qualities in a base that determine whether or not it can provide adequate support are (1) it must be rigid, and (2) it must have a certain suction quality to it that will bind the plaster tightly to it.

That the base should be rigid is self-evident, since any movement in the base will cause the plaster to crack. The metal lath that is often used as a base is quite flexible; this is one of the reasons that a scratch coat must be applied to it—the scratch coat stiffens and increases the rigidity of metal lath.

Suction

And what is *suction?* In plastering, suction is the property a material has of drawing water into itself like a sponge. That is, the base must have a certain amount of porosity. When wet plaster is applied to such a porous base, if the suction is adequate—not too much, not too little—the water in the plaster will be drawn into the base material and, with it, innumerable fine particles of plaster. A plaster sets hard because it forms a dense, interlocking network of fine, needle-shaped crystals. Therefore, when the particles of plaster that have been drawn into the base material set, the crystals that form within the minute crevices and voids on the surface of this base material will hold the base and the plaster tightly together.

Metal Lath and Suction

It is obvious that metal lath exerts no such suction. It is necessary, therefore, to apply a scratch coat of plaster over metal lath to provide the support for the brown and finish coats of plaster that the metal lath cannot provide. For a description of metal lath *see* STUCCO.

Masonry Bases and Suction

Different kinds of masonry have different porosities and thus different amounts of suction. Concrete blocks, most kinds of brick, and most clay tiles have what may be considered an average amount of suction, which means that the plasterer has ample time, using normal plaster mixes, in which to apply the brown coat to the masonry and smooth it out before it begins to set.

Other masonry materials, however, such as some grades of cheap, porous brick, tiles made from soft clay, and all gypsum tiles, are highly porous. When a layer of fresh plaster is applied over one of these high-suction materials, a considerable amount of water is sucked quickly from the plaster. As a result, there is a good chance that the plaster will dry out before it has a chance to set properly. If this should happen, the plaster will be weak and porous.

The plasterer can do one of three things to counteract excessive suction: (1) "kill" the suction by dampening the masonry with water beforehand, (2) make the plaster slightly more watery than normal, knowing that the excess water will be drawn from the plaster, which will leave just enough water in the plaster to allow it to set normally, or (3) deliberately mix the plaster using a higher than normal proportion of sand; since one of the properties of sand is that it can retain a large amount of water, this will counteract the suction and give the plaster time in which to set properly.

Of these three alternatives, attempting to kill the suction by dampening the masonry beforehand doesn't work because it is too difficult to control the amount of water absorbed by the masonry. Either the water will be absorbed too quickly, which will leave the plasterer where he was originally, or the base will become soaked with water, which will prevent the plaster's obtaining a solid grip on the masonry.

If the plasterer should make the plaster more watery than usual, it will be too watery to apply properly, and it will fall off the walls and ceilings because it lacks sufficient cohesiveness.

In practice, therefore, excessive suction is counteracted by mixing more sand into the plaster. This may result in a somewhat weaker plaster, but, if the amount of sand that is added is closely watched, the loss in strength will be negligible and the plaster will hold tight to its base, which will more than compensate for any increase in weakness.

On the other hand, if the masonry should have very poor suction, which happens when the masonry consists of glazed brick or tile or stonework, the plasterer will have to decrease the amount of sand in the plaster in order to increase the strength of the plaster as much as possible.

PREPARATION FOR 3-COAT PLASTERING

Before plaster can be applied to a wall or ceiling, the wall or ceiling must be suitably prepared. In the typical wood-frame house, this means that lathing must first be attached to the studs and ceiling joists. In addition, *grounds* must be installed by the carpenter all around the edges of all doors and windows and along the bottom edges of the walls.

Grounds

Grounds are usually 1 × 2 in. strips of wood, which means the actual size of the wood is ¾ × 1¾ in. The grounds are nailed to the bottoms of the studs by the carpenter as a support to which he will later nail the wood trim that is installed along the bottoms of the walls after the plastering has been completed. Grounds are also installed around door and window frames for the same purpose.

The grounds have another purpose, however. They determine the thickness of the brown coat of plaster. In order that they may serve this function, the grounds must be as straight, level, and/or plumb as the carpenter can make them, since the levelness of the finished plaster surface will to a large extent depend on the accuracy with which the grounds have been installed. The absence of grounds, grounds that weave in and out, or grounds that are not plane relative to each other, will result in an uneven wall surface and in gaps alongside the door and window frames where plaster meets trim—a sure sign of a poorly done job.

Very often in the cheapest construction the builder will not install grounds around the doors and windows at all. He will instead depend on the edges of the door and window jambs to act as guides for the plasterer. (The final wood trim will later be nailed directly to the edges of these jambs as well; *see* DOOR, WOOD; WINDOW.) In this case, the jambs must be wide enough that they come flush with the surface of the finished plaster. The trouble with using the jambs in this way is that the edges of the wood will likely be marked and scored by the plasterer, which will detract from their finished appearance. The wood will also absorb considerable moisture from the wet plaster, which may cause it to shrink and warp.

Ideally, the builder should install temporary door and window jambs that are set in place very accurately. They are removed after the plaster has hardened. Since these temporary jambs are made to the exact size of the finished door and window jambs, the finished jambs can then be installed without any trouble. However, this is rarely done today.

Mixing Plaster

The proportion of sand and cementitious material used to make the plaster for the scratch, brown, and finish coats is shown in Tables 1 and 2. These proportions are based on weight and on the following relationships:

1 part gypsum = 100-lb bag
1 part hydrated lime = two 50-lb bags
1 part lime putty = 1 cu ft
1 part sand = 100 lb

Since it is difficult to weigh out sand directly on most jobs, the weight of the sand is calculated on the basis of the number of shovelfuls it takes to equal 100 lb. A No. 2 shovel is used and 7 shovelfuls of damp sand are assumed to be equal to 100 lb. The shovel must be full but not overloaded. Shovelfuls may not be the most accurate method of calculating the weight of sand, especially since the weight of sand depends very much on how wet it is, but, nevertheless, this measure has proved accurate for most practical purposes, and if the sand is wetter than normal, it means there is less chance of the plaster being oversanded, since a slightly larger proportion of each shovelful will consist of water.

Tables 1 and 2 do not give the proportion of water to be added to a dry mix. The amount of water will range from 7 to 9 gal for each batch of cementitious material and aggregate listed in the tables. The water is not added by quantity, however, but according to the plasticity of the plaster. The plasterer judges the sufficiency of the water content by the feel of the trowel or hoe as it is drawn through the plaster. The amount of water to be added is thus a matter of experience and judgment.

Small quantities of plaster (that is, when only a few bagfuls of plaster are sufficient for the job) are mixed by hand in a mortar box. Half the required amount of sand is spread over the bottom of the box. All the cementitious material is then spread over the sand. Finally, the remainder of the sand is spread on top of the cementitious material. The sand and plaster are then mixed with a hoe until the color of the mixture is the same throughout.

The mixture is then pulled to one end of the box and a gallon or two of water is poured into the opposite end. The sand and plaster are now mixed into the water little by little, using short, chopping, back-and-forth strokes to cut and mix everything together thoroughly. Additional water is added as necessary until all the plaster has been wetted in this way. The plaster is now thoroughly hoed, additional water being added as necessary, until the plasterer is satisfied with the consistency of the plaster. At this point he stops mixing the plaster as overmixing merely stimulates the plaster to begin setting prematurely.

Mixing plaster by machine is much preferred to hand mixing, as the job goes much faster and the plaster has a more uniform consistency. The *plaster mixer* must be in operation all the while that material is being added into it. The required amount of water is poured into the mixer first, followed by half the required amount of sand, all the cementitious material, then the remainder of the sand. The mixer is allowed to run for about 1 min more, and the plaster is then dumped into a mortar box, wheelbarrow, or some other container.

When a plaster mixer is being used, it is wise to be stingy with the water at first, since additional water can always be added if the mortar turns out too stiff. If one were to add too much water at the beginning of the mixing, it would be necessary to add additional sand and cementitious material to increase the consistency of the mix. This may turn into some-

Table 1. Proportions of Cementitious Materials and Aggregates Used for the Base and/or Scratch Coats

	Types of plaster					
Proportion units	Gypsum	Lime		Lime-portland cement		Portland cement
		Dry mix	Putty mix	Dry mix	Putty mix	
1 Part cementitious material =	100 lb gypsum	100 lb hydrated finishing lime (Type S)	1 cu ft lime putty (Type N)	100 lb hydrated finishing lime (Type S) + 94 lb portland cement	Cu ft lime putty (Type N) to 94 lb bags of portland cement	94 lb portland cement
1 Part aggregate =	100 lb sand or 1 cu ft perlite or vermiculite	100 lb sand	100 lb sand	100 lb sand	100 lb sand	100 lb sand or 1 cu ft perlite or vermiculite
Plaster bases	2-coat work					
Gypsum lath	1:2.5					
Gypsum partition tile High suction masonry Medium suction masonry Low suction masonry	1:3 or 1 wood fibered plaster: 1 sand	1:7.5	1:3.5	1:7.5	2:1:9	
Concrete ceilings	Bond plaster					
	3-coat work					
Gypsum lath Metal lath	1:2, 1:3 or both 1:2.5	1:6.75, 1:9	1:3, 1:4	1:7.5, 1:9	1:1:6, 2:1:7	Both 1:3 to 1:5*
Gypsum partition tile High suction masonry Medium suction masonry Low suction masonry	Both 1:3 or both 1 wood fibered plaster: 1 sand					Both 1:3 to 1:5*
Concrete walls and columns	Sc.-Bond plaster, Br.-1:3					

*Up to 10% by weight of dry hydrated lime or up to 25% by volume of lime putty may be added for each part of portland cement as a plasticizer.

Note: Hair or fiber may be added to gypsum plaster scratch coats and should be added as follows (lb fiber per cu yd mortar): Lime: Scratch, 6, Brown, 3.4; Lime-portland cement: Scratch, 6, Brown, 3; Portland cement: Scratch, 4 to 5.

Source: J. H. Callender, Ed., *Time-Saver Standards*, 5th ed., McGraw-Hill, 1974, p. 546.

Table 2. Properties of Cementitious Materials and Aggregates Used for Different Types of Finish Coats

Trowel finishes		
1. Lime putty		: Gypsum gauging plaster
(a) 3		: 1 by volume
		Equivalents
200 lb dry hydrate		: 100 lb gauging plaster
5 cu ft putty		: 100 lb gauging plaster
40 gal putty		: 100 lb gauging plaster
(b) 4		: 1 by volume*
		Equivalents
300 lb dry hydrate		: 100 lb gauging plaster
7.5 cu ft putty		: 100 lb gauging plaster
60 gal putty		: 100 lb gauging plaster
2. Lime putty		: Keene's cement
(a) Medium-hard finish		
50 lb dry hydrate		: 100 lb
		Equivalents
not more than 100 lb putty		: 100 lb Keene's Cement
not more than 1¼ cu ft putty		: 100 lb Keene's Cement
not more than 9 gal putty		: 100 lb Keene's Cement
(b) Hard finish		
25 lb dry hydrate		: 100 lb
	Equivalents	
not more than 50 lb putty		: 100 lb Keene's Cement
not more than ⅝ cu ft putty		: 100 lb Keene's Cement
not more than 4¾ gal putty		: 100 lb Keene's Cement
3. Lime putty		: Portland cement
200 lb dry hydrate		: 94 lb (bag)
		Equivalents
5 cu ft putty		: 94 lb
4. Portland cement		: Sand†
94 lb (1 bag)		: 300 lb‡
5. Gypsum gauging or neat plaster		: Vermiculite fines
100 lb		: 1 cu ft
Float finishes		
1. Lime putty	: Keene's cement	: Sand
2	: 1½	: 4½ by volume
2. Lime putty:	Gypsum gauging plaster	: Sand
1 dry hydrate:	1½	: 2.3 by weight
3. Lime putty	: Portland cement	: Sand
2 dry hydrate	: 1	: 2.5 by weight
4. Lime putty		: Sand
1		: 3 by volume
5. Gypsum neat plaster		: Sand
1		: 2 by weight

Notes: 1. Lime finishes may be applied over lime, gypsum, and portland cement basecoats, other finishes should be applied only to basecoats containing the same cementitious material.

2. A gypsum-vermiculite fines finish should be applied only to gypsum-vermiculite basecoats.

3. Lime equivalents based on Type N hydrated lime.

*Specifications of the Finishing Lime Association of Ohio.

†Finish may be troweled or floated.

‡Lime may be added as a plasticizer in amounts up to 10% by weight of portland cement if dry hydrate or 25% by volume of portland cement if putty.

Source: J. H. Callender, Ed., *Time-Saver Standards*, 5th ed., McGraw-Hill, 1974, p. 547.

thing like what happens when an amateur painter tries to match colors. He adds first this color, then that, then a little more of the first color, until he finally ends up with several unnecessary gallons of paint.

The Scratch Coat

Both gypsum and lime plasters are applied to lath in the same way. Holding a hawk in one hand and a rectangular trowel in the other (see Fig. 2), the plasterer cuts a large batch of plaster from the mortarboard and scoops it up on his hawk by shoveling the hawk and trowel together with the plaster between. The hawk is somewhere between 10 to 14 in. sq, which is large enough to carry several trowelfuls of plaster. Its use makes it unnecessary for the plasterer to return continually to the mortarboard to pick up individual trowelfuls of plaster. The trowel usually used has a steel blade that is 4½ × 11 in. in size.

Picking up a trowelful of plaster from his hawk, the plasterer applies the plaster to the lath with an upward sweeping motion of the trowel across his body. The angle at which the trowel is held to the surface of the lath and the force with which it is pressed against the lath determine how thick the coat of plaster will be. A shallow angle between trowel and lath makes for a thick coat of plaster; a deep angle makes for a thinner coat.

The scratch coat is usually ¼ in. thick, certainly not more than 3/16 in. thick. When applying the plaster to the lath, the plasterer uses just enough force to barely push the plaster completely through the lath, so that the plaster slumps slightly on the opposite side of the openings (see Fig. 3). Excessive force is unnecessary as it will only push an excessive amount of plaster through the openings, and this extra plaster will simply fall off the back side of the lath to the floor. After the plaster has hardened, the fact that it has slumped over the openings will ensure that it is securely *keyed* to the lath.

After a sufficient area of the wall or ceiling has been covered, and before the plaster has a chance to set hard, the plasterer goes back over the plaster to even the surface a bit. He then scores the surface of the plaster with a *scratcher* (see Fig. 2), a small, rake-like tool with which he scribes a series of ⅛-in.-deep grooves in the plaster both horizontally and vertically (*see* STUCCO, Fig. 7).

When gypsum plaster is being used, the plaster will begin to set 2 to 4 hr after it has been mixed with water (the time of set can be controlled at the mill by the manufacturer, or the plasterer can adjust the time of set as described below), and the plasterer can schedule his work accordingly.

When lime plaster is being used, the plaster will take about a day to set. After having smoothed the plaster, the plasterer allows it to set overnight until it has become somewhat firm. He then scratches the surface as described above. There are also lime plasters on the market to which accelerators have been added; these plasters will set in 2 to 6 hr. If he uses one of these plasters, the plasterer can schedule his work just as he would do with a gypsum plaster.

Once the plaster has been scratched, it is left alone to harden completely. For a gypsum plaster, this will take overnight. For a lime plaster, it will take a week. Before the brown coat is applied, as described below, the scratch coat must be hard enough that the plasterer cannot make an indentation in it with his thumbnail. Failure to allow the scratch coat to dry hard may result in its cracking when the brown coat is applied.

The Brown Coat

The brown coat is the thickest of the three coats of plaster. The brown coat gives a plastered wall or ceiling most of its strength and solidity. It also acts as the base for the coat of finish plaster, and it is, therefore, made as plane as possible, with all the corners of the room accurately and neatly formed. For 3-coat work, the brown coat is from ½ to ¾ in. thick, depending on the straightness of the scratch coat and on whether the plaster is made from gypsum or lime. Because lime plaster is weaker than gypsum plaster, the brown coat made from a lime plaster is often ¼ in. thicker than when gypsum plaster is used.

It is important that the brown coat not be thinner than ½ in.

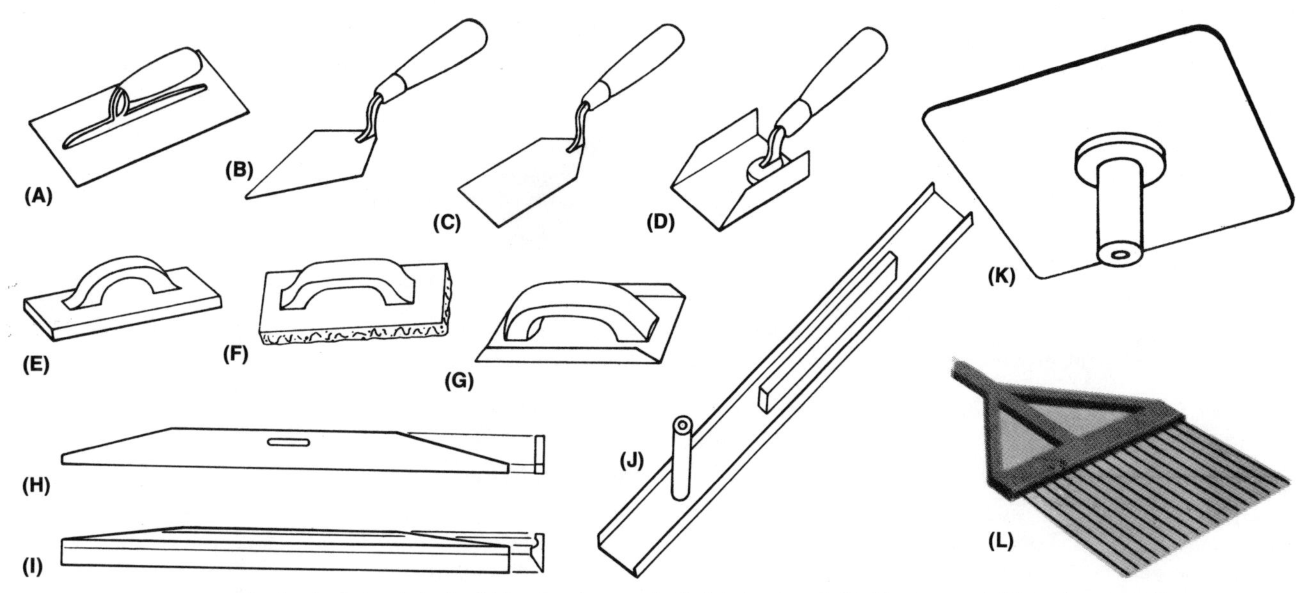

Fig. 2. A plasterer's tools. (A) Rectangular trowel. (B) Pointing trowel. (C) Margin trowel. (D) Angle trowel. (E) Wood float. (F) Sponge float. (G) Angle float. (H) Rod or straightedge. (I) Featheredge. (J) Darby. (K) Hawk. (L) Scratcher.

Studies have shown that if the thickness is reduced to 3/8 in., the probability that the coat will crack increases as much as 60 percent. If the thickness of the brown coat is reduced to 1/4 in., the probability that cracks will occur increases by as much as 82 percent.

But a gypsum brown coat should not be made too much thicker than 1/2 in., either. The thicker the coat of plaster, the heavier and more rigid it becomes. The extra weight may cause the plaster to fall off the wall or ceiling if the adhesion between the scratch and brown coats is not as strong as it should be (or, in 2-coat work, if the adhesion between the base and the brown coat is not as strong as it should be).

Fig. 3. The application of the scratch coat on a metal-lath backing. The plaster is securely keyed to the lath because it slumps over the back of the wire mesh before setting *(U.S. Gypsum Co.)*.

Brown-Coat Application

The plaster is applied using a hawk and trowel and following the same basic technique that has been described for the application of the scratch coat. The brown coat is applied in two layers, the second layer following immediately after the first, a technique called *doubling-up.* Picking up a trowelful of plaster from his hawk, the plasterer applies it quite forcibly in a thin layer upon the scratch coat. The object is to obtain as good adhesion as possible between the scratch and brown coats, and this takes force. A small section of the surface thus having been covered, the plasterer immediately follows with the application of the second layer, the plaster being applied thickly enough to bring it up to the final thickness.

Whether gypsum or lime plaster is used, the plaster must be on the poor side, as shown in Table 1. A mortar that is too rich (that is, with too high a proportion of plaster in the mix) tends to have a sticky quality that makes it difficult for the plasterer to smooth it out evenly. A plaster that is too rich will also shrink an excessive amount as it sets. The plaster may not only crack as it dries out, but the bond that has been established between the scratch and brown coats may also be weakened.

In addition to his trowel, the plasterer uses two other tools, a rod and a *darby,* to help him level and smooth the brown coat. The rod is used to straighten the surface of the brown coat. Any low spots are filled in with additional plaster, and the brown coat is then straightened out again with the rod.

When the plasterer is satisfied with the basic flatness of the surface, he switches to his *darby.* The darby brings the surface of the plaster to the desired degree of planeness, and it also gives the brown coat the desired amount of smoothness. The darby is made of wood and it is about 6 in. wide and from 3½ to 4 ft long. It is used something like an oversize trowel. The plasterer grips it by two handles. He then works the darby back and forth over the surface of the plaster just as he would a trowel, holding it at an angle to the direction of movement and with the leading edge raised slightly off the plaster. The darby smooths and levels the surface of the plaster by pushing the plaster away from high spots and into the low spots.

As the brown coat is rather stiff-working, the plasterer may have difficulty moving the darby back and forth easily and smoothly. When this is the case, he can lubricate the plaster a bit by dashing water upon it with a *browning brush,* which is from 4 to 5 in. wide and 2 in. thick, with bristles about 6 in. long. He must be careful not to throw too much water on the plaster as this may result in a superficial layer of soft, weak plaster to which the finish coat will not adhere. The darby is then carefully worked into the corners of the room until the surfaces abutting the corners are as smooth and level as possible.

The darby, however, cannot get into the actual corners. To define the corners more sharply, the plasterer can use his regular trowel to work over the plaster. More often, he uses a two- or three-sided trowel called an *angle* trowel. The angle trowel can finish a corner very cleanly without undercutting the plaster, which can happen when an ordinary trowel is used.

Another tool the plasterer uses to define the corners is a *featheredge*. This is a straightedge that is 3 to 4 ft long, one side of which tapers to a sharp edge. The plasterer carefully works the featheredge into the corner and then draws it away from the corner to remove any excess mortar. The featheredge is worked into the corner again and is then drawn away against the opposite wall. In this way the plasterer can both define a corner as sharply as possible and straighten it as well, more accurately than by using an angle trowel alone. Actually, the angle trowel and featheredge are used together: the featheredge to straighten and define the corner itself and the angle trowel to remove excess plaster and smooth the wall surface along both sides of the corner.

The final step in the application of the brown coat is to cut the mortar away from around the grounds to a depth of about ⅛ in. This will allow the finish coat to be brought flush to the grounds when it is applied.

Application of Lime Plaster

The basic technique of applying a brown coat that has just been described, and which is called *browning* in the trade, applies equally well to lime plaster. If, however, a lime plaster is used for the brown coat, special care must be taken that the plaster is on the poor side to prevent shrinkage as much as possible. It is also wise to mix into the plaster about ½ lb of fibers per cubic foot of lime putty (or ½ lb of fibers per 100 lb of dry hydrated lime) to enable the plaster to hold itself together while it is setting.

In addition, a brown coat made of lime plaster requires one other step after the darbying has been completed. The day after the brown coat has been applied, while the plaster has set but has not yet hardened, the plasterer must go over the entire surface again with a *devil's float.* This is a trowel with a wooden blade (all trowels with wood blades, of whatever design, are called *floats*) that has one or two nail points projecting from the bottom of the blade by about ⅛ in. When this float is worked over the plaster with a circular motion, the nail points score the plaster much like a scratcher does a scratch coat. The object is to secure better adhesion between the brown coat and the finish coat.

This floating operation is also necessary to compact the plaster, as lime plaster usually shrinks so much overnight that a great many hairline cracks will have appeared. Floating, therefore, also compacts the plaster more solidly and removes the cracks as well as providing a rough surface.

Following the darbying (and floating), the brown coat is left to dry out completely. For a gypsum plaster, this will take a day or two, depending on weather conditions and the thickness of the coat. For a lime plaster, this will take about two weeks.

The Finish Coat

The finish (or *skim*) coat is a 1/16- to ⅛-in.-thick layer of plaster that is applied over the brown coat. It is the plaster you see when you look at a plastered surface. To a very large degree, the finish coat will be only as good as the brown coat, since the brown coat not only provides the solid support that is necessary if the finish coat is not to crack or chip, the brown coat must also have just the right amount of moisture in it when the finish coat is applied; otherwise the two coats will not adhere tightly to each other.

There are two main types of finish coat. First, and by far the most common, is a very smooth, very white, very hard, brittle finish that is known variously as the *cream* coat, *hard* coat, *putty* coat, *smooth* coat, or *white* coat, with putty coat being the most common name. Second, there is a softer, weaker, sandy-textured finish that is known, appropriately enough, as a *sand* finish. It is also called a *float* finish, as it is produced with a float, as described below. Other major differences between the putty-coat finish and the sand finish are in their recipes and in the way they are applied and smoothed.

Putty-Coat Finish

As shown in Table 2, the plaster for a putty coat consists of 3 parts lime putty to 1 part gypsum. The gypsum is of a quality known as *gauging plaster,* which means it has been ground very fine so that it makes a very smooth, lumpless paste. The lime putty provides the plasticity and workability that is necessary if the plasterer is to apply the finish coat to the brown coat smoothly and evenly. The gypsum gives the plaster its strength and quick-setting properties.

When ordinary gauging plaster is used, the finish coat will set in about 2 hr. If the plasterer wants more time than this, because he is plastering a very large room, say, he can add a small amount of retarder to the plaster that will delay the setting time an additional 2 to 4 hr, depending on the amount of retarder added. But if he should add too much retarder to the plaster by mistake, there is a chance that all the water in the finish coat will have evaporated or will have been sucked into the brown coat before it can set completely. As a result, a network of fine cracks will appear over the entire surface of the finish coat after it has dried out completely. To avoid this mischance, instead of adding retarder to the plaster himself, the plasterer can purchase bags of gauging plaster to which the correct amount of retarder has already been added. Different varieties of gauging plaster are sold that will set in 2, 4, or 6 hr, depending on the amount of retarder added. For the method of preparing lime putty, *see* LIME.

But before the plaster is mixed and applied, the brown coat must have just the right amount of moisture in it. Without a special instrument that measures the water content of the brown coat (such instruments exist, but they are expensive), the plasterer must depend on experience when judging the amount of dampness in the brown coat. He looks at the color of the plaster, he runs his hands over it, and experience tells him whether or not the wall is at just the right state of dampness for the putty coat to be applied. His experience must not deceive him, for if the brown coat should turn out to be too dry, too much moisture will be sucked too quickly from the putty coat, the putty coat will be unable to set properly before it dries out, and it will end up having a network of very fine cracks. If the brown coat is too dry, the plasterer must spray the surface with water and then give the brown coat time to absorb the water evenly.

If, on the other hand, the brown coat is too wet (what is referred to in the trade as a *green* wall), the finish coat won't adhere well because the plaster won't be able to get a good grip on the brown coat. In this case the plasterer must wait a bit longer. Once satisfied as to the condition of the brown coat, the plasterer can begin to mix the plaster for the finish coat.

The lime putty and gypsum are mixed on a *finishing board,* which is the same as a mortar board. It is usually 5 ft sq and is placed on a stand or on saw horses to make it convenient for the plasterer to work.

The lime putty is measured out first and placed on the

finishing board in the form of a large ring, with an extra lump of lime putty in the center of the ring.

A measured amount of clean water is now poured into the center of the ring, and a measured amount of gypsum plaster is sifted slowly into the water through a screen. The sifting must be done slowly enough to allow each particle of gypsum to become soaked individually, thus preventing the formation of lumps. If both water and gypsum have been measured out accurately, all the gypsum will combine with all the water, without an excess of either.

Lime putty and gypsum are now mixed together. With a trowel in one hand and a hawk in the other, and working from the inside of the ring out, the plasterer chops, cuts, and stirs the lime putty and gypsum until they are completely mixed. The final mortar should be very stiff. It should also be very smooth and creamy. Once the plasterer is satisfied that he has mixed the lime putty and gypsum completely, he stops working the plaster and thereafter allows it to rest as quietly as possible; for any further working will only accelerate the setting time, which has commenced with the mixing of the gypsum and water.

Application of the Putty Coat

The plaster is applied first in the corners of the room, and to its full thickness (that is, from $^{1}/_{16}$ to $^{1}/_{8}$ in. thick). The plasterer uses a featheredge to define the corners, as already described. A thin layer of plaster is now forcibly applied to the rest of the brown coat to achieve a solid bond between the brown and finish coats. This thin layer is followed by another thin layer that builds up the finish coat to its final thickness. All this work must be done as evenly as possible.

The mortar having been applied (the total area covered at one time will depend on the overall size of the room and the setting time of the mortar), the plasterer now takes a wood float, and working from the corners toward the center of the walls and ceiling, he further straightens, compacts, and smooths the plaster. Floating also removes small depressions in the surface known as *cat faces* that are the result of slight depressions in the brown coat.

This floating operation is called *drawing up* or *laying down* or *tightening* the surface—take your pick. It is often omitted from a plastering job, but it shouldn't be, as it determines just how flat, smooth, and level the final surface will be.

The plaster will have a glossy look to it all the time the plasterer is working it. Once it begins to set, however, it turns dull-looking. The plasterer then begins to give the finish coat its final polish and compaction. Taking in one hand a soft *finishing brush,* which is like a browning brush but with finer, more pliable bristles (or using a sponge instead), and with a rectangular trowel in the other, the plasterer dips the brush (or sponge) in water, brushes the surface of the putty coat with water and follows immediately after with the trowel. He goes over the entire surface with a rapid, sweeping, up-and-down motion, reaching as high as he can, the trowel following close behind the brush (or sponge). The corners of the room are given one final smoothing with an angle trowel, and the entire polishing operation with the brush (or sponge) and trowel is repeated.

When the putty coat has set too hard to be troweled any further, the plasterer brushes the wall down with a wet brush to remove any small dribbles of plaster that may have been left behind.

Sand Finish

The plaster used for a sand finish is very like the plasters used for the scratch and brown coats. As shown in Table 2, the proportions of lime, gypsum, and sand vary according to the working qualities desired in the finished coat. As with a putty-coat finish, additional gypsum may be added to a sand-finish plaster to increase the hardness of the coat, or gypsum may be left out of the plaster entirely, which then becomes an ordinary sand-lime plaster.

The sand used is an especially selected white silica sand that can be obtained dry at building-supply dealers. The sand will have been washed and graded and will have had the very fine particles screened out, as these particles tend only to reduce plaster strength and increase the chances of its cracking.

The degree of sandiness desired in the finish coat can be varied by adding as much or as little sand as desired. To avoid weakening the plaster, the proportion of sand to cementitious material should not, however, exceed 3 parts sand to 1 part cementitious material.

The brown coat should have much more moisture in it than is the case when a putty coat is being applied. Again, it takes experience to judge just how damp the brown coat actually is. More important than the average amount of dampness is whether the dampness is even over the entire area of the brown coat. If it isn't, the areas that are damper will give the sand finish a coarse look, and the areas that are drier will give the sand finish a smoother look.

The brown coat should not be too dry either, as the finish coat will tend to roll under the trowel, and the plasterer will find it difficult to lay the plaster down evenly and smoothly.

The plaster is then left to set overnight. When the plasterer returns the next morning, the plaster will have lost its glossy, wet look, an indication that it has begun to set hard and that it can now be finish-floated. This final floating is done with a float to which a layer of sponge rubber has been glued. Sponge rubber has replaced cork, carpeting, and felt—materials that have been favorites in the past. The rubber is supposed to give the plaster surface an even texture that makes it look like sandpaper. This final floating is not intended to level or compact the plaster. The sponge rubber merely follows the existing surface.

2-COAT PLASTERING

Two coats of plaster are structurally equivalent to three coats whenever the base is rigid enough to support the brown and finish coats without a scratch coat being necessary. In practice, therefore, 2-coat plastering is used whenever the support consists of masonry of any kind or gypsum lath. The only reasons that three coats of plaster may be necessary over any of these base materials is (1) when a local building code requires three coats of plaster regardless of the type of base or (2) when the surface of the masonry is so uneven that three coats of plaster are necessary to level the surface.

The application of the brown coat in 2-coat plastering is much the same as the application of a brown coat over a scratch coat. The plaster is applied in two layers, the first being a thin layer to secure adhesion to the base, the second being thicker to build up the plaster to the height of the grounds. The entire brown coat is straightened and brought flush with the grounds using a rod and darby as has already been described.

The actual plaster mix and the technique by which the plaster is applied will depend on the amount of suction in the base. For brick, tile, and other kinds of masonry that have a normal amount of suction, the plaster usually consists of 3 parts sand to 1 part cementitious material. That is, the plaster is on the poor side. When the masonry has a great deal of suction, the plaster may be made as poor as 4 parts sand to 1 part cementitious material. Gypsum lath has rather poor suction and, therefore, requires a relatively rich plaster mix, one consisting of 2 parts sand to 1 part cementitious material, usually, to make sure that a good bond is established between the lath and the plaster.

Gypsum lath is by far the most common base used for 2-coat plastering in dwellings. In the usual plastering job, it is rare for the brown coat to be leveled with a rod and darby. Instead, the plasterer uses a *slicker*, a flexible length of wood or metal about 3 ft long that resembles a wide yardstick. The plaster is pushed about with the slicker to straighten and level its surface. The final smoothing is done with a float, as usual.

And very often in domestic plastering, the brown coat applied over gypsum lath is not even ½ in. thick, as it should be, but only ¼ in. thick. The intent is, of course, to save as much as possible on the cost of the plastering, at the ultimate expense of the homeowner.

The problem that arises here is that a coat of plaster that is this thin has not the strength to withstand even the small stresses that normally occur in a wood-frame house, such as the seasonal changes in the moisture content of the wood framing that cause the entire house to swell and shrink ever so slightly over the course of a year. As a result, the plaster tends to crack more frequently than it should. Whether any cracks that occur in a particular house are due to insufficient thickness of the plaster can easily be checked by poking a bit into the cracks to see (1) whether the base consists of gypsum lath and, if so, (2) whether the plaster is closer to being ¼ in. thick than ½ in. thick.

Once the brown coat has been completed and allowed to dry hard, the finish coat can be applied in the same way as already described for 3-coat plastering.

MACHINE PLASTERING

The scratch, brown, and finish coats can also be applied by a *plastering machine* (see Fig. 4). This machine consists of an engine-driven worm- or piston-type pump that forces fresh plaster from a hopper, into which the plaster has been dumped, through a hose to the place of application. The plaster is discharged by compressed air from the end of the hose as a spray. The compressed air is obtained from a compressor driven by the same engine that drives the plaster pump. The rate at which plaster is pumped through the hose depends on the speed of the pump. The spray pattern of the discharged plaster, whether fine or coarse, depends on the air pressure.

Machine plastering is done by a crew of five to eight men: one man who operates the nozzle (the *nozzle man* or *gun man*), two to four plasterers who follow after the nozzle man to straighten and smooth the plaster (the *follow-up men, rod men,* or *straighteners*), and two or three men (the *tenders*) who operate the mortar mixer and plastering machine and make sure that everything is available and in place for the nozzle man and follow-up men.

The actual spraying goes so swiftly that the entire job must be planned carefully in advance. The floors and all doors and windows must be protected from flying plaster spray by placing building paper or polyethylene film over them. All the scaffolding that enables the workmen to reach the ceilings and upper parts of the walls must be in place. The plaster, sand, and water must be at hand. Then, once the plastering machine is started up, it can operate continuously until all the plaster has been applied. This is the only way that machine plastering can be done economically.

Fig. 4. A plastering machine *(Gypsum Association)*.

The work schedule must also be planned so that the plastering machine can be shut down about an hour before quitting time so that the pump, hose, and nozzle can be completely disassembled and cleaned thoroughly of any plaster. If this plaster inside the equipment were to set hard, it would, of course, prevent the equipment from working altogether.

The amount of plaster that is actually sprayed on any surface depends on how far the nozzle is from the surface (it is usually 18 to 24 in. away) and by the speed with which the nozzle man moves the nozzle back and forth over the surface. The plaster is discharged from the nozzle in a narrow spray pattern. The nozzle man avoids getting plaster on surfaces where it isn't wanted by controlling the angle at which this spray pattern strikes the surface.

In general, apart from the fact that the plaster is being sprayed on wall and ceiling surfaces, the plaster is applied in the same thicknesses and in the same number of coats as in hand plastering. For 3-coat work on a wire-lath base, the wire lath is first filmed over lightly with a fine spray of plaster. After this plaster has set, the rest of the scratch coat can be sprayed on to the required thickness. For other types of bases, the plaster is sprayed on as thickly as necessary in one pass.

The follow-up men follow quickly after the nozzle man. The plaster must still be rodded and darbied into a plane, smooth coat, and the corners of every room must still be carefully and sharply defined, and all this work must still be done by hand.

The one great advantage of machine-applied plaster is the speed with which the job can be completed. A crew can apply the scratch and brown coats in half the time it takes a crew of plasterers working with hawk and trowel, including the time it takes to set up the equipment and scaffolding, straighten and level the plaster, and clean up afterward. Obviously, this time advantage increases with the amount of surface area that is to be plastered, as in a commercial or industrial structure. The cost advantage of machine plastering is less apparent in the typical wood-frame dwelling, and it disappears altogether if only two or three rooms are to be plastered.

Machine plastering has its disadvantages. It is a very noisy, messy job. The air is filled with a mist of wet plaster that settles on everything. If this film is not washed away immediately and it hardens in place, there is the devil's job removing it afterward. The original covering up of doors and windows and other not-to-be plastered surfaces must be done very carefully, for if there are any openings in the protective coverings, the plaster is sure to find its way through.

The stiff plaster that is used for hand application of the scratch, brown, and finish coats is much too thick to pass successfully through the plastering machine. The plaster must be made much more fluid for this purpose, but the more fluid the plaster, the more it shrinks as it dries, and the more porous and weak it will be when it finally does dry. Machine-applied plaster will be weaker than hand-applied plaster (though it will adhere to its base more tenaciously because of the force with which it has been applied), and there is a greater likelihood that cracks will appear, especially if the plaster has been applied a bit too thickly. In addition, to prevent the plaster's hardening within the equipment whenever the machine must be shut down for a time, a considerable amount of retarder must be added to the plaster, and this, too, may adversely affect the quality of the final plaster coat, as the plaster may dry out before it has a chance to set properly.

Plumbing

Domestic plumbing systems consist basically of three separate piping systems—the cold-water supply, hot-water supply, and drainage. Apart from the fact that the incoming hot and cold water and outgoing wastewater are carried within pipes, and that these pipes are installed by the same person, a plumber, the systems all function independently of one another. The only connection between the hot- and cold-water systems is at the fitting where the water that is to be heated is drawn off from the cold-water supply. (The hot and cold water may later be mixed together by a valve just before being discharged into a fixture, such as a sink or bathtub, but this is a peculiarity of the individual installation and the two systems remain separate and distinct.)

There is definitely no connection whatsoever between the two water-supply systems and the drainage system, except that they may connect to the same fixtures. They are completely separate. They do not even function according to the same hydraulic principles. The water supply circulates under pressure through a closed system of pipes. The waste, on the other hand, is transported to a public sewer or septic tank by gravity flow in a system of pipes that is vented to the atmosphere. The drainage pipes, therefore, are often made from different materials than the water-supply pipes and are assembled with different types of fittings and joints.

DRAINAGE SYSTEM OPERATION

Despite the apparent complexity of Fig. 1, which shows the drainage system of a typical 2-story dwelling, the flow of wastes through such a system is quite simple and direct. Starting at a plumbing fixture—any fixture—the wastewater enters the drainage system through a *trap.* It then flows horizontally a short distance through a *branch line* until it reaches a *waste stack,* which collects the wastewater from one or more branches and conducts the wastewater to a house drain located in the basement or under the house. When excrement and urine are discharged from a water closet into a stack, the stack is called a *soil* stack. (It is the universal custom among plumbers to refer to all the horizontal pipes in a plumbing system, except for the house drain, as *branches* and the vertical pipes as *stacks,* a custom we will adhere to loosely in this article.) Finally, the house drain conducts the wastewater to the *building sewer,* which starts at a point at least 5 ft outside the foundation wall and discharges the wastewater into a public sewer or septic tank.

The flow of wastewater is almost always by gravity. Thus, it is extremely important that both stacks and branches have a minimum number of bends or offsets in them, and that the horizontal branches have a slight pitch, or slope, that allows the wastewater to be carried along by gravity at a good rate of speed. The usual pitch is ¼ in. per ft.

Each branch must be large enough to carry off as quickly as necessary all the wastewater that is likely to be discharged into it without its capacity being taxed. The branches should not be too large, however, as this will not only increase the cost of the installation unnecessarily, but it may also result in water flowing through the pipes so slowly that any solid matter suspended in the water will settle to the bottom of the pipes, which may eventually clog them. Put another way, the pipe diameters should be small enough that the branches are self-scouring but not so small that the water will back up in the pipes.

Since several branches usually empty into one stack, stacks must obviously be larger in diameter than branches. The wastewater discharged into a stack descends within it in a peculiar way. If the interior of the pipe is smooth (as it should be), the water flows down as a thin film that clings against the inside diameter of the pipe. The velocity of the water increases rapidly under the influence of gravity until its rate of descent is counterbalanced by the friction presented by the pipe. In the usual piping installation, the wastewater will have attained its maximum velocity after having fallen about 8 ft, or one story. Projections and roughness within the pipe may interrupt this smooth fall of water for a time and the water may tumble through the stack for a short distance before adhering to the wall of the pipe again.

In the typical 2-story dwelling, therefore, the wastewater reaches the house drain traveling at some considerable speed. If the connection to the house drain is made with a long-turn Y fitting, as it should be, the wastewater, still clinging to the walls of the pipe, will make this turn with little loss in velocity, but, having reached a horizontal pipe, its velocity will drop fairly abruptly, and the water will settle quickly to the bottom of the house drain and continue its journey to the sewer.

The way in which wastewater is discharged into and flows through a waste or soil stack is important because this flow of water has its effect on the air pressures within the drainage system, which will in its turn affect the efficiency of flow. The water rushing down the stack will draw along with it, by friction, the air in the center of the pipe. If the discharge of water into the stack should be considerable, as from a water closet or bathtub, say, this water rushing suddenly into the stack may fall through the pipe in the form of solid slugs of water. These slugs act like pistons, pushing air before them and drawing air after them. The consequence is that the air in the upper part of the drainage system tends to have a negative pressure compared to the outside atmospheric pressure.

And since the flow of water tends to slow down rather abruptly at the bottom of the stack, this down-rushing stream of air becomes compressed within the lower part of the drainage system, where it may build up a considerable positive head of pressure.

Thus, a differential air pressure has been produced within the drainage system. This differential air pressure, if left uncorrected, can interfere seriously with the efficient discharge of wastes from the system. The operation of the drainage system may be compared with water in a drinking straw. As every-

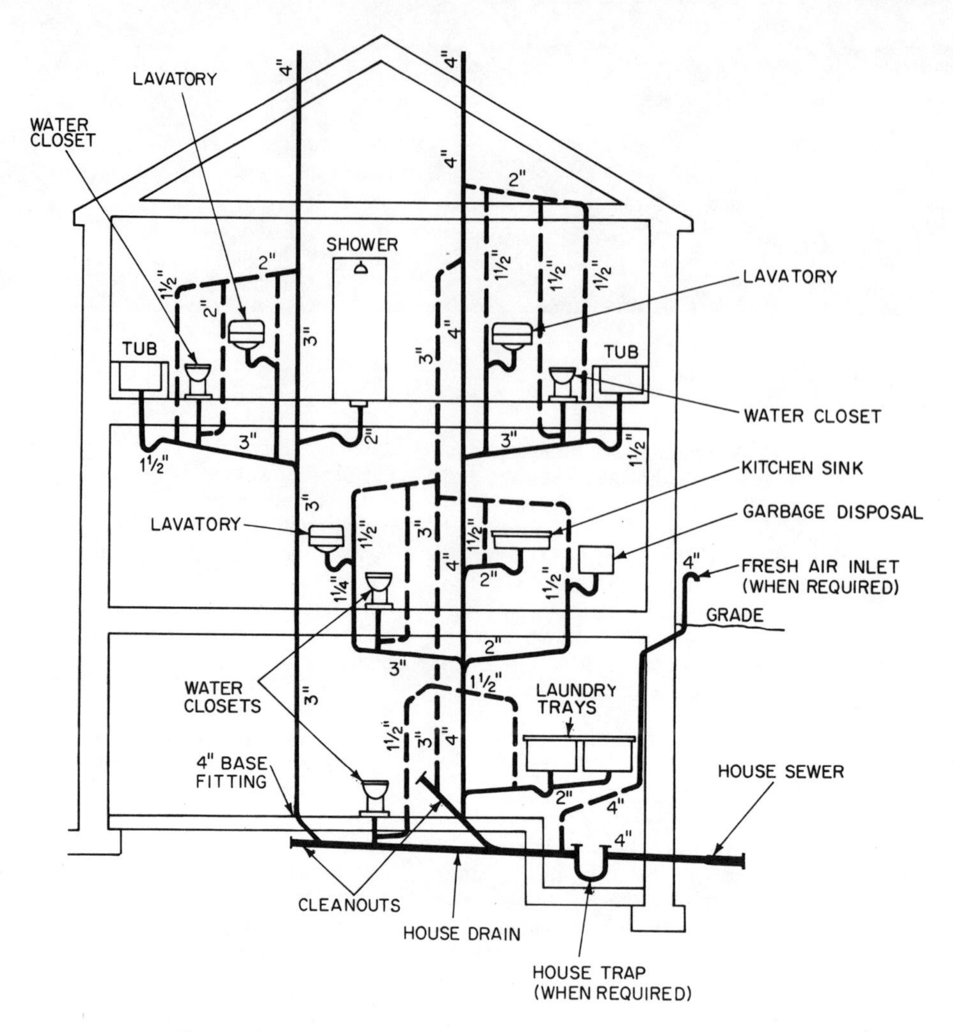

Fig. 1. The drainage system of a typical 2-story residence.

body knows, when you suck water into a straw and then close off the top of the straw with your finger, the water will remain trapped within the straw. The reason is that the air pressures at the two ends of the straw are now unequal, the atmospheric pressure acting against the bottom of the straw being greater. It is this difference in air pressure that prevents the discharge of water from the straw. When you release your finger from the end of the straw, the air pressures are equalized again and the water can run freely out of the straw by gravity flow.

Somewhat the same thing can happen within the upper part of a drainage system because of differential air pressures—if not to the same degree as in a straw. The difference in pressure is sufficient, however, to interfere with the efficient discharge of wastes through the pipes. To equalize the air pressures, therefore, the top of the main waste or soil stack is extended up through the roof of the building, where its open end allows atmospheric pressure into the system. Thereafter, any drop in air pressure within the upper part of the system will draw additional air into the system until the system pressure is again equal to atmospheric pressure.

It is more difficult to deal with the buildup of pressure within the bottom of the stack. How this is accomplished is described later.

It is important that air pressure in the system be equal to atmospheric pressure not only because it allows wastes to flow freely through the pipes, but also because a balanced air pressure is necessary if the traps that are installed at the fixtures are to perform their function of keeping sewer gases out of the house—and this is usually of much greater concern to the occupants of a house.

Traps, Vents, and Sewer Gases

Since domestic drainage systems operate by gravity flow, there cannot be any valves or restrictions within the piping that might interfere with the free flow of wastes. If it were not for the presence of traps at the plumbing fixtures, it would be very easy for sewer gases to make their way into the house, and for cockroaches, water bugs, silverfish, and rats to do likewise. The primary function of traps, therefore, is to prevent these gases and vermin from entering the house.

Sewer gases are products of the decomposition of organic substances within the public sewer or septic tank. The characteristics of these gases are shown in Table 1. Note that six of these eight gases are explosive in relatively low concentrations, four are asphyxiating, three are irritants, and one produces nausea. The odor of these gases is also extremely unpleasant, even in very low concentrations.

It was once believed that dysentery, typhoid, and cholera could be transmitted to human beings by disease organisms that were carried through the pipes by currents of sewer gases, but this is now known to be untrue. And although it is true that the gases can be explosive and/or asphyxiating, the probability that anyone would actually be asphyxiated by them or that the gases would explode if they were to leak into a house is in fact extremely remote. As a practical matter, traps are installed in order to keep the noxious odor of the gases out of the house and to keep vermin out as well.

A great many trap designs have been invented and patented. Only two basic designs are, however, approved for use by most national and local plumbing codes. These designs are the

Table 1. Characteristics of Gases Found in Sewer Air

Gas	Lower explosive limit*	Max. safe concentration*	Physiological effect†	Sp. gr. ref. to air	Btu per cu ft
Carbon monoxide	12.5	0.01	1, 2		
Methane (natural gas)	5.6	See note‡	1	0.63	1000
Hydrogen sulfide	4.3	0.02–0.002	3		
Carbon dioxide	Not explosive	2–3	1		
Gaoline	1.4	1.0	4		
Ammonia	16	0.03	5		
Sulfur dioxide	Not explosive	0.005	5		
Illuminating gas	5.0	0.01	1	0.6	500

*Percentage by weight in mixture with air.
†Classification: (1) asphyxiating; (2) extremely dangerous—odorless, colorless, and subtle, lighter than air; (3) irritant, systemic poisoning; (4) anesthetic—produces headache, nausea, "jag;" (5) respiratory, eye, and mucus irritant.
‡Dangerous when oxygen is displaced more than 8 percent.
Source: H. E. Babbitt, *Plumbing* 3d ed., McGraw-Hill, 1960.

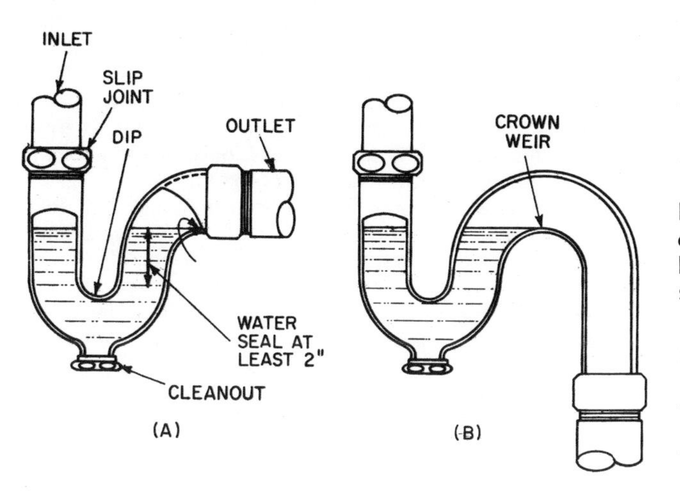

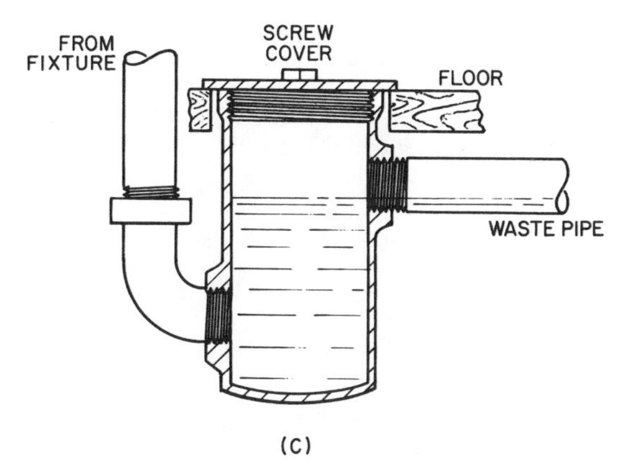

Fig. 2. (A) P trap. (B) S trap. (C) Drum trap. S traps, familiar in older dwellings, are forbidden by most present day plumbing codes.

P trap and the *drum trap* shown in Fig. 2. The P trap is more common. It is found in the drain lines of most lavatories, sinks, and laundry trays. The drum trap is used mainly with bathtubs and shower stalls. Water closets and urinals have their own traps, as described below, and do not require a separate trap in their drain lines.

P Traps

A P trap is simplicity itself. It is merely a length of tubing with a U-shaped bend in it. It is called a P trap because of its supposed resemblance to the letter P. It is also called a ½S trap because there are traps that are S-shaped (see Fig. 2) and the P trap is basically one-half of an S trap as can be readily seen.

The trap is installed as close to the fixture outlet as possible. Wastewater discharged from the fixture rushes through the trap and out into the branch line, leaving a small amount of water trapped behind in the U. This trapped water is called the trap's *seal*. Ordinarily, a trap is so constructed that the depth of the seal is 2 in. That is, the vertical distance between the *dip* and the *crown weir* shown in Fig. 2 is 2 in. In a normally operating drainage system that is adequately vented to the atmosphere, a 2-in. seal is more than sufficient to prevent the passage of sewer gases. Anything less than 1 in. is considered potentially hazardous.

In order to maintain its seal, the trap must have the same air pressure on both sides of it. This balance is ensured by installing a vent line in each branch that is open to the atmosphere. These *back* or *branch vents*, as they are called, are shown in Fig. 1. They are installed in addition to the main stack vent. As long as a trap's vent line remains open, any excessive decrease in system pressure caused by a rush of water through the pipes will be made up for by air rushing into the system through the vent. In the same way, any excessive buildup in system pressure caused by a blockage will be relieved through the vent. In either case, the integrity of the trap seal will be maintained.

Suppose, however, that the trap is not adequately vented because, say, the vent line is clogged shut by an accumulation of grease. As a consequence, the trap might lose its seal because of *siphonage* or excessive *back pressure* within the system. We will discuss siphonage first.

Siphonage

Any inverted U-shaped tube having legs of unequal length can easily be used as a siphon. When the shorter leg is placed in a container of liquid, all that is necessary to initiate the siphoning action is to reduce the air pressure in the longer leg of the U below atmospheric. Anyone who has ever siphoned gasoline from an automobile tank knows this is usually accomplished by sucking on the end of the tube. Once a solid stream of liquid has been drawn through the U and below the level of the liquid in the container, the siphoning action becomes automatic. It will continue until the liquid level in the container falls to the level of the tube opening and air is drawn into the tube, which will break the siphoning action.

In the S trap shown in Fig. 2, a *self-siphoning* action can be initiated when water rushes out of the fixture, through the trap, and into the trap's downward leg. The siphoning action is helped by the fact that the water rushing through the pipes has some momentum to it and because water is cohesive—the drops of water tend to cling together. As long as the downward leg continues to be filled with a solid column of water and this column of water is below the level of the dip, water will flow through the trap until the siphoning action is broken by air reaching the crown weir. The last bit of water at the top of the crown weir falls back into the trap, sometimes filling the trap just at the dip level, sometimes below. If the seal is broken, or is less than 1 in. deep, sewer gases may make their way through the trap and into the house because they usually have a slight amount of natural pressure behind them.

S-shaped traps are particularly prone to siphonage because of the length of their downward legs, which enhances the siphoning action. For this reason, S traps are forbidden by most present-day plumbing codes in favor of P traps, but there are still plenty of S traps remaining in older dwellings. And self-siphonage can still occur in P-shaped traps, though it is more unlikely. With a P-trap installation, self-siphonage can occur if a solid slug of wastewater completely fills the drain line as it flows to the waste or soil stack and if the branch vent line should at the same time be clogged. This may happen even when the piping has been correctly sized for the fixture if the inside diameter of the pipe happens to have been reduced over

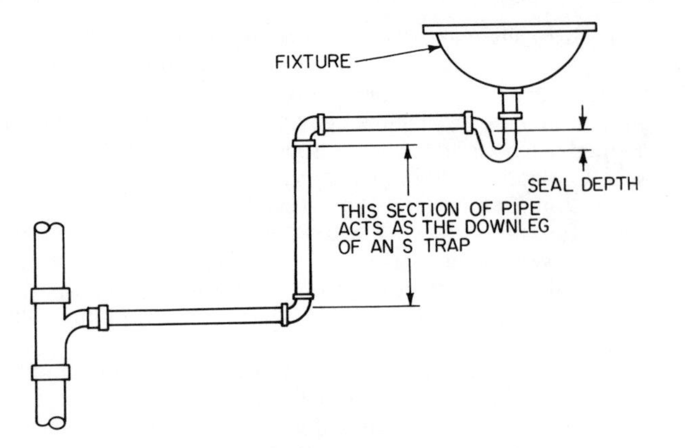

Fig. 3. An incorrectly installed piping system can inadvertently convert a P trap into an S trap.

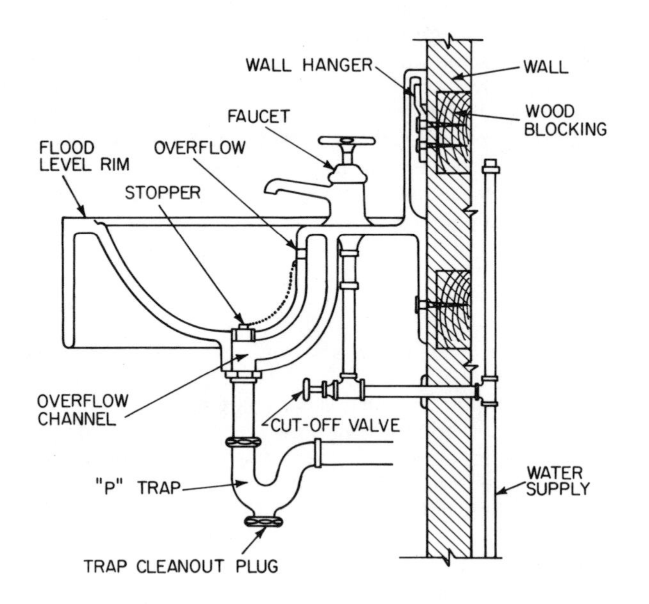

Fig. 4. A typical lavatory installation *(U.S. Dept. of Agriculture).*

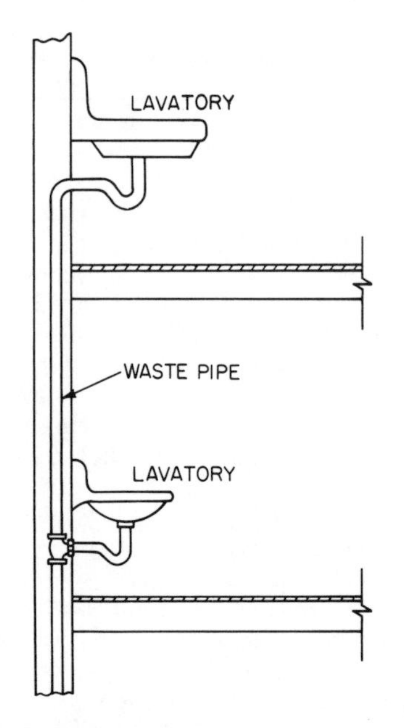

Fig. 5. The water in the lower lavatory trap can be sucked out of the trap by the water rushing through the pipes from the lavatory above. This is called induced siphonage.

a period of time because of an accumulation of grease. If this slug of water should remain intact until the lavatory drains empty, then the water will be siphoned from the trap as described.

Figure 3 shows how an incorrect pipe installation can make an S trap out of a P trap. As shown in the illustration, the drain line has a vertical downleg within the wall that is quite close to the trap. Perhaps the plumber installed the piping this way to avoid some obstacle, or perhaps it was simpler to make the connection to the stack this way. Whatever the reason, a solid column of water running down the vertical pipe can siphon the water out of the trap just as effectively as if an S trap had been installed in the first place.

Fixture manufacturers have done their bit to eliminate self-siphonage by appropriate design. For one thing, the diameters of the outlets on the fixtures are less than the diameters of the traps. This makes it almost impossible for a solid column of water to fill the downward leg of a trap. In addition, lavatories usually have an overflow line built into them that carries any overflow to the drain (see Fig. 4). Whenever the lavatory is drained, air is always sucked through this overflow line into the drain line, where it mixes with the wastewater. The presence of this air further reduces the possibility of a solid column of water filling the downward leg of the trap.

Another type of siphonage is *induced* siphonage, which can occur because of a drop in air pressure within the drainage system. Figure 5 shows how induced siphonage can occur. The illustration shows two fixtures draining into the same waste stack. When wastewater is discharged from the second-floor fixture, the water rushing down the stack past the first-floor branch will create sufficient suction within the first-floor branch line to draw the water from the trap. But for this to happen, the branch vent line must be restricted or blocked.

Siphonage and Vent Lines

It may appear unusual that a suction force strong enough to draw water from a trap can be produced within a piping system. It would appear that 2 in. of water in a trap that is at least 1¼ in. in diameter must surely have enough mass and inertia to resist this suction force. But, in fact, a column of water 2 in. deep has very little resistance to anything at all. The 2-in.-deep seal in a P trap may be compared to an opening in a wall over which a sheet of newspaper has been pasted to keep out the weather. The paper serves well enough when the weather is fine and the breezes light, but it cannot withstand a strong gust of wind.

The same is true of the water seal in a trap. The weight of 2 in. of water in the trap is equal to a pressure of 1.15 oz. This is more than enough to prevent sewer gases at ordinary system pressures from passing through a trap, but 2 in. of water presents no resistance at all when there is an unusually large drop in the pressure within the pipes—a drop in pressure that is almost always due to some deficiency in the venting.

In order to maintain the integrity of the trap seals, it is necessary to prevent any difference in air pressure occurring across the traps. This is accomplished by installing a vent not only for the main stack, as already described, but also for each fixture trap, or for a group of two or three closely related traps. These are the branch vents already mentioned. Figure 6 shows a typical installation. The branch vent admits air at atmospheric pressure to the drainage-system side of the trap, thereby making it impossible for a difference in pressure to occur across the trap, and thus preventing the possibility that siphonage will suck the seal from the trap.

Back Pressure and Running Traps

The increase in air pressure that occurs at the bottom of a waste or soil stack when a fixture discharges wastewater into

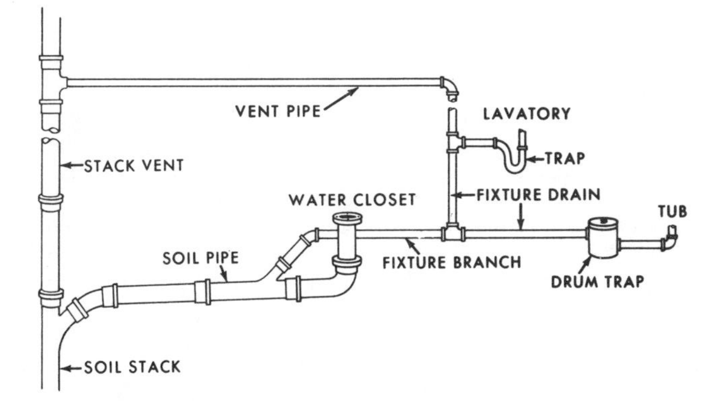

Fig. 6. The piping installation for a group of bathroom fixtures. The vent line assures that the wastes will drain freely from the fixtures and that the trap seals will remain intact *(U.S. Dept. of Agriculture).*

the stack is called *back pressure.* It is not a common problem in 1- or 1½-story dwellings or in 2-story dwellings that do not have full basements. Back pressures are more of a problem in 2-story dwellings with full basements and in the 3-, 4-, and 5-story row houses that are common in cities.

The increase in air pressure at the base of a stack can be considerable if there is no way in which this pressure can relieve itself. It is not uncommon for the back pressure to blow out the trap seals in basement fixtures. The homeowner then wonders where the filth comes from that covers the floor or the bottoms of the laundry trays. The usual suspicion is that the sewer has backed up.

Problems with high pressure at the base of a stack and in a house drain are often aggravated by the U-shaped *house trap* (or *running* trap) that is installed in the house drain line shown in Fig. 1. The purpose of this house trap is exactly the same as that of the fixture traps—to keep sewer gases and vermin out of the house. There is considerable controversy whether house traps are really necessary. Those who are against them say they merely duplicate the function of the fixture traps and add to the cost of the plumbing installation without improving system efficiency at all. Those who are for them say they *do* help to keep sewer gases and vermin out of the house and *do* increase the effectiveness of the fixture traps. The informed consensus seems to be that house traps are unnecessary. The National Plumbing Code, which was until recently the most authoritative of American plumbing codes, recommends that they not be used.

When a house trap is not installed in a house drain, sewer gases are free to enter the drainage system. At the same time, the fact that they are free to enter the drainage system helps equalize the pressures throughout the system.

Also, a municipal sewage system in which all the dwellings connected to the sewers have house traps installed has less capacity than if the traps were absent. The reason is that the gases within the public sewers can build up a back pressure that will interfere with the free flow of sewage through the system. For this reason most municipalities that did not require running traps originally now forbid their installation entirely.

In many other communities, however, house traps have been required for a great many years. A builder constructing a new house in such a community is forced to install a house trap in self-defense, even if the local plumbing code should no longer require them, the reason being the existence of the sewer-gas pressures mentioned above. In a sewage system in which all the dwellings but one have house traps installed, the main vent in this exceptional dwelling will serve as the relief vent for the entire sewage system in that neighborhood. It is also very likely that any excessive system pressures are going to back through the fixture traps installed in this dwelling, which will make life very unpleasant for its inhabitants.

The remedy for excessive back pressures in a drainage system, whether or not a house trap has been installed, is to vent the base of all waste or soil stacks through a separate vent line as shown in Fig. 1. The installation of these vents will relieve any pressure into the top of the main vent (or vents), from where the pressure will be dissipated harmlessly into the atmosphere.

Drum Traps

Drum traps (Fig. 2) are usually 3 to 4 in. in diam. and from 4 to 8 in. deep. Their design allows a greater rate of flow through the pipes than can flow through pipes in which a P trap is installed. Drum traps, therefore, are much more suitable for bathtub and shower stall drain lines. Their principal defect is that sediment and grease are likely to accumulate in the trap. The top of the trap must, therefore, be removable, which means that the top of the trap must be mounted flush with or just below the floor, where its appearance may not be pleasing. In some designs the accumulation of sediment and grease is prevented by having the inlet and outlet lines installed at an angle to the centerline of the trap so that the incoming water swirls through the trap, which effectively scours out any grease or sediment that may have collected.

The comparatively large amount of water held within a drum trap gives it very good antisiphoning qualities. (An *antisiphon trap* is a trap in which such a large amount of water acts as the seal that it is extremely difficult for any differential pressure in the lines to siphon the water from the trap. There are also available types of P trap in which the outlet leg of the trap is made larger than the inlet for this reason.)

Since the position of the drum trap under the floor often makes it difficult if not impossible to install a separate vent line for this trap, a bathtub or shower branch is often vented by what is called a *wet vent.* A typical bathroom installation having such a wet vent is shown in Fig. 6. Note that the bathtub branch connects directly to the lavatory branch, which then empties into the soil stack. The vertical section of the lavatory drain line also acts as the vent for the bathtub branch. This method of venting the bathtub branch is called a wet vent because the wastewater discharging from the lavatory flows through this pipe. No harm is done by connecting the piping in this way. And if the fixtures are installed in such a way that even a wet vent cannot be installed, then the antisiphoning characteristic of a drum trap can be depended upon to maintain the seal within the trap.

WATER CLOSET

Most residential water closets are made of vitreous china, from the same clays used for making fine table china. The complex shape of a water closet requires that it be made from a great many individually molded pieces of clay that are assembled together to form the complete unit. The clay is then coated with a liquid glaze, and the assembly is fired at a temperature of about 2500°F to vitrify the clay. Cheaper residential water closets are made of cast iron coated with porcelain enamel.

Water Closet Operation

That water closets and S traps operate on the same principle should be obvious from an examination of Fig. 7, which shows a *siphon* type of water closet. Both an S trap and a water closet consist basically of a U-shaped passage in which water forms a seal; in the water closet one arm of the U—the bowl itself—is very large. In both an S trap and a water closet, when the downleg fills with water, a self-siphoning action is initiated

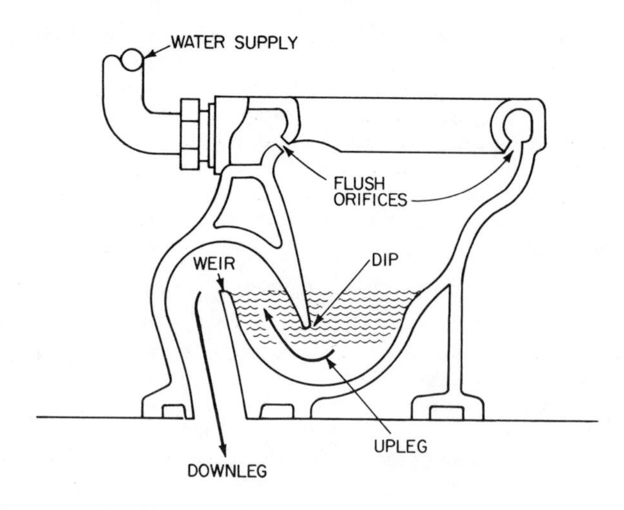

Fig. 7. A siphon-type water closet with a reverse trap.

that draws water from the trap. The big functional difference between an S trap and a water closet is that self-siphonage is deliberately induced in the water closet in order to drain wastewater and its solid waste content from the bowl.

The siphon type of water closet operates as follows. Before it is flushed, the water-closet bowl is filled with water up to the height of the weir. A water closet is usually designed to have a 2½- to 3-in. seal. When the flush is initiated, fresh water enters the bowl through orifices located around its rim. This water washes down the walls of the bowl, thus cleansing them. The initial rush of water also increases the level of water in the bowl. This water overflows the weir and passes into the downleg of the water closet.

This downleg passage is deliberately designed with constrictions and abrupt turns in order to hamper the flow of water. Its cross section is also less than that of the upleg. As a result, the downleg fills completely with water, which initiates the self-siphoning action. The air trapped in the top of the U is carried into the downleg by the solid mass of water that passes into the U from the bowl.

Once self-siphonage has started, water is removed from the bowl at a faster rate than enters it. As a result, all the water that is in the bowl originally—plus any solid wastes—is sucked out. The water level in the bowl continues to decline because of the fast rate of suction until the water level drops below the top of the dip. This breaks the self-siphoning action, the discharge of water from the bowl stops abruptly, and the small amount of water remaining in the upleg falls back into the bowl. The bowl is refilled to its original level by means of an overflow line located in the flush tank, as described in the section below.

The water closet and its associated flush tank are designed to operate as a unit and must be installed as a unit. This is because the rate and volume at which fresh water flows into the bowl will depend on the rate and volume at which wastewater flows out of the bowl. If the flow into the bowl is too rapid or, what amounts to the same thing, if there should be a blockage somewhere in the outlet side, the siphoning action will not be strong enough to flush the solid wastes from the bowl. What will happen in that case will be that the bowl overfills with water and the solid wastes simply remain in the bowl.

If, on the other hand, the flow of fresh water into the bowl should be insufficient, the self-siphoning action will not be initiated at all, or, if it is, it will not sustain itself long enough for all the wastewater and solid wastes to be removed from the bowl.

When a water closet and flush tank are operating properly, the flushing action is strong and positive. A complete flush normally takes from 9 to 15 seconds, depending on the design of the water closet. The first rush of fresh water into and through the bowl will be quite heavy, at the rate of about 30 to 35 gal per min, a rate that quickly declines as the flush tank empties. Since the flushing action occurs for so short a period, only about 3 to 4 gal of water are actually required for a complete flush. The flush tank is designed to hold this quantity of water plus about a gallon more.

Flush Tank

Flush tanks may be mounted low or high. The high-tank installation is practically obsolete. As for low-tank installa-

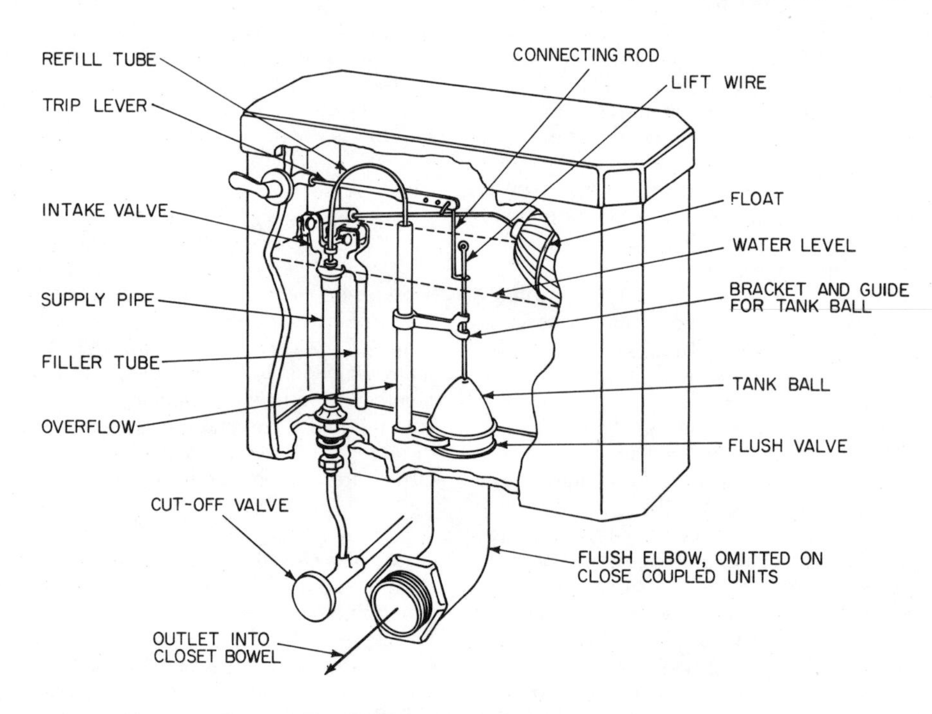

Fig. 8. Typical water-closet flush tank *(U.S. Dept. of Agriculture)*.

tions, the tanks may be either (1) mounted on the wall a few inches above the water closet and connected to it by a pipe, (2) structurally independent of the water closet but resting directly on it, or (3) an integral part of the water closet, the two forming a one-piece combination unit. Whatever the installation, low tanks all operate in exactly the same way. The closer the coupling between water closet and tank, the quieter the flush. On the other hand, the closer the coupling, the more expensive the unit.

The construction of a typical flush tank is shown in Fig. 8. There are innumerable variations of this basic design, each manufacturer having his own preferences and patents; however, most tanks operate on the principle shown in the illustration.

The flush is initiated by tripping the *lever.* This action raises the *rubber ball* off its seat, which allows water to discharge from the tank into the bowl. As the rubber ball is hollow and filled with air, it remains suspended in the water above its seat until the tank has emptied sufficiently for it to descend with the descending water level. Finally it drops close enough to its seat to be sucked back onto it by the outrushing water. Once the ball has reseated itself, the flush terminates and the refilling operation begins.

As the water level drops, the large metal *float* also drops with it. This float is connected to the *water valve* in such a way that, as it drops, the valve opens, which admits fresh water into the tank. The valve will remain open as long as the float is low. As the tank gradually refills with water, the float rises with the water level, and this movement gradually closes the valve again. Note in the illustration that the line discharging fresh water into the tank points downward, with its open end close to the bottom of the tank. This construction reduces the noise of the water rushing into the tank.

There is also a *bypass line* from the valve outlet that discharges into an *overflow pipe,* which empties excess water from the tank into the bowl. In this way the water-closet bowl is refilled with water once the rubber ball has reseated itself. When the float has risen to its original level, both the bowl and the tank will have been refilled with water and the tank will be ready for another flush. This entire refilling operation takes a minute or two to complete. If an attempt is made to reflush the water closet before the tank has been completely refilled, the flush will be incomplete.

PLUMBING INSTALLATION

The complexity of a piping installation, and hence its cost, depends basically on the number and location of the plumbing fixtures. A commercial builder and a couple working up the floor plans for their dream house have two completely different ends in view when deciding on the locations of the bathrooms and kitchen and the positions of the fixtures in these rooms.

A builder usually puts up his houses on speculation. He will, therefore, take care to make all the piping runs as short and direct as possible in order to minimize the plumbing costs, which usually run from 10 to 11 percent of the total construction costs. In a 1-story dwelling he will locate the kitchen and bathroom back-to-back, with all the water supply and drainage lines buried in the wall (called the *stack wall*) that separates these two rooms. An example of such an arrangement is shown in Fig. 9. In a 2-story dwelling, the stack wall will carry all the vent lines and supply and drainage headers, which means the upstairs bathroom and kitchen must share the same stack wall. Such an arrangement is shown in Fig. 10. If there are two bathrooms on the second floor, they will be located back-to-back against the stack wall. The builder will be particularly careful to locate all the water closets as close to the soil stack as possible, both for efficiency of drainage and to minimize the cost of their installation.

A couple is usually more interested in convenience, decorative effect, and modesty. They dislike having a kitchen and bathroom next to each other because they don't want people in the kitchen hearing the noises made by the plumbing. They also prefer to have the upstairs bathrooms separated from each other so that a person in one bathroom will not be able to hear anyone in the other bathroom. The location of the kitchen and of the kitchen fixtures is more likely to be determined by the direction of the morning sun and the view from the kitchen windows than by the lengths of the pipe runs. They don't know and couldn't care less that their layout will require two and

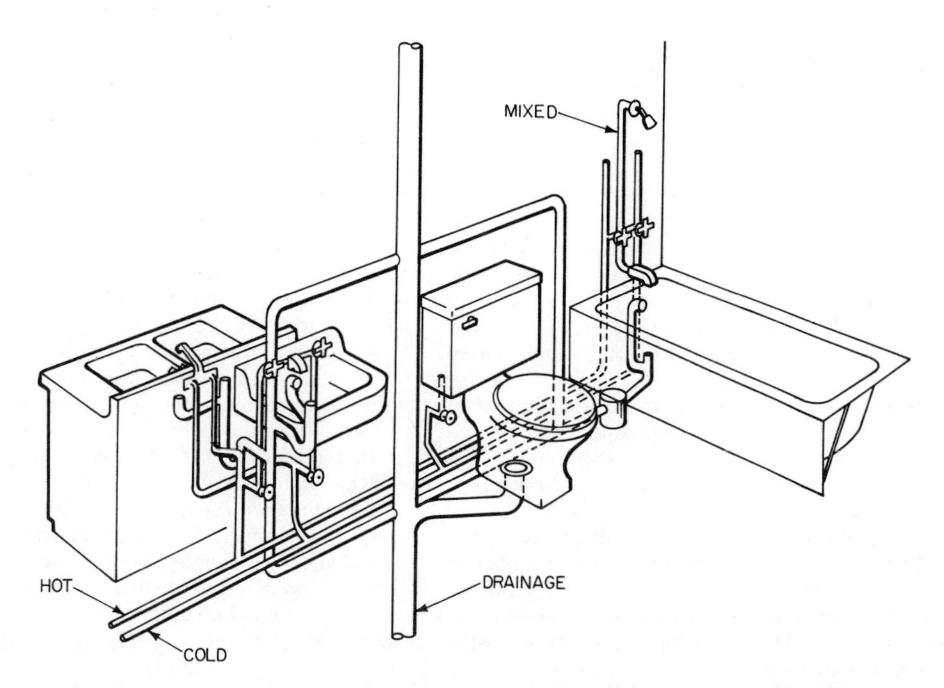

Fig. 9. Plumbing fixtures located back-to-back on opposite sides of a stack wall *(U.S. Dept of Agriculture).*

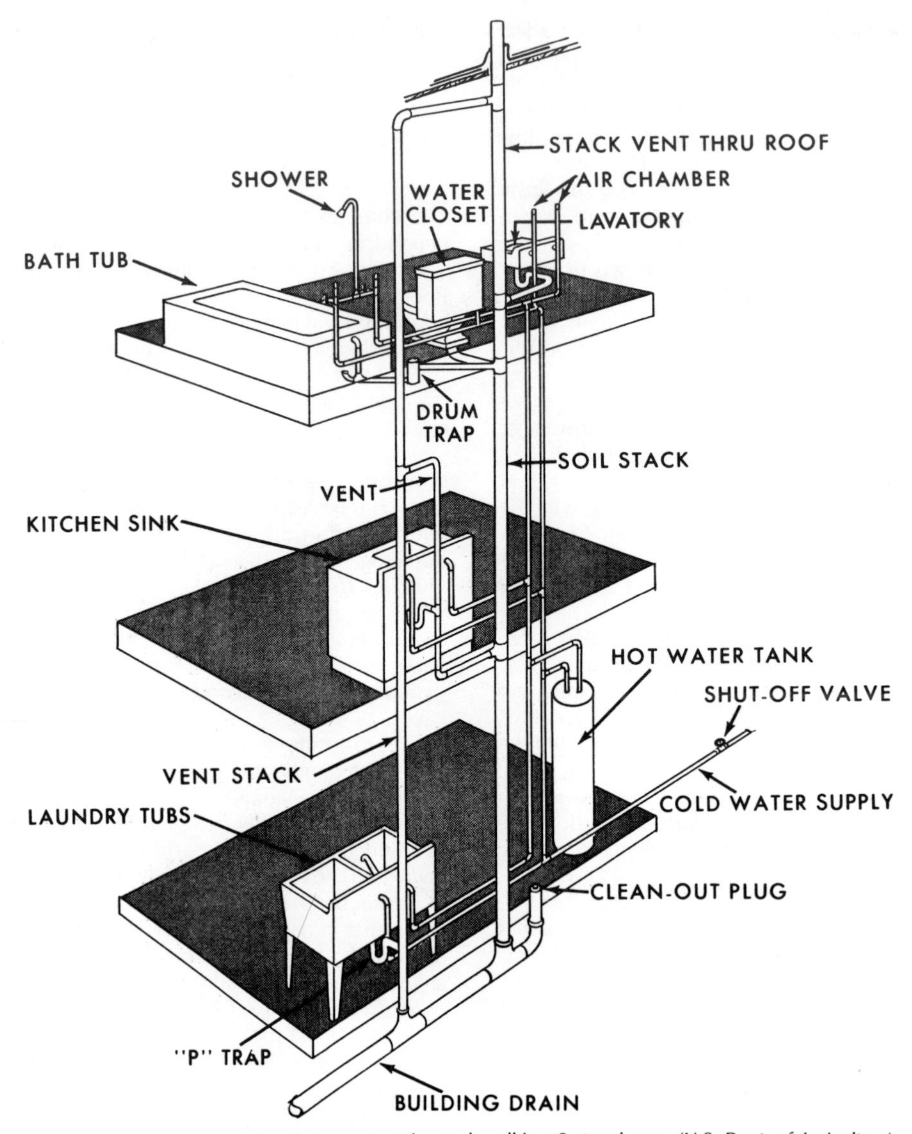

Fig. 10. Plumbing installation against the stack wall in a 2-story house *(U.S. Dept. of Agriculture).*

perhaps three soil stacks instead of only one. Naturally, the cost of their design will be much greater than the cost of the builder's design.

The Plumbing Layout

Before he orders the material for a job, the plumber will draw up a piping layout that shows as accurately as possible the sizes, lengths, and locations of all the pipe runs. In preparing his layout, the plumber must keep certain points in mind regarding the design of drainage systems.

All the pipes (including the vent and water-supply pipes) should be grouped together as much as possible, both for convenience of installation and to minimize the amount of cutting and drilling into the house structure that will be required. Grouping the pipes together will also make it less expensive to break through the wall to find and repair a leak, if a leak should ever develop.

To minimize friction losses in the lines, all the pipe runs should be as direct as possible, with a minimum number of bends and turns. In particular, branch lines should not be offset, either up or down, in order to bypass some obstacle.

All the horizontal branches must slope correctly to assure proper drainage. All pipes up to and including 2 in. in diam. must slope at least ¼ in. per ft of run; pipes 2¼ in. in diam. and larger must slope at least ⅛ in. per ft of run.

All threaded pipes must be cut to length and aligned exactly to avoid stressing their threaded connections, which might cause them to leak. This accuracy of fit is important in the vent lines also, as a crack in a vent line may allow sewer gases into the house.

The plumber must also take into account the inevitable movements that occur in a wood-frame house. These movements must be allowed for if the pipes are not to be stressed and the joints ruptured.

A house moves in different ways for different reasons. If the foundation should settle, the house will settle with it; if the house should settle unevenly, the framework will twist. Soon after a house has been completed, the wood framework will shrink downward as the lumber dries out. Ordinarily the amount of this shrinkage is slight, but if green lumber has been used in the construction, and if platform-type framing has been used, the shrinkage might be considerable (*see* WOOD-FRAME CONSTRUCTION). Finally, there is a once-a-year expansion and contraction of the frame that occurs in climates in which there is a large annual change in the relative humidity. This expansion and contraction is relatively slight and can usually be ignored.

The other two types of movement cannot be ignored. The

plumber must anticipate their occurrence and plan the installation accordingly. With the 1-in.-diam. pipes used for the cold- and hot-water supply he does this by making sure the pipes never run directly between two points without at least one bend in the line to give the line some flexibility of movement; in addition, the pipes should always be loosely supported in such a way that they can move with relative freedom with respect to the structure.

This problem of relative movement does not exist as far as the branch lines are concerned. Usually, the branch lines are so short that the problem will not arise; in any case, traps provide a flexible connection between fixtures and branches that allows for any relative movement in the branches.

The soil stack is another story. It is usually made of heavy cast-iron pipe (though often of copper) that rises in a straight run from the house drain to the top of the roof. Even when the stack does make a bend, the bend does not impart any flexibility to the installation. Flexibility is achieved by leaving the soil stack the single, massive unit it is but making sure that (1) its base is well-supported and (2) it is flexibly supported at each floor level.

Finally, in planning the piping layout, the plumber should avoid locating water-supply pipes in an exterior wall. The pipes might freeze some day, which can cause the pipes to rupture. If the pipe *must* be installed in an exterior wall for any reason, it should be well insulated against the cold.

Pipe Sizes and Fixture Units

In any drainage system the size of the pipes depends almost entirely on the amount of wastewater they are expected to carry. And how is the total quantity of wastewater determined? In principle, by determining the volume and rate of discharge from each of the fixtures. One might expect that an architect or engineer would have available some handy set of formulas that would enable him to take into account the liquid capacity of each fixture installed in a dwelling, the expected usage of each fixture, the rate at which water is discharged from it, and the length of the pipe runs, and come up with the required pipe sizes.

In fact, the number of variables is so large, and the variables vary so much, that trying to design a domestic drainage system on engineering principles is not worth the bother. Instead, pipe sizes are usually determined by reference to a table of *fixture units* contained in the local plumbing code.

All local plumbing codes require that soil branches (i.e., branches carrying the solid wastes from a water closet to the soil stack) be at least 3 in. in diam. and that soil stacks be either 3 or 4 in. in diam. (depending on the local code). Experience has shown that pipes this large are necessary if the liquid and solid wastes discharged from water closets are to pass quickly through the drainage system without blocking the lines or introducing any differential-air-pressure problems. In the selection of the soil pipes, therefore, a plumber has no choice; he must follow the local plumbing code. If there is only one water closet connected to a soil stack, then some plumbing codes require that the soil stack be a minimum of 3 in. in diam., but other plumbing codes require a 4-in. stack. If two or more water closets are connected to a soil stack, then all codes require at least a 4-in. pipe.

As for the branch lines that drain other types of fixtures, a table of *fixture units,* such as Table 2, is consulted to determine the individual pipe sizes. Fixture units are derived as follows. It has been found that the factors that determine the rate at which wastewater drains from a fixture are chiefly the total liquid capacity of the fixture and the size of its outlet orifice. Thus, the typical bathroom lavatory holds about 1.5 gal of water, and it discharges this water through a 1¼-in. outlet at the rate of about 7.5 gal per min. Now, 7.5 gal per min happens to be almost exactly equal to 1 cu ft, which makes this volume a

Table 2. Fixture Units Used to Determine Pipe Sizes in a Plumbing System

Fixture type	Fixture-unit value	Minimum diam. of trap, in.*
1 bathroom group (water closet, lavatory, and bathtub or shower stall)	6, 8†	
Bathtub (with or without overhead shower)‡	2	1½
Bathtub‡	3	2
Bidet	3	1½ (nominal)
Combination sink-and-tray	3	1½
Combination sink-and-tray with food-disposal unit	4	1½ (separate traps)
Drinking fountain	½	1
Dishwasher, domestic*	2	1½
Floor drains§	1	2
Kitchen sink, domestic	2	1½
Kitchen sink, domestic, with food-waste grinder	3	1½
Lavatory¶	1	1¼ (small P.O.)
Lavatory¶	2	1½ (large P.O.)
Laundry tray (1 or 2 compartments)	2	1½
Shower stall, domestic	2	2
Sinks:		
Flushing-rim (with valve)	8	3
Service (trap standard)	3	3
Service (P trap)	2	2
Pot or scullery*	4	1½
Urinal, wall-lip	4	1½
Water closet, tank-operated	4	3 (nominal)
Water closet, valve-operated	8	3

*See American Standard National Plumbing Code (1955), Par. 11.4.3 and 11.4.4 for method of computing unit value of fixtures not listed in this table or for rating of devices with intermittent flows.

†Bathroom group with tank water closet has fixture unit value of 6, with flush-valve water closet 8.

‡A shower head over a bathtub does not increase the fixture value.

§The size of the floor drain is determined by the area of surface water to be drained.

¶Lavatories with 1¼ or 1½-in. trap have the same load value; larger P.O. plugs have greater flow rate.

Source: J. H. Callender, Ed. *Time-Saver Standards,* 5th ed., McGraw-Hill, 1974, p. 885.

convenient unit on which to base a system of measurement. For this reason, 1 cu ft of water flowing from a fixture at the rate of 7.5 gal per min has been made equal to *1 fixture unit.*

The fixture unit having been established in this way, the discharge capacities of other types of fixtures are rated in fixture units by comparing their discharge rates to that of the typical lavatory. Thus, the discharge capacity of the typical bathtub (with or without a shower head) is rated at 2 fixture units, and the discharge capacity of the typical water closet is rated at 3 fixture units. It might be noted, however, that in fact the average bathtub has a discharge rate of about 12 gal per min, which is less than 2 fixture units (2 fixture units = 15 gal per min), and that the average water closet has an average discharge rate of about 18 gal per min, which is less than 3 fixture units (3 fixture units = 22.5 gal per min). That is, fixture-unit ratings tend to be on the conservative side.

The vent lines must be at least 1¼ in. in diam. Pipes this large are required in order to minimize the possibility of their becoming clogged by a buildup of grease. The size of the main vent is usually based on the capacity of the stack to which it is connected. In cold climates the main vent is often increased to 4 in. in diam. as it emerges from the roof in order to prevent its becoming clogged with frost; its construction is described below.

Drainage Pipe and Pipe Fittings

Drainage pipe is made from cast iron, galvanized steel, or wrought iron. Pipes made of brass and lead are also manufactured but they are not used for drainage lines very often, the brass because it is too expensive and the lead not only because it is *very* expensive but also because the art of bending lead pipe and making wiped joints is almost extinct, except among telephone linemen. Of the materials that are used, cast-iron and copper pipes and fittings are preferred for soil and waste stacks and for house drains, with galvanized-steel and wrought-iron pipes being used for vent and branch lines.

Cast iron has the enormous advantage that it is practically impervious to the corrosive effects of water and chemicals. It can be buried in soil or embedded in concrete with no more protection than a hot-dipped coating of coal tar or asphalt, and it will last the life of the house. The principal deficiency of cast-iron piping is its brittleness, which makes it unsuitable for installations subject to vibration, which will cause the pipe to crack.

Galvanized-steel and wrought-iron pipes are easier and cheaper to manufacture in small diameters than is cast-iron pipe. The ductility of these metals also makes it possible for pipes made from them to be cut easily to size and have their ends threaded. These metals are also tough enough to withstand vibration and tensile and torsion stresses that would crack a cast-iron pipe. Neither galvanized steel nor wrought iron is especially resistant to corrosion, however, which limits their use to above-ground locations.

Roughing-In

The actual installation of the pipes, from the terminus of the house drain to the top of the main vent, including all the pipes between, is called *roughing-in.* The only items excluded are the installation of the fixtures and traps. Roughing-in also includes the testing of the installation after its completion.

We have mentioned the importance of making an accurate layout before work is begun. If the plumbing is to be installed without mishap, the drawings prepared by the plumber must show the locations of the doors, windows, and chimney, and the positions of the heating-system pipes or ductwork, kitchen ventilation duct, electrical conduits and panel boxes, floor joists—anything at all that might interfere with the installation of the pipes. If there is interference of any kind, the plumber must consult with the architect or builder to clear up the difficulty before he begins work.

The position of the soil stack is particularly important. Ideally, it should drop vertically in a straight fall to the house drain, and the water closets should be located as close to it as possible. If the location of any other piece of equipment will interfere with the positioning of the soil stack, the soil stack should take precedence. It might be possible to resolve any difficulties by shifting the position of the entire soil stack within the stack wall. But perhaps not, because moving the soil stack would involve moving the water closets and house drain as well, or at least extending their branch lines, all of which will add to the complexity and expense of the installation.

Plumbing is roughed-in after the floor joists and subflooring have been installed but before the interior wall partitions are put up (see Fig. 11). A carpenter should always build partitions to accommodate the piping runs, never the other way around. Since the stack-wall partitions are not meant to be loadbearing structures (not most of the time, anyway), no special difficulties should be encountered in their construction. A stack wall is always built wide enough to completely enclose the soil stack, whatever its size, which may require the wall to be 6 in. or 8 in. wide. The width should not be skimped on because the finished wall surfaces must never touch the piping. There is bound to be some relative movement between the pipes and the wall and this will result in the latter's cracking or bulging. Plaster, for example, invariably cracks whenever it comes into contact with a pipe.

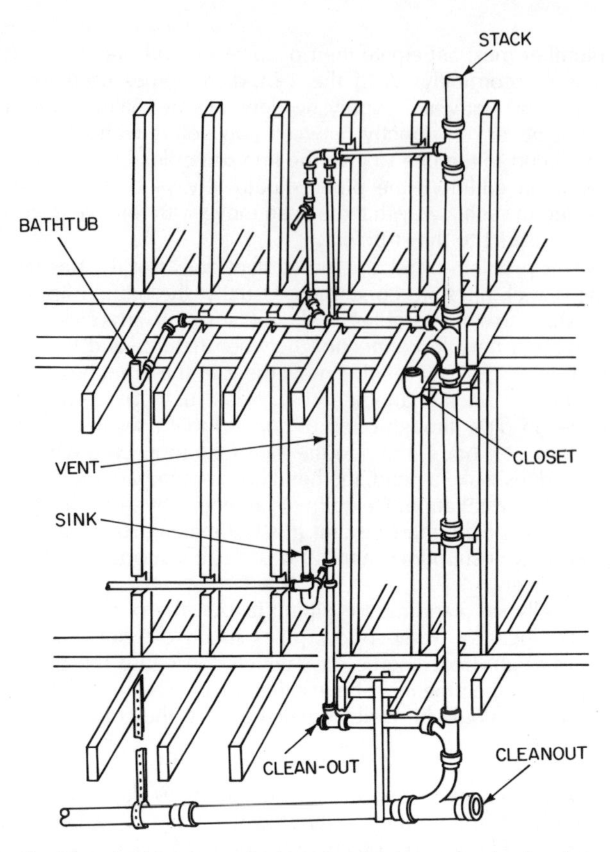

Fig. 11. The plumbing is roughed-in after the house framing has been completed.

Water-Closet Installation

A water closet is connected to the soil stack through a *closet bend,* which is a special fitting having a 90 degree bend, or turn, in the line (see Fig. 12). The long side of the bend is joined to the soil stack. The short side of the fitting projects up through a hole cut in the floor to receive the discharge from the water closet.

The overall dimensions of a closet bend will depend, of course, on the distance between the soil-stack fitting and the water-closet *horn,* which is the name of the opening through which wastes are discharged from a water-closet bowl. All water-closet manufacturers provide roughing-in plans that give the plumber all the information he needs to size and install a closet bend correctly. Closet bends are manufactured in 3- and 4-in. diameters and in various lengths to suit most installations.

To install the water closet, a floor flange is placed over the

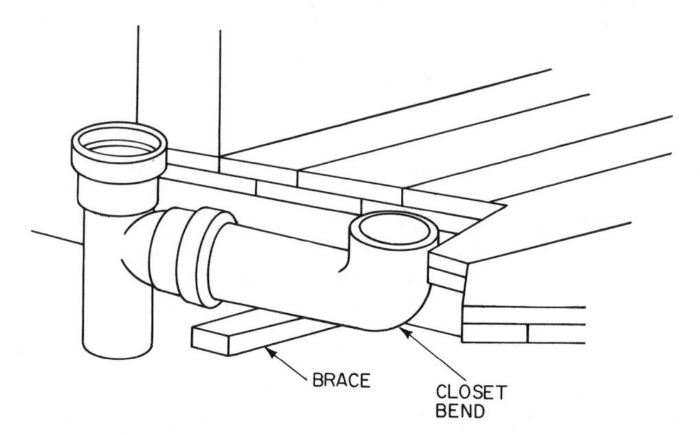

Fig. 12. A closet bend joins a water closet to the soil pipe.

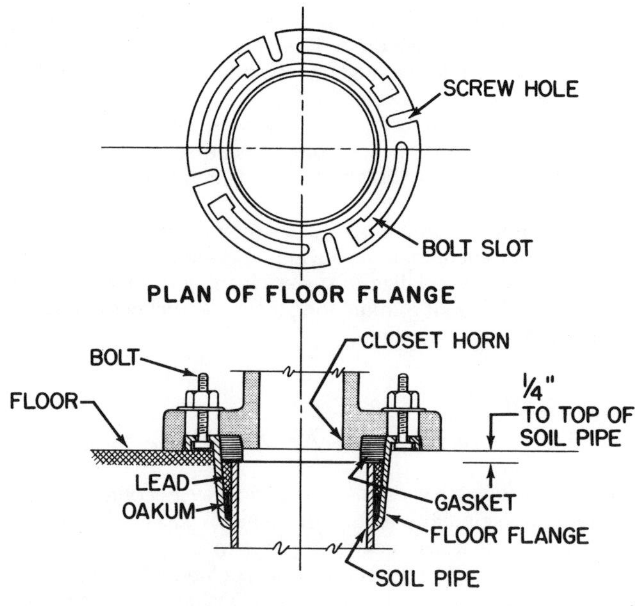

Fig. 13. Typical installation of a water closet to its closet bend *(U.S. Dept. of Agriculture).*

end of the closet bend (see Fig. 13). The joint between the two is then caulked with oakum and molten lead. The water closet is then carefully set down over the flange, and the base of the fixture is checked for levelness. It may be necessary to shim the base to level the water closet and also to ensure the water closet is resting solidly on the floor. The fixture is now lifted off the flange and turned on its side.

The underside of the base is recessed. The plumber fills this recess with a caulking compound, which may consist of either *bowl-setting wax* or an oilless *bowl-setting putty,* in order to seal the joint between the fixture and the flange. Bolts provided by the fixture manufacturer must be installed in the flange, threaded-side-up, before it is caulked in place.

The fixture is now lowered again in position upon the flange. Nuts are threaded onto the bolts. The nuts are drawn down lightly and evenly until the excess caulking compound has been squeezed from the joint. The caulking compound that remains in the joint will provide the necessary watertight seal. Excessive force must not be used to secure the fixture in place. The function of the bolts is merely to keep the fixture from shifting its position for any reason, not to secure a watertight joint.

The installation described above is one of many methods used. Each manufacturer will have his own procedure, which will differ somewhat from the description given above. The general method is much the same, however, regardless of the manufacturer.

Main Vent Installation

In cold climates, the main vent should extend 18 in. above the roof; it can be shorter in more temperate climates, if permitted by the local plumbing code. The pipe should not be fastened rigidly to the roof structure in any way. The gap between it and the roof is, instead, sealed against the weather by flashing (*see* FLASHING).

In cold climates a danger to the normal functioning of the drainage system may arise if that part of the vent exposed to the weather becomes cold enough to condense the water-vapor content of the sewer gases passing through it, and if these water droplets should then freeze to the inner surface of the pipe. If the pipe is small enough and the weather cold enough, the frost may build up sufficiently to block the vent line altogether. If this should happen, the blockage will upset the differential-air balance within the drainage system, with all the consequences to the trap seals that have already been described.

But the extension pipe is unlikely to freeze in the first place, regardless of its length, unless the air temperature drops to 10°F or below for a few days. In localities where the winter temperatures fall this low, then precautions should be taken to prevent the line's freezing.

One method is to make the extension as short as possible. There is no reason why the vent need extend more than 1 in. above a sloped roof, which should keep it from ever freezing. Where the local plumbing code requires the extension be longer, other means must be sought to prevent frost closure of the pipe.

One alternative is to make a double-walled pipe—to surround the vent line with a pipe about an inch larger in diameter that extends down into the attic. The two pipes are sealed together at their upper ends. The air within this double-walled pipe acts as a very effective insulator, and, as this air is continually circulating between attic and pipe, it will be at about attic temperature, which means there is almost no chance that the pipe extension will ever become cold enough to freeze shut.

Another alternative is to increase the diameter of the vent pipe to 4 in., as shown in Fig. 14. It has been found that a pipe this large is unlikely to freeze shut, the reason being that even if a layer of frost should form within it, the frost itself will have a temperature of 32°F regardless of what the pipe temperature is. This being so, most of the warm, moist gases rising through this ring of frost will never become cold enough to condense and freeze; there will always be a free passageway within. But when a large pipe of this kind is installed, it should increase in size at least 1 ft below the roof level.

The only time a very long pipe extension is required by a plumbing code is on a flat roof that is being used as a sundeck, roof garden, or for some similar purpose. In such cases the vent line should extend about 6½ ft above the roof to ensure that no one on the roof will ever be affected by any odors emanating from the end of the pipe.

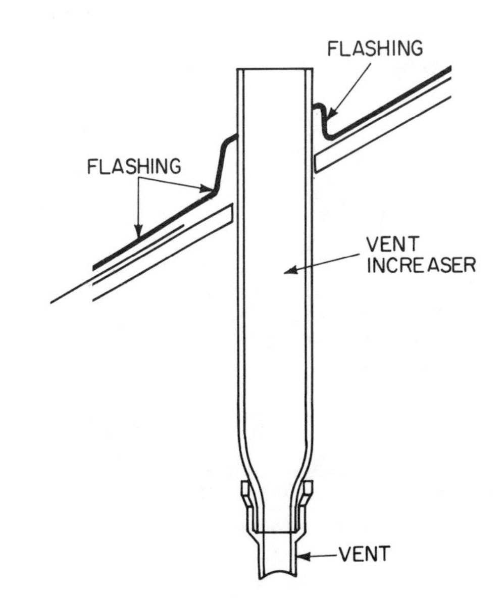

Fig. 14. An "increaser" on the end of a vent line is used very often to prevent frost from closing off the end of the vent line.

Cleanouts

The possibility always exists that some day the drainage system will become blocked. The two most common reasons are that solid matter has become wedged somewhere in the piping, usually at a turn, or that the pipes have become clogged by a deposit of grease. The latter usually happens in the drain line leading from the kitchen sink because many people are in the habit of pouring hot cooking grease into the sink to get rid of it. The grease quickly congeals on the surface of the pipes and sticks there.

Since the possibility of blocked lines exists, it is necessary to install *cleanouts* in the system to enable any blockages to be cleared quickly and conveniently. A cleanout is a Y fitting with two inlets, one of which is closed off by a threaded plug. The plug is made of brass to prevent its corroding tight to the pipe threads. Figure 15 shows typical cleanouts and their method of installation.

One cleanout is always installed in the end of the house drain so that this entire line can be cleaned out if necessary. If the house drain is very long—50 ft or more, another cleanout should be installed in the line just inside the foundation wall to allow the house sewer to be cleaned out also. In addition to these basic locations, a cleanout should also be installed in any horizontal pipe run that is more than 2 ft long and at the base of any waste stack. The kitchen drain line should always have a separate cleanout installed, whether the line runs horizontally for any distance or not.

It is difficult to install cleanouts in the second-floor drainage lines because the piping is located under the flooring and because homeowners object to cleanout fittings projecting above floor level in their living quarters. In any case, in the typical domestic drainage system, only the water-closet branch or lavatory branch will become clogged, and a cleanout is unnecessary for these fixtures. A plumber can always get at a blockage in a water-closet branch by inserting his *auger* through the water-closet bowl; he can get at a blockage in the lavatory branch by removing the trap from the lavatory drain line and inserting his auger into the end of this line.

An *auger* is widely and commonly known as a *snake*. The auger usually kept by homeowners for emergency use consists of a 5- to 25-ft-long length of coiled-steel wire. One end of the auger is inserted into the piping by winding a handle attached to the other end, as if one were cranking up the engine of an old Model T; at the same time the wire is pushed deeper and deeper into the pipe. As the auger twists around and around within the pipe, it scrapes away any grease deposited on the inner surface of the pipe and pushes this grease ahead of it into the soil stack or house drain. If possible, water should be run into the pipe while the auger is being turned in order to flush the loosened grease away.

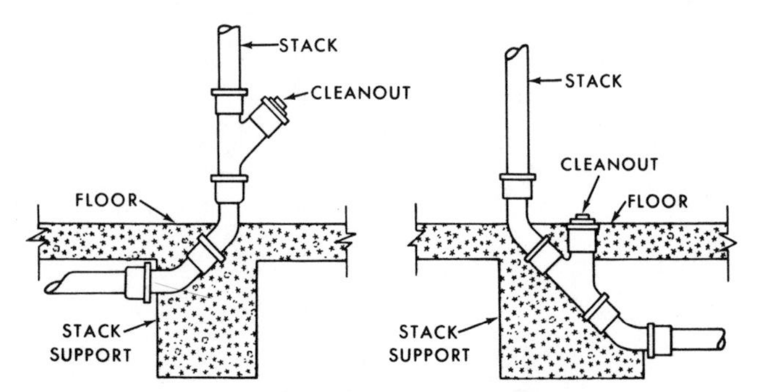

Fig. 15. Cleanouts are required so that soil and waste stacks can be cleaned out if they should become blocked *(U.S. Dept. of Agriculture)*.

Solid wastes such as rags, paper towels, sanitary napkins, or torn-up newspapers (which some people seem habitually to throw into the water-closet bowl) are cleared from a pipe in the same way.

A major blockage in the soil stack or house drain will require heavier equipment than the auger possessed by a homeowner. Plumbers who specialize in sewer cleaning use large, motor-driven augers 100 ft or more long that will quickly clear a 6 in. house drain or sewer line.

INSPECTION AND TESTING

Because a defective drainage system may become a health hazard, municipalities require that a newly installed system be inspected and tested before the system can be used. Most municipalities employ building inspectors who check the plumbing system for adequate workmanship and compliance with the local code. As part of the inspection, the installation is tested for leaks.

Municipal requirements or not, a prudent plumber will test his work in any case, section by section or floor by floor as the pipes are installed. It is to his advantage to do so; otherwise he may find after he has finished the job that he has to disassemble a large section of the completed job in order to replace a cracked fitting located, say, at the base of the soil stack. Or, if the homeowner should discover a leaking pipe after the walls, floors, and ceilings have been finished, the plumber may find himself faced with a lawsuit and the probability of having to correct the fault (plus damage) at his own expense.

The most common method used for testing a drainage system is to fill the system with water and then check all the joints and fittings for evidence of a leak. The drainage system is usually water-tested after the roughing-in has been completed but before the fixtures and traps have been installed, the open ends of all threaded pipes first being capped and rubber test plugs being inserted into the open ends of all cast-iron pipes.

The drainage system should be tested under a pressure of at least 10 ft of water but not more than 40 ft of water. This is achieved by having the end of the soil stack or main vent extend at least 10 ft above that section of the system that is being tested. The pipes are filled with water through the open end of the soil stack or main vent. If the system is watertight, the level of water visible at the end of the open pipe will not change, it must remain unchanged for at least 15 min for the system to be considered watertight. If the water level does drop, then it is presumed a leak exists. The leak must be found and eliminated. The system is then checked again (and again, if necessary) until the plumber or inspector is satisfied that the system is indeed watertight.

After the fixtures and traps have been installed, plumbing codes often require that the complete system be air-tested, even when a water test has been successfully concluded. Because the trap seals can withstand only an inch or two of water pressure, the usual test pressure is equal to 1 in. of water. The system is required to maintain this pressure for at least 15 min. Only then is the system considered completely watertight.

Plywood

Plywood is made from thin sheets of wood called *veneers* or *plies*. These veneers are sliced 1/12, 1/10, 1/8, 1/6, 3/16 or 7/32 in. thick. Several of these plies are assembled together into a panel, with the grain of each ply running at right angles to the grain of the adjacent ply. The plies are then bonded together with a suitable adhesive, the result being a strong, rigid, lightweight panel that is about 1½ times stronger than a sheet of mild steel of the same weight.

Plywood panels are made from both softwoods and hardwoods. Those panels made from softwoods are used structurally in the construction of houses, and those made from hardwoods are used decoratively—as wall paneling, for furniture, and for door paneling, for example—though there is some overlap in usage between the two groups. Each basic kind of plywood is manufactured by different industrial groups, each of which has its own standards, nomenclature, requirements, and markets. The manufacture of softwood plywood panels was for a long time centered in the Pacific Northwest (including British Columbia) but now about 30 percent are made in the South; and hardwood plywood manufacture is spread throughout the southern and midwestern states.

This article will concentrate on the manufacture and grading of the softwood panels that are used in house construction. How these panels are used is described as part of separate articles. For examples *see* FLOOR FRAMING; SHEATHING; SIDING.

Softwood plywood panels are used in house construction mainly as siding, wall and roof sheathing, subflooring and underlayment, and, to a lesser extent, as concrete forms. The panels are usually 4 × 8 ft in size, though panels up to 10 ft in length are also made. The panels range in thickness from 5/16 to 7/8 in. When these panels are nailed and/or glued to the studs, rafters, and joists of a wood-frame house, they greatly increase the strength and rigidity of these structural members.

Manufacture

Table 1 shows how large a variety of softwood species plywood is made from. The largest percentage of plywood, however, is made from Douglas fir or southern pine, both Group 1 (the strongest) woods.

At the mill, the felled logs are cut into lengths of about 8½ ft, the bark is removed, and the *bolts,* as they are called, are placed on huge lathes. Each bolt is rotated against a knife that slices off a thin, continuous sheet of veneer. A bar located just above the knife edge presses against the bolt to prevent splitting or the formation of compression checks, which are fine cracks that tend to form as the wood is "unwound" from the bolt.

The sheet of veneer is then clipped apart, either to form panels of the desired width or to remove large defects in the veneer. The panels are then sent through a drier that reduces the moisture content of the wood to 2 to 6 percent.

If the veneer is to be used as a *face* ply (that is, a veneer that will be exposed to view on the outside of the panel), narrow strips of clear wood trimmed from the continous sheet may be edge-glued together to form a single large sheet; smaller defects in the sheet are removed by a *patching machine* that neatly excises the defects and inserts in their place patches of exactly the same size and thickness. However, this sort of

Table 1. Strength Classification of Softwood Species Used for Plywood Manufactures

Group 1		
Apitong*†	Kerung*†	Pine, Southern
Beech, American	Larch, Western	Loblolly
Birch	Maple, Sugar	Longleaf
Sweet	Pine	Shortleaf
Yellow	Caribbean	Slash
Douglas Fir 1‡	Ocote	Tanoak
Kapur*		
Group 2		
Cedar, Port Orford	Lauan	Pine
Cypress	Almon	Pond
Douglas Fir 2‡	Bagtikan	Red
Fir	Mayapis	Virginia
California Red	Red Lauan	Western White
Grand	Tangile	Spruce
Noble	White Lauan	Red
Pacific Silver	Maple, Black	Sitka
White	Mengkulang*	Sweetgum
Hemlock, Western	Meranti, Red*§	Tamarack
	Mersawa*	Yellow Poplar
Group 3		
Alder, Red	Pine	Redwood
Birch, Paper	Jack	Spruce
Cedar, Alaska	Lodgepole	Black
Fir, Subalpine	Ponderosa	Engelmann
Hemlock, Eastern	Spruce	White
Maple, Bigleaf		
Group 4		**Group 5**
Aspen	Cottonwood	Basswood
Bigtooth	Eastern	Fir, Balsam
Quaking	Black (Western	Poplar, Balsam
Cativo	Poplar)	
Cedar	Pine	
Incense	Eastern White	
Western Red	Sugar	

*Each of these names represents a trade group of woods consisting of a number of closely related species.

†Species from the genus Dipterocarpus are marketed collectively: Apitong if originating in the Philippines; Keruing if originating in Malaysia or Indonesia.

‡Douglas fir from trees grown in the states of Washington, Oregon, California, Idaho, Montana, Wyoming, and the Canadian Provinces of Alberta and British Columbia shall be classed as Douglas fir No. 1. Douglas fir from trees grown in the states of Nevada, Utah, Colorado, Arizona and New Mexico shall be classed as Douglas fir No. 2.

§Red Meranti shall be limited to species having a specific gravity of 0.41 or more based on green volume and oven dry weight.

Source: American Plywood Association.

upgrading is not done except for panels that are to be used decoratively. Veneers of poorer quality that are to be used as inner plies are neither edge-glued nor patched. They are, instead, sorted according to the number and kinds of defects they contain (as described below) and then, with a face ply on either side, are made up into panels.

The sheets of veneer that are to make up a panel pass through a gluing machine where they are coated with an adhesive. The sheets are reassembled into a panel, and they are then pressed together for 2 to 20 minutes, under a pressure of 150 to 300 lb per sq in., depending on how thick the panel is, the type of adhesive being used, and the temperature of the press. A typical temperature is 250°F.

The completed panel is trimmed to its final size and the face plies (of some grades) are sanded smooth, and the panel is inspected for any manufacturing defects and graded according to its overall quality and intended use. The panel is finally marked with a *grade-trademark* stamp that indicates what its quality and intended use are.

PROPERTIES OF PLYWOOD

The essential characteristics of plywood that give it its unique structural properties are (1) the grain of each ply runs crosswise to the grain of the adjacent plies, and (2) the plies on either side of the center ply (i.e., the *core* of the panel) are usually *balanced;* that is, they are mirror images of each other with respect to grain direction, thickness, and strength characteristics.

As a result of this balanced, laminated construction (and in contrast to the properties of ordinary lumber): (1) the strength of a plywood panel is nearly the same along both its length and width, (2) the panel has a much greater degree of dimensional stability, (3) there is far less tendency for the panel to split, warp, or twist, and this construction makes possible in the first place the manufacture of these thin panels in very large sizes. Because a plywood panel consists of an odd number of plies (otherwise its construction couldn't be balanced), the grain of the face plies is always parallel to each other. It is the custom always to have the grain on the face plies running the length of the panel. For the reasons discussed below, this makes the panel stronger in a lengthwise direction than across its width.

Strength Characteristics of Plywood

Because of the fibrous nature of wood, ordinary lumber is about 35 times as strong in a direction parallel to its grain than it is across its grain.

In a plywood panel, the strength of the panel is much more evenly distributed across both its length and width because the grain of the plies runs through the panel in both lengthwise and crosswise directions. This equalization of strength is least apparent in a 3-ply panel because two of the three plies will run the length of the panel, and only one ply runs the width of the panel.

As the number of plies increases, this difference in strength decreases. A 7-ply panel, for example, will have a more equitable distribution of strength than a 5-ply panel, which will have a greater proportion of its strength in its lengthwise direction. Practically speaking, therefore, if one has a choice between two ¾-in.-thick panels, one with five plies and the other with seven, everything else being equal, the decision on which panel to use will depend on whether one is interested in the overall strength and rigidity of the construction or strength in one particular direction.

Dimensional Stability of Plywood

Under ordinary conditions, assuming normal changes in atmospheric humidity, wood will neither expand nor shrink any significant amount in a direction *parallel* to its grain. The same piece of wood will, however, expand and shrink from 3 to 6 percent in a direction *across* its grain. In a 2 × 4, for example, the typical change in size across the 4-in. dimension will amount to something like 0.14 in., or about ⅛ of an inch. In a 6-in.-wide board, the total change will amount to about 0.22 in. across the width of the board, or about ¼ of an inch. Furthermore, in the process of absorbing this moisture, ordinary wood is quite likely to twist or warp or otherwise become distorted because the moisture will not be absorbed evenly throughout the wood. Changes in its moisture content will also cause ordinary wood to split and crack, especially at its ends. (For a discussion of wood and moisture in wood, *see* WOOD.)

These dimensional problems are reduced considerably in a properly manufactured plywood panel: any expansion that might occur in the plies running in one direction because of an increase in their moisture content is firmly resisted by the plies whose grain runs in the other direction. The plies can offer this resistance to expansion because of the strength of the adhesive that binds the plies together. Plywood panels will average about 0.1 percent expansion with normal increases in atmospheric humdity. This amounts to about 3/32 in. over the entire 8-ft length of the panel. An 8-ft-wide wood plank (assuming such a piece of wood existed) would expand about 4 in.

Its cross-ply construction also makes a plywood panel resistant to any changes in its moisture content that would tend to warp or twist it. This is true, however, only when all the plies running in the lengthwise direction are parallel to each other and all the plies running in a crosswise direction are exactly at 90 degree angles to each other. Any misalignment of the grain in either direction will upset the balanced construction of the panel and make it possible for the panel to warp or twist.

Another result of plywood's cross-grained construction is that the plywood panel will not split or crack along its edges. For this reason nails or screws may be driven through the panel as close as ¼ in. to an edge without splitting the wood, which would probably happen with ordinary lumber.

INTERIOR AND EXTERIOR PLYWOODS

At one time, plywood had a very poor reputation as a construction material. To say that something was made of plywood was to say that it was cheaply made and looked it. This may have been true before World War II when most plywoods were bonded together with animal or vegetable glues such as hide and bone glues, blood-albumen glue, and casein glue, none of which is particularly moisture-resistant. Once moisture had worked its way into the glue lines, bacteria and molds could attack the glue, the plies would separate, and all the virtues of plywood disappeared.

But with the introduction of synthetic resin adhesives just before and during World War II, the situation changed completely. Nowadays, when such waterproof, mold- and bacteria-resistant resins as phenol formaldehyde or melamine formaldehyde are used as the adhesive, the glue line is the last thing to fail. It is possible to imagine that, like the grin on the face of the Cheshire Cat in *Alice in Wonderland,* after the wood has disintegrated completely, one might still be left with several paper-thin layers of adhesive suspended in the air.

Indeed, it *is* the adhesive that makes the plywood. Assuming that several different panels have been made from wood of equal quality and thickness, the differences in durability between them will depend almost entirely on the kind of adhesive used to join the plies together.

Plywoods are divided into Exterior and Interior types. The difference between them is due mainly to the adhesive used. Exterior plywoods are made with a *phenol formaldehyde* adhesive. This adhesive is completely impervious to moisture and will last indefinitely under the most extreme climatic conditions. Its only fault is a tendency to stain wood a dark reddish-brown along the glue line, which is an irrelevant complaint in a plywood intended to be used as a construction material.

If the appearance of the panel is important, as it is in some decorative softwood plywoods, then *melamine formaldehyde* can be used as the adhesive. This resin is just as strong and waterproof as phenol formaldehyde; it is more expensive, but it does not stain wood.

A small amount of Interior plywoods, intended for temporary construction work or protected interior installations, is still made from protein glues such as casein, blood-albumen, and soya glues. The overwhelming amount of Interior plywoods are, however, now made from the same phenol formaldehyde resins used in Exterior plywoods. Some Interior plywood is also made with urea formaldehyde resins. These adhesives are cheap, easy to work up and apply, water-soluble (until they have set), and they will maintain the strength of their bond even when they are occasionally, but briefly, soaked with water. They will, however, deteriorate if exposed to moisture for prolonged periods.

The difference in quality between an Interior plywood and an Exterior plywood assembled with the same phenol formaldehyde adhesive is that the surfaces of the Interior veneers are allowed to remain rough, so that the glue lines may not be as strong or as durable as the glue lines in full Exterior plywood. This type of water-resistant Interior plywood may be used even where it is likely to be fully exposed to the weather for a long period of time; that is, where prolonged delays in construction are anticipated before the plywood is finally covered over or otherwise protected from the weather.

APPEARANCE AND ENGINEERED GRADES

Having said all this, we should note that both Interior and Exterior plywoods are separated into two groups—one in which the *appearance* of the plywood is of primary importance and another in which the *structural properties* of the plywood are of primary importance.

The first group is categorized as *Appearance Grade* plywood and the second group as *Engineered Grade* plywood (see Tables 2 and 3). To which of these two groups any particular panel will belong depends not only on the kind of adhesive used in its manufacture but also on (1) the strength and stiffness of the woods used, (2) the appearance of the face plies, and (3) the number and kind of defects in both the inner and face plies, since these defects will affect both the strength and appearance of the panel. (The face plies become the *face* and *back* plies when one of the face plies has a better appearance than the other.)

Strength Groups

The woods used for the manufacture of softwood plywood are placed in one of five strength groups, depending on the inherent strength and stiffness of the wood. This is shown in Table 1, where the strongest woods are those in Group 1, the weakest in Group 5, and the other species are placed somewhere between these two extremes.

The face and inner plies in any particular panel need not belong to the same group or the same species. The inner plies may, for example, consist of a Group 4 or 5 wood, and the face plies may consist of Group 1 or 2 woods. What *is* important is that the strength characteristics of the veneers be balanced across the center core. If this balance is absent, the panel may warp and twist.

The overall strength and stiffness of any given panel is much more dependent on the strength grade of its face plies than on the strength grade of its inner plies. For this reason, when a *group number* is shown on a grade-trademark stamp (such as *Group 1;* see Fig. 1), the group number indicates not only the overall strength and stiffness of the panel, it also indicates the strength of the face plies. When the strengths of the two face plies differ from each other, the strength group of the weaker face ply will be marked on the grade-trademark stamp.

Veneer Grades

The plies are also graded according to the number and kind of defects they contain. These *veneer grades* (see Table 4) are an indication of the surface appearance of the face plies, and they also indicate whether a defect is severe enough to affect the overall durability of the panel. Thus, the lowest veneer grade, Grade D, is not used in Exterior plywoods, even as an inner ply, because of the stringent durability requirements of these plywoods; it can, however, be used in Interior plywoods.

Grade N veneers are completely free of visible defects, except possibly for a few small patches here and there. Grade N is used for the face plies when the panel is to receive a natural finish. Grades A, B, C, and D are progressively poorer looking, with larger and more numerous defects in the veneer.

Appearance Grade Plywoods

Appearance grade plywoods (Table 2) can be identified as to their quality and intended use by examining the grade-trademark stamp that is placed on the back of each panel (Fig. 1). There will be (1) an indication of the veneer grade, or grades, of the face plies (i.e., C-D, A-D, etc.), (2) an indication of the strength classification of the panel (i.e., Group 2, etc.), and (3) the word *Exterior* or *Interior* to indicate the durability or type of the panel. The abbreviation APA on the grade-trademark stamp stands for the American Plywood Association, the trade association the members of which manufacture softwood plywoods. All panels inspected and/or tested by the APA will be identified by the APA grade-trademark stamp.

The stamp will also have the abbreviation PS 1-74 on it, which stands for Product Standard 1, published cooperatively in 1974 by the National Bureau of Standards and the APA. The appearance of PS 1-74 on a panel indicates that the panel was inspected and graded in accordance with the provisions of this product standard.

If the placing of the grade-trademark stamp on one of the face plies will ruin the appearance of a panel, then an *edge stamp* (see Fig. 1) containing the same information will be placed along the edge of the panel.

Engineered Grade Plywoods

As with the appearance grades, Engineered grade plywoods are made using both interior and exterior adhesives. The kind of adhesive used will determine to a large degree the end use of any particular panel. As the appearance of structural plywoods is unimportant, the face plies are not sanded, which adds a bit of thickness to the panels and thus helps to increase their strength slightly.

Most Engineered grade plywoods are classified as C-D, C-C, Structural I, Structural II, or Underlayment, with a subclassification into Interior and Exterior types (see Table 3).

Table 2. Appearance Grades of Plywood[1] *(American Plywood Association).*

Type	Grade Designation (2)	Description and Most Common Uses	Typical Grade-trademarks	Veneer Grade: Face	Veneer Grade: Back	Veneer Grade: Inner Plies	Most Common Thicknesses (inch) (3)					
Interior Type	N-N, N-A, N-B INT-APA	Cabinet quality. For natural finish furniture, cabinet doors, built-ins, etc. Special order items.	NN G1 INT APA PS 1 74 / NA G2 INT APA PS 1 74	N	N,A, or B	C						3/4
Interior Type	N-D-INT-APA	For natural finish paneling. Special order item.	ND G3 INT APA PS 1 74	N	D	D	1/4					
Interior Type	A-A INT-APA	For applications with both sides on view. Built-ins, cabinets, furniture and partitions. Smooth face; suitable for painting.	AA G4 INT APA PS 1 74	A	A	D	1/4		3/8	1/2	5/8	3/4
Interior Type	A-B INT-APA	Use where appearance of one side is less important but two smooth solid surfaces are necessary.	AB G4 INT APA PS 1 74	A	B	D	1/4		3/8	1/2	5/8	3/4
Interior Type	A-D INT-APA	Use where appearance of only one side is important. Paneling, built-ins, shelving, partitions, and flow racks.	A-D GROUP 1 APA INTERIOR PS 1-74 000	A	D	D	1/4		3/8	1/2	5/8	3/4
Interior Type	B-B INT-APA	Utility panel with two smooth sides. Permits circular plugs.	BB G3 INT APA PS 1 74	B	B	D	1/4		3/8	1/2	5/8	3/4
Interior Type	B-D INT-APA	Utility panel with one smooth side. Good for backing, sides of built-ins. Industry: shelving, slip sheets, separator boards and bins.	B-D GROUP 3 APA INTERIOR PS 1-74 000	B	D	D	1/4		3/8	1/2	5/8	3/4
Interior Type	DECORATIVE PANELS—APA	Rough-sawn, brushed, grooved, or striated faces. For paneling, interior accent walls, built-ins, counter facing, displays, and exhibits.	DECORATIVE BD G1 INT APA PS 1 74	C or btr.	D	D		5/16	3/8	1/2	5/8	
Interior Type	PLYRON INT-APA	Hardboard face on both sides. For counter tops, shelving, cabinet doors, flooring. Faces tempered, untempered, smooth, or screened.	PLYRON INT APA PS 1 74			C & D				1/2	5/8	3/4
Exterior Type (7)	A-A EXT-APA	Use where appearance of both sides is important. Fences, built-ins, signs, boats, cabinets, commercial refrigerators, shipping containers, tote boxes, tanks, and ducts. (4)	AA G3 EXT APA PS 1 74	A	A	C	1/4		3/8	1/2	5/8	3/4
Exterior Type (7)	A-B EXT-APA	Use where the appearance of one side is less important. (4)	AB G1 EXT APA PS 1 74	A	B	C	1/4		3/8	1/2	5/8	3/4
Exterior Type (7)	A-C EXT-APA	Use where the appearance of only one side is important. Sidings, soffits, fences, structural uses, boxcar and truck lining, farm buildings. Tanks, trays, commercial refrigerators. (4)	A-C GROUP 1 APA EXTERIOR PS 1-74 000	A	C	C	1/4		3/8	1/2	5/8	3/4
Exterior Type (7)	B-B EXT-APA	Utility panel with solid faces. (4)	BB G1 EXT APA PS 1 74	B	B	C	1/4		3/8	1/2	5/8	3/4
Exterior Type (7)	B-C EXT-APA	Utility panel for farm service and work buildings, boxcar and truck lining, containers, tanks, agricultural equipment. Also as base for exterior coatings for walls, roofs. (4)	B-C GROUP 2 APA EXTERIOR PS 1-74 000	B	C	C	1/4		3/8	1/2	5/8	3/4
Exterior Type (7)	HDO EXT-APA	High Density Overlay plywood. Has a hard, semi-opaque resin-fiber overlay both faces. Abrasion resistant. For concrete forms, cabinets, counter tops, signs and tanks. (4)	HDO AA G1 EXT APA PS 1 74	A or B	A or B	C or C plgd		5/16	3/8	1/2	5/8	3/4
Exterior Type (7)	MDO EXT-APA	Medium Density Overlay with smooth, opaque, resin-fiber overlay one or both panel faces. Highly recommended for siding and other outdoor applications, built-ins, signs, and displays. Ideal base for paint. (4)	MDO BB G4 EXT APA PS 1 74	B	B or C	C		5/16	3/8	1/2	5/8	3/4
Exterior Type (7)	303 SIDING EXT-APA	Proprietary plywood products for exterior siding, fencing, etc. Special surface treatment such as V-groove, channel groove, striated, brushed, rough-sawn. (6)	303 SIDING 16oc GROUP 1 APA EXTERIOR PS 1-74 000	(5)	C	C			3/8	1/2	5/8	
Exterior Type (7)	T 1-11 EXT-APA	Special 303 panel having grooves 1/4" deep, 3/8" wide, spaced 4" or 8" o.c. Other spacing optional. Edges shiplapped. Available unsanded, textured, and MDO. (6)	303 SIDING 16oc T1-11 GROUP 1 APA EXTERIOR PS 1-74 000	C or btr.	C	C					5/8	
Exterior Type (7)	PLYRON EXT-APA	Hardboard faces both sides, tempered, smooth or screened.	PLYRON EXT APA PS 1 74			C				1/2	5/8	3/4
Exterior Type (7)	MARINE EXT-APA	Ideal for boat hulls. Made only with Douglas fir or western larch. Special solid jointed core construction. Subject to special limitations on core gaps and number of face repairs. Also available with HDO or MDO faces.	MARINE AA EXT APA PS 1 74	A or B	A or B	B	1/4		3/8	1/2	5/8	3/4

(1) Sanded both sides except where decorative or other surfaces specified.
(2) Available in Group 1, 2, 3, 4, or 5 unless otherwise noted.
(3) Standard 4x8 panel sizes, other sizes available.
(4) Also available in Structural I (all plies limited to Group 1 species) and Structural II (all plies limited to Group 1, 2, or 3 species).
(5) C or better for 5 plies; C Plugged or better for 3-ply panels.
(6) Stud spacing is shown on grade stamp.
(7) For finishing recommendations, see form V307.

Table 3. Engineered Grades of Plywood *(American Plywood Association)*.

Type	Grade Designation	Description and Most Common Use	Typical Grade-trademarks	Veneer Grade: Face	Veneer Grade: Back	Veneer Grade: Inner Plies	Most Common Thicknesses (inch) (1)					
Interior Type	C-D INT-APA	For wall and roof sheathing, subflooring, industrial uses such as pallets. Also available with intermediate glue or exterior glue. Specify intermediate glue for moderate construction delays; exterior glue for better durability in somewhat longer construction delays, and for treated wood foundations. (2) (10)	C-D 32/16 APA INTERIOR PS 1-74 000	C	D	D		5/16	3/8	1/2	5/8	3/4
Interior Type	STRUCTURAL I C-D INT-APA and STRUCTURAL II C-D INT-APA	Unsanded structural grades where plywood strength properties are of maximum importance: structural diaphragms, box beams, gusset plates, stressed-skin panels, containers, pallet bins. Made only with exterior glue.	STRUCTURAL I C-D 24/0 APA INTERIOR PS 1-74 000 EXTERIOR GLUE	C	C(6)	D(7)	D(7)	5/16	3/8	1/2	5/8	3/4
Interior Type	UNDERLAYMENT INT-APA	For underlayment or combination subfloor-underlayment under resilient floor coverings, carpeting in homes, apartments, mobile homes. Specify exterior glue where moisture may be present, such as bathrooms, utility rooms. Touch-sanded. Also available in tongue and groove. (2) (3) (9)	UNDERLAYMENT GROUP 1 APA INTERIOR PS 1-74 000	C Plugged	D	C(8) & D	1/4		3/8	1/2	5/8	3/4
Interior Type	C-D PLUGGED INT-APA	For built-ins, wall and ceiling tile backing, cable reels, walkways, separator boards. Not a substitute for UNDERLAYMENT as it lacks UNDERLAYMENT's indentation resistance. Touch-sanded. (2) (3) (9)	C-D PLUGGED GROUP 2 APA INTERIOR PS 1-74 000	C Plugged	D	D		5/16	3/8	1/2	5/8	3/4
Interior Type	2·4·1 INT-APA	Combination subfloor-underlayment. Quality base for resilient floor coverings, carpeting, wood strip flooring. Use 2·4·1 with exterior glue in areas subject to moisture. Unsanded or touch-sanded as specified. (2) (5)	2·4·1 GROUP 1 APA INTERIOR PS 1-74 000	C Plugged	D	C & D	(available 1-1/8" or 1-1/4)					
Exterior Type	C-C EXT-APA	Unsanded grade with waterproof bond for subflooring and roof decking, siding on service and farm buildings, crating, pallets, pallet bins, cable reels. (10)	C-C 42/20 APA EXTERIOR PS 1-74 000	C	C	C		5/16	3/8	1/2	5/8	3/4
Exterior Type	STRUCTURAL I C-C EXT-APA and STRUCTURAL II C-C EXT-APA	For engineered applications in construction and industry where full Exterior type panels are required. Unsanded. See (9) for species group requirements.	STRUCTURAL I C-C 32/16 APA EXTERIOR PS 1-74 000	C	C	C		5/16	3/8	1/2	5/8	3/4
Exterior Type	UNDERLAYMENT C-C Plugged EXT-APA C-C PLUGGED EXT-APA	For underlayment or combination subfloor-underlayment under resilient floor coverings where severe moisture conditions may be present, as in balcony decks. Use for tile backing where severe moisture conditions exist. For refrigerated or controlled atmosphere rooms, pallets, fruit pallet bins, reusable cargo containers, tanks and boxcar and truck floors and linings. Touch-sanded. Also available in tongue and groove. (3) (9)	UNDERLAYMENT C-C PLUGGED GROUP 2 APA EXTERIOR PS 1-74 000 C-C PLUGGED GROUP 3 APA EXTERIOR PS 1-74 000	C Plugged	C	C(8)	1/4		3/8	1/2	5/8	3/4
Exterior Type	B-B PLYFORM CLASS I & CLASS II EXT-APA	Concrete form grades with high re-use factor. Sanded both sides. Mill-oiled unless otherwise specified. Special restrictions on species. Also available in HDO. (4)	B-B PLYFORM CLASS I APA EXTERIOR PS 1-74 000	B	B	C					5/8	3/4

(1) Panels are standard 4x8-foot size. Other sizes available.
(2) Also made with exterior or intermediate glue.
(3) Available in Group 1, 2, 3, 4, or 5.
(4) Also available in STRUCTURAL I.
(5) Made only in woods of certain species to conform to APA specifications.
(6) Special improved C grade for structural panels.
(7) Special improved D grade for structural panels.
(8) Special construction to resist indentation from concentrated loads.
(9) Also available in STRUCTURAL I (all plies limited to Group 1 species) and STRUCTURAL II (all plies limited to Group 1, 2, or 3 species).
(10) Made in many different species combinations. Specify by Identification Index.

Table 4. Veneer Grades of Softwood Plywoods *(American Plywood Association)*.

Grade	Description
N	Smooth surface "natural finish" veneer. Select, all heartwood or all sapwood. Free of open defects. Allows not more than 6 repairs, wood only, per 4x8 panel, made parallel to grain and well matched for grain and color.
A	Smooth, paintable. Not more than 18 neatly made repairs, boat, sled, or router type, and parallel to grain, permitted. May be used for natural finish in less demanding applications.
B	Solid surface. Shims, circular repair plugs and tight knots to 1 inch across grain permitted. Some minor splits permitted.
C	Tight knots to 1-1/2 inch. Knotholes to 1 inch across grain and some to 1-1/2 inch if total width of knots and knotholes is within specified limits. Synthetic or wood repairs. Discoloration and sanding defects that do not impair strength permitted. Limited splits allowed.
C Plugged	Improved C veneer with splits limited to 1/8 inch width and knotholes and borer holes limited to 1/4 x 1/2 inch. Admits some broken grain. Synthetic repairs permitted.
D	Knots and knotholes to 2-1/2 inch width across grain and 1/2 inch larger within specified limits. Limited splits are permitted.

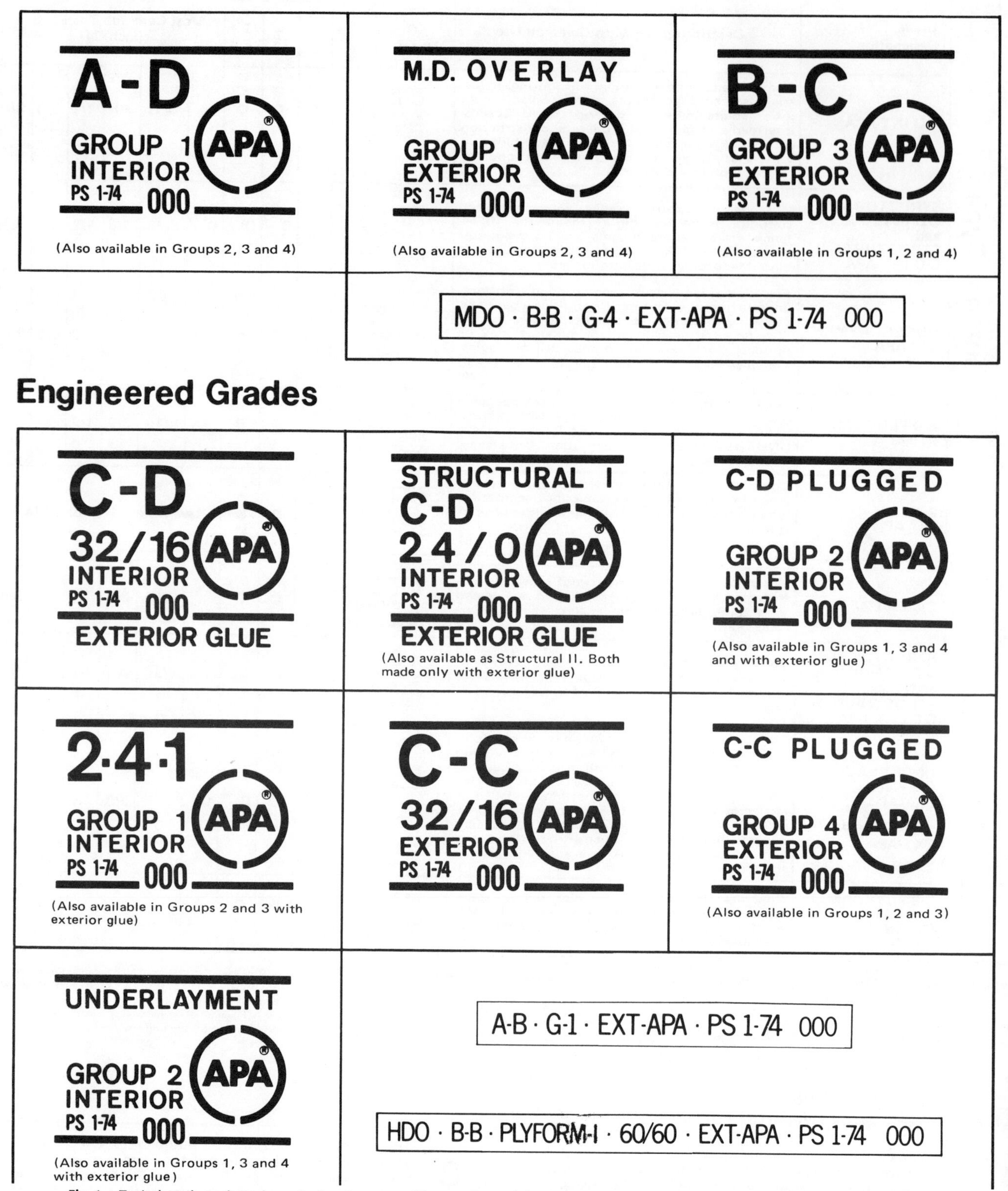

Fig. 1. Typical grade trademarks and edge stamps used by members of the American Plywood Association *(American Plywood Association)*.

Structural I Plywoods

Structural I plywoods are made from Group 1 woods assembled with an exterior glue. When the panels are intended for interior use, the veneer grade of the inner plies may be from woods of Group D. If the panel is intended for exterior use, then Group C or better veneers are required in its construction. Structural I plywoods are used where the maximum possible strength is desired.

Structural II Plywoods

Structural II plywoods are also assembled using an exterior glue, but the woods used for the inner plies may belong to

Groups 2 or 3 as well as to Group 1. As with Structural I panels, Group D veneers may be used if the panel is intended for use indoors, but group C or better veneers are required if the panel is to be used out of doors. Structural II plywoods may be used for the same purposes as Structural I plywoods, but only when strength requirements are not as stringent. Structural II plywoods are not available everywhere, so one would be wise to check on their availability before specifying their use.

Underlayment Plywoods

Underlayment plywoods are used mainly as a combination subfloor-underlayment, or only as underlayment (*see* FLOOR FRAMING). They are especially made to resist surface indentation. Either interior or exterior adhesives may be used to assemble the plies. If an exterior glue is used, the words EXTERIOR GLUE appear on the grade-trademark stamp. The absence of these words indicates that an interior glue used. The word INTERIOR or EXTERIOR by itself indicates the panel is intended either for interior or exterior use, as marked.

Woods of Groups 1 to 5 may be used to manufacture Underlayment panels, and veneer grades C and/or D are used for the face and inner plies. Underlayment panels will also have been touch-sanded to smooth any irregularities in their thickness where they will abut each other. This will prevent wear lines showing through the carpeting or resilient floor tiling that is laid down over the panels.

C-C, C-C Plugged, and C-D Plugged Plywoods

C-C plywood is an unsanded Exterior type that is used primarily for sheathing where the durability requirements are unusually high. C-C Plugged is identical to Exterior Underlayment plywood; it is used as an underlayment in locations where it will be permanently exposed. C-D Plugged is an Interior plywood that is touch-sanded and has the same face appearance as Underlayment but without the indentation-resistant property of Underlayment plywood. It is used for tile backing, counter tops, and other similar uses.

Identification Index Number

Engineered grades of plywood intended to be used as roof sheathing or subflooring (which includes C-D Sheathing, C-D with exterior glue, and most Structural I and II grades) will have an *identification index number* assigned to them (see Table 5). This number consists of two numbers separated by a slash (e.g., 32/16). These numbers indicate the maximum recommended spacing, in inches, for roof rafters or floor joists when a panel of any given thickness is installed over them as roof sheathing or subflooring. The first number indicates what the maximum spacing of the roof rafters must be, in inches, when the panel is to be used as roof sheathing. The second number indicates what the maximum spacing of the floor joists must be, in inches, when the panel is to be used as subflooring. If the second number is a 0, it means that the panel is not strong enough to be used as subflooring. The index numbers assigned to any given panel will depend on the strength characteristics of the wood used and on the total thickness of the panel.

These index numbers are unnecessary or can be ignored when a panel is to be used for wall sheathing because wall sheathing is not stressed in the same way or to the same extent as roof sheathing or subflooring.

Specialty Plywoods

To many people the appearance of ordinary plywood, with its wide bands of aimlessly meandering grain, is not very attractive, and it is not an easy material to paint or finish. The APA has been extremely energetic in finding ways of circumventing these characteristics of plywood in order to extend its use. This circumvention chiefly takes the form of bonding thin films of paper-impregnated plastic to the face plies. This material is usually bonded to both sides of the panel in order to maintain the balance that is necessary if warping or twisting is to be prevented. Either that or the back face of the panel must be constructed in such a way that the panel will maintain the desired balance.

Overlaid Plywood

Two major types of overlaid plywood are produced. Both consist of Exterior plywood to which has been bonded a sheet of paper impregnated with a phenolic resin. For both types, the total thickness of the overlay is 0.012 in. or more. This resin-impregnated sheet may be bonded to one or both sides of the panel. A phenol-formaldehyde adhesive is generally used.

High-Density Overlay. The sheet of paper used for the high-density overlay (HDO) is impregnated with at least 45 percent phenolic resin by weight. The result is a sheet having a very hard, dense, and smooth finish that is completely waterproof and resistant to most chemicals as well. The overlay has a natural translucence to it through which the underlying grain of the wood can be obscurely seen.

High-density overlay was developed originally for use as concrete formwork. It is still used for this purpose, and it gives the surface of the concrete a very smooth, clean look. The advantages of the overlay were so obvious, however, that it is now widely used outdoors wherever a waterproof and extremely durable surface is required, though high-density overlay is seldom used in house construction. Most of the highway signs one sees along the United States Interstate system are made of high-density overlay plywood. (A few of

Table 5. Identification Indexes Used to Select Plywood Panels for Subfloors and Roof Sheathing
(Panels thicker than ⅞ in. shall be identified by group number)

	C-C Exterior C-D Interior			Structural I C-C and C-D Structural II† C-C and C-D	Structural II C-C and C-D
Thickness, in.	Group 1 Group 2*	Group 2 or 3 Group 4*	Group 4	Group 1	Group 2 or 3
5/16	20/0	16/0	12/0	20/0	16/0
3/8	24/0	20/0	16/0	24/0	20/0
1/2	32/16	24/0	24/0	32/16	24/0
5/8	42/20	32/16	30/12	42/20	32/16
3/4	48/24	42/20	36/16	48/24	42/20
7/8		48/24	42/20		48/24

*Panels conforming to special thickness provisions and panel constructions of Paragraph 3.8.6 of PS 1.
†Panels manufactured with Group 1 faces but classified as Structural II by reason of Group 2 or Group 3 inner plies.
Source: American Plywood Association.

the advantages claimed for HDO plywood as a highway-sign material by the APA are that ". . . vandalism costs are drastically reduced . . . resists all forms of deliberate abuse including destructive bending and blows . . . the cross-lamination of plywood limits damage from bullets—they pass through the sign cleanly without shatter or tear." They are indeed signs for our times.)

Medium-Density Overlay. The sheet of paper used for medium-density overlay (MDO) is impregnated with at least 17 percent phenolic resin by weight, which makes the sheet resistant to moisture but not to water vapor. The surface of the sheet has a soft, slightly matte finish that gives it an excellent tooth for paints. Medium-density overlay panels are less expensive than high-density overlay panels, and the panels are used as wall paneling, for siding, and for built-in furniture, when the surfaces of the panels are to be painted.

Textured Plywood

By wire-brushing, sand-blasting, embossing, or sawing the surface of a plywood panel, a wide variety of textures can be achieved. These textures are divided into striations, grooves, rough-sawing, embossing, and relief. Textured plywoods are used indoors as wall paneling and out of doors as siding.

Naturally Finished Plywoods

These are plywoods on which the face ply consists of a decorative softwood veneer. The panel is finished naturally to show off the color and grain of the wood. Woods used for these face veneers include redwood, knotty pine, western and aromatic red cedar, and Philippine mahogany. These panels are used for interior wall paneling.

Roof Framing

Roofs have three purposes, all obvious. They (1) protect the interior of the house against the weather, (2) prevent the escape of heat from the interior during cold weather, and (3) are decorative. All roofs protect and insulate regardless of their type of construction, but the particular shape a roof has also strongly determines the overall appearance of a house. What would a Cape Cod design be without its gabled roof and dormers, a Dutch Colonial without its gambrel roof, or a Victorian mansion without its mansard roof?

The slope, or *pitch*, of a roof is determined primarily by the local climate and by tradition. In areas of heavy snowfall, for example, the pitch is steep to prevent the accumulation of snow on the roof, snow that might eventually find its way into the house as it melts. Relatively steep slopes (that is, roofs having slopes that are at least 25 or 30 degress from the horizontal) are also usual in areas of heavy rainfall because roofing materials such as tiles, slates, wood or asphalt shingles are not waterproof; they merely shed water. Unless a roof having one of these kinds of covering is pitched steeply enough to prevent water or melting snow from making its way past this covering, the water assuredly will make its way into the house some day.

In dry, temperate climates, the pitch of the roof can be very shallow, almost flat, as long as the roof covering is of some impermeable material such as metal or asphalt; but even a flat roof requires some slight amount of pitch to enable water to run off it, for one never knows where or when a crack or pinholes will develop in the roofing material that will allow a puddle of water to leak into the house.

Types of Flat Roof

Flat roofs are basically either *shed* or *lean-to* roofs (see Fig. 1), the difference between them being that a shed roof is usually the main roof of a house and a lean-to roof is part of an extension attached to the main part of the building. Flat roofs may either be dead level or (preferably) have some degree of pitch to them so that rainwater and melted snow can drain away.

Flat roofs are framed in essentially the same way as wood floors (*see* FLOOR FRAMING). A number of ceiling (or roof) joists run from one wall to the opposite wall. These joists are supported by the walls and, perhaps also, by an interior loadbearing partition, in much the same way that floor joists are supported by the foundation walls and a girder. Any openings in the roof are framed about with headers and trimmers just as are openings in a floor.

The ceiling joists on a flat roof must be relatively large and heavy because of (1) the distance they usually span, (2) the weight of the roofing they must carry, and (3) the possibility of snow loads. They must support their own weight as well without sagging.

Alternatively, the roof construction can consist of one or more large beams that span the house to which 2-in.-thick planking is fastened. For a description *see* PLANK-AND-BEAM CONSTRUCTION.

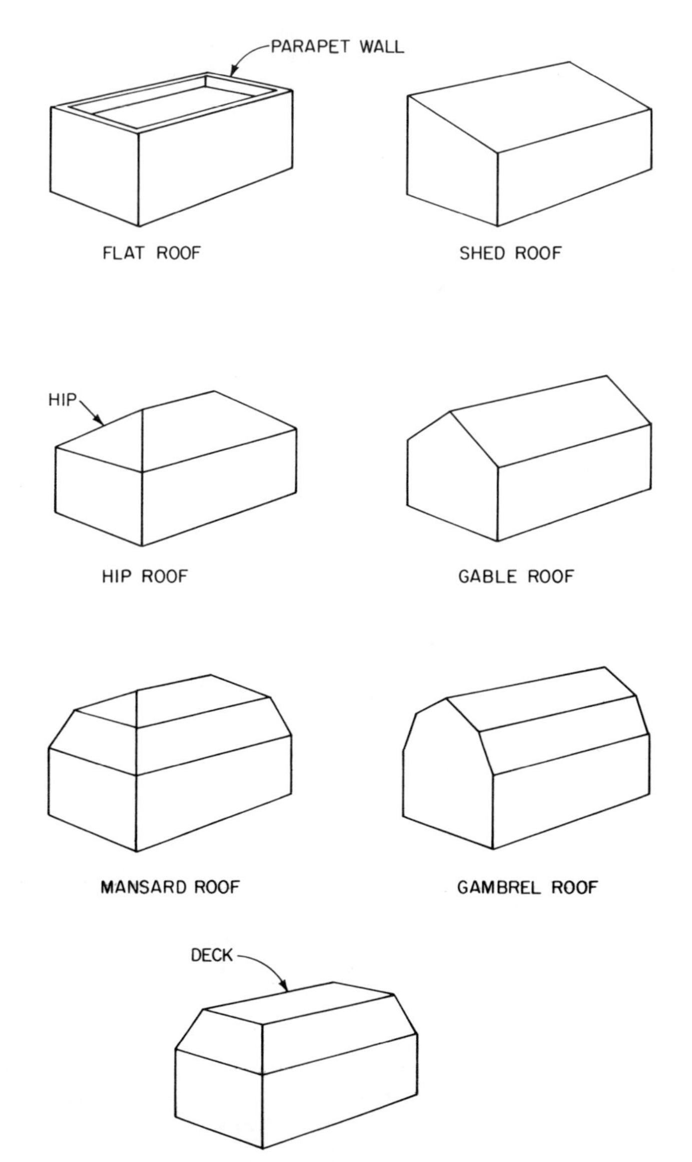

Fig. 1. The principal types of roofs used for dwellings.

Types of Pitched Roof

The principal types of pitched roofs are also shown in Fig. 1. The simplest type is the *gable roof*, which consists of two

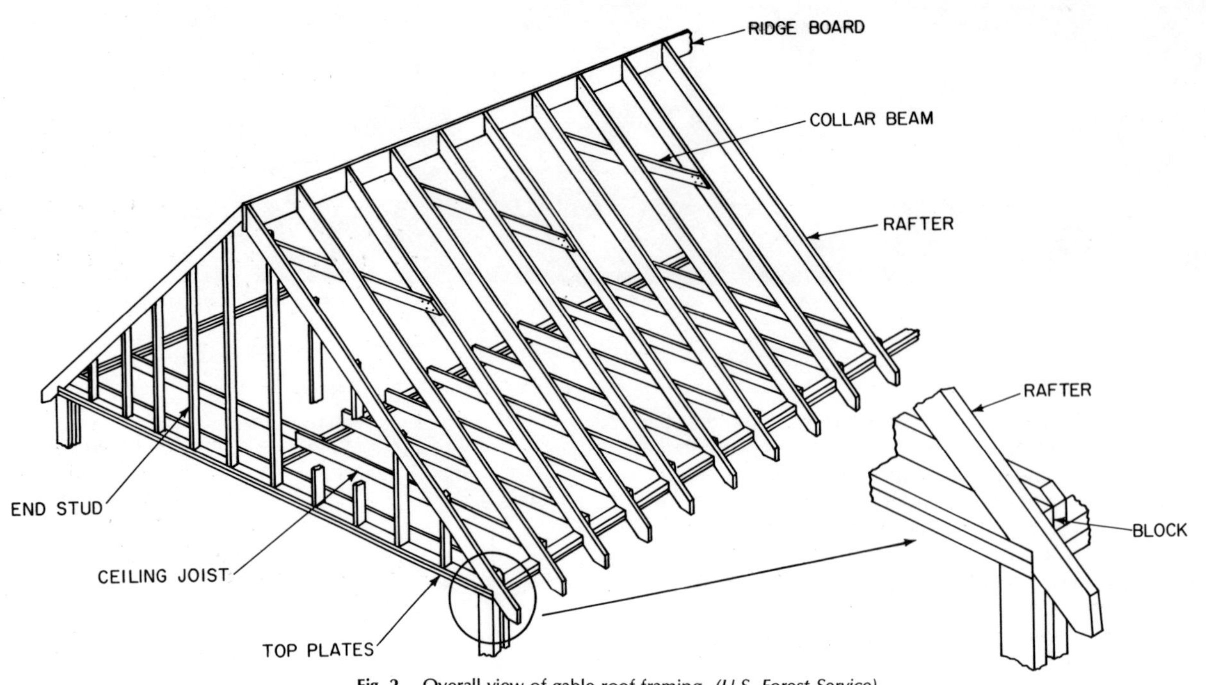

Fig. 2. Overall view of gable roof framing. *(U.S. Forest Service).*

sloping plane surfaces that meet over the center of the house. The two triangular-shaped end walls of the house abutting the roof are the *gables,* or *gable ends,* from which the name of the roof itself is derived.

A *hip roof* has four equally sloped plane surfaces that meet over the center of the house. The "hips" are the four corners at which the plane surfaces meet.

A *gambrel roof* resembles a gable roof in having only two main surfaces that meet at a peak, but it is unlike a gable roof in that each plane surface is bent about halfway up the roof. The point in having a gambrel roof is that it makes available additional storage or living space in the attic without increasing the overall height of the roof. Barns very often have gambrel roofs to enable a large quantity of animal fodder to be stored above the main floor of the barn, the main floor being used to shelter the animals.

A *mansard* (or *French*) *roof* resembles a hip roof in part because it has four main sloping surfaces, and it resembles a gambrel roof in part because each main surface is bent about halfway up the roof. The style was used mainly in large Victorian dwellings and was an adaptation of the French Empire style of architecture that was popular in the middle and late nineteenth century. A mansard roof has the same practical advantages as a gambrel roof in that it allows more of the attic space to be used for storage or for additional living space (the servants had to sleep somewhere).

A *deck roof* is also a nineteenth century roof. It closely remembles a mansard roof except that the top of the roof is flat, like a deck, instead of being sloped.

The structural members of pitched roofs are rafters and ceiling joists (see Fig. 2). The rafters may be thought of as joists installed at an angle. They support the weight of the roofing plus the weight of any snow loads as well; the joists support the weight of a finished ceiling (if any) plus the attic floor loads. The total roof loads thus being divided between the rafters and joists, both can be made from relatively light lumber compared with the size of the joists required for a flat roof covering the same span.

Alternatively, the structural members of a pitched roof can consist of *trusses.* The chief advantage of trusses is that, in addition to being capable of supporting all the roof and ceiling loads, they can support their own weight as well, which enables them to span very wide spaces without center supports being required, as with joists. This, in turn, allows the architect or builder much greater freedom in laying out the rooms in the house. Since they are very rigid structures, they can be completely constructed beforehand and then lifted into place once the walls have been raised.

DESIGN OF A GABLE ROOF

Pitch of the Roof

When one is designing and building a sloped roof, it is convenient if one can express the pitch of the roof in some simple way. There are three such possible methods, as can be seen in Fig. 3.

First, one can express the pitch in degrees from the horizontal. Although this method is convenient for the draftsman who draws up the house plans, it is extremely inconvenient for the carpenter who must lay out and cut the rafters to size, as measurements in degrees works out to very awkward sizes in inches.

Second, the pitch of the roof can be expressed in fractions. Thus, for example, a ¼-pitch roof is one in which the roof rises vertically 6 in. for every 24 in. measured horizontally; and a ⅙-pitch roof is one in which the roof rises vertically 4 in. for every 24 in. measured horizontally. This method, which is also of apparent simplicity, is difficult in practice because the carpenter must continually multiply or divide his measurements by the pitch (i.e., ¼, ⅙, or whatever) in order to find his dimensions in inches.

The third method, and the method actually used, is to express the pitch of the roof as a ratio, or proportion, of the vertical *rise* of the roof per 12 in. of horizontal *run* (see Fig. 4). Thus, to say that a roof has a *rise of 8 in.* or that the roof has a rise of *8 in 12* means that the height of the roof increases 8 in. for every 12 in. of horizontal span. Somewhere on the plans of every house will be a small triangle marked off alongside a side

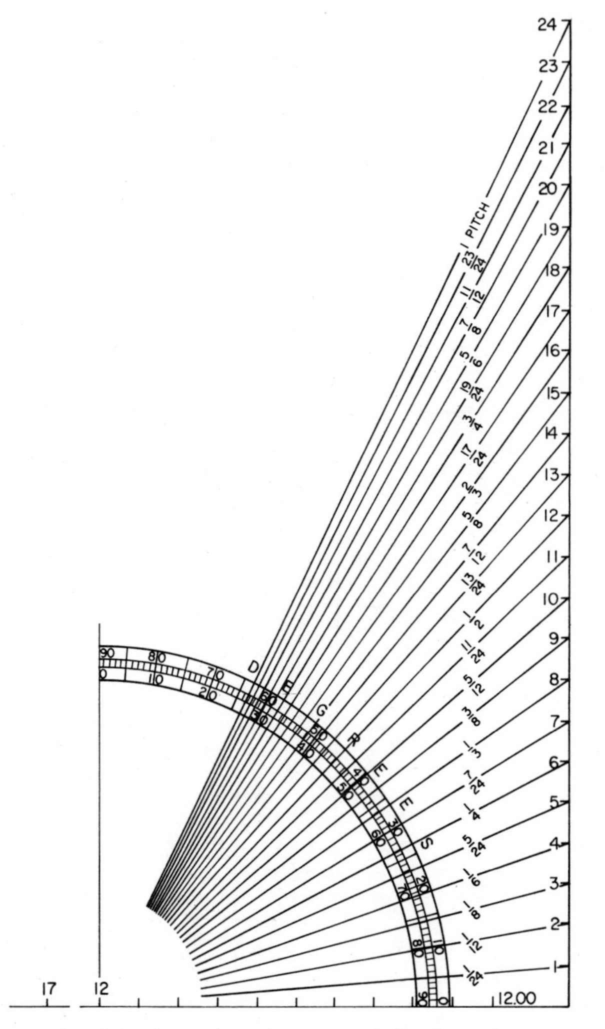

Fig. 3. The pitch of a roof can be expressed directly in degrees, as a fractional proportion, or as the ratio of the span of the roof to its rise.

view of the roof that indicates the ratio of rise to run, thus:

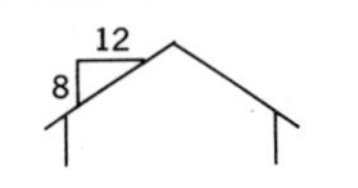

Having these two numbers, as well as the length of the overall roof span, and using his *framing square,* the carpenter can measure out the lengths and cut the angles of all the rafters quickly and accurately.

Rafter Design

In pitched-roof construction, both the rafters and the ceiling joists are designed to resist only bending loads, that is, loads that act directly downward upon them. In Colonial America, the rafters of a house, like the rest of the framework, were made of large, heavy timbers that were shaped to size with an adze. These rafters were from 5 × 5 to 6 × 8 in. in size and spaced 3 to 4 ft apart. They were fitted together carefully at the peak of the roof, and the opposite ends were notched and pinned to the top of the upper wall beams (Fig. 5).

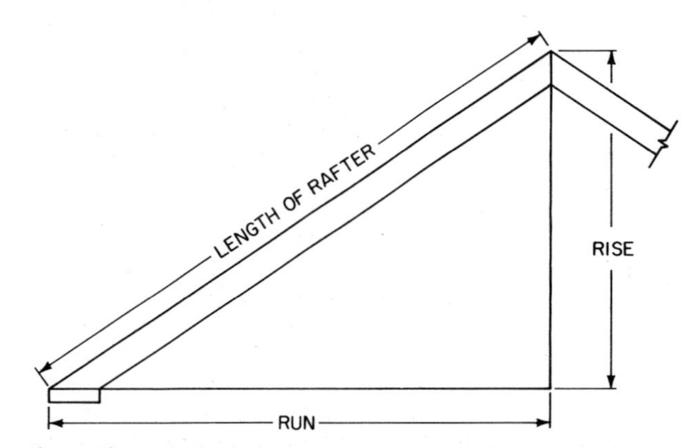

Fig. 4. The pitch of a roof is expressed as a ratio of the rise to one-half the total span. Rise and span form two sides of a right triangle with actual rafter length being the hypotenuse.

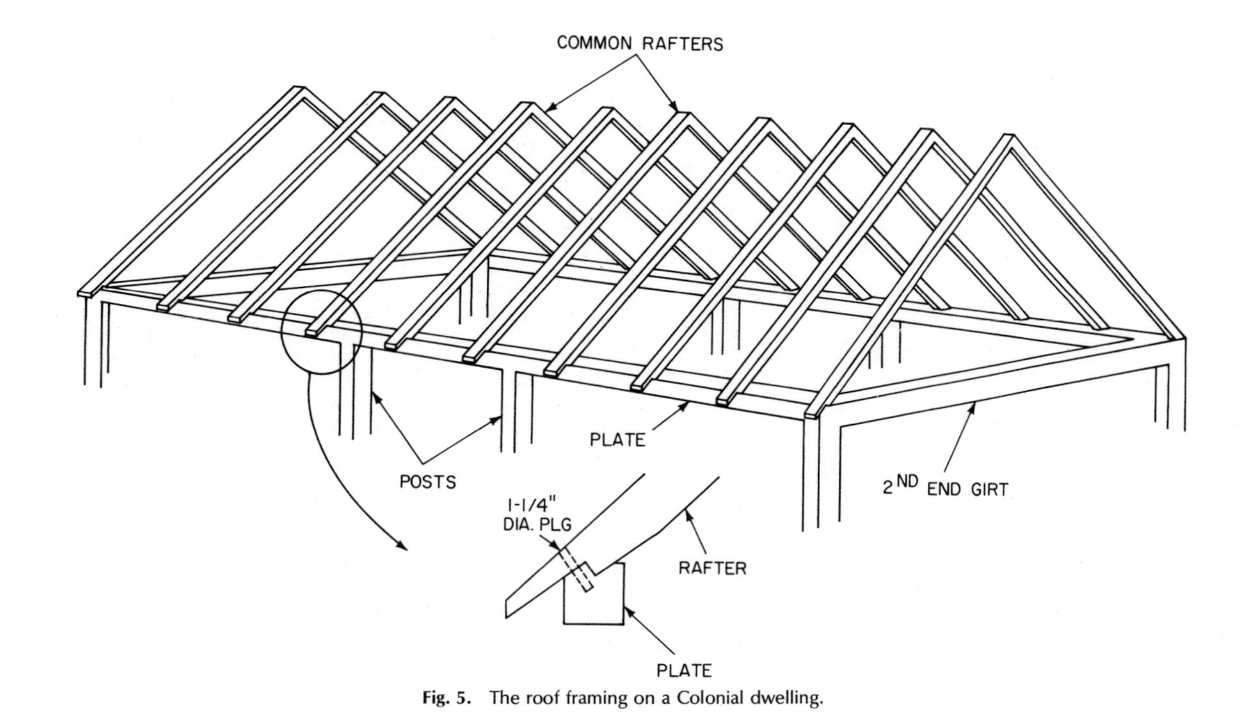

Fig. 5. The roof framing on a Colonial dwelling.

Table 1. Table Used to Find Rafter Size and Spacing for Lumber of Different Strength Characteristics.

LOW OR HIGH SLOPE RAFTERS
20 Lbs. Per Sq. Ft. Live Load
(Supporting Drywall Ceiling)

DESIGN CRITERIA:
Strength - 15 lbs. per sq. ft. dead load plus 20 lbs. per sq. ft. live load determines required fiber stress.
Deflection - For 20 lbs. per sq. ft. live load. Limited to span in inches divided by 240.

RAFTER SIZE (IN)	RAFTER SPACING (IN)	Allowable Extreme Fiber Stress in Bending, "F_b" (psi).										
		300	400	500	600	700	800	900	1000	1100	1200	1300
2x6	12.0	6-7 0.12	7-7 0.19	8-6 0.26	9-4 0.35	10-0 0.44	10-9 0.54	11-5 0.64	12-0 0.75	12-7 0.86	13-2 0.98	13-8 1.11
	13.7	6-2 0.12	7-1 0.18	7-11 0.25	8-8 0.33	9-5 0.41	10-0 0.50	10-8 0.60	11-3 0.70	11-9 0.81	12-4 0.92	12-10 1.04
	16.0	5-8 0.11	6-7 0.16	7-4 0.23	8-1 0.30	8-8 0.38	9-4 0.46	9-10 0.55	10-5 0.65	10-11 0.75	11-5 0.85	11-10 0.96
	19.2	5-2 0.10	6-0 0.15	6-9 0.21	7-4 0.27	7-11 0.35	8-6 0.42	9-0 0.51	9-6 0.59	9-11 0.68	10-5 0.78	10-10 0.88
	24.0	4-8 0.09	5-4 0.13	6-0 0.19	6-7 0.25	7-1 0.31	7-7 0.38	8-1 0.45	8-6 0.53	8-11 0.61	9-4 0.70	9-8 0.78
2x8	12.0	8-8 0.12	10-0 0.19	11-2 0.26	12-3 0.35	13-3 0.44	14-2 0.54	15-0 0.64	15-10 0.75	16-7 0.86	17-4 0.98	18-0 1.11
	13.7	8-1 0.12	9-4 0.18	10-6 0.25	11-6 0.33	12-5 0.41	13-3 0.50	14-0 0.60	14-10 0.70	15-6 0.81	16-3 0.92	16-10 1.04
	16.0	7-6 0.11	8-8 0.16	9-8 0.23	10-7 0.30	11-6 0.38	12-3 0.46	13-0 0.55	13-8 0.65	14-4 0.75	15-0 0.85	15-7 0.96
	19.2	6-10 0.10	7-11 0.15	8-10 0.21	9-8 0.27	10-6 0.35	11-2 0.42	11-10 0.51	12-6 0.59	13-1 0.68	13-8 0.78	14-3 0.88
	24.0	6-2 0.09	7-1 0.13	7-11 0.19	8-8 0.25	9-4 0.31	10-0 0.38	10-7 0.45	11-2 0.53	11-9 0.61	12-3 0.70	12-9 0.78
2x10	12.0	11-1 0.12	12-9 0.19	14-3 0.26	15-8 0.35	16-11 0.44	18-1 0.54	19-2 0.64	20-2 0.75	21-2 0.86	22-1 0.98	23-0 1.11
	13.7	10-4 0.12	11-11 0.18	13-4 0.25	14-8 0.33	15-10 0.41	16-11 0.50	17-11 0.60	18-11 0.70	19-10 0.81	20-8 0.92	21-6 1.04
	16.0	9-7 0.11	11-1 0.16	12-4 0.23	13-6 0.30	14-8 0.38	15-8 0.46	16-7 0.55	17-6 0.65	18-4 0.75	19-2 0.85	19-1 0.96
	19.2	8-9 0.10	10-1 0.15	11-3 0.21	12-4 0.27	13-4 0.35	14-3 0.42	15-2 0.51	15-11 0.59	16-9 0.68	17-6 0.78	18-2 0.88
	24.0	7-10 0.09	9-0 0.13	10-1 0.19	11-1 0.25	11-11 0.31	12-9 0.38	13-6 0.45	14-3 0.53	15-0 0.61	15-8 0.70	16-3 0.78
2x12	12.0	13-5 0.12	15-6 0.19	17-4 0.26	19-0 0.35	20-6 0.44	21-11 0.54	23-3 0.64	24-7 0.75	25-9 0.86	26-11 0.98	28-[illegible] 1.1[illegible]
	13.7	12-7 0.12	14-6 0.18	16-3 0.25	17-9 0.33	19-3 0.41	20-6 0.50	21-9 0.60	23-0 0.70	24-1 0.81	25-2 0.92	26-[illegible] 1.0[illegible]
	16.0	11-8 0.11	13-5 0.16	15-0 0.23	16-6 0.30	17-9 0.38	19-0 0.46	20-2 0.55	21-3 0.65	22-4 0.75	23-3 0.85	24-[illegible] 0.9[illegible]
	19.2	10-8 0.10	12-3 0.15	13-9 0.21	15-0 0.27	16-3 0.35	17-4 0.42	18-5 0.51	19-5 0.59	20-4 0.68	21-3 0.78	22-[illegible] 0.8[illegible]
	24.0	9-6 0.09	11-0 0.13	12-3 0.19	13-5 0.25	14-6 0.31	15-6 0.38	16-6 0.45	17-4 0.53	18-2 0.61	19-0 0.70	19-[illegible] 0.7[illegible]

Note: The required modulus of elasticity, "E", in 1,000,000 pounds per square inch is shown below each span.

Source: National Forest Products Association

Table 1. Table Used to Find Rafter Size and Spacing for Lumber of Different Strength Characteristics. (continued)

RAFTERS: Spans are measured along the horizontal projection and loads are considered as applied on the horizontal projection.

Allowable Extreme Fiber Stress in Bending, "F_b" (psi).											RAFTER	
1400	1500	1600	1700	1800	1900	2000	2100	2200	2400	2700	SPACING (IN)	SIZE (IN)
14-2 1.24	14-8 1.37	15-2 1.51	15-8 1.66	16-1 1.81	16-7 1.96	17-0 2.12	17-5 2.28	17-10 2.44			12.0	
13-3 1.16	13-9 1.29	14-2 1.42	14-8 1.55	15-1 1.69	15-6 1.83	15-11 1.98	16-3 2.13	16-8 2.28	17-5 2.60		13.7	
12-4 1.07	12-9 1.19	13-2 1.31	13-7 1.44	13-11 1.56	14-4 1.70	14-8 1.83	15-1 1.97	15-5 2.11	16-1 2.41		16.0	2x6
11-3 0.98	11-7 1.09	12-0 1.20	12-4 1.31	12-9 1.43	13-1 1.55	13-5 1.67	13-9 1.80	14-1 1.93	14-8 2.20		19.2	
10-0 0.88	10-5 0.97	10-9 1.07	11-1 1.17	11-5 1.28	11-8 1.39	12-0 1.50	12-4 1.61	12-7 1.73	13-2 1.97	13-11 2.35	24.0	
18-9 1.24	19-5 1.37	20-0 1.51	20-8 1.66	21-3 1.81	21-10 1.96	22-4 2.12	22-11 2.28	23-6 2.44			12.0	
17-6 1.16	18-2 1.29	18-9 1.42	19-4 1.55	19-10 1.69	20-5 1.83	20-11 1.98	21-5 2.13	21-11 2.28	22-11 2.60		13.7	
16-3 1.07	16-9 1.19	17-4 1.31	17-10 1.44	18-5 1.56	18-11 1.70	19-5 1.83	19-10 1.97	20-4 2.11	21-3 2.41		16.0	2x8
14-10 0.98	15-4 1.09	15-10 1.20	16-4 1.31	16-9 1.43	17-3 1.55	17-8 1.67	18-2 1.80	18-7 1.93	19-5 2.20		19.2	
13-3 0.88	13-8 0.97	14-2 1.07	14-7 1.17	15-0 1.28	15-5 1.39	15-10 1.50	16-3 1.61	16-7 1.73	17-4 1.97	18-5 2.35	24.0	
23-11 1.24	24-9 1.37	25-6 1.51	26-4 1.66	27-1 1.81	27-10 1.96	28-7 2.12	29-3 2.28	29-11 2.44			12.0	
22-4 1.16	23-2 1.29	23-11 1.42	24-7 1.55	25-4 1.69	26-0 1.83	26-8 1.98	27-4 2.13	28-0 2.28	29-3 2.60		13.7	
20-8 1.07	21-5 1.19	22-1 1.31	22-10 1.44	23-5 1.56	24-1 1.70	24-9 1.83	25-4 1.97	25-11 2.11	27-1 2.41		16.0	2x10
18-11 0.98	19-7 1.09	20-2 1.20	20-10 1.31	21-5 1.43	22-0 1.55	22-7 1.67	23-2 1.80	23-8 1.93	24-9 2.20		19.2	
16-11 0.88	17-6 0.97	18-1 1.07	18-7 1.17	19-2 1.28	19-8 1.39	20-2 1.50	20-8 1.61	21-2 1.73	22-1 1.97	23-5 2.35	24.0	
29-1 1.24	30-1 1.37	31-1 1.51	32-0 1.66	32-11 1.81	33-10 1.96	34-9 2.12	35-7 2.28	36-5 2.44			12.0	
27-2 1.16	28-2 1.29	29-1 1.42	29-11 1.55	30-10 1.69	31-8 1.83	32-6 1.98	33-3 2.13	34-1 2.28	35-7 2.60		13.7	
25-2 1.07	26-0 1.19	26-11 1.31	27-9 1.44	28-6 1.56	29-4 1.70	30-1 1.83	30-10 1.97	31-6 2.11	32-11 2.41		16.0	2x12
23-0 0.98	23-9 1.09	24-7 1.20	25-4 1.31	26-0 1.43	26-9 1.55	27-5 1.67	28-2 1.80	28-9 1.93	30-1 2.20		19.2	
20-6 0.88	21-3 0.97	21-11 1.07	22-8 1.17	23-3 1.28	23-11 1.39	24-7 1.50	25-2 1.61	25-9 1.73	26-11 1.97	28-6 2.35	24.0	

Note: The required modulus of elasticity, "E", in 1,000,000 pounds per square inch is shown below each span.

With the passage of time and the invention of power-driven saws, rafters became smaller and were spaced closer together. Today they are made of nominally 2-in.-thick lumber and are anywhere from 4 to 14 in. deep, depending on the roof loads, the total span, and the spacing between them, which is usually 16, 20, or 24 in. Rafter sizes are selected by consulting a rafter table, such as the one shown in Table 1.

The lower end of each rafter is cut so that it rests flat upon a *wall plate* (or *rafter plate,* depending on the point of view), which usually consists of one or two 2 × 4s nailed securely to the tops of the wall studs (*see* WALL FRAMING).

The upper ends of each pair of rafters are held in position because they bear against and are nailed to a *ridge board,* which forms the peak of a pitched roof (Fig. 2). The ridge board, which is usually made from a length of 1 × 8 in. lumber, is not really necessary, strictly speaking. Its main function is to make sure that the top of the roof forms a straight line. A ridge board should always be installed, however, because it ensures that the peak of the roof *is* straight and because it simplifies the roof construction.

Ceiling-Joist Design

It should be evident from Fig. 2 that the weight of the roofing, plus the weight of the rafters themselves, will cause the rafters to push down and out against the tops of the walls that are supporting them. The walls, therefore, tend to be thrust apart. To counteract this thrust, the tops of the walls are tied together by the ceiling joists, and this is their primary function. The joists do not, however, usually consist of single lengths of lumber that stretch from one wall to the other, not unless the total span is 16 ft or less. To install single pieces of lumber longer than 16 ft would result in their sagging noticeably. To prevent this sagging, the joists would have to be quite large, and this in turn would impose a very heavy dead load upon the walls, not to mention the cost of the lumber. There is, therefore, a distinct economic advantage in having the joists as short as possible.

Usually a *loadbearing partition* is erected that runs down the center of the house (see Fig. 6). Joists that are only half the width of the house can then meet on top of this partition, the joists being either lapped or butted together where they meet.

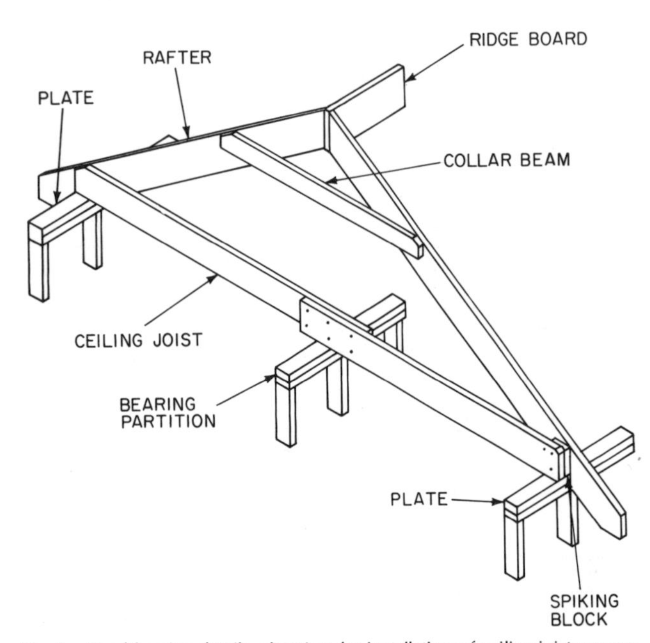

Fig. 6. Roof framing details, showing the installation of ceiling joists over a loadbearing partition and the installation of a collar beam.

It may be noted that one of the disadvantages of having this type of joist construction is that it restricts the builder's freedom in laying out the rooms of the house. The house *must* have that centrally located loadbearing partition. But with such a partition, the joists can be made of reasonably sized, lightweight, and inexpensive lumber that is easy to handle and install.

Ceiling joists are made of nominally 2-in.-thick lumber. The depth of the joists will depend primarily on the span and the weight of the ceiling construction that will be supported by the joists, that is, whether the ceiling is made of plaster, gypsum wallboards, or something else. If, in addition, the attic space is to be used for storage or for living quarters, then the size of the joists must be further increased to support these additional loads. Joist sizes can be selected by referring to a table. *See* also FLOOR FRAMING (Table 1).

Resistance to Wind Loads

The weight of the roofing (which might amount to as much as 8 lb per sq ft for clay tiles) and the weight of any accumulated snow are not the only loads the rafters must support. Of greater importance are the wind loads the rafters must resist; these loads can be of much greater magnitude. In most of the United States, houses must be capable of withstanding wind pressures of 15 lb per sq ft, which is equivalent to a horizontal wind velocity of 75 mi per hr. Along the Gulf coast, where hurricanes are always a possibility, houses must be capable of withstanding wind pressures that are at least twice as great.

For a discussion of the stresses placed upon dwellings by wind loads *see* BUILDING LOADS. It will suffice to note here that these wind loads tend to *lift* the roof off a house, and the steeper the roof, up to a point, the greater this lifting force will be. In the design of a roof, therefore, the different parts of the roof structure must be fastened together in such a way as to resist these lifting forces.

The platform-type wood-frame construction that predominates in the United States is not well designed to resist strong winds (*see* WOOD-FRAME CONSTRUCTION). For one thing, the plates to which the rafters and ceiling joists are nailed are themselves fastened in place by nails that are driven directly downward into the tops of the studs—and nails are at their weakest at resisting tensile forces that tend to pull them directly out of the piece of wood into which they have been driven.

In addition, the nails that hold the rafters and joists to the plates are themselves toe-nailed into the plates at an angle that is too slight to effectively resist the lifting forces imposed by a strong wind.

All in all, the nails in a platform-framed house cannot be trusted to hold the roof down in an extremely strong wind. In the good old days, nailing a roof in place was much less of a problem because the rafters and joists, and the roofing materials as well, were much heavier than they are today. An old slate roof, for example, weighs about 7½ lb per sq ft, compared to the 2 lb per sq ft that present-day asphalt and wood shingles weigh.

All this being so, other measures must be taken to hold the roof and walls together. In most parts of the United States subject to tornadoes or hurricane-force winds, the local building codes require that 18-gauge iron strapping be nailed between the rafters and wall studs as shown in Fig. 7.

Even when iron strapping is not required it would be wise to nail the rafters, joists, and wall plates together using the anchor plates shown in the illustrations of the article FLOOR FRAMING. These anchor plates are more than capable of absorbing any stresses the construction may be subjected to. In addition, the nails holding the anchor plates in place are driven into the wood at right angles to the direction of stress. They are thus at their most effective in helping to resist this stress.

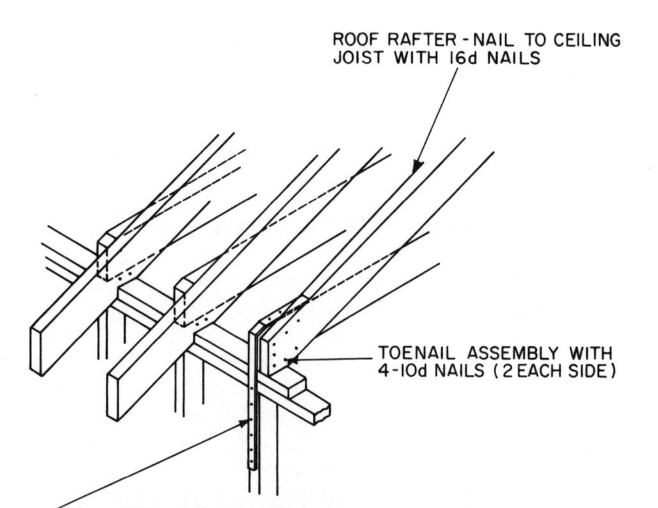

Fig. 7. In high-wind areas, 18-gauge metal strapping is often nailed to the joists and rafters to prevent the roof being lifted off the house.

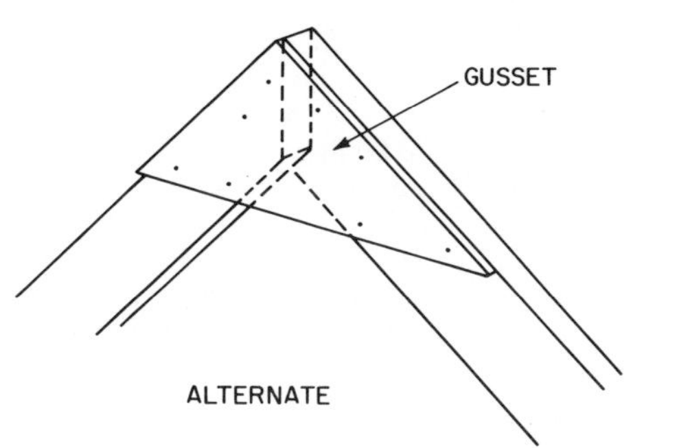

Fig. 8. Installation of a gusset plate to strengthen the rafter installation.

Another weak point in roof construction is at the peak of the roof. Rafters are usually held together at the peak by toe-nailing them directly to each other or to a ridge board. But these nails cannot adequately resist the lifting forces induced by an extremely high wind, and, therefore, other methods must be found to reinforce their juncture at the peak of the roof.

One of these methods is to nail *gusset plates* across each pair of rafters as shown in Fig. 8. These gusset plates are particularly necessary when a ridge board is omitted in the construction.

The most usual method of strengthening the rafter installation is, however, to install *collar beams,* which are usually made from 1 × 6 in. boards, across pairs of rafters as shown in Figs. 2 and 6. Not between every pair of rafters but between every second or third pair, depending on the pitch and span of the rafters and on their distance from one another but, in any case, not more than 4 ft apart.

By tying the rafters together in this way, the collar beams help resist the upward stresses imposed by wind gusts. Collar beams are most effective when they are installed in the upper one-quarter of the height of the roof space.

If attic space is to be finished off as a habitable room, it will be necessary to install collar beams across all the rafters to support the weight of the finished ceiling. In this case, the collar beams should not be made of 1-in.-thick boards but of 2-in.-thick lumber.

DESIGN OF HIP ROOFS

The four corners of a hip roof (see Fig. 9) are defined by the *hip rafters* that run from the corners of the roof to the ridge board at the peak of the roof. If the roof has a square shape, then the four sides of the roof will form a pyramid, the four hip rafters will meet at the peak, and there will not be a ridge board, of course.

The rafters that *do* run between the ridge board and the wall plates are called *common* rafters; all the rafters on a gable roof are common rafters, for example. But as may be seen in Fig. 9, a hip roof also has a great many short rafters that run between the hip rafters and the wall plates. These short rafters are called *jack* rafters.

The hip rafters act sometimes like ridge boards in that their main functions are to help keep the roof lines straight and make it easier to install the jack rafters. That is, they are more of a convenience than anything else. In most hip-roofed dwellings, the hip rafters are made of 2-in.-thich lumber that is about 2 in. deeper than the abutting jack rafters. This additional depth is necessary to give the jack rafters a full bearing where they meet the hip rafters. In addition, the top surface of the hip rafters is often cut away in a double chamfer that provides a flat surface upon which the sheathing can rest (see below), and the lumber must be deep enough for these cuts to be made.

In a small house with a hip roof, collar beams are unnecessary to stiffen the construction, but if the house is more than 20 ft wide, say, and if the roof pitch is not too shallow, as is the case on many hip roofs, then collar beams can be installed across pairs of *common* rafters to help stiffen the overall construction. Ceiling joists are also necessary in a hip-roofed house to counteract the thrusting force of the common rafters.

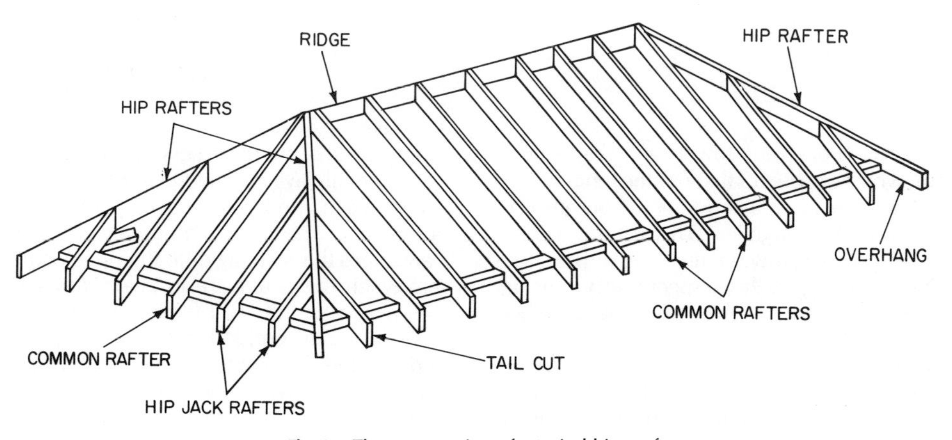

Fig. 9. The construction of a typical hip roof.

In parts of the country subject to tornadoes or hurricane-force winds, the roof must also be anchored to the walls by one of the methods described above for gable roofs. Common rafters meeting at the peak should be tied together with gusset plates when a ridge board is not installed.

DESIGN OF GAMBREL ROOFS

The bent construction of a gambrel roof (see Fig. 10) makes it very vulnerable to downward-acting loads. The upper half of the roof tends to be pushed flat by these loads, which causes the upper rafters to push out against the lower half of the roof. In turn, the lower half of the roof tends to push out against the walls of the house. Each half of the roof must, therefore, be tied together separately, the upper half by collar beams, the lower half by ceiling joists. Ceiling joists are also installed and nailed to the wall plates as already described to resist this thrust.

When the reason for having a gambrel roof is to provide additional living space in the attic, the collar beams can also be thought of as a second set of ceiling joists to which plaster or wallboards can be attached. When this is the case, the collar beams must be designed from the first as if they were joists.

Because of the overall steepness of the roof, the lifting forces induced by a strong wind are of less importance with gambrel roofs. It is more likely that the pressure exerted by a strong wind against the roof will tend to blow the roof sideways off the walls. It is necessary, therefore, if the roof is to resist these forces, to install vertical bracing between the ceiling joists and the lower rafters as is shown in Fig. 10.

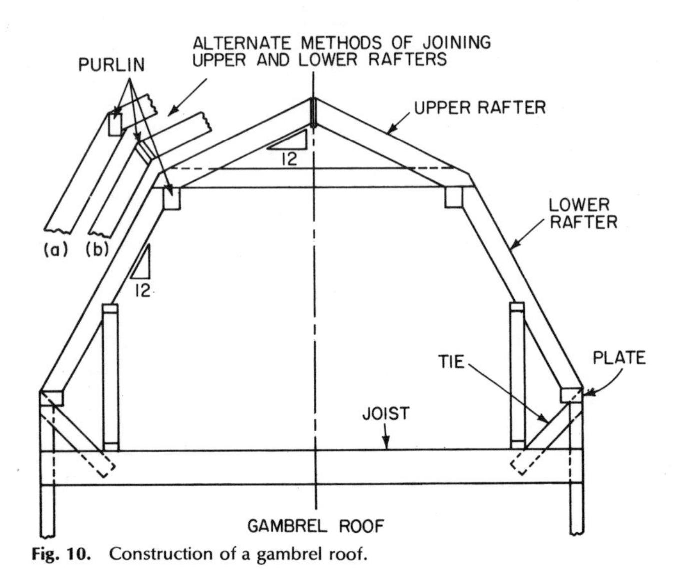

Fig. 10. Construction of a gambrel roof.

Purlins

The juncture between the upper and lower rafters is made at *purlins* (see Fig. 10), which are beams that run the length of the roof. It might appear that the main function of the purlins is to tie the upper and lower rafters together. This is true to some extent since both the upper and lower rafters are, indeed, toe-nailed to the purlins. Actually, in this respect the purlins act much like ridge boards; they are merely a convenient method for fastening the rafters to each other.

Of greater importance, however, the purlins help transmit to the side walls of the house some of the weight of the upper section of the roof, and they also transmit to the side walls some of the pressure exerted by a strong wind. In addition, the purlins help to stiffen the overall roof construction; together with the collar beams, they help the lower half of the roof resist the outward thrust of the upper part of the roof.

DESIGN OF ELL-SHAPED ROOFS

Many houses have extensions built onto the main structure that give the houses an L- or T-shaped look when viewed from above. The roofs of these extensions—or *ells* as they are called—are constructed of joists and rafters in the same way as the main roofs are. But the way in which an ell roof is attached to a main roof will depend on the overall size of the ell and whether or not the attic space in the ell is to be occupied.

Small ells in which the attic space will not be used, such as the roof over a porch or entranceway, are treated as appendages attached to the main roof. The rafters are installed and the sheathing is applied to the main roof before any work at all is done on the roof extension. Then, two 2 × 4s, the purpose of which is to define the slope of the ell, are nailed on top of the main-roof sheathing, and the roof extension is then framed in the same way that the main roof was. Once the ell has been framed, the final roofing material is installed over the entire roof construction at the same time.

This method of constructing an ell can also be used when the ell is as wide as, or almost as wide as, the main roof, if the attic above the ell is not to be used. This is often the situation, for example, when a large garage or carport is attached to a house.

When, however, the attic in an ell will be used, then an opening must be left in the main roof where it is joined by the ell. In this case, *valley* rafters must be installed on the main roof.

Valley Rafters

Valley rafters (see Fig. 11) define the juncture where two roof surfaces meet. Since the pitch of the ell roof is almost always the same as the pitch of the main roof, the angle made by this juncture will be 45 degrees when the house is viewed from above. Geometrically, the valley rafters resemble the hip rafters in both overall length and method of installation.

Unlike the hip rafters, however, the valley rafters must be strong enough to support the weight of the roof above, a weight they will carry almost entirely by themselves. To support this weight, the valley rafters are usually made twice the size of the other rafters. They are made from two 2-in.-thick lengths of lumber that are spiked together, and they must be deep enough to give full bearing to the jack rafters that will be nailed to them.

INSTALLING THE RAFTERS

The general procedure that is followed for the installation of rafters is the same regardless of the type of roof. The common rafters are always installed first. Before the common rafters are set in place, however, the carpenter first measures off the actual rafter spacing on top of one of the wall plates. He then transcribes this spacing onto the ridge board, thus making sure that when the rafters are installed they will be aligned accurately between the wall plates and the ridge board.

For a gable roof, straight rafters are selected for the gable ends (rafters are sometimes slightly warped or bowed) and these *end rafters,* as they are called, are nailed to the ridge board. This is a three-man job. Two of the men will hold the

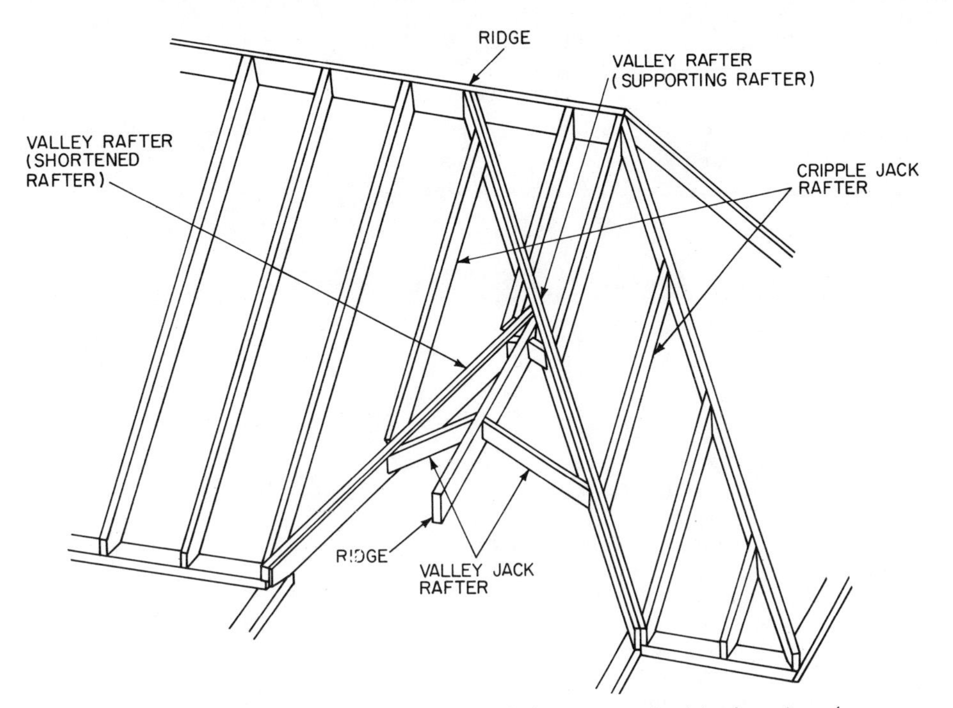

Fig. 11. Construction of a pitched roof where an extension joins the main roof.

rafters and ridge board together in position while the third man nails the rafters to the wall plates. The rafters are then nailed to the ridge board.

The rest of the common rafters can now be installed. Once this job has been completed, the plumbness of the roof is checked (on a gable roof) by dropping a plumb bob from the end of the ridge board to the top of the wall plate below. Any adjustment in the alignment of the ridge board is made and the roof is then braced securely in position. The bracing is removed after the roof sheathing has been installed.

It is also necessary on a gable roof to install wall studs between the wall plates and the end rafters at the gable ends. If the carpenter wishes to, he can lay out these studs using his framing square. More often, he merely marks the positions of the wall studs on the wall plates and then finds the height and angle of cut of each stud directly by setting the stud in position upon the wall plate and then marking off the height and angle of cut of the stud against the end rafter.

Gable roofs usually have false rafters (called *fly* rafters) that extend beyond the house. Fly rafters do not support the roof in any way; they are purely decorative. They must, instead, be supported in position. In jerry-built construction, they are held in place by nails driven through the roof sheathing from above, which is obviously poor practice. In any well-made house the fly rafters will be supported in place more securely. This is usually accomplished by extending both the ridge board and wall plates beyond the house and nailing the fly rafters to these extensions, as well as to the sheathing. There is some danger when this is done that if the wall-plate extensions are not adequately protected against the weather, moisture may seep along them and into the interior construction. For this reason, instead of extending the wall plates, brackets are often installed at the sides of the house to support the fly rafters.

Sometimes fly rafters are made larger and heavier than the actual rafters in order to give a greater feeling of solidity to the appearance of the house. Sometimes, for the same reason, that part of the ridge board that extends beyond the house is also made larger. When this is the case, the enlarged ridge board should extend back into the actual roof 3 or 4 ft so that it may be adequately supported by the rafters to which it is nailed.

Septic Tank System

A septic tank system is a method of disposing of domestic sewage safely and effectively when a municipal sewage system is not available. For most isolated dwellings, the only alternative to a septic tank system is to construct a *cesspool* on the property, a cesspool being simply a pit dug into the ground into which the raw sewage is discharged. Within the cesspool, the organic wastes are partially broken down by bacteria, and the liquid portion of the sewage is gradually absorbed into the surrounding soil, the bacteria along with it. Cesspools will badly pollute any nearby bodies of water, they stink, and they breed disease organisms; for these reasons their construction or use is universally forbidden by local health codes. In the absence of a municipal sewage system, therefore, a septic tank system is the homeowner's only possible method of getting rid of sewage, although there are now coming on the market alternative systems that may eventually replace septic tank systems.

There are two main parts to any septic tank system (see Fig. 1): (1) the *septic tank* itself, which is a large, watertight, sealed container buried in the soil, and (2) an underground *dispersal,* or *distribution,* or *absorption* system (all three terms are synonymous and in widespread use, though *absorption* is the most common term) of some kind that enables the liquid discharged from the septic tank to be safely dispersed (or distributed, or absorbed) into the soil.

The overall design of any septic tank system depends mainly on two factors: (1) the amount of raw sewage that must be treated in the septic tank within a 24-hr period, and (2) the ability of the soil to absorb the treated liquid that is discharged from the septic tank within the same 24-hr period. Some soils are more porous than other soils, some soils are only semi-permeable, and still other soils are completely impermeable and will not absorb any liquid at all. The nature of the soil thus has a very important influence on the overall design of a dispersal system. In comparison, the design of a septic tank is relatively straightforward.

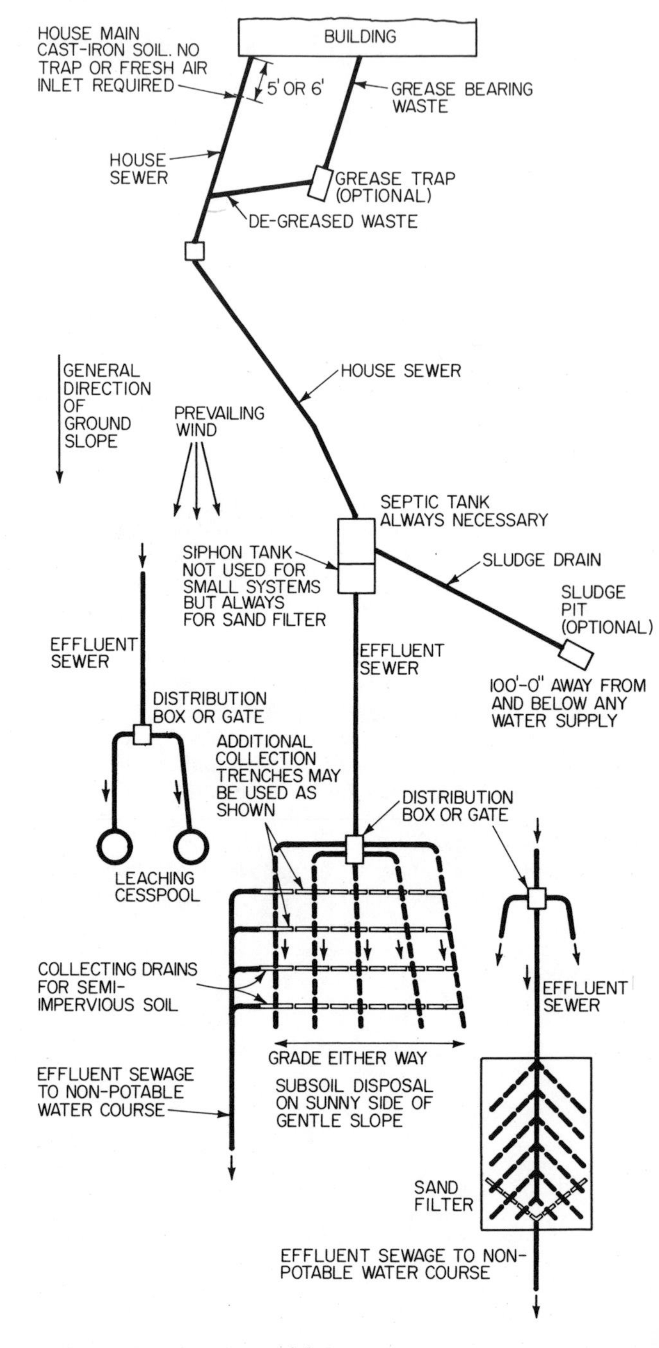

Fig. 1. Septic tank systems and their components.

SYSTEM OPERATION

Septic Tank

Most domestic septic tanks (see Fig. 2) are designed to hold between 500 and 1000 gal of liquid sewage, the actual size of the tank depending mainly on the number of bedrooms or occupants in a house, as described below. This sewage consists of both solid and liquid waste matter. When the sewage is first discharged into the septic tank, that portion of the solid wastes that is capable of disintegrating into small particles in the liquid will do so. Those particles that are heavier than the liquid will slowly settle to the bottom of the tank and become

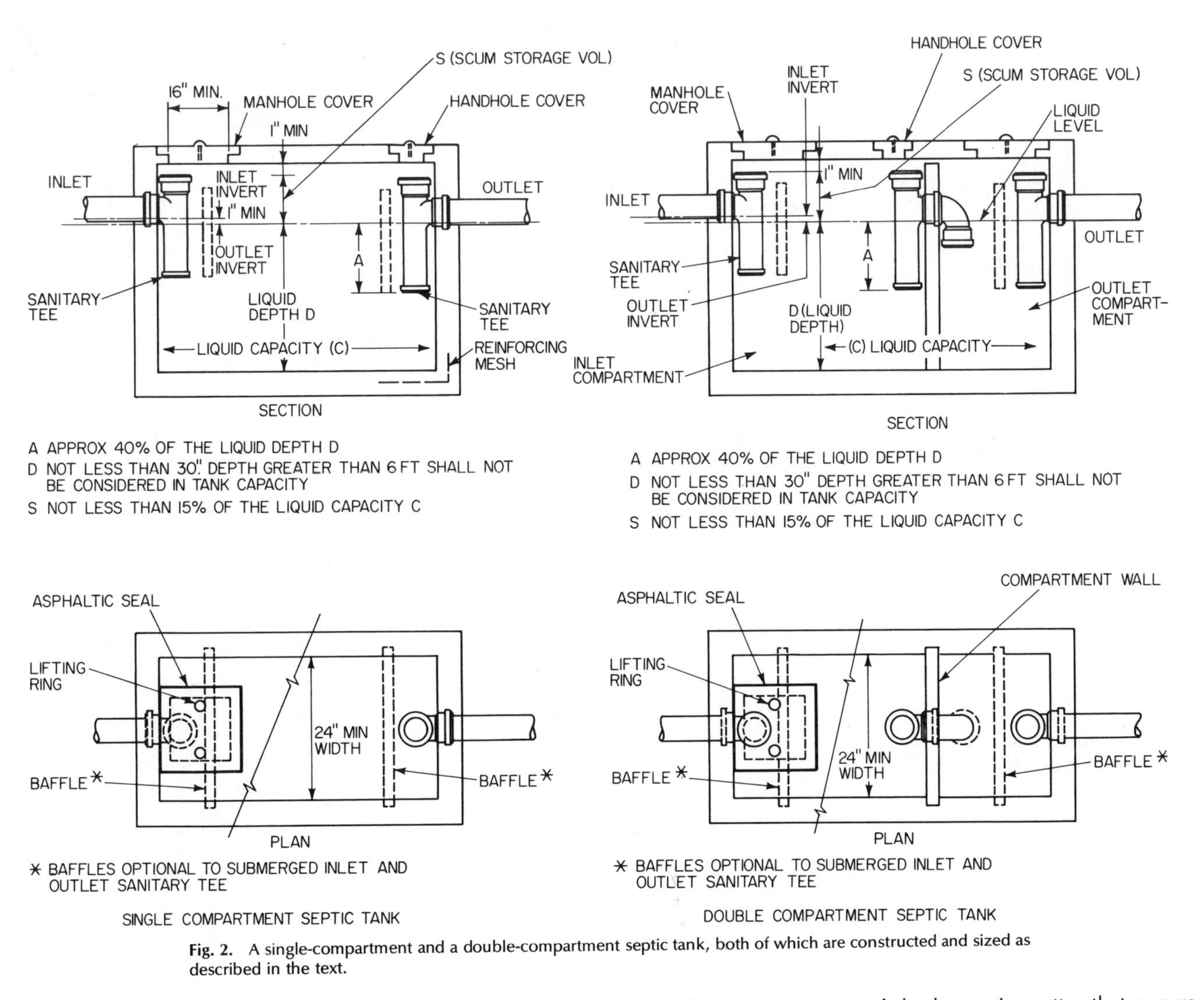

Fig. 2. A single-compartment and a double-compartment septic tank, both of which are constructed and sized as described in the text.

part of a layer of *sludge* lying there. The lighter particles will rise slowly to the top of the liquid and become part of a layer of *scum* floating there.

Living in the sludge are vast numbers of *anaerobic* bacteria, which are bacteria that can live and reproduce in the absence of oxygen. (The layer of scum is very effective at excluding atmospheric oxygen from the sewage.) The bacteria feed upon the fresh sewage, and, as a consequence, the sewage is broken down into chemically simpler substances, which include mainly water plus a number of gases such as methane (also called *swamp gas* or *sewer gas*), carbon dioxide, carbon monoxide, hydrogen, hydrogen sulfide, and sulfur dioxide, plus smaller quantities of other gases. These gases, with the exception of the first four mentioned above, are responsible for the very pungent, putrid stench we associate with decomposing organic matter.

These gases bubble their way up through the liquid and find their way out of the septic tank and into the atmosphere via the house sewer, main soil stack, and main vent in the house, the end of which projects above the roof (for a description *see* PLUMBING). Most of the sewage is decomposed within 24 hr; very little of it actually remains in the tank as solid sludge.

Anyone who has studied elementary biology will recognize the similarity in the way that sewage is broken down by anaerobic bacteria in a septic tank with the digestive process that takes place in the intestines of all living creatures and also with the decomposition of dead organic matter that occurs throughout nature. In a septic tank the end result of this process of digestion, or decomposition, is a relatively small amount of indigestible solid matter that remains behind in the tank as sludge and scum plus a relatively large quantity of *effluent,* a milky white, malodorous, and potentially toxic fluid.

This effluent must now be converted into a clear, odorless, nontoxic liquid. This is accomplished by discharging the effluent into the soil under conditions that will allow the minute particles that remain suspended in the effluent to be filtered out. At the same time, atmospheric oxygen that has made its way into the soil will combine with the anaerobic bacteria in the effluent. The chemical combination of this oxygen with the bacteria (that is, the *oxidation* of the bacteria) will destroy them by converting them into harmless carbon and nitrogen compounds.

Absorption System

In most domestic septic tank systems, the discharged effluent usually enters a system of drain tiles, which is the absorption system (see Fig. 3), with adjacent tiles having slight gaps between them. These gaps allow the effluent to seep out of the dispersal system and leach into the soil.

This kind of absorption system works most effectively when the effluent can spread itself throughout as large a volume of

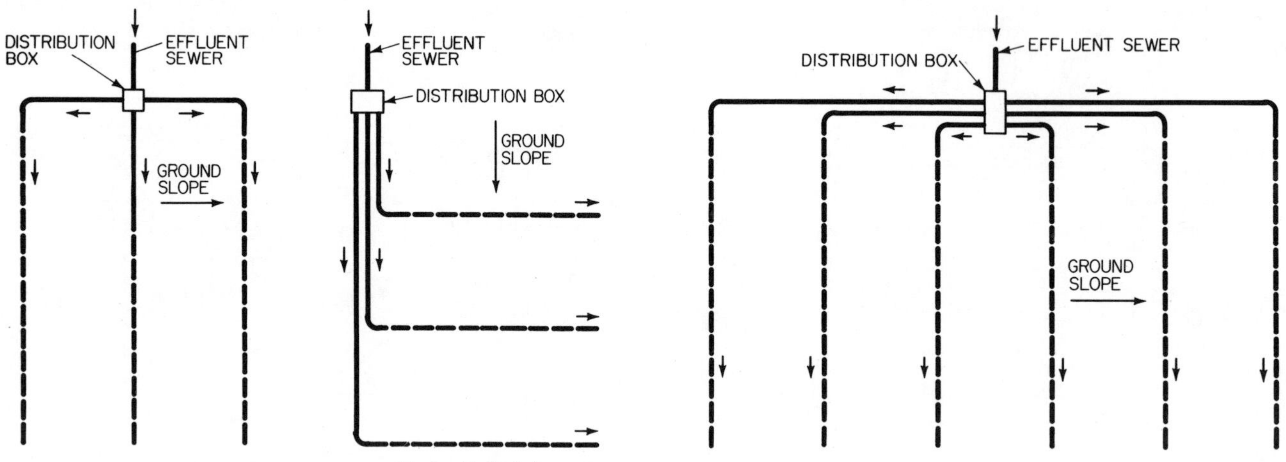

Fig. 3. Typical absorption systems for domestic septic tank systems.

the soil as possible. For this reason, the tiles must be laid dead level, or with only a very slight slope, so that the effluent can run through the entire system and not concentrate in one location. As the effluent seeps out between the gaps, it soaks deeply into the soil, and the solid matter suspended in the effluent is filtered from it by the particles of soil.

Oxygen is rarely found in soil at depths of more than 5 ft. If oxidation of the anaerobic bacteria is to occur, therefore, the absorption system must be installed fairly close to the surface of the earth, which means that the septic tank and all the interconnecting sewer lines must also be located fairly close to the surface of the earth.

In addition, if both the air and the effluent are to make their way deep into the soil, there to meet, the soil must have a sandy or gravelly texture. The more granular the soil, the smaller the absorption system need be, because both the air and the effluent will be able to make their way through the soil with relative ease.

On the other hand, the denser and more clayey the soil, the smaller the voids between the particles and the less efficiently the absorption system will operate. To compensate for this inefficiency, the size of the system must be increased. As soils become increasingly heavy, however, a point is reached at which the soil is so impermeable that it is impossible for either air or effluent to make its way through the soil at all. When this is the case, it will be necessary to install what is called a *sand filter system.*

Sand Filter System

In a sand filter system (see Fig. 4), the soil is removed to a depth of from 5 to 6 ft and layers of coarse sand and gravel totaling 4½ to 5 ft in depth are laid down in its place. After this sand filter system has been installed, a network of *distribution* drain tiles is laid down on top of the bed of sand. The effluent will seep out between the tiles, as already described, and soak into the sand, where it will be both filtered and aerated.

This rejuvenated effluent must now be disposed of in some way; its removal is accomplished by means of a network of *collection* drain tiles that has been installed at the bottom of the sand filter, under the distribution tiles. The system of collection tiles will collect the filtered and oxidized effluent and lead it away to some nearby body of water, to a drainage ditch, or perhaps to a more porous layer of soil nearby. By this time the effluent will be quite harmless and odorless. Although a sand filter is much more expensive to install than the more usual kind of subsoil absorption system, it is exceptionally

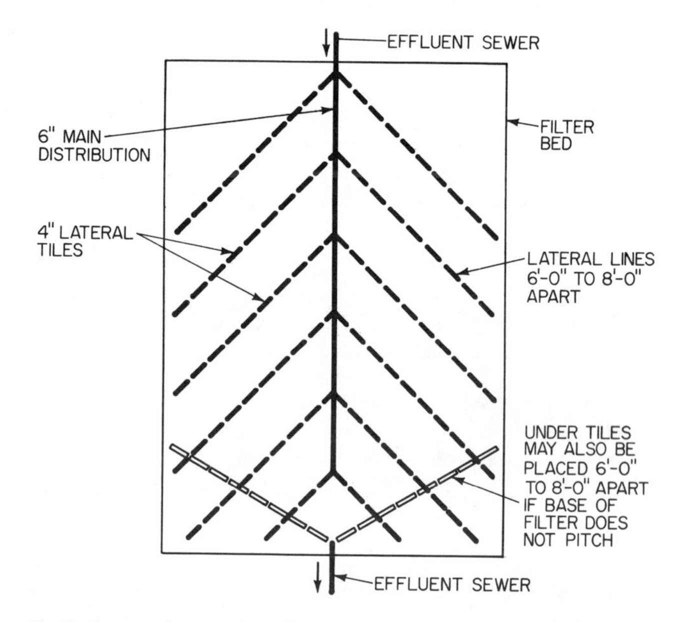

Fig. 4. Layout of a typical sand-filter installation.

efficient, which means that the overall area required for the system can be much smaller.

Very often, when the absorption system must be buried rather deeply in the soil, whether it is a subsoil absorption or sand filter type, vent lines are installed that lead from the drain tiles to the surface of the earth. These vents allow air to circulate through the system, from where it can also make its way into the soil or sand filter.

Seepage Pit

It very often happens that a site is much too small to permit the installation of an absorption system of the required size, even though the soil may be permeable enough to allow the effluent to percolate through it. In this case, a *seepage pit* (or *leaching cesspool,* or *dry well*) must be built in place of an absorption system.

A seepage pit (see Fig. 5) is just that: a circular pit dug into the soil into which the effluent is drained. The pit must be large enough to hold all the effluent that is likely to be discharged from the septic tank in a 24-hr period, and it must be capable

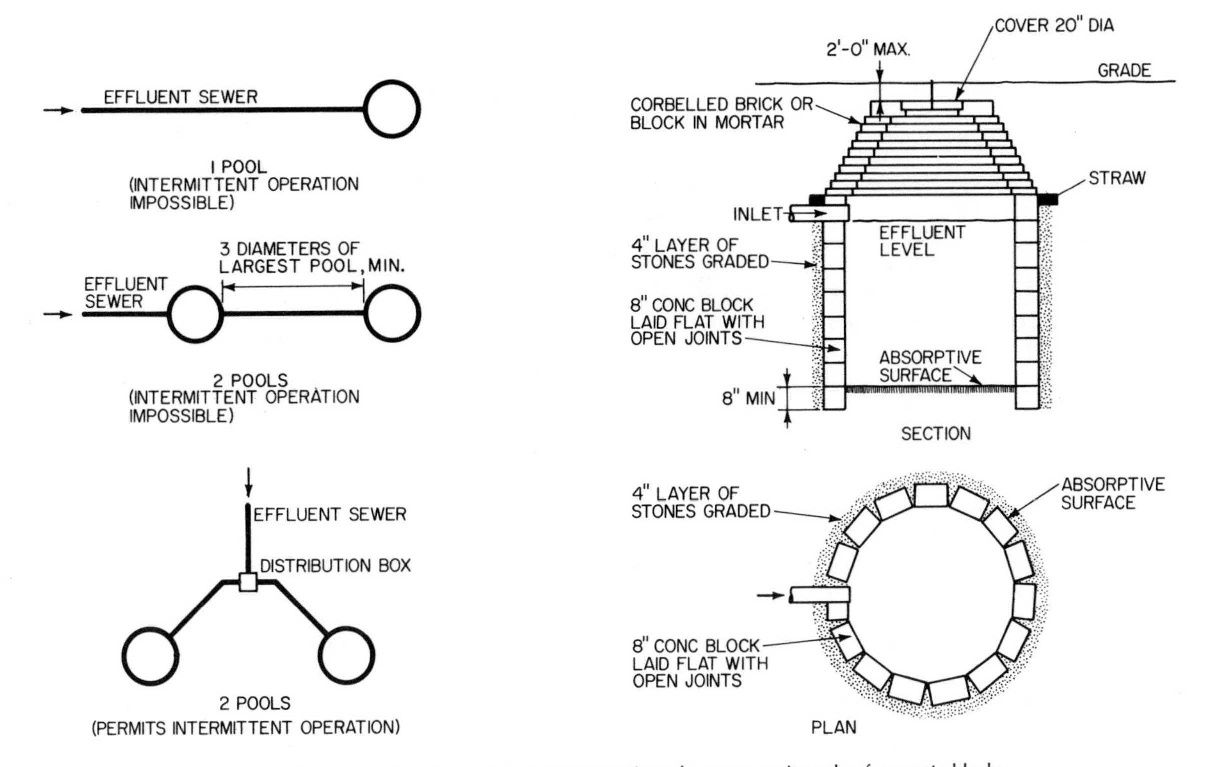

Fig. 5. Layout of seepage pit systems, showing construction of a seepage pit made of concrete blocks.

of allowing all this effluent to seep into the surrounding soil within the same 24-hr period.

A seepage pit is constructed of masonry units, either stone, brick, or concrete blocks, but, whatever the material, the individual masonry units must not be mortared together. Instead, the joints between the units are left open, for it is through these joints that the effluent will seep into the soil.

The soil must, of course, be porous enough to enable the effluent to leach into it at the required rate; otherwise the effluent will simply back up into the septic tank. A seepage pit cannot be dug too deeply into the soil, however, because oxygen must still be able to make its way through the soil to the effluent.

The construction of a seepage pit may also be complicated by the fact that permeable and impermeable layers of soil are mixed together, which will prevent the efficient drainage of the effluent into the soil. Nor can a seepage pit be dug if the groundwater level is too high because, for reasons of health, the effluent should never come in direct contact with groundwater. The local health code usually requires that the bottom of a seepage pit be at least 2 to 4 ft above any groundwater.

If one seepage pit having the required capacity cannot be built on the site, then it will become necessary to build two or three, or more, seepage pits. The distance separating the pits must be sufficient to prevent their leaking into each other. Seepage pits must usually be spaced at least three times their diameter apart.

A seepage pit may also be constructed in tandem with a subsoil absorption system if the soil is not permeable enough to drain away all the effluent that is discharged into the drainage tiles. In this case, a system of collection drainage tiles is installed about 2 ft below the distribution drainage tiles and the partially cleansed effluent is allowed to drain from the collection system into the seepage pit, from whence it can make its way into the soil. In this type of duplex system, the seepage pit can, if necessary, be located a considerable distance from the subsoil disposal system.

Or a seepage pit can be built in tandem with a sand filter. In this case, the seepage pit acts more like a temporary storage tank than as a purification system.

PERCOLATION TEST

Before a septic tank system can be designed for a particular dwelling on a particular building site, the ability of the soil to absorb the estimated amount of sewage that will be produced must be determined. On undeveloped land especially, far from a municipal sewer connection, ignorance of the underlying soil conditions may greatly increase the construction costs, as it may turn out that a much larger or more elaborate septic tank system will be required than was first estimated.

If the soil close to the house is impermeable or rocky, it may be necessary to lay a sewer line to a more suitable location where the absorption system can be installed, and this location may be a considerable distance from the house. It may turn out that the entire site is so stony, with frequent outcrops of rock, that excavating for a septic tank is either impossible or extremely expensive. Or the water table may be so high that the soil cannot be dug into more than a few feet without water being struck. Or the site may turn swampy during rainy weather, which will also make it impossible to install a septic tank system. The existence of any of these conditions can be discovered by boring test holes at the site of the proposed absorption system. In the absence of more definite information, such test borings should be made before the property is purchased.

However, even if none of these conditions exists, a percolation test must still be performed to ascertain the actual permeability of the soil for, as already noted, the size and design of the dispersal system will depend on the soil's permeability.

In principle, a percolation test is simple. One digs or bores a hole into the soil that is from 4 to 12 in. in size and as deep as the drain tile will be laid. Water is poured into this hole until the soil is as a completely soaked as it is ever likely to become during the most prolonged stretch of rainy weather. Finally, the rate at which a measured quantity of water seeps from this hole into this saturated soil is measured, and this result is the percolation rate.

In practice, local regulatory authorities differ in their requirements as to the size of the test hole, the length of time the soil must be presoaked, the meaning of the readings obtained, and who can conduct the test in the first place. In some communities, the percolation test can be performed only by officials of the local department of health or of the buildings department, although other communities will accept test results obtained by a licensed engineer, or by a firm that specializes in the construction of septic tank systems.

In most communities, if it takes the water in a hole longer than 60 min to fall 1 in., then the soil is deemed to be too impermeable for use in a subsoil absorption system, and the potential builder or homeowner will have to consider the installation of a more expensive sand filter system or decide against buying the property.

SEPTIC TANK

Most local regulatory authorities require that a septic tank be located at least 50 ft from any potable water supply (assuming that the house, or a nearby house, obtains its drinking water from a well) and that it not be any closer than 5 ft to the house. It is also sensible to install the septic tank, and the absorption system as well, on the downwind side of the house, in case any future difficulties with the system should result in the exhalation of noxious odors from a broken sewer line or an overflowing septic tank. For the same reasons, the septic tank and the absorption system should be installed on land that slopes down and away from the house.

House Sewer

The connection between the house plumbing system and the inlet of the septic tank is made via a *house sewer* (Fig. 1), which must be made of cast iron for a distance within 5 ft of the house. Beyond this distance, the house sewer can be made either of cast iron (the preferred material), vitreous clay tile, plastic, or concrete. A cast-iron sewer line should be installed no matter how close the septic tank is to the house if there is a potable water supply located within 100 ft of the sewer line or if trees are growing near the sewer line. The great advantage of cast-iron piping under these circumstances is that it is the material least likely to crack, which would allow raw sewage to seep into the soil. Also, roots cannot easily work their way between the pipe joints and thus force breaks in the pipe.

The house sewer is either 6 in. (preferred) or 4 in. in diam. If a 6-in. line is installed, it must slope at the rate of at least 1 in. per 8 ft. If a 4-in. line is installed, it must slope at the rate of at least 1 in. per 4 ft. In northern parts of the United States where the soil is likely to freeze, the house sewer must also be located at least 1½ ft underground to prevent its being damaged by soil heave.

At the same time, because the absorption system must be as close to the surface of the earth as possible, the house sewer cannot be installed too deeply underground. It is, therefore, usually installed just under the first floor of a house so that it can enter the soil as close to the surface of the earth as possible. If the house should have a basement in which a laundry room or sink is located, the high position of the line will create a problem since any wastewater draining from the clothes washer or sink must be able to drain into the house sewer. What is usually done under these circumstances is to install a sump pump that will lift the water to the sewer line.

There is no compelling reason why the septic tank must be located close to the house at all, apart from the cost of laying a sewer line between them. If the land should slope away from the house at the rate of at least 1 in. per 4 ft, the septic tank can be located several hundred feet away from the house, if necessary. All that is required with a long downhill run of pipe is that the final 10 ft or so of pipe not slope more than about 1 in. per 4 ft to prevent having the sewage discharge into the tank at too high a velocity.

Sizing a Septic Tank

The size of a septic tank is determined basically by the number of bedrooms in a house, as shown in Table 1, for it has been discovered that there is a direct relationship between the number of bedrooms in a house and the amount of sewage produced. Most health codes require that the septic tank have a minimum capacity of 500 to 1000 gal, regardless of the number of bedrooms. Since the usual assumption regarding water usage is that each person in a dwelling will on the average use 50 gal of water per day, it may be seen how conservative these requirements are.

Table 1. Determination of Septic Tank Size by Number of Bedrooms and by Use of Garbage Disposals

	Tank size, gal	
No. of bedrooms	Without disposal units	With disposal units
2	500	750
3	600	900
4	750	1,125
For each additional bedroom, add:	180	270

Source: Manas, *National Plumbing Code Handbook*, New York, McGraw-Hill, 1957.

Why, then, is such a large tank required? Because a large tank ensures that freshly discharged sewage will remain in the tank long enough—at least 24 hr—for it to be completely decomposed by the bacteria in the tank. Because a large tank provides a margin of safety when there is a very large but temporary flow of sewage into the tank, as during a large house party, a wedding reception, and so on. Because it has been the experience of most communities since World War II that the per capita consumption of water tends continually to increase. Appliances such as garbage disposal units, automatic dish washers, and automatic clothes washers have come on the market that have greatly increased the per capita water consumption. A great many homeowners also add additional bedrooms and bathrooms to their homes as their fortunes and families wax in size. Where once a household was satisfied with but a single bathroom, two or three are now demanded. And since it would be foolish to dig up and replace a septic tank every few years because of this gradual but constant increase in sewage output, most communities now require that the original installation be made large enough in the first place to accommodate these anticipated increases in sewage output.

Septic Tank Design

A septic tank must have certain characteristics if it is going to be able to handle adequately the sewage that enters it. Basically, the tank must provide the placid and airtight environment that is necessary if the sewage is to be decomposed completely. The flow of liquid through the tank must be slow and steady. The particles of sludge must be allowed to settle quietly to the bottom of the tank where they may rest undisturbed; at the same time, the particles of scum must be allowed to rise to the surface of the liquid.

This tranquil environment is achieved in part by having the sewage enter and leave the tank through *inverts,* or baffles, which are the T-shaped cast-iron fittings shown in Fig. 2. The inverts are integrally cast into the walls of a concrete tank, or, in a steel tank, they are welded to the end walls. The incoming sewage can thus enter the tank with a minimal amount of turbulence and close to the bacteria-rich layer of sludge at the bottom of the tank.

Most tanks are also rectangular in shape, which further helps to induce a calm, unruffled flow of liquid through the tank. Most local regulatory authorities require that the tank have a certain minimum rectangular shape. For example, a tank may be required to be at least 2½ ft wide and 5 ft long, with the bottom of the inlet T at least 4 ft from the bottom of the tank. This latter requirement helps to prevent the incoming sewage from disturbing the layer of sludge lying on the bottom of the tank.

Both the inlet and outlet T's usually extend about 12 in. above the level of the liquid. This extension allows the scum to build up into a very thick layer over a period of time without its either backing up into and clogging the house sewer or entering into and clogging the tiles in the absorption system. Having a thick layer of scum also has a positive function in that it keeps the air at the top of the tank away from the anaerobic bacteria in the liquid. The inverts also allow the gases that are generated by the decomposing sewage to find their way out of the tank past the layer of scum.

Large tanks are often divided by a baffle into two components, as also shown in Fig. 2, their purpose being to further tranquilize the flow of liquid through the tank. In the usual range of tank sizes required for household use, a compartmented tank is unnecessary. An integral baffle does, however, help to strengthen the construction of a tank regardless of its size. Because most of the sludge and scum will accumulate in the first compartment, this compartment is usually required to contain at least two-thirds the total liquid capacity of the tank.

Dosing or Siphon Chamber

During the normal operation of a septic tank, the effluent is discharged from a full tank at exactly the same rate at which raw sewage enters the tank. Every time a water closet is flushed or the dishes are washed after dinner, 5 gal or so of wastewater will enter the tank. As a result, 5 gal or so of effluent will be discharged into the absorption system.

Because of the weak, intermittent nature of this discharge, the soil at the head of the absorption system tends to receive most of the effluent. If the soil should be only semipermeable, it will tend to become saturated with effluent. This soil will never dry out, and the bacteria that collect in it (because of the limited amount of air that can reach the bacteria) will in time cause the soil to become rank and foul-smelling.

Given these circumstances, it would be wise to install a *dosing,* or *siphon chamber,* as part of the original septic tank installation (see Fig. 6). A dosing chamber is a temporary storage tank in which the effluent can collect. At periodic intervals the dosing chamber will empty automatically, the effluent being discharged with some force into the absorption system. As a result, all the soil in which the absorption system is buried will receive an equitable distribution of the effluent. No one area of the soil will have a chance of becoming overloaded with effluent, air will have a chance to get at and destroy all the bacteria, and the filtered and aerated liquid will have time to drain away completely before the next dose of effluent arrives.

The operation of the dosing chamber depends on an *automatic siphon* installed in the chamber (see Fig. 6). Several firms manufacture these devices, which are made of cast iron in patented designs. Figure 6 shows the basic principle. If we assume that effluent enters a dosing chamber from which the liquid has been freshly discharged, the liquid level will begin to rise in the chamber as wastewater is discharged from the house plumbing system. As it does, air is trapped within the straight section of pipe, above the U-shaped trap. As the liquid level continues to rise, the hydraulic pressure exerted on the air column by the liquid will increase until a point is reached at which the hydraulic pressure forces the air down and through the trap. The fluid in the dosing chamber will immediately rush through the automatic siphon and out into the absorption system.

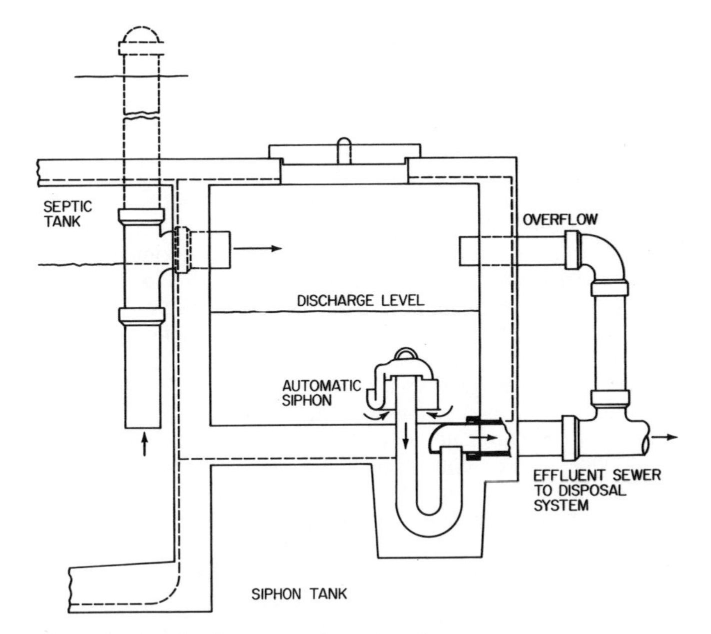

Fig. 6. Dosing chamber as part of a septic tank.

This flow will continue until the liquid level in the dosing chamber falls below the bottom edge of the bell housing. The siphoning action is thereupon broken, the flow of liquid into the absorption system ceases, air is again trapped within the pipe, and the dosing chamber will begin to fill with effluent again. If the siphon should become blocked, then the effluent will flow through the bypass line shown in the illustration and into the absorption system.

Whether or not a dosing chamber is really necessary will depend on the type of soil and the type of dispersal system installed. If the soil is very permeable, a dosing chamber will be unnecessary for most households. But the less permeable the soil, the greater the necessity for a dosing chamber. In addition, as the capacity of the septic tank increases above 1000 gal, a dosing chamber becomes more and more of a necessity. If a sand filter is being installed, a dosing chamber becomes obligatory if the sand filter is to operate efficiently. When a seepage pit is installed, a dosing chamber is unnecessary. To sum up, though dosing chambers are not a necessity in most household installations, they are still very useful devices to have.

No-Nos

Only certain organic substances, chiefly animal foods and wastes consisting of proteins, carbohydrates, and fats decompose easily in a septic tank. A great many other substances that are often unthinkingly disposed of by flushing them down a sink or water closet are not. For this reason, any household that depends on a septic tank for getting rid of its sewage is limited in what it can get rid of via this route.

Cellulose materials, for example, chiefly paper towels and sanitary napkins, but also including such fibrous vegetables as cabbage, lettuce, celery, carrots, and the like, are difficult to decompose. They should never be flushed into the drainage system but collected as solid wastes and disposed of as such. If paper towels and/or sanitary napkins are habitually flushed down a water closet, there is an excellent chance that they will pass through the septic tank and become lodged somewhere in the absorption system, thereupon blocking it. To clear the blockage, it may become necessary to dig up the drain tiles.

It is customary, nowadays, in many communities to install a ¼-in. mesh screen on the outlet invert of the septic tank to prevent the passage of solid matter into the absorption system. Although solid wastes are kept out of the absorption system, they will accumulate as part of the sludge deposit. Their presence in the septic tank will increase the rapidity with which the sludge deposit builds up, and the septic tank will also be more difficult to clean out.

In the ordinary household, the amount of grease and oil that enters the septic tank is too small to be of any consequence, although these substances are very difficult to decompose. Some households, however, are heavy users of cooking oil, or the members may be in the habit of dumping used deep-frying oil into the kitchen sink. Most of this oil and grease ends up as part of the layer of scum in the septic tank, but many small particles do not. These particles are carried by the effluent into the dispersal system, and, over a period of time, they may gradually clog the soil into which the effluent leaches, thus reducing the effectiveness of the absorption system. In such households, a grease trap should be installed in the drain line between the kitchen sink and the house sewer.

Other substances that should never be discharged into a septic tank system include acids, gasoline, caustic compounds such as lye, and disinfectants. All these substances will destroy the anaerobic bacteria, which of course makes the septic tank system inoperative. Sewage will simply pass through the tank unchanged until the bacteria have had a chance to increase in numbers again.

Nor should a septic tank system be used to drain rainwater. If a roof drainage or soil drainage system is connected to the inlet of a septic tank, the rate of flow through the tank during rainy weather will become so great that both the anaerobic bacteria and the sewage will simply be flushed out of the tank and into the dispersal system before decomposition can occur.

The same thing, but to a much lesser extent, happens when a garbage disposal unit and/or an automatic clothes washer and/or automatic dishwasher drain into the septic tank. If any of these appliances is installed in a house, a larger-than-normal-sized tank must be installed in anticipation of larger-than-normal flows of water (see Table 1). Another reason for having a larger septic tank, especially when a garbage disposal unit is installed, is that the unit is used to grind and flush into the plumbing system solid food wastes such as bones and vegetable peelings that would otherwise be collected and disposed of as garbage. As already mentioned, a septic tank is not intended to decompose these substances.

Septic Tank Construction

The walls of a septic tank must be strong enough to sustain the pressure of its contents without cracking and be durable enough to last for a great many years—20 years is not an unreasonable minimum. Most septic tanks are made of reinforced concrete, the walls of the tanks being at least 5 in. thick in the smaller sizes and increasing to 8 or 9 in. in the larger sizes. The walls are heavily reinforced with steel rods to increase their strength. Concrete tanks may either be constructed in place or they may be purchased prefabricated, requiring only a large crane and a crew of men to lower them into holes especially dug for them. A concrete tank, though practically indestructable, can be attacked by some of the gases produced during the decomposition of the sewage. For this reason, to protect the concrete, the inside of the tank above the liquid level should be painted with an asphalt compound.

Steel tanks are also widely used. They are cheaper than concrete tanks because they are easier to fabricate and transport, but the fact that they are made of steel makes them vulnerable to corrosion. The tank must be thickly painted both inside and out to protect the metal against corrosion, but, even so, a steel tank will last on the average only 10 to 15 years.

Septic tanks are also made of such decay-resistant woods as redwood and cedar. Ready-made tanks of these woods are sold knocked-down and ready to assemble. They are much cheaper than either concrete or steel tanks and are said to have a lifetime of 20 years, but since the tanks have not the strength of either concrete or steel, and as the possibility always exists that a seam may open up between the planks because of soil pressures outside the tank or liquid pressures within, most communities forbid their use.

Septic tanks made of plastics are also coming into use. Plastics have the advantage that they are impervious to corrosion or decay of any kind. They are not particularly strong materials, however, and they must be protected against crushing loads. For example, once installed in place, a plastic tank must be filled with water before soil is back-filled around it. Otherwise, the weight of the soil may crush the tank.

ABSORPTION SYSTEMS

Subsoil Absorption System

Sizing the System

The basic question to be answered before a subsoil absorption system can be designed is: What is the permeability of the soil? This question is answered by performing the percolation test described earlier.

The next question to be answered is: For the given rate of seepage, how many square feet of dispersal area are required? Table 2 gives the answer to this question in terms of the number of bedrooms in the dwelling. If, for example, the percolation rate is a 1-in. drop in the water level in 10 minutes, and if the dwelling will have four bedrooms, then Table 2 shows that the number of square feet of dispersal area required is either

100 sq ft × 4 = 400 sq ft, or
120 × 4 = 480 or
135 × 4 = 540 or
165 × 4 = 660

depending on whether or not a garbage disposal unit or an automatic clothes washer, or both, are to be installed in the dwelling. Most health codes assume that if a house doesn't have these appliances when the septic tank system is being built, it will later. They thus throw out the first three columns of Table 2 altogether and make the figures in column 4 the basis for sizing the absorption system.

Construction

A trench 18 to 24 in. wide at the bottom is dug to the required depth (which should, of course, be as shallow as practicable), and a 6-in.-deep layer of gravel, slag, clinkers, crushed stone, or a similar material, from ½ to 2½ in. in size, is laid down at

Table 2. Determination of Size of Subsoil Absorption Systems under Various Conditions

Time in which water falls 1 in., min*	Square feet of trench required for each bedroom			
	Without food-waste unit or automatic washing machine	With food-waste unit	With automatic washing machine	With both food-waste unit and automatic washing machine
2 or less	50	65	75	85
2–3	60	75	85	100
3–4	70	85	95	115
4–5	75	90	105	125
5–10	100	120	135	165
10–15	115	140	160	190
15–30	150	180	205	250
30–45	180	215	245	300
45–60	200	240	275	330

*When it takes longer than 1 hr for the water to fall 1 in., other methods of percolation should be developed.
Source: Manas, *National Plumbing Code Handbook,* New York, McGraw-Hill, 1957.

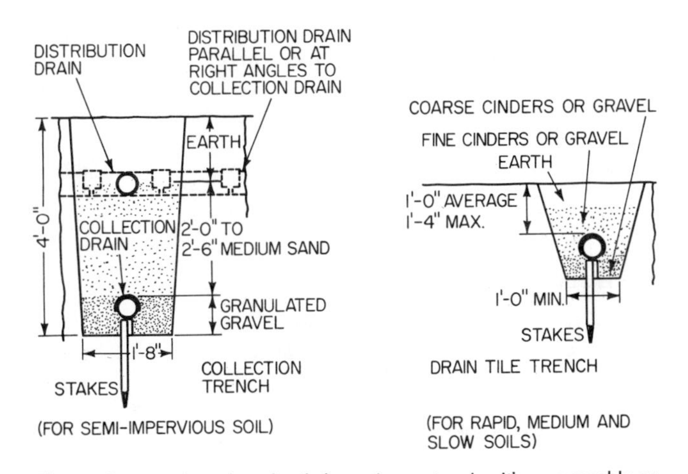

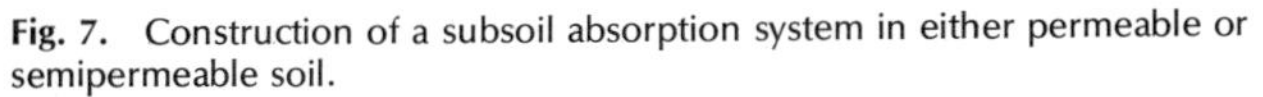

Fig. 7. Construction of a subsoil absorption system in either permeable or semipermeable soil.

the bottom of this trench (see Fig. 7). The drain tiles are then laid on top of this bed. The drain tiles can also be laid directly in the soil, but only when the soil has an exceptionally granular, porous structure through which the effluent can drain freely. Most soils do not have this extreme permeability and require a coarse gravel bed that will enable the effluent to make its way into the soil with greater efficiency.

In the example in which we determined the size of the dispersal area required for a 4-bedroom house, we found that 660 sq ft were required. If the width of the trenches is 18 in., then only 440 *linear* ft of drain tiles need be laid down since 660 sq ft ÷ 18 in. = 440 ft. If the width of the trenches is 24 in., then only 330 *linear* ft of drain tiles are required since 660 sq ft ÷ 24 in. = 330 ft.

No single line of tiles should, however, be longer than 100 ft for the simple reason that it is very unlikely that the effluent will ever travel farther than 100 ft in any absorption system, no matter how level it is. In our example, therefore, if we assume the width of the trenches is 24 in., at least four separate lines of tiles, each at least 82½ ft long, will be necessary in order to dispose of the effluent.

If these lines of tiles are laid out parallel to each other, they must be spaced a certain distance apart to ensure that each will remain capable of draining freely and without interference from the other lines of tiles. Table 3 shows, for example, that for a 24-in.-wide trench, adjacent trenches must be at least 6 ft apart.

The actual layout of the system will depend on how much

Table 3. Determination of Size and Spacing of Subsoil Absorption Trenches

Width of trench at bottom, in.	Recommended depth of trench, in.	Spacing tile lines,* ft	Effective absorption area per lineal foot of trench, sq ft
18	18–30	6.0	1.5
24	18–30	6.0	2.0
30	18–36	7.6	2.5
36	24–36	9.0	3.0

*A greater spacing is desirable where available area permits.
Source: Manas, *National Plumbing Code Handbook,* New York, McGraw-Hill, 1957.

land the homeowner has available and on whether the land is flat or sloped. On flat land, the subsoil absorption system can be laid out in a variety of ways—in parallel lines, in a closed circle or rectangle, in a V shape, or whatever is most convenient or suitable.

When a subsoil absorption system is being installed on flat land, perhaps the single most important point to watch is that the drain tiles are laid as level as possible. If the lines should be laid at a slope, this slope should be very slight, not exceeding (depending on the local code) 1 in. in 24 ft, 1 in. in 36 ft, or 1 in. in 50 ft.

When the land slopes, the levelness of the installation is attained by having the lines of drain tiles follow the contours of the land. Or a cascading system of drainage tile can be installed in which the effluent is led from one level to another through sewer lines. The same requirements regarding the width and depth of the trenches must be met when this type of system is installed.

Distribution Box

All absorption systems having two or more branch lines must have a *distribution box* installed in the sewer line, this box being located at the point where the individual branch lines originate (see Fig. 1). The distribution box is usually made of concrete.

The sewer line carrying the effluent from the septic tank discharges the effluent into the box, about 6 in. from its bottom. The outlet lines are usually about 2 in. lower than the sewer-line inlet. The outlet lines must be exactly level with each other to make sure that each line carries its fair share of effluent to the drainage tiles.

Each branch line can be blocked off by a stop board or gate, the position of which is changed manually. In a 2-outlet box, the homeowner can block off the flow from any given line completely, or in a 4-outlet box, he can control to some degree the proportion of effluent that enters each line. He can let an overworked section of the absorption system "rest" for a while, or he can balance the flow of effluent between the sections of the absorption system. In addition, if any section of the system should ever need to be repaired, the flow to that part of the system can be blocked off until the repairs have been completed.

Seepage Pit System

Sizing a Seepage Pit

Table 4 shows the total surface area, in square feet, required for a seepage pit, given different percolation rates and the installation of different household appliances. If, for example, the percolation rate for a particular soil is a 1-in. drop in the water level in 10 min, and the dwelling will have a garbage disposal unit and a clothes washer, then, according to Table 4, the required surface area of the seepage pit will be 95 sq ft.

Table 4. Surface Area Required for Seepage Pits for Different Percolation Rates and Wastewater Flows

	Absorption area below the inlet—for walls and bottom, sq ft		
Time for water to drop 1 in., min*	Without food-waste unit or washer	With food-waste unit or washer	With food-waste unit and washer
2	30	40	50
3	35	50	60
4	40	55	70
5	45	60	80
10	55	75	95
15	65	90	110
30	110	150	190
60	210	280	360

*Above 60 minutes, seepage pit should not be used.
Source: Manas, *National Plumbing Code Handbook,* New York, McGraw-Hill, 1957.

Local health codes usually require that a seepage pit be at least 4 to 6 ft in diam. For a seepage pit 6 ft in diam., the surface area per foot of depth is 18.85 sq ft. Therefore, 95 sq ft ÷ 18.85 sq ft = 5.04 ft will be the required depth of the seepage pit.

If the percolation rate should be 1 in. in 30 min, a seepage pit 10.08 ft deep would be required. This, of course, is impossible.

What is done in a case like this is to divide the total required depth into two or three interconnected seepage pits. If, for example, two seepage pits are decided upon, each pit need be only about 5 ft deep (that is, about 5 ft below the level of the incoming sewage line).

These calculations assume that the seepage rate will be the same throughout the entire depth of the soil, that is, that the soil will have a homogenous structure. This may not always be the case. Very often one finds that the deeper one digs the less permeable the soil becomes, or layers of soil having different permeabilities may be present. When this is the case, then an average percolation rate must be calculated for the soil.

Sand Filter System

Sizing a Sand Filter

A sand filter system (Fig. 4) requires the same size dispersal area as a subsoil absorption system, if the percolation rate and dwelling size are the same. The only difference between the two is that a sand filter can occupy a much more compact area. If we assume, for example, that 660 sq ft of dispersal area is required for a dwelling (as calculated in our example above), a sand filter can be installed in an area approximately 24 by 28 ft in size, if the distribution tiles are laid parallel to each other and 6 ft apart. In comparison, the 82.5-ft-long drain lines of the subsoil absorption system that has been described above require about 1485 sq ft of area, or an area about 82.5 by 18 ft in size.

Construction

There are two basic kinds of sand filter system—an open system and a closed system (see Fig. 8). A closed system is constructed entirely underground.

In an open system, the sand filter is not covered with topsoil nor are the distribution tiles buried in a layer of gravel. Instead, the distribution drain tiles rest on planks laid on top of the sand, which is in turn exposed to the air and the weather. An open system can become rather smelly and should only be considered if it can be installed somewhere by itself, far from any houses or public thoroughfares. An open system has the advantage, of course, that is somewhat cheaper to construct than a closed system.

SLUDGE REMOVAL

As the layer of sludge builds up higher and higher on the bottom of the septic tank, it must be removed occasionally,

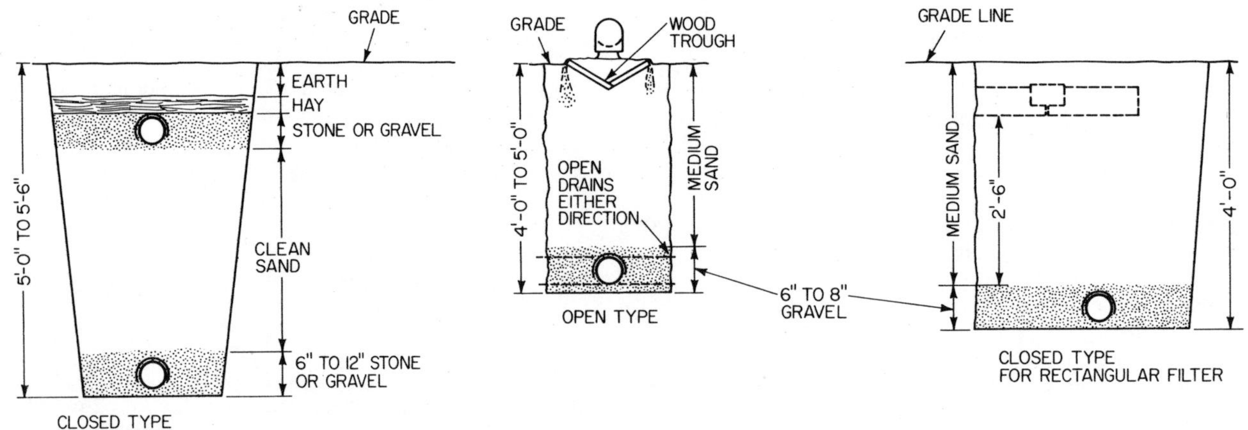

Fig. 8. Construction of open and closed types of sand filter beds.

depending on how rapidly the buildup occurs. If the sludge isn't removed, it will in time reach the height of the outlet line. When this happens, the sludge will pass into the absorption system, where it will clog the drain tile. The entire absorption system will then have to be dug up and cleaned out.

Another reason for cleaning the septic tank at periodic intervals is that the greater the volume of sludge in the tank, the smaller the volume of liquid. As a result, incoming sewage will spend less time within the tank, too little time for it to be completely decomposed. As a result, the effluent will enter the dispersal system only partially decomposed.

To prevent this excessive buildup of sludge, a septic tank should be inspected at least twice a year, during the spring and autumn, say. The depth of the sludge is checked by removing the cover from the top of the tank, lowering a rod into the tank, and noting how far it goes before it touches the layer of sludge. A more accurate method (since the sludge may be too loose for anyone to gauge easily what its height is) is to lower into the tank a length of pipe connected to the inlet of a pump. The pump is started up and the pipe is carefully lowered into the tank, to avoid stirring up the sludge. At some point, as the pipe is being continually lowered, the liquid being discharged by the pump will turn black. This indicates that the top of the sludge layer has been reached. The length of the pipe within the tank is then measured, and this distance is subtracted from the overall depth of the tank. The difference is the depth of the sludge.

If the sludge layer is more than 12 in. deep, it should be removed. How long it takes for this amount of sludge to collect will depend on the capacity of the tank, the size of the house, the living habits of its occupants, and whether they occupy the house continuously all year round. It may, therefore, take less than a year, or it may take several years for 12 in. of sludge to accumulate. To be on the safe side, many local regulatory authorities require that a septic tank be inspected annually.

The sludge must be disposed of. It has no fertilizing value, it smells too bad to be used for landfill or anything else, and it is also a health hazard. The only thing one can do is burn or bury it. The sludge can be run into an isolated trench that has been dug for this purpose. The liquid portion will slowly seep into the soil. The solid matter that remains can be covered over, or it can be allowed to dry; it is then burned. If the tank has been cleaned out by a firm that specializes in this work, they will cart the sludge away to some facility approved by the local health authorities, probably a nearby municipal sewage treatment plant.

Sheathing

Sheathing, often pronounced *sheeting,* consists of a layer of either wood boards, plywood, insulation board, or gypsum wallboard nailed to the exterior framework of a house to stiffen and strengthen the wall and roof construction. Without the stiffening and strength supplied by the sheathing, the walls would tend to *rack* (that is, be twisted out of true by a strong wind), and the same would be true of the roof (*see* BUILDING LOADS; WOOD-FRAME CONSTRUCTION). Roof sheathing also helps to support the weight of the roofing and of any snow loads. Sheathing, therefore, is considered an essential part of the structural framework of a house, always to be installed.

In addition to this basic structural function, sheathing also (1) provides a base to which the exterior finish of a house (that is, the shingles, siding, masonry veneer, or stucco) can be attached, (2) helps to prevent the infiltration of wind into the house through cracks in the exterior finish, and (3) helps somewhat to insulate the interior of the house against extremes in the outside air temperature.

In this article, wall sheathing and roof sheathing will be described separately because of differences in the materials required for the sheathing and because of differences in the way these materials are applied.

WALL SHEATHING

Wood Boards

Board Sizes

The traditional sheathing material, for both walls and roofs, is nominally 1-in.-thick wood boards that are from 6 to 12 in. wide (see Fig. 1). Although boards 12 in. wide can be used, the wider the boards the more they tend to shrink after they have been installed and, therefore, the greater the gaps that may open up between them.

To minimize this shrinkage as much as possible, seasoned lumber should always be used for wall sheathing. Boards

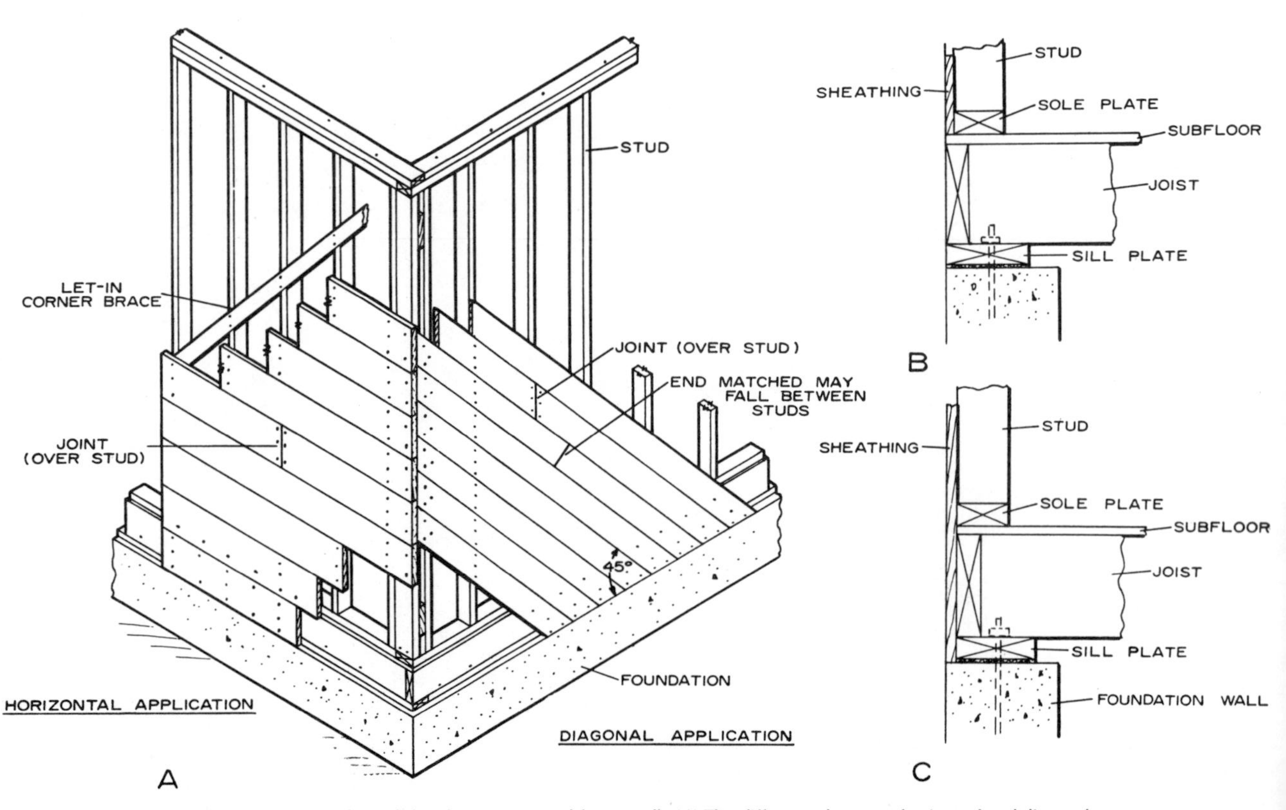

Fig. 1. Application of wood sheathing on a wood-frame wall: (*A*) The difference between horizontal and diagonal application of wood boards; (*B*) sheathing started at the subfloor; (*C*) sheathing started at the sill plate.

should have a maximum moisture content of 19 percent when they are installed. In general, the narrower the boards, the less their overall shrinkage will be. As 6-in.-wide boards tend to shrink least, they are to be preferred for wall siding, but, in fact, 8-in.-wide boards are used more often.

Sheathing boards usually have square edges, but boards the long edges of which are either shiplapped or tongue-and-grooved (the latter also being known as *dressed-and-matched*) are also used, especially in the better class of construction or where weathertight construction is important.

Horizontal Versus Diagonal Installation

Sheathing boards may be installed either horizontally or diagonally, that is, at a 45 degree angle. Regardless of the angle at which the boards are installed, the installation requirements are the same. Boards 8 in. or less in width are nailed to each stud with at least two 8d common or two 7d threaded nails. Boards wider than 8 in. are nailed to each stud using at least three of the above-sized nails at each stud. These are minimum requirements.

The boards must abut against each other tightly. All boards must also begin and end over studs, but the ends of two boards that are adjacent to each other must not begin or end on the same stud.

Sheathing installed horizontally has the advantage over diagonally-installed sheathing that the installation may be completed more quickly and easily (and thus less expensively), and with less waste of lumber. Diagonal sheathing requires that the ends of the boards be cut at 45 degree angles, which takes more time and is wasteful of lumber.

However, instead of cutting the ends of diagonally installed boards at 45 degree angles, *end-matched* boards can be used. These are boards the ends of which have been tongue-and-grooved so that abutting boards fit tightly to each other. This tightness of fit does away with the need for cutting the boards at all, for both horizontally as well as diagonally installed sheathing, since two end-matched boards can be joined anywhere, not only over studs. When end-matched boards are used, however, each board must be long enough that it extends over at least two studs. And even when their ends are matched, no two adjacent boards can have joints within the same stud space.

Differences in ease and expense of installation aside, diagonal sheathing adds far more to the overall strength and stiffness of wall construction than horizontal sheathing does, and for this reason alone it is preferred. In tests conducted as far back as the 1920s, the U.S. Bureau of Standards showed that the stiffness of a diagonally sheathed stud wall was from 4.0 to 7.3 times as great as that of a horizontally sheathed wall and the strength was about 8.0 times as great.

The reason may easily be seen. If you want to stiffen any open framework, not only the framework of a house, it is necessary to add bracing that runs diagonally between the adjacent sides of the frame, as shown in Fig. 2; and the longer this bracing, the stiffer the resultant construction will be. Sheathing installed horizontally cannot compare with diagonal sheathing at all in this respect. No doubt a wood-stud wall is much stiffer and stronger with horizontal sheathing than it would be without any sheathing at all, but the possibility of the frame's being racked has not been eliminated with horizontal sheathing; it has only been reduced somewhat.

Furthermore, with horizontal sheathing, the walls of a house are not tied securely to the foundation. This is especially true when the platform-type construction that predominates in the United States (*see* WOOD-FRAME CONSTRUCTION) is used for the frame. In platform construction (see Fig. 1), the walls are fastened to the foundation by nails driven down to the sill, which is in turn secured to the foundation by anchor bolts. But

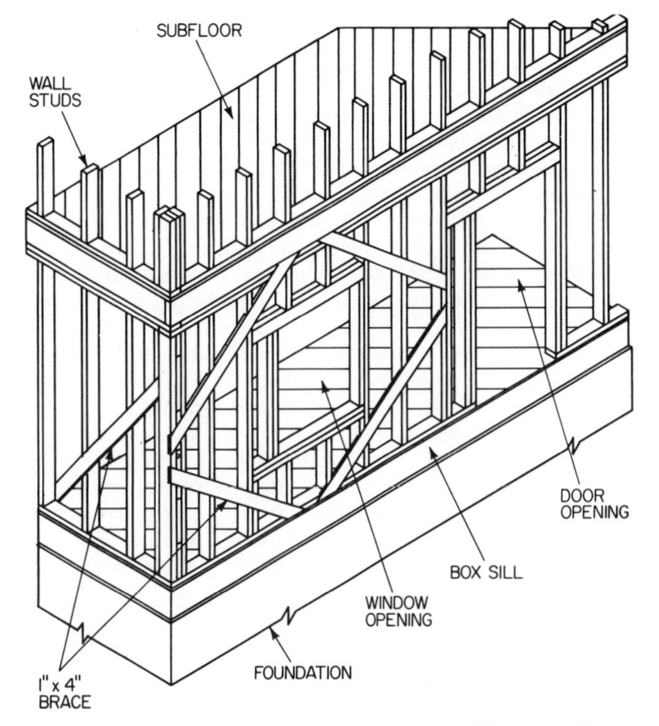

Fig. 2. Let-in bracing, installed on a platform-framed house as shown, greatly stiffens the overall framework of the house.

there is nothing in this construction that will prevent a hurricane-force wind from lifting the entire house off the sill, nails and all—and this has certainly happened to a great many dwellings built along the hurricane-prone coasts of the Gulf of Mexico and the Atlantic Ocean.

Diagonal sheathing, however, when correctly installed as shown in Fig. 1, ties the sill, posts, studs, and wall plates of a wood-frame house together into a single, solid structure. Any racking stresses imposed on the structure will be transmitted by the sheathing directly to the foundation, via the sill.

This is not to say that horizontal sheathing can never be used. Most building codes permit horizontal board sheathing, but only when separate bracing is also installed to stiffen the walls of the dwelling.

Bracing

This bracing usually consists of 1 × 4 in., or wider, lengths of wood that are let into the framework at a 45 degree angle (see Fig. 2). Just enough wood is cut out of the framing members to allow this bracing to lie flush with the outside of the framework.

Bracing is required at all the exterior corners of the house. It must be installed at a 45 degree angle, and it must be at least 8 ft long. Ideally, each brace should run in one length from the top of the post down to the sill. This is often impossible because a door or window is in the way. In such cases, at least three 4-ft-long (and preferably longer) braces must be installed as shown in Fig. 2. Each brace is nailed to each framing member with two 8d nails.

When this kind of let-in bracing has been installed, the overall stiffness of a horizontally sheathed wall can be as much as 4.2 times greater than it would have been without the bracing. But when there are doors and windows in the wall, the bracing only about doubles the strength and stiffness of the construction.

There are two kinds of diagonal wood bracing that are not permitted by some building codes. One type consists of short knee braces that form a K at the corners of a wall. The other type consists of 2 × 4s that are cut into the framework. That is, short lengths of 2 × 4s run from stud to stud at a 45 degree angle, each end of each brace being toenailed to each stud in much the same way that bridging runs from joist to joist in a floor. This type of cross bracing is almost useless in helping to stiffen a wall construction.

The rigidity tests referred to above also showed that plastering will stiffen a wall about as much as bracing or diagonal sheathing does. If one takes into account the fact that the interior surfaces of most walls will in fact be plastered or have gypsum wallboards installed, as well as being sheathed, then the walls will be considerably more resistant to the racking stresses than we have described. These facts have been taken into account in Canada, where horizontal board sheathing is permitted without additional wall bracing being required, as long as the interior surface of the wall is either plastered or covered with gypsum wallboards. But both plaster and gypsum wallboards are brittle materials that will crack or crush quite easily if they are stressed. Once this has happened, their value as wall stiffeners is reduced considerably. Furthermore, though plaster and gypsum wallboards do *stiffen* a wall, they do little to increase the basic *strength* of the construction, as sheathing does.

Plywood

Plywood panels have supplanted wood boards almost entirely as the primary sheathing material (see Fig. 3). Plywood panels have a number of outstanding virtues compared to wood boards. Because they are usually 4 × 8 ft in size, they are easier to install, and they reduce considerably the labor costs of installing sheathing. The large size of the panels reduces to a minimum wind infiltration through the walls. Their inherent stiffness greatly increases the overall strength and stiffness of wood-frame construction. In the stud-wall rigidity tests referred to above, it was found that ¼-in.-thick plywood panels increased the overall stiffness of the construction 4.5 to 7.2 times as much as when horizontal wood boards were used—about the same as when diagonal board sheathing is used. However, the plywood paneling used today for wall sheathing is substantially thicker than the ¼-in.-thick panels used in these tests. It is hardly necessary to add that additional bracing is not needed when plywood panels are installed as sheathing.

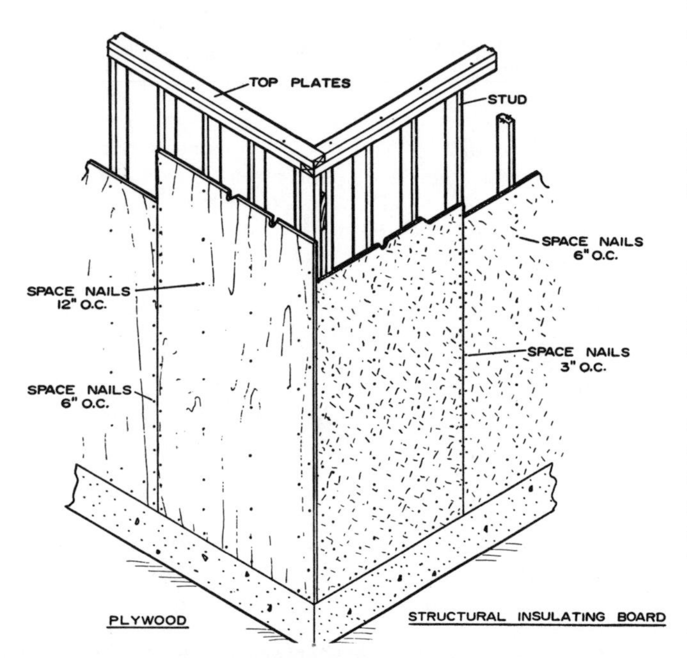

Fig. 3. The vertical installation of plywood panels or insulating-board panels as wall sheathing *(U.S. Forest Service)*.

Plywood Grades and Sizes

The panels recommended for use as wall sheathing are Engineering-grade panels (*see* PLYWOOD). These panels are manufactured as either Interior or Exterior types, depending on whether the adhesive that is used in their manufacture is water-resistant or not.

At one time, the panels recommended for use as sheathing were identified as being of *sheathing* grade. This nomenclature is now obsolete. Instead, the Engineering-grade panels that are recommended for use as wall sheathing are either C-D Interior or C-D Exterior panels. The C-D Interior panels are satisfactory for the great majority of dwellings, but if there is a possibility that the panels will be exposed to the weather for a prolonged period of time before the exterior finish is installed over them, then C-D Interior panels made with an exterior adhesive should be used.

The C-D Exterior plywood need be installed only when it is anticipated that the sheathing will be exposed to the weather for a very prolonged period of time, which would indeed be unusual for wall sheathing.

The most common panel thicknesses used for wall sheathing are 5/16, ⅜, and ½ in. Which thickness is used will depend on the stud spacing and whether or not the exterior wall finish is to be nailed directly to the plywood sheathing or to the wall studs.

In the balmy climates of southern California and the American Southwest, low-cost dwellings are often built without any sheathing at all. Instead, a combination sheathing-and-exterior-finish plywood panel is installed directly to the studs to reduce the overall cost of construction. These combination panels are from ⅜ to ⅝-in.-thick, depending mainly on the stud spacing. They are made of Exterior plywood, and their outside faces are finished off in some decorative texture or overlay.

Installation

Plywood panels are usually installed with their long dimension vertical, although when extra-stiff construction is required (as it may be in some earthquake-prone or hurricane-prone sections of the country), the panels must be installed horizontally. The additional stiffness results when the panels are installed horizontally because the panels extend across and are nailed to more studs.

If the height of a wall is such that the top edge of a panel cannot be nailed to the top plate, then 2 × 4 in. blocking must be toenailed between the studs to provide the necessary edge support for the panel.

When panels are installed horizontally, similar lengths of 2 × 4 in. blocking must be installed between the studs to which the long edges of the panels can be nailed. Panels that are installed horizontally must also have their vertical joints staggered so that two adjacent panels do not begin or end on the same stud.

Wherever the ends of two panels abut, they must be spaced 1/16 in. apart. Wherever the two long edges of two panels abut, they must be spaced ⅛ in. apart. The reason for these gaps is that plywood tends to expand slightly when it absorbs moisture, and the gaps permit this expansion to occur without stressing the construction. If a dwelling is being built in a damp, humid climate, such as along the Gulf Coast, the size of these gaps should be doubled.

Insulation Board

There are three grades of insulation board (or fiberboard) used as wall sheathing. In order of increasing density they are called regular density, high density, and nail base. The high-density material weighs about 25 percent more than the regular-density material. The nail-base material weighs about 5 percent more than the high-density material. We will discuss first the regular-density insulation board and then the other two types.

Regular-Density Insulation Board

Regular-density insulation board is a soft, porous material manufactured as 4 × 8 and 4 × 9 ft panels. Its primary use is as insulation. This material itself is quite weak. However, the large size of these panels, the fact that the thicker panels are stiff enough to be self-supporting, plus the fact that their stiffness is increased when they are nailed to framing members, makes them a quite suitable sheathing material.

For use as sheathing, the panels are either coated or impregnated with asphalt to increase their resistance to moisture. This treatment does in fact make the panels water-resistant, but, more important, it does not affect their permeability to water vapor. That is, any water vapor that enters the wall space can still diffuse out of the wall through the sheathing, thus avoiding condensation problems (*see* CONDENSATION).

These asphalted panels are manufactured in ½ and 25/32 in. thicknesses. The panels have square edges and are ⅛ in. scant all around to compensate for the fact that adjoining panels must be spaced ⅛ in. apart, as the panels will expand by this amount in humid weather.

Both the ½ and 25/32 in. panels are installed the same way (see Fig. 3). The preferred direction of installation is with the long dimension vertical. Large-headed (i.e., ⅜ to 7/16 in. diam.), galvanized roofing nails are used. The point of using roofing nails is, of course, that their large heads allow the nails to grip the panel material more securely than ordinary nails will, although 8d common nails can be used, if necessary.

Whether diagonal bracing will be required will depend on the stud spacing and the thickness of the panels.

If ½-in.-thick panels are being installed, supplementary bracing is required regardless of the stud spacing. This bracing can take the form of the let-in 1 × 4 in. boards described above, or it can consist of ½-in.-thick plywood panels nailed at the corners of the walls.

If 25/32-in.-thick panels are being installed, bracing is not required if the studs are spaced 16 in. apart. Bracing is required, however, if the studs are spaced 24 in. apart. This bracing can take the form of let-in boards or ¾-in.-thick plywood panels.

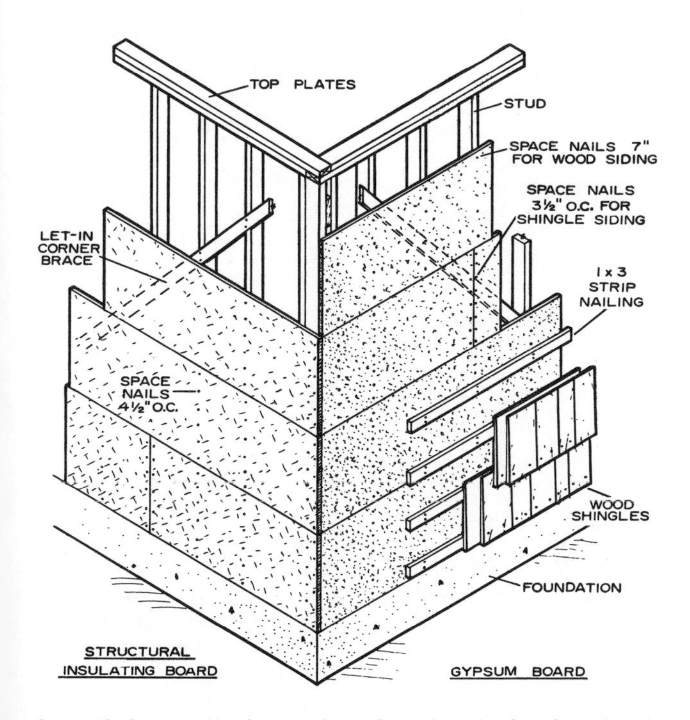

Fig. 4. The horizontal application of 2 × 8 ft panels made of insulation board or gypsum *(U.S. Forest Service)*.

Regular-density insulation board is also manufactured as 2 × 8 ft panels especially for use as wall sheathing (see Fig. 4). The long edges of these panels have V-shaped tongues and grooves that help stiffen the construction and also minimize air infiltration into the building.

Additional wall bracing is always required when 2 × 8 ft panels are installed. As before, this bracing can consist of either let-in boards or plywood panels.

High-Density Insulation Board

High-density insulation board ½ in. thick is manufactured only as 4 × 8 and 4 × 9 ft panels. The panels are installed as already described for regular-density boards, using the same nails and nail spacings.

Nail-Base Insulation Board

Both regular-density and high-density panels are too porous to permit exterior wall finishes to be nailed directly to them. The material has almost no nail-holding ability. This is less true of nail-base insulation board. Wood and asbestos-cement shingles can be nailed directly to this material, but only if special aluminum nails are used.

Nail-base sheathing is manufactured only in ½-in.-thick, 4 × 8 and 4 × 9 panels. The panels are installed using the same procedures and nails, or staples, used for high-density insulation-board sheathing. Additional bracing is required only when the stud spacing is greater than 16 in. on centers.

Gypsum Wallboards

Gypsum wallboards intended for use as sheathing are formulated with emulsified asphalt to increase the moisture resistance of the gypsum, which is otherwise quite poor. To further increase its moisture resistance, the gypsum is covered by a water-repellent paper. The panel is not completely impervious to water vapor, however, which is an advantage as this allows any water vapor that may otherwise be trapped within the wall construction to diffuse out again. The advantages of using gypsum sheathing are that it is cheap, inherently fire-resistant, and easy to install. It can be cut to size merely by scoring the surface with a sharp tool of some kind and then snapping the panel across the score mark.

Gypsum sheathing is manufactured as 4 × 8, 4 × 9, and 2 × 8 ft panels, all of which are ½ in. thick. The 4 × 8 and 4 × 9 ft panels should be installed vertically as shown in Fig. 3, except for the nail spacing. The 2 × 8 ft panels should be installed horizontally as shown in Fig. 4, also except for the nail spacing. The 2 × 8 ft panels also have V-shaped tongues and grooves along their long edges to help stiffen the construction and prevent air infiltration.

When either the 4 × 8 or 4 × 9 ft panels are installed, no additional wall bracing is required. When the 2 × 8 ft panels are installed, however, additional bracing is required. This bracing can consist either of let-in 1 × 4-in. boards or large plywood panels nailed at the outside corners of the walls.

Neither siding nor shingles should be attached directly to gypsum sheathing, as gypsum has no nail-holding ability. Siding must be attached to the wall by nailing it directly to the studs, and shingles must be nailed to *backer boards,* which are in turn nailed directly to the studs. (*See* SHINGLE, WOOD.)

ROOF SHEATHING

Roof sheathing adds greatly to the overall structural integrity of a roof. The rafters that make up the framework of the usual roof are held rigidly in place only by the sheathing. Without this sheathing, the roof construction would not have any stiffness or strength to it at all.

The more extreme the climatic conditions, the more important the roof sheathing becomes, as the sheathing helps support heavy snow loads on the roof, and it greatly stiffens the roof construction against any racking loads imposed by a strong wind. The roof sheathing must also be thick enough to provide a solid support to which the roofing material can be attached, whether it be wood, asbestos, asbestos cement, or slate shingles; sheet-metal or built-up roofing; or clay tiles. For all these reasons, building codes usually permit only wood boards or plywood panels to be used as roof sheathing.

Wood Boards

The general requirements for roof sheathing made of wood boards are much the same as they are for wall sheathing (see Fig. 5). The boards are from 6 to 12 in. wide and nominally 1 in. thick. The edges can be either cut square, shiplapped, or tongue-and-grooved. At least two nails per rafter are required when the boards are 8 in. or less in width; at least three nails are required when the boards are wider than 8 in. Each board must begin and end on a rafter, except when end-matched boards are used. No two adjacent boards can have their ends attached to the same rafter. When the boards are end-matched, no two adjacent boards can have their ends abut over the same rafter space.

As will be noted, there is not much difference in material or installation technique between wall and roof sheathing. If, however, asphalt shingles are to be installed over board sheathing, then other and more stringent requirements must be met. Because of the softness and flexibility of asphalt shingles, the sheathing must be thoroughly seasoned before it is installed. That is, it must have a maximum moisture content of 19 percent; otherwise, as the boards dry out and shrink apart, their movement may cause the shingles to buckle or lift.

Furthermore, there is a strong tendency for moisture to condense under asphalt shingles, and this moisture may make its way into the interior of the house through gaps in the sheathing. For this reason, when asphalt shingles are to be installed, either shiplapped or tongue-and-grooved boards must be used, and one must be sure to install building paper over the sheathing.

Closed Versus Open Spaced Sheathing

Most board sheathing should be nailed together tightly. Traditionally, however, in some parts of the United States, wood shingles are nailed to 1 × 3 or 1 × 4 in. boards that are spaced apart the same distance as the rows of shingles, which will leave a gap of from 1 to 3 in. between the boards (see Fig. 6). The reason given for this practice is that wood shingles may rot unless they have a chance to dry out quickly between rainfalls.

This may be true when wood shingles are installed in the hot, humid climate that exists along the Gulf Coast and in other parts of the South, but it is not true elsewhere in the United States. Where the climate is humid, open-spaced sheathing can be installed when the roof is to be covered with wood shingles. In the rest of the country, open sheathing is unnecessary. Closed sheathing not only helps keep the heat inside a house during the winter, which is important in the Northern states, it also helps to retard the spread of any fire that might occur on a roof; open sheathing does not. (For this reason, most fire insurance companies require closed sheathing.)

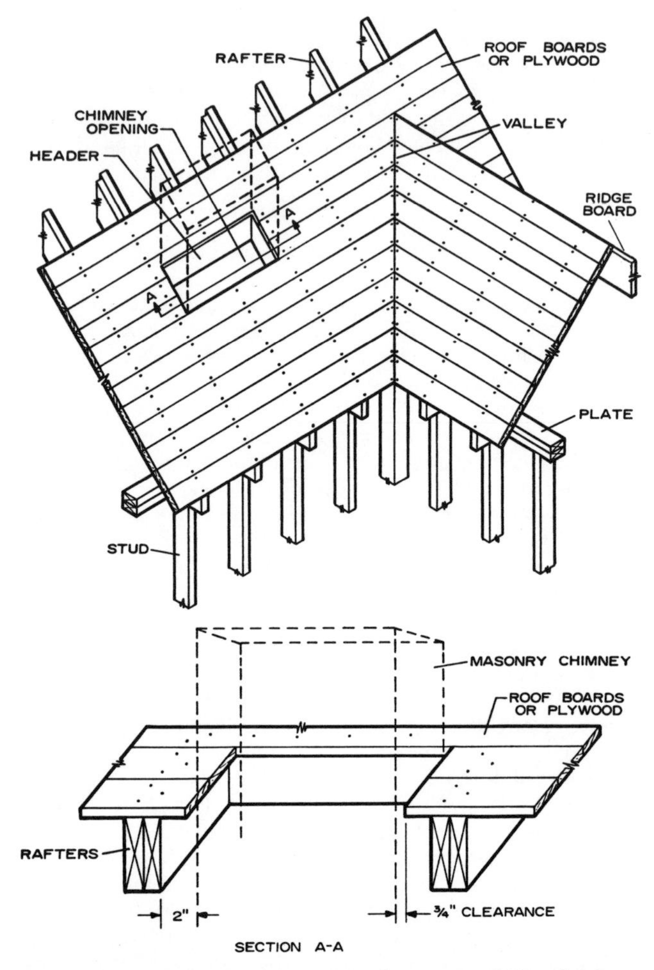

Fig. 5. The installation of wood-board sheathing on a roof. Note that there must be gap between the roof construction and a chimney emerging through the roof *(U.S. Forest Service)*

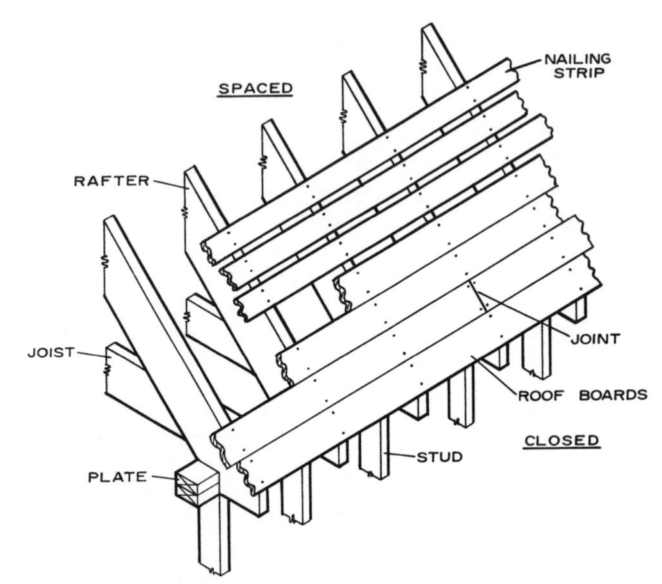

Fig. 6. The difference between open (or spaced) and closed wood sheathing on a roof *(U.S. Forest Service)*.

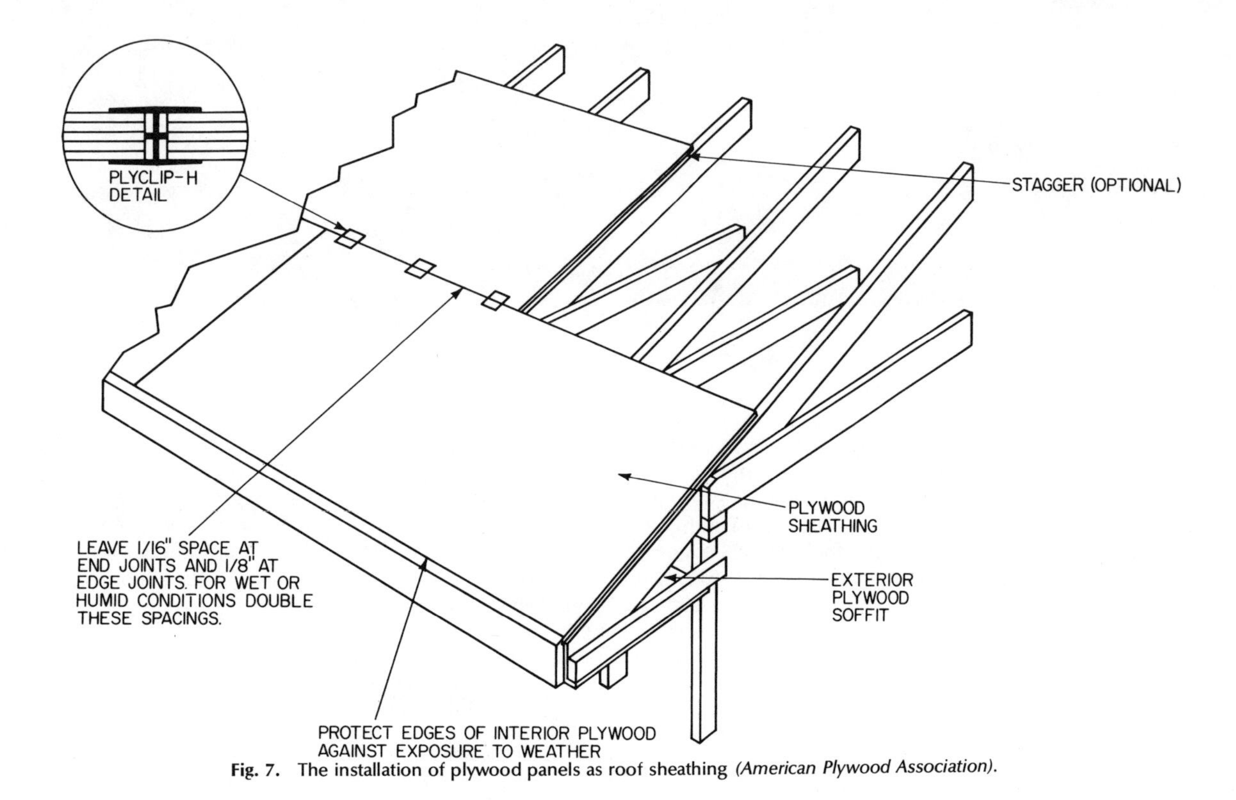

Fig. 7. The installation of plywood panels as roof sheathing *(American Plywood Association)*.

Table 1. Thickness of Paneling Required for a Given Rafter Spacing
Panels thicker than 7/8 inch shall be identified by group

Thickness, in.	C-C Exterior C-D Interior: Group 1 Group 2*	C-C Exterior C-D Interior: Group 2 or 3 Group 4*	C-C Exterior C-D Interior: Group 4	Structural I C-D & C-C Structural II C-D & C-C†: Group 1	Structural II C-D & C-C: Group 2 or 3
5/16	20/0	16/0	12/0	20/0	16/0
3/8	24/0	20/0	16/0	24/0	20/0
1/2	32/16	24/0	24/0	32/16	24/0
5/8	42/20	32/16	30/12	42/20	32/16
3/4	48/24	42/20	36/16	48/24	42/20
7/8		48/24	42/20		48/24

*Panels conforming to special thickness provisions and panel constructions of Paragraph 3.8.6 of PS 1.
†Panels manufactured with Group 1 faces but classified as Structural II by reason of Group 2 or Group 3 inner plies.
Source: American Plywood Association.

Plywood

As with wall sheathing, plywood panels are fast supplanting wood boards for roof sheathing because of the greater strength and stiffness they impart to roof construction, and at a smaller cost (see Fig. 7). Plywood panels 4 × 8 ft in size and ½ in. thick are used as roof sheathing for the usual rafter spacings. The panels are Engineered-grade plywood, either C-D or C-C Interior or Exterior types, depending on the expected moisture conditions. Interior plywoods are satisfactory for the great majority of roofs. Exterior plywood panels should be used whenever the panels will be exposed to the weather for a prolonged period, as, for example, when they will overhang a roof or a cornice, or are exposed on the underside of a carport roof.

The thickness of the panel will depend on the rafter spacing and anticipated roof loads. The American Plywood Association has developed a system for determining how thick the panels must be for different rafter spacings and roof loads. Each panel has marked on it two numbers separated by a slash. The first number indicates, in inches, the maximum rafter spacing over which the panels should be installed. Table 1 shows how this system is used to select panels having the required strength. (*See also* PLYWOOD.)

The panels are installed with their face grain running crosswise to the rafters. Adjacent panels should be staggered by at least one rafter space to prevent nailing the ends of both panels to the same rafters. The long sides of the panels must also be supported. This support is usually provided either by (1) nailing the long edges to 2 × 4 in. blocking installed between the rafters for this purpose, (2) installing panels having tongue-and-grooved edges, or (3) using special H-shaped *plyclips* as shown in Fig. 7. In domestic roof construction there should be at least one of the H-shaped plyclips installed between every two rafters.

The ends of abutting panels should be spaced 1/16 in. apart, and the long edges of abutting panels should be spaced ⅛ in. apart to allow for expansion of the wood during humid weather. In a very humid climate, these gaps should be doubled.

Shingle, Asphalt

An estimated 90 percent of the dwellings in the United States have asphalt shingles installed on their roofs, the reasons undoubtedly being the low initial cost of the shingles and the ease of installing them compared with other types of shingling materials. Besides which, wood shingles, their nearest competitor, are prohibited by many building codes because wood is not a fire-resistant material and asphalt is, to some extent, anyway.

SHINGLE MATERIALS

The manufacture of asphalt shingles begins with an endless sheet of felted material made from rags, wood fibers, or a similar cellulose material. The felt is manufactured by a continuous process that is much like that used to make paper. The same process is also used in the manufacture of asphalt-roll roofing (for a description *see* BUILT-UP ROOFING). The sheet of felt first passes through containers of hot asphalt, where it is thoroughly saturated with the asphalt. Coatings of hot asphalt are then pressed to both sides of the saturated felt, and, finally, granules of crushed slate, granite, or some other mineral are pressed onto one side of the still-hot coating. These granules not only give the final shingles their color, they also protect the shingles against possible mechanical damage, and they also reflect much of the sun's heat, which would otherwise soften the shingles excessively after they have been installed. The completed sheet finally passes between two metal rollers, one of which has knife edges projecting from its surface that cuts the material into individual shingles.

The thickness of the felt core and of the two asphalt coatings determines the final weight of the shingles. The thicker the coatings, the heavier the shingles and the longer-lived they will be. Thus, the weight of an asphalt shingle is an excellent indication of its overall quality and probable lifetime.

The fire resistance of an asphalt shingle is determined more by the ability of the felt to resist fire than by the thickness of the shingle. Underwriters' Laboratories (UL) classifies asphalt shingles as either Class A or Class C, depending on the behavior of the shingles when subjected to a standardized flame test. Shingles made with an organic-felt core—which includes the great majority of asphalt shingles used for dwellings—are given a Class C label, which, in the words of UL Standard 790:

> Includes roof coverings which are effective against light fire exposure. Under such exposures, roof coverings of this class are not readily flammable and do not readily carry or communicate fire; afford at least a slight degree of fire protection to the roof-deck; . . . and may require occasional repairs or renewals in order to maintain their fire-resisting properties.

That does seem to be damning with faint praise. Nevertheless, the asphalt shingles installed on the roofs of dwellings are required by most local building codes to have at least a UL Class C label.

If, instead of rags, wood fibers, or similar organic materials, the felt core of an asphalt shingle is made of a noncombustible material such as asbestos fibers or glass fibers, then the shingle is given a Class A label, which, in the more enthusiastic phraseology of UL Standard 790:

> Includes roof coverings which are effective against severe fire exposures. Under such exposures, roof coverings of this class are not readily flammable and do not carry or communicate fire; afford a fairly high degree of fire protection of the roof-deck; . . . and do not require frequent repairs in order to maintain their fire-resisting properties.

In other words, not only do Class A asphalt shingles have much greater fire resistance, but the last part of the sentence quoted above tells us that an inorganic-felted asphalt shingle has a much longer lifetime than an organic-felted asphalt shingle.

Shingle Weights

The weight of asphalt shingles is calculated on the basis of their installed weight *per square,* which is equal to 100 sq ft. Thus, for example, shingles whose weight is given as 235 lb per square will weigh a total of 235 lb when they are installed over 100 sq ft of roof surface in accordance with the manufacturer's instructions.

The 3-tab *square-butt strip shingles* shown in Table 1 are by far the most popular types of asphalt shingle. When the strips are installed on a roof, the slots in the strips give the illusion that smaller, individual shingles have been laid. This type of strip shingle is made almost entirely in one size, 12 × 36 in., and in a wide range of colors and weights, the weights ranging from 145 lb per square to over 325 lb per square. Most shingle manufacturers guarantee the lifetimes of their shingles, when correctly installed, according to weight, approximately as follows (the weights being based on shingles with organic cores):

Weight per square, lb	*Lifetime, yr*
325–290	25
240–235	20
235–220	15
195 and less	10 and less

One should always consult a manufacturer's catalog for specific lifetime guarantees.

Shingle Types

The *hexagonal-tab* strip shingles shown in Table 1 are also made in 2-tab and 3-tab patterns. The strips are usually 11⅓ × 36 in. in size. They are available in weights of 195 lb per square and less, and usually have a 10-year life. They are used mainly for resurfacing an existing asphalt-shingled roof.

The 16-in.-square *diagonal shingles* shown in Table 1 are

Table 1. Asphalt Roofing Shingles Available in Most Areas of the United States *(Asphalt Roofing Manufacturers Association).*

	SHINGLE TYPE	APPROXIMATE SHIPPING WEIGHT, LB	BUNDLES PER SQUARE	SIZE, IN.		SHINGLES PER SQUARE	EXPOSURE, IN.
				LENGTH	WIDTH		
STRIP SHINGLES	LAMINATED WOOD APPEARANCE MORE THEN ONE THICKNESS	285 TO 390	4 OR 5	36 OR 40	11-1/2 TO 15	67 TO 90	4 TO 6
STRIP SHINGLES	WOOD APPEARANCE SINGLE THICKNESS	250 TO 350	3 OR 4	36 OR 40	12 OR 12.25	75 TO 90	4 TO 5-1/8
STRIP SHINGLES	SELF-SEALING	205 TO 325	3 OR 4	36	12 OR 12.25	78 OR 80	5 OR 5-1/8
STRIP SHINGLES	SELF-SEALING NO CUT OUT	215 TO 290	3 OR 4	36 OR 36-1/4	12 OR 12.25	78 TO 81	5
STRIP SHINGLES	2 AND 3 TAB HEXAGONAL	195	3	36	11-1/3	86	5
INDIVIDUAL	STAPLE LOCK	145	2	16	16	80	—
GIANT INDIVIDUAL	AMERICAN	330	4	12	16	226	5
GIANT INDIVIDUAL	DUTCH LAP	165	2	16	12	113	10
LOCK DOWN	BASIC DESIGN	180 TO 250	3 OR 4	20 TO 22-1/2	18 TO 22-1/4	72 TO 120	—

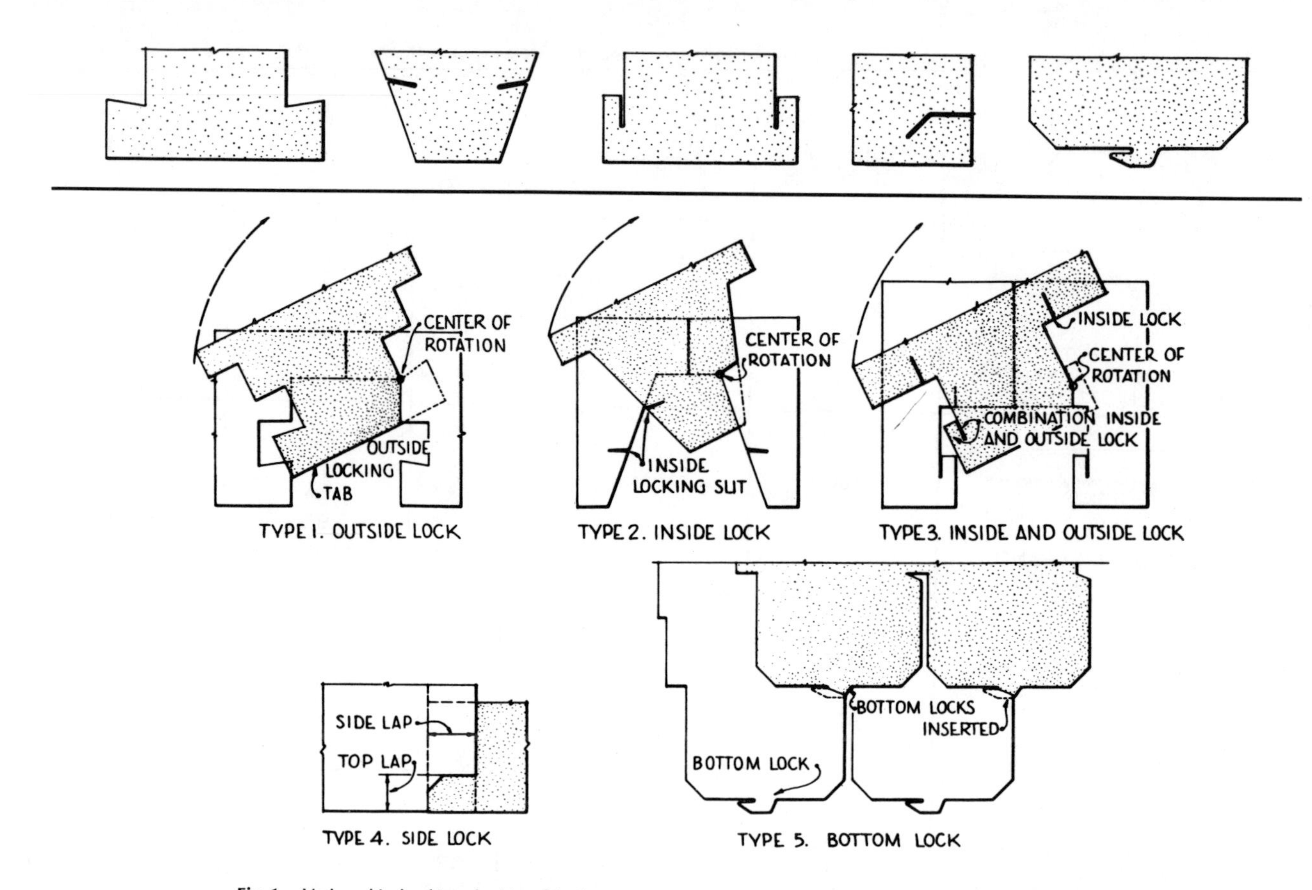

Fig. 1. Various kinds of interlocking shingles and how they lock together to prevent their being lifted off a roof by the wind *(Asphalt Roofing Manufacturers Association).*

also intended primarily for resurfacing a roof. The shingles weigh approximately 145 lb per square and have a 10-year life.

The 12 × 16 in. *Dutch-lap shingles* shown in Table 1 are usually available in two weights, 175 and 150 lb per square. Both have 10-year lives. When Dutch-lap shingles are laid with their long dimension horizontal, they are installed according to what is called the *Dutch method.* When the same shingles are installed with their long dimension vertical, they are installed according to what is called the *American method.*

Dutch-lap shingles are also known as *giant individual shingles.* They are installed according to the Dutch method primarily when an existing asphalt-shingled roof is being resurfaced. The same shingles when installed according to the American method can be used either for new roofs or for resurfacing old roofs.

Shingles are also cut with tabs and slots in a variety of patterns (see Fig. 1). These shingles are designed to lock together in such a way as to prevent their being lifted off a roof by a strong wind. The shingles are made in weights up to 250 lb per square and usually have a 15-year life. They, too, are intended primarily for resurfacing an existing roof.

Wind-Resistant Shingles

The shallower the slope of a roof, the greater the tendency the wind has to get under the shingles and lift them off the roof surface. If the wind is strong enough, and if the shingles have not been securely attached to the roof in the first place, some of the shingles may be torn entirely loose.

The wind may even tear loose the asphalt shingles installed on the downwind side of a roof, much to the homeowner's exasperation and astonishment. The reason is that a strong wind blowing over the downwind side of a sloping roof tends to suck the shingles off the roof (see BUILDING LOADS). Indeed, this suction force may be greater than the direct wind pressure blowing against the windward side of the roof and result in greater damage to the roof.

Strong winds and heavy rainfalls often go together. Even when a wind is not blowing hard enough to tear the shingles off a roof, the shingles may still be lifted up enough for the wind to drive the rain underneath them, which may make it possible for the rain to find its way into the house.

Shingles for Shallow Slopes

For this reason, at one time asphalt shingles were not recommended for roofs that sloped less than 4 in 12, that is, when the vertical rise of the roof was less than 4 in. for every 12 in. of horizontal span. Square-butt strip shingles—and these shingles *only*—are now approved for use on roofs that slope as little as 2 in 12, if special precautions are taken before and during their installation to prevent their being either lifted by the wind or allowing water to make its way under them.

For one thing, the shingles must be heavy enough that they will tend to lie flat no matter how strong the wind may be blowing. For low-sloping roofs, the shingles should weigh at least 235 lb per square, and the heavier the shingles are, the better. The shingles should also be installed in such a way as to provide *triple coverage,* which means that the shingles must overlap each other so that three layers of material cover the entire surface of the roof. To provide this triple coverage, square-butt strip shingles that measure 15 × 36 in. are especially manufactured for installation on low-sloping roofs.

Finally, the roofer must either place dabs of quick-setting asphalt cement under the shingle tabs as each shingle is nailed in place, or he must use shingles on which dabs of asphalt cement have already been applied by the manufacturer. Figure

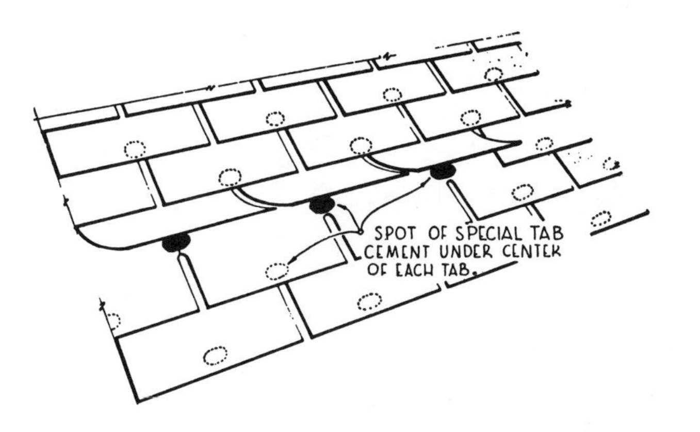

Fig. 2. The shingle tabs can be cemented down to prevent their being lifted off a roof *(Asphalt Roofing Manufacturers Association).*

2 shows an example of such a self-sealing shingle, with the dabs of asphalt cement running across the shingle just above the tab slots. Such self-sealing shingles are pretty much standard in the United States today. Once these self-sealing shingles have been nailed in place on a roof, the heat of the sun plus the weight of the shingles above pressing down upon them will cause the shingles to adhere to each other.

Underwriters' Laboratories has a special "Wind Resistant" label for asphalt shingles that have either self-sealing adhesive applied or that are designed with interlocking tabs. Shingles so labeled are guaranteed not to lift off a roof when they are subjected to a direct wind pressure equal to a wind velocity of 60 mi per hr. However, the adhesive tends to dry out over a period of time and lose its adhesive properties. Shingle manufacturers, therefore, do not guarantee the wind-resistant properties of these shingles for more than 5 years.

SHINGLE INSTALLATION

Roof Preparation

Before asphalt shingles can be installed, the roof must be suitably prepared, which means mainly that steps must be taken to ensure that any water that does find its way under the shingles for any reason will not find its way into the house.

The roof rafters are usually covered by *sheathing,* to which the shingles are nailed. Sheathing consists either of wood boards or of plywood panels (for a description of their installation *see* SHEATHING). Some roofing materials, wood shingles in particular, can be installed on a roof on which the wood-board sheathing has large gaps between the boards (*see* SHINGLE, WOOD). Asphalt shingles should never be installed on a roof having this "open" type of wood-board construction. When asphalt shingles are to be installed, the sheathing must present a solid surface upon which the shingles can rest and into which the roofing nails can be driven.

Underlayment

If the sheathing is satisfactory, *underlayment* is laid down over the sheathing. Underlayment consists of 36-in.-wide rolls of asphalt-saturated felt that weighs about 15 lb per square. This underlayment is installed over the sheathing in overlapping layers as shown in Fig. 3.

The primary function of the underlayment is to prevent water that has made its way past the shingles from making its way into the interior of the house through gaps in the sheathing. The underlayment is waterproof, but it is at the same time permeable to *water vapor,* which means that any water vapor in the attic or within an enclosed attic-roof construction can find its way out of the house through the underlayment. This is a great advantage as it helps prevent the condensation of this water vapor on the underside of the sheathing, which may result in the sheathing's rotting away within a short time (*see* CONDENSATION). The attic should be ventilated in any case, summer and winter, to prevent buildup of water vapor.

Underlayment is especially necessary when asphalt shingles are installed over wood-board sheathing made from southern pine, as this lumber contains resinous compounds that will attack the asphalt and cause the shingles to deteriorate. The underlayment acts as a barrier that keeps the shingles and the sheathing apart, thus prolonging the life of the shingles.

Drip Edges

Drip edges are usually installed along the eave (the bottom edge) and rakes (the sides) of a roof as shown in Fig. 3. The drip

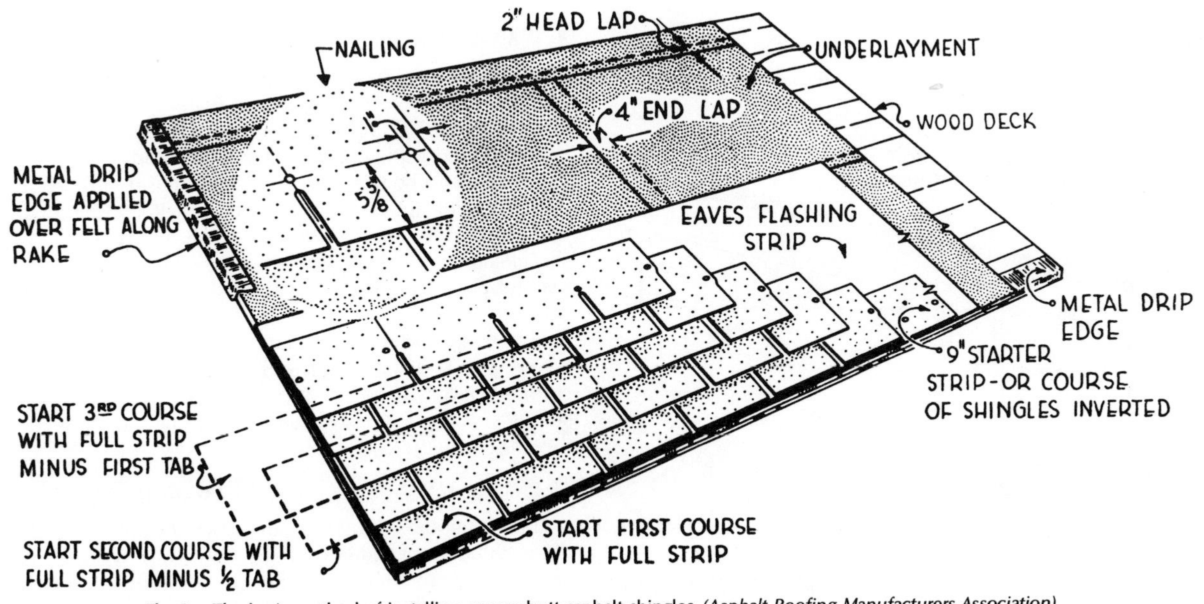

Fig. 3. The basic method of installing square-butt asphalt shingles *(Asphalt Roofing Manufacturers Association).*

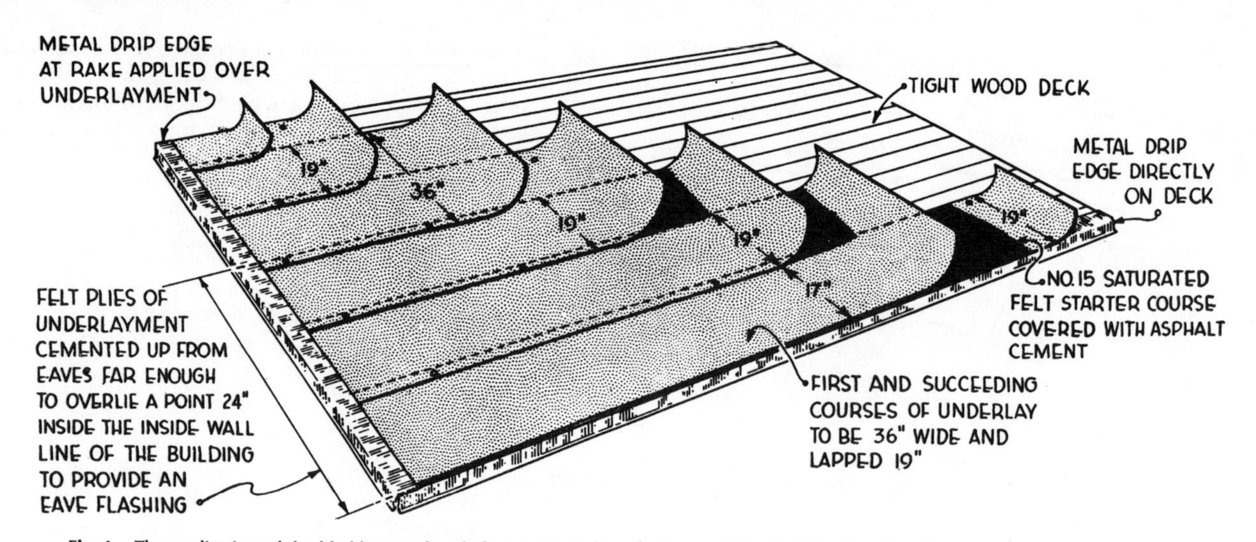

Fig. 4. The application of doubled layers of underlayment on a low sloping roof *(Asphalt Roofing Manufacturers Association).*

edges allow rainwater running down the roof to fall clear of the construction. Without the drip edges, it would be possible for this water to soak into the sheathing at the eave and rakes and then be carried back under the shingles a considerable distance by capillary action. As shown in Fig. 3, the underlayment is installed on top of the drip edge at the eave and under the drip edges at the rakes to facilitate the drainage of water.

How many layers of underlayment are installed depends on the slope of the roof. When the slope is 4 in 12 and more, a single layer of underlayment is sufficient. When the slope is between 2 in 12 and 4 in 12, a double layer of underlayment should be installed as shown in Fig. 4.

Eaves Flashing

Eaves flashing is laid down along the bottom edge of a roof whenever there is a chance that snow will collect on the roof during the winter. If the roof has an overhang that extends beyond the interior wall line of the house, any snow resting on this overhang is less likely to melt than is the snow resting on the roof located above the interior of the house. As a result, when the snow higher up on the roof does melt, the water will be trapped at the overhang by the unmelted snow, which is called an *ice dam*. The water will back up on the roof, work its way under the shingles, and, except for the presence of the underlayment, would very likely find its way into the house.

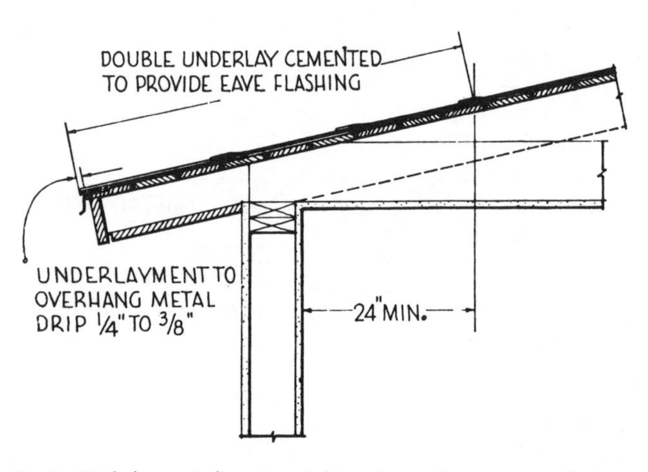

Fig. 5. Underlayment plies cemented together to form an eaves flashing *(Asphalt Roofing Manufacturers Association).*

If the slope of the roof is 4 in 12 or more, the eaves flashing can consist of another 36-in.-wide layer of underlayment installed from the eave to a point at least 12 in. inside the interior wall line of the house (see Fig. 3). If the roof overhang is so great that a shingle width of 36-in.-wide underlayment isn't wide enough and another width of underlayment must be laid down, then the two widths must overlap beyond the interior wall line of the house. This is to prevent backed-up water from finding its way through the overlap and into the house.

When the slope of the roof is less than 4 in 12, the roof will already have a doubled layer of underlayment installed (Fig. 4). In addition, the overlapping layers that extend from the eave to a point at least 24 in. inside the interior wall line of the house (see Fig. 5) must be cemented together with plastic asphalt cement. The cement must be applied at the rate of 2 gal per 100 sq ft of eaves flashing. The cement must be applied as completely and as evenly as possible to ensure that it does in fact form a watertight seal between the two layers of underlayment.

(In addition to all this, before the shingles are installed, it will be necessary to install flashing around the chimney, dormers, roof vents, and any other roof joints; *see* FLASHING.)

Shingle Overlap

If the shingles are to shed water effectively, they must overlap. The greater this overlap, the more watertight the construction. On new roofs, double exposure of the shingles is normally required (and triple exposure for low-sloping roofs). The shingles that are to be used on new construction must be selected, therefore, for their ability to provide this double or triple coverage. This requirement effectively limits one's choice of shingles to square-butt strip shingles, some types of interlocking shingles, and giant individual shingles that are laid down according to the American method.

When an old roof is being resurfaced, however, single coverage is usually adequate. In this case, hexagonal-tab strip shingles, large diagonal shingles, some types of interlocking shingles, and giant individual shingles laid down according to the Dutch method will prove satisfactory.

Installation of Square-Butt Strip Shingles

In the remainder of this article we will describe in detail only the installation of square-butt strip shingles, as the great majority of new asphalt-shingled roofs use this type of shingle. But as all types of asphalt shingle are installed using this same basic procedure, it would make a tedious and unnecessarily long article to repeat over and over what has already been said regarding the installation of square-butt shingles. Therefore, the installation of these other types of shingle will be left to the illustrations.

Laying Out the Shingles

A new roof having been prepared as described above, the first step in the actual installation of the shingles is to plan their position on the roof. Square-butt strip shingles are usually laid down with a 5-in. exposure (see Fig. 6) as already described, which means that only 5 in. of the shingle's total 12-in. width will be exposed to the weather. The rest of the shingle will be covered by the courses of shingles laid after it. The 5-in. exposure will provide double coverage plus an additional overlap (called a *headlap*) of 2 in.

If the roof surface is completely free of projections or breaks, such as a chimney, the shingles can be laid with the standard 5-in. exposure. But if there is a large chimney on the roof, or

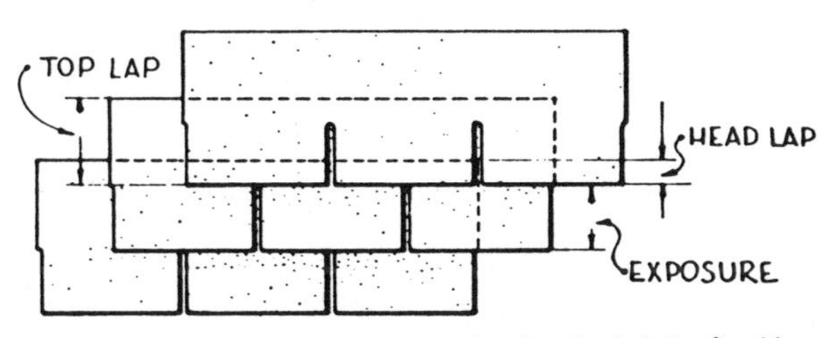

Fig. 6. Terms used in the laying of asphalt shingles *(Asphalt Roofing Manufacturers Association).*

some other break, then for the sake of a neat appearance it may be necessary to increase or decrease the exposure slightly in order to align the butt ends of the shingles with the bottom of the chimney.

Starter Course

The first course of shingles laid down along the eave of the roof is the *starter course*. It consists of a doubled layer of shingles that provide the proper support for the second course of shingles and that back up the tab slots of the first course, which would otherwise expose the underlayment underneath. There are several methods of laying down this starter course.

In one, a row of *wood* shingles is first nailed to the sheathing, the shingles extending over the eave by about ½ in. to ensure that water running off the roof falls clear of the eave and does not double back under the shingles. (The shingles at the rakes of the roof are similarly extended beyond the edge of the roof by about ½ in. as each course is laid down.) A course of asphalt shingles is then laid down on top of the wood shingles. Wood shingles are used because they are stiff enough to support the weight of the asphalt shingles above them, in spite of the fact that they are projecting over the eave.

Another method of installing the starter course is to first lay down a strip of mineral-surfaced asphalt roll roofing that weighs at least 55 lb per square. A course of regular shingles is then laid down on top of this strip of roll roofing. The advantage of using a single strip of roll roofing is that it has no gaps through which water might find its way into the house. The roofing is usually from 9 to 18 in. wide, although on a shallow roof in a high-wind area it might be 36 in. wide. The strip is usually nailed in place, but in a high-wind area it might also be sealed to the underlayment with plastic asphalt cement.

The method preferred by the Asphalt Roofing Manufacturers Association is to cut shingles lengthwise to remove the tabs. These strips are then laid down along the eave, and the line of adhesive on the strips will assure that the starter course is properly sealed down (see Fig. 3). The first course can then be installed on top of the starter course.

Laying Subsequent Courses

Before the second course of shingles is laid down, two roofers, one at each end of the roof, carefully measure the desired exposure from the butt end of the starter course. They then snap a chalk line between the two measured points. The straightness of the second course thus being assured, they proceed to lay it. This procedure is followed with all the courses until the top of the roof is reached. The vertical alignment of the tabs is, of course, controlled either by the edge of the roof or by the vertical chalk line (or lines) snapped down earlier.

The first shingle strip installed in each course is laid down along the edge of the roof, or, in the case of a hip roof, against a vertical chalk line. This shingle is trimmed by one-half, one-third, or two-thirds, as necessary, to maintain the correct tab alignment for that course. Once this first shingle strip has been cut and nailed in place, the remaining shingle strips in the same course will follow along automatically to maintain the desired tab spacing.

Nails and Nailing

The nails used to fasten asphalt shingles are corrosion-resistant roofing nails, usually made of aluminum or hot-dipped galvanized steel, although copper is the preferred material. The nail shanks are 10 to 12 gauge in size, and the heads are from ⅜ to 7/16 in. in diam. The shanks may be smooth, but threaded or barbed shanks are preferred because of their greater gripping ability. The nails should be long enough to penetrate the sheathing. In new construction, the usual nail length is 1¼ in.

A minimum of four nails per strip is usually recommended, but using six nails per strip is better, especially in high-wind areas. The nails should be located about ⅝ in. above the butt line of the succeeding course of shingles when the exposure is the standard 5 in; that is, the nails will be hidden by about a ⅝-in. overlap of the shingles that will be laid on top of the shingle being nailed. This technique is often called *blind nailing*.

When shingles of the succeeding course hide the nails of the preceding course with this ⅝-in. overlap, it means that each shingle will actually be held in place not by four or six nails but by eight or twelve nails, the reason being that each nail will penetrate two layers of shingles, not one. If the nails were to be located much higher on the shingles, say more than 1 in. above the butt line of the succeeding course, with the standard 5-in. exposure, each shingle would then be held in place by only four or six nails.

When the roof slope is less than 4 in 12, a dab of plastic roofing cement should be placed under each tab of each strip, as shown in Fig. 2, if the shingle doesn't already have adhesive applied. The cement can be obtained in different consistencies for brushing, troweling, or for use with a caulking gun. The cement should be applied immediately after the shingle has been nailed in place.

In those parts of the United States classified as medium-wind areas (see Fig. 7), dabs of cement should be applied to the shingles on all roofs that slope less than 4 in 12, and on all roofs that slope 4 in 12 or more if the shingles weigh less than 275 lb per square.

In those parts of the United States classified as high-wind areas, all the shingles on all roofs, regardless of their slope, should have cement applied under the shingle tabs.

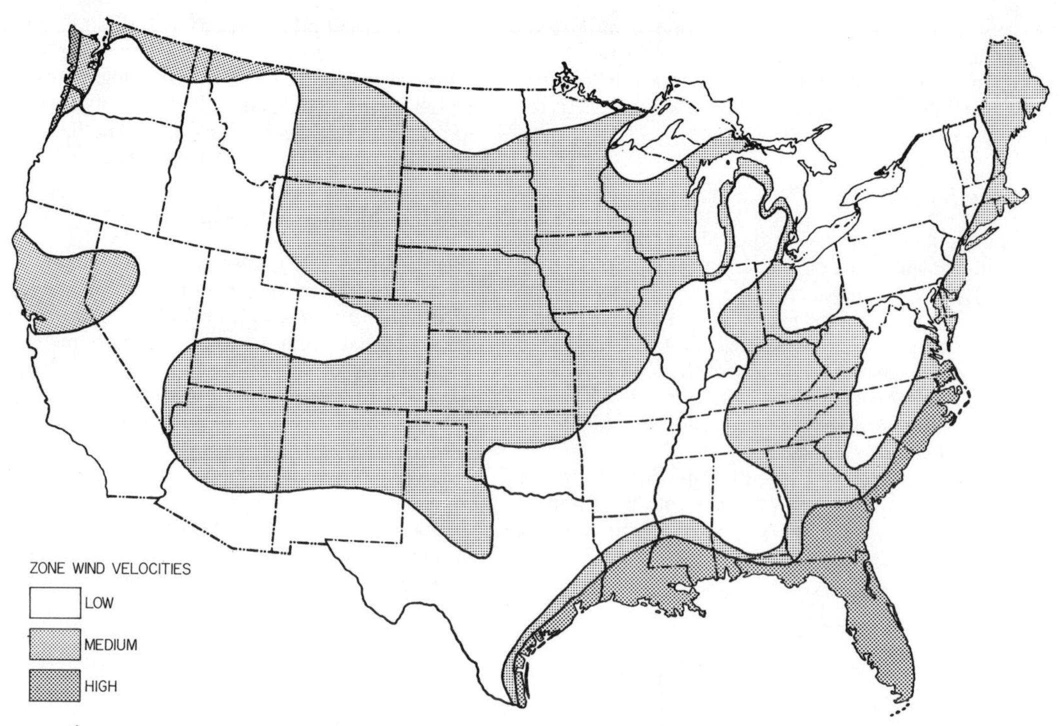

Fig. 7. Low, medium, and high wind-velocity areas in the United States, insofar as they affect the installation of asphalt shingles *(U.S. Federal Housing Administration).*

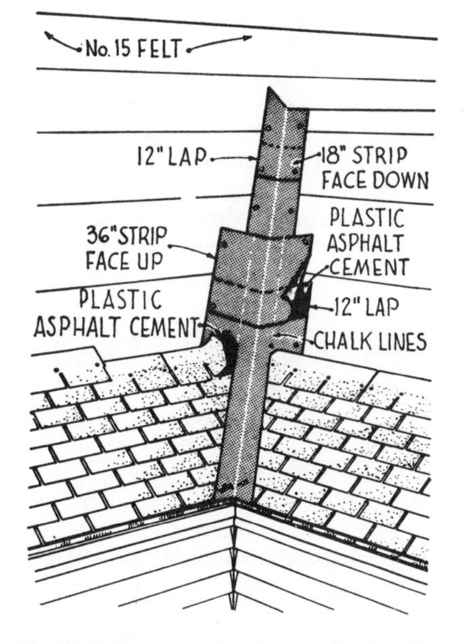

Fig. 8. Open valley flashing using mineral-surfaced roll roofing *(Asphalt Roofing Manufacturers Association).*

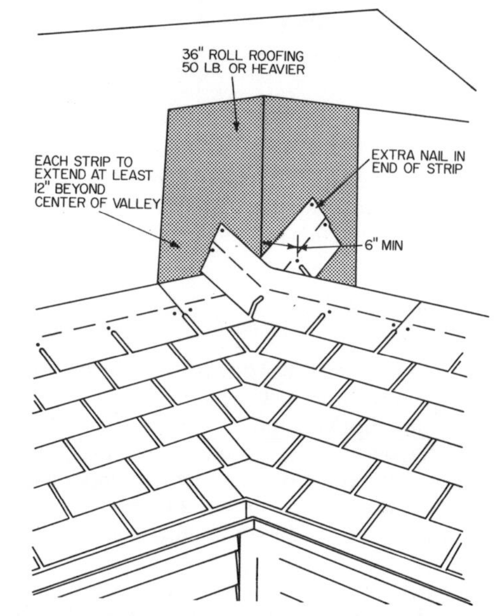

Fig. 9. Closed valley flashing using woven strip shingles *(Asphalt Roofing Manufacturers Association).*

Valley Installation

Where two roof surfaces meet, they form a valley. Rainwater running down the two roof surfaces tends to collect in this valley. This juncture between the roof surfaces is also very vulnerable to leaks and care must be taken to make the construction watertight. This is accomplished by installing flashing down the center of the valley (*see* FLASHING).

As far as asphalt shingles are concerned, there are two methods by which the shingles can be installed at a valley. They can either stop alongside the valley, leaving an open channel down which the rainwater can run (Fig. 8), or they can be laid through the valley (Fig. 9), which means that the rainwater will run over them on its way down the roof. The first type of shingle installation results in an *open* valley; the second results in a *closed* valley.

Hip and Ridge Shingles

The roof surfaces that meet at the ridge of a peaked roof, and at the hips of a hip roof, must also be protected against leaks by flashing. The flashing consists of the underlayment that will have already been laid down as described above. After the surfaces of the roof have been shingled right to the edges of the ridge and hips, these junctures are finished off by nailing across them overlapping strips of 9 × 12 in. shingling material. These strips are bent in half and nailed over the juncture like saddles.

Shingle, Wood

Shingles, whether made of wood, asphalt, or asbestos-cement (the three most popular shingling materials), keep the interior of a house dry by shedding the water that falls on the roof. Shingles do not make a roof watertight, as built-up roofing and metal roofing do. All they do is present a barrier to running water, a barrier that depends primarily on the slope of the roof for its effectiveness. It is the steepness of a roof that ensures rainwater will run off the shingles before it has a chance to work its way around the gaps between the shingles and perhaps find its way into the house.

The steeper the slope of a roof, the less chance there is that water will either soak into the shingles or work its way around them. The shallower the slope, the greater the chance. When a roof slopes less than 3 in 12 (that is, when the roof rises vertically 3 in. for every 12 in. of horizontal span; this is equal to an angle of about 14 degrees from the horizontal) shingles cannot be installed on a roof. When the slope is between 3 in 12 and 5 in 12, the shingler must take special precautions during the construction to ensure that rain cannot work its way around the shingles; these precautions are described below. It is only when a roof slopes 5 in 12 and more that special precautions become unnecessary.

Not only is the watertightness of the construction dependent on the slope of the roof, but so is the lifetime of the shingles; for the shallower the slope, the less quickly will wet shingles dry out. Shingles subjected to prolonged wetting tend to decay more quickly. For example, when a roof slope is 4 in 12, split shingles made from red cedar have an expected lifetime of at least 40 years. When the roof slope is equal to 8 in 12, the same shingles will have a lifetime of at least 60 years. And when these shingles are used as wall siding, their lifetime is indefinitely long—they will outlast the house. (But to give one perspective on this point, we should note that slate shingles, clay tiles, and copper are all expected to last for hundreds of years when used as roofing, and they usually do.)

Another reason for having a reasonably steep roof when wood shingles are installed is that it becomes much more difficult for a strong wind to drive rain or snow underneath the shingles, from where it may make its way into the interior of the house. To make it even more difficult for wind-driven rain or snow to make its way into a house, roof shingles are usually required to have at least *triple coverage,* which means that at least three layers of shingles must cover the entire surface of the roof. How this is accomplished is described below.

Because vertical surfaces such as the walls of a house are so much more efficient at shedding water, they are considered adequately protected when the shingles have only double coverage, as described below.

SHINGLING MATERIAL

The Romans made shingles from oak. The shingles had a reputed lifetime of about 50 years. Oak continued to be used for shingles throughout Europe until the gradual deforestation of the countryside forced a switch to slate shingles and clay tiles.

In Colonial America, the abundance of oak made wood shingles popular for both roofing and wall siding from the start. Even today wood shingles are a popular building material in New England.

The virtues of white cedar as a shingling material were discovered early in American history (the logs split neatly and evenly into shingles, and the shingles were long-lasting), and this wood continued to hold first place in the hearts of American shinglers until the settlement of the West Coast made apparent the superior virtues of western red cedar and redwood as shingles. These two woods (and cypress, also) have the inestimable advantage over white cedar that they are all extremely resistant to decay, a property they owe to certain natural oils within them. As roofs are often wet, and as this wetness tends to persist on the underside of shingles, which leads to the decay of the wood, resistance to decay is a very necessary quality in a wood shingle.

Fig. 1. Hand-splitting shakes from a log using a mallet and a froe *(Red Cedar Shingle & Handsplit Shake Bureau).*

Other qualities are also important. The grain of the wood must be close and even if shingles are to be split smoothly from the log. The wood should have a relatively low coefficient of expansion when wet, and it must be rather impermeable to water in the first place. Finally, the wood must be durable enough to resist the constant wetting and drying, heating and freezing cycles it will be subjected to. Western red cedar, redwood, and cypress all have these qualities but western red cedar (*Thuja plicata*) has them to an outstanding degree. This wood is by far the preferred shingling material in the United States. ·

Western red cedar is, like the redwood, an exceptionally tall, slow-growing conifer with a long trunk that rises straight and true from the ground. Because the tree is so slow-growing, the grain has a very close, uniform texture that is largely free of knots or other defects. Western red cedar also has an exceptionally low coefficient of expansion when it is soaked with water (about 3 percent) that helps prevent it from checking or splitting.

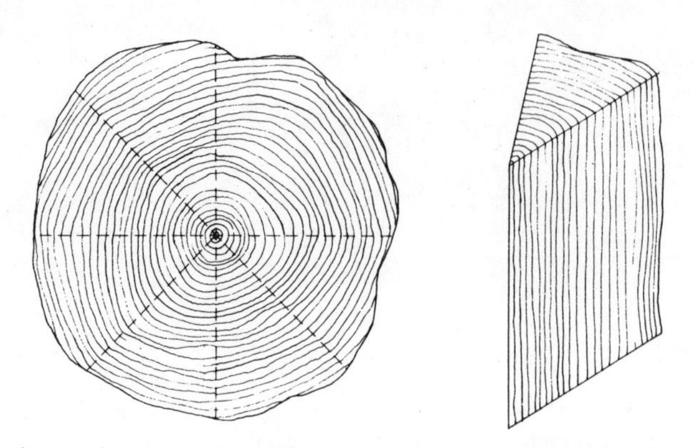

Fig. 2. A log must first be cut into wedges in order to expose as much of the edge grain in the wood as possible *(Smith, Materials of Construction, McGraw-Hill).*

Shingle Manufacture

In the old days, shingles were split off a log by a man using a *froe* (a large knifeblade attached to an offset handle) and a wooden mallet (see Fig. 1). When a shingle is split off a log by a blade in this way, the wood tends to cleave between the fibers that make up the grain of the wood. It is because the fibers remain intact that the wood is able to shed water so effectively.

When shingles are sawed from a log, however, the teeth of the saw cut through the wood fibers, which exposes the entire cross section of the wood to moisture. For this reason, sawed wood is much less weather-resistant then split wood. Thus, for example, if a split shingle will last for 60 years on a roof having an 8-in-12 slope, a sawed shingle of the same wood installed on the same roof will last only 35 years.

The logs from which shingles are to be sawed are first cut into 16, 18, or 24-in. sections, these dimensions being the lengths of the finished shingles. The sections are then quartered and requartered as shown in Fig. 2 before being passed through a saw. As each strip of shingle is sawed off the log, a sawyer trims its sides square with the *butt* (i.e., the thick bottom edge) of the shingle. If there are any defects in the wood, these are also trimmed out, which usually converts one wide shingle into two narrower shingles. The sawyer then grades the shingles as to their quality and places them in separate piles according to grade.

Grading Shingles

The best shingles have certain desirable characteristics. The wood will be 100 percent *heartwood,* as the heartwood of a tree is more resistant to decay than the sapwood. The wood will be entirely *clear,* that is, free of knots, splits, or any other

Table 1. The grades and sizes of red-cedar shingles *(Red Cedar Shingle & Handsplit Shake Bureau).*

CERTIGRADE RED CEDAR SHINGLES

Grade	Length	Thickness (at Butt)	No. of Courses Per Bundle	Bdls/Cartons Per Square	Description
No. 1 BLUE LABEL	16" (Fivex) 18" (Perfections) 24" (Royals)	.40" .45" .50"	20/20 18/18 13/14	4 bdls. 4 bdls. 4 bdls.	The premium grade of shingles for roofs and sidewalls. These top-grade shingles are 100% heartwood, 100% clear and 100% edge-grain.
No. 2 RED LABEL	16" (Fivex) 18" (Perfections) 24" (Royals)	.40" .45" .50"	20/20 18/18 13/14	4 bdls. 4 bdls. 4 bdls.	A good grade for many applications. Not less than 10" clear on 16" shingles, 11" clear on 18" shingles and 15" clear on 24" shingles. Flat grain and limited sapwood are permitted in this grade.
No. 3 BLACK LABEL	16" (Fivex) 18" (Perfections) 24" (Royals)	.40" .45" .50"	20/20 18/18 13/14	4 bdls. 4 bdls. 4 bdls.	A utility grade for economy applications and secondary buildings. Not less than 6" clear on 16" and 18" shingles, 10" clear on 24" shingles.
No. 1 or No. 2 REBUTTED-REJOINTED	16" (Fivex) 18" (Perfections) 24" (Royals)	.40" .45" .50"	33/33 28/28 13/14	1 carton 1 carton 4 bdls.	Same specifications as above but machine trimmed for exactly parallel edges with butts sawn at precise right angles. Used for sidewall application where tightly fitting joints between shingles are desired. Also available with smooth sanded face.
No. 4 UNDER-COURSING	16" (Fivex) 18" (Perfections)	.40" .45"	14/14 or 20/20 14/14 or 18/18	2 bdls. 2 bdls. 2 bdls. 2 bdls.	A utility grade for undercoursing on double-coursed sidewall applications or for interior accent walls.

CERTIGROOVE GROOVED RED CEDAR SIDEWALL SHAKES

Grade	Length	Thickness (at Butt)	No. of Courses Per Bdl/Carton	Bdls/Cartons Per Square	Description
No. 1	16" (Fivex) 18" (Perfections) 24" (Royals)	.40" .45" .50"	33/33 28/28 13/14	1 carton 1 carton 4 bdls.	Same specifications as rebutted-rejointed shingles, except that shingle face has been given grain-like grooves. Natural color, or variety of factory-applied colors. Also in 4-ft. and 8-ft. panels.

LENGTH AND THICKNESS	Approximate coverage of one square (4 bundles) of shingles based on following weather exposures																									
	3½"	4"	4½"	5"	5½"	6"	6½"	7"	7½"	8"	8½"	9"	9½"	10"	10½"	11"	11½"	12"	12½"	13"	13½"	14"	14½"	15"	15½"	16"
16" x 5/2"	70	80	90	100*	110	120	130	140	150I	160	170	180	190	200	210	220	230	240†	....	....	....	....	....	....	....	....
18" x 5/2¼"	....	72½	81½	90½	100*	109	118	127	136	145½	154½I	163½	172½	181½	191	200	209	218	227	236	245½	254½†	....	....	....	....
24" x 4/2"	....	....	....	....	....	80	86½	93	100*	106½	113	120	126½	133	140	146½	153I	160	166½	173	180	186½	193	200	206½	213†

NOTES: * Maximum exposure recommended for roofs. I Maximum exposure recommended for single-coursing on sidewalls. † Maximum exposure recommended for double-coursing on sidewalls.

defects that will allow water through the shingles or that might in time cause the shingle to split. The shingles will also consist of 100 percent *edge grain;* that is, the grain runs perpendicular to the face of the shingle, as shown in Fig. 2. Edge-grained shingles are preferred because the wood is much less likely to warp or cup then when the wood contains flat grain, which is grain that runs parallel to the face of a shingle. (And we might also note that thick shingles are much less likely to warp than thin shingles, and that narrow shingles are much less likely to warp than wide shingles, points that should be kept in mind when selecting shingles.)

Shingles made of red cedar are graded as to quality by the Red Cedar Shingle & Handsplit Shake Bureau, an industry-sponsored inspection organization. Shingles inspected and graded by this organization (see Table 1) can be identified by a label attached to the bundle the shingles are packed in.

INSTALLATION ON ROOFS

Roof Preparation

Roof shingles are applied over *sheathing,* which consists of wood boards or plywood panels that have been installed over the roof rafters (for a description *see* SHEATHING). If the sheathing consists of plywood panels, the sheathing must of necessity be *closed;* that is, the panels will form a solid surface over the entire roof to which the shingles can be nailed.

When wood boards are installed, however, the builder has the option of nailing these boards tightly together (i.e., *closed* sheathing) or of spacing them several inches apart (i.e., *open,* or *spaced,* sheathing; see Fig. 3). There is a difference of opinion as to which is preferable. Proponents of open sheathing say that air must be allowed to circulate around the undersides of the shingles; otherwise they will quickly rot. Proponents of closed sheathing say this is nonsense; that, except for certain very humid sections of the United States, especially along the Gulf Coast, as long as an attic is well ventilated, there is little chance of the shingles staying wet; that, besides, closed sheathing helps to keep a house warmer in wintertime.

In fact, there doesn't seem to be much difference in the weatherability of a shingle roof whether open or closed sheathing is installed. Local custom is the best guide. However, in those parts of the United States where it is likely that wind-driven rain or snow will be blown under the shingles, solid sheathing should be installed. In addition, regardless of the section of the country, attics should always be well ventilated to prevent any condensation of water vapor on either the shingles or sheathing.

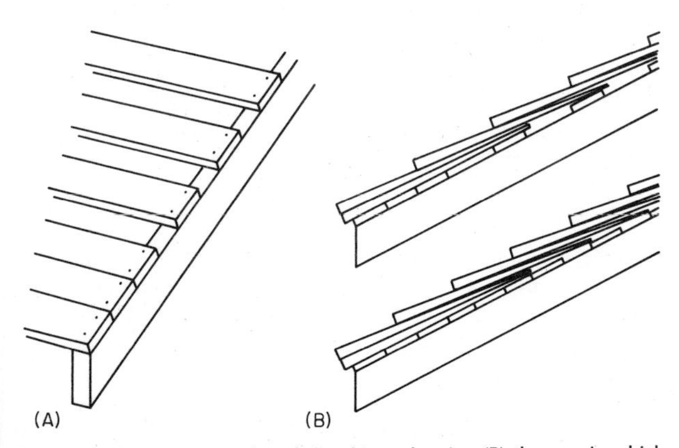

Fig. 3. (A) Open or spaced roof sheathing, showing (B) the way in which shingles are nailed to the sheathing boards.

When spaced sheathing is being installed, 1 × 4 or 1 × 6 in. boards are installed (*see* SHEATHING). The center-to-center distance between two adjacent boards must be the same as the exposure between two adjacent courses of shingles, as described below. Spaced boards should never be spaced more than 10 in. apart, center to center, in any case.

Underlayment and Eaves Flashing

Some roofing materials such as asphalt shingles are nailed over a layer of *underlayment,* which consists of 15 or 30-lb asphalt-saturated felt laid down over the sheathing (*see* SHINGLE, ASPHALT). The purpose of this underlayment is to keep water from entering the house, if the water should make its way past the roofing material. Underlayment of this kind should *never* be installed under wood shingles, as its presence will definitely result in persistently damp shingles, which will cause them to rot all the quicker.

The only exception to this rule, and it is only a partial exception, is when shingles are installed on a roof overhang in a section of the country where it snows every winter. Under these circumstances, whenever there is a snowfall, the snow on that part of the roof directly over the interior of the house tends to melt more quickly than the snow resting on the overhanging part of the roof. As a result, an *ice dam* (see Fig. 4) will form on the overhang, and this ice dam will trap the melted snow running down the roof. The melted snow then backs up under the shingles, and it may find its way into the house.

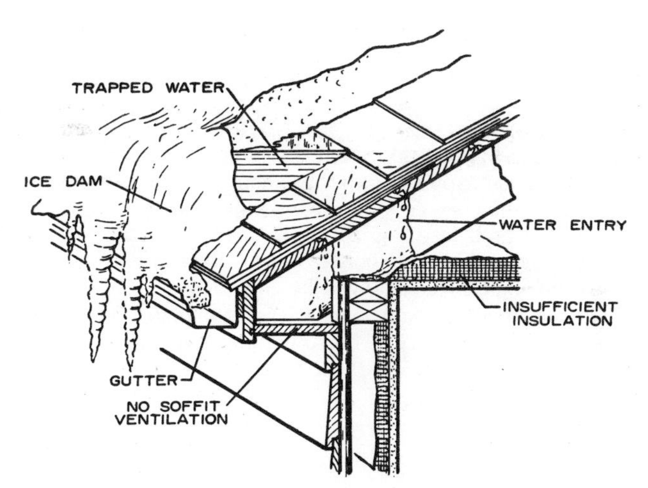

Fig. 4. An ice dam along the eave of a roof may allow melted snow to back up under shingles and enter the house *(U.S. Forest Service).*

To prevent this penetration of melted snow into the house, *eaves flashing* is installed along the overhang. The flashing consists of a double layer of 15-lb asphalt-saturated felt laid down on top of the sheathing (*see* SHINGLE, ASPHALT, Fig. 5).

If the roof slope is 4 in 12 or more, the flashing should extend at least 36 in. inside the interior wall line of the house. If the roof slope is less than 4 in 12, the double layer of eaves flashing must extend 36 in. inside the interior wall line of the house, and plastic asphalt cement must be applied between the two layers of felt to form a seal. The cement should be applied at the rate of 2 gal per 100 sq ft of roof.

Roof Shingle Installation

Shingle Exposure

On a roof having a slope of 5 in 12 or more, wood shingles must have triple coverage. That is, if a shingle is 16 in. long,

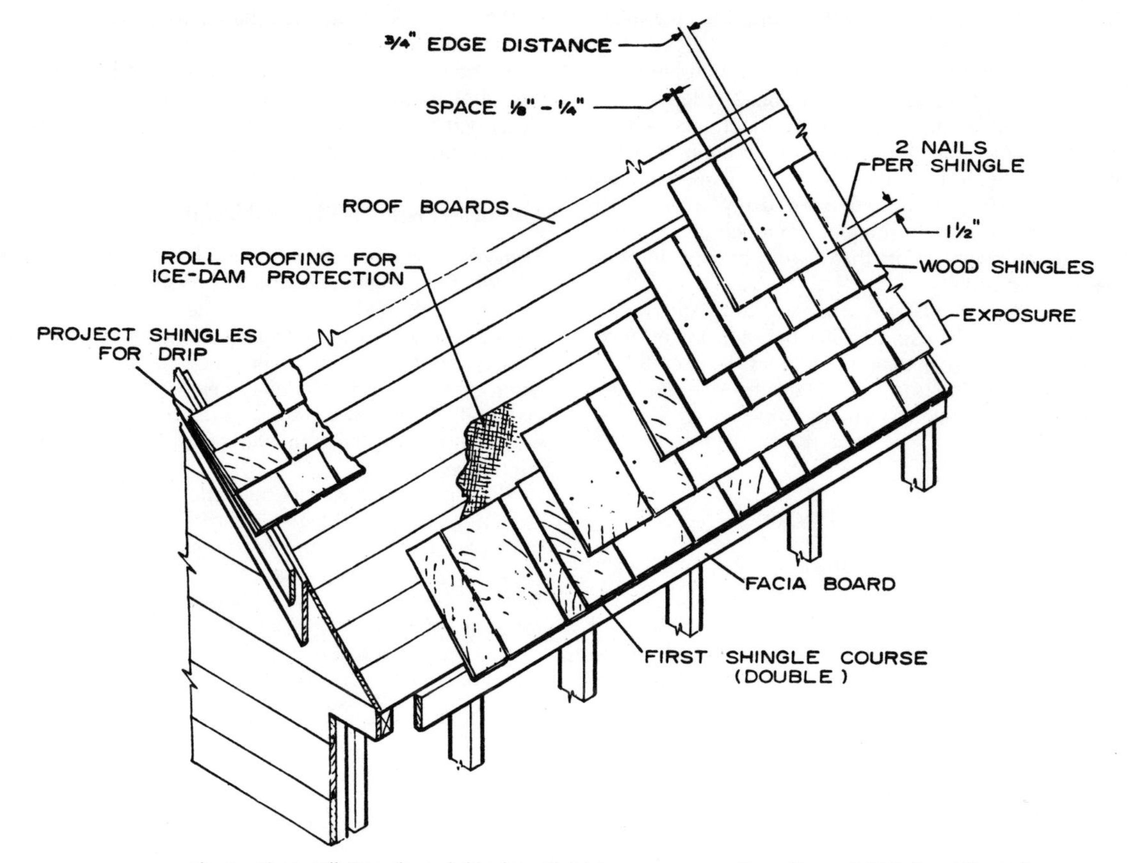

Fig. 5. The installation of wood shingles with triple coverage over the entire roof *(U.S. Forest Service)*.

only 5 in. of the shingle, which is a bit less than one-third its length, can be exposed to the weather; the remainder of the shingle will be covered by the two shingle courses that will be installed over it (see Fig. 5). Since the entire roof will be constructed in this overlapping manner, there will be a triple layer of shingles installed over the entire roof.

For the three standard shingle lengths manufactured, three standard exposures result when the shingles are laid with triple exposure:

Shingle length, in.	*Exposure, in.*
16	5
18	5½
24	7½

When bundles of shingles are being packed at the mill, each bundle contains a sufficient number of shingles to cover 25 sq ft of roof, assuming the shingles will be laid with a standard exposure. Knowing the total area of a roof, therefore, and assuming a standard exposure, one can easily calculate the total number of bundles that will be required on any roof.

Shingles need not, however, be installed with a standard exposure. A standard exposure can always be increased (but slightly) or reduced, if desired, so that the spacing of the courses will present an even appearance over the entire expanse of a roof.

The exposure is sometimes dependent on the overall size of the roof, since a small exposure looks better on a small roof, and vice-versa; and also because there may be dormers or a large chimney projecting from the roof. When this is the case, the shingle exposure must be worked out in such a way that the butt ends of the courses will be aligned with the bottoms of the dormers or chimney.

Although there is no limit to how much a standard exposure may be reduced, there is a limit to how much an exposure can be increased, for one must still maintain at least triple coverage of the shingles. The following formula is used to work out the maximum exposure:

Maximum exposure = (shingle length − 1 in.)/3.

Shingle Exposures on Shallow-Sloped Roofs

When a roof slope is between 4 in 12 and 5 in 12, the overlap of the shingles must be increased to increase the watertightness of the construction. Therefore, the maximum shingle exposures must be reduced as follows:

Shingle length, in.	*Exposure, in.*
16	4½
18	5
24	6¾

When a roof slope is between 3 in 12 and 4 in 12, the overlap of the shingles must be increased even more, until there is quadruple coverage over the entire roof. With quadruple coverage, the maximum exposures become:

Shingle length, in.	*Exposure, in.*
16	3¾
18	4¼
24	5¾

These maximum exposures can be twiddled with slightly, if necessary, to achieve an even-looking appearance in the courses.

Laying the First Course

The exposure having been decided upon, the first course of shingles can now be laid down along the eave. The butt ends of the first course must overhang the eave by 1 to 1½ in. so that the shingles form a *water drip,* which will prevent water running off the roof from backing up under the shingles. There should be a similar 1 in. overhang at the *rakes,* or sides, of the roof.

In the usual construction, this first course of shingles is always *doubled;* that is, two layers of shingles are installed, one directly on top of the other (see Fig. 5). In the best construction, the first course is *tripled.* Water always tends to soak deep into the shingles installed along the lower edge of a roof and doubling or tripling the shingles makes it more difficult for water to soak through the shingles and back up under them by capillary action.

When the first course is laid, the shingles in the bottom layer are spaced ⅛ to ¼ in. apart, depending on their widths, as the wider a shingle the more it tends to expand when wet. When the top layer of shingles is laid down over the bottom layer, the joints in the top layer must be staggered in such a way that they are at least 1½ in. away from the joints in the bottom layer.

If a third layer of shingles is being installed, then the joints between the shingles in this third layer must be located at least 1½ in. away from the joints in both the first and second layers. Above all, the joints in all three layers should not be aligned with each other in order to prevent rainwater finding its way through the roofing, but, unfortunately, this cannot always be avoided.

The shingles in the bottom layer (and second layer also, if one is installed) may be No. 3 Grade. The top layer, however, must consist of No. 1 Grade, as must all the shingles in all the succeeding courses. Shingles used for roofing should always be of No. 1 Grade. No. 1 Grade shingles will last so much longer than No. 2 Grade shingles that it is very much a false economy to install anything other than No. 1 Grade shingles on any roof.

Laying the Succeeding Courses

If a roof is straight, without any intersecting roof surfaces, the shingles can be laid from either end of the roof as is convenient for the shingler. Each course of shingles should begin with either a narrow or a wide shingle to emphasize the staggered appearance of the shingles, and to make sure that sufficient distance is maintained between the joints in adjacent courses. When two roof surfaces meet at a valley, as described below, each course should be laid beginning at this valley.

Nailing

Two nails, and only two nails, are used to fasten down a shingle, regardless of its width. The nails are positioned so they will be covered by the shingles in the course above by about 1 in., and they are at least ¾ in. from the sides of the shingles. The nails should be driven until the nail heads just make contact with the shingle. The nail heads definitely should not crush the wood.

Nails should preferably be made of a corrosion-resistant metal such as copper or bronze, but aluminum and hot-dipped zinc-coated nails are satisfactory, cheaper, and used more often. Plain steel or iron nails should never be used, even if they have been galvanized by an electrical plating process, as the protection afforded a nail by this process is poor. The coating is likely to crack through, and if it does the metal will quickly corrode. If an entire roof has been installed using this type of nail, it is possible that the overall life of the roofing will be reduced considerably.

Installation at Valleys

Where two roof surfaces intersect, they form a valley. During a heavy rainstorm, a considerable volume of water may run through the valley and any overflow will certainly find its way under the shingles located alongside the valley. The roof at the valley juncture must, therefore, be made completely watertight. For methods of protecting the underlying roof structure from water *see* FLASHING. In this article we will describe only the methods of installing wood shingles in such a valley.

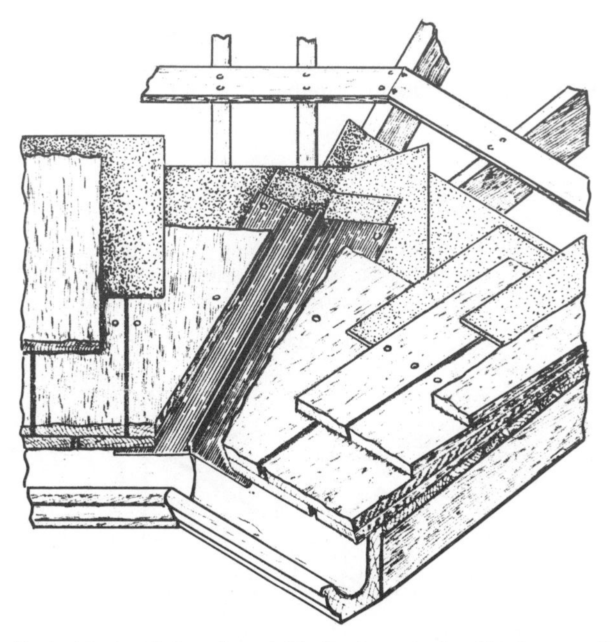

Fig. 6. The installation of wood shingles at a roof valley. This is open-valley construction, in which a gap runs down the center of the valley through which rainwater can run. *(Red Cedar Shingle & Handsplit Shake Bureau.)*

There are two such methods: (1) *open-valley* construction in which a gap runs down the center of the valley alongside, which the shingles are laid (see Fig. 6), and (2) *closed-valley* construction, in which the shingles run through the valley without interruption. Open-valley construction is usually preferred as it is both simpler and cheaper than a closed valley. An open valley is also less difficult to make watertight.

Hip and Ridge Shingles

The hips of a hip roof (*see* ROOF FRAMING) are covered by two overlapping rows of shingles that are assembled and fastened to the roof in much the same way as the shingles on the main part of the roof. The same size shingles are used, and the shingles will have the same exposure as those on the main roof. The hip shingles are usually as wide as the exposure is deep, but in any case they should be at least 5 in. wide. Before the shingling begins, the shingler usually picks through the bundles of shingles to find shingles that are about 1 in. wider than the desired width. Shingle manufacturers also make up prefabricated pairs of hip shingles for the most common roof slopes. They need only be nailed in place on the roof.

The hip shingles should be installed in pairs, with each succeeding pair of shingles overlapping each other in opposite directions. This method of laying the shingles is called a modified Boston lap.

The peak of the roof is covered with pairs of overlapping shingles that are assembled and nailed to the roof in the same way that has just been described for the hips. If there is a prevailing wind direction, the shingles should be installed with their butt ends facing away from the wind.

As with the hip shingles, the shingles installed along the peak should have the same exposure as the shingles on the main part of the roof and be as wide as the exposure is deep, as long as they are at least 5 in. wide. Strip flashing must also be installed between each pair of shingles as already described to prevent water entering the gap that exists at the roof juncture.

SIDING CONSTRUCTION

Shingles are used for wall siding as well as for roofs. The shingles can be applied to a wall by either of two methods, which are called *single coursing* and *double coursing.*

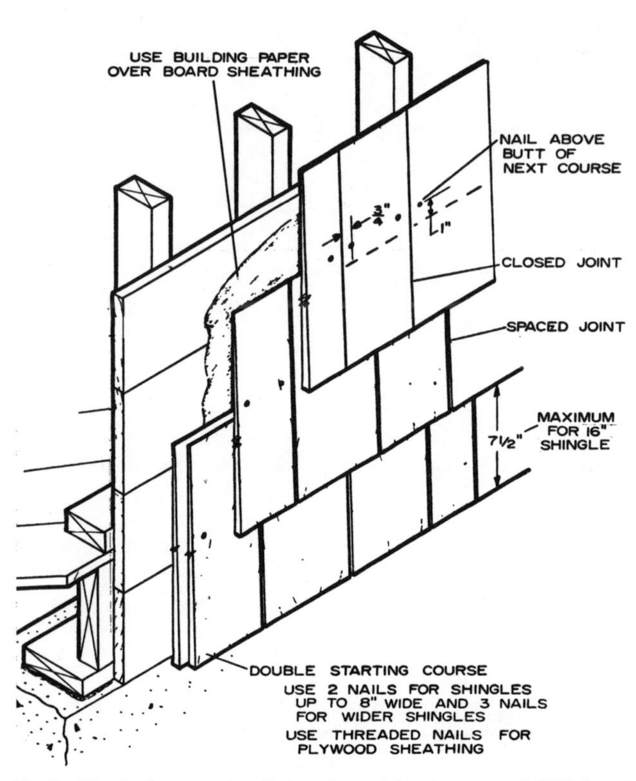

Fig. 7. The single-course installation of wood shingles on a wall *(U.S. Forest Service).*

In single-course construction (see Fig. 7), the shingles are installed using double coverage and following the same general construction techniques used to install roof shingles. That is, the shingles will overlap each other in such a way that a double layer of shingles covers the entire wall surface. Standard wall exposures for single-course shingle walls are as follows:

Shingle Length, in.	*Exposure, in.*
16	7½
18	8½
24	11½

In double-course construction (see Fig. 8), two layers of shingles are installed in each course, which are called the *undercourse* and the *outer course.* A lower grade of shingle (either No. 3 Grade or Undercourse Grade) is used for the undercoursing while No. 1 Grade shingles are used for the outer course. Since each course is doubled, the rows of shingles can be spaced much further apart than is the case with single-course construction. As a result, double-course siding is cheaper to install than single-course siding, despite the fact that more material is required for double-coursing. And the fact that the butt ends of double-course rows stand a bit farther

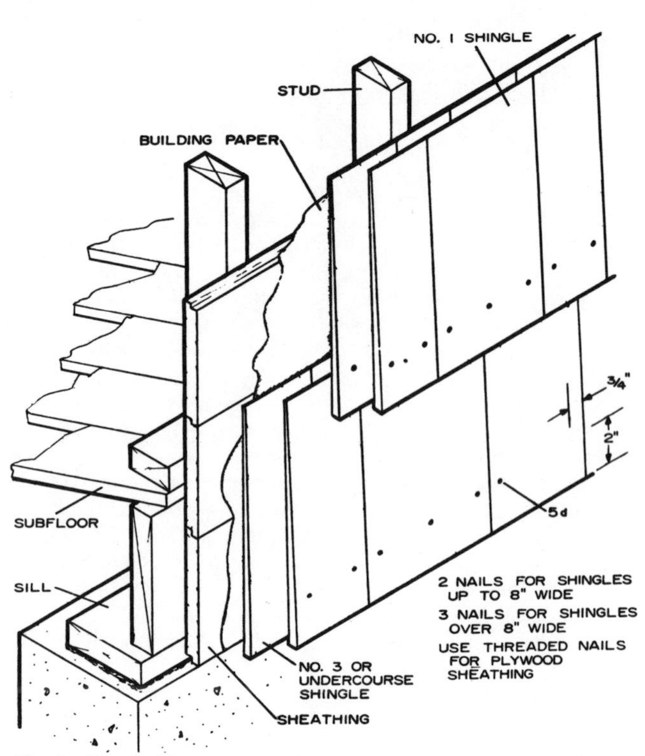

Fig. 8. The double-course installation of wood shingles on a wall *(U.S. Forest Service).*

away from the wall gives the courses a heavier, more emphatic shadow line that many homeowners find attractive.

The standard exposures used with double-course shingle walls are as follows:

Shingle length, in.	*Maximum exposure, in.*
16	12
18	14
24	16

Wall Preparation

If wood-board sheathing has been installed, the boards must be protected against moisture (which would cause the wood to swell), and the entire wall surface must be protected against air infiltration by the installation of building paper over the sheathing (*see* SHEATHING). The building paper is nailed over the sheathing using only the minimum number of nails required to hold the building paper in place until the shingles have been nailed in place.

Building paper has the great advantage that, although it is moisture-resistant, it will still allow water vapor to pass freely through the wall construction. This is important as it prevents the buildup of condensation within the wall construction, with potentially disastrous consequences (*see* CONDENSATION). It is a great mistake to use instead of building paper an impermeable barrier of any kind, such as polyethylene film, to cover the sheathing.

If either plywood, insulation board (i.e., fiberboard), or gypsum wallboard has been installed as the sheathing, the use of building paper is unnecessary.

Apart from the installation of building paper over wood-board sheathing, the only other special wall preparation required is the installation of flashing around all openings (*see* FLASHING).

Before the shingles are installed, their exact exposure must be determined. For appearance's sake, the courses must all have the same exposure, and their butt ends must also be aligned with the tops and bottoms of all windows, with the tops of all doors, and with the bottoms of any wall overhangs.

The exposure, therefore, must be carefully worked out with a tape measure. Sometimes slight adjustments in the exposure of individual courses are necessary to assure alignment with a door or window. A 1-in. overhang at the bottom of the wall must also be allowed for. This overhang will act as a water drip that prevents rainwater from soaking into the construction behind. The shingler usually begins his computations using the maximum permissible exposure and then adjusting this exposure downward as necessary.

Once a suitable exposure has been worked out, the height of each course is transferred to a *story pole,* which is a 1 × 1 or 2 × 2 in. stick on which is marked the exact height of each course. The story pole is as long as the wall is high. The story pole is set against the corners of the wall, the height marks are transferred to the wall, and chalk lines are then snapped along the wall to establish the height of each course.

Single-Course Construction

The first course along the bottom of a wall is doubled or tripled, just as with the first course of roof shingles. The shingles in the first undercourse (or first two undercourses) are spaced 1/8 to 1/4 in. apart. The shingles in the outer course may be spaced apart or butted tightly together, depending on the effect desired.

Corrosion-resistant shingling nails 3d or 4d in size are used. The nails should be long enough that they penetrate completely through the wood-board or plywood sheathing, if either of these two materials has been installed as the sheathing. If the sheathing is made of plywood, the nails should also have threaded shanks. The nails are located 1 in. above the butt line of the next course to be laid and 3/4 in. away from the sides of each shingle. If a shingle is 8 in. or more in width, a third nail should be nailed at the center of the shingle.

The rest of the courses are installed as already described for roof shingles. The joints in any given course of shingles must be located at least 1½ in. away from the joints in the course directly below it.

Double-Course Construction

In any given course of double-course shingles, the butt ends of the undercourse are located about ½ in. above the butt ends of the outer course. The simplest method of achieving this spacing is to use a strip of shiplap siding as a guide. The undercourse is nailed in place first, using a single 3d nail, or a staple, located at the top of the shingle.

The outer course is then nailed in place using two nails for each shingle, the nails being located at the *bottom* of the shingle, about 2 in. above the butt and 3/4 in. away from each side. If the shingle is 8 in. or more in width, then the nails should be spaced about 4 in. apart; as many nails as necessary should be used. Thus, it will be noted that one of the distinctive features of double-course wall shingling is that the nails are exposed. The nails should be small-headed, corrosion-resistant, and 5d in size.

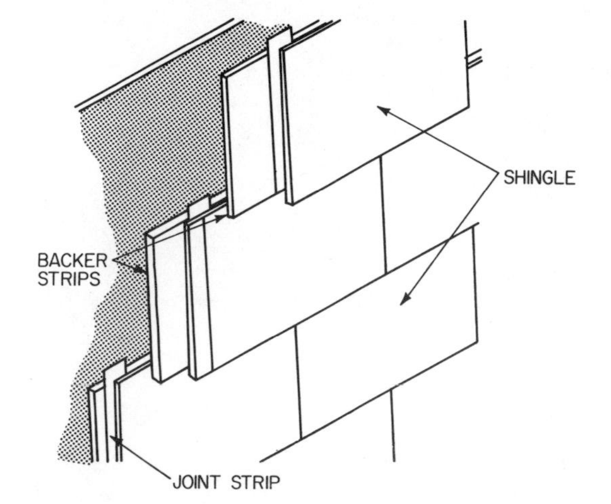

Fig. 9. Backer strips made of insulation board can be used instead of shingles as the undercourse in double-course installation.

The undercoursing on the first course of shingles is doubled to build up the butt line. This first course should be located so that the butt ends of the outer course will extend about 1 in. below the top of the foundation wall to form a water drip. The joints between the shingles in adjacent courses must be located at least 1½ in. away from each other, as already described.

An alternative to the use of undercoursing shingles is *backer strips* (see Fig. 9). These are made of insulation board. They are manufactured in 3/8 and 5/16-in. thicknesses, they are 4 ft long, and they are either 13½ or 15½ in. wide, for use with 12 and 14 in. exposures, respectively. To install the backer strips, one nails them to each stud with a single 8d, corrosion-resistant nail located 3/4 in. from the bottom of the strip. The butt ends of the shingles should extend ½ in. below the bottom edge of the backer strips.

Siding

Siding, by which one usually means *wood* siding, is a method of applying an *exterior wall finish* to a wood-frame house. The purpose of any exterior wall finish is to protect the structural framework from the weather, to prevent dust and wind from entering the house or heat from leaving it, and to give the house an attractive appearance. In the United States, siding has traditionally consisted of lengths of wood applied horizontally (usually) and in an overlapping manner to the studs or wall sheathing (see Fig. 1). Wood siding of this type is, in fact, as characteristically American in its way as a Thanksgiving Day turkey.

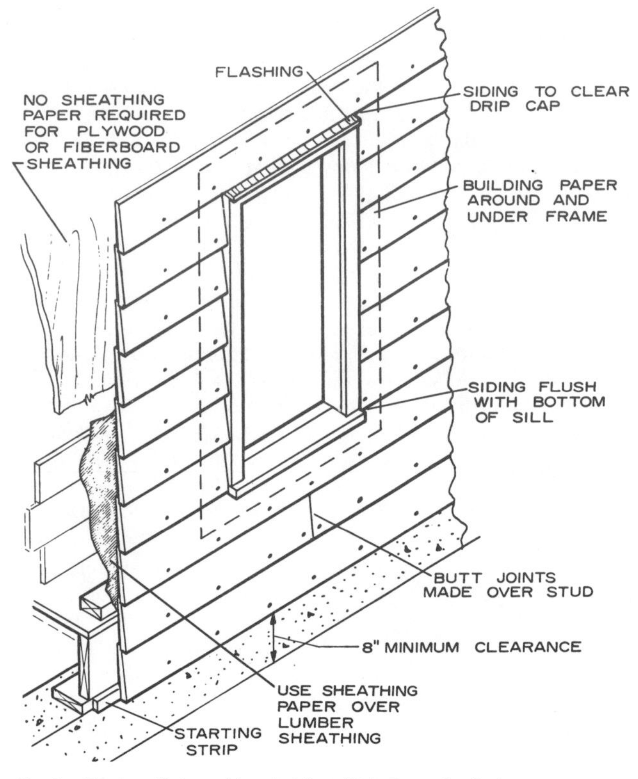

Fig. 1. The installation of bevel siding *(U.S. Forest Service)*.

A great many materials other than wood are now used for siding, including aluminum and steel, plywood, hardboard, asbestos cement, and polyvinyl chloride (or "vinyl"). Most of these materials are designed in imitation of wood siding, and they are installed to resemble a wood-siding installation. These materials do, however, have properties that are often improvements on wood, but, since these properties, good or bad, can only be appreciated by reference to wood siding, wood siding will be described first.

WOOD SIDING

Because wood siding is exposed to the weather, the wood used must have excellent weather-resistant properties. In particular, the wood must be capable of withstanding constant changes in its moisture content without warping, splitting, or decaying. The cedars—eastern white and western red—as well as cypress, redwood, and the white pines are particularly suitable for siding because they all have a straight, close grain that limits the amount of warpage and splitting that is likely to occur and because they also have excellent decay-resistant properties. Such woods as ponderosa pine, southern yellow pine, Douglas fir, and hemlock are also used for siding, but these woods should definitely be considered second-best choices.

If the siding is to be stained or left to weather naturally, then a clear grade of one of the preferred woods mentioned above should be selected; that is, the wood shouldn't have knots, pitch pockets, or other defects (for lumber grading *see* LUMBER). But if the wood is to be painted, then a lower grade of wood can be used as long as its defects are the kind that are easily hidden by a coat of paint, such as small tight knots, for example.

Paint adheres best to edge-grained wood, and, therefore, edge-grained wood should always be selected in preference to flat-grained wood when a siding is to be painted. Indeed, edge-grained wood should always be selected in any case because the wood will warp less, split less, and shrink much less across the width of the siding than flat-grained wood.

To prevent as much as possible the wood's shrinking after it has been installed, which would tend to pull the nails loose or force the wood to split, the wood should be kiln-dried or air-dried to the average moisture content of the air in the locality in which it is to be installed. In most parts of the United States, the average moisture content of the air is usually between 10 and 12 percent; in the Southwestern states, it is usually between 8 and 9 percent. In addition, narrow boards tend to have a smaller total shrinkage than wide boards, and this fact should also be taken into account when selecting a siding material, particularly if there are wide swings in the humidity level in the locality over the course of a year.

Bevel Siding

Bevel siding (Fig. 1) is also known as *bungalow* siding and *Colonial* siding when it is 8 in. or more in width and as *clapboard* when it is narrower. Bevel siding is also known as *lap* siding and as *weatherboarding*. Clapboards as such are no longer manufactured, though the name lingers romantically on. They were made by sawing lengthwise through a log to obtain long tapered boards that were from 6 to 8 in. wide. The boards were then installed in an overlapping manner so that 4 to 5 in. of each board was exposed to the weather.

Today, bevel siding is manufactured by first planing a board smooth and to size and then sawing the board diagonally along its length, which results in two lengths of tapered siding, one side of which is smooth-surfaced while the other side is rough-sawed. If the siding is to be painted, the smooth side is usually left exposed. If the siding is to be stained, or left to weather naturally, then the sawed side is usually left exposed as the stain penetrates more deeply into a roughly sawed surface, and it also results in a more attractive looking finish.

Bevel siding is made in nominal widths of 4, 5, 6, 8, 10, and 12 in., which means the actual widths are 3½, 4½, 5½, 7¼, 9¼, and 11¼ in. The thin edge of the board is usually 3⁄16 in. thick, regardless of the board's overall width, which means the thick edge of the board will vary in thickness according to the overall width of the board. Thus, for example, if a board is nominally 4 in. wide, its thick edge will be about 7⁄16 in., but if the nominal width is 12 in., then the thick edge will be about ⅝ or ¾ in. (These 12-in.-wide boards are sometimes referred to as *Anzac* siding.)

The two distinguishing characteristics of bevel siding are its tapered cross section and its square-cut edges, both of which are a consequence of its method of manufacture. These features allow the siding to be installed by laying one board partially on top of another, the amount of overlap being determined by the requirements of the installation, as described below.

Drop Siding

Drop siding (see Fig. 2), which is also widely known as *novelty* siding and *rustic* siding, differs from bevel siding basically in having parallel sides, though the face side may be planed to a bevel finish, and in having either tongue-and-grooved or shiplapped edges. Drop siding is nominally 1 in. thick (actual thickness being ¾ in.) and from 4 to 12 in. wide, though most drop siding manufactured is nominally 6 and 8 in. wide. The tongue-and-grooved edges are also called *matched* edges because the convex tongue matches and fits into the concave groove of another board; and, by extension, drop siding having matched edges is called *matched siding.* Shiplapped edges are also called *rabbeted* edges, and, by the same kind of analogous thinking, drop siding with rabbeted edges is called *rabbeted siding.*

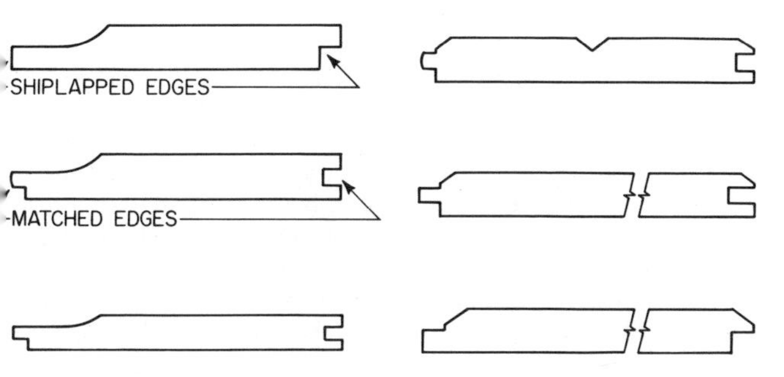

Fig. 2. Typical patterns of drop siding that are widely available.

Drop siding is made from clear grades of lumber when the wood is to be stained or left to weather naturally, but drop siding is more often made in lower grades of lumber because it is usually painted. Drop siding is sometimes installed on low-cost housing as a combination sheathing-and-exterior-finish, separate sheathing being omitted entirely in the wall construction. Drop siding can be used as a substitute sheathing in this way because its one flat side allows it to be nailed tight against the studs, and its matched or shiplapped edges make the construction reasonably windtight and watertight.

However, the installation of drop siding as a combination sheathing-and-exterior finish is strongly to be deprecated in any house. But if, to cut costs, drop siding is used for this dual purpose, the wall construction should first be braced, but, even so, this construction should be used only for garages, barns, summer cottages, and similar lightly constructed structures. (For a description of the bracing *see* SHEATHING; WALL FRAMING.)

Dolly Varden Siding

Mention should be made of *Dolly Varden* siding, which is a mixture of bevel and drop siding. It is drop siding because it has shiplapped edges and a flat back; it is bevel siding because the exposed face is beveled. In appearance it resembles true bevel siding and it is often used in place of true bevel siding. It is installed, however, like any other drop siding.

WOOD-SIDING INSTALLATION

Siding can be installed either horizontally or vertically—or at a diagonal if a homeowner or architect thinks it makes an interesting effect—but siding is usually installed horizontally, and this is the type of installation we will describe first. It is assumed that sheathing has been installed over the studs (*see* SHEATHING), and that building paper has been nailed over the sheathing (if necessary) to help prevent the infiltration of air and moisture into the house.

It is also assumed that the door and window frames have been installed. (*See* DOOR, WOOD; WINDOW.)

Installation of Bevel Siding

Before bevel siding is installed, it is necessary to make a detailed layout of the exterior walls in order to determine what the exposure of the siding will be. The intent is to install the bottom edge of one course of siding level with the bottoms of the window sills, and to install the bottom edge of another course of siding level with the tops of the window dripcaps, all the courses of siding in between having the same exposure. This is necessary if the overall installation is to have a neat, tidy look (see Fig. 1). It would also be nice if the same exposure could be achieved for the courses of siding that will be installed between the foundation and the bottom of the window sills, and between the tops of the window dripcaps and the under-side of the roof.

Using his rule, the carpenter marks the relevant wall measurements on a sheet of paper and then calculates the possible siding exposures to see if he can fit the siding in such a way that the courses alongside the windows and those at the top and bottom of the wall will have the same exposure. He also has to take into account the fact that all bevel siding must overlap by a certain minimum amount if the construction is to be watertight. For example, the minimum overlap for nominally 4-in.-wide siding is 1 in., and for nominally 12-in.-wide siding, it is 1½ or 2 in., the other widths of siding having minimum overlaps that range between these two extremes.

In practice, it is usually found that the exposure between the tops and bottoms of the windows can be accurately adjusted but that the exposures for the siding installed above and below the windows will differ from this exposure by ¼ to ½ in. per course—differences that will not, however, be apparent to a casual observer.

When bevel is being installed on a wall, another strip of wood about 3⁄16 in. thick called a *cant strip* is sometimes nailed along the bottom of the sheathing, underneath the first course of siding, to give the first course the required amount of flare. This bottom course of siding should extend at least 1 in. below the top of the foundation to keep moisture out of this juncture (see Fig. 1). The bottom of the siding is then beveled to form a water drip.

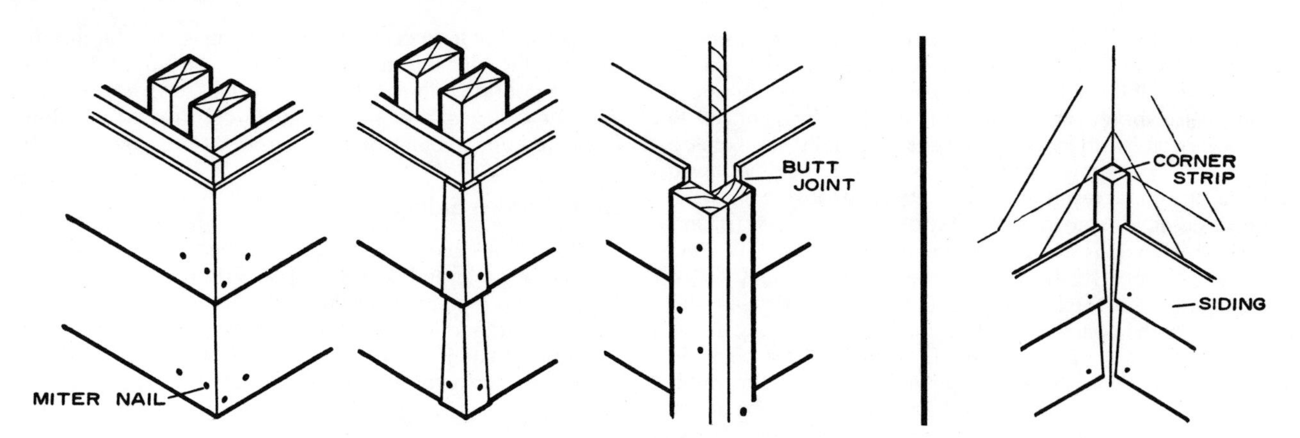

Fig. 3. (Left) Three methods of joining the ends of siding at the external corner of a house, and (right) the way that an interior corner is made *(U.S. Forest Service)*.

Making Joints in Siding

The first course of siding having been installed, the carpenters (at least two are required) mark off the height of the second course and then snap a chalk line against the sheathing to guide them in positioning the second course of siding accurately. The siding is then nailed in place; and so on to the top of the wall.

For the sake of appearance, as few joints as possible should be made in the siding as it is installed. Before the installation begins, therefore, the siding should be sorted out by lengths and the longest pieces saved for the long stretches of wall that lie below and above the windows. The shorter pieces are then used to fill in the spaces between the doors, windows, and corners.

When a joint does have to be made in a course, the preferred method of making it is to cut the two pieces of siding at an angle to the face of the siding. When the cut is carefully made, the joint will be barely visible. A joint of this kind should always be located over a stud, so that both ends of the joint can be securely nailed in place. If joints are required in two adjacent courses, they should not occur over the same stud but be staggered.

Where a strip of siding abuts a door or window frame or a corner board (see below), the juncture should be made as neatly and as tightly as possible. A gap might allow moisture to enter the wall, besides being unsightly and a sure sign of poor workmanship. The overall length of the siding must be measured carefully and the wood then cut squarely so that it is slightly on the long side. The cut edge is then trimmed neatly and accurately with a plane until the siding is the desired length. The juncture is finally sealed tight by caulking compound applied by the painter before the siding is painted.

Corner Construction

There are three methods by which joints can be made at the external (outside) corners of a dwelling. *Metal corner pieces* can be installed, *corner boards* can be installed, or the abutting edges of the siding can be *mitered* together. All three methods are shown in Fig. 3.

Mitering two abutting lengths of siding together makes the neatest, cleanest, and most attractive-looking job, if the joints are well made and if they do not open up subsequently because of the effects of the weather on the wood. The mating edges should fit closely together along the entire width of the siding, and each strip of siding should be nailed securely into the studs or sheathing behind. As an additional precautionary measure, if building paper hasn't been installed over the sheathing (and sometimes it is not required, especially if the sheathing consists of plywood panels), a long strip of building paper should be folded against the corner to keep any moisture that may make its way past a gap in the siding from entering the house.

Unfortunately, mitered corners tend to open up, even when they are well made, because of the stresses imposed on the wood by changes in the humidity. This is particularly true in the South, where the summers are hot and humid. And even when siding with mitered corners is installed in a temperate, dry climate, the corners take time to make and they are, therefore, expensive. Their use, therefore, is usually limited to more expensively made dwellings.

For a more ordinary class of construction, especially in the South, prefabricated corners made of aluminum or galvanized steel are used to cover the ends of two pieces of siding where they meet at an external corner. These corner pieces give a neat, finished look to the installation. The ends of the siding need not be finished off neatly at all; they need only overlap roughly and without the edge of one board projecting beyond the side of the other board. The metal corner pieces are then nailed over the juncture.

Corner boards are lengths of wood, edge-butted together, that are nailed to the external corners of a dwelling. The ends of the siding then butt up against them. Corner boards are usually made from one 3 × 1⅛ in. board and one 4 × 1⅛ in. board nailed together to form a corner having equal widths (see Fig. 3). Narrower boards than this can be used to make up the corner—they give a somewhat more elegant look to the house—as long as the boards are not so narrow that they cannot be nailed securely to the wall studs behind. Once the corner boards have been installed, the siding is butted against them.

Lengths of siding that meet at an internal corner can also be mitered together, as described above for external corners, but corner boards are used more often, except that on an internal corner a corner board actually consists of a square stick of lumber—1¼ or 1⅜ in square—that is simply nailed into the corner (see Fig. 3). The siding then butts against this strip of wood.

Nails and Nailing

Each strip of bevel siding is attached to the wall by a single nail driven into each stud behind the siding as shown in Fig. 4. The nail is located at the bottom of the siding, which allows the wood to expand and contract freely, up and down, with changes in its moisture content. Care must be taken that the nails are not placed so close to the bottom edge of the siding

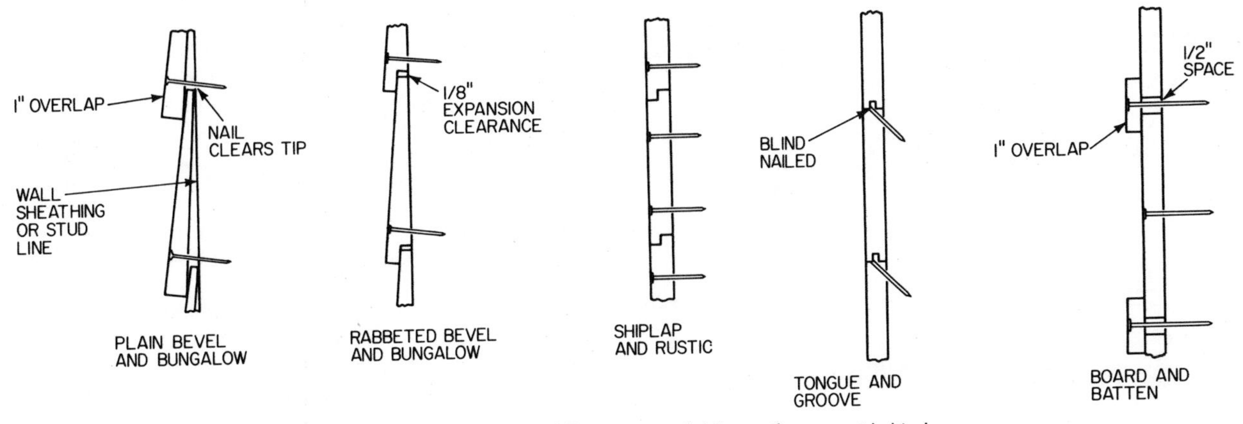

Fig. 4. Recommended methods of nailing different types of siding to the support behind.

that the top edge of the siding underneath it is also nailed to the stud.

Installation of Drop Siding

As we have mentioned, drop siding is quite often installed on a wall without a layer of sheathing being installed first. In effect, the siding acts as both the exterior finish and as sheathing. This is very poor practice and is usually done on low-cost houses on which the costs are being watched very closely—too closely, in fact. If combined siding and sheathing has been installed in a purportedly more expensive dwelling, it is a sign that the house is not worth the price being asked for it.

The problem with this combination siding and sheathing is that the siding is incapable of bracing the wall construction against the racking loads imposed by a high wind; which is the primary structural function of sheathing. If the sheathing has been omitted, therefore, the corners of the walls must be strengthed by installing diagonal bracing (for a description *see* SHEATHING). Failure to install either sheathing or diagonal bracing will very likely result in cracked plastering or wallboards, walls that are out of plumb, doors and windows that jam, and leaks everywhere.

Sheathing has another purpose; to help keep a house warm in the winter. In a warm climate sheathing may indeed be unnecessary for this purpose, and a builder might very well think he can get along without sheathing. As long as the walls are adequately braced, then sheathing may indeed be omitted. But these considerations do not hold in a climate where the winters are cold, in which case sheathing should always be installed, since the cost of installation will in the long run more than pay for itself in reduced fuel costs.

As for the installation of the drop siding itself, it differs from the installation of bevel siding mainly in that the exposure of each strip of siding is fixed by its matched or shiplapped edges. The siding will have an exposure that cannot be altered. For nominal 6 in. drop siding, this exposure will be 5¼ in. For nominal 8 in. drop siding, it will be 7¼ in.

In the cheapest type of drop-siding installation, the siding is completely installed before any of the door or window frames have been set in place. The siding is rough-trimmed to size at the door and window openings, and the frames are then set in place over the siding. Since drop siding usually has a pattern of some kind planed into one side, the result will be gaps between the siding and the door and window frames through which rain and wind can enter the house, unless the junctures are thoroughly caulked. Furthermore, when this method of construction is followed, it is impossible to install flashing at the tops of the doors and windows, with the further result that rain washing down the wall will make its way under the frames and into the house, unless these junctures are thoroughly caulked also.

In a somewhat better class of construction, the siding is installed as described above, but the siding abuts strips of wood that have been nailed all around the door and window openings. These weather strips are as thick as the siding, and, together with the door and window frames, against which they press, they form seals that will keep rain and wind out of the house.

In the best class of construction, the door and window frames are installed first, with proper flashing being installed around the frames (*see* FLASHING). The drop siding is then installed, which means the siding will have to be accurately cut and fitted around the frames. If the drop siding is to look well, this work must be done with the same care as when bevel siding is installed.

Installation of Vertical Siding

Installing siding vertically originated as a type of exterior finish for barns, sheds, and similar utilitarian structures. As adapted for dwellings, either drop siding or ordinary wood boards can be used.

If sheathing has not been installed, the walls must be braced to stiffen the construction. (*see* SHEATHING; WALL FRAMING.) Lengths of 2 × 4 in. blocking are then nailed horizontally between each pair of studs, the blocking being spaced about 24 in. apart. The siding will be attached to this blocking. Before the siding is installed, however, building paper must be applied over the entire wall to keep wind and dust out of the house.

When either 1-in.-thick wood boards or ½-in. thick plywood panels have been installed as sheathing, the installation of vertical siding is much simplified as the siding can be nailed directly to the sheathing; the 2 × 4 in. blocking is unnecessary. But when the sheathing consists of either insulation boards or gypsum wallboards, then the 2 × 4 in. blocking will still be required as these sheathing materials have no nail-holding ability.

Boards and Battens

Boards and battens (see Fig. 5) are a popular exterior wall finish. The boards usually consist of nominally 1-in.-thick 4-, 5-, or 6-in.-wide lumber that is nailed either to the sheathing or to 2 × 4 in. blocking. The battens consist of 1 × 2 in. or 1 × 3 in. wood strips nailed over the joints between the boards, thus covering the joints and ensuring a watertight construction. Whatever the widths of the boards (almost any width can be used, this being a matter of taste), they should be long enough

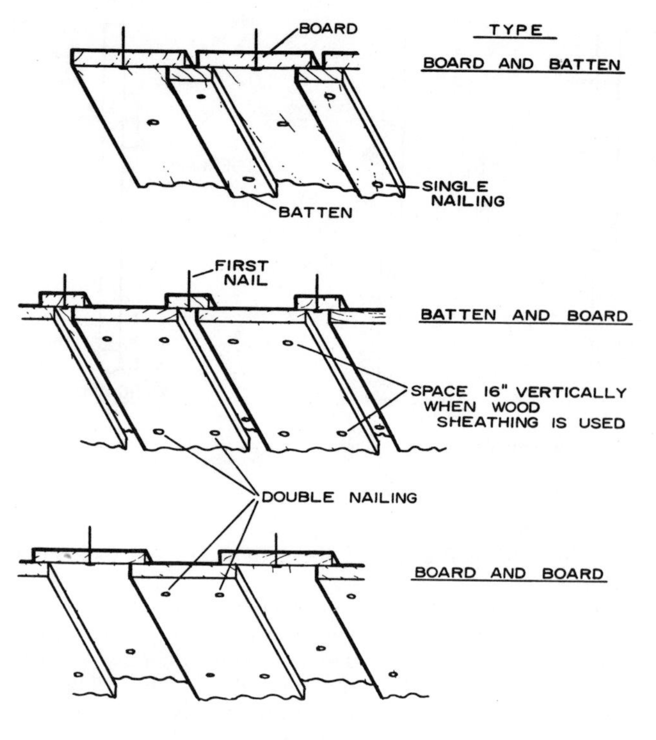

Fig. 5. Different types of vertical sidings *(U.S. Forest Service)*.

that they run the height of the wall in one piece. Redwood, red cedar, and cypress are the woods used most often for board-and-batten construction as they are both attractive and exceptionally decay-resistant. The boards can be finished smoothly or rough-sawed, and either painted, stained, or left to weather naturally.

In the best class of board-and-batten construction, the abutting edges of the boards are covered with caulking compound before they are butted tightly together and nailed to the wall. The backs of the battens are then coated with the same caulking compound before they, too, are nailed in place.

In a variation of this technique, with boards having matched edges, a ribbon of the caulking compound is run into the grooved edge of each board before it is installed. This technique results in a very tight and strong construction.

In the more usual construction, however, regardless of whether square-edged or matched boards are being used, the caulking compound is omitted entirely between the boards and, to obtain a watertight seal, the carpenter depends instead on caulking compound spread on the backs of the battens. A gap of ⅛ to ¼ in. is left between the boards (if they have square edges, that is) when the battens are coated with caulking compound in this way.

PLYWOOD SIDING

Plywood panels especially manufactured for use as siding are very popular for the good and sufficient reasons that they are easy and inexpensive to install and, once installed, provide a strong, stiff, and watertight construction. A wide range of attractive finishes that suit present-day tastes is available, perhaps more so than the traditional bevel- and drop-siding styles.

Plywood panels intended as siding are manufactured in 4 ft widths and 8-, 9-, and 10-ft lengths, and in thicknesses that range from 5⁄16 to ¾ in. The panels are Exterior grade, which means that a completely waterproof adhesive has been used in their construction. The panels, therefore, are completely unaffected by weather. A wide variety of plywood grades, types, and groups are available (see Fig. 6). (For a description of them, *see also* PLYWOOD.)

As far as this article is concerned, plywood panels that are manufactured specifically for use as decorative exterior siding are categorized as 303 Siding. They are manufactured in a large variety of surface finishes and textures, some of which are shown in Fig. 7.

Most 303 Sidings can be stained or left to weather naturally. When panels are to be painted, they are covered with a thin sheet of plastic-saturated paper called Medium Density Overlay (MDO). The result is a surface that is completely impermeable to water and that has a *tooth* that makes an excellent base for paints. (For a description of MDO plywoods *see* PLYWOOD.)

Plywood siding is usually, though not necessarily, installed with the long dimension of the panel running vertically, the long edges of the panels being located over studs while the short edges are located over the sill plate and top plates of the wall construction.

Plywood Bevel and Lap Siding

For those who prefer a house of more traditional appearance, bevel and lap sidings made of plywood are available (see Fig. 6). The bevel siding has a tapered cross section, and the lap siding has parallel sides. Both types of siding are manufactured in widths of 12, 16, and 24 in., both are available in lengths up to 16 ft, and both are available in thicknesses that range from ⅜ in. to ¾ in. (For the bevel sidings, these thicknesses are the thicknesses through the butt edge of the sidings.) As these plywood sidings are usually intended to be painted, they are surfaced with MDO, although sidings are available with 303 surfaces. Both the bevel and lap sidings are installed in much the same way as already described for bevel siding. The overlap between any two courses can be as little as ¾ in. or as much as necessary to achieve an even exposure over the entire height of a wall.

METAL SIDING

Siding made of sheet aluminum and sheet steel is manufactured that simulates the appearance of wood bevel siding and vertical wood board-and-batten siding. Metal has the advantages over wood in being longer-lasting, lighter in weight, and impervious to attack by molds or termites. However, in the thicknesses in which it is used, metal siding (especially aluminum) is not very strong and it is easily dented if mishandled. A hailstorm can sometimes ruin the appearance of metal siding entirely. Most manufacturers guarantee their products for 20 to 30 years.

Another disadvantage of metal siding is that it may build up a large charge of static electricity if it is not adequately grounded. This, in turn, makes the building vulnerable to lightning. The builder must make sure, therefore, that the siding is grounded, which is simple enough to do, requiring only a cable connection between the siding and a water-supply pipe.

Metal siding may also absorb a terrific amount of heat during a hot summer day, heat which is then radiated into the building, just as happens with parked automobiles during a hot summer day. This is why metal siding is usually painted white—to reflect the sun's heat as much as possible. In addition, siding manufacturers, aware of the efficiency with which their products can radiate heat into a house, usually sell them

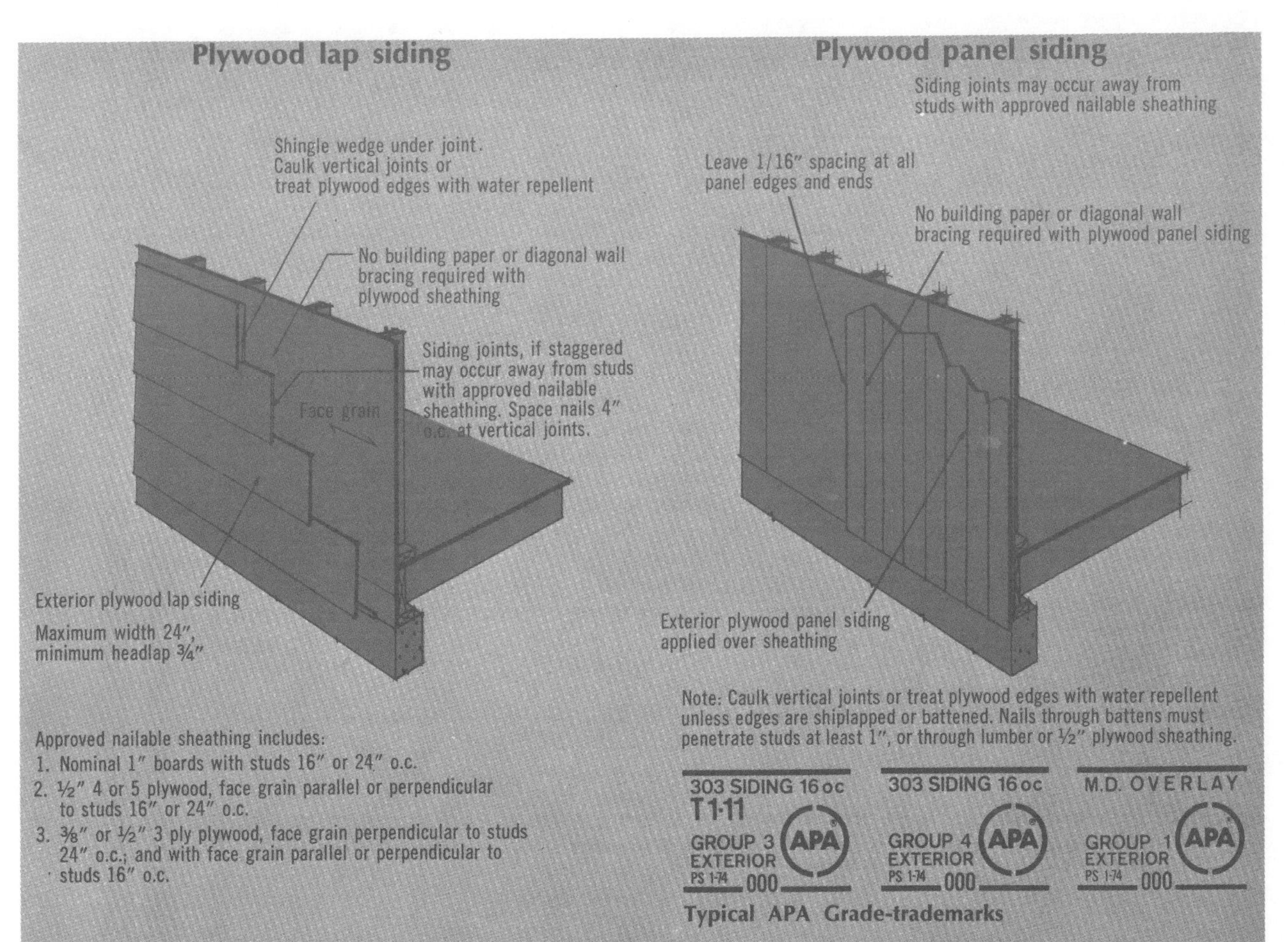

Exterior plywood siding over sheathing

(Recommendations apply to all species groups)

Plywood Siding			Max. Stud Spacing (in.)		Nail Size (Use nonstaining box, siding or casing nails)	Nail Spacing (in.)	
Type	Description	Nominal Thickness (in.)	Face Grain Vertical	Face Grain Horizontal		Panel Edges	Intermediate
Panel Siding	MDO EXT-APA	5/16	16 (d)	24	6d for panels ½" thick or less; 8d for thicker panels. (a)	6	12
		⅜	16 (d)	24			
		½ & thicker	24	24			
	303-16 o.c. Siding EXT-APA	5/16 & thicker	16 (d)	24			
	303-24 o.c. Siding EXT-APA	7/16 & thicker	24	24			
Lap Siding	MDO EXT-APA	5/16	–	16(b)	6d for siding ⅜" thick or less; 8d for thicker siding. (c)	4" @ vertical butt joints; one nail per stud along bottom edge.	8" @ each stud, if siding wider than 12".
		⅜	–	16(b)			
		½ & thicker	–	24			
	303-16 o.c. Siding EXT-APA	5/16 or ⅜	–	16(b)			
	303-16 o.c. Siding EXT-APA 303-24 o.c. Siding EXT-APA	7/16 & thicker	–	24			

(a) Use next regular nail size when sheathing (other than plywood or lumber) is thicker than 1/2".
(b) May be 24" with plywood or lumber sheathing.
(c) Use next larger nail size when sheathing is other than plywood or lumber, and nail only into framing.
(d) May be 24" with approved nailable sheathing, if panel is also nailed 12" o.c. between studs.

Fig. 6. The installation of plywood panel siding and lap siding over sheathed walls *(American Plywood Association)*.

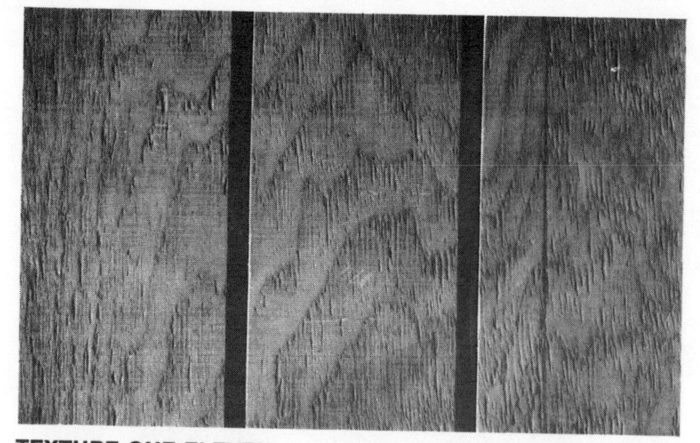

TEXTURE ONE-ELEVEN

CHANNEL GROOVE

REVERSE BOARD & BATTEN

ROUGH SAWN

KERFED

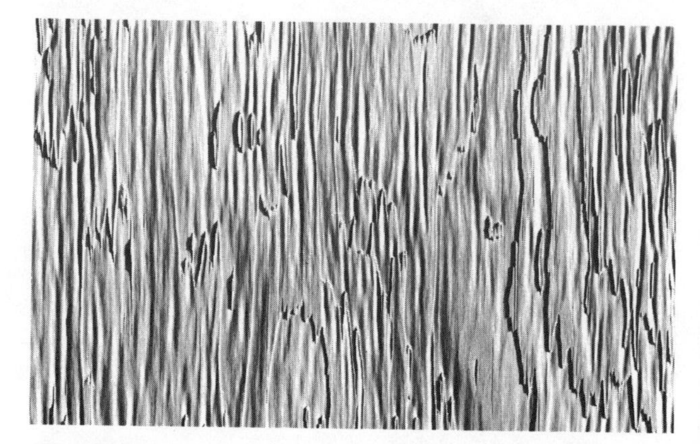

BRUSHED

FINE LINE

STRIATED

Fig. 7. Some of the surface textures available in plywood exterior siding *(American Plywood Association)*.

with an insulated backing, the insulation consisting of a ⅜-in.-thick layer of insulation board or polystyrene. Whether or not this insulation is adequate for a particular dwelling can only be determined by making a heat-gain calculation for the house.

During the winter, the opposite phenomenon occurs; that is, the metal siding tends to radiate the heat within the house rapidly to the outdoors. Again, this transfer of heat can be forestalled somewhat by insulating the siding. Thus, in addition to the summertime heat-gain calculations, wintertime heat-loss calculations must be made.

Simulated wood bevel siding made of both steel and aluminum is manufactured in 12½ ft lengths and in widths that provide 4-, 5-, or 8-in. exposures. The lengths of siding are attached to the sheathing by nails driven through elongated holes punched in the top edges of the siding. The long edges of the siding are also formed in such a way that adjacent strips will overlap each other and be locked together, as shown in Fig. 8.

The lengths of siding are not nailed tightly to the wall for a good reason: metal expands and contracts considerably with changes in its temperature. The nails must be carefully located at the center of the slots and not be driven home. Instead, the nail heads should project a short distance away from the wall so that in effect the siding is hanging loosely from the nails. The siding is thus free to expand and contract.

The ends of the siding butt against metal trim that is nailed around the door and window frames and at all the corners. A gap of 1/16 in. must always be left between this trim and the lengths of siding to allow the siding to expand in hot weather. If a joint is necessary in a course of siding, the joint is made by overlapping the two pieces of siding by about ½ in.

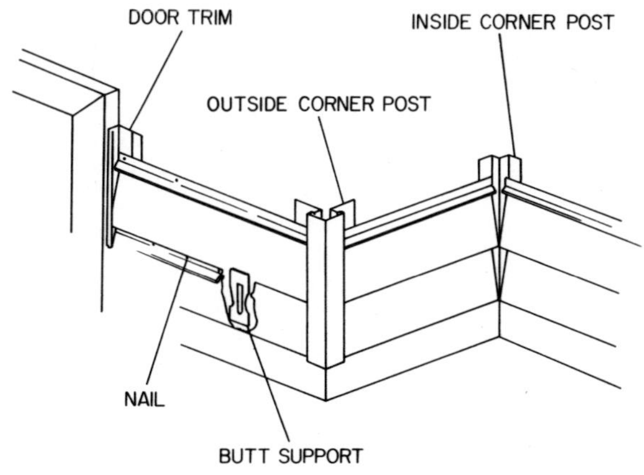

Fig. 8. The installation of metal siding.

POLYVINYL CHLORIDE SIDING

Polyvinyl chloride does not rust, corrode, warp, dent, rot, peel, crack, burn, or blister. It is unaffected by acids, alkalies, or organic solvents of any kind. It can be manufactured with a color built into it so that it never requires painting. Dirt washes off with a spray of water. Its only fault is that it tends to turn brittle in very cold weather, and it may then crack if struck a sharp blow. Otherwise, polyvinyl chloride is an apparently indestructible material—the ideal siding—although manufacturers will guarantee their sidings for only 20 years. (Why is this?)

The virtues of polyvinyl chloride as a siding material are so self-evident that the manufacturers of competing materials have hastened to provide their products with a film of polyvinyl chloride so they, too, may claim for their products the same claims made by the manufacturers of "solid" polyvinyl chloride siding. ("Solid" polyvinyl chloride siding is all of 0.045 in. thick; the films applied by the manufacturers of competing products to their siding are from 0.0015 to 0.0025 in. thick, which is certainly thick enough, provided the film remains unbroken.)

As a siding material, polyvinyl chloride does have one unfortunate characteristic: this is a tendency to expand and contract with changes in its temperature, just as metal sidings do, and this characteristic has determined the way in which the siding has been designed and the way it is installed. Polyvinyl chloride siding is, in fact, manufactured according to the same principles as metal sidings are, and it is installed in much the same way. That is, the top edge of the siding has elongated holes punched into it, and the siding is hung from a wall by nails driven through the midpoints of these holes.

Special molded polyvinyl chloride trim is used around the door and window frames and at the corners of a building. The lengths of siding, however, should neither butt against this trim nor be fastened to it. Instead, the trim is so designed that the ends of the siding slide under and are hidden by the trim, which enables the siding to expand and contract without hindrance. (The siding will expand about ¼ in. per 4 ft of length in hot weather.) If a joint must be made in a course of siding, the joint is made by overlapping the two strips by 1 to 1½ in.

Unlike metals, polyvinyl chloride does not transmit heat. On the other hand, it does not act as an insulator either. The manufacturers of polyvinyl chloride siding provide ⅛-in.-thick strips of insulation that can be applied to the back of the siding to give the siding a bit of insulating value. The effectiveness of this insulation can only be determined in any given dwelling by making the appropriate heat-gain and/or heat-loss calculations.

Slab-on-Grade

A slab-on-grade (or *slab-on-ground*) floor is just that—a thin (4-in. minimum thickness) slab of concrete that is poured either directly upon the surface of the earth or upon a suitably prepared gravel or crushed-rock base that has been spread upon the surface of the earth. The concrete is leveled, troweled smooth, and, after it has set, it becomes the ground floor of a dwelling.

The advantages of a properly constructed concrete slab are many. It is a much simpler and less expensive method of building the floor of a dwelling compared with the traditional wood joist floor (*see* FLOOR FRAMING), and it thus lends itself to mass-production methods of house construction. It is much used by large developers for these reasons, although its use is not limited to developers by any means. A slab-on-grade floor eliminates the expense of having to excavate for a partial or full basement or, in a basementless house, the expense of constructing a foundation to support a traditional wood joist floor above the ground. Once the slab has hardened, it becomes a very strong, rigid structure that is impervious to termites, fungi, or fire, and it also provides a flat, solid base upon which carpeting, resilient flooring of any kind, or a traditional wood floor can be laid.

The first concrete-slab floors were constructed rather naively and without much thought being given to the potential defects of this type of construction. Thus, very often the soil upon which the slab was to be poured was simply leveled, and the concrete was then poured directly upon the soil. After the house had been completed, if there was moisture in the soil, this moisture migrated through the concrete and into the house. As a result, the floor felt cold and the rooms felt continually damp. In addition, if the finish flooring consisted of an impermeable material such as asphalt or vinyl tiles laid in mastic, the moisture that had made its way up through the slab would be trapped between the mastic and the concrete, with the result that the mastic would lose its grip on the concrete and the floor tiles would come loose.

There is nothing inherently wrong in pouring a concrete slab directly upon the ground—*if* the soil is stable, well-drained, and reasonably dry. If it isn't, then troubles such as those described above are likely to occur. In addition, if the soil is a clayey or silty type that expands when wet, not only is it likely that moisture will migrate through the slab and into the house, but it is also likely that the expansion of the soil that occurs as it absorbs water will crack the slab. (*See also* FOUNDATION.)

Sometimes a builder gives little or no thought to proper soil drainage. He reasons that, after all, concrete is impermeable to water, isn't it, and how much pressure can low-lying water exert against a solid concrete slab, anyway? Although the hydrostatic pressure exerted by water under a slab may be trivial, it is there and it is persistent, and the water will eventually find its way into the house through any cracks or porosity in the slab; or else the slab will crack because the water will have washed away the soil under the slab. A large, 4-in.-thick slab of unreinforced concrete has very little tensile strength. Even if the slab remains intact, water will still soak into the concrete and cause the problems described above.

Another problem that may occur in northern climates is an exceptionally cold floor during the winter, especially along the perimeter of the house. What happens is that, as the ground temperature drops during the fall and winter, heat is drawn from the slab into the soil, with the result that the perimeter of the floor becomes much colder than the center of the floor.

Still another problem that occurs with slabs that have been poured directly upon the soil was at first thought to be another consequence of moisture migrating through the slab: This is the existence of a wet floor during hot, humid weather. During hot weather, the slab tends to be cooler than the air. As a result, if there is an excess amount of water vapor in the air it will condense on the slab. This is particularly likely to happen when the slab is covered by such impermeable flooring materials as linoleum or vinyl or asphalt tiles, which will prevent the condensate from migrating into and through the slab to the soil below. (For a detailed discussion, *see* CONDENSATION.)

All the problems we have described have given concrete slabs-on-grade a bad name. There are any number of people who automatically reject a house with a slab floor; they would not be caught dead in one. Nevertheless, there is nothing inherently wrong with a slab-on-grade floor—as long as the builder takes, or has taken, the trouble to site the house carefully and build the slab properly.

The basic elements of a well-constructed slab-on-grade floor are (1) a well-drained site with the surface of the soil graded away from the house, (2) the installation of a base course of gravel or crushed rock that will prevent the transmission of moisture into the slab via capillary attraction, (3) the installation of a drainage system under the slab to carry away any excess water, (4) the installation of a vapor barrier under the slab to prevent the migration of water vapor into the slab from the soil below, and (5) the insulation of the slab to prevent the transfer of heat out of the house during cold weather. The slab itself must be well made and reinforced, if necessary, so that it will be able to support the loads imposed on it without cracking.

Not all these items are required in every slab (see Table 1). It depends on the local climate and on the soil conditions on the site, as described in detail below.

Soils and Soil Drainage

Whether or not a concrete slab can, or should, be built on a particular site in the first place will depend on the nature of the soil on the site. The best possible soils on which to build a slab are gravelly soils that have a coarse, well-graded texture and are well-drained and packed together solidly. A coarse sandy soil is equally satisfactory if it is packed down solidly and is well drained, and if there is not too great a proportion of silt or clay mixed in with it. On soils such as these a slab can be poured directly.

There is a second group of soils that are not as good.

Table 1. How Soil and Moisture Conditions Influence the Design of a Slab-on-Grade Floor

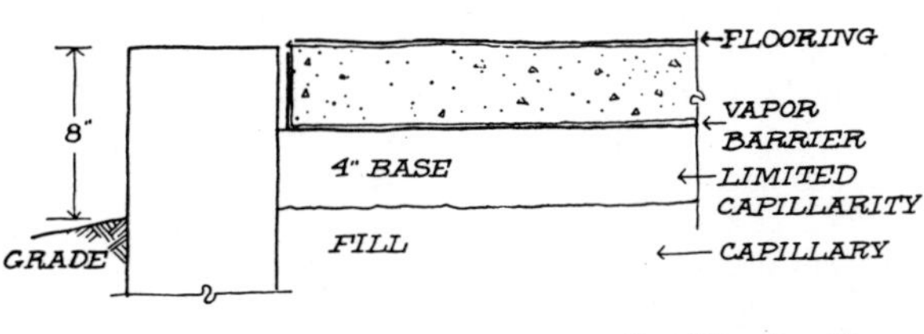

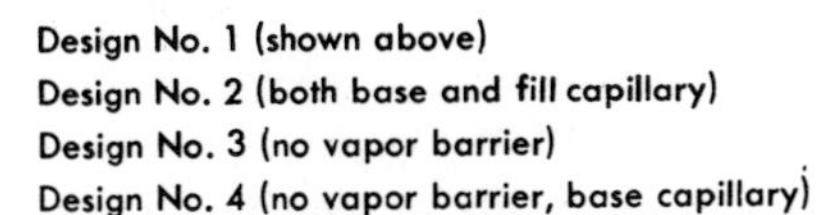

Construction Item	Design No. 1	Design No. 2	Design No. 3	Design No. 4
Vapor Barrier	Provide	Provide	No vapor barrier. Separator as noted [1]	No vapor barrier. Separator as noted [1]
Base of Limited Capillarity Material	Provide [2]	None	Provide [2]	None
Fill or Base	Capillary	Capillary	Capillary	Capillary
Flooring Material	Group A or B	Group A Group B only as noted [3]	Group A Group B not allowed	Group A only as noted [4] Group B not allowed

[1] Provide as listed under the subhead, "Separator"
[2] A duct or plenum system should have a 4 in. base material of limited capillarity under the entire system.
[3] To determine if Group B flooring may be used with Design No. 2:
(a) determine type of soil.
(b) from table for capillary rise of water in various soils, determine figure which applies to this soil.
(c) if the water table for the site is at a distance below the ground surface greater than this figure, Group B floorings may be used.
[4] To determine if Group A flooring may be used with Design No. 4:
(a) and (b) same as in Note 3.
(c) if the water-table for the site is at a distance below the ground surface greater than this figure, Group A flooring may be used.

Source: J. H. Callender, ed., *Time-Saver Standards*, 5th ed., McGraw-Hill, 1974.

Capillarity figures

Capillary water does not rise above the water table more than the following height in these soils

Gravel	0.0 Ft.
Coarse Sand	2.6 Ft.
Fine Sand	7.5 Ft.
Silt	11.5 Ft.
Clay	11.5 Ft.

Flooring types

Group A: Asphalt tile, rubber tile, vinyl-asbestos tile, flexible vinyl tile (unbacked).

Group B: Cork tile, linoleum, felt or fabric backed flooring compositions, wood block.

Gravelly soils are missing from this group entirely, the sands are finer grained, and they are mixed with larger quantities of silt or clay. The finer the particles of sand and the greater the admixture of silt or clay, the poorer a base the soil will be for a concrete slab. A slab may still be poured on such a soil, if the soil is well-drained and reasonably dry. However, the bearing strength of these soils is not as good as the soils in the preferred group, and, therefore, it will be necessary either to reinforce the slab with welded-wire fabric to increase its strength or construct piers under the slab to give it greater support.

There is a third group of soils that are plainly incapable of supporting the weight of a concrete slab plus the weight of the house above. These soils consist of fine silts and clays, organic soils such as peats, and very fine, loose sandy soils that will compress under the weight of a slab and house. The volumes of all these soils are very much affected by changes in their water content. When they are soaked with water, they expand; when the water drains away, they contract again. Clayey soils in particular are highly plastic and will shrink and expand greatly with changes in their water content.

The only ways in which a concrete slab can be built on one of these problem soils are (1) build the slab so heavily reinforced with steel bars that it becomes a self-supporting "raft" that is capable of "floating" on top of the soil, or (2) build a self-supporting slab that rests entirely on piers, pilings, or grade beams. In both of these cases, the slab is still poured directly upon the soil, but the soil is actually a kind of temporary form that remains in place after the concrete has hardened. It is not expected to support the slab at all, and it doesn't. As both of these methods of constructing a slab are quite expensive, it makes the construction of a concrete slab-on-grade floor rather pointless.

This rather simple picture of adequate soils versus inadequate soils is greatly complicated by how well any particular site is drained of its excess water—both underground water and any surface water that may soak into the soil. Good drainage is essential in even the most gravelly and porous of soils; otherwise it may still be impossible to construct a slab upon such a soil. As for those soils that are on the borderline of acceptability, the adequacy of their drainage will make all the difference in the world as to whether or not they will make a suitable foundation for a slab.

Surface Drainage

A slab should not be built on any site where the surface water tends naturally to collect or run through, which means mainly on the side of a hill or in a valley or hollow. Surface drainage is achieved most simply by constructing the house on a site where the surface of the earth slopes away from the slab in every direction. A slope of 1 in 25 is considered minimal, and this slope should be maintained for a distance of at least 25 ft from the site. If this type of natural drainage is impossible because the site is located on the side of a hill or too close to adjoining property, then some kind of drainage system will have to be built that will divert the surface water around the site.

Underground Drainage

There are two basic conditions that can exist in regard to underground water—either the water table is so high that a definite hydrostatic pressure will be exerted against the bottom of a slab or the water table, while high, will be too low to exert such a direct pressure. In the latter case, the problem will be one of moisture rising through the soil and into the slab via capillary attraction.

When the water table is so high that a definite hydrostatic pressure exists on the underside of the slab, then it simply isn't possible to build a slab-on-grade floor on the site (and you probably couldn't get a bank to finance construction of such a floor in the first place). The reason should be obvious enough. The constant head of pressure acting against the slab will quickly force water through cracks or porosity in the slab, thus flooding its top. The water will also wash away the soil under the slab. If the slab has not been reinforced, once it has lost this support, it will collapse of its own weight.

If a person is absolutely determined for some reason to build a house with a slab floor on such a water-soaked soil and nowhere else—perhaps because of advice received from his astrologer—then either the slab must be so heavily reinforced that it is structurally self-supporting and/or a complex drainage system must be built under the proposed slab that will drain away the water as fast as it collects.

It is, however, far more common that the water table will be some distance below grade level. When this is the case, the problem then becomes one of draining away any excess water that may soak into the soil during a period of heavy rainfall, if we assume that the surface drainage is otherwise adequate. In this case, a drainage system under the slab will still be necessary.

Capillarity and Base Courses

Even when water drainage is no problem, most soils except the most gravelly contain enough moisture to create a condition of perpetual dampness on the underside of a concrete slab, dampness that will have migrated through the soil from the water table below by capillary attraction and that will migrate through the slab because of the same capillary attraction. The result will be all the problems with dampness that were described at the beginning of this article. Therefore, regardless of whether a drainage system has been installed under the slab, additional steps must be taken to break the capillarity that exists within the soil.

This is accomplished by laying down on top of the soil a base course of gravel, crushed rock, cinders, crushed slag, or a similar material (see Fig. 1). A base course of this kind can break the capillarity because the voids that exist between the pieces of gravel, et al., are too large to sustain the capillary movement of the moisture.

And the thicker the base course the better. A base course that is 4 in. thick should be considered a minimum, and one that is 6 to 8 in. thick is much better. The material should be washed, graded so it contains a full range of particle sizes, and screened to eliminate particles larger than 2 in. and smaller than ¼ in. The material definitely should not contain any sulfates (which attack concrete), organic matter such as sticks or roots, or building rubble such as brickbats. The base course should be compacted solidly together after it has been installed to make sure that it will not settle after the house has been completed.

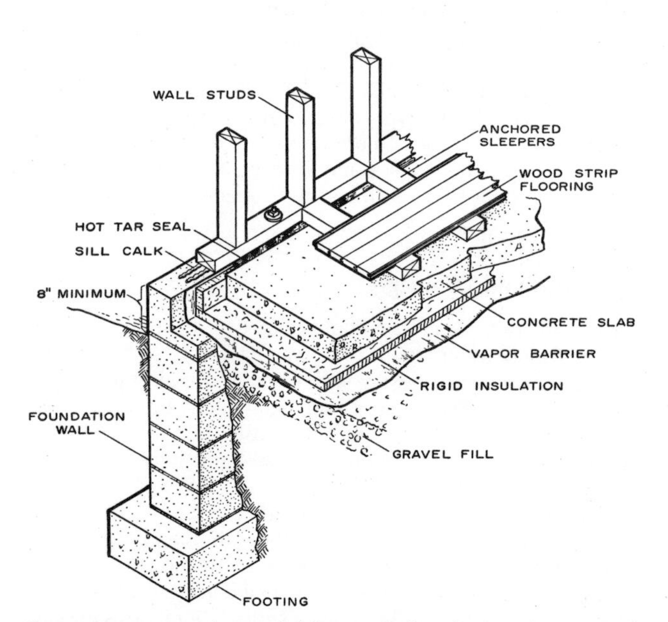

Fig. 1. Construction of a typical slab-on-grade floor that is resting on gravel fill. The floor and foundation walls are completely independent of each other, structurally speaking *(U.S. Forest Service).*

Vapor Barrier

A distinction must be kept in mind between liquid *water* passing through a concrete slab because of capillary attraction and water *vapor* passing through the same slab because of a difference in the water-vapor pressure across the slab.

Water vapor is a gas, like air, and like all gases it always travels from an area of high pressure toward an area of low pressure. In most houses with a slab-on-grade floor, the area of high vapor pressure is under the slab, and the area of low vapor pressure is on top of the slab. Therefore, water vapor will migrate through the slab and into the house. The fact that the water vapor must pass through a solid slab of concrete to get inside the house means nothing. Once on top of the slab, the water vapor condenses upon the floor, where its presence will excite surprise and wonder in the houseowner, not to mention the feelings aroused when this water soaks away the finish on a wood floor, and perhaps warps the wood itself. If a vinyl-tile floor has been laid down, the condensate will loosen the tiles. If the condition is a persistent one, the condensate may rot the wood mouldings that are usually installed along the bottoms of the walls, and the legs of any wood furniture resting on the floor will rot as well. That part of the water vapor that enters the air results in a damp, clammy-feeling atmosphere.

The way to stop this migration of water vapor through a concrete slab is to install a *vapor barrier,* which is also known as a *waterproof membrane,* or *damp course,* under the slab (see Fig. 1). A vapor barrier will not only prevent water vapor from passing through a concrete slab, it will also prevent the passage of liquid water, and it will protect the concrete itself against any sulfates that may be in the soil. The vapor barrier will also keep moisture away from any insulation that may be installed alongside the slab. And, finally, it will keep moisture away from any heating coils or warm-air ducts that may be embedded within the slab.

At one time the usual method of installing a vapor barrier required, first, the placing of a 2-in.-thick concrete slab, which was troweled flat and allowed to harden. An ⅛-in.-thick layer of hot asphalt or coal tar was then spread on top of this base. After this coating had hardened, another 3 in. of concrete was placed on top of the bituminous coating to complete the slab.

Alternatively, after the 2-in.-thick base slab had been placed, two layers of 15-lb asphalt-saturated roofing felt were mopped together and to the base slab as well, just as is done when a built-up roof is installed (see BUILT-UP ROOFING). The top layer of concrete was then placed.

These waterproof membranes are effective but they are also expensive to install. Since the introduction of plastic-film vapor barriers made of polyethylene and polyvinyl chloride, the usual method of installing a vapor barrier has been to spread one of these plastic films over the soil or base course before the concrete is poured.

If polyethylene film (the preferred material) is not available, other materials can be used instead. These include 45-lb or 55-lb asphalt-saturated roll roofing, which is spread over the soil or base course. The edges of adjacent sheets should overlap by at least 4 in. The overlapped edges are then sealed together with asphalt cement.

There is also *duplex paper,* which consists of a thin layer of asphalt laminated between two sheets of kraft paper. Rolls of duplex paper are spread over the soil or base course in much the same way that roll roofing is.

But polyethylene film is by far the most commonly used

vapor barrier, and we shall concentrate on its method of installation. The film is manufactured in 6-, 12-, and 18-ft-wide rolls. The film should be as wide as possible in order to minimize the number of overlapping joints required. Polyethylene film 4 mils thick (i.e., 0.004 in. thick) is usually recommended, but this thickness should be considered a minimum, especially if the material used for the base course has sharp edges that might puncture the film while it is being laid in place or while the concrete is being placed. Polyethylene film that is 6 mils thick is much better.

The film should be installed right to the top of the foundation wall so that water vapor cannot enter the slab by going around alongside the foundation. The film will thus be able to protect as well any insulation that is installed under or alongside the slab, as described below.

All overlapping edges of the film should overlap each other by at least 6 in. If the film is being laid directly on the soil, these overlapping edges must then be sealed with an adhesive recommended by the manufacturer of the film to prevent water making its way between the overlapped edges by capillary attraction. But when the film is laid down over a base course, this type of edge sealing is unnecessary; the weight of the concrete pressing down upon the overlapped edges will be sufficient to maintain the seal.

Problems may also arise where plumbing and electrical lines pass through the slab, as the film must be split at these points. To maintain the seal, the film must be carefully cut around the lines and the cut edges then taped tightly to the pipes. In general, an occasional puncture in the film is not a serious problem because the amount of water vapor that will be able to pass through a small tear will be inconsequential. If, however, there should be a large gap in the film where it is pierced by a pipe, then a considerable amount of water or water vapor might make its way alongside the pipe and into the house.

Slab Insulation

Concrete slabs do not lose their heat in quite the same way that the walls and roof of a house do, which is through a combination of radiation, convection, and conduction. A concrete slab loses its heat mainly by conduction, that is, by direct contact with the soil.

The rate at which heat will be lost will depend mainly on the difference in temperature across the slab. The temperature of soil a few feet below ground level tends to remain relatively constant over the course of a year. In the northernmost parts of the United States, this temperature averages about 40°F. In the southernmost parts it averages about 70°F. Assuming the temperature inside a heated house is 70°F, then during the winter months the temperature differential that might exist across a slab in the northern states will be something around 30°F. Contrast this with the temperature differential that will exist across the walls and roof of the same house during the winter, which will be something on the order of 70°F or 80°F. Obviously, it is more important that the walls and the roof be insulated than it is for the floor slab.

Nevertheless, it is a common complaint of people who live in a house with a concrete-slab floor that the floor feels cold. There are two reasons for this. First, a concrete floor has a very high *heat capacity*, which means that an enormous quantity of heat must soak into the slab before its temperature will be raised to anywhere near the room temperature. If the heating system is started up in October, say, it might take until the middle of November before enough heat has soaked into the slab to take off its summer chill.

The second reason that a concrete-slab floor tends to feel cold during cold weather is due to *edge loss*. That is, the description given above of how heat soaks into a concrete-slab floor during the winter is only partially accurate. More accurately, during the winter months a considerable amount of heat *is* lost through a concrete-slab floor, but this heat loss is concentrated almost entirely within a 3-ft-wide strip all around the perimeter of the floor. It is this part of a concrete-slab floor that feels coldest during the winter. The heat loss occurs because the edge of the slab is close to the surface of the earth. This soil is much more affected by cold weather than the soil directly under the house. Since the temperature differential will be greatest along the perimeter of the floor, it is the perimeter of the floor that loses more heat and, consequently, feels colder.

In order to prevent this edge loss, insulation must be installed completely around the perimeter of the slab. If heat-loss calculations show that the ground is too cold for this insulation to be completely effective, it may then be necessary to install a perimeter heating system within the slab as well.

Insulating Materials

Any material used to insulate the perimeter of the slab must have certain properties if it is to be effective. It must, of course, be an excellent insulator, but it must also be unaffected by dampness, frost, fungi, or termites; it must not absorb water, and it must be strong enough to resist the crushing loads that will be imposed on it. Few materials are suitable, and, in fact, the choice is limited to four: foamed glass, glass fiber, foamed polystyrene, and foamed polyurethane (*see* INSULATION).

All these materials are manufactured as rigid panels that are soaked in hot asphalt to seal them against moisture, since moisture would reduce their insulating value. In addition, moisture tends to dissolve the adhesive that binds together the fibers of glass-fiber insulation.

All these materials are manufactured in slabs that are from ½ in. to 5 in. thick, depending on the material. The thicker the material, the greater its insulating value, of course.

SLAB CONSTRUCTION

This section will describe only the peculiarities of constructing a slab-on-grade. For the preparation and placing of a batch of concrete, *see* CONCRETE.

A concrete slab can be built in any of several different ways: (1) with or without internal reinforcement, (2) integrally with the foundation to form a single unit (see Fig. 2) or separately from the foundation, (3) with the slab supported by the soil or

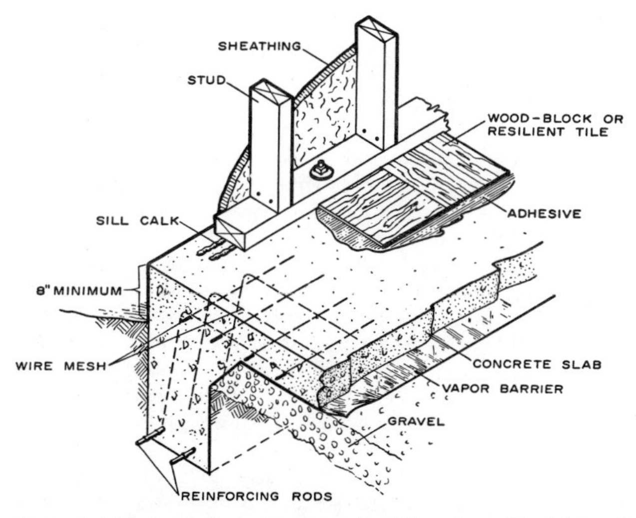

Fig. 2. A slab-on-grade floor can consist of a reinforced, combined slab and foundation wall *(U.S. Forest Service).*

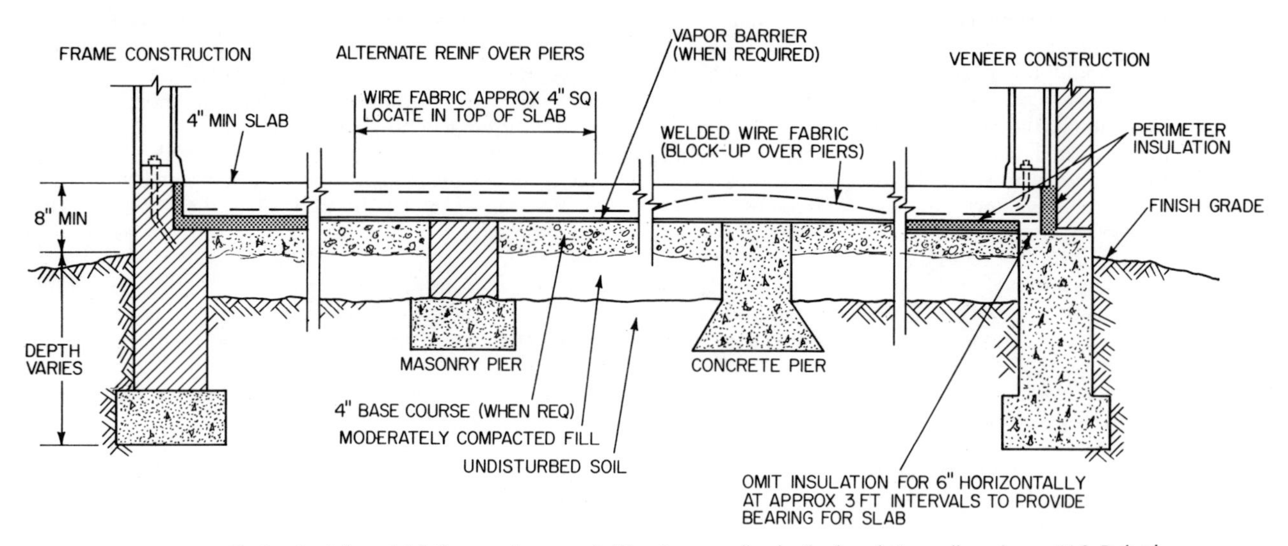

Fig. 3. A reinforced slab floor can be supported by piers as well as by the foundation wall, as shown *(U.S. Federal Housing Authority)*.

base course on which it rests, or (4) by piers (see Fig. 3), piling, or grade beams, or by a combination of these supports.

If the soil is well-drained and has sufficient bearing capacity to support the weight of the slab without settling, then the slab can be placed directly on the soil without any internal reinforcement being required at all. If, in addition, the climate is mild and the soil never freezes, then the foundation walls and the slab can be placed as a single, monolithic unit, as shown in Fig. 2.

On the other hand, when the climate is cold enough to freeze the soil, the foundation should extend below the frost line to prevent its being shifted out of position by soil heave. The foundation and the slab should in addition be placed separately to allow for any movement between the two that is the result of soil heave. The edge of the slab must not rest either in part or wholly upon the foundation because if there should be some slight relative movement between the two, the slab is likely to crack.

If there is a chance, no matter how slight, that the soil may settle because of the weight of the slab and house pressing upon it, or that the soil may swell if it becomes soaked with water, then both the slab and the foundation must be placed separately. The sides of the slab must also in this case be able to shift its position relative to the foundation without being stressed, which will likely result in its cracking. That is, the edges of the slab must not be locked to the foundation in any way.

Furthermore, if there is reason to doubt the ability of the soil to support the weight of the slab and house, or if an analysis of the soil shows that the soil is likely to swell if its moisture content increases, then the slab must be reinforced to increase its rigidity. The typical, 4-in.-thick, unreinforced concrete slab cannot really be depended upon to sustain its own weight without cracking, nor is it strong enough to resist any external stresses that may be imposed upon it. In this case, the slab should be reinforced with steel rods or welded-wire fabric.

This reinforcement accomplishes two things. It greatly increases the tensile strength of the slab, enabling the slab to be used over soils having a relatively poor bearing capacity, and it greatly reduces the amount of shrinkage that will occur as the slab dries out.

Fresh concrete shrinks because it is always mixed with an excess amount of water. This excess water is required to give the concrete enough plasticity to be worked and placed. As the concrete hardens, this excess water evaporates away, and the slab will shrink. The amount of shrinkage is slight, but it is nevertheless sufficient to result in cracks. To prevent cracking, the usual limit in the size of an unreinforced 4-in.-thick slab is about 30 ft square. But when the same slab is reinforced with welded-wire fabric, the slab can be as large as 75 ft square.

A concrete slab may also crack because of the stresses set up at the interior corners of an L-shaped or T-shaped slab as the concrete shrinks. Cracks that originate at these points of stress then spread throughout the slab. Reinforcing the slab cannot eliminate this type of cracking, but it can limit both the extent and size of the cracks. One method of controlling the cracking in an unreinforced slab is to build control joints into the slab that originate at the interior corners. Control (or contraction) joints are commonly seen in concrete sidewalks; they divide the sidewalk into squares. Their purpose is to create weak points within the slab along which the slab will crack, in preference to its cracking in a random manner.

The purpose for which reinforcement is installed in any given slab will determine its location within the slab. If, for example, the purpose of the reinforcement is primarily to increase the rigidity of the slab, then the reinforcement will be located about 1 in. from the bottom of the slab. If the purpose of the reinforcement is to limit or prevent cracking, then it will be located about 1 in. from the top of the slab. If necessary, two layers of reinforcement can be embedded in a slab—one at the bottom of the slab to increase its rigidity, another at the top to limit cracking around pipe openings and interior corners.

The height of the forms above grade into which the concrete will be placed will, of course, determine the overall thickness of the slab, as well as its flatness. In the usual slab-on-grade construction, the foundation is placed first. The foundation then serves as the form. The top of the foundation acts as a *screed* as well, a screed being a straightedge that is used to determine the overall thickness and flatness of a slab. To level the top surface of the slab, a *screed board,* which is a very long straightedge, is laid across the two opposite sides of the foundation and pushed against the freshly poured concrete with a zig-zag motion.

This method of placing concrete means, of course, that the top of the foundation and the top of the slab will be flush with each other. Since most local building codes require that (1) the top of the slab be at least 8 in. above finish grade (a requirement meant to ensure the slab remains dry in rainy weather) and (2) the wood framework of the house also be at least 8 in.

above grade level (a requirement meant to ensure that the wood framework in contact with the slab will also remain dry), using the foundation in this way to establish the height of the slab is a very convenient method of constructing a floor slab.

As for the placing of the concrete itself, the most desirable method of doing so is all at once, so that the slab becomes a single, monolithic structure. This will maximize the strength of the slab, and it also does away for the need for expansion joints, which are required when (1) two separately poured slabs of concrete abut each other and (2) when an unreinforced slab is more than 30 ft on a side.

If vertical insulation has been installed all around the inside of the foundation, the top of this insulation should be about 1 in. below the top of the slab-to-be. To achieve this, boards 1 in. deep and 1 in. wide are placed on top of the insulation. After the slab has hardened, the boards are removed and the gap is filled with hot asphalt or tar.

Exactly the same technique is used around any pipe openings in the slab. The pipes are wrapped about with strips of cloth before the concrete is poured. Once the concrete has set, the cloths are removed. The gaps thus created are filled with the same hot asphalt or tar that is used to fill the other gaps in the slab. This method of sealing the pipes has the added advantage that it allows the pipes to expand and contract with changes in their temperature without stressing them enough that they crack.

Although a monolithic slab is always preferred, it is not always possible to pour a slab monolithically since it is difficult to make and use a screed board that is more than 10 ft long. Therefore, practically, 10 ft becomes the maximum width that any slab can be. When a slab must be wider than 10 ft, it will be necessary to install a 2 × 4 in. screed down the center of the house and pour the slab in two sections. After one-half the slab has been poured, screeded, and troweled smooth, the screed is removed and the other half of the slab is then poured. To prevent a gap opening up where the two slabs abut, a 1-in.-square strip of wood is nailed along the edge of the screed. After the concrete has hardened and the screed has been removed, a 1-in.-deep and 1-in.-wide gap will remain where the two slabs abut. This gap is then filled with hot asphalt or coal tar to seal the juncture. Whenever it is necessary to place concrete in sections in this way, one should always try to locate the gap under a future partition. The gap will later be covered by the partition and the finish flooring can be installed without worrying about any irregularities in the base underneath.

One final point that should be mentioned is that heating coils or warm-air heating ducts are often embedded in a concrete floor slab, both to provide radiant heat for the house and also to warm up the floor itself; this also prevents the condensation problems described above.

All such coils or ductwork must be surrounded on all sides by at least 1 or 2 in. of solid concrete (the actual thickness will depend on the local building code). Whenever ducts or heating coils are embedded in a slab, expansion cracks are likely to occur in the slab because of temperature differences produced by the heating system. The slab, therefore, must be reinforced to resist such cracking. In addition, if a vapor barrier has not been installed as described above, a *water stop,* which is a vapor barrier under another name, must be installed under the coils and ducts to prevent moisture making its way up through the concrete and into the heating system, if there should be any system joints embedded in the slab. In any heating system that has been embedded in a floor slab, therefore, particular care must always be taken that any joints are tightly sealed to avoid this problem.

Most local building codes also require that the bottoms of any warm-air heating ducts embedded in a concrete slab be at least 2 in. above finish-grade level to reduce the possibility of moisture making its way into the ducts.

Stucco

Stucco is a mortar-like substance made from portland cement and/or lime, plus sand and water, that is applied to the exterior of a building as a weather-resistant covering. Stucco is thus closely related to concrete, mortar, and plaster (for descriptions of them *see* CONCRETE; MORTAR; PLASTER). Insofar as its preparation is concerned, stucco is indistinguishable from ordinary mortar; insofar as its application is concerned, stucco is indistinguishable from plaster. In fact, when stucco is applied to the interior surfaces of a building, it is defined as a *plaster* finish.

What differences do exist between stucco and concrete, mortar, and plaster occur chiefly because stucco is applied in a thin layer to the exterior of a building where it is exposed to frequent and extreme changes in its moisture content and temperature, as well as to dimensional changes in its support. Any of these conditions may stress the stucco and cause it to crack. Since the primary purpose of stucco is to provide a weather-resistant covering, any condition that might result in cracks must be anticipated and either prevented or neutralized. It is this circumstance that accounts for what differences do exist between stucco and its cousins.

Lime versus Portland-Cement Stuccoes

Anyone who has traveled extensively outside the United States and who has an eye for building construction will have noticed how widespread the use of stucco is. Stuccoing is without doubt the preeminent method of finishing the exterior of a building. It is used throughout the world except for regions such as northern Russia, Scandinavia, and Canada where there are extensive forestlands and lumber is both cheap and abundant. In most parts of the world where stucco isn't used, brick masonry is. That wood continues to be the principal material used for exterior coverings throughout the United States seems, in contrast, rather anomalous.

Stuccoes made from lime have been used for thousands of years. Lime's virtues are many. It is cheap, widely available, easy to work and apply, and it hardens (if slowly) into a very durable and weather-resistant covering (*see* LIME). As long as the stucco is applied to a very stable support (such as a thick-walled masonry building), as long as care is taken to slake the lime thoroughly and to cure it for a period of several weeks or months before using it, and as long as the stucco is applied in several thin coats (each of which is allowed to shrink down and harden before the next coat is applied), if the final coat of stucco is painted or whitewashed every few years, the result is a coating that will last for a hundred years and more. This is how lime-based stuccoes were prepared and applied for thousands of years and still are in most temperate parts of the world.

One characteristic of a lime stucco that helps improve its longevity is that, although it is strong, it is not as strong or as rigid as the support to which it is usually applied, which is either brick, concrete blocks, or stone. On masonry supports such as these, the stucco can be applied directly. The stucco adheres to its support because (1) the rough surface of the support enables the stucco to get a solid mechanical grip, and (2) the support usually has a suction effect on the wet stucco that draws the stucco tightly to it.

More basic, however, the fact that the support is usually stronger and dimensionally more stable than the stucco is extremely important in helping to maintain the integrity of the stucco. If the support were weaker than the stucco, the stucco would eventually fail, no matter how inherently strong it may be or how well it may have been applied to its support.

Once the stucco has hardened, if some slight differential movement should occur between it and its support (and such movements are normal), the stucco, being weaker, will adapt itself to this movement by forming a multitude of very fine, hairline cracks. In this way the stucco relieves the stresses that have been set up within it without either cracking visibly or coming away from its support. In fact, one of the reasons that stucco must be continually painted or whitewashed is to seal these minute cracks and thus prolong the life of the stucco.

This kind of situation may be satisfactory in a temperate climate such as exists around the Mediterranean, in the Caribbean, and throughout most of Europe and the United States, but in harsher, more northerly climates where cold, driving rains and freezing-thawing cycles are facts of life, lime stuccoes quickly fail. The hairline cracks open up, water soaks into both the stucco and its support, and the stucco ends by cracking badly and separating from its support. Although it is not impossible to use lime stuccoes in a northern climate, the stucco must be carefully maintained and repainted frequently.

Once portland cement had been invented (by Joseph Aspdin in 1824), the use of lime stuccoes could not help but decline. To early enthusiasts, stuccoes made with portland cement seemed to have advantages that completely outclassed those of lime stuccoes. Cement stuccoes were (and are) much harder, stronger, denser, and more durable than lime stuccoes, they were (and are) exceptionally weather-resistant; and, what may have been of greatest importance, a coat of cement-based stucco will set hard in 2 to 3 hr, which allows a crew of plasterers to apply three coats of stucco to a wall within the space of 2 weeks instead of the months often required for lime stucco. Economically, this savings in time is no small advantage. In addition, a cement stucco is ideal for interior locations that will receive hard usage or are perpetually damp, such as garages, basements, steam rooms, and bathrooms. Finally, their great durability does away with the necessity to paint the walls once every year or two. The principal disadvantage that cement stuccoes seemed to have were that they were much stiffer working and much more difficult to apply than lime stuccoes.

However, the advantages of an all-portland-cement stucco have proved in practice to be largely illusionary. The very strength, hardness, and rigidity of the cement stuccoes work to their disadvantage. Where lime can adapt itself to differential wall movements by developing a network of fine cracks, cement stuccoes cannot. They are much too strong. They can only relieve the stresses imposed upon them by cracking, and the strength of the stucco ensures that these cracks will be few but large rather than minute and pervasive.

Once any crack has appeared, unless it is quickly tended to, the virtues of a cement stucco disappear altogether because moisture can now make its way between the stucco and its support. But because cement stucco is so dense, the moisture cannot easily make its way out of the wall again. Although the outer section of the stucco dries quickly, the inner section does not. The result of this differential drying is that the stucco bends away from its support. Adhesion is lost, and a complete failure of the stucco coating follows within a relatively short time.

There are two possible solutions to this problem. Either modify the all-cement stucco formula or modify the way in which the stucco is applied to a wall. In Europe the solution chosen has been to add lime to the portland cement. This modifies, but not excessively, the virtues of an all-cement stucco. The mixture of lime and cement results in a stucco that is harder and stronger than an all-lime stucco and sets as quickly as an all-cement stucco does. At the same time, the presence of the lime makes the stucco more pliant and yielding under stress. In addition, the lime gives a much-needed plasticity to the stucco, which enables the plasterers to work the stucco with ease. (For comparable experiences with "weak" and "strong" mortars *see* MORTAR.)

In the United States the solution has been to keep the all-cement stucco but modify the method of application. This decision has been influenced by the fact that the support more often than not consists of a wood-framed wall. Now, wood is a notoriously unstable material, dimensionally speaking. It shrinks and expands readily with changes in its moisture content, and much more so across its grain than parallel with its grain (*see* WOOD). As it happens, the predominant method of framing a house in the United States is *platform framing,* in which a significant fraction of the total height of a house consists of wood laid edgewise. As a result, the typical 1-story wood-frame house will expand and contract up and down something like ½ in. over the course of a year because of normal seasonal changes in the humidity. (*See* WALL FRAMING; WOOD-FRAME CONSTRUCTION.)

It would be fatal to attempt to apply stucco directly to such a support. The solution, therefore, has been to isolate the stucco as much as possible from the wood frame; to treat the stucco, in fact, exactly as if it were a thin slab of reinforced concrete that is structurally independent of the frame.

To accomplish this, a metal reinforcement consisting of *expanded metal lath* or *wire lath* (see below) is first attached to the wall studs; the studs are dimensionally quite stable. (For the reasons *see* LUMBER; WOOD.) The stucco is then troweled over the reinforcement, surrounding it completely. The result is an exceptionally strong and rigid slab of reinforced stucco. The reinforcement not only strengthens the stucco, it also compels the stucco to resist any tendency it may have to expand or contract, which would cause it to crack.

To prevent cracks occurring because of any differential movement between the stucco and the structure of the house, the stucco is isolated from the structure by a layer of waterproof building paper. In short, the stucco is supported entirely by its metal reinforcement. The reinforcement is in turn supported on the wall studs by nails or staples. To further isolate the stucco slab from any external stresses, *control joints* (see below) are installed in the construction wherever there is a chance that differential movements between different parts of the house's structure might affect the stucco.

However, when the support consists of traditional masonry construction (brick or concrete blocks usually), the stucco may be applied directly to the masonry without any reinforcement being required, just as is done in European construction.

STUCCO PROPORTIONING AND MIXING

The cementitious materials used for stucco—that is, the portland cements and limes—are described elsewhere, (*see* CONCRETE; LIME).

It is the general rule when preparing a stucco mortar that the proportions of cementitious material to the aggregate—that is, the sand—be in the ratio of 1:3 cementitious material/sand (see Table 1 A–F). This rule applies whether the cementitious

Table 1. Recommended Proportions for Stucco Mixes

A. Portland Cement Mortar

Portland cement, bag	Sand, no. 2 shovels	Lime, bag	Mason cement, bag	Water, gal
1	18	½		9
1	18		½	9

B. Portland Cement Mortar with Lightweight Aggregate

Portland cement, bag	Perlite	Vermiculite	Lime, bag	Mason cement, bag	Water, gal
1	¾ bag 3 cu ft		½		10
1	¾ bag 3 cu ft			½	10
1		¾ bag 3 cu ft	½		11
1		¾ bag 3 cu ft		½	11

C. Portland Cement Finish

Portland cement, bag	Sand, no. 2 shovels	Lime, bag	Water, gal	Silica sand, bag
½	21	2	17	
½		2	17	3

D. Recommended Stucco Mixes*

Group	Portland cement	Lime	Masonry cement	Plastic cement	Sand
C	1	0–¼			3–4
C	1		1		6–7½
C	1	¼–½			4–6
L	1	½–1¼			4½–9
L			1		3–4
F	1	1¼–2			5–10
P	1			1	6–10
P				1	3–4

*Proportions are given in parts by volume.

E. Stucco Mix Selection Guide

Stucco base	Recommended groups for stucco coats		
	Scratch	Brown*	Finish†
Low absorption (placed concrete, dense clay brick)	C, P	C, L, P	L, F, P
High absorption (concrete unit masonry, clay brick, and structural tile)	L, P	L, P	L, F, P
Metal reinforcement‡	C, P	C, L, P	L, F, P

*Use as base coat in 2-coat work over concrete and masonry bases.
†Finish coat may be factory-prepared "stucco finish."
‡Over any type of supporting construction.

(continued)

Table 1. Recommended Proportions for Stucco Mixes (continued)
F. Recommended Curing Schedule for Stucco Finishes

Application*	Damp curing, hr†	Total setting time‡
Scratch coat	12–24	At least 48 hr (between coats)
Brown coat	12–24	At least 7 days (between coats)
Finish coat	12–24	At least 48 hr (after application

*In 2-coat work, requirements for base and finish coats are the same as for brown and finish coats, respectively.
†Damp curing should be delayed until scratch and brown coats are sufficiently set to prevent erosion, and for 12 to 24 hr for finish coat.
‡At least 50°F temperature should be maintained during and after application.

material is all-cement, all-lime, or any mixture in between. Thus, for example, one common stucco mixture consists of 1:1:6: cement/lime/sand; the overall proportion between the cementitious materials and the sand remains 1:3.

Practically speaking, in order to obtain particular qualities in a given stucco, it is more usual to use mixtures of portland cement and lime rather than have either an all-cement or an all-lime stucco. At one extreme a small proportion of lime will be added to an all-cement stucco to improve its plasticity, and at the other extreme a small proportion of cement will be added to an all-lime stucco to increase its strength. The following proportions of portland cement to lime are widely used:

For a strong, waterproof base coat—1:¼ cement/lime.
For a general-purpose stucco—1:¼ to 1:½ cement/lime.
For the finish coat—1:½ to 1:1 cement/lime.

Admixtures

Admixtures are sometimes added to a stucco to impart specific properties to the stucco. In this sense, the small amount of lime added to an all-cement stucco is often considered merely as an admixture that helps improve the workability of the stucco, since an all-cement stucco is notoriously difficult to apply to a wall. Vegetable or glass fibers may be added to a stucco to increase both its plasticity and cohesiveness. These fibers are anywhere from ½ in. to 2 in. long and have been cleaned of any grease, oils, or dirt. The usual amount added to a stucco mix is ½ lb per 1 cu ft (i.e., one bag) of cementitious material.

Sand

For any given batch of stucco, once the basic strength and working characteristics of the stucco have been determined by the proportions of cement and lime in the stucco, the actual quality of that batch of stucco will depend primarily upon the quality of the sand. The importance of having a clean, well-graded sand that is free of any clay, loam, or vegetable matter cannot be overemphasized.

A clean sand free of extraneous matter is important because it allows the *cement paste* (the mixture of cementitious material and water) to coat the particles of sand completely and bind them together tightly. This is necessary if the stucco is to develop its full strength.

In addition, well-graded particles that range smoothly and evenly in size from about ⅛ in. down to about 0.006 in. are essential if all the potentially desirable qualities of the stucco are to be realized. A stucco made with such a well-graded sand is as dense and weather-resistant as it is possible for a stucco to be. The stucco is also easy to work, the shrinkage that normally occurs during setting and the expansion that normally occurs because of moisture penetration are minimized, and the overall strength of the stucco is at the same time maximized (for any given cement/lime ratio, that is).

A poor sand, on the other hand, will ruin a batch of stucco. Take, for example, a stucco in which the sand particles are too coarse. In such a stucco the voids (or spaces) between particles will be very large. If the correct amount of water were added to the mix, the stucco would be extremely hard to work, and it would tend to curl off a wall. To improve the plasticity and cohesiveness of the stucco, an additional amount of water must be added to the mix. Although this extra water improves the plasticity, it does so by diluting the cement paste. The consequences are that as the stucco dries out it will shrink excessively and crack badly. Since the sand particles are not bound together at all well by the weakened cement paste, the result is also a weak stucco.

At the other extreme, assume a batch of stucco in which the sand particles are too fine. Again an excessive amount of water is required, this time because of the large surface area presented by the sand particles. That is, excess water is required to extend the cement paste so that all the particles will be coated by the paste. A stucco with an excess of fine particles is extremely easy to work and apply, but, again, as the stucco dries out it will shrink excessively and crack badly. It, too is a weak stucco.

When, however, the sand is well-graded, the voids between the largest particles will be filled with somewhat smaller-sized particles, and the voids between these smaller particles will be filled by even smaller particles, and so on. As a result, a minimal amount of water is required to make the cement paste, which is just what is necessary to maximize the strength of the stucco. The wide range of particle sizes ensures that the stucco is easy to work while the close packing of the particles ensures that the shrinkage, and consequent cracking, will be at a miminum. (For example, in one study, a stucco made with fine sand shrank 11 percent as it set while a stucco made with a graded sand shrank only 4 percent.)

Mixing the Stucco

Stucco may be mixed either by hand or with a mechanical mixer. Hand mixing is hard work and only a small amount of stucco can be prepared at one time, about 1 cu yard at most. A mechanical mixer should always be used, except for the smallest jobs.

BASE SUPPORT

Masonry Base

In traditional stucco construction, the stucco is applied directly to a masonry base of some kind (see Fig. 1). This support usually consists of a brick or concrete-block wall, although walls made of poured-in-place concrete and structural clay tiles are not uncommon, especially in commercial and industrial construction. In this traditional type of construction, the stucco is held to the masonry primarily by the mechanical key that exists between them because of the surface roughness of the masonry. In addition, the masonry will usually have a suction quality to it that enables it to draw the fresh, plastic stucco deep into its surface. The effect of this suction is to increase the mechanical grip between the masonry and stucco.

Whether or not any given masonry base is rough-textured enough to hold a layer of stucco will be obvious at a glance. Whether the masonry also has sufficient suction is tested by spraying water upon its surface. If the water soaks into the masonry immediately, good; if it lies on the surface of the masonry, forming droplets of water that run down the wall, that's bad.

Masonry can also have too much suction. Soft, crumbly bricks or concrete blocks made with a lightweight aggregate

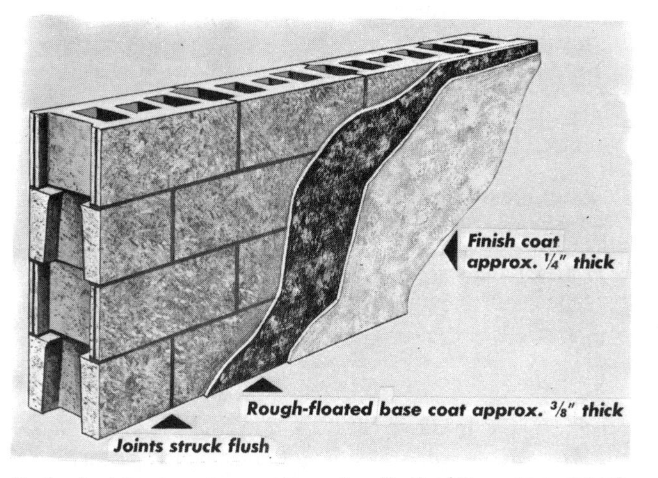

Fig. 1. Applying stucco to a masonry surface *(Portland Cement Association).*

tend to have an excessive amount of suction. Excessive suction will draw the water out of fresh stucco too rapidly. The stucco will stiffen quickly and become difficult to work. It will very likely dry out so quickly that it won't have time to cure properly, and it will, therefore, be much weaker than it would be otherwise. If a masonry support does have too much suction, this suction must be reduced before the stucco is applied. This is accomplished by spraying the masonry lightly with water. The masonry should appear slightly damp, not soaked.

Some types of masonry will have an uneven amount of suction. This occurs particularly in walls made of old hand-burnt bricks or glazed concrete blocks. When this is the case, the layers of stucco that will be applied over the masonry will dry out unevenly. As a result, after the stuccoing has been completed, it will be noticed that the stucco has a mottled appearance. Furthermore, because the adhesion of the stucco to the masonry varies, being stronger at some sections of the wall than at others, cracking is likely to occur, since those sections that have good adhesion will grip the stucco securely and resist any tendency the stucco may have to shrink as it dries, while in those sections with poor adhesion, the stucco will pull away from the wall as it dries out. The stresses set up along the boundaries between these different sections will cause the stucco to crack.

A support must have other properties in addition to good adhesion if it is to make a satisfactory base for stucco. Of equal importance is rigidity. The support must be rigid enough to resist without cracking any bending loads that might be exerted against it. Obviously, any solidly built masonry structure is very unlikely to bend or flex or crack (though it may expand under certain circumstances; see "Control Joint" below), which makes masonry an ideal base for the thin but very rigid layers of stucco that are applied over it.

Uneven settlement of the house is another problem, as settlement may cause even the thickest masonry wall to crack. To prevent settlement cracks, a newly built house with masonry walls must be allowed to "rest" for several months after it has been completed so that any settlement that is going to take place has time to take place. Only then should the stucco be applied. (But time is money to most builders, and they simply cannot afford to tie up their capital in a half-finished house in this way, though it is very much to the long-term advantage of the homeowner that this be done.)

Any masonry base to which stucco cannot be directly applied, either because the base is not rough enough or because it lacks sufficient suction, and where it would be too expensive either to roughen or clean the masonry, can always have metal reinforcement (lathing) attached to the wall to which the stucco will adhere, as described below for wood-frame houses.

Wood-Frame Base

In the first place, as has already been mentioned, a wood-frame structure is not stable enough dimensionally to support a rigid layer of stucco. The wood frame expands and shrinks in response to normal changes in atmospheric humidity. These dimensional changes cannot be prevented, and, if a stucco coating were to be applied directly to the wood, the stucco would inevitably either bulge away from the base or crack, or both. This is particularly true of platform-framed houses, but much less true of balloon-framed houses. Anyone who is contemplating having a stucco finish applied to a wood-frame house must, therefore, first make sure that the house has been balloon-framed (*see* WOOD-FRAME CONSTRUCTION).

It also helps enormously if the climate is dry and there is a minimal annual change in the humidity. If the climate is dry and unchanging, then dimensional problems are unlikely to be a problem even with a wood-frame house. For this reason stucco finishes are much more practical in the desert climate of the Southwest from west Texas to southern California than anywhere else in the United States; which does not, we hasten to say, make stuccoing impractical in the rest of the United States.

But even a balloon-framed house must be adequately braced against any twisting or racking loads that may be imposed upon it by high winds, earthquake shocks, or by settling of the foundation. If the stucco is to be long-lived, it is necessary that the framework be as rigid as possible and that the footings be large enough to support the house against settlement. (For a discussion of what constitutes adequate wall bracing *see* SHEATHING; WALL FRAMING; for a discussion of stable foundations *see* FOUNDATION.)

Sheathed versus Unsheathed Wood Frames

It is strongly recommended that a wood-framed house always be sheathed to stiffen the structure and to minimize any racking or twisting movements (see Fig. 2). In the South sheathing is often thought of as an unnecessary frill, as if it were only a kind of insulation that might be required in the North but is certainly an extravagance in the South. In fact, the insulating value of ordinary sheathing is insignificant. It should instead be thought of as an essential part of the structural framework of a house,

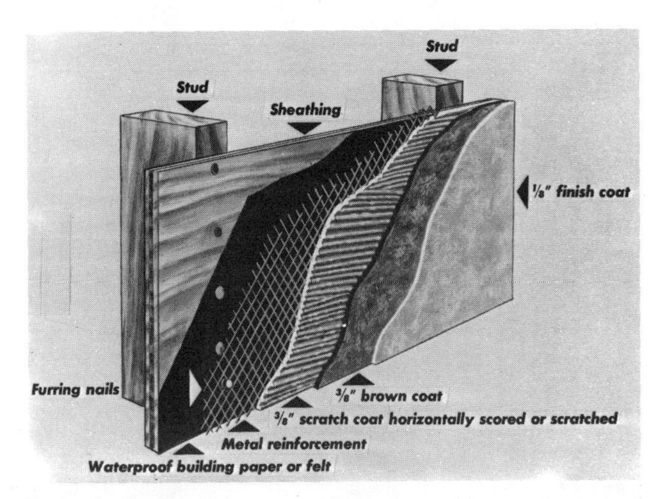

Fig. 2. Application of stucco to a sheathed, wood-frame wall *(Portland Cement Association).*

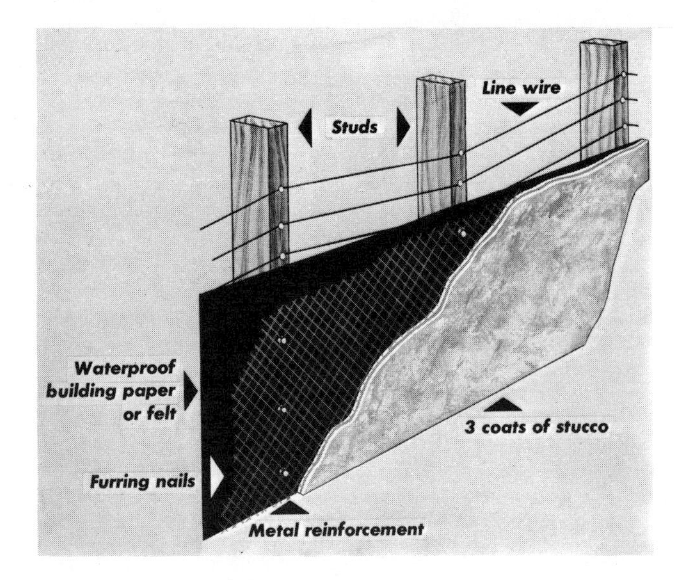

Fig. 3. Application of stucco to a wood-frame wall without sheathing *(Portland Cement Association).*

without which the house is much more flimsily constructed.

However, for economic reasons sheathing is often omitted in low-cost houses built in the South, and when this is the case special steps must be taken to support the stucco adequately (see Fig. 3). Basically, the wall studs must be spaced a maximum distance of 16 in. apart, and enough time must elapse between the completion of the framing and the application of the stucco that the moisture content of the wood has reached a state of equilibrium with the moisture content of the atmosphere. The walls of an unsheathed house must be braced (*see* SHEATHING).

These essentials having been complied with, *line wire,* which is 18 gauge (or heavier) galvanized-steel wire, is wrapped completely around the house. Starting at the bottom of the walls, the plasterers wrap the wire around and around the house, spacing adjacent lengths of wire 6 to 8 in. apart, nailing or stapling the wire to every fifth stud.

Once the house has been "wrapped" in this way, the line wire is stretched taut by raising or lowering the strands between the points of attachment and then nailing or stapling these strands to intermediate studs. The line wire thus provides a support that will keep the backing paper, which is applied over it, from sagging when the stucco is in turn applied against the backing paper. In no sense, however, should the line wire be thought of as strengthening or stiffening the house structure itself.

Waterproof *backing paper* (or 15-lb asphalt-saturated roofing felt) is now nailed or stapled to the studs. The purpose of backing paper (or roofing felt) is to isolate the wet stucco from the wood framework of the house, thus avoiding soaking the wood with water. The backing paper will also help support the stucco as it is being applied, and, later on, if the hardened stucco should crack, the backing paper will keep rain out of the interior of the house.

Backing paper should also be applied over sheathing. Once the backing paper has been installed, the metal reinforcement can be nailed or stapled to the walls.

Metal Reinforcement

The two main types of metal reinforcement used to support stucco are *expanded metal lath* and *wire lath* (see Fig. 4). Expanded metal lath originates as a sheet of lightweight metal. The sheet is first inserted into a press in which dies cut a pattern of slits into the metal. The sheet is then gripped on opposite sides and pulled apart, or expanded, with the result that the slits open up into the diamond-shaped patterns shown in Fig. 4.

Wire lath is made by weaving or welding steel wires

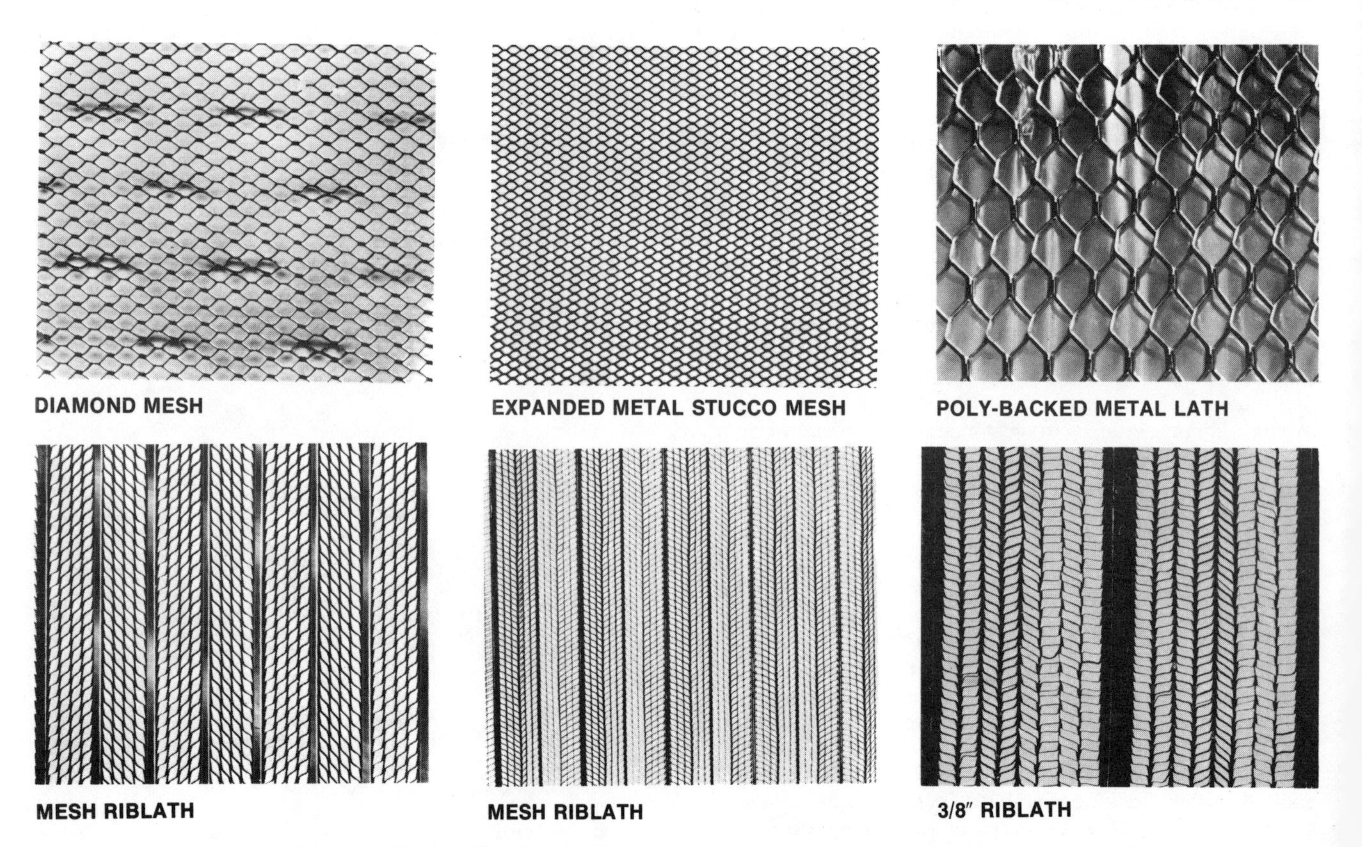

Fig. 4. Different types of metal lath used for stuccoing *(U.S. Gypsum Co.).*

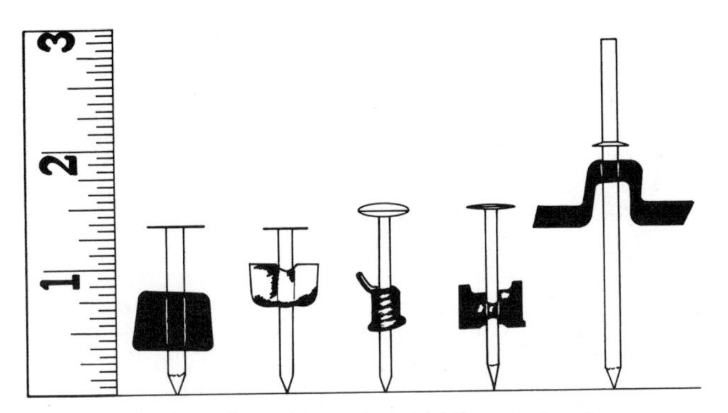

Fig. 5. Self-furring nails used to attach metal lath to a support.

together in such a way as to form the desired fabric pattern.

If the metal reinforcement is to be completely embedded in the stucco, as it should be, it is essential that the reinforcement stand away from the backing paper by at least ¼ in., with ⅜ to ½ in. being preferred. This spacing was obtained at one time by nailing thin strips of wood furring over the backing paper, the reinforcement then being nailed or stapled to the furring.

Furring strips have since been replaced by *self-furring nails,* which consist of galvanized roofing nails inserted through spacers of some type (see Fig. 5). The use of self-furring nails makes it possible to install the metal reinforcement much more conveniently. (By using masonry nails, this metal reinforcement can even be attached directly to concrete or brick.)

To save further on the time and expense of installing metal reinforcement, manufacturers of these products have begun pressing dimples into their products, which enables the reinforcement to stand away from the base support by itself. This type of reinforcement is called *self-furring* reinforcement.

To reduce the installation costs even further, some types of self-furring reinforcement are manufactured with backing paper already attached to one side, the entire assembly then being nailed to the studs at one time. Unlike the first two kinds of reinforcement described above, which are manufactured in rolls, the paper-backed types are manufactured as flat sheets. Paper-backed reinforcement is used mainly when the stucco is applied by machine, as described below.

Flashing

An essential preliminary to stuccoing is making sure that adequate flashing is installed wherever moisture might make its way past the edges of the stucco. If moisture should get behind the stucco, it would certainly rust away the metal reinforcement as well as weaken the bond that exists between the stucco and its base support. In freezing weather, if moisture has collected under the stucco, it will expand and forcibly push the stucco away from its support as it freezes.

To prevent moisture penetration, therefore, flashing must be installed over all door and window tops, under all door and window sills, under chimney and parapet capping, and at the junctures between roofs and chimneys and between roofs and parapets. On walls, if the top edge of the stucco is not protected by an overhanging eave, flashing must be installed along the top of the wall as well. (For a description of flashings and their installation *see* FLASHING.)

STUCCO APPLICATION

Stucco is applied to its support in two or three layers in almost the identical way that plaster is applied to an interior wall. (For a detailed description of this technique *see* PLASTERING.)

Stucco is applied over masonry in only two coats, whether or not metal reinforcement is used to strengthen and stiffen the construction. These two coats are called the *base coat* and the *finish coat* (see Fig. 1). Together, they should be at least ⅝ in. thick. The purpose of the base coat is to provide a flat support for the finish coat, a support to which the finish coat can adhere tightly. The finish coat is, of course, the visible layer of stucco.

Stucco is applied over a wood-frame base in three coats, which are called the *scratch coat, brown coat,* and *finish coat* (see Fig. 2). Together, these three coats form a layer the minimum thickness of which should be ⅞ in.

The scratch coat is always embedded in a metal reinforcement of some kind. The reinforcement gives the scratch coat the strength and stiffness it must have to support the brown and finish coats. The brown coat increases the overall thickness of the stucco coating, thus adding to the strength and rigidity of the stucco. The brown coat also provides a suitable base to which the finish coat will adhere. And, as in 2-coat work, the finish coat presents a decorative appearance.

The layers of stucco can be applied either by traditional hand methods or by machine. We will describe first the hand application of stucco, then the machine application. But however the stucco is applied, the final straightening and smoothing of the layers, as well as the overall quality of the construction, will depend primarily on the knowledge and experience of the plasterer and on the skill with which he uses his trowel and float to smooth and level each layer (*see* PLASTERING). Machine application is basically a method for hurrying up the job. It does little or nothing to replace, correct, or cover up poor workmanship. Machine application is not a substitute for a good mechanic.

Three-Coat Stuccoing

Scratch Coat

The first, or scratch, coat must be applied with sufficient force that the space between the support and the backside of the reinforcement is filled completely with stucco (see Fig. 6). The

Fig. 6. Applying the scratch coat on metal lath reinforcement *(Portland Cement Association).*

reinforcement must be completely embedded in the stucco. If it isn't, moisture within the air pockets surrounding the reinforcement will eventually corrode the metal, which will in turn seriously weaken the stucco.

The plasterer tries to arrange his work schedule so that the stucco is always being applied on the shady side of the house. By following the sun around a house in this way, he avoids having the stucco dry out prematurely. He does not attempt to smooth out the scratch coat although it must, of course, be applied in an even layer. Most building codes require that the scratch coat be at least $\frac{3}{8}$ in. thick, although in the best quality work it should be $\frac{1}{2}$ in. thick.

Once the stucco has begun to set, it is *scratched*, or scored horizontally, with a hand tool that resembles a child's rake (see Fig. 7). Scoring provides a mechanical key that will help the brown coat to get a good grip on the scratch coat.

Fig. 7. Scratching the scratch coat *(Portland Cement Association)*.

The scratch coat having been applied, it is *damp-cured* for at least 48 hr. That is, a fine spray of water is directed against the stucco at frequent enough intervals to keep the surface damp. Damp curing is a necessity if the stucco is to achieve its full strength, which it will do over the course of several days, as long as it doesn't dry out. Failure to damp-cure the stucco will result in its being weak and porous.

After the damp curing has been completed, the scratch coat is allowed to dry out completely, which will take about 5 days, depending on the weather. During this drying-out period, the stucco will shrink down to its final size, which is necessary if the brown coat is to obtain a secure grip to it. In addition, the stucco having dried out, the plasterer is in a position where he can dampen the scratch coat evenly, thus assuring himself that the scratch coat will have the same suction over its entire surface.

Brown Coat

The brown coat is also required to be at least $\frac{3}{8}$ in. thick, but, like the scratch coat, in the best quality work it should be $\frac{1}{2}$ in. thick. The final appearance and overall quality of the job will depend greatly on the skill with which the brown coat is applied. Any imperfections in the brown coat will adversely affect both the adhesion and appearance of the finish coat, as scratches, score marks, and other defects in the brown coat will show through to the finish coat. It is necessary, therefore, that the brown coat be applied as carefully as possible.

The stucco must be applied with considerable force if it is to bond tightly to the scratch coat. Once the stucco has been applied, it is carefully leveled with a *rod* (a 4 to 8-ft-long straightedge having hand grips) and smoothed with a *darby* (a 4 to 8-ft-long float-like tool, also with hand grips). No attempt is made to smooth the surface of the brown coat excessively. It is, instead, left flat but rather rough to give the finish coat something it can grip. The plasterer tries to cover and smooth an entire wall surface at one time, to avoid leaving lap or joint marks where wet stucco meets dried-out stucco. Such lap marks will show right through the finish coat.

As soon as the stucco begins to set, it is damp-cured for 48 to 72 hr and then left to dry out completely, which will take about 5 days. The finish coat can now be applied.

Finish Coat

The finish coat must be at least $\frac{1}{8}$ in. thick. The ingredients for the finish coat are usually purchased premixed by the bag rather than mixed together on the job. Purchasing premixed ingredients has several advantages. Most important, the plasterer is assured that the proportions of the ingredients in the mix will be consistent from batch to batch, which in turn will give him assurance that the finish coat will present a uniform appearance over the entire wall area. If he tries to mix the ingredients himself, he may run into problems with sand size and grading, in determining the correct amount of water to add to the mix, as well as problems in proportioning the sand, cement, lime, and any admixtures. Colors are especially difficult to mix accurately and slight miscalculations in the amount of coloring matter added to different batches of stucco can have a disasterous effect on the final appearance of the job. By using a factory-prepared mix, the plasterer not only avoids all these difficulties, he is assured of a uniform product that will have the color, texture, and physical characteristics he wants in the finish coat.

The finish coat is applied starting at the top of the wall and working down. The finish coat must at the same time be applied rapidly and continuously to avoid lap marks that may occur because the stucco has dried out. A fast job is usually achieved by hiring as many plasterers as are necessary, all of whom work continuously around the walls of the building until the entire finish coat has been completely applied. Once the stucco has been applied, it is damp-cured for 24 hr, and the job may be considered finished.

Figure 8 shows some of the kinds of decorative finishes that can be applied. As one might imagine, the application of any of these finishes requires skill and is not the sort of job to be entrusted to the beginner. Even the simplest and most common type of finish—a smooth-surfaced but rough-textured *sand finish*—requires skill to apply.

The sand finish is obtained using a wood float, the blade of which is covered with a sheet of rubber, plastic foam, or a piece of carpeting. Different materials produce slightly different effects. The float is worked lightly over the surface of the still-wet stucco to smooth and flatten it and at the same time expose the grains of sand lying just under the surface of the stucco. The skill comes in manipulating the float so that the texture is the same over the entire wall area. Care must be taken to avoid overworking the stucco, as this tends to build up a surface layer of very fine but weak cement paste that will spall or flake away within a few weeks or months.

Another very common and attractive type of finish is called *pebble dash*. The finish coat is first worked into a smooth surface. Small stone pebbles, or some similar material, about $\frac{1}{8}$ in. in size, are loaded into the hopper of a small hand-held

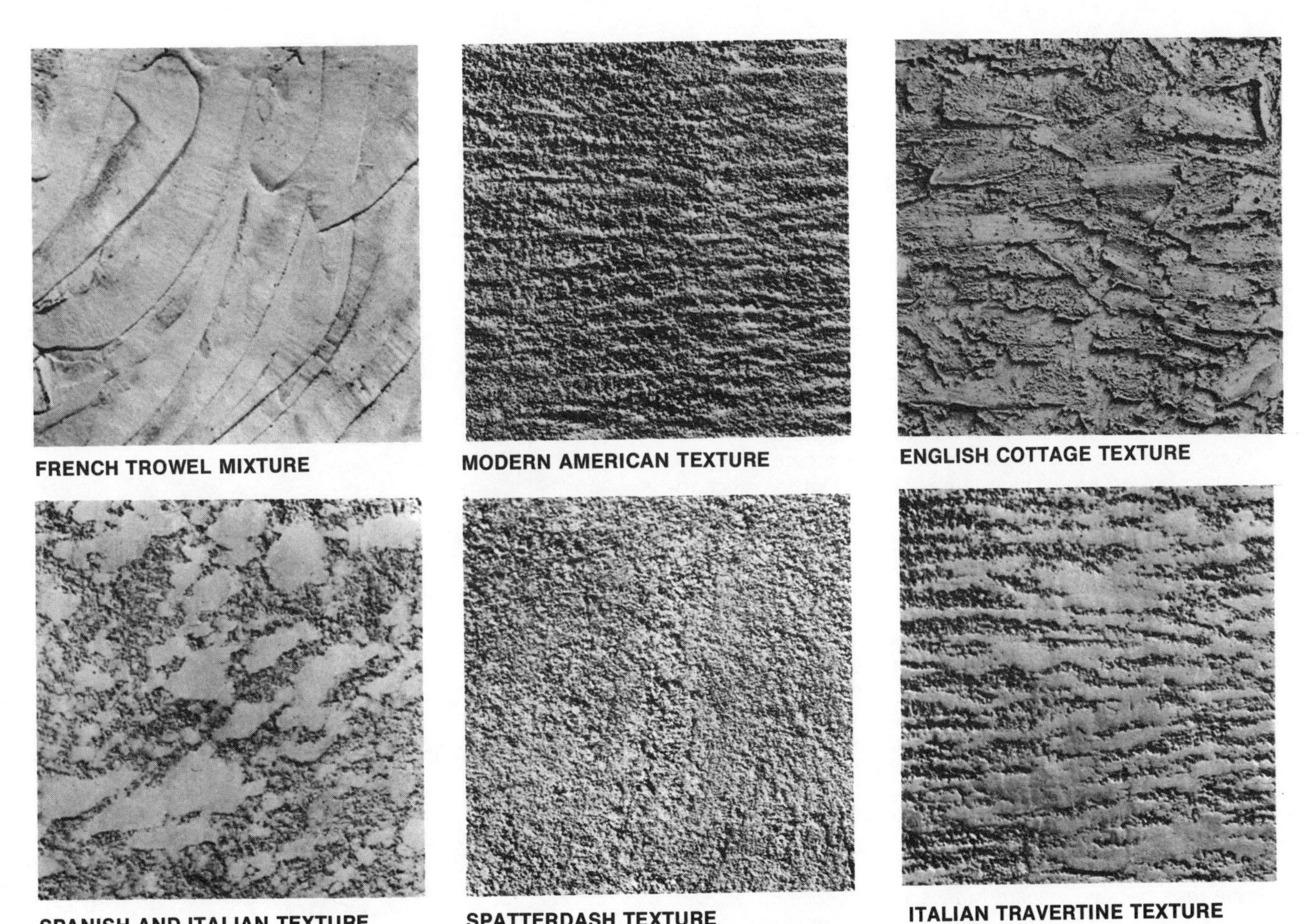

Fig. 8. Some of the attractive surface finishes possible in stucco *(Portland Cement Association)*.

machine and blown from the machine against the still-wet stucco (see Fig. 9). The result is a very attractive granular finish that has the added virtue that the particles provide focal points around which minute cracks can form that will relieve shrinkage stresses in the stucco. If one looks closely at a pebble-dash finish, one can see these tiny cracks around the particles. The cracks are invisible to the casual glance, they do not allow moisture through the wall, and, by having relieved the stresses in the finish coat, the particles help prevent the formation of large, disfiguring cracks.

Two-Coat Stuccoing

In 2-coat stuccoing, the base coat acts as a combined scratch-brown coat. The required minimum overall thickness of both base and finish coats is ⅝ in.

It is worth reemphasizing that the suction in the masonry must be adequate and the same over the entire surface of the support. If the support lacks these characteristics, and if it would be too difficult or expensive to refinish the masonry properly, then backing paper and metal reinforcement should be applied over the masonry before the stucco is applied as has been described above for a wood-frame wall.

As much force as possible must be used when the base coat is applied to masonry so that the bond between the two is as tight as possible. If only normal pressures were used to apply the base coat, a thin film of air would be trapped between the masonry and the stucco. This film of air must be forced out if a good bond is to be obtained.

Some masonry surfaces are so uneven that the thickness of the base coat applied over them will vary greatly, which will lead to shrinkage cracks as the stucco dries out. When this is

Fig. 9. Application of a pebble-dash finish, using a mechanical blower *(Portland Cement Association)*.

the case, it will first be necessary to apply a kind of quasi-scratch coat to the masonry in order to straighten out the wall surface. Once this preliminary coat has dried, the usual base coat can be applied.

The base coat having been applied, it must be damp-cured for at least 48 hr and then left to dry completely, which will take around 5 days. The finish coat can then be applied as described above for 3-coat work.

Machine Application

The machine application of stucco is much more common nowadays than hand application. Economically, the reason is plain. A crew of men using a plastering machine can apply and smooth in a couple of hours the same amount of stucco it would take the same crew an entire day to apply and smooth by hand. (For the technique of machine application *see* PLASTERING.) Insofar as the quality of the work is concerned, the great advantage of machine application is that the stucco is thrown against the support with such force that the bond between the stucco and the support is much stronger than it is when stucco is applied by hand; furthermore, the stucco is compacted together much more densely, which makes it stronger and less subject to shrinkage, and this allows it to be applied in a thicker coat. The finish coat can also be applied more uniformly by machine than by hand, and the finished job will thus have a more even texture and color.

However, in order to realize all these advantages, the stucco must be mixed together completely and thoroughly, the nozzle from which the stucco is discharged must be skillfully manipulated, and the temptation to apply the stucco too thickly must be resisted. The ease with which the stucco can be applied has led to a school of thought that believes it is better in 3-coat work to apply the scratch and brown coats at the same time because a more homogenous, and thus stronger, coat of stucco will result. But the thicker a coat of wet stucco, the more it will shrink as it dries out and the greater the probability that it will crack. Furthermore, because the stucco must be more watery when it is applied by machine—in order to reduce as much as possible the friction between the stucco and the pump and hose, applying the stucco too thickly often causes the stucco to sag, when the stucco doesn't fall off its support altogether. Indeed, adding a bit too much water to the mix is the greatest potential vice of machine-applied stucco. The stucco may flow easily but the final result is likely to be excessive shrinkage and cracking and lower strength.

Machine application is particularly advantageous for applying the finish coat. A variety of surface textures can be obtained by varying the air pressure, pump speed, or the distance of the nozzle from the wall, or all three. The consistency of the mix can also be varied to obtain a desired texture, though the mixture for a finish coat is usually thicker than it is for the scratch and brown coats.

Machine application is also advantageous when the stucco is to be colored. Colored stucco is applied in two light layers. During the first application the entire wall is lightly sprayed until the desired color has been obtained. The desired texture is then obtained during the second application.

CONTROL JOINT

A stucco wall may crack for any number of reasons. There are shrinkage stresses, which have already been described. Pipes may pass through the wall, and these pipes may restrain the normal shrinkage of the stucco in such a way that the stucco

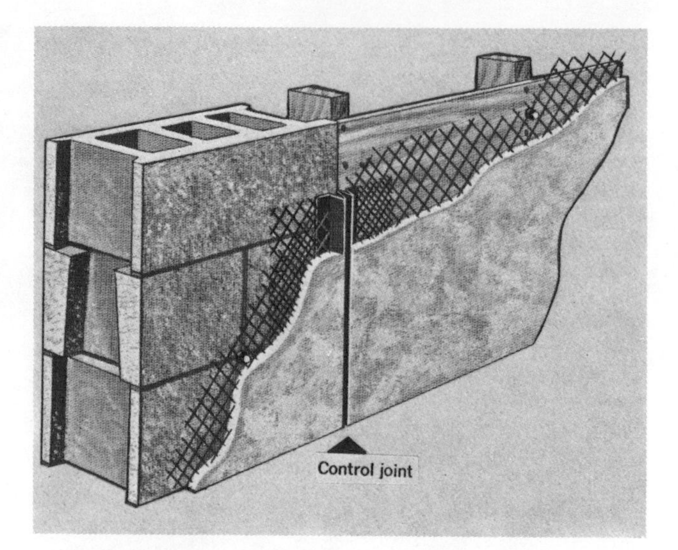

Fig. 10. Installation of a control joint in a wall where two dissimilar materials meet *(Portland Cement Association).*

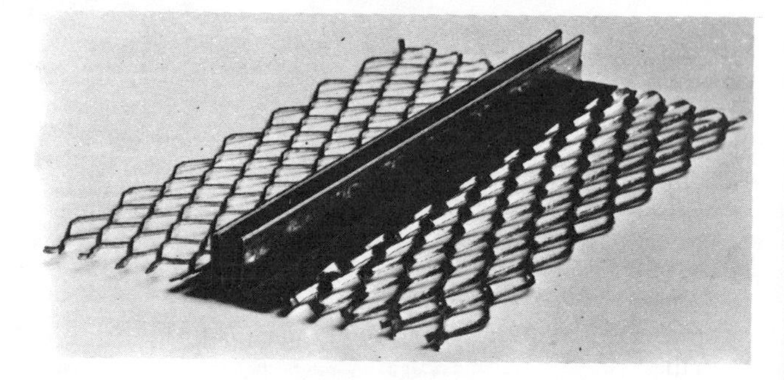

Fig. 11. The accordion construction of a control joint allows it to absorb any expansion in the walls to which it is attached *(Portland Cement Association).*

becomes stressed and cracks. There may be differential movements in the underlying wall structure because two dissimilar materials abut each other, materials that expand and contract at different rates. A long masonry wall may expand enough in hot weather to crack the overlying stucco.

Under any of these circumstances, a control joint installed in the wall (see Fig. 10) will limit the total amount of movement that can take place in the stucco. Cracking is thus either prevented or very much reduced. A control joint is usually required when the total surface area of a stuccoed wall exceeds 150 sq ft. A control joint is also required when stucco has been applied directly in front of an expansion joint in the underlying masonry construction (*see* CONCRETE-BLOCK CONSTRUCTION), or directly in front of any construction where two dissimilar materials meet each other, or wherever a wall meets a loadbearing column or pilaster.

As shown in Fig. 11, a control joint works because its accordion-like construction enables it to absorb differential movements. Whenever a control joint is installed, therefore, it must be installed in such a way that its ability to move differentially is not interfered with. Basically, this means that both sides of the control joint must not be attached to the same piece of metal reinforcement or to the same structure. Instead, there must always be a gap in the construction underneath, a gap that is bridged only by the control joint.

Wall Framing

This article describes the traditional method of constructing wood-frame walls in which the principal structural members are 2 × 4 in. studs. Both exterior walls and interior partitions are described since the method of constructing both is the same. For a description of masonry walls *see* BRICK-MASONRY CONSTRUCTION; CONCRETE-BLOCK CONSTRUCTION.

In traditional Japanese construction, the walls do not carry any loads at all. Instead, the entire weight of the roof is supported by large posts placed at fairly frequent intervals around the perimeter of the house as well as down the center of the house. These supports allow the exterior walls to be made of light wooden frames that are covered with sheets of translucent paper. The frames can be removed entirely from the sides of the house, if desired.

In most of the rest of the world, walls are of more substantial construction since they, and not posts, are expected to support the weight of the roof and, in a 2-story house, the weight of the upper floor as well. In the typical American wood-frame house these loads are carried by 2 × 4 in. studs spaced 16 or 24 in. apart around the perimeter of the building.

In the heavy-timbered houses that had been built up to, say, 150 years ago (*see* WOOD-FRAME CONSTRUCTION), the studs did not carry any structural loads at all; their sole purpose was to support the weight of the exterior and interior wall finishes. The studs, therefore, could be built rather lightly and spaced widely apart.

But as the heavy timbers used originally in dwellings gradually decreased in size, the loads carried by the studs gradually increased, and they were spaced closer and closer together to support these loads. Furthermore, as plastered walls gradually supplanted the wood paneling that had been the common interior wall finish, strips of wood lathing, which were usually cut into 48-in. lengths, were nailed to the studs to support the plaster. To increase the stiffness of the construction, the lathing was staggered. To allow the 48-in. lathing to be staggered, the stud spacing became standardized at its present 16-in. dimension.

When gypsum-wallboard and plywood-panel manufacturers first began to supply their products to builders, it was only natural that they should adopt a panel size—which is 4 × 8 ft—that would allow their products to be used with the standard 16-in. stud spacing. And so most studs have continued to be spaced 16 in. apart even though wood lath is practically obsolete as a support for plastered walls.

Stud Design

Structurally speaking, a stud is merely a long column, the purpose of which is to support loads acting directly downward upon it. In the typical 1-story dwelling, a stud spacing of 24 in. on centers is usually considered adequate to support the downward-acting loads imposed by the roof, plus any snow or wind loads. For a 2-story dwelling, a maximum stud spacing of 16 in. on centers is specified for the first-floor studs, while a maximum stud spacing of 24 in. is satisfactory for the second-floor studs. For a 3-story dwelling, the first-floor studs should be 2 × 6 in. in size, the studs being spaced a maximum of 16 in. apart on centers, with the second and third floors having the same stud size and stud spacing required for the typical 2-story dwelling.

We have mentioned ceiling height. Most 1- and 2-family dwellings built today are sized according to the dimensions of the different paneling materials that are available. Since the standard panel size is usually 4 × 8 ft, most houses are built to make economic use of panels having these dimensions, which is the basic reason that most dwellings built today have 8-ft-high ceilings. (Second-floor ceilings are usually required by local building codes to be at least 7 ft 6 in. high. If a ceiling should slope because a pitched roof is directly overhead, then at least one-half the floor area should have a 7 ft 6 in. ceiling above it.)

BALLOON FRAMING

There are two methods in use today by which the framework of a house is constructed: platform framing (see Fig. 1) and balloon framing (see Fig. 2). Of the two, balloon framing is the older method and it provides the more solid construction. Platform framing is, however, by far the most widely used

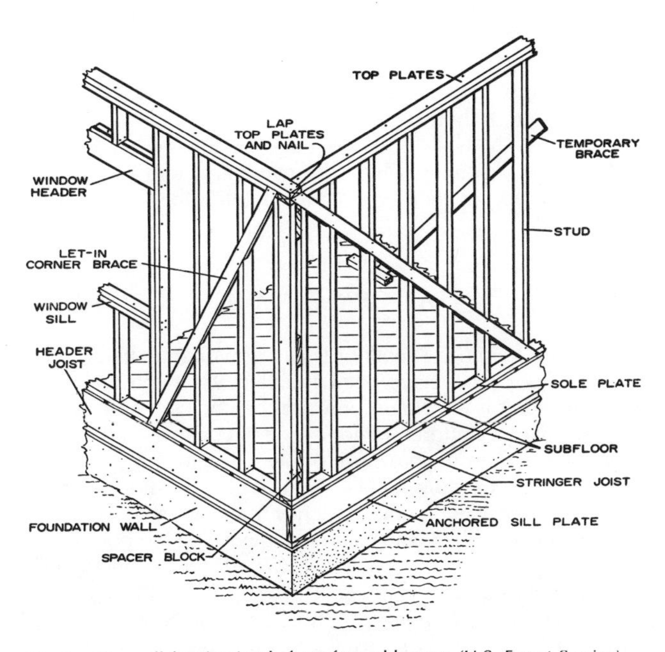

Fig. 1. The wall framing in platform-framed houses *(U.S. Forest Service)*.

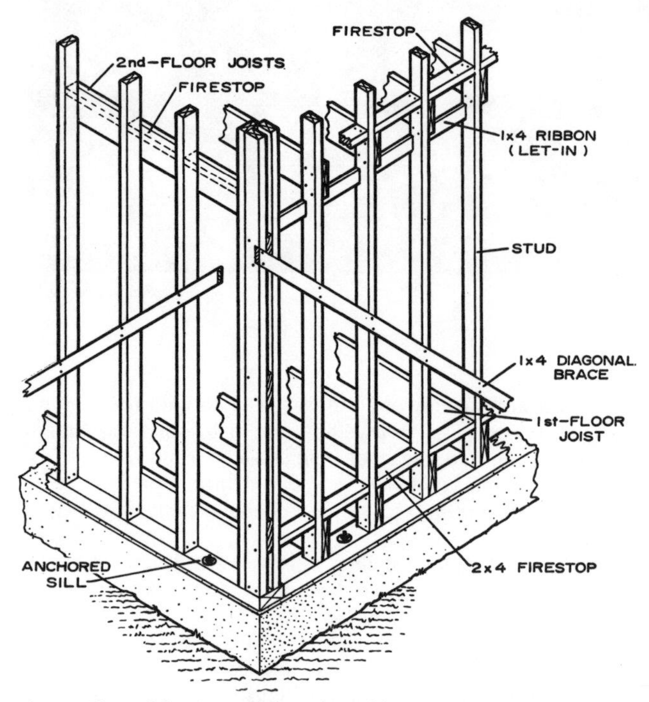

Fig. 2. The wall framing in balloon-framed houses *(U.S. Forest Service).*

method of constructing wood-frame houses in the United States since it is a much faster, and thus less expensive, way of putting up a house. (For a description of both methods *see* WOOD-FRAME CONSTRUCTION.)

In balloon framing, the studs rest on a *sill* and are nailed to the joists as shown in Fig. 2. These studs are installed before the floor is constructed. The builder can construct the walls either by raising each stud in place individually or by building the entire wall flat on the ground, complete with the top plate and the doors and windows framed in. He then raises the wall to an upright position on the sill, aligns and plumbs the wall, braces it in position, and finally drives in the nails that join the lower ends of the studs to the joists.

The first method is laborious and is perhaps best suited to an enthusiast building a house single-handed. The second method, while faster, can be difficult if there is no convenient place nearby that is flat and on which the carpenters can lay out the studs and nail them together. For a 2-story house, raising the studs individually is still the most convenient way of constructing the walls since it is not often that there is a large, flat area nearby upon which the carpenters can lay out the walls, not to mention raising them after they have been assembled.

PLATFORM FRAMING

In platform framing, a subfloor is laid down immediately after the first-floor joists have been installed, *then* the walls are constructed. It is the almost universal practice to construct the entire wall of a platform-framed house flat on this subfloor and then raise this wall into position as a single unit. It is this convenience of construction that is the platform-framed house's greatest, and perhaps, only virtue. Each wall is raised, aligned and plumbed, and then held in position temporarily by 1 × 6 in. or 1 × 8 in. braces that are nailed to the subfloor via 2 × 4 in blocks as shown in Fig. 1.

As may also be seen in Fig. 1, the bottom of the wall consists of a 2 × 4 called a *sole plate,* which is nailed through the subflooring into the joists below. Although this method of joining the walls and floor together appears to be solid enough—after all, the weight of the roof presses downward, doesn't it?—in fact, this construction by itself offers very little resistance to the lifting forces imposed by a high wind blowing against the house or to the pushing forces imposed by the same wind blowing directly against the walls (*see* BUILDING LOADS). In fact, in those parts of the country where hurricanes occur, some local building codes require that iron straps be placed under the walls and that the wall studs be nailed to these straps as shown in Fig. 3.

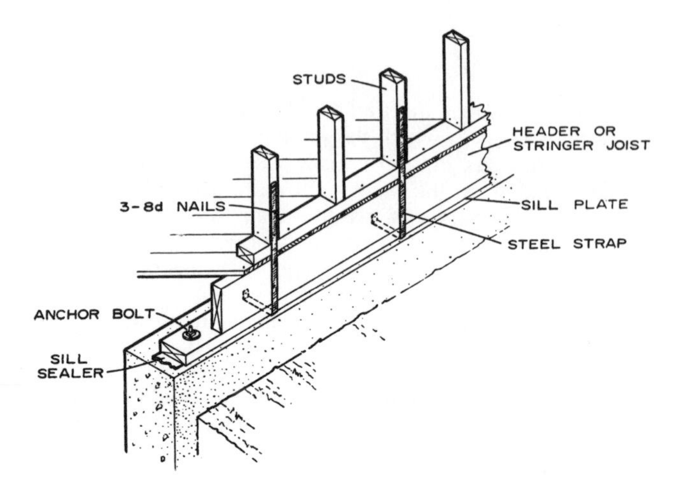

Fig. 3. Iron straps are used in some hurricane-prone areas of the country to tie the walls of a platform-framed house securely to its foundation *(U.S. Forest Service).*

Top Plate

Both balloon-framed and platform-framed houses have a *top plate* that joins the upper ends of the studs together. The top plate also provides a support for the roof rafters and ceiling joists, to which they are also nailed (*see* ROOF FRAMING). In low-cost dwellings, the top plate may consist of a single 2 × 4 attached to the studs by nails driven directly downward into the ends of the studs.

More expensively built dwellings will have a top plate that consists of two 2 × 4s nailed together, the lower plate first being nailed to the studs as described above with the upper plate then being spiked to the lower plate. The joints between mating lengths of 2 × 4 should be staggered.

Another method of constructing the top plate is to lay two 2 × 4s edgewise on top of the studs and then toenail each 2 × 4 into the studs below. Where the top plates meet at the corner of a house, they overlap as shown in Fig. 4.

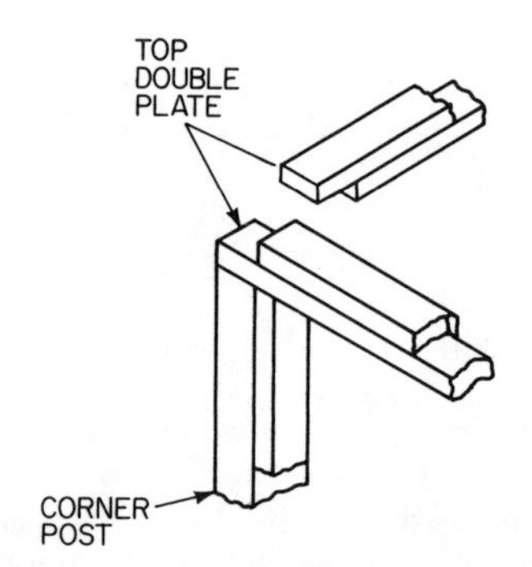

Fig. 4. The overlapping construction of doubled top plates where they meet at a corner post.

Corner Post

As far as the downward-acting dead loads of a house are concerned, there is no reason at all why a house should have large corner posts; ordinary 2 × 4s would be quite satisfactory. But the corner posts have other functions as well, functions that require they be larger, stiffer, and stronger than ordinary wall studs.

The main structural function of the corner posts is to tie the walls together. Both sole plates and top plates (or the sills and top plates in a balloon-framed house) must be joined securely together at the corner posts if the walls are to resist the racking loads imposed by a high wind, which would otherwise tend to blow the walls flat, or at least twist them out of true. In addition, should the walls not be joined together securely at the corner posts, the entire wall construction would be considerably less stiff than it should be.

In very old houses, the corner posts are usually solid 4 × 6 in. or 4 × 4 in. timbers. Solid lumber of this size is seldom used today because of its cost, although it does provide a simple method of constructing strong corner posts. If 4 × 6s are used, a 2 × 4 is nailed along one side as shown in Fig. 5A. The result is a post that provides a solid nailing surface for both the inside and outside wall finishes. When a 4 × 4 is used, two 1 × 2s are nailed to it as shown in Fig. 5B.

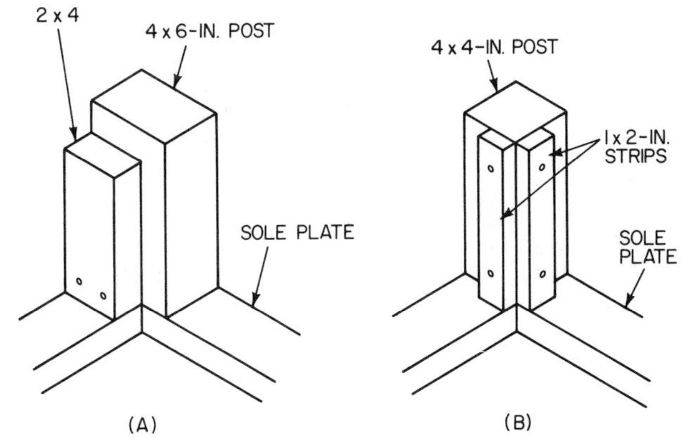

Fig. 5. Two methods of using large lumber for corner posts.

It is far more usual nowadays to build up the corner posts out of several 2 × 4s. When this is done, the lumber selected must be as straight as possible, since a considerable number of nails will be driven into the wood. Figure 6A shows what is perhaps the simplest method; the danger is that the 2 × 4 into which the others are nailed may split because of the number of nails.

Figure 6B shows a method of building up a corner post in which 2 × 4 in. blocks spaced 3 to 4 ft apart are used to build up the thickness of the post. This type of construction provides the large nailing surfaces that are necessary in order to attach the interior finish to the studs, as well as to attach 4 × 8 ft panels that are being used as sheathing and wallboards.

Figure 6C shows the least desirable method of constructing a corner post, although it is used by many builders. Because of the way the 2 × 4s are nailed together, one of the inner surfaces of the corner post is ½ in. shy of being level with the inner surfaces of the wall studs with which it is supposedly aligned. To build out the surface of the post, a special ½-in.-thick strip of wood must be nailed to the post.

Wall Openings

To make room for the doors and windows, parts of the wall studs must, of course, be cut away. But when parts of the studs

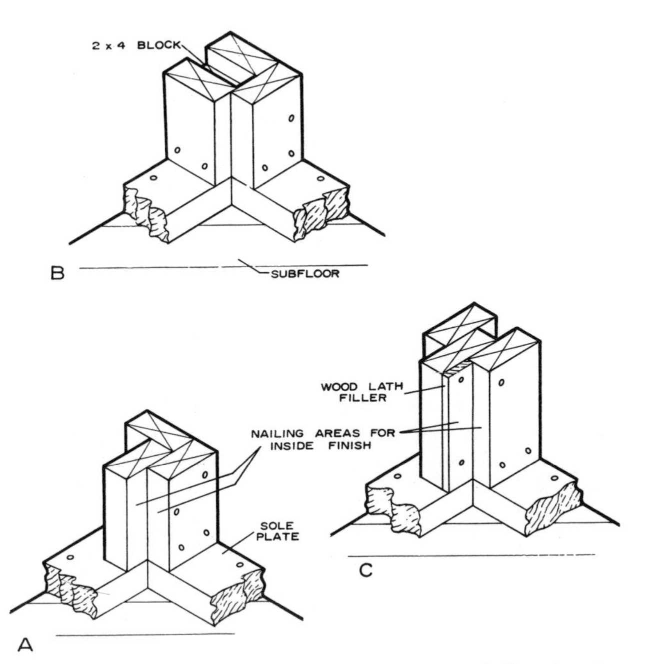

Fig. 6. Three different methods by which corner posts are built up from 2 × 4 in. lumber. The pros and cons of each method are discussed in the text *(U.S. Forest Service)*.

are removed in this way, some method must be found of supporting the cut ends of the studs so that the studs can continue to support the building loads. In loadbearing walls and partitions, this support is provided by a *header,* which consists (usually) of two lengths of 2-in.-wide lumber laid edgewise across the top of the opening (see Fig. 7). The ends of the shortened *cripple* studs, as they are called, are spiked to this header and to the top plate. The header is in turn sup-

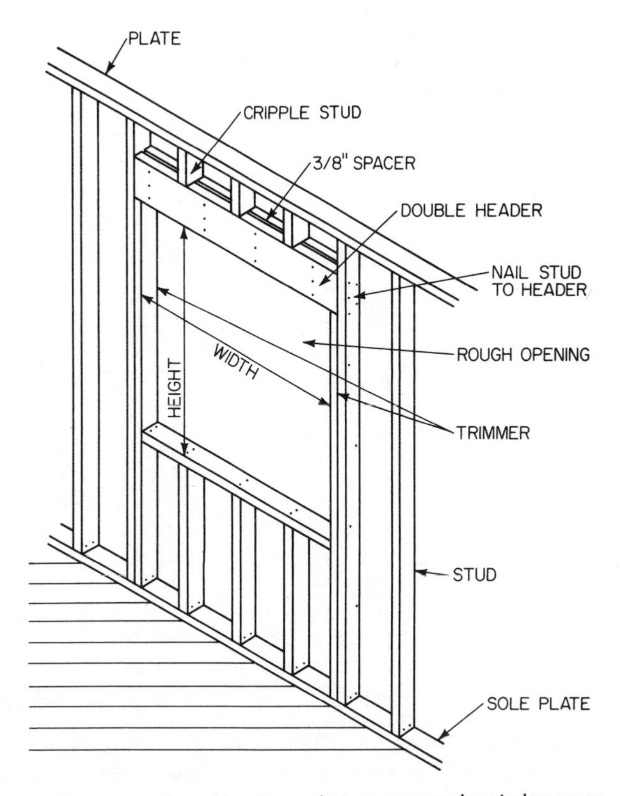

Fig. 7. The construction of headers and trimmers around a window opening. Doors are framed in a similar way.

ported by *trimmers*, which are 2 × 4 in. studs located against the sides of the opening.

These trimmers help share the load imposed by the header upon the studs. In addition, they provide backing to which the finish trim that surrounds the doors and windows is nailed. The trimmers also provide backing to which plaster grounds, lathing, and wallboards can be nailed. On doors, especially, having a doubled 2 × 4 on the lock side of the door provides a solid support the door can slam against. The doubled 2 × 4 on the hinge side of the door provides material into which the hinge can be fastened to prevent the door's sagging, particularly when it is large and heavy (see DOOR).

The most important structural requirement of a header is that it must be strong—it must carry the loads imposed on it without bending. For this reason, most headers are installed with the lumber laid on edge, since the narrowest dimension of a piece of lumber is always the side that is most resistant to bending loads.

For wall openings up to 4 ft in width, two 2 × 4s set on edge (and with ½-in.-side strips of wood nailed between the 2 × 4s to space them out to the 3½-in. dimension of the cripple studs) are sufficient. For longer spans, the size of the double headers typically increases as follows:

Maximum span, ft	*Header size, two of each*
0–4	2 × 4
4–6	2 × 6
6–8	2 × 8
8–10	2 × 10
10–12	2 × 12

The wider the opening and the greater the thickness of the header, the greater also is the potential amount of wood shrinkage. Excessive shrinkage through the header may result in plaster cracks alongside the opening. In addition, the use of large-sized lumber increases the cost of constructing the header. For these reasons, *box* headers have been developed. A box header consists of ½ in.-thick plywood panels that are nailed and glued to both sides of a 2 × 4 in. framework built above the opening. The construction is shown in Fig. 8. Box headers are stiff and strong, and they don't shrink.

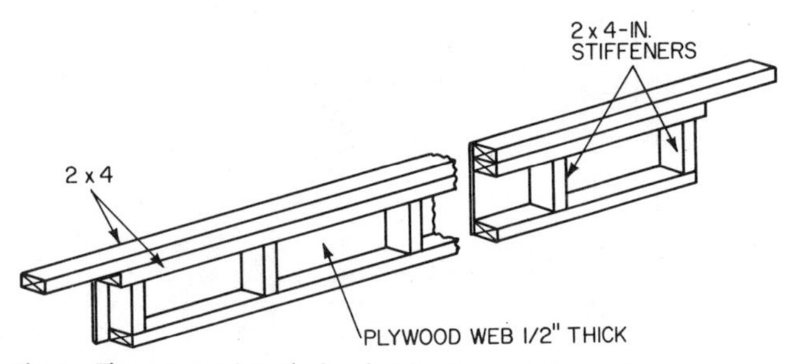

Fig. 8. The construction of a box header above a wall opening.

INTERIOR PARTITIONS

Interior partitions can be divided into those that are *loadbearing*, which means they support the weight of the floor or roof above, or they are *nonloadbearing*, which means they support only their own weight. There is a difference in the construction of partitions for platform-framed and balloon-framed houses; in what follows, the construction for platform-framed houses is described first, the construction for balloon-framed houses being described after.

Nonloadbearing Partitions

A nonloadbearing partition can run either parallel to or across the direction of the floor joists. Since the ceiling joists above the partition are adequately supported already, the construction of the partition can be skimped on somewhat. The basic construction of the partition is identical to that of the exterior walls. That is, a sole plate is nailed to the subfloor, the studs are toenailed to this sole plate, and the tops of the studs are joined together by a top plate.

In a nonloadbearing partition, no harm is done, and some cost is saved, if 2 × 3 in. studs are used instead of the more common 2 × 4 in. studs. In addition, the stud spacing can be 24 in. on centers instead of 16 in. on centers. As with exterior walls, the entire partition, including any door openings, can be built flat upon the subfloor and then raised into position for the final nailing in place.

The top of the partition must be secured in place. If the partition is located directly underneath a joist, a 1 × 6 in. board must first be nailed to the underside of the joist to provide backing for the ceiling lathing or wallboards that will be installed later (see Fig. 9). The top plate of the partition is then nailed to the joist by nails driven through the 1 × 6 in. board.

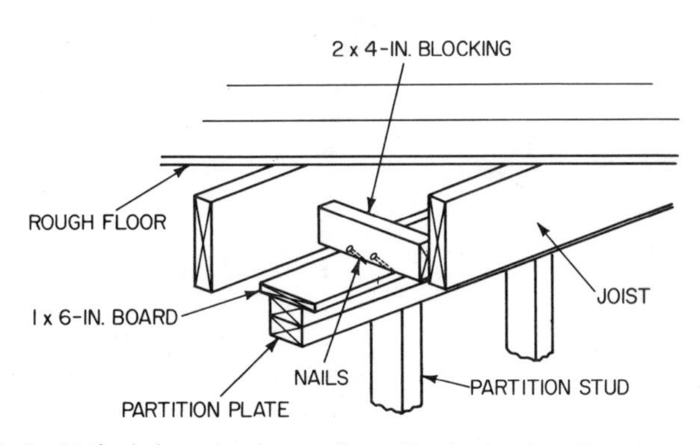

Fig. 9. Method of securing the top of a partition in place by nailing it to a 2 × 4 in. blocking spaced between the ceiling joists. The 1 × 6 in. board on top of the partition is used to support the interior wall finish.

If the top of the partition is located between two joists, 2 × 4 in. blocking spaced 3 to 4 ft apart is nailed between the joists. The 1 × 6 in. board mentioned above is then nailed to the underside of this blocking, and the top plate is nailed into the blocking through the board (see Fig. 9).

If there is an opening in the partition, a single 2 × 3 or 2 × 4 will more than suffice as a header above the opening, with a single 2 × 3 or 2 × 4 acting as a trimmer on either side of the opening. The studs also provide backing into which the door trim, lathing, and wallboards can be nailed.

Loadbearing Partitions

A loadbearing partition usually runs crosswise to the direction of the joists. It serves the same function as the girder that is installed under the first-floor joists; that is, it supports the second-floor or ceiling joists spanning the house. The partition is, therefore, constructed as solidly as the exterior walls. The studs are 2 × 4 in. in size, they are spaced 16 in. apart, and the top plate will consist of two 2 × 4s spiked together. Any openings in a loadbearing partition are constructed in the same way and to the same standard of strength as an opening in an exterior wall.

The most solid construction is obtained when the first-floor loadbearing partition is located directly above the first-floor girder. The second-floor or ceiling joists will then be solidly supported by a system of framing members that is supported ultimately by the foundations of the house.

Partitions in Balloon-Framed Houses

What has been described above is the construction of partitions in platform-framed houses. These partitions, although perfectly adequate for platform-framed houses, are not at all satisfactory in balloon-framed houses. (For the reasons, see WOOD-FRAME CONSTRUCTION.) In a balloon-framed house, the interior loadbearing partitions must be constructed in such a way that as little lumber as possible is installed on edge, in order to reduce to a minimum any possible shrinkage in the construction.

In a balloon-framed house, loadbearing partitions must be built as shown in Fig. 10. (Nonloadbearing partitions may be built as described above for platform-framed houses.) The first-floor partition studs will rest on the first-floor girder and be nailed to the first-floor joists. A 2 × 4 in. top plate is then nailed to the tops of the first-floor studs to hold them in alignment. The second-floor joists or ceiling joists then rest on this top plate.

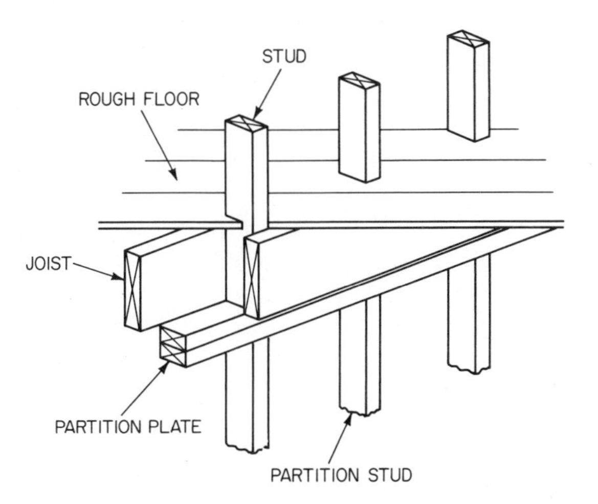

Fig. 10. Method of constructing a loadbearing partition in a balloon-framed house.

The construction of a second-floor loadbearing partition to support the ceiling joists is identical to the construction of the first-floor partition. Note especially that the only lumber installed on edge in the entire construction are the two 2 × 4 in. plates, which total 3 in. in depth. Whatever shrinkage does occur in this construction can occur only in this 3 in. thickness of lumber, and the amount of shrinkage involved will, of course, be trivial.

Before the subflooring is installed, 2 × 4 in. blocking must be nailed between the joists alongside the partition studs to provide backing to which the subflooring can be nailed. This blocking also acts as fire stops; and it provides backing to which the baseboard trim, plaster grounds, and/or wallboards can be nailed as well.

This entire construction obviously requires more time to install and a greater amount of material than platform-framed partitions and, therefore, it is not likely to be found in houses built to ordinary standards of quality.

Partition Corners

Where two partitions intersect, or where a partition meets an exterior wall, they must be attached together securely. If they aren't, plaster cracks will appear in the walls where the two walls work against each other.

Figure 11 (A and B) shows the two main kinds of construction used. Two things should be noted. First, normal stud spacing must always be maintained in the exterior wall, although there is no harm in shifting an exterior stud an inch or two in order to align the wall stud with the partition in order to construct the corner shown in Fig. 11A. Second, backing must be provided behind the partition to which lathing or wallboards can be nailed. Both parts of the illustration show how such backing can be installed.

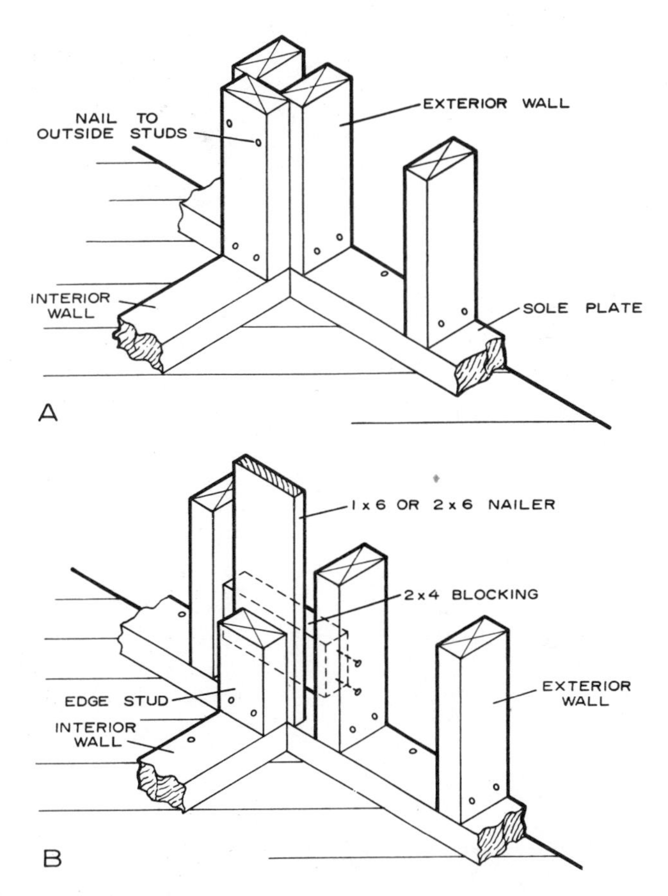

Fig. 11. The methods by which interior wall partitions abut an exterior wall: (*A*) The partition meets the exterior wall at a wall stud; (*B*) the partition meets the wall between two studs *(U.S. Forest Service)*.

WALL BRACING

Bracing may or may not be built into the corners of the exterior walls, depending on whether the walls are sheathed (*see* SHEATHING). If the walls are left unbraced, then they should certainly be sheathed with diagonally installed wood boards or with plywood paneling. (For a description of this process *see* SHEATHING.)

Bracing stiffens the wall construction and increases its ability to resist racking and twisting loads. Bracing also counteracts any tendency the walls may have to warp or twist out of true. Bracing can be either of two types.

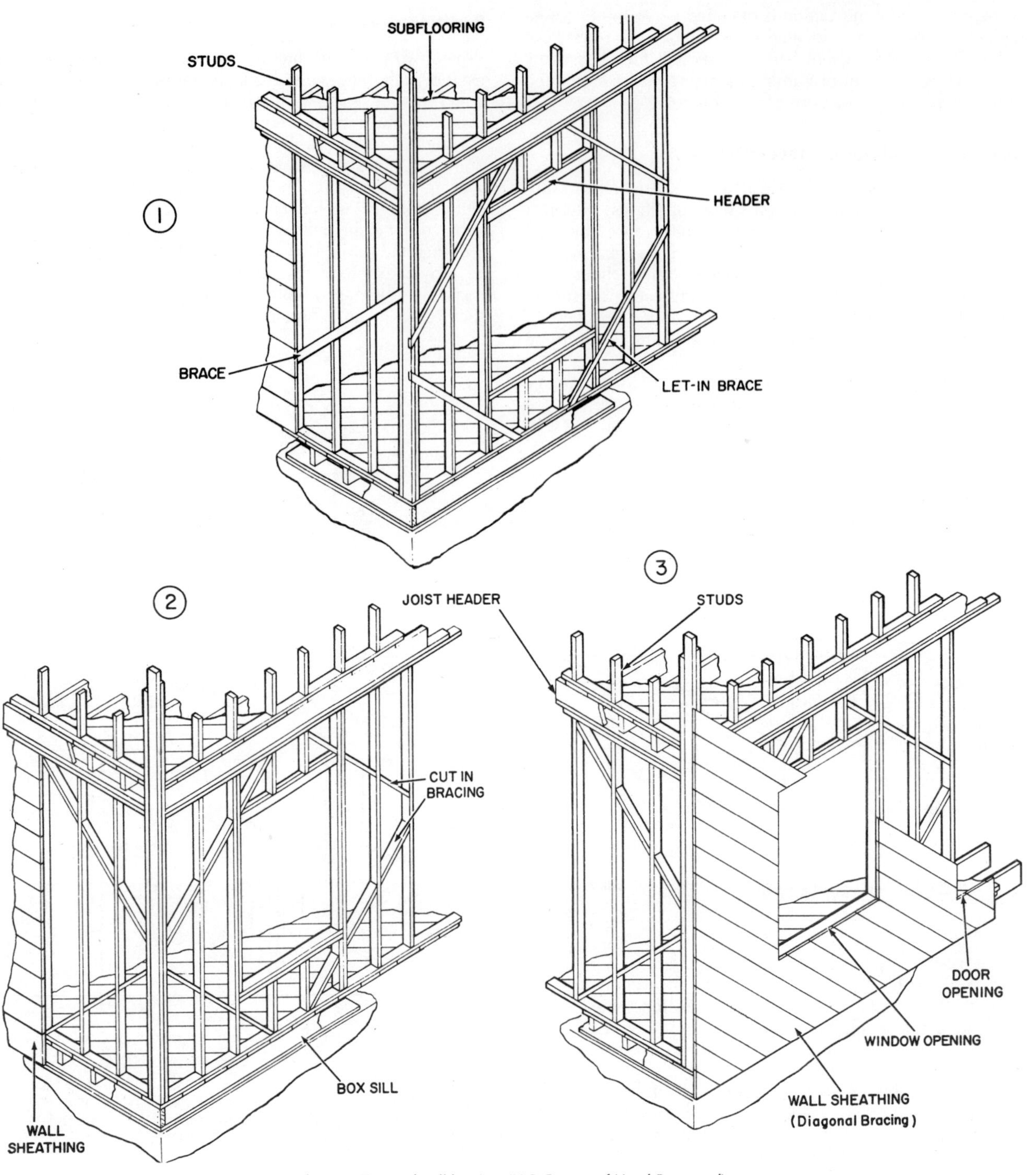

Fig. 12. Types of wall bracing *(U.S. Bureau of Naval Personnel)*.

Let-in bracing (see Fig. 12) consists of 1 × 4 in. or 1 × 6 in. boards that run from the top plate to the sole plate or sill. Connecting a brace to a sill is much the better construction as the brace will then tie the entire wall firmly to the foundation, thus greatly increasing the building's overall resistance to any racking or twisting loads.

To *let in* the bracing, the board is laid against the studs at a 45 degree angle, pencil lines are drawn on the studs to mark the location of the board, and the studs are then cut out as deep as the brace is thick. The brace is then nailed to each stud as shown in Fig. 12. The brace should be made from one length of lumber. Using short, disconnected pieces of lumber will obviate the entire point of installing the bracing.

Cut-in bracing consists of pieces of 2 × 4 cut at angles that allow the pieces to fit tightly between the studs. The 2 × 4s are installed in a line that runs from the top of the corner post to the sole plate or sill, as shown in Fig. 12. The construction is not as good as let-in bracing.

Window

Windows have two obvious practical functions: they admit light and they admit fresh air into a dwelling. Windows also have the important, though less apparent, function of enabling the occupants of a dwelling to make their way quickly from it in case of a fire or some other emergency; thus, whatever other qualities windows may have, they should be easy to open and large enough to crawl through, especially on the second floor of a dwelling. Windows also have an important psychological function in that they allow the occupants to look outdoors. Few people indeed would feel comfortable in a well-lighted, well-ventilated house that was also windowless.

Most building codes require that each habitable room in a dwelling be naturally lighted and ventilated by at least one window that opens to the outdoors. The glass area of this window is usually required to be at least one-tenth the floor area of the room, with the further proviso that the glass area of the window be at least 10 sq ft in size, except in bathrooms, where 3 sq ft is acceptable.

These figures should, however, be considered absolute minimums. In fact, if any room is to be considered adequately lighted, its window area should be at least 20 percent of the floor area.

The building code may also specify that, when the window is fully open, the open area be at least 4 percent of the floor area. Other codes may require that at least 50 percent of the glass area of the window be capable of being opened—which amounts to about the same thing. A building code may also allow habitable rooms without windows if the rooms are ventilated artificially by some approved system or if the rooms are air-conditioned, but, as we mentioned, few people would be happy to live in such rooms continuously.

Design Requirements for Windows

The sizing and placement of windows for maximum daylight illumination is a complicated business in which the position of the house on its site, the size and location of the rooms within the house, the summertime and wintertime positions of the sun in the sky, the average amount of cloud cover throughout the year, the characteristics of the glass used, and many other factors must be taken into account, not to mention such aesthetic considerations as the overall appearance of the house and whether there are views that the occupants might want to look out upon. When for some reason maximum daylight illumination is important to a potential homeowner, then he should consult a lighting engineer before deciding finally on the design of his house.

Very roughly, however, the higher the windows are in a wall, the farther will daylight penetrate into a room. The lower the windows, the better will the floor area near the window be lighted. The larger a window opening is in proportion to the wall in which it is located, the more evenly will the light illuminate the room; in this respect, one large window is much more effective than many smaller windows. In a cold climate, however, there is an insoluble conflict between the requirements for maximum daytime illumination and those for efficient room heating. It is impossible to maximize both at the same time except at the cost of very high fuel bills.

How well a house is ventilated during the summer months also depends on window size and placement—and the best window placement for maximizing the illumination may very well conflict with the requirements for efficient ventilation; one may very well have to make a number of unpleasant choices. To induce the maximum flow of air through a house, the windows should be capable of being opened completely—100 percent. There should also be more windows on the downwind side of the house than on the windward side (if the wind usually blows from the same direction), as this draws air through the house more effectively. Effective ventilation also depends on the angle at which the air enters the house. During the summer, when one wants to enjoy the cooling effects of a breeze, the windows should direct the air down toward the floor.

During the winter, however, if windows are opened at all, they should be capable of directing the air up toward the ceiling in order to reduce drafts as much as possible. In winter, however, the big problem is not one of maintaining a suitable draft but of preventing any draft at all. When the windows are closed, they should be completely closed. But in cold weather outside cold air tends insistently to find its way through the smallest cracks in the windows. To reduce this unwanted air infiltration as much as possible, the window sash must fit very closely against the frame.

Most windows installed in northern climates should have weather stripping installed around the sash to reduce air infiltration as much as possible, since air infiltration past the windows is one of the principal causes of heat loss from a house, which adds greatly to the cost of heating the house. But, the windows (and the doors as well) must not be sealed too perfectly because some air infiltration is essential if the inhabitants are to have a constant supply of fresh air to breathe and if the heating unit is to be supplied with enough fresh air to allow it to operate at maximum efficiency.

TYPES OF WINDOWS

Structurally, there are two main parts to any window. There is, first, the *window frame,* which is securely fastened to the wall of the house. There is, second, a more lightly built, movable framework held within the window frame that in turn holds the panes of glass in place; these are the *window sash.*

Double-Hung Window

In a general way, all windows can be classified according to the way in which the sash move within the frame (see Fig. 1).

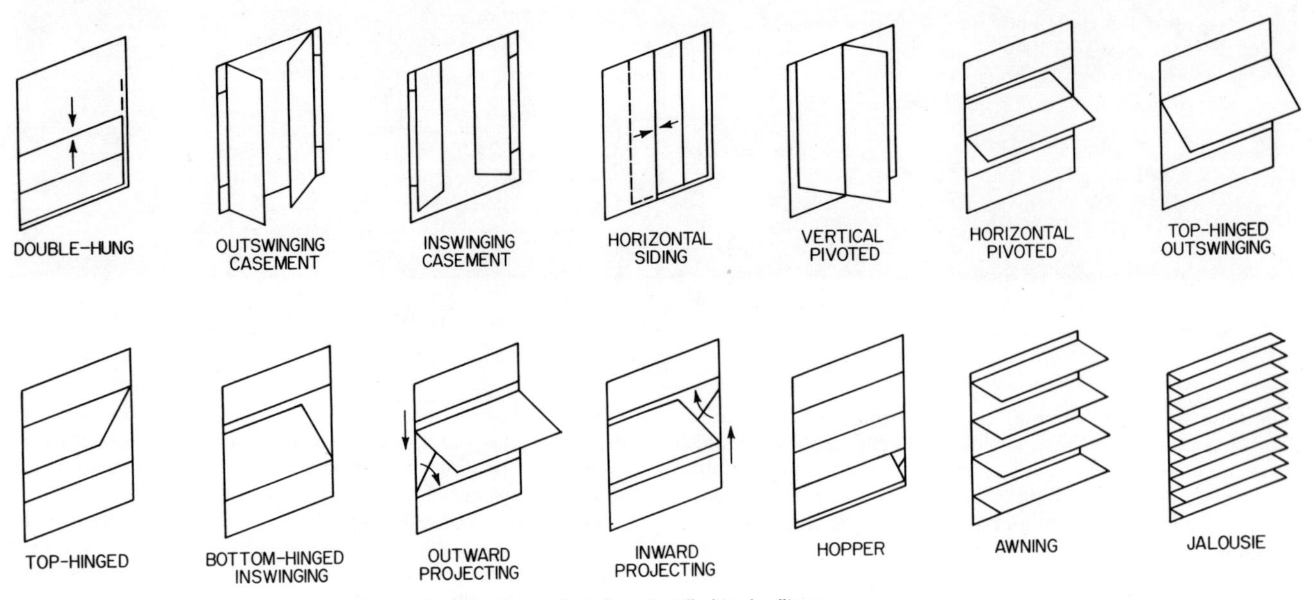

Fig. 1. Common types of windows installed in dwellings.

In one group of windows, the sash are held in place within grooves, or tracks, and the windows open and close as the sash slide back and forth within these grooves or tracks. The sash in *double-hung* windows slide vertically; the sash in *sliding* windows slide horizontally.

Double-hung windows are undoubtedly the most familiar type of window to most people. The sash slide up and down within the frame, their weight being counterbalanced by sash weights or spring balances of one design or another. Because of their construction, as described below, they are the most practical kind of window to install in a cold climate because they can easily be made weathertight. Their principal functional defect is that they can be opened a maximum of only 50 percent, which limits the amount of ventilation they can provide. Furthermore, when they are opened, they cannot direct the incoming air either up or down. And, unless the sash are easily removable, which is possible in some modern designs, it is very difficult to clean the outsides of the windows.

A sliding window has the advantage that its long dimension is horizontal. It can thus be installed along the top of a wall where a double-hung window cannot be fitted. In a bedroom, for example, it can provide both light and air, at the same time assuring privacy. A counterbalancing mechanism is not required for a sliding window. Thus, it is usually much less expensive than a double-hung window of the same size. Sometimes the sash ride on nylon rollers; sometimes the sash are suspended from overhead tracks.

Casement Windows

The sash of a second group of windows are held in their frames by hinges attached between the frame and one edge of the sash. In a *casement* window, the hinges are located along one of the vertical edges of the frame, and the sash can thus be swung either in or out like a door. In an *awning* window, the hinges are located along the top horizontal edge of the frame, and the sash can thus be swung up and out from the frame. In a *hopper* window, the hinges are located along the bottom horizontal edge of the frame, and the sash can thus be swung down and in from the frame.

Casement windows are made to swing either in or out. The outswinging type is more popular because the construction can more easily be made watertight than the inswinging type. The fact that the sash swings out, however, makes it impossible to install a screen or storm sash on the outside of the window. They must, instead, be installed on the inside of the window. When screens or storm sash are installed in an outswinging window, therefore, the sash must be operated by a crank located at the bottom of the window frame. Either that or it must be possible to slide a section of the screen back so that someone can reach through and open the sash manually.

This sort of problem does not arise when a casement window swings inward. But, when the window is open, the sash extend into the room where they may interfere with draperies or shades, and it is always possible that someone will come along and knock into the open sash.

Both inswinging and outswinging casement windows can be opened fully to allow 100 percent ventilation. Outswinging sash can also be attached to a frame with *extension hinges* that allow the sash to be swung out and away from the window frame when the window is opened. By reaching an arm in the space between the window frame and the opened sash, a homeowner can easily clean the outside of the window.

Although an entire window can consist of several awning or hopper sash installed together within the same window frame, it is more usual in dwellings to install only a single awning sash at the top of a window or a single hopper sash at the bottom of a window. There is also a type of combination window that is very common in which a single awning sash is installed at the top of a window and a single hopper sash is installed at the bottom of the same window. Windows of this design can be ordered from stock from most window manufacturers.

The great advantage of awning and hopper windows is that they allow fresh air to enter a house while at the same time they prevent rain from entering. They are thus especially useful in warm climates where it rains frequently. Since both types of window can also direct incoming air up toward the ceiling, they can be installed as ventilators that avoid drafts that run along the floor. Both can be completely opened to allow 100 percent ventilation. Extension hinges can be installed on awning and hopper sash that enable the outside of a window to be cleaned from inside the house when the sash are fully open.

Jalousies are a type of hinged window that resembles a venetian blind. The sash consists of glass slats 3 to 8 in. wide that are fastened together in a metal framework. By turning a crank, all the slats can be made to swing open simultaneously. Jalousies are neither airtight nor raintight when they are fully closed. They are limited, therefore, to warm climates and to locations where this deficiency is relatively unimportant. Their principal virtues are that they allow the maximum amount of air circulation through a house and that the slats can be opened to direct the air up or down as desired.

Projecting Window

In a third major group of windows, the sash are held in place by pivots located midway along opposite sides of the frame. This group of windows is known as *projecting, pivoted,* or *ventilator* windows. When the window is opened, the sash swing, or pivot, in such a way that they project beyond both sides of the window frame at the same time. This mode of operation makes it impossible to install screens or storm sash in the window frame, and when the sash are open they tend to interfere with draperies and shades. People also knock into them. It is easy, however, to clean both sides of the window. Projecting sash are used almost entirely in large commercial and industrial structures.

Picture Window

There is, finally, a fourth group: *fixed* windows in which the sash are immovable. Store windows are the principal commercial example of this type and *picture* windows the principal domestic example. As in commercial shop windows, the glass in a picture window is often installed directly in the frame, instead of in a sash that is in turn inserted into the frame. This method of construction makes it more difficult to install or remove the glass, if this should ever become necessary.

It is, of course, impossible to ventilate a room when a picture window is installed in the wall. Picture windows are, therefore, often used in combination with casement or pivoted windows, which are installed alongside it; or metal ventilating units may be installed in the wall below the window to provide ventilation.

WINDOW CONSTRUCTION

Almost all windows in 1- and 2-family dwellings are made of wood. Metal windows are used mostly in apartment windows, for which local fire-code requirements are more strict. Metal windows may be more fire-resistant than wood windows, but wood is the traditional window material, for which it has many advantages. Wood has better insulating properties than metal, there being much less heat transfer through a window frame made of wood than through one made of metal. There is very little tendency for condensation to form on wood compared with metal (excessive condensation on a metal window frame can result in the decay of the surrounding woodwork). Wood can be painted or stained to suit the style of the house or the owner's taste, but metal, especially aluminum, cannot.

Because windows are exposed to the weather, the wood used in their construction must be able to resist warping, shrinkage, and decay as much as possible. White pine has been the traditional window material, especially in the eastern United States. Nowadays, however, ponderosa pine is by far the main wood used, especially for mass-produced stock-sized windows, as it is easily worked and glued and has a light, uniform color that allows the wood to be stained as well as painted. Other woods used include sugar pine, Douglas fir, cypress, redwood, and cedar. The wood must be straight-grained and all-heartwood, free of loose knots, excessive checks, or other defects.

The great majority of windows installed in houses are, and have nearly always been, stock windows. That is, the windows are mass-produced in stock sizes and in stock designs. The lumber is first kiln-dried to a moisture content of between 6 and 12 percent. After it has been cut to size, the wood is treated with a water-repellent wood preservative. The window is then assembled, including installation of the glass in the sash and the sash in the frame, and all the required hardware is attached. The assembled window is given a primer coat of paint or stain, and it is ready for shipment. All the carpenter on the job need worry about when he installs the window is that it hangs plumb and that the sash operate smoothly within their frame, which may require some slight adjustments.

Stock windows are manufactured in accordance with Standard Specification IS-2, published by the National Woodwork Manufacturers Association. In particular, stock windows are sized according to modular standards (*see* MODULAR CONSTRUCTION). That is, the window sizes are based on a 4-in. module, and they can be ordered in a range of heights and widths that increase in size in intervals of 4 in. An architect who could not find a stock window in a suitable size and style to suit the house he has in mind would really be finicky.

Nevertheless, in the remainder of this article we shall describe in detail only the construction and installation of double-hung and casement windows of traditional design and construction rather than attempt to describe the variety of stock designs that are available at the present time. In the first place, windows of traditional design are installed in millions of older houses, even if they are no longer being manufactured. Second, each manufacturer of stock windows tries to carve out a niche for himself using patented mechanisms and unique construction materials and methods. We could not possibly begin to describe all the variations. Nor is there need to. All windows must solve the same problems of unwanted air infiltration, weathertightness, and smooth sash operation. Once the reader understands how these problems have been solved in windows of traditional design, he will have no problem in applying this knowledge to the window designs of today.

Double-Hung Windows

The most elaborate type of traditional double-hung window is of *box-frame* construction, so-called because the sides of the frame form boxes, inside of which sash weights are suspended. This is the design we shall describe first.

Window Frame

The window frame (see Fig. 2) consists basically of four pieces of wood—two side *jambs,* a top jamb (or *head*), and a bottom *sill.* When the window sash are attached to sash weights via cords or chains that run through pulleys located at the tops of the side jambs, the side jambs are called *pulley stiles* and the head is called a *yoke.*

The pulley stiles and yoke are made of nominally 1-in.-thick lumber. A groove is cut along the center of each pulley stile and a strip of wood called a *parting strip* inserted into this groove divides the pulley stiles in two. The purposes of the parting strips are to separate the top and bottom sash from each other, and they also form one edge of the tracks in which the sash slide up and down. Because of the wear they are subjected to, the parting strips and pulley stiles are often made from yellow pine, a tough, hard wood; or the pulley stiles may be faced with strips of aluminum or galvanized steel.

The outside edge of each pulley stile is rabbeted to a *blind stop* (or *outside rough casing*). The blind stop has two functions: (1) it forms the second edge of the track in which the upper sash slide, and (2) it encloses the weight pocket in which the sash weights are located. Thus, the blind stops must be wide enough to form the 2-in.-wide weight pocket and to overlap the framing stud by at least 1 in. As shown in Fig. 2, the blind stops abut and are flush with the wall sheathing. The gap

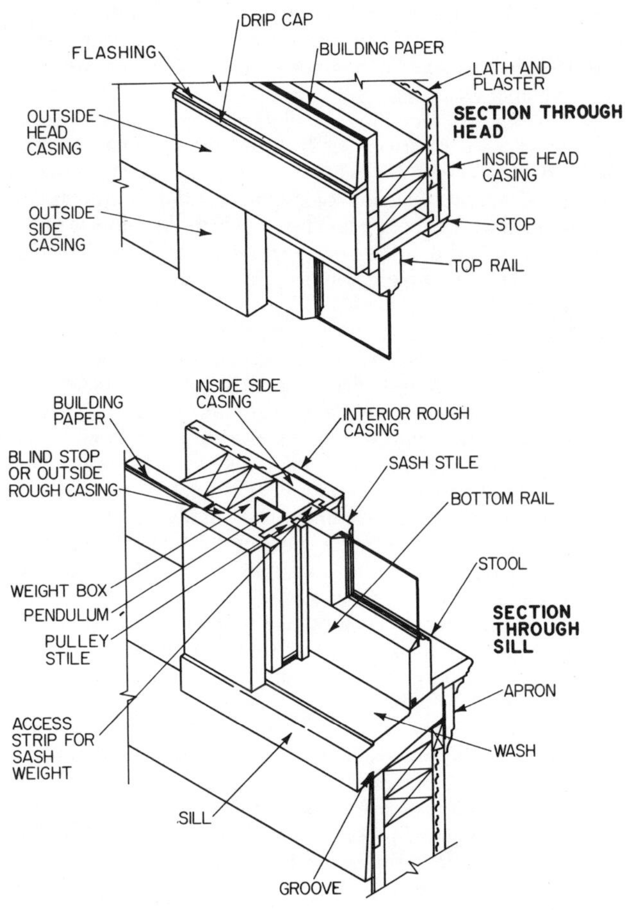

Fig. 2. A double-hung window of traditional box-frame construction.

in the construction between them is covered by building paper to keep moisture from making its way through the gap.

The top of the window frame is also finished off by a blind stop that is rabbeted to the yoke and abuts the sheathing. As there is no need for a weight pocket at the top of the window, the blind stop is made narrower. The space between the top of the yoke and the 2 × 4s above it need be no wider than is necessary to enable the carpenter to level the window accurately within the rough framing. This space is usually about ½ in. deep.

Similar pieces of wood—the *interior rough casings*—form the other sides of the weight pockets and the top of the window frame. The interior rough casings abut the plaster or wallboard. If the walls are to be plastered, then the interior rough casings also form the *grounds* that are used by the plasterer as a guide to bring the plaster out to its final thickness and to level the plaster off smoothly (*see* PLASTERING).

The *sill* is made of nominally 2-in.-thick wood. The distinctive characteristic of the sill is its slope, the purpose of which is to drain away quickly any water that falls on it. Figure 2 shows two steps cut into the top surface of the sill. The outer step forms a seat for a screen or storm sash, and it thus helps make the installation more weathertight. The inner step forms an obstruction that helps prevent wind-driven rain from forcing its way under the bottom of the lower sash.

The outside of the window frame is completed by two *outside side casings* and an *outside head casing*. These casings should be the same size and design as the casings that surround the exterior door frames of the house. If decorative mouldings are cut into or nailed to the outside casings, these same mouldings should be used for every other door and window in the house. Whatever the appearance of these outside casings, their practical function is to cover the gaps that exist between the sheathing and the blind stops, thus further preventing the penetration of moisture into the house. The outside casings are set back slightly from the edges of the blind stops to form a step all around the window frame in which a screen or storm sash can fit.

The interior of the window frame is completed in the same way by two *inside side casings* and an *inside head casing*. These must be of the same wood and same design used for the rest of the interior wood trim. The practical purpose of these casings is to hide the construction gaps between the plaster (or wallboard) and the rough casings. A thin molding called a *stop bead* is nailed all around the inside of the window frame. The stop bead hides the gap between the rough and inside casings. The stop bead also forms the second edge of the track in which the bottom sash slide up and down. In the best construction, the stop beads are fastened to the inside casings with round-headed screws, which are set in metal cups to prevent the screw heads marring the wood. This method of attachment allows the sash to be removed and replaced easily without damaging the wood.

Window Sash

The sash for a double-hung window fit within the window frame as shown in Fig. 2. The framework of each sash consists of four pieces of wood—two side *stiles* and a top and bottom *rail*. The edges of this framework are rabbeted as shown in Fig. 3 to enable panes of glass, or *lights,* as they are usually called, to be fitted into the sash.

When a window has several lights installed in each sash, the lights are separated by and fit into thin pieces of wood called *muntins*.

In double-hung windows the sash are hung in pairs and slide past each other as shown in Fig. 2. The *meeting rails* are often cut in a bevel. When the window is closed, the meeting rails form an airtight seal; when the window is opened, the sash can move away from each other without binding or interference. One of the functions of the crescent-shaped window lock that is usually located on the meeting rails is to draw the meeting rails together to increase the tightness of the seal.

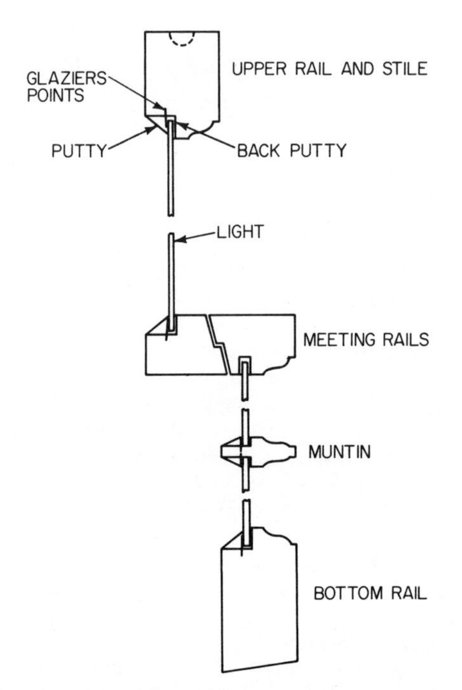

Fig. 3. The sash in a double-hung window fit within the window frame as shown.

Sash Weights

Double-hung sash are normally about ⅛ in. narrower than the framework in which they fit. This clearance enables them to slide up and down without binding or excessive friction. The sash are counterbalanced by *sash weights*. Sash weights not only make it easy for a person to move the sash up or down, they also make it possible for the sash to remain at intermediate positions without falling of their own weight.

Sash weights are made of cast iron. They are usually 1½ in. in diam. and are made in different lengths to provide different amounts of counterbalancing for different size sash. The weights move up and down in the weight pockets already described. To prevent the weights knocking into each other, a thin strip of wood called a *pendulum* is inserted into the weight pocket to divide the pocket in two.

Mullions

When two double-hung windows are placed side-by-side within the same wall opening, they are separated by a *mullion*, which is a hollow box that contains the sash weights for both windows. Thus, the mullion must be at least 6 in. wide. The weight pocket within the mullion is divided into four separate shaftways by pendulums inserted into the weight pocket.

Simplified Box-Frame Construction

Cold air can infiltrate a house through gaps in the window frame with as much facility as it can pass through gaps in the window sash. When a window frame fits poorly in the wall, having tightly sealed or weather-stripped sash is pointless.

The box-frame construction that has been described above is extremely good at preventing the infiltration of cold air past the window frame because the double-casing type of construction effectively seals off any gaps through which outside air might enter the house. For this reason, the box frame may also be said to have excellent insulating properties. The construction also, of course, prevents the penetration of wind-driven rain into the house.

However, the complexity of construction required is unnecessary in more temperate climates. Besides, it is expensive. The windows in a great many dwellings, therefore, are made according to what is called *simplified box-frame* construction, in which the interior rough casings are omitted altogether and the blind stops (or outside rough casings) are reduced merely to strips of wood nailed to the pulley stiles. This construction is shown in Fig. 4.

Note that this construction results in a large gap between the blind stops and the rough framing, a gap that is supposed to be closed by the outside casings that are nailed over the sheathing. How effectively this gap is sealed depends on the skill of the carpenter and on whether or not the outside casings are caulked all around their edges, as they should be. Caulking is often omitted, although most carpenters do wrap building paper all around the rough framing to seal off this framing from moisture. Though the simplified type of construction is less expensive, it also affords much less protection against air infiltration, and it also has less insulating value.

Cottage-Frame Construction

In the least expensive type of box-frame construction, which is often called *cottage-frame* construction, the blind stops are omitted altogether and the outside casings are brought forward until they extend into the window frame as shown in Fig. 5. The casings thus replace the blind stops in providing an edge for the track in which the upper sash slides. By doing so, the step that had been cut into the outside casings that had provided a support for a screen or storm sash is eliminated. Although a screen and storm sash can still be installed in the window, there is now much less assurance that insects or cold air will be prevented from passing through the window.

Cottage-frame construction also uses thinner lumber than the other types of window construction described. The window frame, therefore, is much more likely to warp and allow cold air into the house.

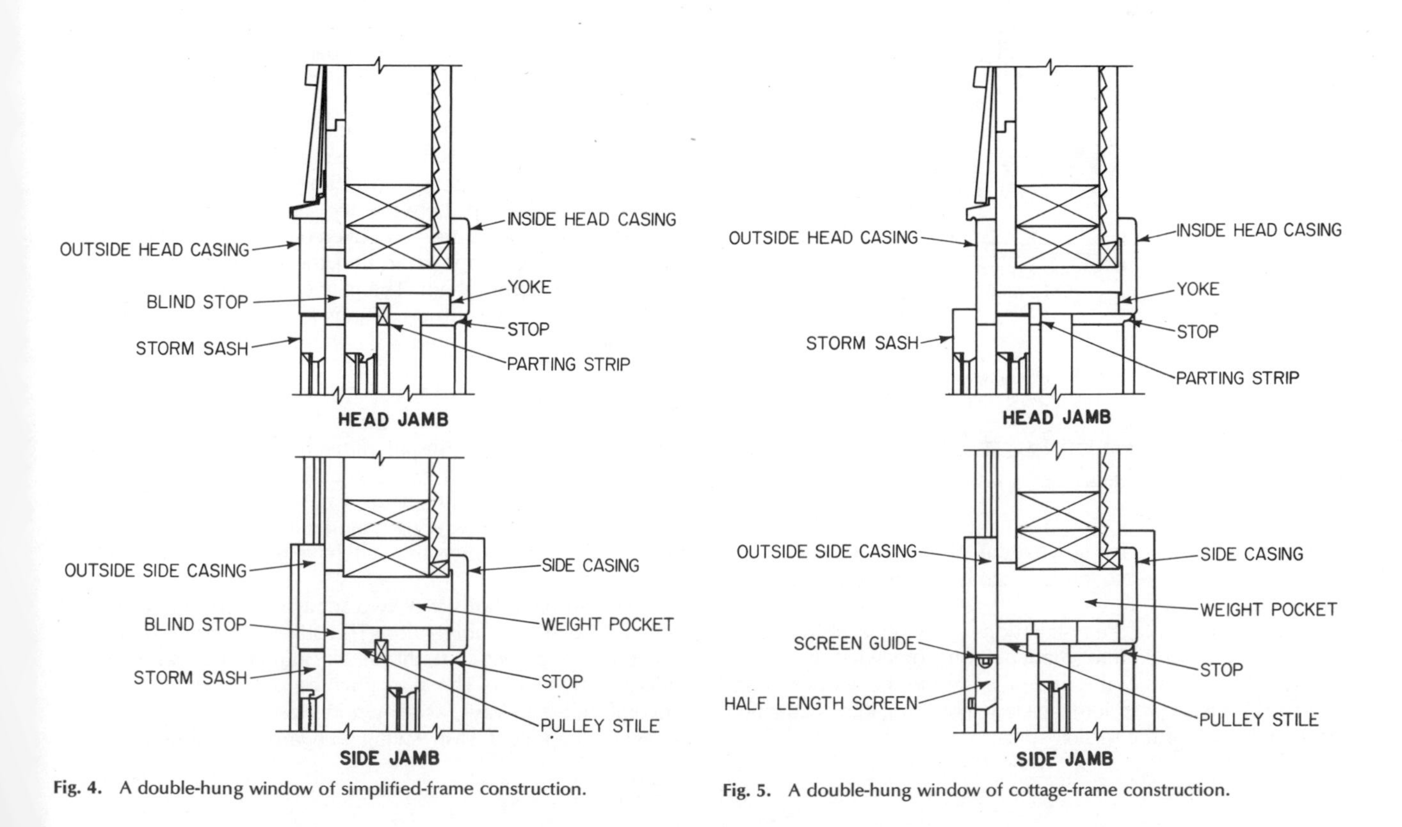

Fig. 4. A double-hung window of simplified-frame construction.

Fig. 5. A double-hung window of cottage-frame construction.

Sash Balances

After all that has been said, it is time to say that double-hung windows incorporating sash weights are no longer manufactured. This type of window is too complex in construction, and thus too expensive to manufacture, compared with newer designs of double-hung window in which sash *balances* have replaced sash weights. These sash balances can be divided into torsion-spring and tension-spring types; there are also compression-type sashes.

A *tension-spring balance* incorporates a wind-up type of coiled spring such as are used to operate alarm clocks, children's toys, and tape measures. The spring is mounted in a case that is installed at the top of the pulley stile, one spring balance being installed at each pulley stile. Attached to the spring is a long cord or a flat steel ribbon, the end of which is fastened to the side of the sash. The tension of the coil spring is thus able to counterbalance the weight of the sash. This spring tension also enables a partially opened sash to remain where it has been placed.

A *torsion-spring balance* operates because of the twist in a helically coiled spring (see Fig. 6). There are two such balances for each sash. Each spring is enclosed within a metal cylinder located against the jamb; usually the side of the sash is cut away to accommodate the cylinder.

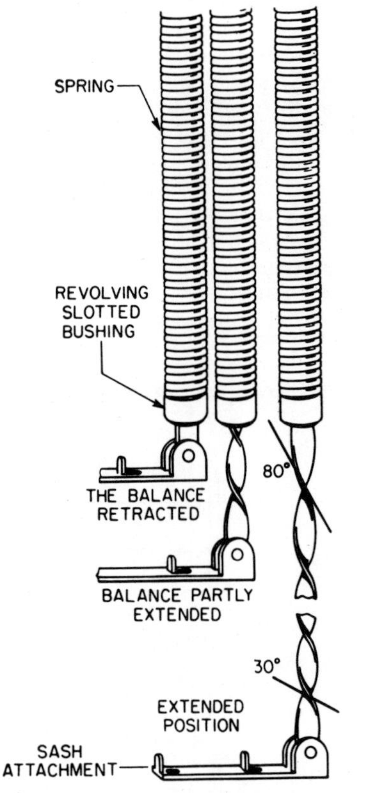

Fig. 6. The operation of a torsion-spring balance.

The top end of the spring is fixed in place. The bottom end is free to twist. Attached to this free end is a bushing with a slot in its center.

The second part of the installation is a long, flat strip of steel that has been twisted into a helical form, somewhat like a drill bit. The bottom end of this ribbon is fixed to the window sash. The top end is inserted into the slotted bushing that is attached to the free end of the torsion spring.

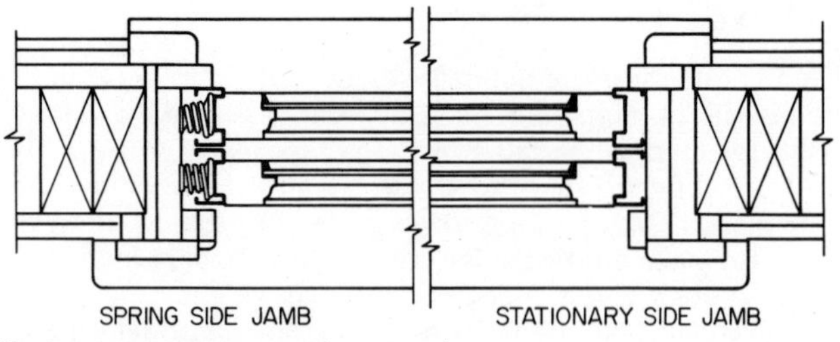

Fig. 7. A compression, or take-out, sash.

When the sash is opened, the helical-shaped steel ribbon is forced through the slotted bushing. The helical ribbon forces the bushing to rotate, which forces the spring to twist, which increases the torsion on the spring. It is this torsion that counterbalances the weight of the sash and enables the sash to remain wherever it has been placed.

The third type of sash balance, which is known as a *compression sash,* or as a *take-out sash,* consists of a thin metal or plastic liner that makes up one of the two tracks within which the sash slides up and down (see Fig. 7). This liner is forced against the side of the sash by springs. To install the sash, it is necessary to compress these springs. Once the sash is in place, the spring tension pressing against the liner holds the sash in place, in whatever position it may be placed. The same spring tension also produces a very weathertight seal along both edges of the sash. Another advantage of this type of sash installation is that the sash can readily be removed from the window frame whenever it is necessary to wash the window.

Casement Windows

The construction of casement windows, whether the sash swing in or out, is remarkably similar to the construction of exterior doors (*see* DOOR). They can, in fact, be considered as miniature doors in both construction and operation. The window frame consists of four pieces of wood—two side *jambs,* a top *head,* and a bottom *sill* (see Figs. 8 and 9). As there are no sash weights, there is no need for sash pockets. Casement windows, therefore, have a comparatively larger glass area than double-hung windows of the same frame size, and they are also less expensive.

The outside of the window frame is finished off by two side casings and a head casing. The wood is usually 1⅛ in. thick (but thinner in less expensive dwellings) and as wide as necessary to cover the gaps between the sheathing, the jambs, and the head. These gaps are also covered over by building paper, the paper being wrapped around the framing studs to keep moisture away from them. The inside casings are constructed in the same way. The inside casings will also, of course, match the style of the other interior wood trim.

Outswinging Sash

Sash that swing out are installed much more often in casement windows than sash that swing in as it is much easier to make the installation watertight, as may be seen in Fig. 8. Once the sash in an outward-swinging window are closed, the sash rest tightly within the rabbets that will have been cut all around the window frame.

The sash often consist of two separate units that swing out together, like a double door. The stiles where the two sash meet are a potential point of air infiltration and must be adequately sealed. The gap may be closed in any one of several different ways, as shown in Fig. 10.

When casement sash swing outward, it is impossible to

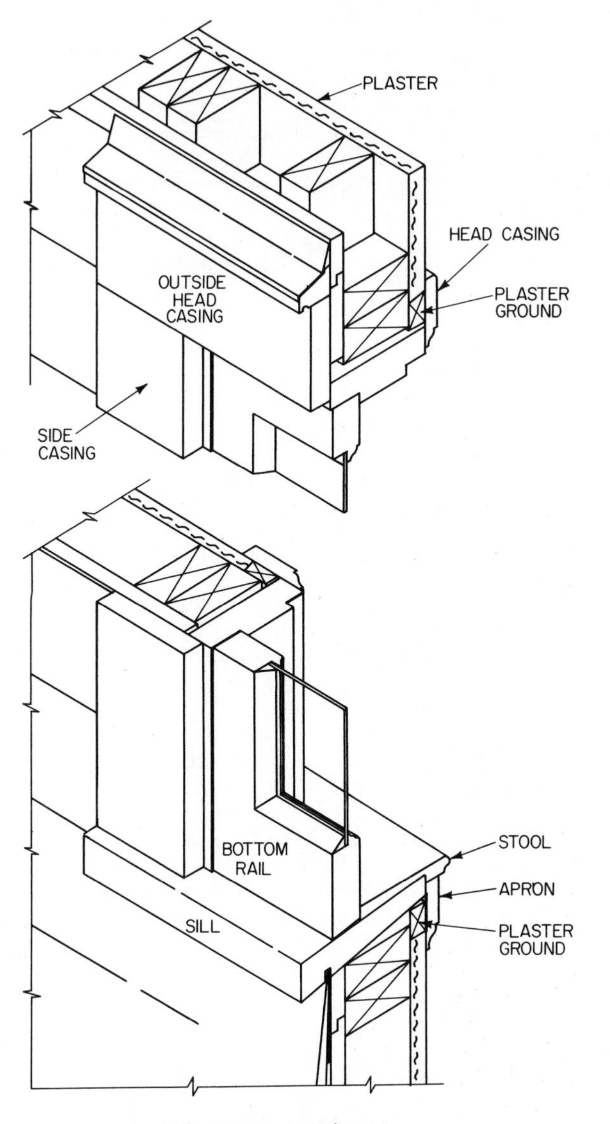

Fig. 8. An outward-swinging casement window.

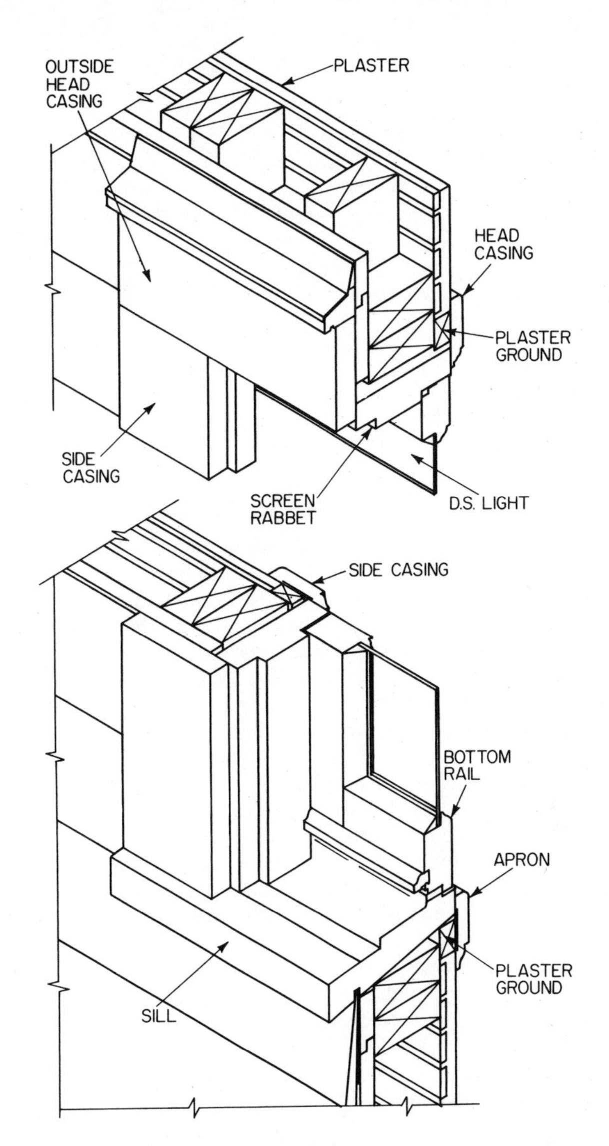

Fig. 9. An inward-opening casement window.

install a screen or storm sash on the outside of the window. But if a screen or storm sash were to be installed on the inner side of the window, it would be impossible to open or close the sash manually. Outswinging sash, therefore, are usually provided with a crank that is geared to sash operators that are attached to the bottoms of the sash. Turning the crank handle makes the sash swing out and in. For high windows, it may still be necessary to install a manually operated sash lock at the tops of the sash to make sure the sash are seated tightly within the frame.

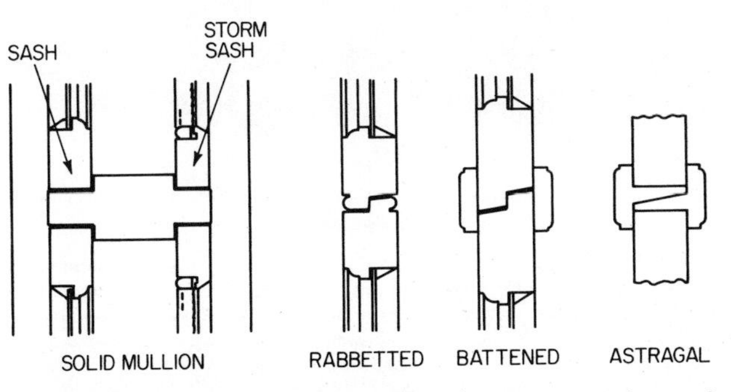

Fig. 10. Double sash meeting at the center of a casement window can be made airtight in any of several different ways.

Inswinging Sash

With sash that swing inward, screens and storm sash can be installed on the outside of the window and the window sash can then be opened and closed by hand. But inswinging sash are more difficult to seal adequately, especially along their lower edges, unless pains are taken in the construction. The sill must not only be rabbeted to provide a seat for the sash, a step must also be cut into the top surface of the sill to obstruct any wind-driven rain that might otherwise be driven under the sash. In addition, a drip moulding should be nailed to the bottom of the sash as shown in Fig. 9. Providing these details are done carefully, one need have no fear that rain will make its way past the window and into the house.

Both awning and hopper windows, being basically nothing more than casement windows in which the hinges are located along the top and bottom of the sash, respectively, are constructed in exactly the same way as the casement windows that have just been described.

Wood

Wood is without doubt the most widely used building material in the United States. There is an abundant supply of lumber available everywhere, and a large (and expensive) labor force that is familiar with the techniques required to construct houses out of lumber. As a building material, wood has the virtues of strength, lightness, and durability. Timbers, boards, planks, and framing lumber are easily fastened together. Wood is unaffected by airborne pollutants and most chemicals. Finally, wood weathers attractively, or it can be painted or stained. As a structural material wood has one major fault—it absorbs and gives off water with changes in atmospheric humidity; as it does, the overall dimensions of the wood change slightly, and this can result in damage to a house if the builder has neglected to take this characteristic of wood into account when building the house.

All wood used for construction purposes comes from *exogenous* trees, which are trees that increase in diameter by the growth of additional material next to the bark. Exogenous trees are, in turn, divided into the familiar *conifers* (evergreens, or needle-leaved trees) and *deciduous* (broad-leaved) trees. In the building trades, conifers are more commonly known as *softwoods* and deciduous trees as *hardwoods*. Softwoods are not always softer than hardwoods, though as a general rule they are. Softwoods, however, have characteristics that make them especially useful as construction lumber while hardwoods are especially useful as interior trim and flooring.

Characteristics of Wood

Certain gross characteristics of wood are known to everyone. Saw a tree trunk through and examine the wood. What is immediately obvious is the concentric growth rings (see Fig. 1). These rings are a measure of both the age of the tree and its rate of growth. There is one ring for each season's growth, although in some species of wood the annual growth rings are difficult to find. At the very center of the trunk, in the bull's-eye of the rings, is the *pith,* a dark brown ring perhaps a ¼ in. or less in diam. The pith is what remains of the sapling from which the tree grew originally. The entire tree is sheathed by bark, which, like a skin, protects the inner parts of the tree from insects, molds, and extremes of temperature.

Peel away the rough, corky outer bark from a tree and one sees the *bast,* the inner bark, which can be pulled away from the trunk in long, fibrous shreads to expose a thin, bright green layer of cells called the *cambium.* All the growth that occurs in the tree occurs within the cambium. Most trees grow from the cambium at a rate of 0.10 to 0.25 in. per year; rarely, a tree might grow more than 0.5 in. a year.

Observing the cross section of the trunk, we note that the outer portion of the trunk is lighter in color than the inner portion. This lighter-colored wood is the *sapwood,* so-called because the tissues of the wood are filled with sap, a watery fluid that carries foodstuffs from the roots of the tree to its leaves. The layer of sapwood is usually 1½ to 2 in. thick, though in some species it may be 6 in. thick and in other species only two or three rings thick.

The darker inner portion of the trunk is called the *heartwood.* As the tree has grown, this part of the wood has gradually ceased to transport the sap. The cells have become clogged with gums and resins, and the role of the heartwood has become the purely passive one of providing mechanical support. Some people believe the heartwood is inherently stronger than the sapwood. This is not true.

Chemically, wood consists of cellulose, lignin, and various extractives. About 60 percent of the weight of wood consists of *cellulose,* an organic substance whose molecules are joined together to form long, hollow, helical-shaped fibers that form the basic cells of the wood. It is these cellulose fibers that give wood its strength, elasticity, and toughness.

The fibers of each species of wood differ from those of other species, but, in general, the fibers of softwoods and hardwood have family resemblances. Softwood fibers are from ⅛ to ⅓ in. in length; hardwood fibers are much shorter, being about 1/25 in. in length. In all woods, the fibers are from 20 to 50 times longer than they are thick, with the fibers of hardwoods being

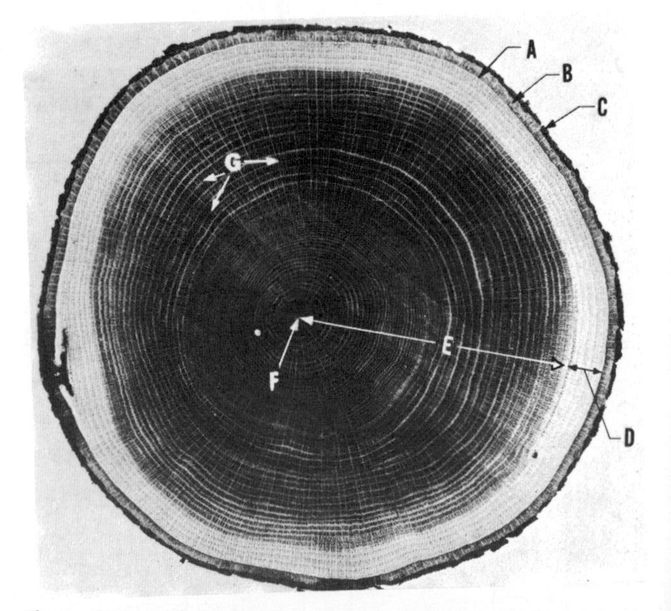

Fig. 1. A tree trunk. (*A*) *Cambium layer* is inside the inner bark; wood and bark cells originate in the cambium layer. (*B*) *Inner back* carries food from leaves to all growing parts of tree. (*C*) *Outer bark* or *corky layer* is composed of dry dead tissue; it protects the tree against external injury. (*D*) *Sapwood* is the light-colored wood beneath the back; it carries sap from roots to leaves. (*E*) *Heartwood* is formed by a gradual change in the softwood; it is inactive and gives tree mechanical support. (*F*) *Pith* is soft tissue about which the first wood growth takes place. (*G*) Wood *rays* connect the various layers from pith to bark and transfer food between them. *(U.S. Forest Service).*

somewhat stumpier than those of softwoods. In softwoods, also, the ends of the fibers taper and are closed at each end. In hardwoods, the ends of the fibers are open and abut one another. The fibers form long tubes, extraordinarily long considering their diameter, through which the sap travels. In addition, hardwoods have special, larger-diameter fibers called *vessels* that also carry sap through the trunk. Softwoods have *resin tubes,* which are nothing more than voids between groups of cells through which the resin travels. Both softwoods and hardwoods also have fibers called *rays, wood rays,* or *medullary rays,* that travel through the trunk from pith to bark. These rays serve to transport the sap radially through the sapwood.

The fibers of cellulose are bonded tightly together by the *lignin,* which acts as a cement. Perhaps 20 to 25 percent of the weight of wood consists of lignin, different species having differing amounts. Another 1 percent, or less, of the wood consists of *minerals.* When wood is consumed completely by fire, the ash that remains is this mineral matter.

Extractives also make up about 1 percent or less of the wood. Extractives include resins, gums, pigments, tannin, oils, fats, and other substances that are responsible for the color, odor, taste, and resistance to decay that distinguishes one species of wood from another. The extractives also influence to some extent the resistance of a wood to water, and thus the way in which the wood shrinks or swells in response to changes in atmospheric humidity. These substances are called extractives because they can be dissolved in water, or alcohol, benzene, acetone, or ether, and thus extracted from the wood.

The remaining substances in wood consist of starches, which provide the living cells with their source of energy and the cambium with the raw materials out of which the cambium manufactures additional cellulose and lignin.

The Growth Rings

A great deal can be told about the life history of a tree by observing the number and relative sizes of its growth rings. Trees grow more slowly in dry years than in wet years, for example. Each species also has its typical growing pattern which is reflected in the general appearance of the rings. It does not take much experience to learn to identify different woods by noting the characteristic appearance of these growth rings.

Regardless of species, when we examine a growth ring closely, we see that it is made up of two parts. The inner portion (i.e., closest to the pith) is light in color. Under a microscope (see Fig. 2) we can see that the fibers have relatively large central cavities and relatively thin walls. About 10 percent of the cross-sectional area of these fibers consists of wall. This portion of the annual ring was formed during the spring, a period of very vigorous growth for the tree. It is, consequently, called *springwood,* or *early wood.*

The outer portion of the growth ring is much darker in color (usually). It was formed during the summer months, a period of slower growth and is, therefore, called *summerwood,* or *late wood.* Under a microscope, we can see that the fibers of summerwood have relatively thick walls and very small central cavities. About 90 percent of the cross-sectional area of summerwood consists of fiber wall. In sum, summerwood is considerably denser and harder than springwood. It follows that a wood in which the annual rings are close together and in which there is a high proportion of summerwood is stronger than wood in which the rings are spaced wider apart and in which there is a higher proportion of springwood. The number of annual rings per inch is, therefore, often used to estimate the density and, thus, the strength of wood, since the closer the rings, the greater the proportion of summerwood and, therefore, the stronger the wood. (*See also* LUMBER.)

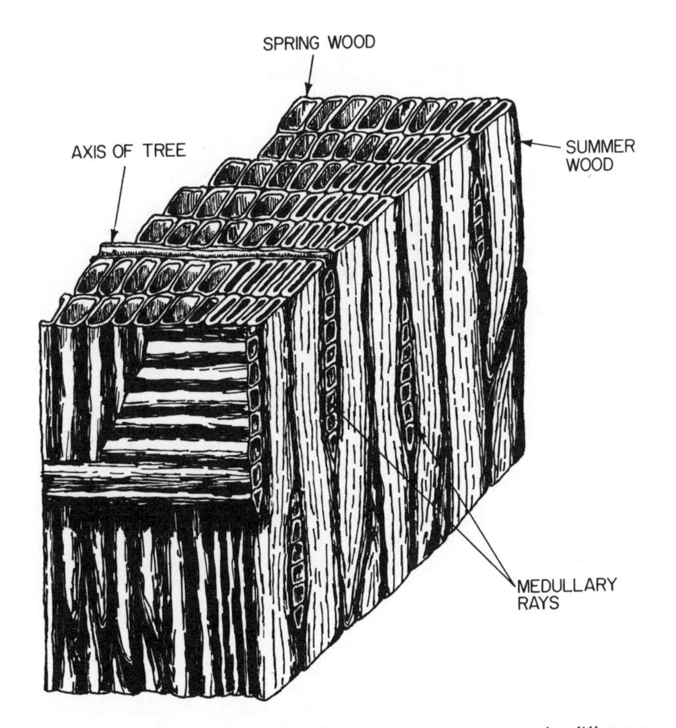

Fig. 2. Microscopic cross section through a conifer, showing the difference between springwood and summerwood.

Wood Shrinkage

A freshly felled tree contains a considerable quantity of water. The wood is "green." In some species, this water content will increase the overall weight of the wood by two or three times its bone-dry weight. This water is of two kinds. First, there is the water that has been absorbed by the cellulose fibers. Second, there is the water contained within the hollow pores and channels of the wood. This second kind of water is called *free* water. It has absolutely no influence on the strength or shrinkage characteristics of the wood. It merely adds to the weight of the wood. When green wood first begins to dry out, it is this free water that evaporates first. The free water closest to the surface of the wood evaporates away first. The remainder of the free water then gradually migrates to the surface of the wood until it, too, has evaporated away.

When the last of the free water has disappeared, but while the wood fibers themselves are still saturated with water, the wood is said to be at its *fiber saturation point.* For most green woods, 26 to 30 percent of its total weight consists of this water, compared to the bone-dry weight of the wood. Once this absorbed water begins to evaporate away, the strength of the wood begins to increase gradually and in proportion to the decrease in its moisture content.

As the fibers lose their water, the wood also begins to shrink. This shrinkage is also proportional to the amount of water lost. If the wood has a normal water content of 30 percent (normal in the sense that at a constant relative humidity of 100 percent the wood will contain 30 percent water by weight), then, as the wood loses half this absorbed water, it will shrink about half its total possible amount. Another way of saying this is to say that for every 1 percent loss in water content below the fiber saturation point, wood will shrink about $\frac{1}{30}$th of its total possible shrinkage. And vice versa, because for every 1 percent increase in its moisture content, wood will swell about $\frac{1}{30}$th of its total possible expansion.

Fortunately for builders and homeowners, wood does not shrink the same amount in every direction. Wood shrinks very little—about 1 percent—longitudinally, that is, in a direction

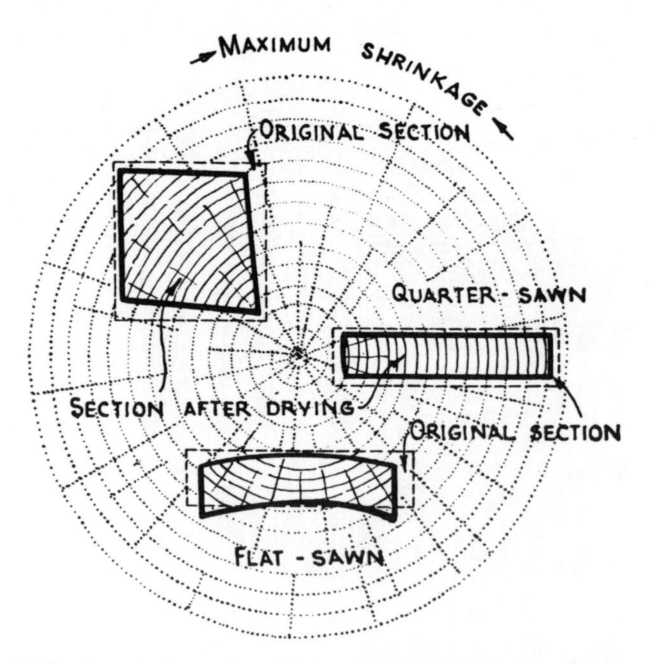

Fig. 3. Green lumber cut from different parts of a log will shrink and warp in different ways because wood does not shrink the same amount in every direction.

parallel to its grain. For all practical purposes, therefore, one can assume that the length of a piece of lumber will remain constant during the lifetime of a dwelling.

Wood shrinks considerably more in a radial direction, that is, in a direction from pith to bark (see Fig. 3). Wood shrinks most of all tangentially, that is, in a direction parallel to its annual rings. If we assume that a piece of wood will shrink a unit length of 1.00 longitudinally, then the wood will shrink radially and tangentially as follows: Longitudinally—1.00; radially—50 to 60 times 1.00; tangentially—100 to 120 times 1.00.

The reason for the large amount of radial and tangential shrinkage is not hard to see. Radially, there are the annual rings—springwood, summerwood, springwood, summerwood, etc.—in order through a cross section of the wood. The fibers of springwood have thin walls and large cavities. Therefore, even when the springwood loses its water, there is not much material there in the first place capable of shrinking. The summerwood, however, with its much thicker fiber walls, can and does shrink considerably more. But since the summerwood usually makes up less than half the total volume of wood in an annual ring, the total amount of possible shrinkage is limited.

Tangentially, however, we have a situation in which there are many long bands of summerwood running through the wood from one side to the other. When this summerwood shrinks, the result can be a considerable amount of shrinkage. The resistance offered by the springwood to this shrinkage is minimal.

Wood Seasoning

All lumber used for home construction is seasoned before being sold; at least, it *should* be. *Seasoning* means giving the wood time to dry out naturally immediately after it has been cut into rough lumber. At the sawmill, the rough-cut lumber is stored, loosely stacked, in open sheds, so that air is free to circulate around each piece of lumber but the lumber is still protected from rain or the heat of the sun. Once the lumber has dried to the desired moisture content, it is finished to size, or it may be dried further in a kiln before being finished to size (see LUMBER). Generally speaking, lumber intended for general construction is merely air-dried. The better grades of lumber or wood intended for specialized uses such as mouldings, cabinet work, or flooring will be kiln-dried (see Table 1).

The main reason wood is seasoned is to enable the wood to shrink approximately to its final size under controlled conditions. Wood warps as it dries. Warping is difficult to prevent, but it can be minimized if the wood is seasoned carefully; careful seasoning will also reduce the amount of checking that will occur.

A *check* is a separation of the wood fibers parallel to the grain and across the annual rings. Checks usually occur at the ends of a piece of lumber because the wood tends to dry out more rapidly at the ends, which stresses the wood fibers there more than anywhere else. Mills often paint the ends of cut lumber to reduce the rate at which moisture evaporates from the ends, thus reducing the rate of shrinkage and the amount of

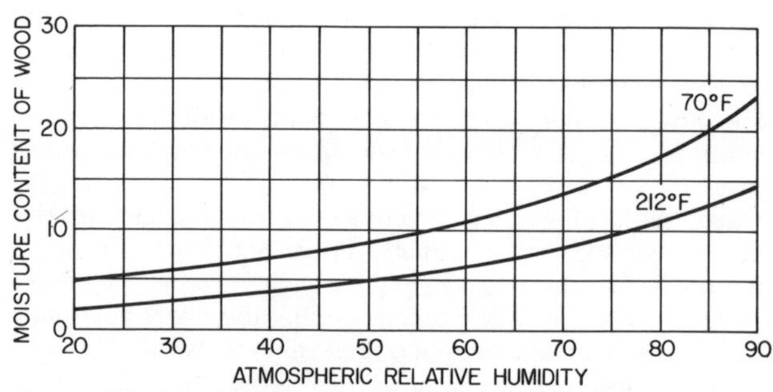

Fig. 4. The relationship between long-term relative humidity in a locality and the long-term moisture content of wood *(U.S. Forest Service)*.

Table I. Maximum Recommended Moisture Contents of Woods Intended for Specific Uses in Different Parts of the United States

	Moisture content (percentage of weight of oven-dry wood)					
	Dry southwestern states		Damp southern coastal states		Remainder of the United States	
Use of lumber	Average	Individual pieces	Average	Individual pieces	Average	Individual pieces
Interior finish woodwork and softwood flooring	6	4–9	11	8–13	8	5–10
Hardwood flooring	6	5–8	10	9–12	7	6–9
Siding, exterior trim, sheathing, and framing	9	7–12	12	9–14	12	9–14

Source: U.S. Forest Service.

checking that occurs. Checks may also so appear along the sides of a very large timber because of the greater difference in moisture content that exists between the surfaces of the timber and its interior.

Kiln drying enables the mill to reduce the moisture content of the wood to a lower value than can be achieved by air drying alone in most parts of the country. Kiln drying is also used to speed up the seasoning process, and it also kills any insects or fungi that are in the wood.

Although it is possible to reduce the water content of a piece of lumber almost to zero by kiln drying it, there is no point in doing so. The wood needs to be air-dried or kiln-dried only to the point where its moisture content is in equilibrium with the normal relative humidity levels prevailing in a locality (see Fig. 4). Let us assume that this relative humidity level is around 50 percent. This means that wood that has been air-dried in this locality will have a moisture content that is about 15 percent by weight. To attempt to kiln-dry the wood below this 15 percent figure only means that once the wood is out of the kiln it will quickly absorb moisture from the atmosphere until its moisture content is in equilibrium with the moisture content of the atmosphere. Since it is impossible to dry all the pieces of lumber in a batch to exactly the same moisture content, it is common practice to specify the allowable moisture content in terms of maximum moisture content. That is, a maximum moisture content of 19 percent means the lumber will have an average moisture content of 15 percent.

If we assume that the lumber used in a house has been dried to the desired moisture equilibrium level before it is installed, it will thereafter shrink or expand very little. Rapid but temporary changes in the relative humidity—as high as 95 percent during a period of very hot, rainy weather and perhaps as low as 30 percent during a period of hot, windy weather—will hardly affect the wood at all dimensionally. What dimensional changes do occur will be seasonal in nature because of annual changes in the climate from, say, long, cold winters when the building is heated to hot, dry summers when the building is open.

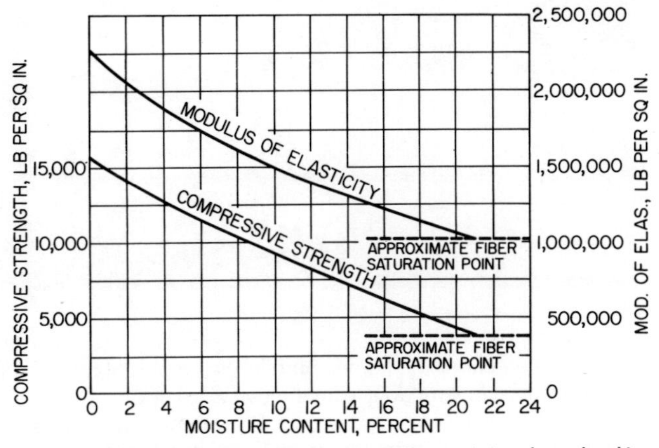

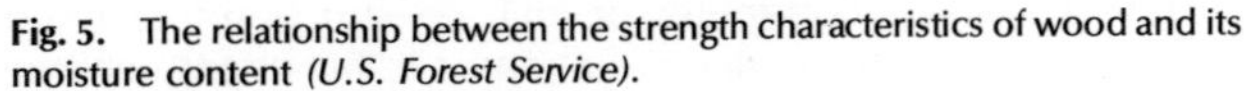

Fig. 5. The relationship between the strength characteristics of wood and its moisture content *(U.S. Forest Service).*

Dimensional problems occur with wood mainly when a house has been built using green lumber or partially seasoned lumber. Then the floor joists may shrink as much as ½ in. until they reach their final size, and the other framing lumber may shrink proportionately, thereby causing unanticipated damage to doors, windows, and interior finish.

Lumber is tested for its moisture content using an electrical-resistance moisture meter in which the lumber is touched by two prongs. The resistance to current flow from one prong to the other is dependent on the moisture content of the wood.

One tremendous advantage of using seasoned wood that has not been mentioned so far is that the strength of wood increases greatly as it loses its moisture (see Fig. 5). As a result, lumber that has been properly seasoned will have substantially increased resistance to bending and crushing loads. The wood will also be harder, more resistant to indentations, and it will be more difficult for nails and other fastenings to be installed in it.

Wood-Frame Construction

Over 80 percent of the homes built in the United States are of wood-frame construction, sometimes called *frame* construction or *light* frame construction, the latter term to distinguish it from the *heavy* wood-frame, or *mill,* construction techniques used to build factories and warehouses in nineteenth-century America before the invention of steel and reinforced concrete.

As for the remaining 20 percent, even though their exterior walls may be made of stone, brick, or some other masonry material, their interior structural framework is still made largely of wood. In particular, the main structural difference between a house with solid masonry walls and a house with wood-framed walls is that the floor joists in the masonry house rest in niches let into the masonry instead of being supported by and nailed to wood studs. Otherwise, the ceilings, interior partitions, and roof of both a masonry house and an all-wood-framed house are constructed in much the same way. Therefore, most of what is said in this article concerning wood-frame construction will apply also to masonry houses. (*See also* BRICK-MASONRY CONSTRUCTION; CONCRETE-BLOCK CONSTRUCTION.)

This article, then, will (1) describe in general terms the three main methods that are used to build the structural framework of a house entirely out of framing lumber, and (2) give the reader some idea of their comparative virtues and shortcomings. (For more detailed descriptions *see* FLOOR FRAMING; ROOF FRAMING; WALL FRAMING.)

One of these framing methods—*braced-frame construction*—is almost obsolete, though it is still used somewhat in the New England states. It will be described first and in detail because it is an excellent way of introducing the two framing methods that came after and are now in common use—the *balloon frame* and the *platform frame*—of which the platform frame is by far the most widely used.

BRACED-FRAME CONSTRUCTION

Braced-frame construction, which is also known as *eastern, combination,* or *barn* framing, is descended from the half-timbered framing we associate with Tudor England but was, in fact, used throughout northern Europe during most of the Middle Ages. This method of constructing houses was introduced to America by the first colonists, who merely transplanted to the New World what they were familiar with in seventeenth-century England (see Fig. 1). The methods and materials used were exactly the same, even to the oak timbers

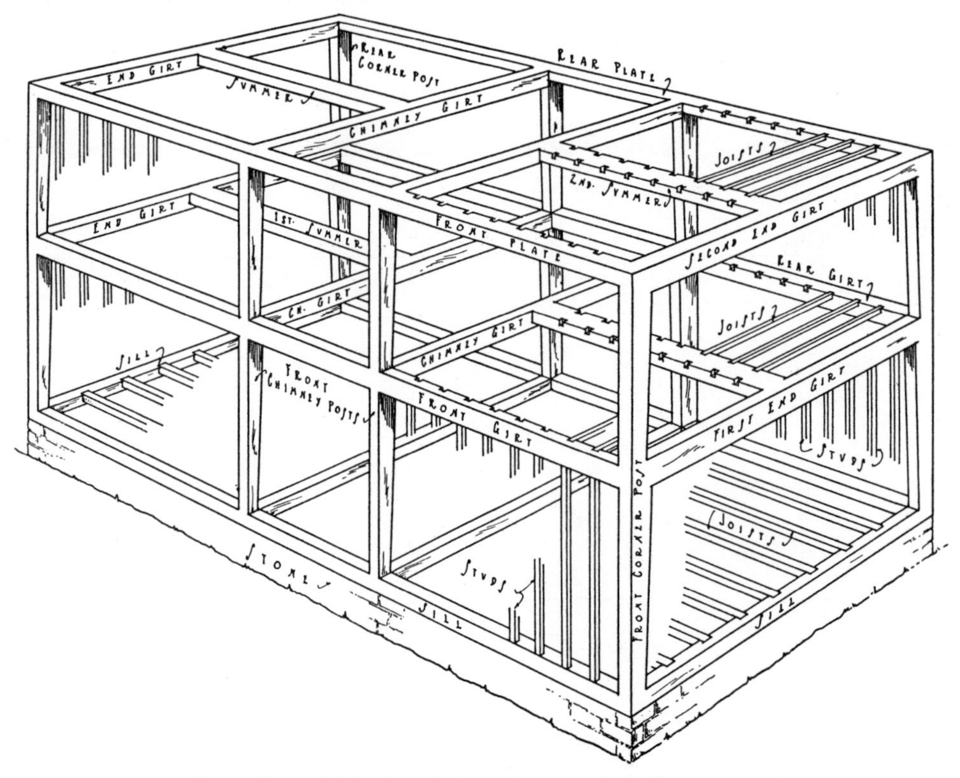

Fig. 1. The typical framing of a house built during Colonial times.

used for the framing. They built the houses the way they did because of tradition, cheap labor, a plentiful supply of timber, and because nails were scarce and expensive.

Everything was large, heavy, substantial. The timbers were larger than they had to be because it was laborious to saw large oak logs into smaller planks, and because the builders had little idea of the actual strength of wood. Braces were often let into the corners of the timbers for additional support. The result was an exceptionally stiff, rigid frame that could withstand any load or stress that man or the weather could possibly place upon it. Today we would say the houses were overbuilt; but many of them are still standing, as sound today as the day they were built, 300 years ago.

During the course of the nineteenth century, the invention of power-driven saws made it easier to manufacture smaller, more easily handled pieces of lumber. The increased availability of cheap, mass-produced nails (nails had been made by convict labor, the way automobile license plates are made today) made it possible to join these smaller timbers together efficiently. There gradually evolved the modified type of braced-frame construction we shall now describe (see Fig. 2).

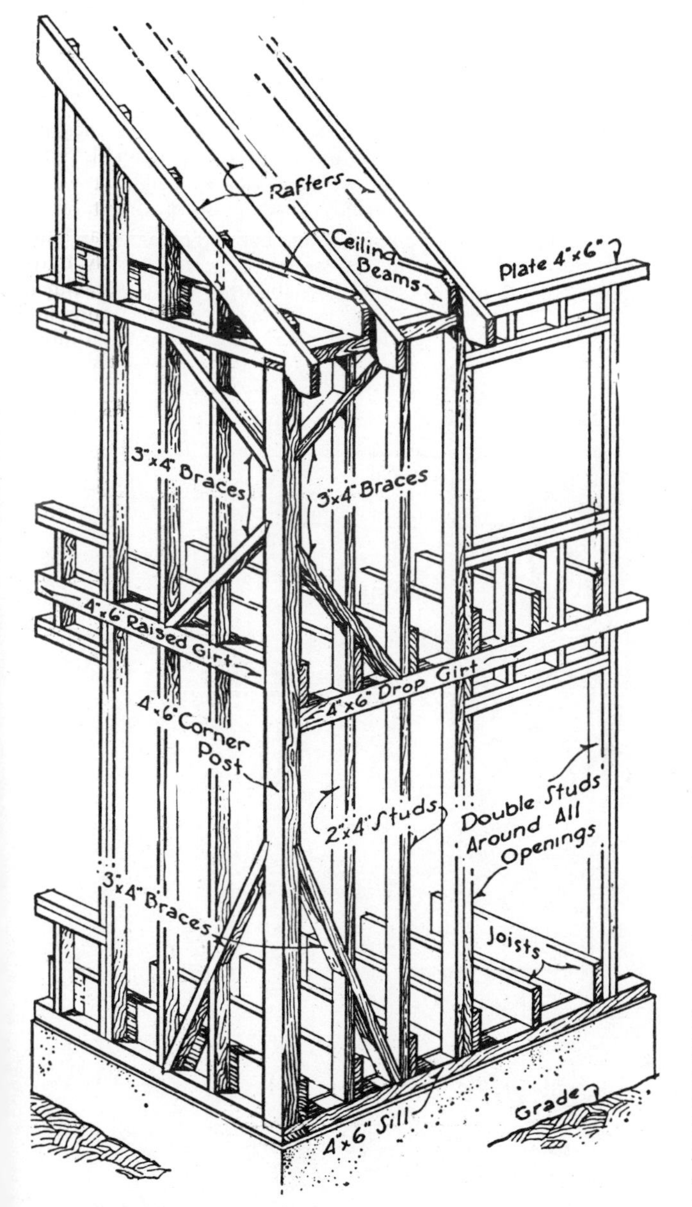

Fig. 2. Braced-frame construction.

Sills

The structural framework of the house rests on a sill, which is a 4 x 6 in. length of lumber that is laid directly on the foundation of the house. It is the function of the sill to transfer the weight of the building, plus the stresses imposed by high winds, to the foundation. The sill is set dead level in a bed of mortar (since its flatness will determine how level the rest of the superstructure will be), and, once the mortar has set, the sill is fastened securely in place by anchor bolts that had been built into the foundation for this purpose (see Fig. 3). The sills are half-lapped together at the corners of the building and fastened together with spikes.

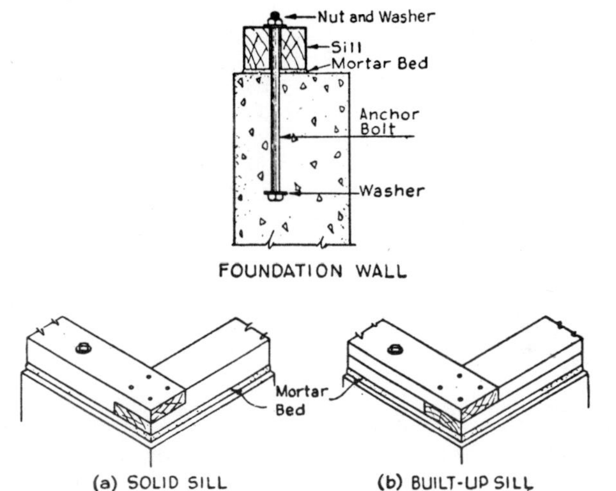

Fig. 3. The sill is anchored securely to the foundation. Where sills meet at the corner of a house, they are half-lapped and spiked together.

Posts

The corner posts are anywhere from 4 × 6 in. to 6 × 8 in. in size. Each extends from the sill to the roof rafters in one continuous length. The posts are placed in position on top of the sill, spiked securely to it, plumbed true in both directions, and held upright by temporary braces until the girts and top plates have been fastened to them.

Girts

Girts are the timbers that run horizontally from one post to another at the second-floor level of the house. In the original braced-frame construction (Fig. 1), the girts were large enough to support the entire weight of the second-floor joists and all their loads by themselves, but their present reduced size now requires that the wall studs help carry these floor loads. Today, the main function of the girts is to support the floor joists, to tie the frame together laterally, and to absorb any stresses transmitted to them by the corner bracing.

Girts have the same dimensions as the corner posts and, ideally, should be attached to the posts with mortise-and-tenon joints. But today labor is expensive and less skilled, and time flies. Today, the girts are usually supported by blocks of wood fastened to the posts by spikes, or by steel angle plates attached to the posts with lag screws. One pair of girts running parallel to each other on opposite sides of the building help support the second-floor joists; they are known as *dropped* girts. The other pair of girts are at the same level as the joists, and they are called *raised* girts.

Top Plate

This horizontal framing member runs from one corner post to another to tie the frame together at the top of the house. It usually has the same dimensions as the sill. Structurally, it also supports and takes the thrust of the roof rafters.

Instead of using a solid piece of lumber for the top plate, sometimes a 2-in.-thick piece of lumber is laid on top of the wall studs and spiked directly into the ends of the studs. Another piece of 2-in.-thick lumber is then spiked to the first piece of lumber. An advantage to this latter method of construction is that less warping is likely to occur when two 2-in.-thick pieces of lumber are spiked together than when a single 4-in.-thick piece of lumber is used. Adjoining top plates are lapped and spiked together over the corner posts just as the sills are.

Braces

The junctures between posts and girts, and between posts, sill, and plate, must be stiffened by bracing if the frame is to withstand the racking stresses imposed by high winds. This bracing usually consists of 4 × 4 in. lumber. There are two kinds of bracing: pairs of knee braces can be installed at each corner (this is the older method), or longer single lengths of lumber that extend at least three stud spaces from the corner posts can be installed at each corner. Both kinds of bracing are set at 45 degree angles.

Joists

In the modified braced-frame construction we are describing, the studs and joists play a larger structural role than they did originally, since the dimensions of the posts, sill, girts, and top plate are much reduced from what they were a hundred years ago and more. Most joists installed today are 2 in. thick and anywhere from 6 to 14 in. deep, depending on their overall span and on the floor loads they are expected to support.

At the first-floor level, the ends of the joists can either rest on the sill or hang from the sill in metal hangers (*see* FLOOR FRAMING). If the joists are to rest on the sill, then each is spiked both to the sill and the adjacent stud.

At the second-floor level, the joists rest on the dropped girt, and, again, each is spiked both to the girt and to the adjacent stud.

At the attic-floor level, two 1 × 8 in. boards called *ledger boards* are set into the studs just below the top plates located on opposite sides of the house. The studs are sometimes cut away to enable the ledger boards to lie flush with the inner sides of the studs. The attic-floor joists then rest on the ledger boards and are spiked securely to the adjacent studs.

The joists are usually set 16 in. apart from each other on centers, but they may be set as close together as 12 in. if the floor loads are expected to be exceptionally heavy, or as much as 24 in. apart if the floor loads are expected to be exceptionally light. The actual size and spacing of the joists will depend on the span of the joists and on the loads they are expected to carry.

Studs

Whatever the spacing of the joists, the studs must have exactly the same spacing; otherwise the framing will become quite complicated and unnecessarily expensive to install. Structurally, the function of the studs is to transmit the floor loads from the second-floor or ceiling joists and the rafters to the sill, and thence to the foundation. The studs are usually 2 × 4 in. in size, though 6-in.-wide studs are often used in a wall if waste

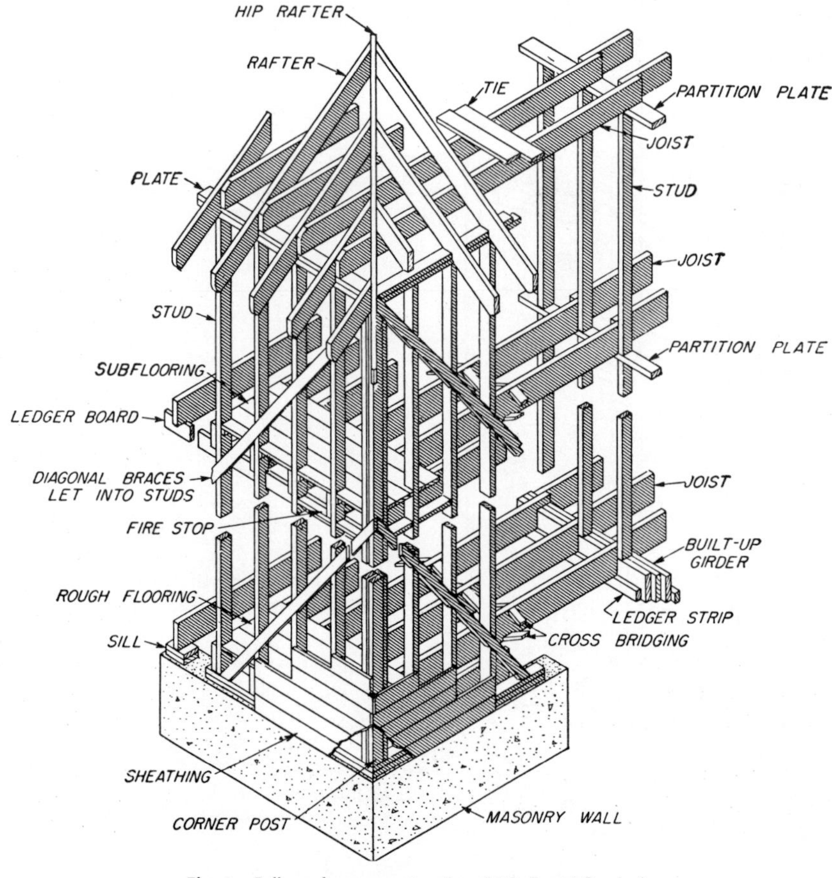

Fig. 4. Balloon-frame construction *(U.S. Forest Service).*

pipes or the ducts for warm-air heating systems will run inside the wall. The studs are spiked to the sill, girts, plate, and, as already mentioned, to the joists.

BALLOON-FRAME CONSTRUCTION

Balloon framing (see Fig. 4) differs from braced-frame construction mainly in that the girts are absent. Instead, the studs run straight up to the top plate; this is balloon framing's distinguishing characteristic. A second major difference is that 2 × 4 in. lumber is used almost entirely.

Balloon-frame construction is much lighter and much more likely to be racked by a high wind than braced-frame construction. Stiffness is added to the framing by *sheathing,* which in balloon-frame construction is considered an integral part of the framing, always to be installed (*see* SHEATHING).

Sill

The sill is usually made of a 2 × 6 in. piece of lumber. As in braced-frame construction, the sill is laid in a bed of mortar, leveled dead flat, and then fastened securely to the foundation with anchor bolts (see Fig. 5).

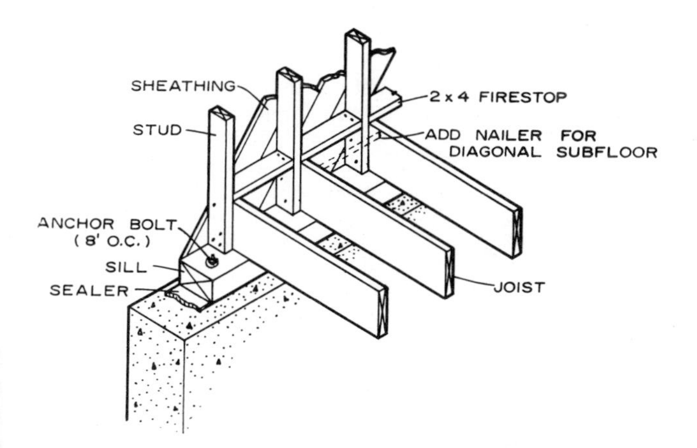

Fig. 5. The installation of the sill in balloon-frame construction and the installation of first-floor joists and studs at the sill *(U.S. Forest Service).*

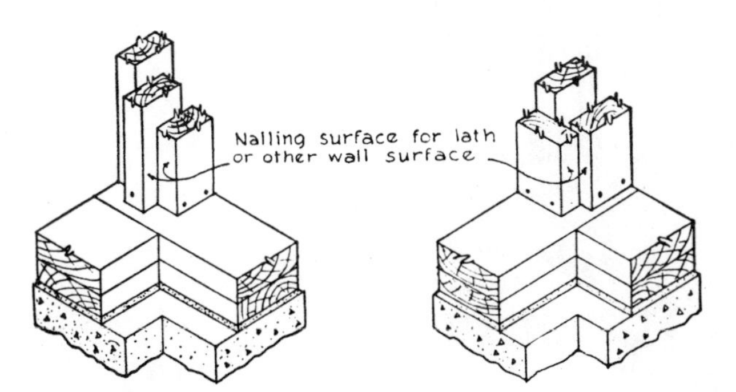

Fig. 6. The construction of the corner posts in balloon-frame construction.

Posts

Once the sill is in place, the corner posts are raised into position. The procedure is the same as for braced-frame construction. In balloon framing, however, the posts are usually built up from three 2 × 4s spiked together to form one solid unit, the individual 2 × 4s being so arranged that the interior surfaces of the post form a right-angled corner to which the interior wall finish can be fastened (see Fig. 6).

Joists

Joists are made of 2-in.-thick lumber and are as deep as required to support the expected floor loads without sagging. The joists are usually from 6 to 14 in. deep. As in braced-frame construction, they are usually set 16 in. apart, on centers, though this distance may actually be anywhere from 12 to 24 in., depending on the span and the anticipated floor loads. The first-floor joists are placed on top of the sill and spiked to the studs (Fig. 5).

Studs

The studs are made from 2 × 4s and, as already mentioned, they are as long as the building is high. The studs, stiffened and tied together into a single structural unit by the sheathing, support the entire weight of the building above the first-floor joists. The studs are raised individually, set plumb, and then spiked into the sill and the adjacent joists. They must have the same spacing as the joists.

Ribband

At the second-floor joist level, two *ribbands,* or "ribbons" as they are often called, are set into the studs, on opposite sides of the house (see Fig. 7), to support the second-floor joists. Each ribband is a 1 × 6 in. or 1 × 8 in. board. The joists are laid on top of the ribbands and then spiked securely to the adjacent studs. Another pair of ribbands are let into the studs at the ceiling-joist level in the same way to support the ceiling joists.

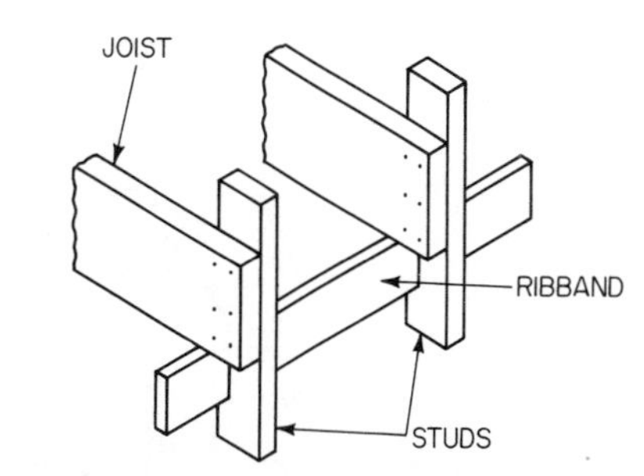

Fig. 7. Ribbands, let into the studs at the second-floor and attic-floor levels of a house, support the joists. The joists are also nailed to the studs.

Plate

The tops of the studs are fastened together at a top plate in the same way as has been described for braced-frame construction. That is, a 2 × 4 in. length of lumber is laid on top of the studs and spikes are then driven through the plate and into the ends of the studs. Another length of 2 × 4 is then laid on top of the first, and the two are spiked together to complete the plate. The 2 × 4s are half-lapped and spiked together where they meet over the corner posts.

Sheathing

Sheathing is necessary in balloon-frame construction to add rigidity to the framework. Any wind loads that would otherwise rack the framework are now simply transmitted to the sill. Either diagonally installed wood boards or plywood panels are satisfactory as sheathing; *see* SHEATHING.

Bracing

Although bracing is rarely installed in balloon-frame construction when sheathing is installed, there is no reason why the frame cannot be braced to increase its stiffness, especially in high-wind areas. Bracing also helps the frame resist the lateral loads imposed by earthquakes, if the house is being built in an earthquake-prone part of the country. Long 1 × 6 in. boards are let into the studs, corner posts, and sill at a 45 degree angle. If doors or windows interfere with the run of the bracing, shorter knee braces can be let into the frame instead. (For a description *see* WALL FRAMING.)

Fire Stops

If there should be a fire in the basement of a balloon-framed house, the spaces between the studs would act like long flues, which would conduct the fire rapidly to the top of the house. To forestall the spread of a fire, these flue-like passageways must be blocked off. Lengths of 2 × 4s called *fire stops* are fitted between the studs for this purpose.

In a 2-story dwelling there should be two fire stops within each stud space; one at the sill (see Fig. 5) and one at the second-floor ribband.

PLATFORM-FRAME CONSTRUCTION

Platform framing, also called *western* framing, is the most widely used method for constructing houses in the United States (see Fig. 8). Most of the custom-built ranch houses and split-level houses put up by builders and almost all the mass-produced houses built in large developments use platform framing. The distinguishing characteristic of platform framing is that each floor of a house is constructed as a platform that is largely independent of the floors built above or below it.

Briefly, the first-floor joists are installed and a subfloor is laid down on the joists. The subfloor becomes a platform on which the rest of the first-floor framing is installed. Once the first floor has been completed, the second-floor joists are installed, another subfloor is laid down, and the second-floor framing is built on top of this subfloor. And so on.

As far as a builder is concerned, platform framing is the quickest and most economical way there is of constructing a house. Once the subfloor is in place, the walls can be rapidly assembled on top of the subfloor and then raised into position with a minimum of effort; ideally, the plumbers, electricians, and other trades can go about their respective businesses on the first floor while the carpenters are busy laying down the second floor.

Sill

The sill consists of a 2-in.-thick length of lumber that is installed on the foundation with anchor bolts in the same way as described above for balloon-frame construction.

Posts

The corner posts are also built up out of several lengths of 2 × 4 as described above for balloon framing. As in balloon fram-

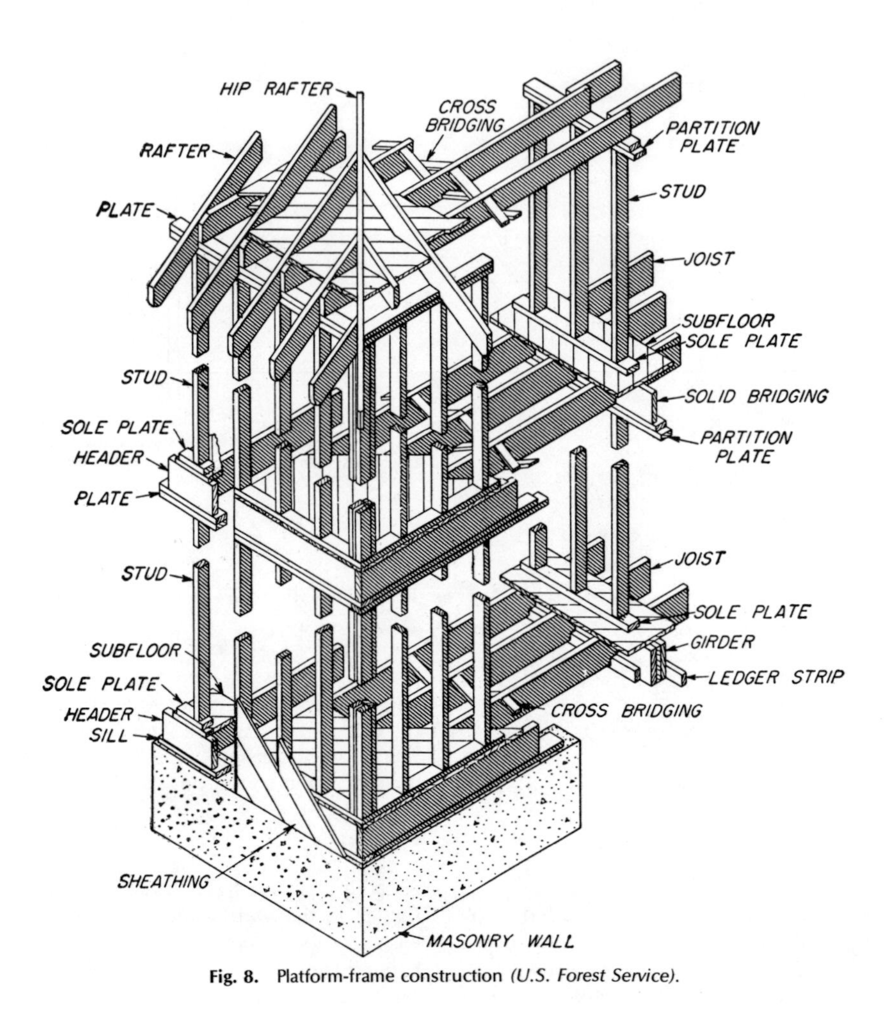

Fig. 8. Platform-frame construction *(U.S. Forest Service)*.

ing, the posts *should* run the full height of the building. It is more usual, however, for a builder to install posts that are only 1-story high.

Header

A header is a structural member used only in platform-frame construction (see Fig. 9). The header has the same dimensions as the joists, and it is set on edge on the sill and across the ends of the joists and runs completely around the periphery of the house. Once the sill has been installed, the header is placed on top of it and the two are spiked together. Together, they form what is called a *box sill.*

Joists

Everything that has been said regarding the sizing and spacing of joists in balloon-frame construction applies also to the joists in platform-frame construction. In platform framing, the joists are set on top of the sill, abutting the headers on opposite sides of the house, and they are then fastened to the headers by spikes driven through the headers.

Subfloor

The subflooring is then laid down on top of the joists and headers. The subflooring will extend to the very edges of the headers all around the periphery of the floor.

Sole Plate

The sole plate, or *shoe,* is another structural member that is found only in platform framing (Fig. 9). It is usually the same size as the studs, and it is spiked in place on the subflooring as a foundation for the studs. A sole plate is also laid down wherever there is to be an interior partition in the house.

Studs

As already mentioned, the studs are all 1-story high and they rest on the sole plate, to which they are spiked. Everything said above regarding the sizes and spacing of studs for balloon framing applies also to the studs in platform framing.

Top Plate

Two 2 × 4s are nailed to the tops of the studs and to each other to form a top plate as described above for balloon framing. In platform framing, the first-floor top plate also acts as the second-floor sill. The second-floor header is spiked to it and the second-floor joists rest on it just as described above for the first-floor construction.

Sheathing

Sheathing is even more essential in platform-framed houses than it is in balloon-framed houses as it is the only tie between floors, especially if the corner posts are only 1-story high.

Some form of bracing is also required in addition to the sheathing to help stiffen the framework. The bracing should be installed in the same way as described above for balloon-framed houses. (*See also* SHEATHING.)

Fire Stops

Separate fire stops are not as essential in platform-framed houses as they are in balloon-framed houses because the side plates that are installed at each floor reduce the chances that a fire will be able to spread through a house via the stud spaces.

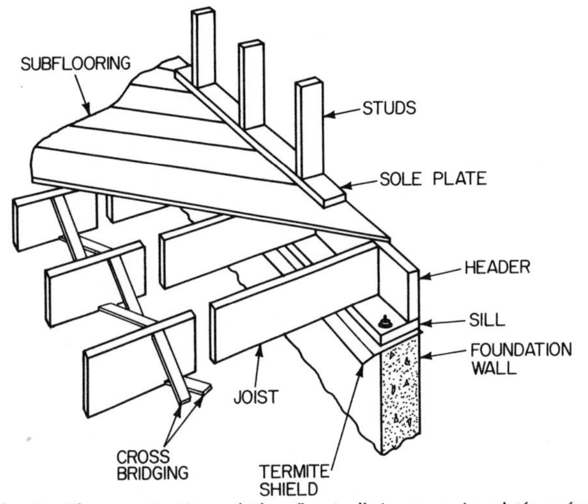

Fig. 9. The construction of the floor/wall juncture in platform-frame construction.

BALLOON FRAMING VERSUS PLATFORM FRAMING

Economically, the advantages are all with platform framing. It takes more time, more effort, and thus more money, to construct a balloon frame than it does a platform frame. A balloon-framed house is superior structurally, however, because the studs and joists overlap at each floor level and are fastened together by spikes that are driven perpendicularly through the lumber. As long as a sufficient number of nails are used at each joint, the result is an exceptionally rigid framework that will withstand any but the severest stresses, such as hurricane-force winds, for example, or a strong earthquake, without being racked.

A platform-framed house, however, is basically, as we have said, merely a number of platforms built one on top of the other. It is, therefore, extremely important that the framing members and sheathing be tied solidly together with a sufficient number of nails. A platform-framed house properly nailed together into a single structural entity will be more than capable of resisting any external forces imposed upon it. But if the nailing is inadequate, the structural adequacy of the house may be compromised.

Another reason for preferring balloon framing over platform framing is the negligible amount of wood shrinkage that takes place in balloon framing. As wood shrinks, the amount of shrinkage that occurs parallel to the grain of the lumber is trivial while the amount of shrinkage that can occur across the width and breadth of the same piece of lumber can be considerable. The framework of a house the joists of which are made of green lumber can shrink as much as 1 in., for example, before the wood has dried out to the point where its moisture content is in equilibrium with the outside relative humidity. (For a discussion of how and why lumber shrinks, *see* WOOD.)

As may be seen in Fig. 4, there is very little lumber in the framework of a balloon-framed house that is likely to shrink. What lumber is likely to shrink amounts actually to the thicknesses of the sill and the top plate. For this reason balloon framing should be used whenever a house is to have masonry walls.

In a platform-framed house, however, as shown in Figs. 8 and 9, the joists, the header, the top plate, and the sole plate will all shrink down somewhat as the lumber dries out to its final size. Only the length of the studs will remain unchanged. Practically, what all this shrinkage means is that a house cannot be platform-framed if the exterior finish is to consist of masonry veneer or stucco because these materials will not shrink down with the wood framing. If a builder tries to put masonry and platform framing together, the masonry will stay where it is while the wood will not. As a consequence, joints in the framework will open up wherever they are free to do so. Where the wood is restrained, it may split and warp to relieve the stresses placed upon it. Sometimes on a house with a stucco finish, the shrinkage of the wood crushes the stucco along the first-floor level of the house, and large pieces of the stucco will fall off the walls.

Index